The Rangelands of Libya

The Rangelands of Libya

Gustave Gintzburger and Slim Saïdi

CABI

CABI is a trading name of CAB International

CABI
Nosworthy Way
Wallingford
Oxfordshire OX10 8DE
UK

Tel: +44 (0)1491 832111
E-mail: info@cabi.org
Website: www.cabi.org

CABI
200 Portland Street
Boston
MA 02114
USA

Tel: +1 (617)682-9015
E-mail: cabi-nao@cabi.org

A catalogue record for this book is available from the British Library, London.

ISBN-13: 9781800627130 (hardback)
9781800627147 (ePDF)
9781800627154 (ePub)

DOI: 10.1079/9781800627154.0000

Commissioning Editor: Ward Cooper
Editorial Assistant: Emma McCann
Production Editor: James Bishop

Typeset by Straive, Pondicherry, India

Contents

About the Authors

Gus Gintzburger, a botanist/range ecologist, has lived and worked on arid rangelands from Morocco to the Middle East and Central Asia for the past 50 years. His first visit to Libya was in 1966, hitchhiking from Tunisia to the Middle East. He returned to Libya in 1975 as an FAO rangeland expert, surveying the rangelands and systems from west to east and deep into the desert, working with Libyan colleagues on rangeland development projects up until 2010.

Slim Saïdi is a rangeland ecologist/geomatician specializing in RS and GIS research and developments with *The Maison de la Géographie* in Montpellier since 1993. He worked for many years in the field of rangeland ecology at the CEPE/CNRS (*Centre d'Etudes Phytosociologiques et Ecologiques*). He currently teams up with the CIRAD (*Centre de Coopération Internationale en Recherche Agronomique pour le Développement* Montpellier) on rangeland desertification and restoration of Mediterranean arid zones of north and sub-Saharan Africa, developing spatial modelling and mapping applications for agricultural forecasting in Africa, Asia and Latin America.

في موسم جز الصوف في ليبيا يتغنى الرعاة بأهازيج عديدة وهم يقومون بجز وجمع الصوف، تترجم تمنياتهم لأنفسهم ولأغناهم بسقوط الأمطار وتحسن المراعي ووفرة ألإنتاج منها
هذه الأهزوجة باللغة الدارجة :

يجيبلك الله يا ضان ......ربيع بغو وسنين دايرة

وهي دعاء في صورة أهزوجة تتمنى أن يرزق قطيع الغنم بموسم ربيع ممتاز وأن يتكرر ذلك على مدى السنين

During the sheep-shearing season in Libya, shepherds sing various melodies as they shear and gather wool, expressing their wishes for themselves and their livestock for rain to fall, rangelands to improve, and production to flourish. Among these traditional tunes is the following:

'May God grant you, O lamb ! . . . a spring of abundant pastures for many many years'

It is a prayer in the form of a folk song, wishing for the flock to be blessed with an excellent spring season and for this blessing to be repeated over the years.

(original text and translation from Dr Manam Saaed, Benghazi)

Foreword

In Tribute to a Remarkable Devotion and Passion
for the Libyan Rangelands

The establishment, in 1972, of the Agricultural Research Centre ushered in a new era for Libyan agriculture and brought much excitement among agriculturists. However, the challenges ahead of the fledging centre looked insurmountable. We had to start from almost zero in several areas: building a credible research and technological cadre and facilities and developing a vision for the future. We had to rely on expatriate expertise while building our national capacities. We received help from several quarters, but the breakthrough that put the new centre's development on an assured footing was signing a novel agreement with the United Nations Food and Agriculture Organization. The organization was to assist us in recruiting experts in crucial research areas, training young researchers and preparing them for further postgraduate studies.

The process of assembling our team was a gradual one. Still, we were fortunate to bring together a group of individuals with diverse backgrounds and talents, all under the capable leadership of the affable Dr P. Huxley, who we remember with deep gratitude and fondness. It did not take long for the FAO team and the ARC staff to find their rhythm, working in perfect harmony. The various experts, including Gus Gintzburger, made indelible contributions to the early development of the centre, both in terms of their professional achievements and the human connections they forged and maintained throughout their tenure. Their collaborative efforts, characterized by mutual respect and shared objectives, were pivotal to the centre's success.

I met Dr Gintzburger for the first time on 15 August 1975. I found him very amiable and self-assured while briefly discussing his scientific and professional background. Later, at the customary welcome meeting of recruits, he enthusiastically expanded on his previous work with a touch of pleasant humour. But the real surprise came when he revealed that he set foot in Libya for the first time in August 1966, coming from Tunisia. With a light backpack, he hitchhiked his way, visiting places and meeting people, culminating in a dream-fulfilling visit to the renowned rock paintings at Akakus, a place that one of his high-school teachers marvelled about. He emphasizes that his previous trips to southern Tunisia and Algeria prepared him well for the Libyan trip and that he remained, as he put it, 'fascinated with these regions, the history, landscapes, vegetation, wildlife and the people'. The man made an excellent impression, especially on the young researchers, and he became known to them simply as Gus, as he demanded.

He did not waste time and promptly started, with his national counterpart, criss-crossing the vast areas of the country and camping out in the wilderness for days, meanwhile collecting rangeland plants, developing a herbarium, mapping vegetation, and organizing and running an international collection of forage seeds from arid rangeland vegetation. No wonder, as he claims, he exhausted two Toyota Land Cruisers in less than five years! His time with the ARC came to a formal end, but the daunting task of evaluating the amassed Libyan collection remained to be completed. And here is when the WA Department of Agriculture went to the rescue in a fruitful collaborative endeavour involving Australian scientists and technicians, highly skilled Australian agronomists, and a resolute Libyan team of agricultural engineers.

Years of hard work has culminated in the soon-to-be-published volume entitled *The Rangelands of Libya*, by Gus Gintzburger and Slim Saïdi. This publication, made possible by the generous support of CABI, is a testament to the significant impact of Gus's work on establishing agricultural research in Libya. It adds a new landmark to CABI's long and valuable academic publishing endeavour and is a legacy of Gus's dedication and love for the rangelands. The sheer volume of the book and massive supporting documentation is a testimony to the outstanding achievement of the two authors and the individuals and organizations who made this possible. According to the experts, the publication of the book 'not only fills a significant void in the existing literature but also serves as a cornerstone for future research endeavours in Mediterranean arid rangeland management and conservation'.

Indeed, it has been a remarkable journey; luckily, there has always been the loving care of Monique, Gus's wife, and the understanding of their children, one of whom was born in Tripoli – no escape from association with Libya!

Dr Mohamed Sidiki Zehni
BSc (Alex.), MSc (Calif.), PhD (Cantab.)
Founder of the Libyan Agricultural Research Centre
June 2024

Preface

Libya, spanning 1.75 million km^2, stands as Africa's fourth-largest country by land area. Despite its extensive 1900 km coastline along the Mediterranean Sea, the prevailing climate is arid, with arid and hyper-arid environments covering roughly 95% of the territory. Densely vegetated areas represent a mere 1% of the land, primarily clustered in El-Jabal El-Akhdar region of north-eastern Cyrenaica. The rangelands occupy approximately 7.7% of the nation's total area, primarily situated in the northern regions along the coast, desert oases, and bare mountainous areas in the middle and south of the desert, alongside agricultural areas utilized for livestock grazing and occasional rainfed cereal cropping. Notwithstanding the arid climate and formidable environmental conditions, rangelands have historically served as the primary habitat for most of the population, significantly influencing their livelihoods over time.

Old records, such as ancient Egyptian temple inscriptions and papyri dating back to the Pharaonic era and earlier, depict the Libyan lands west of Egypt as verdant rangelands abundant with livestock. These records also highlight Egypt's dependence on Libyan exports of olive oil, grain and livestock products through trade and barter during peacetime and times of war. The Greek historian Herodotus, who purportedly visited Libya in the 5th century BC, attributed the arrival of Greeks to eastern Libya (Cyrenaica) to the region's renowned rangelands and plentiful livestock and wildlife. Rangeland activities and the trade of animal products remained primary sources of income for most of Libya's inhabitants throughout successive eras until the discovery of oil in the mid-20th century. Even today, rangelands and pastoral activities continue to represent a significant source of earnings for numerous Libyan households.

In the context of the present book, although I have not had the opportunity to become well acquainted with the second author, Dr Slim Saïdi, an arid zones ecologist expert in remote sensing/GIS, I have a significant scientific connection with the first author, Dr Gustave Gintzburger, commonly known in the wider region as Gus. My initial encounter with him was approximately 20 years ago while working on the rangelands of the Marmarica Plateau in the far north-east of Libya, bordering Egypt. At that time, I had perused several publications and reports authored by Gus concerning Libyan rangelands, which greatly aided my work. Gus's publications, along with those of the late French rangeland scientist Henry Noël Le Houérou, solidified my path in studying and understanding rangeland systems in Libya. The first material I read from Gus was a highly informative technical report prepared in 1977, submitted under the auspices of the FAO/ARC to the Libyan Ministry of Agriculture: *Survey of the present situation and production of the Libyan rangelands*. Since then, I have actively sought and studied his publications. Last year (2023), I acquired a copy of his *Rangelands of the Arid and Semi-arid Zones in Uzbekistan*, published in 2003 by CIRAD & ICARDA. I consider him, along with the Australian rangeland scientist David J. Tongway, among my mentors, whose publications and books I have had the honour of reading. Apart from minor differences in viewpoints concerning the evaluation of certain concepts related to restoration and rehabilitation methods for degraded rangelands, which may stem from generational differences in approach, I find myself largely in agreement with Gus's ideas and perspectives. I learn from him, enjoy reading his work and benefit greatly from his comments, and occasionally engage in discussions with him on social media as he resides in Western Australia and I am based in North Africa in Cyrenaica, Eastern Libya.

The study of vegetation in Libya has a rich history dating back centuries, marked by notable scientific explorations. The first modern scientific exploration was conducted by the European scientist Lemaire in 1703, who embarked on a quest to uncover the historical extinct plant *Silphium* in Cyrenaica. Over subsequent centuries, European scientists continued to contribute to the botanical knowledge of various regions across Libya. However, it was not until 1931 that the first focused study on rangelands took place, led by Italian scientist Maugini in Cyrenaica, followed by American scientist Barry C. Park's basic study of Libyan rangelands in 1957.

The latter half of the 20th century witnessed increased attention to rangelands due to their recognized importance as a vital component of the country's economy. This prompted thorough research conducted by contracted international institutions and scientists, and later by Libyan researchers, aimed at understanding and maximizing the benefits derived from these rangelands. However, despite these efforts, further studies are warranted to deepen our understanding of rangeland ecosystems. This need is particularly evident considering socio-economic shifts over the past seven decades resulting from oil exploration and production in Libya, which have significantly influenced land-use practices and perspectives. Despite the attention rangelands have received, they still require further investigation, especially regarding their current status, dynamics, potential, degree of degradation, and the development of appropriate methods and techniques for their rehabilitation. Addressing these aspects is crucial for ensuring the sustainable management and preservation of Libyan's rangelands in the face of evolving environmental and socio-economic factors.

In addition to his extensive contributions to Central Asia, the Middle East, North Africa and Australia, Gus stands out as a key figure in the study and assessment of Libyan rangelands, along with the implementation of conservation and rehabilitation techniques across various regions of the country. Libya holds particular significance in his career; it is where he dedicated substantial efforts to research and engage with its landscapes, climate and communities. His involvement in Libya spanned from his initial visit in 1966 until his departure in 2010. A summary of his professional journey is:

- Rangeland scientist position at the National Algerian Agriculture Research Institute (1970–1971) discovering the Algerian steppe, vegetation and agricultural systems
- Presented a PhD in Plant Computer Modelling at Strasbourg University (1971–1975)
- Rangeland scientist with the FAO team embedded with the Libyan Agricultural Research Centre, Tripoli (1975–1980)
- Position with the WA Department of Agriculture (Perth, Australia), identifying and evaluating the Libyan native forage seed collection, particularly annual medics (*Medicago* sp.) (1980–1985)
- Research scientist with INRA, Montpellier, France (1985–1992) working on forage and Mediterranean livestock systems
- Programme leader for PFLP (pasture, forage and livestock) at ICARDA, based in Aleppo (1992–1999)
- Research Director at INRA-CIRAD Montpellier (2000–2005) working on Central Asian rangelands
- Retired: consultancy work in rangeland development in Qatar, Abu Dhabi, Georgia, Central Asia and Northern China (2005–2007)
- Returned to Libya to contribute to the Rangeland Development Project (Suani, 2007–2010)

Throughout his career, Gus has produced a wealth of scientific works and publications. These include chapters in 7 books, 22 refereed journal papers, 5 non-refereed journal papers (original results), 31 contributions to congresses, symposia, workshops etc., 2 authored books, 85 detailed technical reports on Libyan rangelands, 9 technical reports on rangelands in other regions of the world, participation in 11 other projects related to rangeland development in various arid regions, and supervision of 21 Master's students and 3 doctoral students in the field of rangelands.

The present book aims to bridge a significant portion of the information gap concerning rangelands in Libya and potentially extend its relevance to the broader North African region. Covering a diverse array of topics related to rangelands, it delves into environmental aspects such as climate, soil composition, landforms, botany, vegetation patterns and production, rainfed cereal cropping, livestock systems, water sources, wildlife, and agro-ecological zoning. Additionally, it explores social dimensions, including social relationships, customs and traditions, human–environment interactions, cultural practices, traditional grazing methods, and local production systems. Furthermore, the book provides a comprehensive overview spanning more than four decades, incorporating accumulated experience, empirical observations and scientific experiments. It documents the evolving landscape of rangeland management practices and societal changes and offers critical insights into the complex dynamics at play. Special attention is given to discussing rangeland degradation/desertification and recovery, and their underlying causes, and suggesting strategies for sustainable restoration and rehabilitation, specifically of the Libyan rangelands.

Supported by a robust framework of data analysis and illustrated with a plethora of visual aids, including photographs, tables, graphs and maps, the book offers a synthesized narrative reflective of the authors' extensive fieldwork and scholarly engagement in North Africa over a prolonged period.

The subject matter addressed in this book stands as one of the paramount environmental concerns in Libya and the adjacent arid countries. Its publication not only fills a significant void in the existing literature but also serves as a cornerstone for future research endeavours in rangeland management and conservation.

As I extend my congratulations to the authors, it is pertinent to commend the rangeland scientist community for publishing this seminal work. I express my heartfelt gratitude to all individuals and organizations who played a role in the conception, execution and dissemination of this exceptional book. In North Africa, particularly in Libya, we are privileged to benefit from the dedicated efforts of the authors who have invested about 12 years of their lives in producing this comprehensive compendium on rangelands. Their scholarly endeavour encapsulates a wealth of knowledge and expertise, making it an indispensable resource for practitioners, researchers, decision makers and landowners in rangeland investigation, rehabilitation and management.

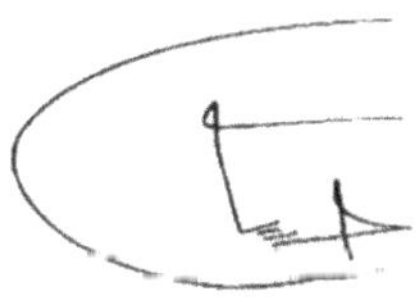

Dr Manam W.B. Saaed
Landscape and Rangeland Ecologist
Associate Professor, Dean of Scientific Affairs for the Faculty of Arts and Sciences (Al-Abyar branch)
Benghazi University
March 2024

مقدمة

تمتد أراضي ليبيا على مساحة تبلغ 1.75 مليون كيلومتر مربع، مما يجعلها رابع أكبر دولة في إفريقيا من حيث المساحة. وعلى الرغم من الساحل الطويل الذي يمتد لمسافة 1900 كيلومتر على طول البحر المتوسط، فإن المناخ السائد فيها يعتبر جافاً في الغالب، حيث تغطي البيئات الجافة ما يقرب من 95% من أراضي البلاد. تمثل المناطق ذات الغطاء النباتي الكثيف نسبةً ضئيلةً جداً من مساحة البلاد لا تتجاوز 1%، وتتواجد أساساً في منطقة الجبل الأخضر، شمال شرق ليبيا. وتشغل المراعي حوالي 7.7% من المساحة الإجمالية للبلاد، وتنتشر بشكل رئيس في المناطق الشمالية على طول الساحل، وفي الدواخل حول الواحات الصحراوية، والمناطق الجبلية في وسط وجنوب الصحراء، يضاف لها المناطق الزراعية المستخدمة لرعي الماشية ومناطق الزراعات المطرية. وبالرغم من المناخ الجاف والظروف البيئية القاسية، كانت أراضي المراعي هي الموطن الأساسي لمعظم سكان البلاد، وهي التي تحكمت في سبل معيشتهم وبشكل كبير جداً على مر العصور.

السجلات التاريخية، مثل النقوش الهيروغليفية المصرية القديمة والبرديات التي يعود تاريخها إلى عصر الأسرات الفرعونية وما قبلها، تصور الأراضي الليبية غرب مصر كمراعي مليئة بالماشية. تسلط هذه السجلات الضوء أيضاً على اعتماد مصر على صادرات ليبيا من الزيت والحبوب ومنتجات الماشية في توفير جزء من احتياجات المصريين، من خلال التجارة والمقايضة في فترات السلم، واخذها كغنائم في أوقات الحرب. كما يعزو الكاتب الإغريقي هيرودوتس سبب وصول الإغريق إلى شرق ليبيا (برقة)، في القرن السابع قبل الميلاد، إلى شهرة الإقليم بالمراعي الجيدة، وأنها تعج بالماشية والحياة البرية. لقد ظل النشاط الرعوي وتجارة منتجات الحيوانات مصدر دخل رئيس لمعظم سكان ليبيا على مر العصور، حتى اكتشاف النفط في منتصف القرن العشرين. وحتى اليوم، لا تزال المراعي والأنشطة المرتبطة بها، تمثل مصدراً هاماً لدخل العديد من الأسر الليبية.

في سياق الكتاب الحالي، "*The Rangeland of Libya*"، فعلى الرغم من عدم حصولي على فرصة لتوثيق علاقة وثيقة مع المؤلف الثاني "Dr. Slim Saïdi" الخبير في علم البيئات الجافة ونظم الاستشعار عن بعد ونظم المعلومات المكانية، إلاّ أنني أمتلك اتصالاً علمياً مهماً مع المؤلف الأول "Dr. Gustave Gintzburger"، الشهير باسم "جاس". لقد كانت معرفتي الأولى به تقريبًا قبل عشرين عاماً، أثناء عملي في مراعي هضبة مارماريكا شمال شرق ليبيا، قرب الحدود الليبية المصرية. في ذلك الوقت، قرأت العديد من المنشورات والتقارير التي كتبها جاس حول المراعي الليبية، مما ساعدني كثيراً في عملي. ساهمت منشورات جاس، جنباً إلى جنب مع منشورات عالم المراعي الفرنسي "Henry-Noël Le Houérou" في ترسيخ مساري العلمي في دراسة وفهم أنظمة المراعي في ليبيا. أول مادة علمية قرأتها لجاس كانت تقريراً فنياً ذو معلومات غنية، كان قد أعده في عام 1977، وقُدِم تحت رعاية منظمة الأغذية والزراعة (FAO) ومركز البحوث الزراعية الليبي (ARC) إلى وزارة الزراعة الليبية تحت عنوان " Survey of the present situation and production of the Libyan rangelands". منذ ذلك الحين، سعيت بكثير من الإهتمام للاطلاع على العديد من منشوراته. في العام الماضي (2023)، حصلت على نسخة ورقية من كتابه " Rangelands of the Arid and Semi-arid Zones in Uzbekistan"، الذي نشر في عام 2003 من قبل CIRAD & ICARDA. أنني أعتبر جاس، مع عالم المراعي الأسترالي " David J. Tongway"، من أساتذتي في مجال دراسة المراعي، الذين يشرفني قراءة منشوراتهم وكتبهم. وبصرف النظر عن الاختلافات الطفيفة في وجهات النظر بشأن تقييم مفاهيم معينة متعلقة بأساليب استعادة وتأهيل المراعي المتدهورة، والتي قد تنبع من الاختلافات بين الأجيال في النهج وطريقة التفكير، إلا أنني أجد نفسي في معظم الأحيان متفقاً تماماً مع أفكار وآراء جاس. أتعلم منه، وأستمتع بقراءة منشوراته، وأستفيد كثيراً من تعليقاته، وأشارك أحياناً في مناقشات مباشرة معه عبر منصات وسائل التواصل الاجتماعي، حيث يقيم هو في نصف الكرة الجنوبي في غرب أستراليا، وأنا مقيم هنا في شمال إفريقيا في برقة، شرق ليبيا.

دراسات الغطاء النباتي في ليبيا لها تاريخ طويل يعود إلى قرون مضت، مميز بالاستكشافات العلمية الهامة. أُجريت أول استكشافات علمية حديثة بواسطة العالم الأوروبي "Lemaire" في عام 1703، الذي شرع في مهمة لتتبع احتمالية استمرار وجود النبات التاريخي المنقرض "السيلفيوم" في برقة. خلال القرون التالية، استمر العديد من العلماء الأوروبيين في المساهمة في دراسات الغطاء النباتي في مناطق مختلفة من ليبيا. ومع ذلك، لم يكن حتى عام 1931 أي دراسة على المراعي في ليبيا، حين قام العالم الإيطالي "Maugini" في تلك السنة بدراسة الغطاء النباتي في أراضي المراعي في برقة، تليها دراسة أساسية للمراعي في عموم ليبيا من قبل العالم الأمريكي "Barry C. Park" في عام 1957.

خلال النصف الثاني من القرن العشرين شهدت ليبيا زيادة في الاهتمام بالمراعي نظراً لاعتبارها أحد دعائم الاقتصاد الوطني. وقد نتج عن هذا الاهتمام إجراء دراسات موسعة بالتعاون مع مؤسسات وعلماء دوليين، وفي وقت لاحق من قبل علماء ليبيين، بهدف فهم النظم الرعوية، ومحاولة المحافظة عليها وإعادة تأهيلها وتنميتها، وتعظيم الفوائد المستمدة منها. وعلى الرغم من كل الجهود التي بذلت، فإننا لا زلنا بحاجة إلى إجراء المزيد من الدراسات لتعميق فهمنا لنظم المراعي في ليبيا. تبرز هذه الحاجة جلية نظراً للتحولات الاجتماعية والاقتصادية للمجتمع على مدى العقود السبعة الماضية، نتيجة لاستكشاف وإنتاج النفط، والتي أثرت بشكل كبير على ممارسات استخدام الأراضي، ومنها مناطق المراعي. لذلك، فنحن لا نزال نحتاج لمزيداً من الدراسات، خاصة فيما يتعلق بتقييم الوضع الحالي للمناطق والنظم الرعوية، وفهم ديناميكياتها، وتحديد إمكانياتها بدقة، ومعرفة درجة تدهورها، وتطوير الأساليب والتقنيات المناسبة لإعادة تأهيلها. تعتبر معالجة هذه الجوانب أمراً حاسماً لضمان إدارة وحفظ المراعي الليبية بطريقة مستدامة، في مواجهة التغيرات البيئية والاقتصادية والاجتماعية المتسارعة خلال العقود الأخيرة.

علاوة على إسهاماته الواسعة في آسيا الوسطى والشرق الأوسط وشمال إفريقيا وأستراليا، يبرز جاس كعالم مهم في دراسة وتقييم المراعي الليبية. فبالإضافة إلى تنفيذ تقنيات الحفظ والتأهيل عبر مناطق مختلفة من البلاد، قدم جاس جهوداً كبيرة في مجالات البحث والدراسة، وطرق التعامل مع مراعيها ومناخها ومجتمعاتها المحلية. استمرت مشاركته في ليبيا على فترات متقطعة بدءاً من زيارته الأولى في عام 1966 حتى آخر زيارة له في عام 2010. وتتضمن النقاط التالية ملخصًا لأهم محطات حياته العلمية والتطبيقية المختلفة:

- عمل كعالم بالأراضي الرعوية في المعهد الوطني الجزائري للبحوث الزراعية (1971-1970) حيث تم إجراء دراسات للسهول الجزائرية وأنظمة الغطاءات النباتية الطبيعية والزراعية بها.
- قدم رسالة دكتوراه في نمذجة الغطاءات النباتية باستخدام الحاسوب في جامعة ستراسبورغ، فرنسا (1975-1971).
- عمل كعالم للأراضي الرعوية ضمن فريق منظمة الأغذية والزراعة (FAO) الذي عمل مع مركز البحوث الزراعية الليبي في طرابلس، ليبيا (1980-1975).
- عمل مع إدارة الزراعة في ولاية أستراليا الغربية (بيرث، أستراليا)، حيث تم تعريف مجموعة البذور الأصلية الليبية للنباتات البرية الرعوية وتقييمها، بشكل خاص الأنواع النباتية الحولية مثل بعض أنواع المديك (*Medicago* spp.) (1985-1980).
- شغل وظيفة عالم أبحاث في المعهد الوطني للبحوث الزراعية (INRA)، مونبلييه-فرنسا (1992-1985). حيث عمل على أنظمة الأعلاف والماشية في منطقة البحر المتوسط.
- عمل كقائد لبرنامج المراعي والأعلاف والماشية (PFLP)، في منظمة إيكاردا (ICARDA) بمقرها في حلب-سوريا (1999-1992).
- عمل مديراً للبحوث في منظمة INRA-CIRAD بمونبلييه-فرنسا (2005-2000) حيث عمل على المراعي في آسيا الوسطى.
- تقاعد سنة 2005، وعمل كاستشاري في تطوير المراعي في قطر، أبو ظبي، جورجيا، آسيا الوسطى، وشمال الصين (2005-2007).
- عاد إلى ليبيا للمساهمة في مشروع تطوير المراعي (2010-2007).

خلال مسيرته المهنية الطويلة، قدم جاس مجموعة غنية من الأعمال العلمية والمنشورات. تشمل هذه الأعمال تقريباً فصولاً كاملةً في سبع كتب مختلفة، 22 ورقة بحثية محكمة، 5 ورقات بحثية غير محكمة (نتائج أصلية)، 31 مساهمة في مؤتمرات علمية وندوات وورش عمل، تأليف كتابين، 85 تقريرًا تقنيًا مفصلًا حول المراعي الليبية، تسع تقارير تقنية حول المراعي في مناطق أخرى من العالم، مشاركة في 11 مشروعًا آخر متعلق بتطوير المراعي في عدة مناطق جافة مختلفة حول العالم، وأشراف على 21 طالب ماجستير وثلاث طلاب دكتوراه في مجال المراعي.

أما الكتاب الحالي فهو يغطي جزء كبير من الفجوة المعرفية المتعلقة بالمراعي في ليبيا، وتمتد أهميته إلى منطقة شمال إفريقيا بشكل أوسع. يتناول هذا الكتاب مجموعة متنوعة من الموضوعات المتعلقة بالمراعي مثل المناخ، التربة، الأشكال الأرضية، الغطاء النباتي، أنماط الإنتاج في مناطق المراعي، الزراعات المطرية، نظم تربية الماشية، مصادر المياه، الحياة البرية، والتخطيط الزراعي البيئي. بالإضافة إلى ذلك، يستكشف الكتاب بعض أبعاد وجوانب المجتمعات المحلية بأراضي المراعي الليبيه، بما في ذلك العلاقات الاجتماعية، والعادات والتقاليد، والتفاعل بين الإنسان والبيئة، والممارسات الثقافية، وطرق الرعي التقليدية، وأنظمة الإنتاج المحلية.

علاوة على ذلك، يقدم الكتاب نظرة شاملة تمتد لأكثر من أربعة عقود، ويدمج التجارب العملية والتطبيقية مع الملاحظات العلمية والنظرية. الكتاب يوثق أيضاً التطور الذي شهدته ممارسات إدارة المراعي والتغيرات الاجتماعية، مما يقدم رؤى حاسمة في الديناميكيات المعقدة التي تؤثر على أراضي المراعي الليبية. كما أولى المؤلفون اهتماماً خاصاً لمناقشة مشاكل التدهور والتصحر في أراضي المراعي، والأسباب الكامنة وراء ذلك، وتقديم تصور عن استراتيجيات مستدامة لإعادة تأهيل المناطق المتدهورة مصممة خصيصاً للمراعي الليبية.

مدعوماً بتحاليل بيانية قوية، ومزوداً بمجموعة كبيرة من الوسائط المرئية، بما في ذلك عدداً كبيراً من الصور والجداول والرسوم البيانية والخرائط، يقدم الكتاب سرداً عميقاً ورؤية مستفيضةً، تعكس العمل الميداني والمشاركة العلمية الواسعة للمؤلفين في ليبيا وشمال أفريقيا على مدى فترات زمنية طويلة. تعتبر الموضوعات التي يناقشها هذا الكتاب واحدة من أبرز القضايا البيئية في ليبيا والدول الجافة الأخرى خاصة المجاورة. إن طباعة ونشر هذا الكتاب لا يملأ الفجوة الكبيرة الناتجة عن نقص مثل هذه المعلومات عن المراعي في ليبيا وشمال أفريقيا فحسب، بل يعتبر أيضاً حجر زاوية للجهود البحثية المستقبلية في الحفاظ على المراعي وإعادة تأهيلها وإدارتها بشكل مستدام

إنني إذ أوجه التهاني الحارة للمؤلفين، يجدر بي تقديم التهاني أيضاً لجميع علماء المراعي في المنطقة وحول العالم على نشر هذا العمل المهم. وأعرب بشكل شخصي عن امتناني العميق لكل الأفراد والمنظمات الذين ساهموا بطريقة أو بأخرى في استكمال ونشر هذا الكتاب. ففي شمال أفريقيا، وخاصة في ليبيا، نحن محظوظون للاستفادة من الجهود المخلصة للمؤلفين، الذين استثمروا حوالي اثنتي عشرة عاماً من حياتهم في إنتاج هذا الموسوعة الشاملة حول المراعي الليبية، والذي يعتبر ثروة من المعرفة والخبرة، مما يجعله مورداً لا غنى عنه للممارسين والباحثين وأصحاب القرار، وملاك الأراضي الرعوية على حد سواء.

د. منعم وافي براني سعيد
أستاذ مشارك
مختص بإيكولوجي اللاندسكيب والمراعي
عميد الشؤون العلمية، كلية الآداب والعلوم-الأبيار، جامعة بنغازي، ليبيا
الأحد، 24 مارس، 2024

Acknowledgements

Special recognition and gratitude are due to the late François de Villaret (*Si Abderrahman*), the late Dr Henry-Noël Le Houérou, the late Dr Peter Huxley, the late Dr Clive Francis, Dr Mohamed Zehni, Dr Abubaker Maddur, Dr Ali Nefzaoui, and Dr Manam Saaed.

To the many Libyan, Algerian, Tunisian and Middle East friends and colleagues who shared their knowledge and always offered their warm hospitality: Abdamotalab Gashoot, Abdelhafiz Rojbani, Abdulhakeem Naffati, Abdullah Beit-el-Mal, Abdulrazzak Al-Sharif, Abdulsalam Albukhari, Abubaker Blahou, Adel Mahmoud, Adnan Jabril, Adnan Sbeita, Ali Agosh, Ali Amel, Ali El-Gadi, Ali El-Taiaf, Ali Gwasem, Ali Khalifa, Ali Madi, Ali Salem Ibrahim Shreidi, Ali Sultan Nezdef, Anwar Mousa, Atayari Amar Abusatta, Ayad Hamid, Azeden Akramy, Basher el-Ouashy, Dareef Omrani, Driss Haddouche, Fahim Ghassali, Faik Bahhady, Fauzi Marhei, George Arab, Hajj Al-Arabi bin Saleh, Hamdi Zurqani, Hasan Bashaga, Hedi H'diden, Hussein el-Sudani, Hüseyin Firincioglu, Khalid El Fadli, Lutfi Bakkar, Mahfouz Abu-Zanat, Mahmood Eswayah, Mohamed Abdalkarem Alashab, Mohamed AbuHadra, Mohamed Ammar Kalel, Mohamed Ansari, Mohamed Asid, Mohamed Snam, Mohsen Abolagras, Mustafa Badawi, Nabil Bathikha, Omar Jally, Rafa Essokne, Ramadan Rabah, Saddeg Tamer, Salem Essaker, Samir Badawi, Suleiman Belkheir, Tahar Telahique, Yakoub El-Barasi, Yousef Bengharsa, and the late Ali Abozeid, Ali Ben Ramadan, Ali Shilmani, Aïssa Abdelguerfi, Basher Joudah, Haj Sassi, Loutfi Boulos, Mohamed Al-Idrissi, Milad Sklehieh, Mohamed el-Fitory, Mohamed Saleh Alshhibi, Mohamed Salem el Bargathy, and many other unknown friendly Libyan farmers and shepherds.

To friends and colleagues we were working with in Libya and Western Australia: Ann Hamblin, Andrew Watson, Dragomir Dumancic, Geoffroy Collins, George Woolston, Gilbert Webb, Jeremy Wallace, John Allen, Kevin Foster, Mario d'Antuono, Mike Ewing, Norman Halse, Peter Skinner, Stephen Hill and Warren Standing.

To Scott Christiansen, Tom Nordblom and Euan Thomson for the many field trips and fruitful discussions about West Asia, North Africa and Central Asia while working together at ICARDA.

To botanists and French colleagues who shared their deep knowledge of the Mediterranean arid regions: Cyrille Chatelain, Alain Dobignard, Bernard Faye, Claude Lemmel and Édouard Le Floc'h.

We are thankful for the financial support from Guillaume Duteurtre (CIRAD-Selmet), Mrs Sandrine Dury (CIRAD, Méditerranée, Moyen-Orient et Pays des Balkans), Stephen Hill (Kimseed) and Kim Sanders (PTY LTD); and Dr David Holthouse, for the manuscript preparation before submission to CABI.

Our countless thanks to Michael Major, Padma and Joe who steadily and heroically sorted out and organized our bibliography and more.

Finally, to our fathers and mothers who taught us, from early age, to observe, learn, understand and respect all countries and their people, and, most important, to our wives and families who stoically endured our long Libyan endeavour.

1

Introduction

My Libyan Journey

Gus Gintzburger

August 1966, early morning. At the Tunisian border post at Ras-ej-Jdeir, I step into Libya for the first time, hitchhiking from Tunis, going east around the Mediterranean Sea with a detour to southern Libya, to Sebha/Ghat, and reach the *Tadrart Akakus*. At high school in France, one of my teachers, who took part in several French archaeological missions in south-western Libya in the 1950s, showed us photos and drawn colour replicas of the famous rock paintings in the Tassili and petroglyphs from Wadi Matkendush. My long-standing teenage dream! I am well prepared for this trip to Libya having done previous hitchhiking trips to southern Tunisia and southern Algeria. I remain fascinated with these regions, the history, landscapes, vegetation, wildlife and people. As a kid born in Tunisia, my French father, a photographer by trade who did the WWII Tunisian campaign, amateur botanist, geologist and herpetologist, used to take me with him on long field trips to central and southern Tunisia. I learnt a lot from him, to observe, listen, memorize and respect.

For this new and long Libyan trip from Tunis to Egypt and further, I carry a light backpack and a blanket. Soon, a Libyan teacher driving his brand-new Peugeot 403 gave me a lift to Zuara. After waiting for a few hours along the sabkha, I get another ride to Tripoli, reaching it by 6 pm. I spend my first night sleeping wrapped in my blanket on the hard pavement in front of the small Georgimpopoli police station. The police staff kindly shared a few cups of sweet tea and a watermelon. Early next morning, after a quick breakfast of black tea with Carnation milk and bread, I was on my way, walking through Tripoli centre towards Sidi-el-Mesri and Tajoura.

That was the beginning of my discovery of Libya – slowly hitchhiking along the coast – Homs, Misurata, Taworga, Bu-Gren – I soon realized that going south to Sebha/Ghat would be extenuating with the summer heat. I then decide to go further east towards Cyrenaica. And here comes the steppe – salty glaring flats along the narrow road dipping into Sirte, Nofilia and Marble Arch (I slept in it).

On the road again. Hot, agonizing scorching weather – in the 40s – waiting for a welcoming driver to carry me further; patiently eating hot tinned tuna fish and Tunisian *Shakshuka* with dry bread, waiting for the next hike under the glaring and searing sun. Now in the bin of a police truck with workers sharing their rock-hard biscuits and with sand under my teeth, gulping tepid water from my aluminium water bottle, burnt by the sun and the wind, covered with yellow dust. Ajdabiya, Soluk, Benghazi. At last,

DOI: 10.1079/9781800627154.0001

some green landscape and this red soil – Tokra – on the seashore, where I slept in a cave crawling with spiders. No *shakshuka* nor bread nor milk left, I collect some periwinkles and crabs, boil them in seawater and grudgingly eat them. Salty, salty! Then El-Marj – still with many collapsed buildings from the 1963 earthquake. Beida, Shahat, Cyrene. I visit the museum and the antique ruins. On, on! Derna, Tamimi, Tobruk – roasted for hours on the bare rocky plateau at the gate of the Tobruk WWII French military cemetery, sleeping on the roadside. At 4.30 am, a friendly police truck stops, and I jump in the bin. I am now used to it. Rocks as far as my reddened eyes can see – bare desert, hot, hot days! Bardia, Musaid, and reaching the border post at Sollum at night. Sollum is not a good place to spend the night in 1966, this lonely foreigner having to suffer a thorough search of my poor belongings and multiple and insistent requests for *baksheesh* from the local authorities! I go on – Egypt to Luxor – flying with great difficulty to Lebanon, hitchhiking my way back to France through Syria, Turkey, Soviet Bulgaria, Romania and Hungary, Austria and Germany. Finally, back to France after two gruelling months on the road, and happy and ready to do it again!

After two years at Strasbourg University where I did a Master's degree in plant biology, I went, in 1970, as a young scientist working with the National Algerian Agricultural Research Institute, collecting herbarium specimens and ecological data on the high plateau in Ouled Nail country (Tadmit – Djelfa – Laghouat) with Fr François De Villaret. While working there I meet with Dr Henry-Noël Le Houérou and join him during extraordinary field trips to the Hodna, learning from him, in a few days, more than I learned during all my time at university. Back to France, I compile my PhD in Strasbourg (1975) in plant computer modelling – far away from the steppe? Not much, as Henry-Noël unexpectedly called me from Rome offering me a position with FAO in Tripoli to work with the Libyan Agricultural Research Centre (Sidi-el-Mesri) as a young rangeland expert. No hesitation, despite having gained a solid researcher's position with the French National Research Centre (CNRS).

On, on! To Tripoli again in August 1975, driving past, nine years later, the Georgimpopoli police station and being now stationed at the ARC (Sidi El-Mesri). Soon equipped with a Toyota Land Cruiser, I start criss-crossing and camping out in the Jeffara plain, the Jebel Nefusa, the Hamada, deep into Sirte area, the Jebel Al-Akhdar to Tobruk-Bardia with ARC Engineer M. Alidrissi, technicians H. Hidden and H. El-Sudani. We collect rangeland plants, develop a herbarium and a pastoretum at Tajura ARC farm, map vegetation, and organize and run an international collection of forage seeds from arid rangeland vegetation. This was possible with the unfailing support from Dr M. Zehni (ARC Director) and Dr P. Huxley (FAO TF LIB 6 project leader). I exhausted two Toyotas in five years! We collected over 3000 native seed samples with ecological site data that became the LMEP (Libyan Medic Evaluation Project).

After my time with FAO, I was contracted (1980–85) by the WA Department of Agriculture (Perth, Australia) to evaluate this Libyan collection, especially annual medics (*Medicago* sp.) with Dr C. Francis and with a dedicated team of Libyan agricultural engineers – A. Sbeita, M. Fituri, A. Rojbani, and P. Skinner, a highly skilled Australian technician agronomist. During this time, I set up the Libyan Medic Computer Data Base (LMDB) covering the detailed ecological descriptors and agronomic evaluation of some 2800 ecotypes, starting seed production of the most productive and early-flowering ecotypes in Australia and in western Libya. In three years of incessant work with my Libyan team, we shuttled between Australian and Libyan sites taking advantage of opposed seasons in the southern and northern hemisphere.

From there, I left Australia and returned to France, taking a research position with INRA France in Montpellier (1985–1992). Boring work and a grim time! I left INRA in mid-1992 for an ICARDA position as Program Leader PFLP (pasture, forage and livestock) based in Aleppo (1992–1999). Fantastic opportunities! I developed collaborative work with national programmes on rangelands from Morocco to Central Asia/north-west Pakistan. I learned lots about Central Asian rangelands – another teenage dream come true!

Back to France (2000–2005) at INRA-CIRAD Montpellier as Research Director, concentrating on Uzbekistan. I authored a book, *Rangelands of the Arid and Semi-Arid Zones in Uzbekistan*, published in 2003. I travelled and worked extensively on rangelands in Turkmenistan, Kazakhstan, Tajikistan, and trekked in northern China (Taklamakan to Beijing).

Mid-2005: retired to Western Australia, did rangeland consultancies in Qatar and Georgia, and again in Central Asia for Fauna and Flora International. I was again contracted by the WA Department of Agriculture to work episodically with the Rangeland Development Project (Suani-Libyan Ministry of Agriculture) from 2007 to the end of 2010. I was pleased to reconnect with Libya and friends there from west to east. Due to the war, the project stopped.

Back home in Western Australia, I watched in horror the events, keeping as much as possible in touch with friends in Tripoli and Benghazi and still hopeful that I could go back there. 'No way,' my Libyan friends told me repeatedly.

Often reviewing my thousands of Libyan photos, recalling sites I visited, a project slowly germinated to write a book about Libyan rangelands as a token of gratitude to the many Libyan friends and colleagues who, over so many years, came with me on bumpy, dusty, never-ending tracks, sweated or shivered when camping in remote places, cooked so many *am'bakbaka* and kilos of *macarona* with sandy *terfes*, shared the barely drinkable salty water of forgotten wells and litres of sweet tea flavoured with *chengoura* or *zaater*, ate much sand and bread with the matchless tinned tuna fish with fresh onions, sweated to free our 4x4, bogged down in soft sand, jolted on endless rocky tracks, and got immobilized for days during a sandstorm in Soluq.

This book is for the many Libyan shepherds and farmers, the ARC and RDP colleagues who offered their simple and warm hospitality, taught me the names of the places, of plants and wildlife, took me to strange lost antique sites, shared some of their deep knowledge and history of their fascinating and rough country.

To all of them, greetings and deep respect!

2

The Libyan Rangelands under Siege

Rangelands are all land used by free-roaming livestock or under the guidance of shepherds. Rangelands are usually too arid, too rocky or too sandy to offer any prospect of reliable and sustainable cultivation. They are located between the sown and the desert.

The sown are rainfed or irrigated cropped land where livestock do not graze freely or wander around, simply to avoid damaging crops or fruit trees belonging to farmers. Crop residues may be accessed by livestock during a short period of the year, usually after harvest. This is usually during summer and under agreement or contract between farmers and livestock owners.

The desert is mostly devoid of vegetation and water resources. It is commonly accepted that it lies below the 100 mm mean annual rainfall (MAR) isohyet (Le Houérou, 1959, 1969). It offers little to shepherds and their livestock, except during short periods in winter and early spring when annual vegetation sprouts.

In between the sown and the desert, the rangelands are a vast treeless expanse usually called the steppe or *Al-Suhub* in North Africa and further east. Though receiving sparse rainfall, allowing a meagre cover of native perennial and annual plants, the steppe cannot be cropped on a regular basis due to inter- and intra-annual rainfall variability, poor soils, rocky outcrops, saline depressions, or deep sand dunes. This is the livestock grazing domain on land officially controlled by the government, though more often, practically managed by local tribes and communities, and custodians, according to ancestral laws.

In Libya, rangelands cover most areas from west to east receiving a mean annual rainfall (MAR) between 50 mm and 500 mm a year, even with forests and matorral in the highest rainfall zones in the east. Little cropping occurs below the 200 mm MAR isohyet with few exceptions in large wadi and non-saline depressions receiving substantial run-off (wadi Sof-el-Jin, wadi Mimoun, wadi Sassu, and tributaries, south of the Jebel Al-Akhdar, etc.) where sporadic rainfed barley crops are attempted despite facing many failures. Widely spaced large rainfed olive tree plantations are common above the 300 mm MAR extending into the 200 mm MAR isohyet (The Dahar, Tarhuna, Msalata, Khoms to Misurata) on sandy-loam soils. In lower rainfall, down to 150 mm MAR, farmers from the Jebel Nafusa, from Nalut to Yeffren, practise an ancestral way to grow individual olive and other fruit trees on terraces in small water catchments. In even lower rainfall areas in the west, olive, figs and almond groves colonize the wadi beds as at wadi Ghirza, collecting large amounts of water from vast barren surrounding watersheds.

Above 300–400 mm, in the east (Jebel Al-Akhdar), cropping and fruit trees dot and expand onto matorral, garrigues and forested

DOI: 10.1079/9781800627154.0002

areas, slowly but irremediably encroached on and nibbled away by ever-escalating urbanization, quarries, roads, tracks and charcoal manufacturing.

Over the years, and since the mid-1960s, due to population and livestock numbers steadily escalating, modern ways of cultivation and transportation have become increasingly available; the pressure being on the best rangelands first and then on more marginal steppe, exploding all over northern Libya, with risky, rainfed barley cropping creeping into lower-rainfall zones. The availability of irrigation from deep wells and artesian wells leads to more irrigated cropping and fruit tree projects planted on new land, even in the lowest rainfall zones such as wadi Sof-el-Jin, wadi Mimoun, wadi Sassu, Bey-el-Kebir, etc. Furthermore, water and supplementary feed trucked to flocks became the norm all year long and all over the most remote rangelands, once only seasonally grazing native vegetation. The pressure and ensuing rangeland degradation and desertification affected mostly the 100–500 mm MAR zones (Fig. 2.1), not only in Libya but also all over North Africa to the Middle East and further east (Gintzburger, 1996).

After a few years of extraordinary and expensive rangeland and agricultural project developments (1975–1990), Libyan rangelands, from the margin of the desert to the forested highest rainfall zones in the east, are now under siege, slowly and steadily being depleted, continuously used and abused with little sustainable and necessary action to control, stop or reverse this dramatic degradation trend recognized by all.

This book is a contribution to depicting the Libyan rangeland conditions and evolution witnessed by the authors from 1975 to 2010.

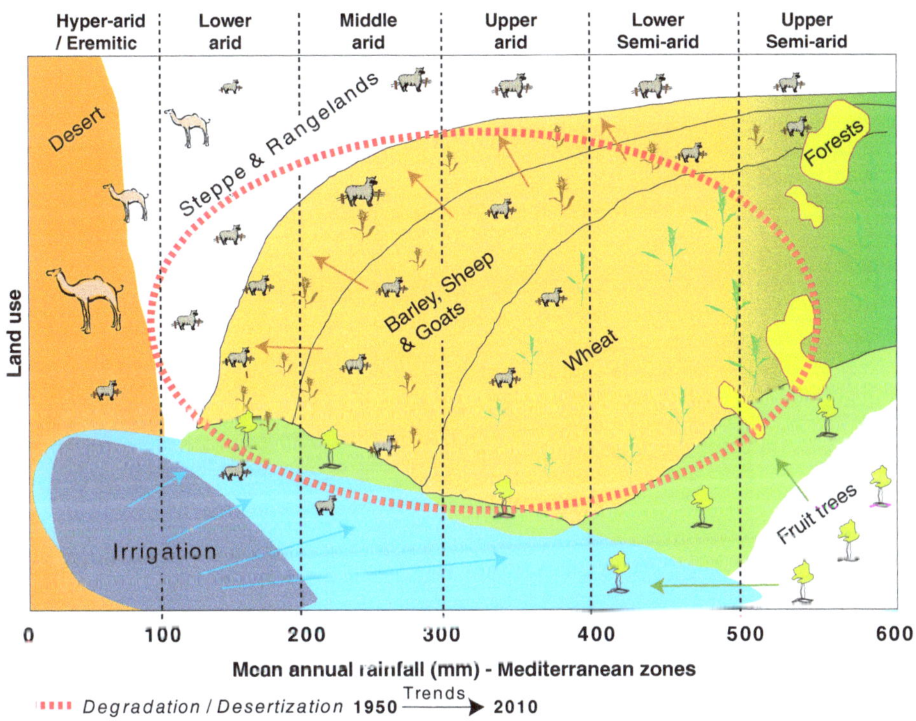

Fig. 2.1. Land-use trends in west Asia and North Africa. (From Gintzburger, 1996)

References

Gintzburger, G. (1996) *PFLP Annual Report.* International Center for Agricultural Research in the Dry Areas, Aleppo.

Le Houérou, H.-N. (1959) *Recherches écologiques et floristiques sur la végétation de la Tunisie méridionale.* Vols 1-2, N° 6, 510 p. + maps and annexes. PhD dissertation, Université Alger Institut de Recherches Sahariennes, Algiers.

Le Houérou, H.-N. (1969) *La Végétation de la Tunisie Steppique (Avec Référence aux Végétations Analogues d'Algérie, de Libye et du Maroc)*, Vols 1-2, N° 42/5, 622 p. + maps and annexes. Annales de l'Institut National de la Recherche Agronomique de Tunisie, Tunis.

3

Climate, Bioclimates and Droughts

Abstract

After summarizing the seasons perceived by Libyan herdsmen, we statistically and spatially analysed the Libyan climatic data available. They show differences in Mediterranean climatic conditions, especially in terms of spring rainfalls between the north-west related to the Maghreb and the north-east attached to the Mashreq, separated by the Sirte region where the Sahara reaches the Mediterranean Sea. Reviewing the climatic factors impacting the native vegetation and rainfed cropping, we develop a bioclimatic synthesis based upon the mean annual rainfall controlling the native plant distribution and crops, and the mean daily minimum temperatures of the coldest month limiting their winter growth. Statistical analysis from annual rainfall data from 1945 to 2010 differentiates wet and dry years, drought length and trends at 14 northern meteorological stations. It indicates that current climate change may not affect all stations. Paleoclimatology and archaeology provide proof that the region was not immune to long-past drastic climatic oscillations.

Introduction

Climate is the description and analysis of long-term meteorological data (rainfall, temperatures, solar radiation, wind, evapotranspiration etc.). Bioclimates define the thresholds of climatic data and variability on living organisms, and how climatic data impact the native vegetation distribution and agricultural activities.

The Climate and Seasons Perceived and Memorized by Herders, Farmers and Desert Dwellers

Long before recording climatological data, local populations observed, named and defined the seasonal events that rule their survival and life. In arid Australia, detailed seasonal calendars have long been used by aboriginal tribes in their own local languages. They help them to characterize the evolution of the vegetation, and the time of availability of food and material essential to their daily life in extremely difficult arid and tropical environments (https://www.csiro.au/en/research/indigenous-science/Indigenous-knowledge/Calendars (accessed 8/8/23)).

Unsurprisingly, similar seasonal calendars were developed, memorized and shared for the same practical reasons by native populations, herders, farmers and desert dwellers living on arid lands. They have long learned to distinguish climatic events and periods that are distinct from the conventional Gregorian calendar divided into four seasons. Their seasonal calendar is functional and guides year-long agricultural activities. It brings useful recurring climatic information

DOI: 10.1079/9781800627154.0003

collected over centuries. An example is the Ghilan's agricultural calendar transmitted by shepherds and farmers in southern (Medenine) and central (Sidi Bouzid) Tunisia, into the Jeffara plain and possibly further east. With information collected from western cameleers, shepherds and farmers, and with assistance from Engineer D. Amrani and Dr A. Maddur, we decided to reveal it before it is forgotten.

The Ghilan's pastoral calendar

In North Africa and further east, Arab and Berber herdsmen roaming the steppe and arid rangelands have developed from experience their own practical and seasonal calendar. It describes the expected weather and impact on humans, animals and vegetation during the year, especially in central Tunisia (Sbeitla, Sidi Bouzid) and southern Tunisia (Medenine, Douz, Tataouine, Ksar Ghilane). It is known as Ghilan's calendar. The legend reports that Ghilan was a cameleer from the *Awlad Ghaylan* and possibly the Durayd tribes from northern Tunisia. Ghilan was leading his flock between Sbeitla and Medenine according to the seasons.

The main periods of Ghilan's pastoral calendar

Ghilan's calendar names and describes the weather during specific periods of the year observed by herdsmen living on rangelands. It names and dates the successive periods in dialectal Arabic. It is transmitted by oral tradition and extends to the Berber population from North Africa with Amazigh names. The calendar starts early in winter when the sun gets to its lowest daily elevation in the sky and the days are the shortest in the northern hemisphere. This is the time of the winter solstice, 21 December in the Gregorian calendar. However, for them, in southern Tunisia and Libya, the winter starts at the end of November.

Beginning of winter: 29 November

***Illayali ilbidh* = the white nights: 25 December–13 January**
It refers to clear skies without clouds during long nights, hence a very cold period with cold nights with possible frost and morning dew. Day temperatures remain at the lowest. Farmers know that no seeding must be performed during this period as the soil is too cold preventing germination, and irrigating crops or fruit trees is useless as vegetation is dormant.

***Ilyali issoud* = the black nights: 14 January–2 February**
This is a period when heavy clouds accumulate. The weather is still cold during the day, but nights are milder.
Then starts a period called *Al-Azara* when winter and spring are fighting.

***Guerit il azara* = 'shivering of bachelors': 3–13 February**
The weather during this period is unstable like, they say, young bachelors' moods. It may change from one extreme to another but remains cold and possibly wet.

***Guerit il anz* = 'goats shivering': 14–19 February**
During this usually wet and windy period, the livestock suffers much. The goats, which are very vulnerable to cold, wet and wind, are particularly affected, with kids and lambs needing protection from harsh and wet weather and kept close to the tent in a *zeriba* or even inside the tent for the youngest ones.

***Jamrit il houa* = 'ember of air': 20–26 February**
During this period, the cold air fades away gradually and days and nights become milder.

***Jamrit il ma'a* = 'ember of water': 27 February**
This period lasts for one day when the sea and wadi waters are warming up.

Beginning of spring: 27 February

After *Al-Azara* is *dabashk*, when the spring is taking over from the cold, with vegetation growing vigorously, rangelands flowering and crops growing fast. This is the time of *ashaba* or *eushba* when plenty of green annual forage is available on rangelands.

***Jamrit ittrab* = 'ember of soil': 6–10 March**
This period comes immediately after the beginning of the spring season, when the soil starts to warm up and spring rains may occur. The spring rains secure abundant feed on pastures and rich crops. Farmers consider March rains as gold. By this time, herdsmen put away their *wazra* and winter cloth and may spend nights out with their

flocks, away from the camp, sleeping rough, and benefiting from night grazing.

***Ilhoussoum* (no meaningful translation): 10–17 March**

This period is characterized by windy days facilitating crops, trees and native vegetation pollination. The name of this period apparently refers to *Al-Hassum*, an event mentioned in the Holy Qur'an in the verses 6 and 7 of Surat *Al-Haqqah*. It recounts: 'And as for 'Âd [A biblical tribe], they were destroyed by a furious, bitter wind which Allah unleashed on them non-stop for seven nights and eight days, so that you would have seen its people lying dead like trunks of uprooted palm trees'.

By mid-April, they often refer to *Qara Hayyan*, which is the period when livestock and new-born are secured and will prosper. This is also the time for starting cereal harvesting.

Beginning of summer: 30 May; summer solstice: 21 June

***Oussou or Osso*: 25 July–2 September**

This is a summer period which is characterized by very high temperatures. People from Zuara call *aoussou* the very early summer when the weather is still pleasant and when early morning swimming in calm sea is popular. After July, grain harvest is too late as grain will shed easily. Livestock move on to and graze cereal crop residues and dry plants (*guech*) on rangelands while they slowly move to perennial pastures on the hills and Jebel as the summer proceeds.

Beginning of autumn: 30 August; autumn equinox: 22 September

Farmers start ploughing after the first autumn rain by mid-October and start seeding cereals when there is enough water in the soil and the soil is still warm. Autumn rains may be coming late. Cereal seeding past mid-November into December is ill-advised as cold weather may occur, with soil getting colder, preventing germination and early growth.

Ghilan's pastoral calendar is unique

While it may not be recognized in eastern Libya, it synthesizes centuries of skilled and practical observations of climatological events in the arid zones of southern Tunisia and western Libya. Even with modern powerful computer tools, it would be difficult to reach the same accuracy as it would necessitate analysing huge amounts of climatic data over centuries, which we do not have and never will.

To further explore the Libyan climate, we have to rely on recorded meteorological data over approximately the past century.

Climatic Data Availability and Limitations

Libya is an arid country with about 93–95% of the national territory strictly desert, receiving less than 100 mm/year. This is south of a line running approximately between Nalut and Ghadames in the west to Bu Green – south Sirtc, about 30–40 km inland from the coast and extending to Ajdabiya – Mekhili and Bardiayah-Musaid in the east.

We concentrate our climatic study north of the above-mentioned line receiving a mean average winter rainfall close to or above the 50 mm isohyet where rangelands are dominant agricultural activities. The Mediterranean climate rules the rangeland vegetation cycle and agricultural activities (Fig. 3.1).

We will not consider the desert dotted with oases and modern irrigation farming projects that makes little overall contribution to rangelands.

The most comprehensive past study of the Libyan climate and bioclimates was published by Le Houérou (1984). We will refer to it in this chapter as he had access to meteorological data from the Italian period and later, up to 1980, while he was working with FAO at Sidi-el-Mesri (ARC – Tripoli). PhDs and papers were later published by Libyan colleagues (Ageena *et al.*, 2013a, 2013b; Elfadli, 2009) (Albukhari, Benghazi, 2022, personal communication).

Data availability

Climatic data are collected and currently stored by the Libyan National Meteorological Centre

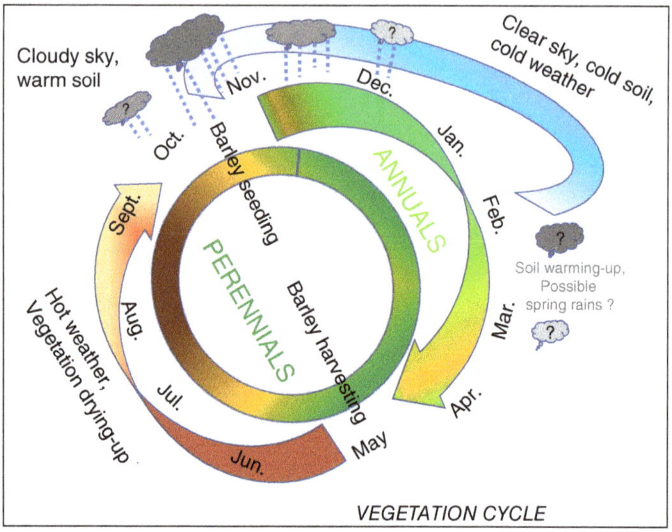

Fig. 3.1. Rangeland vegetation cycle in relation to Mediterranean climatic conditions.

(LNMC). Elfadli (2009) reports that precipitation records started in Tripoli and Benghazi by the Italian and French consulates as early as 1879–81 (Fantoli, 1952, 1967, 1968). The meteorological network expanded during the Italian occupation including 66 complete stations and 214 sub-stations monitored by police or military. By the mid-1960s, the meteorological network operated 45 full stations from the old network and 255 rain gauges located at police stations all over the country.

Most climatic data were consigned on paper. The LNMC embarked on digitizing all data. By 2009 (Elfadli, 2009), monthly temperature data from 1915–1949 were partly digitized while rainfall data for 66 stations were not yet completed for the same period. Daily and monthly rainfall data from 1956 to 2008, and the longest rainfall monthly data series (since 1879–81) from Tripoli and Benghazi, were completed, too, but unfortunately not available to us.

By 2000, the LNMC supported by the World Meteorological Organisation (WMO of the UN) upgraded its meteorological network with 19 automatic weather stations (AWS) receiving weather parameters in real time at the climate department in Tripoli. However, most new AWSs are located along the coast or deep into the desert with the exception of Nalut, Yeffren and one on the southern Hamada airfield (N 29.30 E 12.57). All LNMC meteorological data are now managed under CLISys (Climate Data Management System), a new climatic data management and application system using Oracle Database.

We accessed Libyan meteorological data series from earlier publications (Le Houérou, 1984) and some from Libyan colleagues. These data may vary slightly from one to another in terms of calculated MAR and temperatures according to the recorded period considered. It does not affect the general Libyan climatic and bioclimatic pictures. Rainfall and other climatic data recorded at police stations on remote stations on rangelands often appear to be incomplete and sometimes doubtful.

Limitations in meteorological data collection, validity and interpretations

While data were collected on long time series on standard meteorological stations, some series are incomplete due to equipment breaking down or being replaced, local unrest and wars. Station equipment evolved over years and replacement or upgrading occurred. We do not have records of these events and what equipment was either changed or upgraded at different stations. Furthermore, over the years, several stations or rain gauges once installed on open space according to international meteorological specifications have seen their surroundings changing with urban areas fast expanding, new buildings encroaching or added close by, and trees planted close to the station. As an example, the Al-Assah standard wooden weather station was established in the early 1970s and operated in an open space about 50 m west of a small police two-room building. Over the years, this police station (N 32.83 E 11.62) expanded, with new buildings added and *Eucalyptus* trees planted around. By 2007, five metre-high gargaresh stone walls were enclosing the police radio tower and the weather station. No doubt temperatures and rainfall data recorded at Assah for several years may have become progressively irrelevant to the local environment. Similar microclimate alterations may have happened with other weather stations rendering some data interpretation doubtful. In our opinion, extreme caution regarding temperature and rainfall recording must be considered past 1980.

Wind

Precipitations are brought by autumn/winter winds coming from the north-west, blowing over the sea. They bring rainfall to Libya, first along the coast and then inland to the Jebel Nafusa and the Jebel Al-Akhdar.

The wind from the sea is called *bahri* and is mostly prevalent on the coast in the west (Zuara, Tripoli, Misurata) and Sirte to Ajdabiya, barely penetrating inland. While it is stronger during winter months (6–10 km/h), the average annual wind speed ranges from 7 km/h to 9 km/h on the western coast. Inland, in the steppe zone of the Jeffara plain (Aziziya), the Jebel Nafusa and on northern Hamada (Nalut, Giado, Zintan, Yefren, Tarhuna, Mizda, Bani Walid to Ghadames), the average annual wind speed climbs to 12–17 km/h.

In the east, near Benghazi, Benina has a low average annual wind speed (3.7 km/h), a suitable place for an airport. Further east, climbing onto the plateau, from Soluq to Al Abyar, El Marj, Beida, the wind is a common feature blowing at an average annual speed around 18 km/h. Shahat benefits from a lower average annual wind speed of about 13 km/h but with high seasonal variations from 7 km/h on average in July/August to nearly an average 22 km/h in December.

Remarkable is the *ghibli*, a powerful hot, dry, dusty desert wind blowing from the Sahara in spring/early summer. It usually blows strongly for a couple of days with hurricane wind speed reaching 80–100 km/h, lifting large amounts of dust and abrasive sand particles. During *ghibli*, the air temperatures rise swiftly by 15–25°C, reaching 40°C and above in a few hours, with air humidity dropping to 5–10%, devastating rainfed cereal crops in a few hours, destroying flowering, grain filling or fruit formation. Crops affected by *ghibli* are usually lost. In nearby Tunisia, this hot desert wind is named *sirocco* while in Egypt it is referred to as *khamsin*, reputedly sometimes, but rarely, blowing for up to 50 days. These hot, windy episodes are extremely oppressive and inconvenience humans and livestock.

Severe dust storms and sandstorms reached the coastal zones, as was the case in January 2009 at Soluq, Gumerah region (Photo 3.1), obscuring daylight and making navigation difficult, with visibility dropping to a few metres. This dust storm lasted five days with wind gusts reaching 80–100 km/h. Another strong dust storm episode affected all of northern Libya in April 2022.

Precipitation

Mean annual rainfall (MAR)

The highest long-term mean annual rainfall (500–550 mm) is recorded in the eastern regions of Libya on the uppermost elevation of the Jebel Al-Akhdar (Shahat – Beida). In the west, the top of the Jebel Nafusa (Gharyan to Jendouba) receives up to 300–350 mm.

Valuable rangelands benefit from MAR close to or above 100 mm, which is the upper limit of the Sahara desert mostly devoid of vegetation. Stations in the Sahara (Sebha, Murzuk, Ghat, Zella, Awjilah, Jarabub, Kufra etc.) receive little or no rain most years.

Photo 3.1. Dust storm at Soluq, Gumerah region (January 2009, 11.50 am).

A crude rainfall/climatic map of north-western Libya was published by Fantoli (1939) and later used by the British administration for the development of Tripolitania (Rowland, 1945). Le Houérou (1984) hand-drew the isohyet map of northern Libya from MAR collected from 57 stations over periods 10 to 101 years up to 1980. His isohyet map was based on his deep field experience and knowledge of plant ecology in relation to rainfall (Le Houérou, 1959a, 1969). He clearly identified the arid 150 mm MAR isohyet extending from the Tunisian border to Giosh, Bir Ayyad – Bir-el-Ghanem to Aziziya in the eastern Jeffara plain.

Isohyet map of northern Libya

Le Houérou's rainfall map remains the reference confirmed with minor refinements by using state-of-the-art GIS geospatial methods with most recent rainfall records available up to 2010. To this effect, we used the ANUSPLIN software (Australian National University, 2013) and the climate databases compiled by the Global Historical Climatology Network (GHCN), FAO and WMO (Fick and Hijmans, 2017). ANUSPLIN interpolates on a 1 km^2 resolution grid, noisy multivariate data, in our case, the average monthly rainfall data, latitude, longitude and elevation as independent variables and distance to the coast per station. The updated rainfall map of northern Libya is presented (Map 3.1).

Annual and seasonal rainfall variability

The total annual and seasonal rainfall variability are key factors affecting plant growth on rangeland and rainfed cropping systems.

Annual rainfall variability

MAR variability is expressed by the MAR coefficient of variation. It is the ratio of the MAR standard deviation of a long series of annual rainfall records by the calculated average long-term MAR at each station. It shows that the MAR variability of a station is inversely proportional to its MAR (Table 3.1; Fig. 3.2).

It confirms the relationship between MAR and MAR coefficient of variation (Le Houérou, 1959a, 1969, 1984) also valid for Morocco, Algeria and Tunisia (Table 3.2).

This is also confirmed with the ratio MAR MAX / MAR min recorded that is greater than

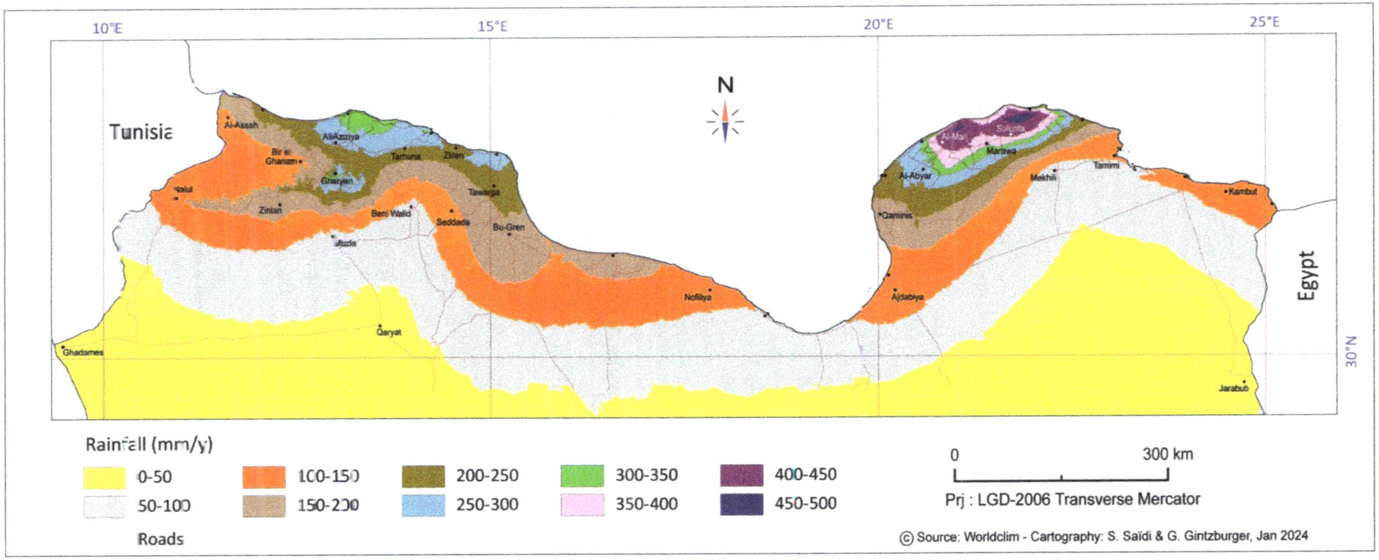

Map 3.1. Isohyet map of the northern part of Libya.

Table 3.1. Stations' MAR 1945-2010. (Raw rainfall data provided by Albukhari, Benghazi, 2022, personal communication)

Station (recorded period)	MAR (mm)	MAR STDev (mm)	MAR coef. of variation (%)	MAR min	MAR MAX	MAR MAX / MAR min
Kufra (1946–2010)	2	3	182	0.1	14.0	140
Jarabub (1949–2009)	18	20	108	0.2	90.0	450
Hon (1948–2010)	31	28	90	0.3	161.0	537
Jalo (1949–2010)	10	11	111	0,1	50	503
Ghadames (1945–2009)	34	31	91	0.4	204.0	510
Gharyat (1968–2009)	55	38	68	8.4	151.0	18
Mizda (1980–2005)	63	42	67	0.7	141.0	201
Nalut (1945–2010)	148	81	56	37.0	569.0	15
Ajdabiya (1949–2010)	149	62	42	25.0	351.0	14
Tobruk (1985–2009)	174	54	31	71.0	265.0	4
Sirte (1946–2010)	185	83	45	6.7	429.0	64
Zuara (1945–2010)	226	89	39	40.0	453.0	11
Surman (1975–2009)	257	92	36	111.0	454.0	4
Derna (1946–2010)	264	80	30	124.0	479.0	4
Benina (1945–2010)	265	82	31	105.0	464.0	4
Yefren (1983–2010)	265	92	35	87.0	491.0	6
Zawia (1957–2009)	265	107	41	68.0	573.0	8
Misurata (1945–2010)	270	81	30	77.1	462.0	6
Tripoli (1945–2010)	276	96	35	49.0	567.0	12
Khoms (1991–2007)	285	104	36	111.0	518.0	5
Gharyan (1955–2009)	348	142	41	106.0	755.0	7
Beida (1979–2010)	546	118	22	312.0	827.0	3
Shahat (1946–2010)	555	127	23	283.0	963.0	3

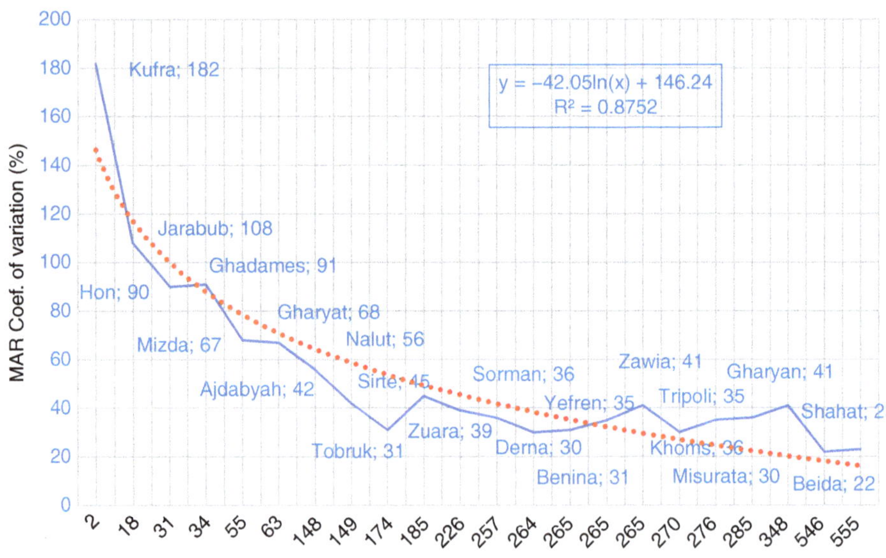

Fig. 3.2. Relationship between MAR and MAR coefficient of variation from 22 Libyan stations.

100 in the Saharan stations below 100 mm, quite high (10–15) on the coastal zone (Zuara, Tripoli, Misurata, Sirte, Ajdabiya) and dropping to 10–3 inland in the 250–550 mm MAR.

Seasonal rainfall variability

We do not have enough detailed raw rainfall data to evaluate the seasonal rainfall variability. It is anticipated that it is in the same order as the MAR variability. However, it appears that localized events of heavy spring rainfalls are quite erratic with catastrophic spring wadi flows common in the Jebel Nafusa (May 1976, Tiji-Nalut, 1996, near Bir Tutta, etc). These unpredictable spring rainfalls determine the success of rainfed cropping and good years for *terfas* harvesting.

Seasonal rainfall distribution

Precipitations in Mediterranean regions occur mostly during autumn (mid-Oct., Nov., Dec.) and winter (Jan., Feb., early Mar.). The opening rains start from mid-October to mid-November and extend into January to mid-March, allowing plant growth if those daily temperatures are mild. For crops and rangelands, the difference occurs with anticipated late spring rains (mid-March, April, May and early June) granting seed filling. The summer (sometimes starting as early as mid-May), extending through June, July, August and part of September, is dry as in all Mediterranean climates.

For the sake of simplification, and due to the lack of available data, we present the overall percentage of MAR seasonal distribution on 26 stations (Table 3.3) from northern Libya, from the Tunisian to the Egyptian borders.

It confirms that most precipitations occur in Libya during the autumn/winter months, followed by little spring rains in the west. Looking at Figs 3.3. and 3.4 by stations arranged from west to east along a longitude gradient, the west receives 76–90% autumn and winter rainfall and some spring rains, while the east receives most of its precipitation in autumn/winter (90–95%) and little or no spring rain. There are variations due to elevation and distance to the sea. As there are no summer precipitations, the difference between the western and eastern regions is that the former benefits from more frequent spring precipitations (+20–6% of MAR) while the east receives less spring rain (8.3–3.9% of MAR) (Fig. 3.4). These western spring precipitations are important for appropriately finishing seed filling, despite overall western lower MAR than eastern MAR.

It illustrates the climatic differences between the west of Libya receiving winter/autumn and spring precipitations as in Morocco, Algeria and Tunisia (Le Goff, 1985; Le Houérou, 1984), and eastern Libya that receives mostly autumn/winter precipitations with negligible spring rainfall, as is the case in the Middle East and west Asia (Le Houérou, 2004, 2005a, b). The two distinct Libyan rainfall zones are separated by the Sirte region where the Sahara reaches the coastline between Bu-Green and Ajdabiya.

Table 3.2. MAR and MAR coefficient of variation in North Africa. (From Le Houérou, 1984)

MAR (mm)	MAR coef. of variation
600–400	20–30%
400–200	30–50%
200–100	40–70%
100–50	60–100%
< 50	100% and more

Other Precipitatons

Dew and fog

Dew as water droplets is common on crops and fruit trees in the evening and early morning in the El-Marj plain. Experiments were conducted

Table 3.3. Overall percentage of MAR seasonal distribution on 26 stations from northern Libya

	Autumn	Winter	Spring	Summer
Average	42.5	44.8	8.5	4.1
STDev	5.0	5.1	4.3	2.3
Seasonal rainfall coef. of variation	11.8	11.5	50.9	54.9

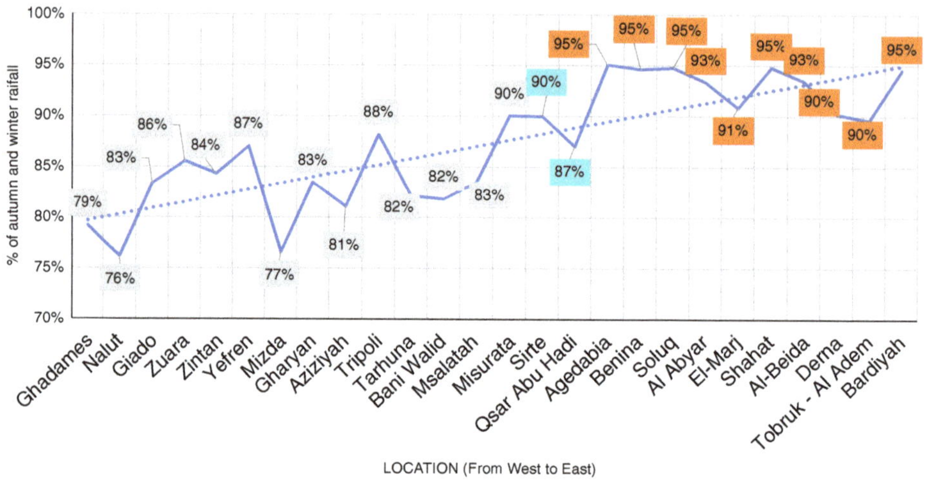

Fig. 3.3. Percentage MAR of autumn and winter rainfall in northern Libya from west to east along a longitude gradient.

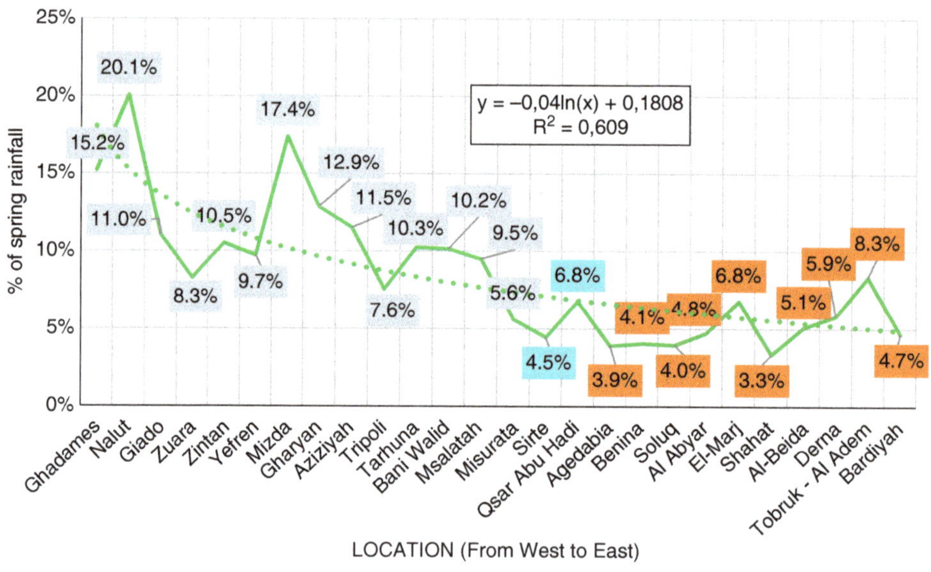

Fig. 3.4. Percentage MAR of spring rainfall in northern Libya from west to east along a longitude gradient.

by the FAO-ARC team at the ARC station El-Marj in the mid-1970s to harvest dew with different implements to supplement rainfall for fruit trees. Fog is usual in autumn/winter on the highest part of the Jebel Al-Akhdar. It allows the vigorous development of lichens (*Ramalina* sp. and *Teloschistes* sp.) on *Juniperus phoenicea*.

Snow

Snowfalls are uncommon in Libya during winter. They occur only on high elevations from the Jebel Al-Akhdar (as in the winters of 2008 and 2022) and exceptionally on the Jebel Nafusa (as in the winter of 1955/56).

Potential Evapotranspiration

Annual potential evapotranspiration (ETP^A)

The annual potential evapotranspiration is computed using the relation between the MAR and (t) the mean annual temperature (Le Houérou, 2005a, b, Le Houérou *et al.*, 1975; UNESCO, 1977). They are presented in Fig. 3.5 (Le Houérou, Montpellier, 2004, personal communication). Quite logically, the milder coastal stations have the lowest annual evapotranspiration with 1100–1300 mm/year, increasing to 1500–1800 mm/year in the dry steppe, and up to 2500 mm in the desert. While these values could be useful for managing irrigated lands, they are of not much use for rangelands vegetation or rainfed crops as they are all far above the MAR.

Seasonal potential evapotranspiration (ETP^S)

As the rangelands and rainfed crops rely only on seasonal cold-season winter rainfalls, the ETP^S is important (Maps 3.2 and 3.3). Roughly, we expect the native vegetation to be active when the soil-water reserve minus the ETP^S is positive. When the ETP^S is greater than the received rainfall and the soil-water reserve, we expect no or little growth from the native vegetation.

Air humidity

Air humidity follows the pattern of evapotranspiration in relation to daily temperatures, wind and solar radiation. Along the coast, less than 10–20 km inland, the ambient humidity ranges between 60% and 70% with a summer minimum. Inland, in the arid and semi-arid zones, it varies between 50% and 60% with maximum during the cool winter. Further south, on the Hamada, south Sirte, and south Jebel Al-Akhdar, it fluctuates between 40% and 50% during the year.

During the brief period of *ghibli*, the air humidity may drop as low as 5–10% with dire consequences for crops and native vegetation.

Air Temperatures

The mean annual temperature of a Mediterranean station makes little sense and impact on native vegetation and crops. More important are the winter temperatures that regulate native plants and crops growth when water is available in the soil, providing that the air and/or soil

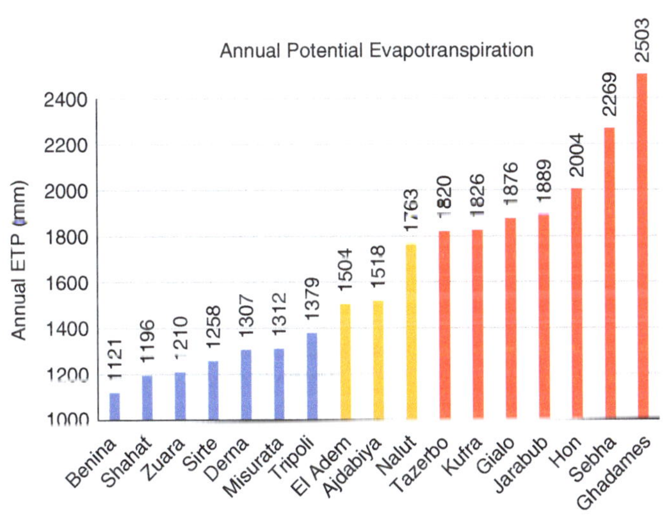

Fig. 3.5. Annual potential evapotranspiration (ETP^A) (*blue* = coastal stations; *orange* = steppe stations; red = desert stations). (After Le Houérou, Montpellier, 2004, personal communication).

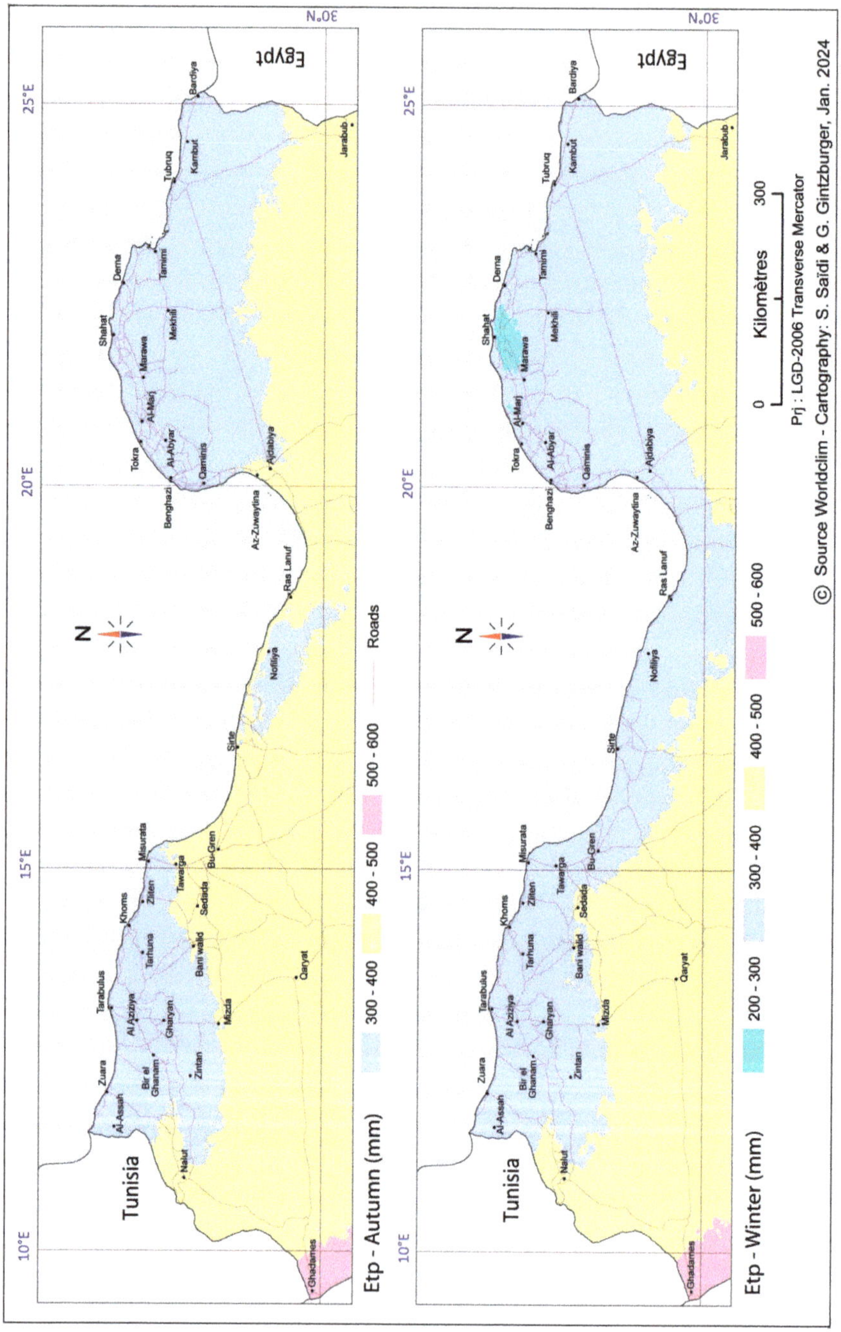

Map 3.2. Seasonal potential evapotranspiration for autumn and winter. (After Fick and Hijmans, 2017)

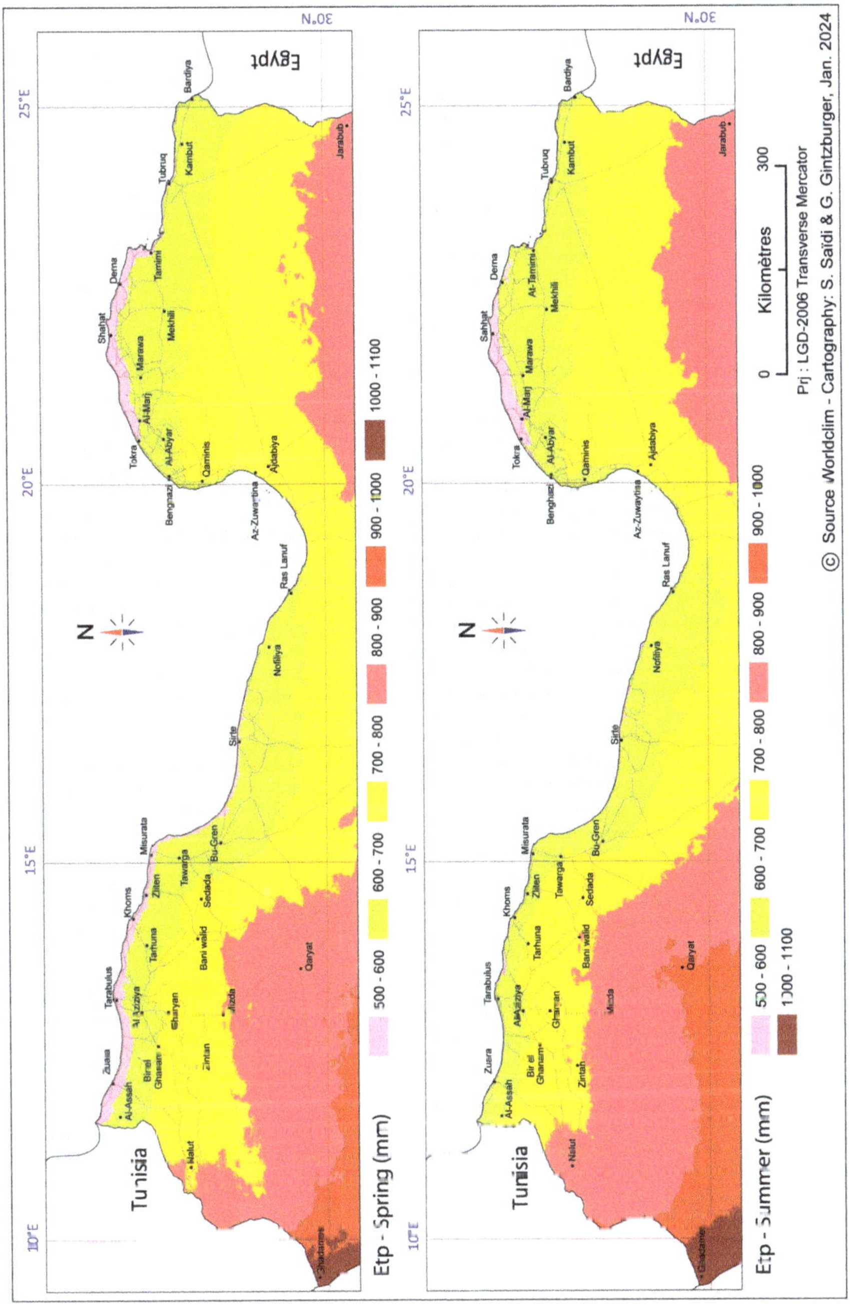

Map 3.3. Seasonal potential evapotranspiration for spring and summer. (After Fick and Hijmans, 2017)

temperature are above the 'zero growth' temperature. When below this 'zero growth' temperature, which can be different between plants, no plant growth is possible.

Winter temperatures

While rainfall occurs during the autumn/winter period, plant germination and growth is possible providing mean minimum and mean maximum temperatures permit and water is available in the soil.

Mean daily minimum temperatures (m) of the coldest month

Winter temperatures must not fall below the 'zero growth', usually around 3–5°C stopping all plant growth. This rarely occurs in the west and Sirte region, while it is common on the high elevations (Shahat – Beida) of the Jebel Al-Akhdar.

Le Houérou (1984) indicates the following 'm' limits valid for all North Africa (from Morocco to Libya) in relation to plant growth. The accepted thresholds are:

- m > 9°C: very warm winters; no frost occurring; most tropical crops (banana, sugarcane) and citrus possible with irrigation;
- 9 > m > 7°C: warm winters; some rare frosts possible; most native plants will grow during winter as well as citrus.

The above two zones are essentially coastal, where vegetables, alfalfa and berseem are grown all year long under irrigation.

- 7 > m > 5°C: mild winters; a few days (5–10) of light frosts; growth of native vegetation reduced to 10–30 days in December/January; cultivation of citrus is risky as well as Australian *Acacia* sp. plantations; most favourable zone for olive tree plantations.
- 5 > m > 3°C: temperate winters with 10–20 days of light frosts; growth of native vegetation reduced during most of December/January. Citrus, *Ceratonia siliqua* plantations most unsafe; olive plantations are still possible providing they are sheltered (east Jebel Nafusa); favourable for fruit trees (apples, pears, plums etc.) requiring some winter chill for flowering (El-Marj, Gharyan, Yefren).
- 3 > m > −1°C: cool to cold winters, 20–40 days of possible light frosts; native plants growth reduced from November to February; *Eucalyptus*, Australian *Acacia* and spineless *Opuntia* plantations will fail. *Ceratonia siliqua* will struggle to survive as well as olive trees and date palms. Cold-hardy *Pistacia vera* plantations are still suitable (Gdama – Gharyan).

Mild frosts are often recorded in the desert, east and south of the Jebel Nafusa, and south of the Jebel Al-Akhdar in the lower-rainfall zones (< 200 mm MAR). The coastal line of the Jebel Al-Akhdar from Tokra to Ras-Hilal and all along the western and Sirte coast benefit from warm winters without frost.

Mean daily maximum temperatures (M′) of the coldest month

Despite low winter temperatures, the mean daily maximum temperature of the coldest months (M′) is important during winter when M′ may climb by 10–20°C above the minimum daily temperature, allowing plant and tree growth for a few hours providing that water is available in the soil.

Le Houérou *et al.* (1979) differentiate M′ climatic sub-variant thresholds within each area defined by 'm' as follows:

- M′ < 10°C = micro-emerothermal: little plant growth during daytime winter temperatures
- 10 < M′ < 12°C = meso-emerothermal: some plant growth during daytime winter temperatures

Out of 24 stations distributed from west to east, and from the coast to deep inland (Ghadames), M′ ranges from 12°C to 19°C (Fig. 3.6) belonging to the following sub-variants:

- 12 < M′ < 15°C = macro-emerothermal – sites mostly located in high elevations;
- M′ > 15°C = mega-emerothermal – sites mostly located on coastal zones or in desert.

Summer temperatures

The mean maximum temperature (M) of the hottest month (July/August) stays mild (25–30°C) close to the sea and even on the high elevation of the Jebel Al-Akhdar. It climbs above 35°C inland in the west, in the centre of the Jeffara plain

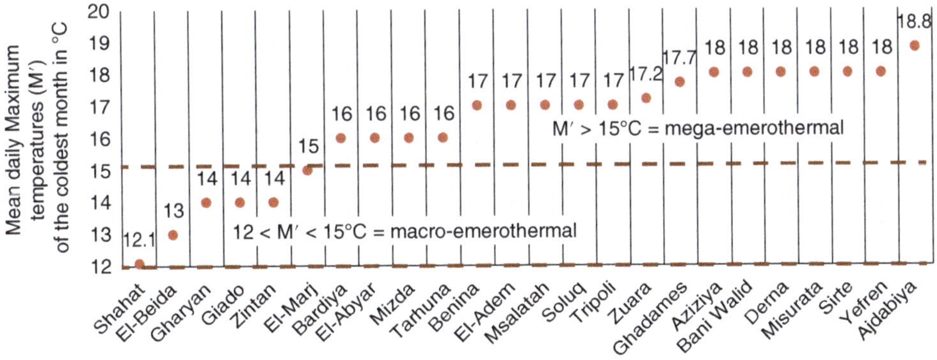

Fig. 3.6. Mean daily maximum temperatures (M′) of the coldest month for 24 meteorological stations from northern Libya.

(Aziziya, Bir-el-Ghanem, El-Watiya), and above 40°C on the Hamada (Sinaween, Derj, Ghadames) and in the desert (Murzuk, Sebha, Ghat) due to the continentality of the stations. In general, the summers are hotter in the west than in the east at comparable latitudes, elevations and continentality. We may explain this by the fact that the east has a larger coastal shore and highlands exposed to westerlies bringing rainfall, with large areas covered with forest and matorral compared to the predominantly barren west.

During the summer, due to lack of water in the soil, the native vegetation is at rest. High summer temperatures coupled with water deficit negatively impact native perennial vegetation (*Artemisia herba-alba*, *Retama raetam*, *Genista saharae*, *Anthyllis henoniana*, *Arthrophytum* sp., *Salsola* sp., *Atriplex* sp. etc.) that shed new leaves and even younger stem to reduce evapotranspiration (Fahn, 1964; Fahn and Sarnat, 1963; Orshan, 1972). Libyan herders know this very well and graze as much as possible the non-aromatic vegetation before it sheds spring leaves and young stems: there is nothing left to graze during the sizzling summer days before moving onto dry *Artemisia herba-alba*, *Thymus* sp. etc. Feeding their animals with barley grain, stubbles and concentrates, herdsmen must wait for the first autumn rain and new rangeland growth.

Thermal continentality

Thermal continentality (Debrach, 1953; Emberger, 1930, 1955) is defined by the 'M-m' values for each station. It is the temperature amplitude between the mean daily maximum temperatures (M) of the hottest month and the mean daily minimum temperatures (m) of the coldest month. In Fig. 3.7, we represent the thermal continentality as a function of the shortest distance to the sea of 52 meteorological stations. It ranges from 20°C to 40°C as a logarithmic function of the distance to the sea.

The coastal stations within 50 km of the sea have a thermal continentality ranging from 19°C to 30°C with variations observed for some stations in the centre of the Jeffara plain (Bengashir, Aziziya, Bir-el-Ghanem). Inland stations more than 50–75 km from the sea have a thermal continentality between 30°C and 40°C. The shorter the distance to the sea is, the lowest is the thermal amplitude buffered by the proximity of the sea. The further from the sea, the higher the thermal amplitude, which reaches extremes, as in the case of inland and desert stations. The higher the thermal continentality, the more stress is exerted on the native vegetation and crops.

Map 3.4 shows the spatial distribution of the mean maximum of the warmest month (M) and mean minimum of the coldest month (m).

Climate Shift Observed at Meteorological Stations

We remain cautious about writing about climate change in Libya as the climatological data we could access only cover the period from 1945 to 2010. With what we do have available, we could

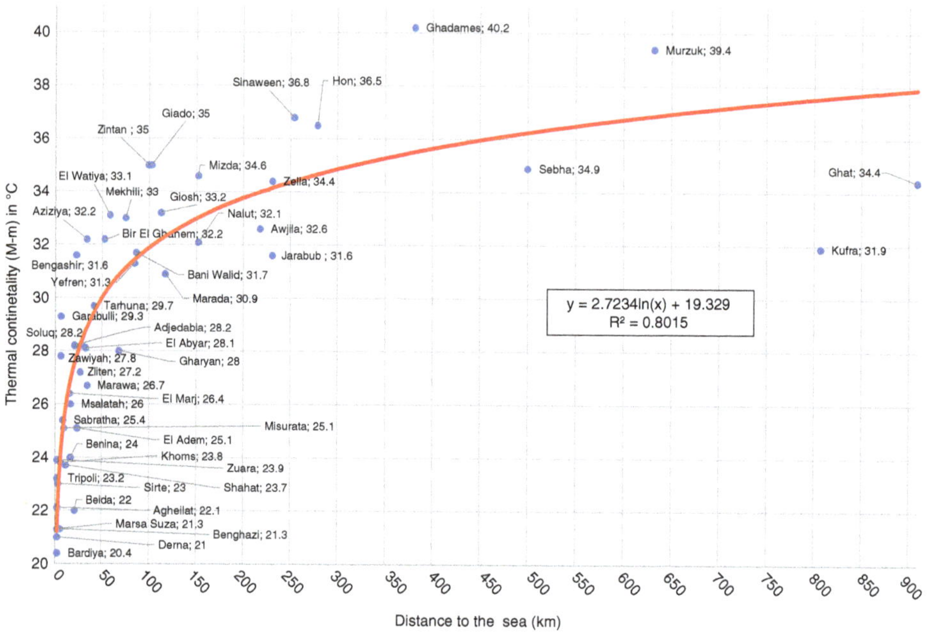

Fig. 3.7. Thermal continentality (M-m, in °C) of meteorological stations as a function of the shortest distance to the sea (km).

observe positive and negative trends, or none, that need to be confirmed with longer climatic series.

Temperature changes: real or artefact due to urbanization post mid-1970s?

El-Tantawi (2005) reports an average increase in mean annual temperatures in Libya of about 0.07°C/decade with two major warm periods, 1910–1945 (0.14°C/decade) and 1976–2000 (0.17°C/decade). However, these results appear to simply be a report from global meteorological compilations of Jones *et al.* (1999, 2000). When computing long-term trends of mean annual temperature from actual meteorological data from the Libyan Meteorological Department (Tripoli), El-Tantawi found positive and statistically significant trends during 1946–2000 for Ajdabiya, Kufra, Ghadames, Nalut, Sebha, Tripoli city and airport, Zuara, Benina and Shahat, weakly significant for Sirte and Jalo and not significant for Derna, Jarabub and Misurata. He further identified different trends during 1946–1975, significant for Tripoli city (+0.38°C/decade), Misurata (–0.35°C/decade) and Benina (+0.23°C/decade) and not significant on all other stations with –0.18°C/decade (Jalo and Ghadames) and from 0.01°C to 0.08°C/decade for all other stations. When processing the mean annual temperature data from 1976 to 2000, all trends were significantly positive for all the above-mentioned stations with +0.23°C to +0.78°C/decade, except for Benina and Tripoli airport. He further examined the mean annual minimum temperatures during 1946–2000 with no significant trends during 1946–1975, but significant positive trends mainly during 1976–2000 ranging from +0.18°C to +0.40°C/decade on all stations.

As for the mean annual mean maximum temperatures (1946–2000), they were poor or not significant on all stations.

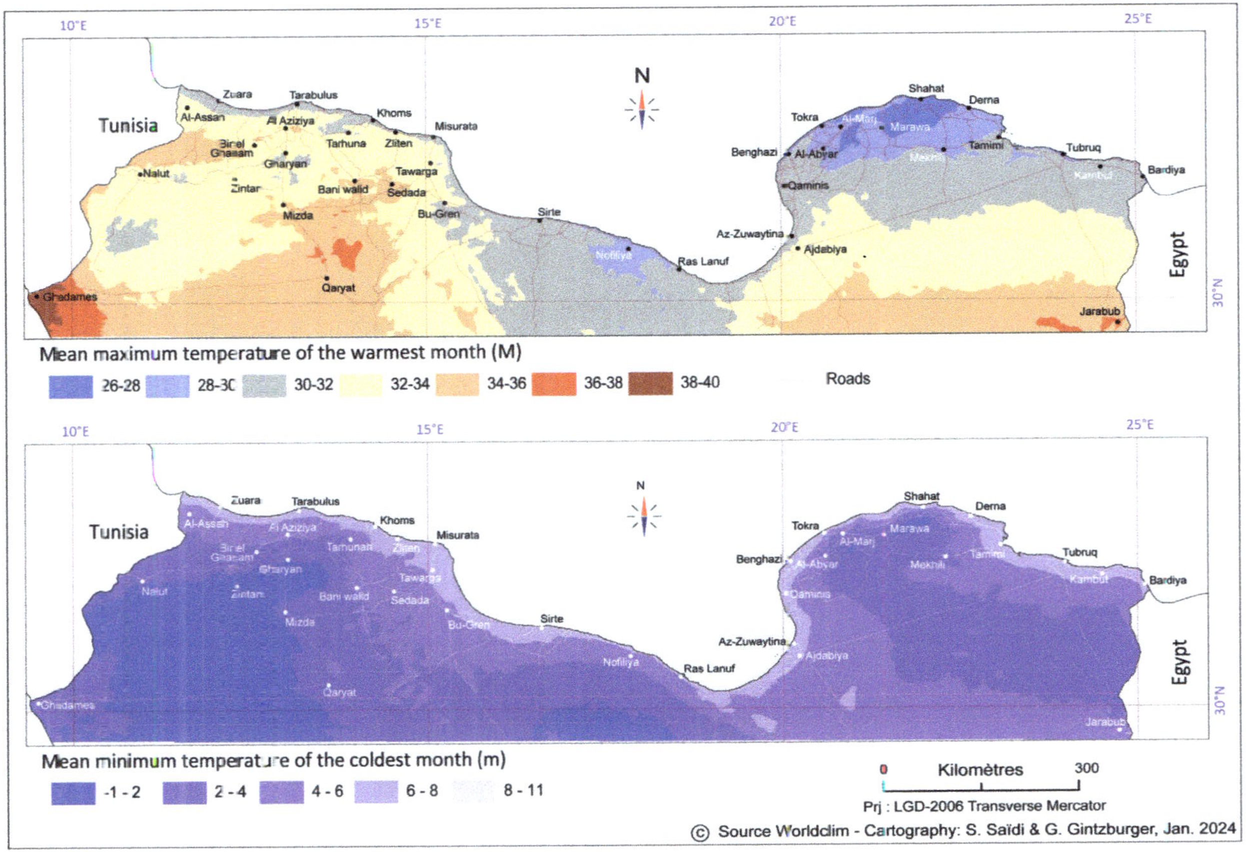

Map 3.4. Distribution of the mean maximum of the warmest month (M) and mean minimum of the coldest month (m). (After Fick and Hijmans, 2017)

Ageena *et al.* (2013a, b) studied the changes affecting the daily and monthly minimum temperatures, and maximum and mean average temperatures during the period 1945–2009 based on Libyan meteorological data. They report a significant increase of average minimum temperatures at a rate of 0.032°C/year, citing a considerable faster increase of the global mean temperature of 0.74°C ∓ 0.18°C over the last 100 years. Looking at their Fig. 3.4 (Ageena *et al.*, 2013b), it is remarkable that during the period 1945–1977on six stations the annual mean minima stays variable but stable, while the recorded mean minimum temperatures climbed significantly from 1978 to 2009 on all Libyan stations. This annual temperature increase (1979–2009) is also doubtfully reported on temperature evolution on the CRU TS Version 4.03 Google Earth Grid Interface on half-degree cells (https://crudata.uea.ac.uk/cru/data/hrg/cru_ts_4.03/ge/ [accessed 3/1/24]) based on Harris *et al.* (2014) on Figs 3.8, 3.9 and 3.10.

What would be a possible explanation of these positive temperature trends identified during the 1976–2000 period?

It raises the issue of artefacts and biases in recording climatological data: Is it real or could it be due to replacement of measuring instruments by the mid-1970s or simply weather stations being gradually surrounded by galloping urban developments, new buildings, asphalt roads, concrete walls etc., all due to the economic and population boom post-1969? El-Tantawi (2005) briefly mentioned 'urban heat islands' that locally alter the climatic conditions citing the increase of urban population in Libya from about 69% in 1980 to 88% in 2000. The World Bank data (https://data.worldbank.org/indicator/SP.URB.TOTL.IN.ZS?end=2021&locations=LY&start=1960&view=chart [accessed 3/1/24]) show an increase of the Libyan urban population from about 27% in 1960 to 46% in 1969 and 74% in 1984, stabilizing after that to about 80%; hence, a very rapid Libyan urbanization

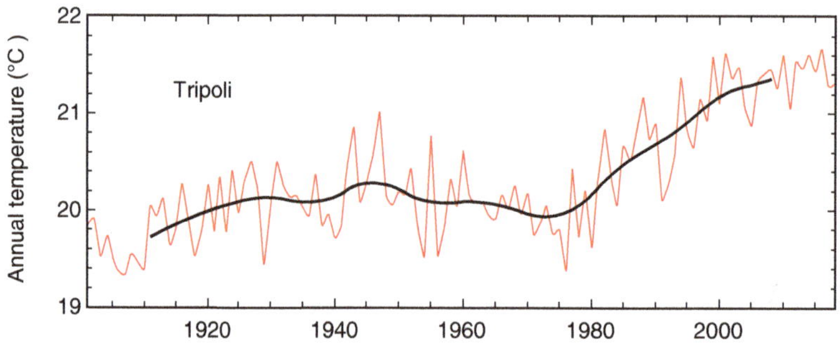

Fig. 3.8. Evolution of mean annual temperature near Tripoli (on grid 32.75 N, 13.75 E), 1900–2010.

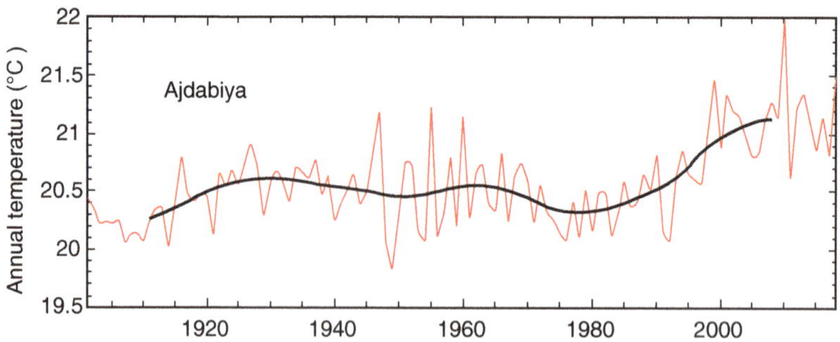

Fig. 3.9. Evolution of mean annual temperature near Ajdabiya (on grid 30.75 N, 20.25 E), 1900–2010.

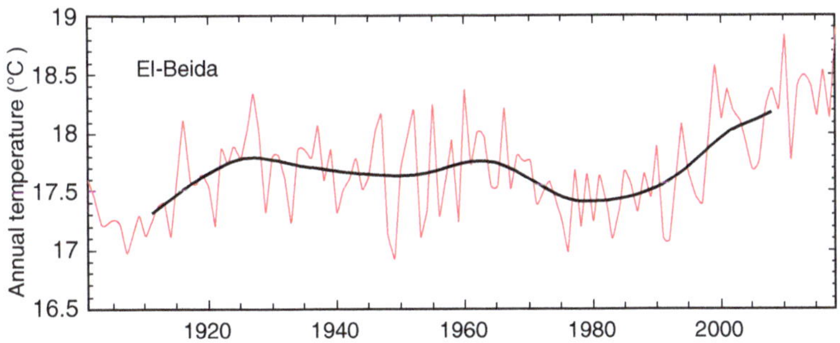

Fig. 3.10. Evolution of mean annual temperature near El-Beida (on grid 32.75 N, 21.75 E), 1901–2018.

from the mid-1960s to 1985. However, Ageena *et al.* (2013a, b) report no convincing relationship between mean annual minimum and the rate of population growth, and provide no information on urban expansion close to the meteorological stations studied. The fact that El-Tantawi (2005) identifies positive trends of the mean annual minimum temperatures on all stations from 1976 to 2000 could only be due to microclimate alteration from urbanization around the meteorological stations.

It remains that the temperature trends identified on Libyan meteorological stations raises the issue of compiling reliable data from various origins without a critical approach and proper field checking of the history of the weather stations.

Bioclimatic Synthesis

A bioclimatic synthesis was attempted by different methods developed by Emberger (1930, 1955) with his Q2 Aridity Index refined by Le Houérou (1969, 1984).

The most important bioclimatic limits

We prefer using the two main easily accessible climatic parameters that control plant growth, i.e. the MAR and the mean daily minimum temperatures (m) of the coldest month. These two climatic parameters are commonly available and do not require complicated calculations. It is usually represented on a bioclimagram plotting the position of meteorological station on a graph with 'm' *in abscissa* and MAR in ordinate (Fig. 3.11).

Le Houérou (1959b, 1969, 1984) gives the following MAR bioclimatic limits (Tables 3.4., 3.5 and 3.6).

These are combined with the mean daily minimum temperatures (m) of the coldest month (Table 3.5).

Therefore, we have a two-dimension matrix of bioclimate by 'm' with the position of 61 Libyan meteorological stations (Table 3.6 and Fig. 3.11).

This matrix could further be refined with a third dimension, which is the daily maximum temperatures (M′) of the coldest month.

Bioclimate and Plant Distribution

Vegetation and cropping limits

In terms of rainfall, the justification of the main limits adopted by Le Houérou are:

On flat land without water harvesting or cropping in depressions or wadis:

- rainfed wheat requires a minimum of 300 MAR to allow for a reliable grain harvest;
- rainfed barley requires a minimum of 200 MAR to allow for a reliable grain harvest;
- this lower limit is often extended in 175 MAR sites and depressions targeting the best rangelands vegetation (*Artemisia herba-alba*).

Rainfed cereal cropping below these MAR limits or unpredictable drought year will most of the time result in crop failure.

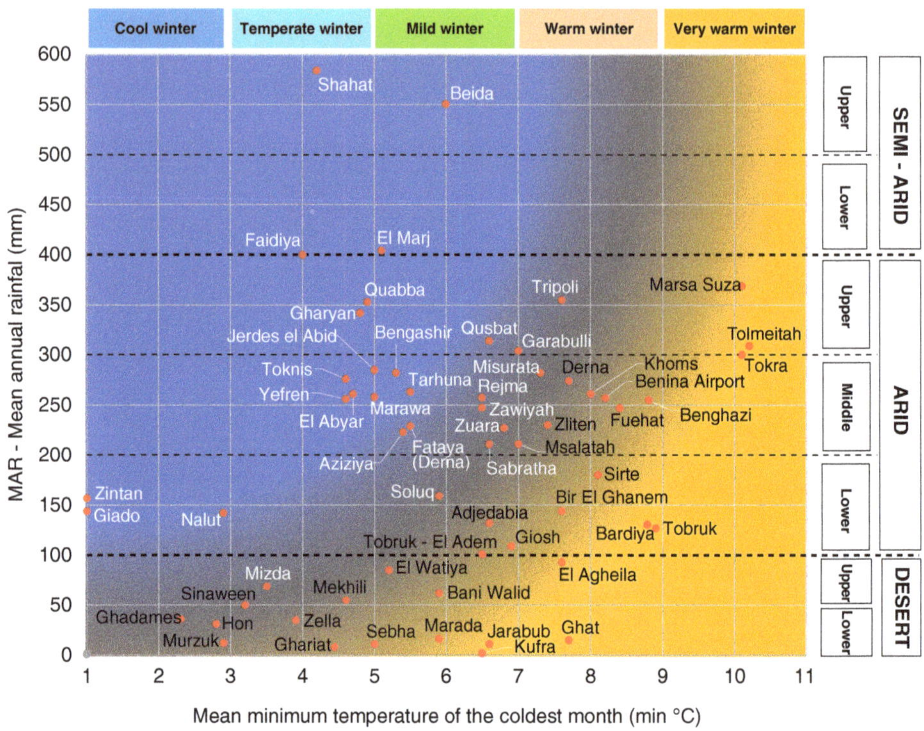

Fig. 3.11. Bioclimagram of Libyan meteorological stations.

Table 3.4. Bioclimate as a function of MAR.

Range of MAR (mm/year)	Bioclimate	Sub-level	
500 < MAR < 600	SEMI-ARID (SA)	Upper	SA-U
400 < MAR < 500		Lower	SA-L
300 < MAR < 400	ARID (A)	Upper	A-U
200 < MAR < 300		Middle	A-M
100 < MAR < 200		Lower	A-L
50 < MAR < 100	SAHARAN (S or Eremitic)	Upper	S-U
MAR < 50		Lower	S-L

Table 3.5. Bioclimate variant according to mean daily minimum temperatures (m) of the coldest month.

m > 9°C	Very warm winters	VWW
9 > m > 7°C	Warm winters	WW
7 > m > 5°C	Mild winters	MW
5 > m > 3°C	Temperate winters	TW
3 > m > −1°C	Cool winters	CW

The 100 mm MAR is recognized as the upper limit of the Sahara. Below 100 mm MAR, native vegetation tends to be of the 'contracted' type where the Saharan vegetation survives only in small depressions and in wadis collecting extra water. The flats around remain mostly bare or with sparse 'diffuse' native vegetation, a common feature on the Hamada. Any vegetation destroyed below the 100 mm MAR will never recover. As an example, the tracks left in 1941/42 by WWII tanks and army vehicles that destroyed the poor perennial vegetation (*remeth, Arthrophytum scoparium*) south of Jebel Al-Akhdar and Mekhili were still visible in 1975/76. Here was no

Table 3.6. Matrix of Libyan bioclimates (MAR x 'm') for 61 Libyan meteorological stations.

		MAR (mm/year)						
		S-L (0–50)	S-U (50–100)	A-L (100–200)	A-M (200–300)	A-U (300–400)	SA-L (400–500)	SA-U (500–600)
Mean daily minimum temperatures (m) of the coldest month	**VWW (m > 9°C)**	n.a.	n.a.	Bardiya, Tobruk	Benghazi	Marsa-Suza, Tholmeitah, Tokra	n.a.	n.a.
	WW (9 > m > 7°C)	Ghat	Agheila	Sirte, Bir-el-Ghanem, Giosh	Misurata, Derna, Homs, Benina, Fuehat, Zliten, Msalata	Tripoli, Garabulli	n.a.	n.a.
	MW (7 > m > 5°C)	Marada, Jarabub, Kufra, Awjilah	Watiya, Bani Walid	Soluq, Ajdabiya, El-Adem	Bengashir, Tarhuna, Rejma, Zawiya, Fataya, Zuara, Aziziya, Sabrata	Qusbat	El-Marj	Beida
	TW (5 > m > 3°C)	Sinaween, Gharyat, Sebha, Zella	Mizda, Mekhili	Giado, Nalut	Jerdes el-Abid, Toknis, El-Abyar, Marawa, Yefren	Gharyan, Quabba	Faidiya	Shahat
	CW (3 > m > −1°C)	Ghadames, Hon, Murzuk	n.a.	Zintan	n.a.	n.a.	n.a.	n.a.

vegetation regeneration on these tracks 30 years later. Damage done by oil exploration vehicles on native vegetation in the desert remains visible for many years if not forever. In short, if desert vegetation is destroyed, there is very little likelihood of it being restored.

The main Libyan phytocoenosis in relation to MAR

In arid and semi-arid environments, rainfall is the main limiting factor determining plant phytocoenosis, plant diversity and distribution. Soil types are another crucial element as they modulate the water stored in soil profiles.

The list of Libyan plants by bioclimate is not exhaustive; it only is intended to give an indication of the main species present in each bioclimatic zone. Some species may be present across several nearby bioclimates according to micro-environmental conditions. We do not review the phytocoenosis present on saline soils as they are azonal and independent from rainfall.

The upper and lower semi-arid bioclimatic zone (400 < MAR < 600 mm)

It is an area on the rise of the Jebel Al-Akhdar (Cyrenaica) from El-Marj to Shahat-Beida. It does not occur in western Libya.

The dominant vegetation consists of clear forests (> 7–10 m tall trees), garrigues/matorral (scrub < 6–7 m high) and small shrubs and perennials on karst landforms. Main plant species are:

- Trees: *Pinus halepensis, Juniperus phoenicea, Cupressus sempervirens, Quercus calliprinos, Ceratonia siliqua*
- Garrigue/matorral: *Arbutus pavarii, Olea sylvestris, Laurus nobilis, Viburnum tinus, Myrtus communis, Phyllyrea media, Rhamnus lycioides*
- Small shrubs and perennials: *Ballota* sp., *Marrubium* sp., *Phlomis* sp., *Thymus* sp., *Tamus communis, Smilax aspera, Erica* sp., *Rosmarinus officinalis, Cistus* sp., *Coronilla glauca, Genista acanthoclada, Lonicera* sp., *Asparagus acutifolius, Arum cyrenaicum*
- Grasses: *Lolium perenne, Brachypodium ramosum, Cynosurus* sp., *Phalaris* sp., *Oryzopsis holciformis*

The upper-arid bioclimatic zone (300 < MAR < 400 mm)

This is the area where rainfed wheat cropping can still be economically viable but becoming uncertain with MAR close to the lower limit of 300 mm.

IN TRIPOLITANIA – WESTERN LIBYA. This is mostly found in the eastern part of the Jebel Nafusa (Yefren – Gharyan – Msalata – Qusbat, up to the coast near Homs) where the hills are covered with *Stipa tenacissima* (*Halfa*), a remnant of forest of *Pinus halepensis* (and possibly *Juniperus phoenicea*) eradicated over centuries. This is still witnessed by the presence of *Globularia alypum* and rare *Oryzopsis miliacea*, both common companions of *Pinus halepensis* forests in North Africa. It should also be noted that *Stipa tenacissima* is a strict North African western species that disappears east of Zliten – Daphnia and is not present in Sirte and the Jebel Al-Akhdar.

Associated perennial species to *Stipa tenacissima* formations are: *Globularia alypum, Rosmarinus officinalis, Thymus hirtus, Genista microcephala, Helianthemum ruficomum, Fumana thymifolia, Teucrium fruticans, Periploca laevigata, Lavandula* sp., *Dactylis glomerata*, and a prostrate *Juniperus oxycedrus* ssp. *macrocarpa* (on the coast near Sidi Benhur – Naqqazah).

IN CYRENAICA. This is the typical vegetation from most of the plateau of Cyrenaica with a garrigue/matorral of dominant low and dense bushes of *Juniperus phoenicea* and *Pistacia lentiscus*, with companion species such as *Periploca laevigata, Cistus* sp., *Globularia alypum, Satureja* sp., *Helichrysum* sp., *Phagnalon* sp., *Cyclamen rholfsianum, Corydothymus capitatus, Achillea santolina, Sanguisorba minor, Oryzopsis miliacea* and the inconspicuous geocarpic *Factorovskaya aschersoniana* on *Terra rossa* in the Benghazi plain. Remarkable is the unique degraded stage of the above garrigue on the colder part of the Jebel Al-Akhdar plateau with a low hemispherical shrub belonging to the eastern Mediterranean species *Sarcopoterium spinosum* (most common in Jordan, Lebanon, southern Syria and Palestine). Other associated species in this formation are: *Phlomis floccosa, Marrubium vulgare, Ballota hirsuta, Trisetaria macrochaeta, Stipa capensis, Cynosurus capitatus, Hordeum murinum, Elymus* sp., mostly unpalatable plants on very degraded rangeland, and a valuable small forage species such as *Poa bulbosa* in the cool subvariant bioclimate.

The middle-arid bioclimatic zone (200 < MAR < 300 mm)

This is the area where rainfed barley cropping can still be economically viable but becoming uncertain with MAR close to the lower limit of 200 mm.

IN TRIPOLITANIA – WESTERN LIBYA.

- In the sandy Jeffara plain (north Bir-el-Ghanem to Reqdalin – Jmeil – Agheilat – south Sorman to Aziziya), *Stipagrostis pungens* dominates with associated *Retama raetam, Genista saharae, Echiochilon fruticosum, Teucrium polium, Matthiola longipetala, Plantago albicans, Helianthemum sessiliflorum, Koeleria* sp., *Haplophyllum tuberculatum, lotus pusillus. Artemisia campestris* fast colonizes post-cultural sandy areas.
- In the Jebel Nafusa, large hilly areas are still covered with *Stipa tenacissima* and associated species as in the upper-arid bioclimatic zone, but *Halfa* have disappeared on the western part of the Jebel from Yeffren to the Tunisian border with diminishing rainfall and rangeland degradation by industrial harvesting, overuse and overgrazing. This phytocoenosis displays small perennials and annuals such as *Helianthemum* sp., *Matthiola fruticulosa, Dactylis hispanica, Hyparrhenia hirta, Hippocrepis scabra, Launaea resedifolia, Phagnalon saxatile.*, with *Erodium guttatum, Erodium glaucophyllum* and *Lygeum spartum* appearing on gypseous soils. Rare large *Pistacia atlantica* trees dot the degraded *Halfa* landscape (Gharyan – Gdama). Most uncommon is *Caralluma europaea* (Sherif *et al.*, 2020), the only Libyan succulent cactus-like plant hiding amongst the *Halfa* tufts. *Caralluma* is also reported present in Marmarica (Derna to Bardiya coast).

IN CYRENAICA. The garrigue extending to the south is further degraded with bushes of *Pistacia lentiscus* getting smaller, and most other shrubs and trees have disappeared due to water deficit, overall overuse, overgrazing and fuelwood harvesting. Worth noticing are the remnant clear forests with remarkable *Juniperus phoenicea* trees still dotting the rugged landscape from Jardas-el-Abid – Taknis, Marawa to south of Gendoula. The last isolated wind-battered *Juniperus* trees can still be seen about 15–20 km south of Gendoula where what would be the 200 mm isohyet.

Adjoining to the south is the lower-arid bioclimatic zone (100 < MAR < 200 mm), with large extents of low shrubs of *Artemisia herba-alba* steppe on loamy soils and *Noaea mucronata* (a mostly irano-turanian species), and *Arthrophytum scoparium* interspersed with a carpet of *Poa bulbosa. Anabasis* sp. appears on sandy sites.

The lower-arid bioclimatic zone (100 < MAR < 200 mm)

This is the typical domain of the steppe with low shrubs.

IN TRIPOLITANIA – WESTERN LIBYA.

- In the sandy/sandy-loam Jeffara plain: a good example is near wadi Abu-Shayba – Bir-el-Ghanem, on the ex-sheep project on sandy-loamy flats *Rhanterium suaveolens* (now nearly all disappeared) used to cover large areas up to Bir-Tuta to Aziziya and expanding to Qasr-el-haj. *Rhanterium suaveolens*, a threatened species in Libya, is an exclusive western species also present in southern Tunisia to Sirte, but not in eastern Libyan rangelands. Common associated species with this vegetation type are *Argyrolobium uniflorum, Polygonum equisetiforme, Stipa lagascae, Stipa parviflora, Helianthemum lippii, Nolletia chrysocomoides*, and abundant annual *Medicago* sp. (*M. truncatula* ssp. *tricycla*) and small grasses (*Cenchrus ciliaris, Cutandia* sp., *Schismus* sp., *Sphenopus* sp.). These excellent rangelands were totally obliterated as persistently targeted for rainfed barley cropping and irrigation projects.
- On the flat land above the Al-Assah *sebkha*, *Salsola vermiculata* ssp. *brevifolia* is most common with *Salsola* sp., *Argyrolobium uniflorum, Stipa lagascae* with carpets of annual *Medicago truncatula* var. *tricycla* and *Lotus* sp.

- On the flats with heavier soils below the northern foothills of the Jebel Nafusa, wadi flats (Bir Ayyad – wadi Ethel with MAR close to 100–125 mm) host *Haloxylon scoparium* and *Artemisia herba-alba* with a cohort of *Vella annua, Herniaria fontanesii* and *Anabasis articulata*. Shallow rocky soils bear *Gymnocarpos decander, Helianthemum kahiricum* and *Atractylis serratuloides*. Remnant overgrazed bushes of *Periploca laevigata* dot the wadian and creek lines descending from the Jebel Nafusa.

IN EASTERN SIRTE (AJDABIYA), SOUTHERN JEBEL AL-AKHDAR AND MARMARICA (TOBRUK TO BARDIYA) AT THE LOWER END OF THIS LOWER-ARID BIOCLIMATIC ZONE.

- On wind-swept flat sandy soils, most common phytocoenosis are colonized with *Thymelaea microphylla, Thymelaea hirsuta, Arthrophytum schmittianum* and *Pituranthos tortuosus*. The occurrence of *Traganum nudatum, Bassia muricata* on sandy soils indicates a strong Saharan influence.
- On shallow and loamy soils, *Haloxylon scoparium, Artemisia herba-alba, Rhus tripartita, Anabasis articulata, Anabasis oropediorum, Suaeda* sp., *Limonium* sp., *Reaumuria vermiculata. Noaea mucronata* is also present in the east in the cool part of this bioclimate.

The Saharan (desert) bioclimatic zone (MAR ≤ 100 mm)

Rainfed barley cropping is exceptional in wadis with large water harvesting systems operating such as at wadi Merdum, Bani Walid, Sof-el-Jin, Zamzam, Bey-el-Kebir and Ghirza.

- In western Jeffara plain, in the low sand dunes south El-Assah (Bir-Shabunyah – North Takut – Tiji), and south Nalut towards Derj-Sinaween, *Calligonum comosum* and *Ephedra alata* are present.
- South of Jebel Nafusa, close to the 100 mm MAR, the low shrubs/low-cover steppe on shallow loamy soils is dominated with *Anthyllis henoniana* and *Arthrophytum scoparium* (Nalut – Haraba).
- Further south-east (wadi Bey) and Sirte area, *Anthyllis henoniana* is missing. *Haloxylon scoparium, Gymnocarpos decander, Anastatica hierochuntica, Helianthemum kahiricum, Salsola tetragona, Salsola tetrandra, Salsola sieberi, Anabasis articulata* and *Fagonia* sp. are dominant on shallow rocky soils. *Traganum nudatum, Bassia muricata* and *Cornulaca monacantha* colonize shallow sand veils. These species are common on the southern desertic slopes of the Jebel Al-Akhdar and in Marmarica.
- Worth noting is *Astragalus spinosus* (wadi Bey – Bu-N'gem) and *Cymbopogon shoenanthus* (south Buerat-el-Hsun), present in the Sirte area on gypseous soils.
- In wadis only in the west and uncommon in the western part of Sirte (south of Buerat-el-Hsun), *Acacia tortilis* and the rare *Cocculus pendulus* climber (in warm bioclimate subvariant – Beni Walid, Seddada, Zorzor) and *Pistacia atlantica* (in cool bioclimate subvariant – Mizda) are present though persistently targeted by illicit charcoal manufacturing. In these western wadis, common associated species are *Panicum turgidum, Pennisetum dichotomum, Zilla spinosa, Moricandia nitens, Spergularia tomentosa, Anvillea garcinii, Zygophyllum album, Lasiurus hirsutus, Digitaria nodosa, Calotropis procera, Neurada procumbens*.
- *Acacia tortilis* is not present in the southern Jebel Al-Akhdar or Marmarica. Rare *Pistacia atlantica* are noted near Daryanah (on the coast 30 km north-east of Benghazi).
- In the west, further into the MAR with less than 50 mm, *Farsetia aegyptiaca, Anvillea garcini, Oudneya africana, Aristida obtusa* and extremely drought-resistant annual *Medicago* (*Medicago laciniata, Medicago laciniata var. brachyacantha*). Trees and bushes of *Tamarix* sp. thrive south of El-Watiya on the Hamada (wadi Ghirza, wadi Arandel) and deep into the desert close to oases.

Rainfall Trends and Drought Occurrences in Libya

Climatic changes, alteration, cold, wet or dry periods during prehistoric times can be deduced with different methods of studying fossils, sediment deposits, pollen, and rock engravings or paintings. Climatic changes during history are difficult to assess due to the lack of recorded climatic data older than a century at the most.

Paleoclimatology

Climate change over the past millenaries is witnessed with presence of millions of years-old fossilized/petrified forest in the Libyan desert (Ubari, Uweinat), near Nalut-Cabao in Cretaceous Albian sandstone formations (–100 million years old), and prehistoric rocks engraving of wildlife in Libya (see Chapter 14). Records of past vegetation changes are not available though it could be approached with palynology research in southern Tunisia (Damblon and Vanden Berghen, 1993; Schulz *et al.*, 1995, 2002) and studies on prehistoric sites at Haua Fteah and other caves in the Jebel Al-Akhdar (Douka *et al.*, 2014; Graeme *et al.*, 2009; Jacobs *et al.*, 2017).

Paleoclimatology reconstructs by tephrochronology[1] high-resolution flood history and fire activity in *sebkha* sediments in southern Tunisia (Lakhdar *et al.*, 2006; Marquer *et al.*, 2008). Marquer *et al.*, (2008) combined a multidisciplinary physical and biological approach (sediments colouring, morphoscopy, composition, charcoal dating, palynology) with chronology calibration using volcanic sediments from the Tyrrhenian eruptions historically dated, i.e. Vesuvius 24–26 Oct. 79 CE[2], Vesuvius 17 Dec. 1631 and Stromboli 11 Sept. 1930. They identified four major wet/dry cycles during the last two millenaries, *c.*550–950 CE, 950–1300, 1300–1570 and 1570–1870, with overall seven major wet cycles. The wettest periods were during 1000–1500 AD and the Little Ice Age (1400–1850), with six wet/dry episodes and only one earlier, between 600 and 700 CE. None was recorded earlier during the late Roman-Byzantine period (300–500 CE) that appeared to be consistently warm and possibly dry, similar to the current 19th–20th centuries' climatic conditions.

Historic climatology

Aridity and periodic droughts have long been reported in nearby Tunisia between 1846 and 1867 (Doumet-Adanson, 1888; Dupuy, 1939; Guérin, 1862; Kralik, 1855; Mayet, 1886; Raynaud, 1853) between 1918 and 1927 (Dupuy, 1939), between 1945 and 1949 (Fantoli, 1952, 1967, 1968; Foret, 1958) and later (Floret *et al.*, 1986). Libya was not immune to past droughts and will face future droughts.

Climate change is a global trend and remains a source of potential immense social and economic damage. Unfortunately, due to a lack of long-term reliable data and often partial analyses, this finding remains unclear at the Libyan level. In these environments, where desertification takes an increasing dimension, the rainfall trends are not always easily perceived and often remain controversial.

The temporal evolution of total annual precipitation recorded from 1945 to 2010 on 14 stations are analysed in an attempt to identify drought occurrences. These time series remains the only official rainfall series provided by the Libyan National Meteorological Office (El-Fadli and Albukhari, Tripoli, 2021, personal communication). Unfortunately, we could not access the raw meteorological data back to 1920 published by Fantoli (1952, 1967, 1968).

Rainfall trends analysis

The rainfall evolution (approx. 1945–2010) is assessed using a graphical analysis of the raw series smoothed with four-year moving average with the wet years appearing on the graphs (Figs 3.12 to 3.16) in pale blue and the dry years in orange. These smoothed averages reduce the amplitude of the inter-annual fluctuations and better highlight trends.

The dry and wet sequences commonly are shorter than four years for most stations. We also observe that the moving averages of the annual rainfall of Ajdabiya, Jarabub, Benina, Derna, Gharyan, Tripoli, Zuara, Sirte, Nalut, Misurata and Hon display a lack of trend until the 1970s when there is a net cumulative rainfall with an average duration of 19 years (Std = ±4). The smoothed curve of the moving averages no longer exceeds the average for all the stations, with the exception of the Saharan stations of Ghadames and Jalo over an average duration of 15 years (Std =±4). This deficit appears significant as the rainfall curve remains clearly and durably below the median indicating a downward trend. The graphic analysis of the majority of the stations displays variations in rainfall

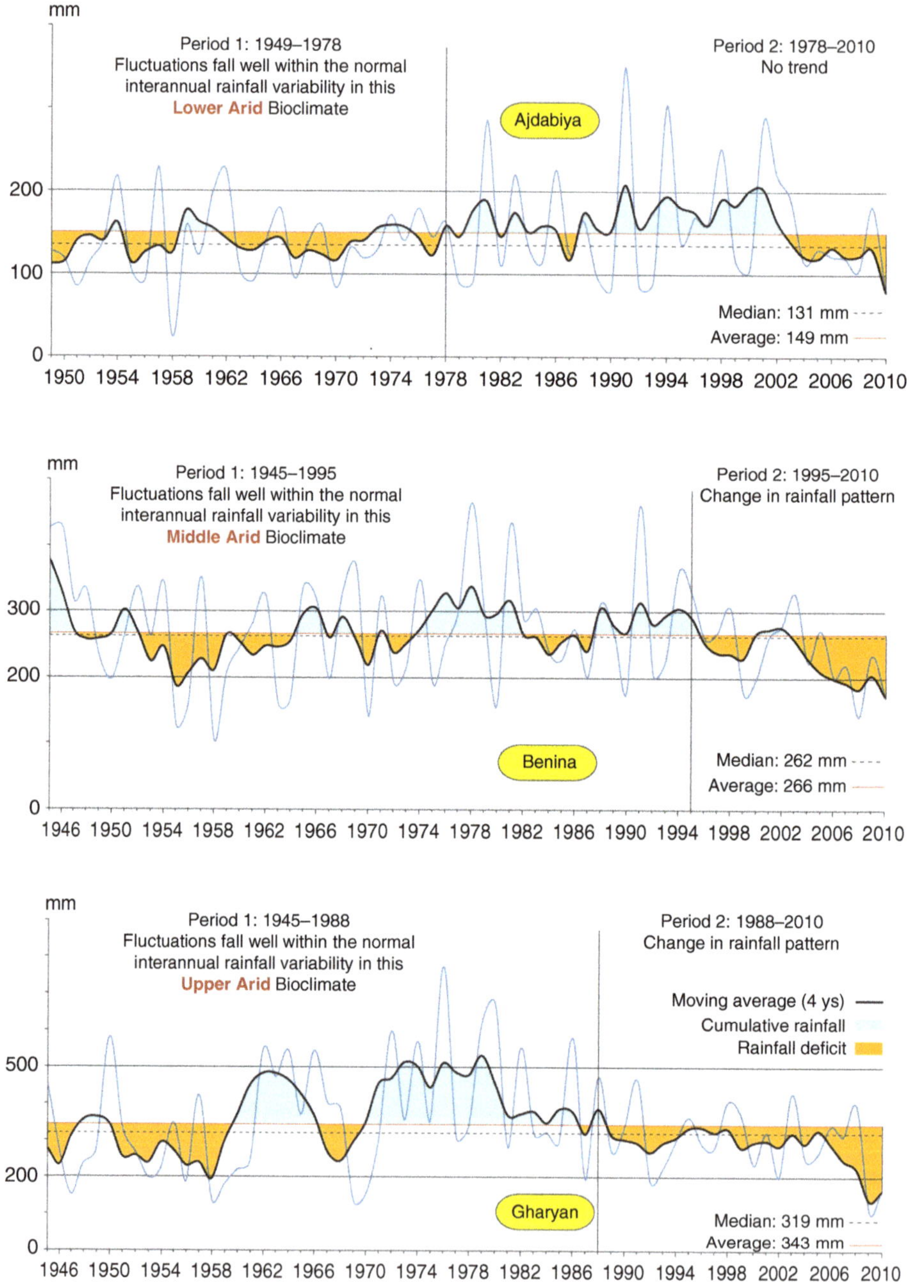

Fig. 3.12. Four-year moving average for total annual rainfall at Ajdabiya, Benina and Gharyan.

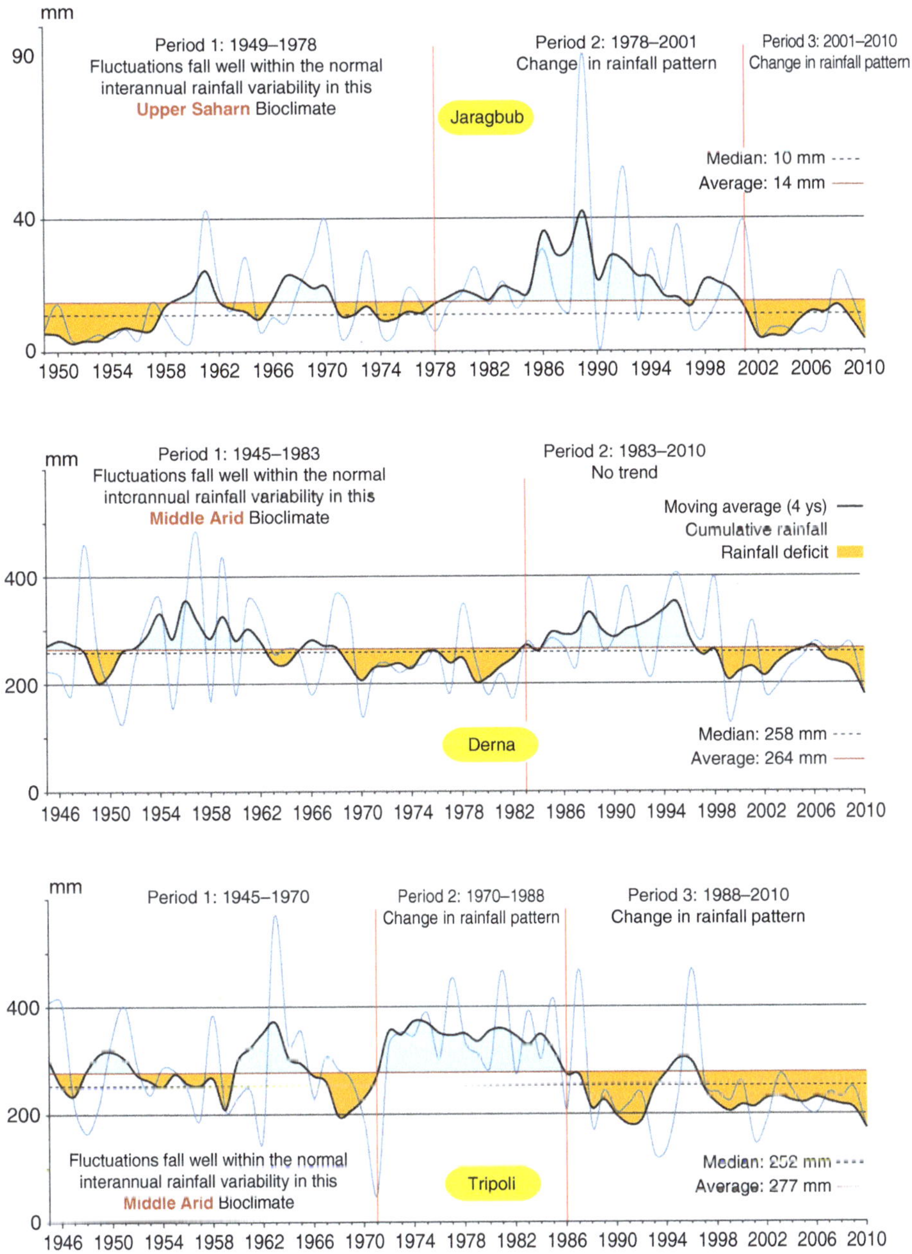

Fig. 3.13. Four-year moving average for total annual rainfall at Jarabub, Derna and Tripoli.

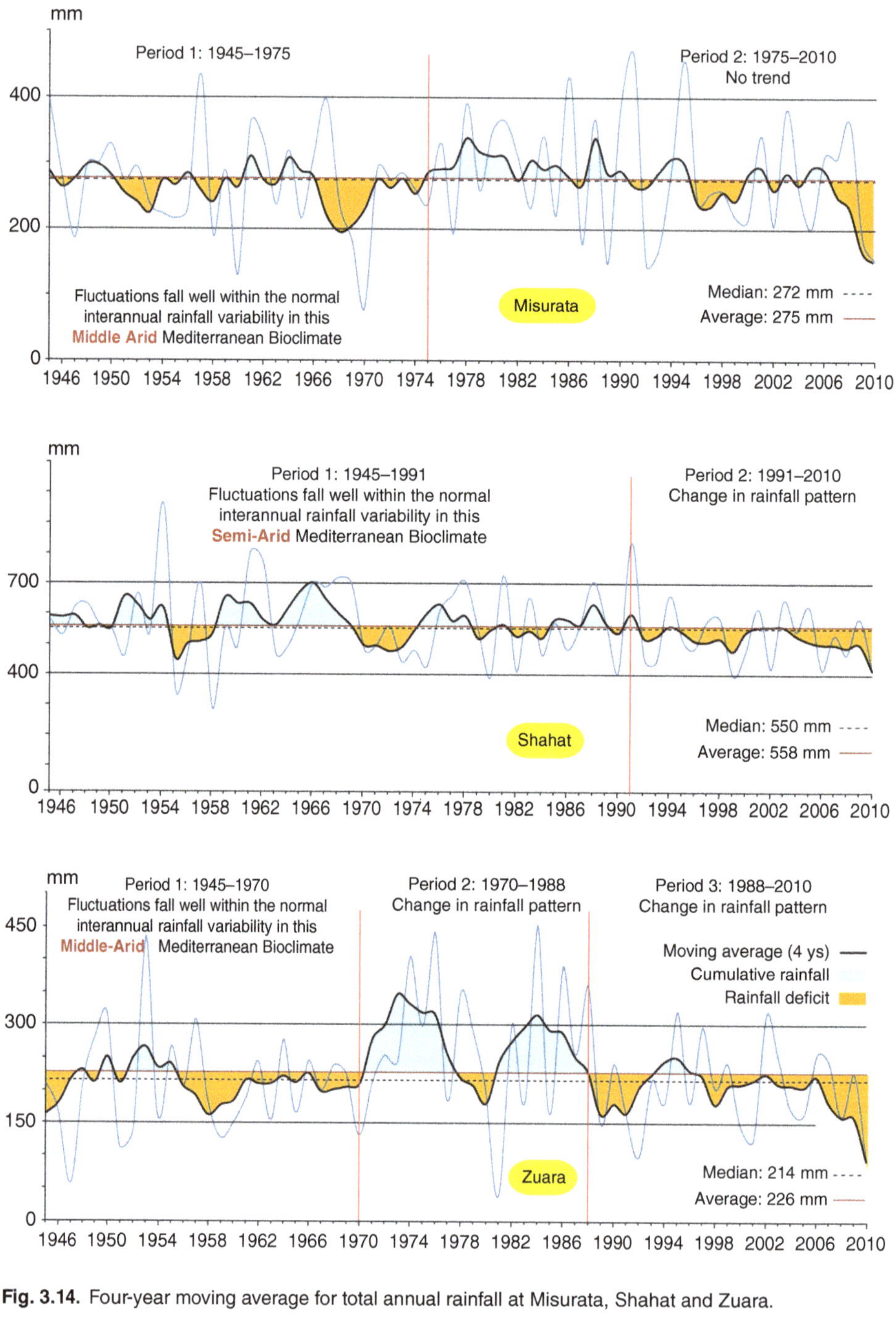

Fig. 3.14. Four-year moving average for total annual rainfall at Misurata, Shahat and Zuara.

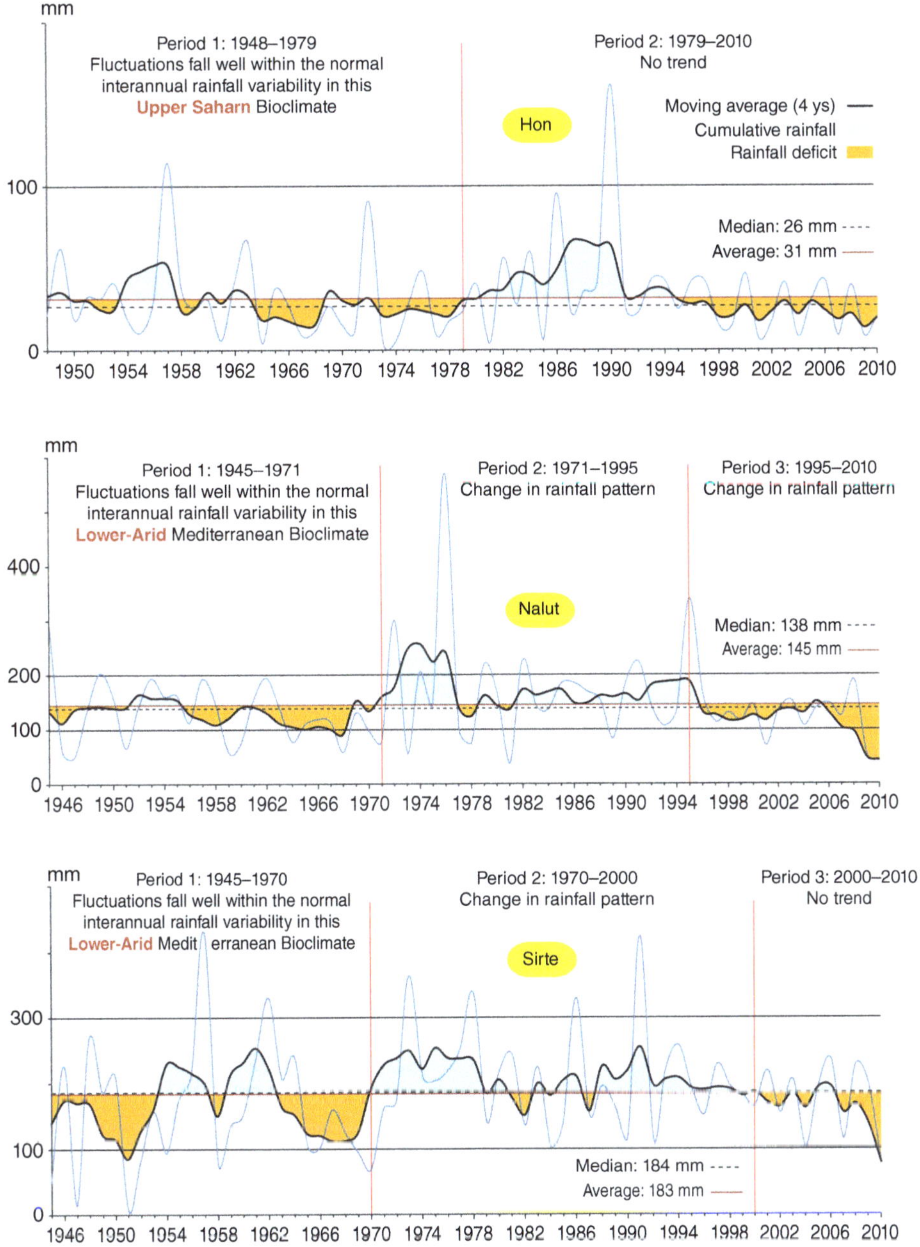

Fig. 3.15. Four-year moving average for total annual rainfall at Hon, Nalut and Sirte.

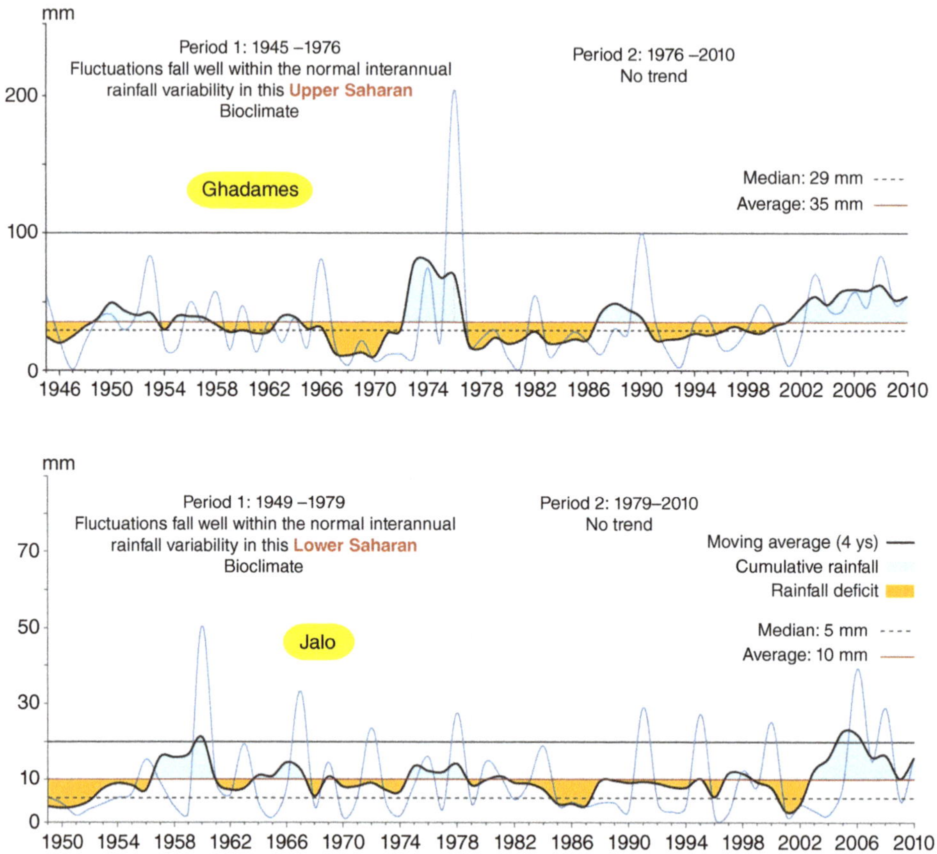

Fig. 3.16. Four-year moving average for total annual rainfall at Ghadames and Jalo.

amount and distribution over years indicating a possible drying trend.

Defining wet or dry years

To define a wet or dry year, we may use the quantiles, the standard deviations or the percentage of annual rainfall from the median. In this study, we used the latter. A year is considered normal if the difference from the average value of annual rainfall time series does not exceed 10%. A year will be considered dry when the difference is below 10% and wet beyond (Hirche, 2007; Le Goff, 1985). The 10% interval remains questionable as it is relatively low in absolute terms so that normal years are not the majority. However, it has the advantage of a better discrimination between dry and wet years in this arid environment where even a small reduction in yearly rainfall may lead to a drought period (Table 3.7).

The number of dry years is higher than the wet years for all stations. This difference is, on average, 32% (Std = ±5) under Saharan bioclimate, 19% (±3) under hyper-arid bioclimate, 11% (±8) under lower-arid bioclimate, 10% (±4) under average arid bioclimate, 13% under higher arid bioclimate and 7% under semi-arid bioclimate.

Assessing the duration of dry sequences by frequency class makes it possible to understand the perception of the dry year depending on whether it follows another dry or a wet year (Rognon, 1996). The dry sequence generally swings between one and four years depending on the bioclimatic variant (Table 3.8). However, these can vary between 6 and 12 years for the Saharan and pre-Saharan stations.

Table 3.7. Number of years with wet, normal and dry sequences (% of year) at 14 meteorological stations (1945–2010).

Bioclimate	Stations	(year) Wet	Normal	Dry	Total
Semi-arid	**Shahat**	**21**	**19**	**26**	**66**
	% year	*32*	*29*	*39*	
Upper Arid	**Gharyan**	**21**	**15**	**30**	**66**
	% year	*32*	*23*	*45*	
Middle Arid	**Benina**	**25**	**11**	**30**	**66**
	% year	*38*	*17*	*45*	
	Derna	**18**	**23**	**25**	**66**
	% year	*27*	*35*	*38*	
	Misurata	**24**	**15**	**27**	**66**
	% year	*36*	*23*	*41*	
	Tripoli	**20**	**15**	**31**	**66**
	% year	*30*	*23*	*47*	
	Zuara	**22**	**15**	**29**	**66**
	% year	*33*	*23*	*44*	
Lower Arid	**Ajdabiya**	**19**	**11**	**32**	**62**
	% year	*30*	*18*	*52*	
	Nalut	**20**	**20**	**26**	**66**
	% year	*30*	*20*	*39*	
	Sirte	**27**	**10**	**29**	**66**
	% year	*41*	*15*	*44*	
Hyper-arid	**Ghadames**	**22**	**8**	**36**	**66**
	% year	*33*	*12*	*55*	
	Hon	**24**	**5**	**34**	**63**
	% year	*38*	*8*	*54*	
Sahara	**Jarabub**	**21**	**3**	**38**	**62**
	% year	*34*	*5*	*61*	
	Jalo	**18**	**4**	**40**	**62**
	% year	*29*	*6*	*65*	

Statistical analysis

We use the z-score transformation test (Falissard, 1998) to compare the average values of two selected periods to the z-score transformation table with a 0.05 error probability. It allows interpreting curve differences (Figs 3.12 to 3.16) between periods of annual rainfall for each meteorological stations affected with a dry spell trend between 1945 and 2010:

$$Z = |\mu 1 - \mu 2| / \sqrt{[(\sigma^2 1/n1) + (\sigma^2 2/n2)]}$$

where Z = z-score transformation test
$\mu 1$ and $\mu 2$ = rainfall averages for the two periods considered
$\sigma^2 1$ and $\sigma^2 2$ = variances of two samples
n1 and n2 = number of years of measurements

The choice of the periods compared is determined by the change in the shape of the smoothed moving average curve (Figs 3.12 to 3.16). The z-score transformation results are presented in Table 3.9.

Climatic trends at selected stations

The climatic trends of selected meteorological stations are detailed in Tables 3.7, 3.8, 3.9 and Figs 3.12 to 3.16.

AJDABIYA. During the years 1949–2010, 52% are dry years compared with 30% wet years. Dry sequences vary between 1 and 5 years, with an average value of 2 years, with two sequences of 4 successive dry years and only one sequence of 5 successive dry years. The average rainfall recorded at the station increased by 15% between period 1 and period 2. The z-score transformation value is smaller than 1.96 (5% risk threshold)

Table 3.8. Frequency of dry sequences (1945–2010).

Bioclimate	Stations	Nb of year sequence	Frequency of dry sequences	Nb of years observed
Semi-arid	**Shahat**	1	13	13
		2	5	10
		3	1	3
	Average		**1.36**	
Upper Arid	**Gharyan**	1	7	7
		2	3	6
		3	3	9
		4	2	8
	Average		**2**	
Middle Arid	**Benina**	1	7	7
		2	6	12
		3	2	6
		5	1	5
	Average		**1.9**	
	Derna	1	8	8
		2	4	8
		3	3	9
	Average		**1.7**	
	Misurata	1	14	14
		2	3	6
		3	1	3
		4	1	4
	Average		**1.4**	
	Tripoli	1	4	4
		2	4	8
		3	1	3
Lower Arid	**Ajdabiya**	1	7	7
		2	6	12
		4	2	8
		5	1	5
	Average		**2**	
	Nalut	1	11	11
		2	5	10
		5	1	5
	Average		**1.52**	
	Sirte	1	11	11
		2	2	4
		3	1	3
		4	1	4
		7	1	7
	Average		**1.8**	
Upper Saharan	**Ghadames**	1	5	5
		2	3	6
		3	3	6
		5	1	5
		7	2	14
	Average		**2.6**	
	Hon	1	12	12
		2	5	10
		3	2	6
		6	1	6
	Average		**1.7**	

		4	1	4
		5	1	5
		7	1	7
	Average		**2.6**	
	Zuara	-	13	13
		2	6	12
		4	1	4
	Average		**1.45**	
Lower Saharan	**Jarabub**	1	6	6
		2	4	8
		3	2	6
		6	1	6
		12	1	12
	Average		**1.7**	
	Jalo	1	7	7
		2	3	6
		3	3	9
		5	1	5
		6	1	6
		7	1	7
	Average		**2.5**	

Table 3.9. Z-score transformation trend test on 14 meteorological stations.

		Periods observed			Average values (mm)			Nb years observed			Z-score test		
Bioclimate	Stations	P1	P2	P3	P1	P2	P3	P1	P2	P3	*P1 & P2*	*P2 & P3*	
Semi-arid	Shahat	1945–1994	1994–2010		574.8	517.3		47	19		2.10		Trend
Upper Arid	Gharyan	1945–1988	1988–2010		367	294.6		44	22		2.37		Trend
Middle Arid	Benina	1945–1995	1995–2010		274	235		51	15		2.18		Trend
	Derna	1945–1983	1983–2010		258.4	271.7		39	27		0.70		No trend
	Misurata	1945–1995	1995–2010		266.9	281.6		31	35		0.72		No trend
	Tripoli	1945–1971	1971–1986	1986–2010	276	343	235	27	15	24	2.58	4.52	Trend
	Zuara	1945–1970	1970–1988	1988–2010	209	280	203	26	18	22	2.45	2.73	Trend
Lower Arid	Ajdabiya	1949–1978	1978–2010		139	159		30	32		1.35		No trend
	Nalut	1945–1971	1971–1995	1995–2010	128	179	122	27	24	15	2.07	2.35	Trend
	Sirt	1945–1970	1970–2000	2000–2010	159	207	173	26	30	10	2.08	1.56	Trend
Upper Saharan	Ghadames	1945–1976	1976–2010		36.1	33.3		32	34		0.37		No trend
	Hon	1948–1979	1979–2010		28.9	34		31	31		0.74		No trend
Lower Saharan	Jarabub	1949–1978	1978–2001	2001–2010	10.6	21.9	6.5	30	23	9	2.52	3.41	Trend
	Jalo	1949–1979	1979–2010		9.4	9.9		31	31		0.18		No trend

indicating that the difference between the two average rainfall values for the two periods considered is not significant. According to this result, the rainfall fluctuations observed at Ajdabiya station does not show a significant trend after 1978 despite the cumulative rainfall appearing increasing and durably above the average curve. However, after 2001 the rainfall from this meteorological station has significantly decreased.

BENINA. During the years 1945–2010, the dry years represent 45% compared to 38% wet years. Dry sequences vary between 1 and 5 years with an average value of 1.9 years. Only one sequence of 5 successive dry years is recorded. The average rainfall at Benina station decreased by 14% between period 1 and period 2. The z-score transformation value is greater than 1.96 (5% risk threshold) indicating a statistically significant dryer trend after 1995.

GHARYAN. During the years 1945–2010, dry years represent 45% compared to 32% wet years. Dry sequences vary between 1 and 4 years with an average value of 2 years with three sequences of successive dry years and two sequences of 4 successive dry years. The average rainfall recorded by the station decreased by 20% between period 1 and period 2. The z-score transformation value is greater than 1.96 (5% risk threshold) indicating that the difference between the two average rainfall values for the two periods considered is significant. Gharyan station shows a convincing dry climate trend post-1988.

JARABUB. In this Saharan station, during the years 1949–2010, the dry years cover 61% compared to 34% for the wet years. The dry sequences vary between 1 and 12 years with an average value of 1.7 years, with one sequence of 6 successive dry years and one sequence of 12 successive dry years. The average rainfall recorded by the station increased by 106% between period 1 and period 2 and it has decreased by 71% between period 2 and period 3. The z-score transformation value is greater than 1.96 (5% risk threshold) indicating that the difference between the two average rainfall values for the two periods considered (period 1 vs. period 2 and period 2 vs. period 3) is significant. According to this result, the rainfall fluctuations observed at Jarabub station show a significant wetter trend after 1978, followed by a statistically significant dryer trend after 2001.

DERNA. During the years 1945–2010, dry years represent 38% compared to 27% wet years. However, the dry sequences vary between 1 and 3 years with an average value of 1.7 years, with three sequences of dry years. The average rainfall recorded by the station increased by 5% between period 1 and period 2. The z-score transformation value is smaller than 1.96 (5% risk threshold) between period 1 and period 2 indicating that the difference between the two average rainfall values for the two periods considered is not significant. According to this result, the Derna precipitation did not show a statistically significant trend after 1983.

TRIPOLI-AIRPORT. During the years 1945–1986, the dry years represent 47% compared to 30% wet years. Dry sequences vary between 1 and 7 years with an average value of 2.6 years and only one sequence of 5 successive dry years and only one sequence of 7 successive dry years. The average rainfall recorded by the station increased by 24% between period 1 and period 2 and it has decreased by 31% between period 2 and period 3. The z-score transformation value is greater than 1.96 (5% risk threshold) indicating that the difference between the two average rainfall values for the two periods considered (period 1 vs. period 2, and period 2 vs. period 3) is significant. According to this result, the rainfall fluctuations observed at Tripoli station show a significant wetter trend after 1971, followed by a dry trend after 1986. The rainfall time series of this last period did show a statistically significant dry trend.

MISURATA. During the years 1945–2010, dry years cover 41% compared to 36% wet years. The dry sequences vary between 1 and 4 years with an average value of 1.1 years, and only one sequence of 4 successive dry years. The average rainfall recorded by the station increased by 6% between period 1 and period 2. The z-score transformation value is smaller than 1.96 (5% risk threshold) between period 1 and period 2, hence not significant. The rainfall oscillations observed at Misurata fall well within the normal interannual rainfall variability in this lower-arid Mediterranean region.

SHAHAT. During the years 1945–2010, the dry years represent 39% compared to 32% wet years. Dry sequences vary between 1 and 3 years, with an average value of 1.36, with only one sequence of 3 successive dry years. The average rainfall recorded by the station decreased by 10% between period 1 and period 2. The z-score transformation value is greater than 1.96 (5% risk threshold) indicating that the difference between the two averages rainfall values for the two periods considered is significant. Hence, Shahat station displays a proven dryer climate post-1991.

ZUARA. During the years 1945–1988, the dry years represent 44% compared to 33% wet years. Dry sequences vary between 1 and 4 years, with an average value of 1.45 years, and only one sequence of 4 successive dry years. The average rainfall recorded by the station increased by 34% between period 1 and period 2 and decreased by 28% between period 2 and period 3. The z-score transformation value is greater than 1.96 (5% risk threshold) indicating that the difference between the two average rainfall values for the two periods considered (period 1 vs. period 2, and period 2 vs. period 3) is significant. According to this result, the rainfall fluctuations observed at Zuara station show a significant trend after 1970. However, after 1988 the rainfall decreased. The time series for this latter period did show a statistically significant dry trend.

HON. During the years 1948–2010, the dry years represent 54% compared to 38% wet years. The dry sequences vary between 1 and 6 years with an average value of 1.7 years, and one sequence of 4 successive dry years and one sequence of 6 successive dry years. The average rainfall recorded by the station increased by 18% between period 1 and period 2. The z-score transformation value is smaller than 1.96 (5% risk threshold) between period 1 and period 2 indicating that the difference between the two average rainfall values for the two periods considered is not significant. The rainfall time series at Hon does not show a statistically significant trend and fall well within the normal inter-annual rainfall variability in this Saharan bioclimate.

NALUT. From 1945 to 1995, the dry years represented 39% compared to 30% wet years. Dry sequences vary between 1 and 5 years with an average value of 1.52 years, and only one sequence of 5 successive dry years. The average rainfall recorded by the station increased by 39% between period 1 and period 2 and decreased by 32% between period 2 and period 3. The z-score transformation value is greater than 1.96 (5% risk threshold) indicating that the difference between the two average rainfall values for the two periods considered (period 1 vs. period 2, and period 2 vs. period 3) is significant, indicating that the rainfall fluctuations observed to Nalut station show a significant wetter trend after 1971. However, after 1995 the rainfall curve remains clearly and durably below the median indicating a drying trend.

SIRTE. During the period 1945–2000, the dry years represented 44% compared with 41% wet years. Dry sequences vary between 1 and 7 years with an average value of 1.8 years, and one sequence of 4 successive dry years and one sequence of 7 successive dry years. The average rainfall recorded by the station increased by 30% between period 1 and period 2 and decreased by 16% between period 2 and period 3. The z-score transformation value is greater than 1.96 (5% risk threshold) between period 1 and period 2 indicating that the difference between the two average rainfall values for the two periods considered is significant. However, it is smaller than 1.96 (5% risk threshold) between period 2 and period 3 indicating that the difference between the two average rainfall values for the two periods considered is not significant. Therefore, Sirte station displays a convincing wet climate trend after 1970 until 2000. However, after this year, the rainfall level observed has significantly decreased but does not show a statistically significant trend.

GHADAMES. From 1945 to 2010, the dry years represented 55% compared to 33% wet years. Dry sequences vary between 1 and 7 years, with an average value of 2.6 years and one sequence of 5 successive dry years and two sequences of 7 successive dry years. The average rainfall recorded by the station decreased by 8% between period 1 and period 2. The z-score transformation value is smaller than 1.96 (5% risk threshold) indicating that the difference between the two average rainfall values between period 1 and period 2 is not significant indicating that the rainfall time series of Ghadames did not show a statistically significant trend. The rainfall fluctuations observed from this meteorological station fall well within the normal interannual rainfall variability of the Saharan bioclimate. However, after 2001, the rainfall amount recorded increased but without indicating a statistically significant trend.

JALO. From 1949 to 2010, the dry years represented 65% compared to 29% wet years. The dry sequences vary between 1 and 7 years with an average value of 2.5 years, and with one sequence of 5 successive dry years, one sequence of 6 successive dry years and one sequence of 7 successive dry years. The average rainfall recorded at this station increased by 5% between period 1 and period 2. The z-score transformation value is smaller than 1.96 (5% risk threshold) indicating that the difference between the two average rainfall values for the two periods considered is not significant. According to this result, the time series of Jalo precipitation did not exhibit a statistically significant trend after 1979. The rainfall fluctuations observed from this meteorological station fall well within the normal inter-annual rainfall variability of the Saharan bioclimate. However, after 2001, the rainfall observed increased but did not show a statistically significant trend.

The cartographic location of the selected stations reveals that the drying trend increases as one moves towards the Libyan coast (Fig. 3.16).

Discussion and Conclusions

In our presentation, we focus on northern Libya, from the Mediterranean coast to the 29–30° latitude to the south. The climate is mostly arid Mediterranean with dominant scarce and irregular winter rainfall. We clearly distinguish two different rainfall regimes between western and eastern Libya with more random spring storms in the west. In the east, rainfall is concentrated during cold autumn/winter seasons as is common in the Middle East. The limit between the two winter rainfall regimes is in the Gulf of Sirte (approx. 16–20° longitude east) where the Sahara desert reaches the Mediterranean sea. It demonstrates that Libya is at the carrefour of North Africa and the Middle East, not only in terms of climatology but also in terms of phytogeography and history, between Tripolitania of the Roman world to the west and Pentapolis of the Greek world to the east. The extreme aridity of the Sahara and the deep south, though not studied in detail, impacts the vegetation with intrusion from tropical elements at the southern margin of the coastal zones.

Long-term MAR is an important factor affecting vegetation distribution and diversity. As in all arid zones, inter-annual and seasonal variability increase with aridity. Combination of long-term MAR with the mean daily minimum temperatures (m) of the coldest month (and with the mean daily maximum temperatures (M′) of the coldest month) defines the Libyan bioclimates that better explain the vegetation distribution and diversity.

The question of climate change affecting Libya remains open although diminishing rainfall and longer periods of droughts appear to be affecting several Libyan stations. Available temperature data do not display a convincing warming trend as there could be bias due to the evolving urban environment around old meteorological stations. This may be elucidated in the future with the new 19 automatic weather stations (AWSs) installed by the LNMC.

The majority of the stations show trends of wet years followed by a decrease in rainfall with prolonged periods of drought. From the two Saharan stations (Jalo and Jarabub), only Jarabub displays a drying trend starting since 2001. On the other hand, the Jalo station, even if the statistical tests show no drying trend, records longer dry sequences and a very high coefficient of variation. As for Ghadames and Hon, also located in the hyper-arid Saharan bioclimate, they do not show any proven trend, despite their high rainfall coefficient of variation and a high frequency of dry sequences particularly remarkable in Ghadames.

From the three stations of the lower-arid (Ajdabiya, Nalut and Sirte), only Ajdabiya does

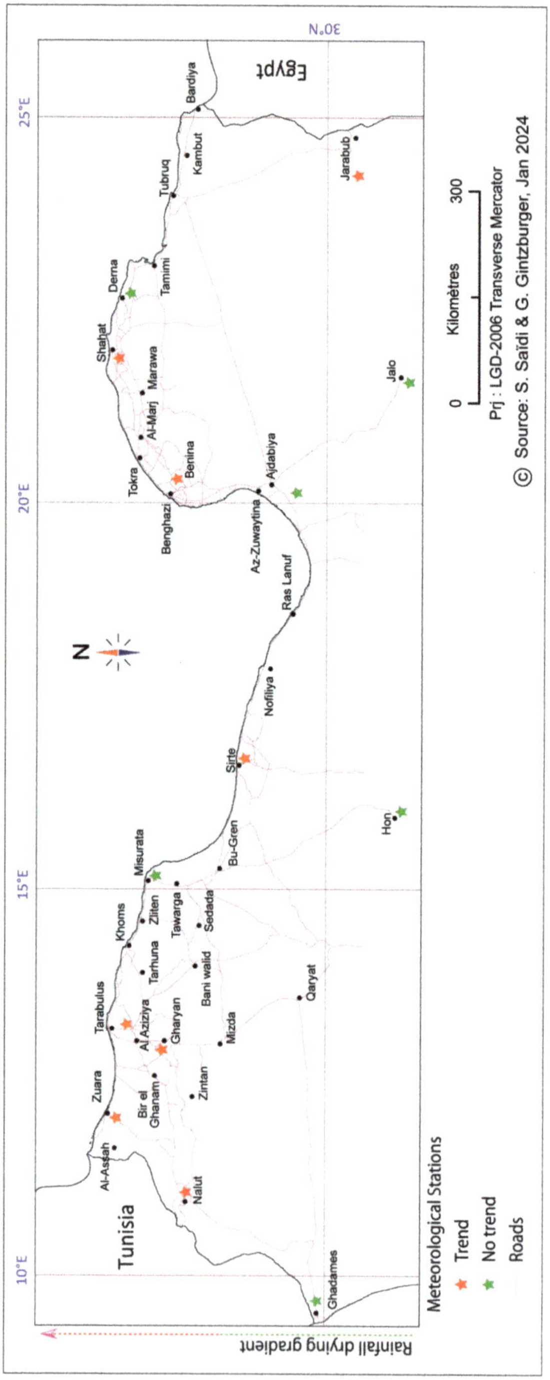

Map 3.5. Long term (1945–2010) rainfall data station location and drought trend.

not show a proven drying trend. However, this latter station shows a remarkable drop in annual rainfall from 2000 to 2010. Benina, Tripoli airport and Zuara station, three out of five stations in the middle-arid bioclimate, show a dryer MAR trend. This trend appears between 1980 and 2010. This may partly explain why previous studies based on series not exceeding the 1980s missed this trend. Finally, the Shahat station in the semi-arid and the Gharyan station in the upper-arid, both show a proven drying trend, mainly during the decades from 1990 to 2010.

Assessment of these time series shows that in these regions, rainfall, a preeminent factor in climatic characterization, has indeed undergone a clear drying trend over the past 30 years. Of the 14 stations examined, several of them show clear trends with a recurrence of non-periodic dry and wet sequences often exceeding four to five years. The longest dry sequence is observed at Jarabub with 12 consecutive dry years between 1949 and 1960. Ranging from the lower-arid to the upper-arid through the semi-arid as well as the Saharan (Jarabub), these stations display a marked drying up since the 1980s, both with a reduction in the total amount of annual rainfalls and with increasing dry sequences. It needs to be confirmed with a tighter network of stations located in the lower-arid to the semi-arid bioclimates and with longer observation series.

While our observations on Libyan meteorological data barely cover the past 80 years, palaeo-climatological studies in nearby Tunisia clearly reveal long periods of climate changes with long-lasting wet/dry cycles over the past 2000 years. Climate change is not new. In any case, it is better to be prepared, as Le Houérou's early premonition (1993) warned in its conclusion that 'Unless strong and determined action is taken by the year 2025, disaster is inescapable', and that was 29 years ago.

Notes

[1] Tephrochronology is a geochronological technique that identifies discrete layers in sediment deposits.

[2] CE = Christian era

References

Ageena, I., MacDonald, N. and Morse, A.P. (2013a) Variability of maximum and mean average temperature across Libya (1945–2009). *Theoretical and Applied Climatology* 117(3–4), 549–563. DOI: 10.1007/s00704-013-1012-z

Ageena, I., MacDonald, N. and Morse, A.P. (2013b) Variability of minimum temperature across Libya (1945–2009). *International Journal of Climatology* 33(3), 641–653. DOI: 10.1002/joc.3452

Australian National University (2013) *ANUSPLIN*. Available at: https://fennerschool.anu.edu.au/research/products/anusplin (accessed 28 August 2024).

Damblon, F. and Vanden Berghen, C. (1993) Etude paléo-écologique (pollen et macrorestes) d'un dépôt tourbeux dans l'île de Djerba, Tunisie méridionale. *Palynosciences* 2, 157–172.

Debrach, J. (1953) Notes sur les climats du Maroc Occidental. *Maroc Méditerranée* 342, 1–14.

Douka, K., Jacobs, Z., Lane, C., Grün, R., Farr, L. *et al.* (2014) The chronostratigraphy of the Haua Fteah cave (Cyrenaica, northeast Libya). *Journal of Human Evolution* 66, 39–63. DOI: 10.1016/j.jhevol.2013.10.001

Doumet-Adanson, N. (1888) *Rapport sur une mission botanique exécutée en 1884 dans la région saharienne, au nord des grands chotts et dans les îles de la côte orientale*. Imprimerie Nationale, Paris.

Dupuy, A. (1939) *Les colonies françaises: La Tunisie*. Laurens éditeur, Paris.

Elfadli, K.I. (2009) *Precipitation Data of Libya*. Libyan National Meteorological Center (LNMC).

El-Tantawi, A.M.M. (2005) Climate change in Libya and desertification of Jifara Plain using geographical information system and remote sensing techniques. PhD dissertation, Johannes Gutenberg Universität, Mainz, Germany.

Emberger, L. (1930) Sur une formule applicable en Géographie Botanique. *Comptes Rendus de l'Académie des Sciences de Paris*, 189, 389–390.

Emberger, L. (1955) Une classification biogéographique des climats. *Naturalia Monspeliaca* 7, 3–43.
Fahn, A. (1964) Some anatomical adaptation of desert plants. *Phytomorphology* 14, 93–102.
Fahn, A. and Sarnat, C. (1963) Xylem structure and annual rhythm of development in trees and shrubs of the desert. IV. Shrubs. *Bulletin of the Research Council of Israel*, 11D4, 198–209.
Falissard, B. (1998) *Comprendre et Utiliser les Statistiques dans les Sciences de la Vie*, 2nd edn. Elsevier Masson, Paris.
Fantoli, A. (1939) *Elementi preliminary dela pluviometria Libyca l'agricultura coloniale 1*. L'Agricultura Coloniale, Florence, Italy.
Fantoli, A. (1952) *Le Pioggie Della Libia Con Particolare Riguardo Alle Zone Di Avvaloramento*. A cura dell'Ufficio studi del Ministero A.I, Rome.
Fantoli, A. (1967) *Contributo alla climatologia della Tripolitania. Parte 1. Riassunto dei Risultati e tabelle meteorologiche e pluvionetriche*. Ministero degli Affari Esteri. Cooperazione scientifica e tecnica; Ministero degli Affari Esteri., Rome.
Fantoli, A. (1968) *Contributo alla climatologia della Cirenaica – Riassunto dei risultati e tabelle meteorologiche e pluviometriche*. Ministero degli affari esteri. Cooperazione scientifica e tecnica.
Fick, S.E. and Hijmans, R.J. (2017) WorldClim 2: new 1-km spatial resolution climate surfaces for global land areas. *International Journal of Climatology* 37(12), 4302–4315. DOI: 10.1002/joc.5086
Floret, C., Le Floc'h, É. and Pontanier, R. (1986) La désertisation en Tunisie présaharienne. *Revue des Mondes Musulmans et de la Méditerranée* 41(1), 291–326. DOI: 10.3406/remmm.1986.2464
Foret, C. (1958) Le cheptel tunisien et les problèmes d'ensemble de son alimentation. *Terre de Tunisie* 4(1), 1–30.
Graeme, B., Antoniadou, A., Barton, H., Brooks, I., Candy, I. (2009) The Cyrenaican Prehistory Project 2009: the third season of investigations of the Haua Fteah cave and its landscape, and further results from the 2007–2008 fieldwork. *Libyan Studies* 40, 55–94.
Guérin, V. (1862) *Voyage archéologique dans la Régence de Tunis, Plon éditeur* (Vols 1 and 2). Henri Plon, Paris.
Harris, I., Jones, P.D., Osborn, T.J. and Lister, D.H. (2014) Updated high-resolution grids of monthly climatic observations – the CRU TS3.10 Dataset. *International Journal of Climatology* 34, 623–642. DOI: 10.1002/joc.3711
Hirche, A. (2007) Évolution de la pluviosité annuelle dans quelques stations arides algériennes. *Sécheresse* 18(4), 314–320.
Jacobs, S., Li, B., Farr, L., Hill, E., Hunt, C. (2017) The chronostratigraphy of the Haua Fteah cave (Cyrenaica, northeast Libya) – optical dating of early human occupation during Marine Isotope Stages 4, 5 and 6. *Journal of Human Evolution* 105. DOI: 10.1016/j.jhevol.2017.01.008
Jones, P.D., New, M., Parker, D.E., Martin, S. and Rigor, I.G. (1999) Surface air temperature and its changes over the past 150 years. *Reviews of Geophysics* 37, 173–199.
Jones, P.D., Parker, D.E., Osborn, T.J. and Briffa, K.R. (2000) *Global and Hemispheric Temperature Anomalies: Land and Marine Instrumental Records (1850–2015)*. Environmental System Science Data Infrastructure for a Virtual Ecosystem (ESS-DIVE) (United States); Carbon Dioxide Information Analysis Center (CDIAC), Oak Ridge National Laboratory (ORNL), Oak Ridge, Tennessee. DOI: 10.3334/CDIAC/cli.002
Kralik, L. (1855) Lettre (de Tunisie) à M. Webb. *Bulletin de la Société Botanique de France* 1 and 2, 23–27, 116–120.
Lakhdar, R., Soussi, M., Ben Ismail, M.H. and M'Rabet, A. (2006) A Mediterranean Holocene restricted coastal lagoon under arid climate: case of the sedimentary record of the Sabkha Boujmel (SE Tunisia). *Palaeogeography, Palaeoclimatology, Palaeoecology* 241(2), 177–191. DOI: 10.1016/j.palaeo.2006.02.014
Le Goff, Y. (1985) Variabilité interannuelle des précipitations au Maroc. In: *Actes de la Conférence 'Sécheresse, gestion des eaux et production alimentaire'*. Agadir, Maroc.
Le Houérou, H.-N. (1959a) Ecologie, phytosociologie et productivité de l'olivier en Tunisie méridionale. *Bulletin du Service de la Carte Phytogéographique*, CNRS B (IV), 7–72.
Le Houérou, H.-N. (1959b) *Recherches écologiques et floristiques sur la végétation de la Tunisie méridionale*. PhD dissertation, Université Alger Institut de Recherches Sahariennes, Algiers.
Le Houérou, H.-N. (1969) *La Végétation de la Tunisie Steppique (Avec Référence aux Végétations Analogues d'Algérie, de Libye et du Maroc)* (Vols 1–2). 42–45. Annales de l'Institut National de la Recherche Agronomique de Tunisie, Tunis.
Le Houérou, H.-N. (1984) An outline of the bioclimatology of Libya. *Actualités Botaniques* 131(2–4), 157–178. DOI: 10.1080/01811789.1984.10826657

Le Houérou, H.-N. (1993) Changements climatiques et désertisation. *Sécheresse* 4(2), 95–111.
Le Houérou, H.-N. (2004) An agro-bioclimatic classification of the isoclimatic Mediterranean zones. *Arid Land Research and Management* 18(4), 1–47.
Le Houérou, H.-N. (2005a) *Atlas of Climatic Diagrams for the Mediterranean Isoclimatic Zones Volume 1.* Copymania, Montpellier, France.
Le Houérou, H.-N. (2005b) *The Isoclimatic Mediterranean Biomes: Bioclimatology, Diversity and Phytogeography* (Vol. 2). Copymania, Montpellier, France.
Le Houérou, H.-N., Claudin, J., Haywood, M. and Donadieu, P. (1975) *Etude phyto-écologique du Hodna (Algérie).* (Technical Report No. 3). FAO, Division Ags, Dp/Alg/66/509, Rome.
Le Houérou, H.-N., Claudin, J. and Pouget, M. (1979) Etudes bioclimatiques des steppes algériennes. *Bulletin Société Histoire Naturelle Afriques du Nord* 68 (3–4), 33–74.
Marquer, L., Pomel, S., Abichou, A., Schulz, E., Kaniewski, D. and Campo, E. (2008) Late Holocene high resolution palaeoclimatic reconstruction inferred from Sebkha Mhabeul, southeast Tunisia. *Quaternary Research* 70(2), 240–250. DOI: 10.1016/j.yqres.2008.06.002
Mayet, V. (1886) *Voyage dans le sud de la Tunisie.* Boehm et fils.
Orshan, G. (1972) Morphological and physiological plasticity in relation to drought. In: McKell, C.M., Blasdell, J.P. and Goodin, J.R. (eds) *Wildland Shrubs, Their Biology and Utilization: An International Symposium.* Logan, Utah, July 1971 pp. 245–254.
Raynaud, E. (1853) *Description de la Régence de Tunis.* Imprimerie Impériale, Paris.
Rognon, P. (1996) Sécheresse et aridité: leur impact sur la désertification au Maghreb. *Sécheresse* 7, 287–297.
Rowland, J.W. (1945) *Survey of Land Resources in Tripolitania.* British Administration, Department of Agriculture.
Schulz, E., Abichou, A., Hachicha, T., Pomel, S., Salzmann, U. and Zouari, K. (2002) Sebkhas as ecological archives and the vegetation and landscape history of southern Tunisia during the last two millennia. *Journal of African Earth Sciences* 34(3–4), 223–229. DOI: 10.1016/S0899-5362(02)00021-0
Schulz, E., Smykatz-Kloss, W., Abichou, H., Fromm, R., Pomel, S. *et al.* (1995) Zaderg environmental history of the semidesert region in southern Tunisia. *Geologisches Zentralblatt, Paläont. Teil I*, 3–4, 423–440.
Sherif, A.S., Mahklouf, M.H. and Betelmal, A.G. (2020) Floristic study and species diversity of Msallata-Garaboulli Province in Libya. *Journal of Advanced Botany and Zoology* 7–3, 1–18. DOI: 10.33258/bioex.v2i2.232
UNESCO (1977) Map of the World Distribution of Arid Regions. *Technical Notes* 7.

4

Geomorphology and Landforms

Introduction

Geomorphology and landforms condition the vegetation spatial distribution – more so in arid environments where water is a limiting factor as well as the action of water and wind erosion. Mediterranean arid zones landscapes, geomorphology and landforms present nearly always the same features, mostly due to past and present water erosion during erratic winter episodes.

The northern Libyan landscapes (Jebel Nafusa, Jeffara plain, wadi zones south of Misurata to Beni Walid, Sirte region, southern Jebel Al-Akhdar) are predominantly affected by water and wind erosion and deposits. The southern Jebel Nafusa, the Hamada, Sirte, and southern Jebel Al-Akhdar are affected by wind erosion (transport and deposit).

Water impact on landforms

Water erosion can be violent when strong rainfall strikes the denuded landscape. Rainfall as little as 10–20 mm in a few hours will trigger large amounts of water running down watershed slopes, gathering in sudden walls of water in wadis, transporting huge volumes of rocks, colluvium and alluvium to the lower land (Photo 4.1). With these floods, large temporary lakes may appear in the desert (south Nalut – Derj) with fine sediment accumulating in large endoreic depressions as south Jebel Al-Akhdar in the *Balat* region.

Wind erosion and deposits

Wind erosion and deposits are common features all over the Libyan steppe and more so in the desert. The Jeffara plain is covered with mostly sand deposits fixed by vegetation of *sbot* (*Stipagrostis pungens*). The *reg* and the *erg* on the Hamada and further south, in Sirte and southern Jebel Al-Akhdar, are continuously battered by wind moving fine and coarse sand when wind blows at more than 15–20 km/h (Photo 4.2).

Fine sand and aeolian loam deposits cover large, cultivated areas between Zintan and Al-Asabah. Thick loess quaternary deposits are present on the northern slope of the Jebel Nafusa between Gharyan and Tarhuna. Violent dust storms often occur when the southern hot *Gibli* wind blasts the landscapes.

The terraces landforms in the Mediterranean arid and semi-arid zones

In arid and semi-arid zones, in the Mediterranean region of the old World, in North Africa, the Middle East and further to Central Asia, the

DOI: 10.1079/9781800627154.0004

Photo 4.1. Winter torrential wadi flow (Bir Tuta bridge, N 32.21 E 12.88, 4 Oct. 1994).

Photo 4.2. *Reg* landscape submitted to wind erosion on the Hamada (N 30.70 E 11.88, April 2009).

terraces systems (Fig. 4.1) are similar, with a succession of three (sometimes four) levels of interfluves or terraces (Gintzburger, 2009) capped with hardpan, caliche or crusts. These are the result of long-past climate changes (Le Houérou, 1997), European glaciations (Günz, Mindel, Würm I, Würm II and Würm III) from peri-glaciations episodes in North Africa, and warmer inter-glaciations, alternating long millenaries of cold/wet-humid years when soluble salts (Carbonate $CaCO_3$, Sulphate $CaSO_4 2H_2O$) leached through soil profiles and then percolated

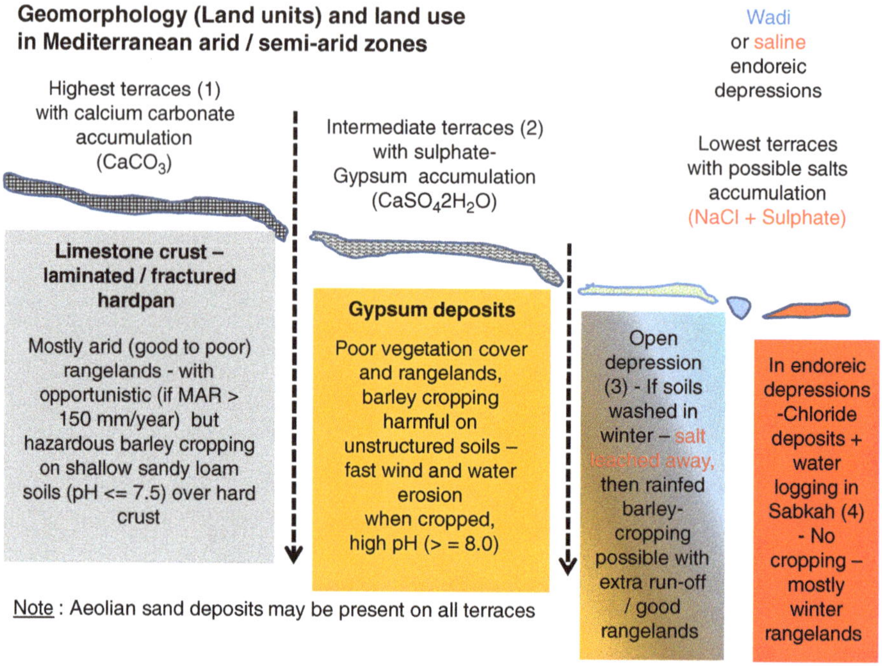

Fig. 4.1. Schematic of the Mediterranean arid/semi-arid terraces in western Libya, Sirte and southern slope of Jebel Al-Akhdar.

back to the soil surface during the warmer long periods, forming hard crusts (see Chapter 5). In Libya, these capped terraces are common on the northern and southern foothills of the Jebel Nafusa, yet not much in the Jeffara plain, and extending onto the Hamada, in Sirte and south of the Jebel Al-Akhdar.

The upper calcareous terraces (1) are capped with a hardpan, sometimes up to a metre thick and more, that sounds hollow. It is laminated in long ribbons of calcareous white/salmon-coloured concretions and nodules. It looks like a geological formation when exposed by water erosion along the edge of interfluves and gullies. However, it is easily recognized as it often caps softer underlayers of marls or colluvial material. This hard crust may be also impregnated with gypsum.

This second lower terrace (2) covers a considerable extent of the Jeffara plain to the Sirte region. It is mostly gypseous, hardened or pulverulent, sometimes with calcareous nodules in the lower horizons.

The third terrace (3) is mostly flat and made of fine alluvium (loam and/or clay) transported over long distance and spread by wadis in depressions (*gar'a* in the west or *balta* in the east) and flats (*feida*). These open depressions, when flooded, clean up from soluble salts (sodium chloride and sulphates) ending up in saline *sebkha*.

The last terrace (4) is the base level where highly soluble salts (NaCl) leach into endoreic muddy saline depressions (*sebkha*). They are often flooded in winter. As summer progresses, they dry up and are covered with pulverulent gypsum on the periphery and sodium chloride efflorescence in the middle.

Sand veils and veneer often cover the terraces at all levels.

The landforms and landscapes of northern Libya

The north-western region (Nafusa escarpments and Jeffara plain)

This is typified in northern western Libya on the Jebel Nafusa where wadis drain into the sandy

Jeffara plain. The Jebel Nafusa elevation varies from Nalut (about 600 m) along the Tunisian border, climbs to around Gharyan (800–900m), drops gently through Tarhuna (about 400 m), Msallata (about 250 m) to Homs on the seashore.

The northern slopes of Jebel Nafusa are a series of monoclinal alternations of hard limestones, soft gypseous marls and crumbling sandstones (Gray, 1971). They all belong to geological formations from the mid-Cretaceous, especially the Turonian (hard dolomitic limestone from Nalut plateau with strata of marls), the Cenomanian (Calcareous sandy tuffs, sandstones tuffs and multicoloured clay) and the deep Albian (compact marly limestone sandwiched with sandy grey-green marls), the last two exposed below Giado escarpments. Worth mentioning are the volcanic features (*burkan*) with phonolites, porphyry and basalt lava fields – (post-Ecocene/Quarternary) along north-west/south-east geological faults and dotting the landscape from east of Kikla fault, Bir Tuta, Qawasim (Kaf-el-Tekut), Gharyan and east of Mizda on the Hamada towards Beni-Walid.

The thick monoclinal layers from the Jebel Nafusa form a series of cuestas gently sloping to the south (Fig. 4.2). To the north, it ends with multiple cliffs and spurs with overhanging hard and soft layers abruptly dropping 250–300 m into the Jeffara plain. The cliffs are dissected with deep canyons and valleys (*foum*), bucketing rocks, scree, colluviums and alluviums, onto lower glacis or interfluves on long distances to alluvial depressions (*gar'a* or *feida*) and faraway lower saline *sebkha*. In the valleys, from Gharyan to Tarhuna, thick loess deposits are frequent.

The upper glacis, closest to the Jebel foothills, usually bear a calcareous caliche (or hardpan). Further down, the second glacis is capped with a calcareous crust still, but more often with a gypseous crust. These hardpan, caliche and crust are the results of long-past climatic pedogenesis. The glacis are dissected by wadis and deep ravines.

Endoreic wadi courses (wadi Nalut, wadi Mzerah, wadi Ethel, wadi Er-Rabt, wadi Gharyan, wadi Hira, wadi Abu-Shayba reaching Bir-el-Ghanem, etc.), flowing from the northern slopes of the Jebel Nafusa, disappear deep under the Jeffara plain sands. Small dry artesian white mounds, surrounded with gypseous crust aureoles, speckle the Jeffara plain south of Sorman.

Rare and short coastal wadis reach the sea between Tripoli and Homs: wadi Megenin irregularly flowing north to Tripoli, and wadi Tareglat, wadi Kaam and wadi Lebda between Tripoli and Homs, descending from the Msallata hills.

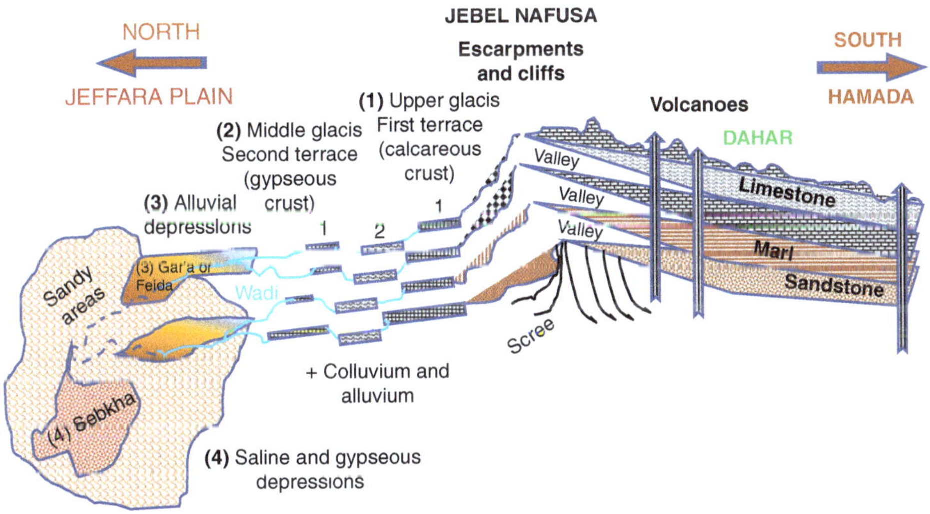

Fig. 4.2. Landforms transect from south to north, from Hamada to Jeffara plain.

South of the Nafusa: The Hamada-el-Hamra (the Hamada)

South of the Jebel main escarpment starts the *Dahar*. The top Cretaceous limestone tableland with gently rolling country extends south of Giado – Zintan – Yefren – Gharyan – to south Tarhuna. This strip of land, 10–50 km wide with thick *ard hamri* pinkish sandy-loam soils (a few metres over limestone), still allows sustainable rainfed arboriculture (olives, figs and almond trees) and rainfed barley cropping.

Further south, as rainfall drops drastically, begins the Western Hamada, south of Nalut, Zintan, Mizda – vast extents of desertic Palaeocene limestone rocky plateau slopping to the south, with immense bare rocky, gravelly (*reg*), and sandy flats and sand dunes (*erg*, with isolated *gara*), low ridges and scarps broadly oriented west–east; the plateau, with elevation ranging from 400 m to 600 m, is intersected with small endoreic wadis and gullies, and sometimes bears scarce contracted Saharan vegetation in small pans covering a few hundred square metres. The desertic Hamada extends further south from the latitude of Ghadames – Derj – Ash-Schwarif – for another 200 km to the Ubari sand sea (*erg*, *ramla* or *edeyen*).

East of Mizda – As-Schwarif, further south of Tarhuna – Beni Walid, the Eastern Hamada, has a rugged landscape dissected with large wadis flowing south-west–north-east. These are wadi Beni-Walid, wadi Meimun, wadi Sassu (flowing into the powerful Sof-el-Jin), and wadi Zamzam (with several modern artesian wells), all parts of the Solul-el-Khoder region, sporadically draining eastward into the large Taworga coastal *sebkha* that has a vast sweet-water artesian lake close to the seashore.

They are adjoining the western Sirte region.

The Sirte middle zone

This is where the Sahara reaches the Mediterranean Sea from Bu-Gren to Sirte, El-Agheila and Ajdabiya. Most are upper Cretaceous-Eocene deposits. This is the antique Syrtica region.

The western Sirte region tableland drops from 200 m to 300 m in the west to the sea level to the east. It is largely tertiary limestone and gypseous marls (Palaeocene to Miocene) plunging under quaternary deposits along all the coastlines. This desert landscape presents numerous geologic faults running north-west to south-east. The landscape is nearly flat with numerous dry channels, eroded gullies and some *reg* extending south, deep inland. The large watershed of wadi Bey-el-Kebir ends in the coastal *sebkha* east of Buirat-Al-Hasun. Many short wadis run northward from the tableland to the coastline peppered with numerous muddy, salty and gypseous white, glaring *sebkha*, surrounded with white limestone sand.

Past Ras-Lanuf and El-Agheila, (the antic limit between western Libya (Tripolitania) and eastern Libya (Cyrenaica), the landscape changes to an even country on solid limestone and marls without geologic fault. This is also where the Sahara proper reaches the Mediterranean Sea. Quaternary white oolitic sand and consolidated dunes border the coastline.

Reaching the Ajdabiya trough, long strings of large *sebkha* and quaternary white sand extend up to Zuwetina on the coast. Inland, south of Ajdabiya, along the road to Kufra, whitish fixed sandy desert country spreads out to El-Beda settlement and further south to the deep desert of Gialo, sinking into the flat Serir.

The Jebel Al-Akhdar

The Jebel Al-Akhdar covers the two antique Greek-Roman regions of Cyrenaica (western Jebel Al-Akhdar) and Marmarica (east of Jebel Al-Akhdar).

CYRENAICA. North of Ajdabiya, towards Soluk, starts the Cyrenaica platform with a gently undulating dusty plain extending from El-Magrun to Soluk; the first rolling Miocene hills of eastern Libya begin extending to the red karstic Benghazi plain. This low, curved coastal plain (elevation 40–80 masl) wedges north-east from Soluk towards Daryanah, Tokra and ending in Ad-Dirsiyah against the first escarpment. The wadi Qattara, collecting water on the second escarpment, cuts the first escarpment flowing westward to the salt flats south of Benghazi.

This first limestone escarpment (Cretaceous to Eocene), locally called *lusaita* or *al-Wasaita*, raises to 150–200 masl, encloses the eastern part of the Soluk plain and the Benghazi plain, curls as a crescent pointing eastward, encloses the El-Marj plain (elevation 250–350 masl), extending as far as Marsa-Suza and Ras Hillal, and fades away near Derna. The northern edge of the first escarpment is dissected with numerous short gorges and deep wadis flowing into the sea. The El-Marj plain is a large polje (about 50,000 ha), a common karstic formation on Mediterranean limestone country, dotted with dolines, sinks and caves, receiving water from the second escarpment.

The second limestone escarpment called *al-zahir* is the domain of *al-ghaba* (the forest), raises east of El-Abyar (300 m elevation) to Jardas-Al-Abid (600 m elevation) onto a rugged tableland of fractured geology, deep wadis, canyons and karst forested landscape. From El-Marj, the road climbs this second ridge to Taknis (450 m), continuing the top of the tableland through the jugged canyons of the wadi El Kuf, to El-Beida-Shahat (600 m) on the northern edge of the limestone ridge, to Labraq (700 m) and descending to Guba (550 m) and Derna on the seashore. Again, many wadis originate on the top of this second escarpment and descend sometimes onto the first escarpment and further north to the sea. Water collected on watersheds from the highest rainfall zone of Libya mostly disappears into the many karts' sinkholes and caves, reappearing in wells and springs along the ridges (Shahat) and onto the lower land.

The southern forest-matorral-covered limestone slope of the Jebel rises eastward to Marawah (450 m), Gendula (650 m), Soluntah (780 m), Faidiya (750 m), reaching the Al-Qayqab polje (750 m) adjoining Labraq.

About 20 km south of Gendula-Soluntah, the top tableland gently slopes into steppe country where *Juniperus* trees disappear fast, still with the last rainfed barley crops in wadi and the last dolines. Wadis flow southward into the desert, along geologic faults oriented north-west/south-east, and spreading into large depressions (*balta*) south of Mekhili (180 m).

Further north-east, on the eastern side of the gorges of wadi Derna, we are entering Marmarica.

MARMARICA. Marmarica adjoins Cyrenaica to the east. It extends approximately from east of Derna to the Gulf of Bombah-Tamimi. This region is locally called *Al-Batnan*. Then comes the *Al-Dafna* stretching to Bardia-Sollum and further into Egypt to Marsa-Matruh.

From Derna, eastward, we enter the large Merleb Garab red soils polje (approx. 13,000 ha, 200 m, approximately corresponding to the El Marj polje elevation) cut by the gorges of wadi Chalig and wadi El-Hmesa running from the west, turning north-east into the sea at Marsa-el-Chabta. South of the polje, the rainfall again drops radically when we reach Martuba (300 m elevation) with a steppic hilly country regularly sloping down eastward to the Gulf of Bumbah. Ten to fifteen kilometres south of Martuba, the large catchment of wadi Maalegh now flows straight west–east into the Gulf of Bumbah. Further south, geologic faults in the limestone and marls run north-west–south-east, collecting water from many small wadis into wadi Es-Sceggaa/wadi Tamimi, ending at Tamimi.

Not far east from Tamimi, from Ain-el-Ulaymah along the coast, starts the Mio-Pliocene outcrops of sandy limestone, sandstone, shale, sand, white marls ridges running parallel to the coast to Tobruk. East of Tobruk is the elongated polje (Kambut) with large, eroded interfluves parallel to the coast. The first scarp (60 m) is the location of the actual Tobruk harbour. The second scarp, only 8–10 km south of Tobruk town, reaches 130–140 m elevation followed by a third minor scarp (120 m) another 15–17 km south near Adam. The three ridges merge south of Kambut and drop abruptly into the sea at Bardia-Sollum, at the border with Egypt.

Conclusions

Cretaceous limestone is predominant in the geology of northern Libya with the Jebel Nafusa rise and the karst of the Jebel Al-Akhdar extending into Marmarica. The Sirte region results from marine transgression millions of years ago. These landscapes were altered by long-past peri-glaciations and alternating humid-dry eras. Most important in understanding the vegetation ecology is the formation of paleo-terraces capped with calcareous and gypseous crust covering large areas of the steppe and the pre-desert, common features of all the arid Mediterranean regions. Saline coastal *sebkhas* and endoreic depressions dot the landscape. The sand coming from nearby oriental *erg* is continuously blown by westerlies over the north-west and the Sirte region where it remains an important environmental feature. Wind and water erosion keep altering landscapes and soils, certainly at a very slow pace, nonetheless aggravated by current relentless land clearing and overgrazing.

Climatic conditions modelled the soils of northern arid Libya and the native vegetation cover we are now going to explore.

References

Gintzburger, G. (2009) *Landforms and Geomorphology on Libyan Rangelands – Terraces.* Range Development Projects (Suani) Training Session.

Gray, C. (1971) Symposium on the geology of Libya. Faculty of Sciences, University of Libya, Tripoli.

Le Houérou, H.-N. (1997) Climate, flora and fauna changes in the Sahara over the past 500 million years. *Journal of Arid Environments* 37(4), 619–647. DOI: 10.1006/jare.1997.0315

5

The Libyan Soils

Abstract

Soil classification is complex and requires examination of deep soil profiles and specialized expertise. Based on research and experience in neighbouring southern Tunisia, we provide a simple identification of the main arid Libyan soil types including aspect, colour, texture, structure and soil surface with information about associated native vegetation. Our classification focuses on the topsoil and the upper horizon supporting the native vegetation or rainfed crops. While our soil type description does not follow the complex US, European or Soviet classifications, it remains easily accessible to rangeland ecologists and land developers. Based on our classification and field experience, we produced a synthetic soil map of northern Libya. It provides for easy inference and soils identification from vegetation type, ecological group, plant association, or presence of specific indicator plants.

Introduction

Soil identification and potential have been the focus of communities and land developers for agricultural purposes since time immemorial. With increasing expectations from fast-expanding local populations for higher agricultural system production, scientists focused their research on soil types, texture, structure, chemical composition and agricultural value. It was only by the mid-19th century that they turned their attention to soil surveys and mapping from arid regions from North Africa and Central Asia. They soon discovered the agricultural potential of some arid-zone soils, and their relationship to the native rangeland vegetation. However, it was not without dissent between the different soil research schools from the Soviet Union, Europe and USA, with their observations and experience not always suitable and relevant to the Libyan soils and climatic conditions. Soil scientists working in North Africa developed their own soil-type classification especially fitting the Libyan arid zones.

Mapping the Libyan Soils

Soil surveys, research, and mapping in Libya has a long history dating back to the early 1950s (Elaalem *et al.*, 2021; Rowland, 1945, Zurqani, 2021). By the early to mid-1970s, the European soil classification was used for agricultural development purposes in the western and Sirte zones (GEFLI, 1972a, b, c, d; 1974a, b; 1977). At the same time, a Soviet team of 500 staff from Uzbekistan and Turkmenistan, all well aware of steppic and arid-zone soils, used the Russian soil classification system for the most comprehensive Libyan soil mapping. They carried

DOI: 10.1079/9781800627154.0005

out detailed field surveys supported with extensive soil sampling and chemical analysis (Belobrov, 2015; Selkhozpromexport, 1980a, e, f; Sishov and Kharin, 1980) mapping the whole of northern Libya, the Jeffara plain, part of the Jebel Nafusa, the Sirte region, and the north-east region. These were published by the Selkhozpromexport team at 1:200,000 scale (Selkhozpromexport, 1980b, c, d), then simplified and published at 1:1,000,000 scale (Selkhozpromexport, 1984a, b). These Soviet maps were field checked, extremely detailed and accurate.

By 1990, the US soil classification was adopted by Libyan soil scientists and foreign contractors with many local pedological maps produced for agricultural development at large scale (1:20,000 to 1:50,000). The small-scale Soviets' pedological maps covering most of northern Libya were subsequently converted to the US soil classification; however, without ground-truthing. These Soviet maps were then redrawn by ACSAD (Arab Centre for the Studies of Arid Zones and Dry Lands) and the GWA (General Water Authority) at 1:250,000 and 1:1,000,000 scale, developed out of a complex combination of digital elevation model, geology, lithography and pedology (ACSAD-GWA, 2005). This northern-region new soil map was laboratory-interpreted and synthetized by ACSAD, and later by a Dutch team from the Soviet maps. Unfortunately, ACSAD and the Dutch complicated legends associated with the US soil classification poorly reflect the field reality. Other small-scale soil and land-use mapping followed (FAO, 2006), adding little to what was already known from the Libyan soil types and distribution.

Which Soil Classification to Use for Libyan Rangelands?

The USA (USDA, 1999) and FAO (FAO *et al.*, 1998; FAO Soils, 2015) soils classification systems, while comprehensive and universally accepted, are too complex and difficult to adhere to for most rangeland ecologists unless specializing in soil sciences. By contrast, the Russian soil classification (Stolbovoi, 2002) is relevant to the Libyan environment as generally referring to steppic soils and arid rangeland from Central Asia (Gintzburger *et al.*, 2003).

For the purpose of simplification, our classification intends to bring soil observations that most field botanists and environmentalists can refer to. Our classification focuses on the topsoil and the upper horizon supporting the native vegetation or rainfed crops and, most of the time, without having to dig a deep soil profile. We use simple terms that would assist most colleagues in identifying soil types during rangeland surveys and field trips. We refrain from using the American or the FAO soil classifications used these days in Libya (Elaalem *et al.*, 2021), which we find difficult to use for non-soil scientists. For our practical fieldwork on Libyan rangelands, we adopted the soil descriptions and classification published in nearby central and southern Tunisia (Belkhodja *et al.*, 1973; Le Houérou, 1959, 1960, 1969; Novikoff, 1961). They are entirely relevant to the Libyan soil types occurring from semi-arid to Saharan environments. It has the advantage of being pragmatic, easily and rapidly recognized by rangeland ecologists. We give a summary and description of their findings that readily suits the Libyan conditions.

The Soils on Libyan Rangelands

The predominant soil types present on the rangelands of the northern part of Libya belong to:

- Developed soils

Terra Rossa and brown-chestnut soils: always heavy textures – clay soils, exclusively found in the Jebel Al-Akhdar resulting from alteration/dissolution of limestone bedrocks in the highest-rainfall zones and, the latter, under dense matorral vegetation cover. They evolved to degraded shallow lithosols under water erosion and overgrazing.

Steppic soils: with a range of texture ranging from sand to clay, developing under a low vegetation cover, hence with a low organic content with slow evolution due to low rainfall. These soils evolve towards lithosols by the action of water and wind erosion intensified by overgrazing.

- Young soils
    - Non climatic soils;
    - Raw climatic soils – desertic soils.
- Paleo-soils: as exposed or buried hardpan calcareous and gypseous crusts from long-past climatic era (see Chapter 4). These are

shallow soils hosting sparse xerophytic vegetation.

- Azonal halomorphic soils: independent from rainfall, hence azonal. They are located in endoreic depressions close to the seashore and inland. They are rich in soluble salts (sodium chloride (NaCl) and calcium sulphate dihydrate ($CaSO_4.2H_2O$)).

Developed soils

Terra rossa and brown-chestnut soils

These red Mediterranean[1] and brown-chestnut soils have a clay texture and a lumpy structure. They occur in mosaics exclusively in the Jebel Al-Akhdar on the highest-rainfall zones (MAR ± 300–450 mm) under a matorral/Phrygana vegetation type (*Quercus calliprinos, Pistacia lentiscus, Ceratonia siliqua, Juniperus phoenicea*, etc.).

The *Terra rossa* (Photo 5.1) soils are associated with karst formations.

Karst are Mediterranean landscapes that develop on calcareous or dolomitic bedrock (Miocene to the late Pleistocene – quaternary) by alteration, dissolution of calcium carbonate over thousands of years by rainfall, free running and/or stagnant water. It creates characteristic landforms as chaos of altered bedrock, lapiaz, ridges and pavements with sharp edges, sinkholes, depressions (dolines), caves and subterranean water courses. Typical karsts are exclusively present on the highest-rainfall zones of the Jebel Al-Akhdar from El-Marj, wadi Kuf, Beida – Shahat to Derna.

Terra rossa results from mineral alteration (rubefaction) due to long-lasting alternating warm/humid periods followed by dry episodes. Iron freed during the desiccation episodes gives it a bright red colour. Alumina and silica during these processes recombine into fine loam-clay material (kaolinite and illite (alumina-silicate)) accumulating with organic matter (OM) in the karst's cracks and pockets. Besides the red colour, the clay content may be between 50% and 60% and 30–35% loam with about 1–3% organic matter (Omar, 2023). These soils have a low water reserve and are quite dry. Water infiltration is fast when the *Terra rossa* is dry, but soon turns to sticky mud when wet. These heavy red soils, common in Benghazi and El-Marj plain, have been used for ages for rainfed cropping and fruit trees (figs, olives, almonds, vines etc.) and rainfed cereal cropping.

Photo 5.1. Karst formation (Lapiaz) with *Terra rossa* (Labraq, N 32.795 E 21.993).

On matorral sites with dense vegetation cover in the highest-rainfall zones, the upper horizon (first 5–10 cm) has an organic matter content as high as 8–10% giving a brown (reddish)-chestnut colour (Photos 5.2 and 5.3). The top layer structure of these soils is gritty/clotted, turning polyhedric and lumpy in the sublayers including calcium carbonate nodules, and filamentous concretions witnessed by the presence of calcicole plants (*Cistus* sp., *Sanguisorba minor, Sarcopoterium spinosissimum*). Climatic alteration and leaching result in some profiles being totally free of calcium carbonate with pH slightly acidic, and hosting calcifuge plants (*Erica* sp., *Arbutus pavarii*).

Rendzina[2] and lithosols

This soil has only a thin (5–20 cm) organic horizon (5–10% OM) sitting directly on the fractured

Photo 5.2. Mosaics of *Terra rossa* and brown-chestnut soils under dense matorral (*Quercus calliprinos, Pistacia lentiscus, Ceratonia siliqua, Juniperus phoenicea*, etc.) – N 32.872 E 21.971.

Photo 5.3. Degraded *Terra rossa* soils accumulated behind a small water harvesting terrace (Gendula, N 32.542 E 21.564, south Jebel Al-Akhdar) with patches of brown-chestnut soils under shrubs.

limestone bedrock. The organic horizon is brown under matorral vegetation to dark grey under Aleppo pine, with a fine polyhedric texture (loamy clay) abundantly colonized with roots and rootlets. The transition to the geological substrate is quite abrupt and clear-cut, with hard fractured limestone, or softer marl and fine colluvium may be exposed. The rendzinas are dry soils with little water-holding capacity. Cultivation or overgrazing removes the organic matter by water erosion, leading to lithosols with the fractured geological substrate exposed.

Steppic soils

With rainfall diminishing in adjacent southern zones, with low vegetation cover, the soil organic matter content decreases due to soil erosion, relentless land clearing, rainfed cropping and grazing. It gives room to brown sub-arid soils, brown sub-desertic soils, lithosols and grey sub-desertic soils (often named *Sierozem* (*Sensu lato*) in the Russian soil classification).

The transition from one steppic soil type to another is not always clearly discernible without examining soil profiles.

Brown sub-arid soils

They have a loamy texture and occur mostly under 150–250 mm MAR on slopes and hilly country of the Jebel Nafusa (Photo 5.4) covered with *Stipa tenacissima*, *Genista microcephala* (Yefren to Homs). They often evolve towards skeletal lithosols on steep slopes and on overgrazed sites.

Special mention should be made of these brown sub-arid soils in northern Libya that are covered with:

- *Ard Hamri* (in the Dahar region on the southern slope of Jebel Nafusa – Zintan) which are pinkish sandy-loam traditionally planted with rainfed fruit trees and occasionally cropped with rainfed barley; and
- localized thick silty to loamy aeolian loess: yellowish to brown deposits (in the Jebel Nafusa – Gharyan to Tarhuna-Msalata) (see 'Soils issued from wind transport' below).

The brown sub-arid soils are also present in patches on the southern slopes of the Jebel Al-Akhdar; they appear under a residual clear vegetation of degraded stands of *Juniperus phoenicea*. The superficial horizons have a yellowish-brown colour, an organic content of 1–2% with fluffy calcareous nodules in deeper horizons. They have a higher organic matter content than the next brown sub-desertic soils. These sites are usually used for grazing and rarely cropped due to rocks, vegetation and land clearing costs.

Brown sub-desertic soils

They occur in the 150–200 mm MAR zones on all topography but are also encountered down to the 100–150 mm MAR, on large rangeland flats (Photo 5.5) and along wadi terraces. Their organic content is low, in the range 0.5–1.0%. They have a loamy texture with stones from fractured calcareous crust, carbonate nodules and

Photo 5.4. Brown sub-arid soils (Tarhuna, Msallata to Homs, N 32.496 E 13.615) with *Stipa tenacissima*.

Photo 5.5. Brown sub-desertic soils profile with horizon of white limestone crust at 40–50 cm (plateau between Misurata and wadi Sassu, N 32.202 E 14.812).

concretions in the profile below the crust. They remain an easy target for opportunistic and risky rainfed barley cropping on flat country and due to low fertility, low water holding capacity and marginal rainfall zones where they occur.

These brown sub-desertic soils could also be located, though rarely, into more arid sites vegetated with *Rhanterium suaveolens – Argyrolobium uniflorum* on reddish-brown loam with carbonate nodules covered with sand veil (Gintzburger and Alidrissi, 1976). In this case, rainfed cropping using disc plough or even minimum tillage implements is regularly leading to failure, not only due to low rainfall and higher inter-annual and intra-annual rainfall variability, but mainly due to repeated cultivation stimulating wind erosion removing the superficial sandy-loam veil, exposing the deeper hard loamy horizon (Fig. 5.1; Photo 5.6).

South of the Jebel Al-Akhdar, brown sub-desertic soils are present on large extent allocated to rainfed barley cropping with water harvesting and soils conservation operations (Photo 5.7 and 5.8).

At their most southern extension limit, brown sub-desertic soils gradually give way to grey sub-desertic soils.

Grey sub-desertic soils

They have a lower organic matter content than the brown sub-desertic soils. They are loamy and characterized by a greyish colour due to less free iron giving the reddish hue and less organic matter giving the brown shade. The presence of sublayers with calcareous (or gypseous) accumulation or even overcalcareous and gypsum paleo-soil crusts result in a soil surface with abundant fractured stones. They occur close or below the 150 mm MAR on large glacis and flat areas covered with sparse *Arthrophytum scoparium*, *Artemisia herba-alba* and *Anthyllis henoniana* as found south of the Jebel Nafusa (Haraba – south Zintan) on the northern fringe of the Hamada (Mizda – Seddada), the Sirte region to Ajdabiya and extending to Marmarica (Tobruk – Bardiya). Large flocks of small ruminants graze these areas (Photo 5.9).

A note on crusted topsoil

On loamy soils or brown and grey soils on degraded rangelands, the topsoil is often covered with a thin (10–15 mm) lamellar crust (Photo 5.10) resulting from lack of organic matter, poor soil structure, compounded by rainfall impact and

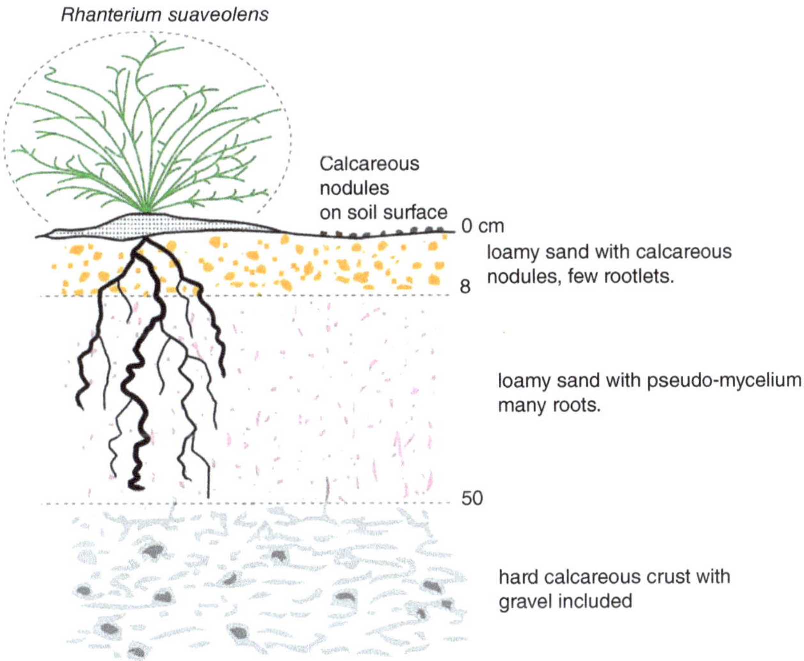

Fig. 5.1. Soil profile of a degraded *Rhanterium suaveolens* (RS) on sand veils due to endless cropping at Bir-el-Ghanem (Gintzburger and Alidrissi, 1976).

Photo 5.6. Sand veil and topsoil removed, exposing reddish-brown loam soil hard sublayers following repeated cropping (Bir-el-Ghanem, N 32.315 E 12.639).

Photo 5.7. Brown sub-desertic soils assigned to rainfed cropping (between Mafsel and Bir Shaweb, N 32.432 E 21.512, South Gendula, Jebel Al-Akhdar).

Photo 5.8. Brown sub-desertic soils (Mafsel, N 32.421 E 21.517) with the most southern *Juniperus phoenicea* trees (MAR ± 175–200 mm, south Gendula, Jebel Al-Akhdar).

running water battering the loam. It may also be colonized with biogenic crusts made of lichens, fungi and cyanobacteria, a common occurrence on heavy soils in arid and semi-arid Central Asian (Orlovsky *et al.*, 2004) and Tunisian (Schulz *et al.*, 2014) rangelands with *Catapyrenium lacinulatum, Heppia lutosa, Collema* sp., *Diploschistes diacapis* and *Fulgensia fulgens*. This thin but compact topsoil crust prevents water infiltration and emergence from native plants after germination. Trampling and recurring cultivation of these crusted soils will only accelerate water erosion. Techniques (Gintzburger *et al.*, 1986a, b; Gintzburger and Skinner, 1985a, b, c) to trigger rangeland vegetation regeneration on such crusted topsoil exist and have proved efficient using native rangeland plants such as *Artemisia herba-alba* and annual Medicago (see Chapter 16).

Young soils (not developed)

Non climatic soils

They result from soil material being reshuffled and transported either by water or wind.

Photo 5.9. Grey sub-desertic soils on the northern edge of the Hamada (South El-Haraba, N 31.720 E 11.580) (Author's own photo)

Photo 5.10. Crusted topsoil on loam on degraded *Artemisia-herba-alba – Haloxylon scoparium* vegetation type (wadi Sassu).

Soils issued from accumulation due to water erosion

These soils are issued from water transport: either as colluvium (lateral transport on slopes) and on short distances on terraces and glacis (Photo 5.11), or alluvium and eluvium (longitudinal transport on longer distance into wadi, flood plains and depressions).

COLLUVIUM. They are composed of coarse material (scree, rock, stones, gravels) creating glacis from overhanging cuesta and peaks. Slopes may vary from very steep to a few degrees.

Photo 5.11. Coarse colluvium and alluvium on wadi terraces (N 32.047 E 20.733), wadi Maghur, south Jebel Al-Akhdar).

The longer the distance spreading the colluvium, the finer the soil textures are.

ALLUVIUM AND ELUVIUM. Reshuffled fine sediments (loam and clay) are transported over long distances and form young soils on large flats and flood plains. Once rich productive lowland rangelands, these flats, sometimes covered with a sand veneer, have been turned into rainfed cereal cropping land or large development projects in the Jeffara plain after clearing native vegetation (*Ziziphus lotus* Nebkha, *Rhanterium suaveolens*, *Artemisia herba-alba*, etc.).

South of Karuba and Mekhili (south Jebel Al-Akhdar), the *balat* are large endorheic flat loamy-clay depressions in which drainage systems and wadis (wadi Samalus and others) have spread their fine sediments over millenaries to these days. Some of the largest *balat* (Balta Az-Zalaq (about 6000 ha), Balta Abu-Khays (2000 ha), Balta Barir (1000 ha) etc.) are opportunistically cropped with rainfed barley. Long water harvesting trenches are dug across *balat* (Balta Muhammad, Balta Lawlab al Ḥalq, Balta al Lubaydah, Balta Mani, Balta al Atash, Balta Mazatah, Balta al Milaz, Balta Minqar al Gharnuq, etc.) to collect and keep water long into the summer for livestock watering and encampments.

Alluvium and eluvium result from seasonal and powerful flooding when winter rainfall collected in the denuded catchments rushes down Jebel and slopes. A coarse matrix of gravels and rocks with little fine sediments and little organic matter are transported along wadis descending from the northern slope of the Nafusa (Bir Tuta, Bir Ayyad) with *Pennisetum setaceum*, *Cenchrus ciliaris*, *Hyparrhenia hirta* and *Cynodon dactylon*. If not revamped or disturbed by renewed catastrophic water erosion, they may slowly evolve towards brown sub-arid or sub-desertic soils according to the rainfall. These are most common in the Jebel Nafusa wadis and adjacent northern and southern foothills. They also extend below the Nafusa northern foothill into the Jeffara plain onto large water-spreading areas much richer in fine/loamy sediments once covered with *Artemisia herba-alba* (wadi Ethel, southern part of wadi Abu-Shayba), targeted and cleared for large-scale rainfed barley cropping or irrigation projects (Bir-el-Ghanem).

These soils expanding on huge areas have various vegetation types from *Artemisia herba-alba* on large loamy flats or piedmont glacis on weak slopes (Bir Tuta, Oushtata, and south Jebel Nafusa) to *Retama raetam* – *Ziziphus lotus* in large wadis (wadi Sof-el-Jin, Photo 5.12) and large flats

Photo 5.12. Large wadi sediment flats used for barley rainfed cropping (wadi Sof-el-Jin, N 31.864 E 14.070) with *Ziziphus lotus* Nebkha).

(wadi Abu-Shayba, wadi el-Hay, wadi Diga). These soils are common targets for large rainfed barley cropping due to appropriate average annual rainfall, usually 150–175 mm (wadis, flood plains and depressions in Solul-el-Khoder), but rarely for rainfed wheat that requires in Libya at least a minimum 300 mm MAR.

In the more arid northern fringe of the Hamada and in the southern arid parts of the Jebel Al-Akhdar, they transition towards grey sub-desertic soils on large gently sloping areas with fine-textured soils covered with *Artemisia herba-alba* and/or *Arthrophytum scoparium* devoted to sheep and goats' summer/autumn grazing.

Soils issued from wind transport

INLAND LOW SAND DUNES. Deep fixed yellowish-pinkish sand deposits (up to 3–10 m thick) cover vast gently rolling areas of central Jeffara (Photo 5.13). These sand dunes are mostly fixed and colonized with *Stipagrostis pungens* and *Retama raetam* (from El-Watyah – south Sorman to north Bir-el-Ghanem – Aziziya) to the east.

Rainfall permitting along the western coastal zone, private irrigated farms are expanding fast, drawing on limited underground water resources. These fixed sandy sites are traditionally planted with olives, almonds, figs and palm trees. Inland (Jmeil – Ajaylat – Regdalin), they are opportunistically cropped with barley and fast reversing to an *Artemisia campestris – Thymelaea hirsuta* post-cultural vegetation. Repeated ploughing with disc plough of this loose sand induces wind erosion that is difficult to control.

A large mobile sand dune patch (about 23,000 ha) covers the land south of Sidi Benhur (south-east Tajura, Photo 5.14). As these dunes are threatening the surrounding irrigated farms, dunes fixation projects were implemented with forestry plantations of *Acacia cyanophylla* and *Eucalyptus* sp., with asphalt sprays or grids of dry material issued from *Stipagrostis pungens*. Wood harvested from these trees is transformed into charcoal for the local market.

A triangular large area of higher and more mobile dunes expands on more arid zones (MAR < 150 mm) close to the Tunisian border, between Takut – Tiji – Giosh and Al-Assah – Beshiul, Shabunyah to El-Watiya – Bir-el-Hamra, where *Ephedra alata* and *Calligonum* sp. are more common among *Stipagrostis pungens*.

LOAMY SAND ON CALCAREOUS CRUST. Issued from loam and fine sand deposits brought from north-western winds blowing over the Jeffara, these shallow soil formations occur on the northern foothill on the Jebel Nafusa (from Takut to Bir-el-Ghanem). They cover large areas with buried or exposed calcareous crust with various vegetation of *Artemisia herba-alba*, *Anthyllis henoniana*, and *Arthrophytum scoparium*, with patches of *Rhanterium suaveolens* etc. These areas are exclusively devoted to grazing.

Photo 5.13. Inland fixed sandy areas (south Sorman – track to Al-Assah through Bir Shajarah (N 32.494 E 12.117)).

Photo 5.14. Asphalt spray to fix mobile sand dunes (*c.*1976) before afforestation project with *Eucalyptus* and *Acacia saligna* (Sidi Benhur, N 32.730 E 13.573), south-east Tajura).

COASTAL SAND DUNES. White sand deposits all along the coast result from sea gusts. It could be low, mobile white sand dunes close to the seashore. It often covers fossilized sand dunes and Gargaresh stones formations, the latter quarried since antiquity for building material (Sabratha, Oea, Leptis Magna). In the Sirte region, these white coastal sands are common, again close to the coast, bordering *sebkha* and covering gypsum layers. They become less common along the rocky coast of Jebel Al-Akhdar and Marmarica.

LOESS DEPOSITS. They form thick (up to 20–30 m) quaternary homogeneous aeolian deposits on slopes of the Jebel Nafusa extending to Homs (most abundant near Tarhuna-Shiarchara, much thinner extending to Gharyan – Takut). The loess yellowish matrix is undifferentiated mostly carbonate-gypseous fine loam and very fine sand. It is compact and biocrusted on the topsoil when dry and becomes slippery and sticky when wet. As their surface is crusted, they are often subject to strong water erosion leading to gully erosion (road approaching Tarhuna – deep gullies and bad lands) covered with sparse *Stipa tenacissima*, *Rosmarinus officinalis* on slopes and *Lygeum spartum* in the bottom of the gullies. The thick stable loess formations have long been used to dig troglodyte homes (*dawamis*, sing. *damus*) by the Nafusi population.

SAND VEIL/VENEER. This is a common occurrence on all landscapes in the western zone, Sirte (Photo 5.15) and south Jebel Al-Akhdar on all substrate (rocks, gravels, various soil types). The mobile sand veneer may be from a few centimetres to 1–2 m thick and covered with *Stipagrostis pungens*, *Rhanterium suaveolens* (Bir-el-Ghanem) or *Arthrophytum schmittianum* (south Al-Assah, Jred – east of Bir Ougradia). *Lygeum spartum* steppe would grow on a thinner sand veil covering gypseous substratum (Magdar-es-Sbot, East Al-Assah).

MICRO-DUNES. These result from shallow (a few centimetres to half a metre) pyramidal accumulation of mobile sand (Photo 5.16) behind bushes (*Ziziphus lotus*, *Lycium* sp., *Arthrophytum schmittianum*, *Anthyllis henoniana* etc.). The sand is often colonized with annuals (*Shismus barbatus*, *Cutandia dichotoma*, annual *Medicago* sp., etc.).

NEBKHA. These are mounds (0.5–15 m diameter (Photo 5.17) of aeolian material accumulated by plants such as *Ziziphus lotus* (Bir Saman, Jbebina), *Nitraria retusa*, *Limoniastrum guyonianum*, *Tamarix* sp., *Atriplex halimus*, and various Salsolaceae (at Hebiliyah, Taworga, Kararim, Sirte coast). The ones from spiny *Ziziphus lotus* and dense *Atriplex halimus* accumulate organic matter and nitrogen as they serve as refuge to a rich wildlife (birds, rodents, snakes, foxes, jackals, etc.).

Raw climatic soils

These are shallow raw soils often with little, or totally devoid of, vegetation. They do not evolve due to unremitting fracturing action of

Photo 5.15. Sand veneer with *Stipagrostis pungens* and *Rhus tripartita* (El-Hisha, N 31.321 E 15.664), east Bu-Green).

Photo 5.16. Micro-sand dunes accumulating behind *Lycium* sp. and *Ziziphus lotus* bush (near Bir Beshiul, N 32.501 E 11.612, south-west Al-Assah).

Photo 5.17. Sand accumulated in a large *nebkha* by a bush of *Ziziphus lotus* between Jbebina and Bir Saman (N 32.909 E 11.576, Al-Assah).

alternating extreme temperatures, incessant wind removal and deposit in small depressions accumulating fine soil benefiting from rare run-off.

REG – DESERTIC RAW SOILS. The Hamada *reg* formations occur below 100 mm MAR. They result from wind erosion and deflation exposing geological rock layers to extreme alternating temperatures causing slow rock shattering. Hamada surfaces are rather flat, covered with fractured slabs, rocks, stones, gravel, over vast desertic areas (*Reg*, Photos 5.18, 5.19 and 5.20). In between

Photo 5.18. The *reg* on the Hamada, north of Girzah (N 30.971 E 14.546)

Photo 5.19. *Reg* on Hamada at wadi Nefad (N 31.533 E 14.715 – Seddada).

slabs and rocks, a thin grey desertic soil without organic matter may evolve very slowly. Rare bushes of *Arthrophytum scoparium*, *Gymnocarpos decander*, *Salsola tetragona* and *Salsola tetrandra* may find a niche in-between rocks. A contracted vegetation type may occur in shallow depressions of a few hundred square meters hosting clumps of rare *Aristida obtusa*, *Aristida plumosa*, *Medicago laciniata* var. *brachyacantha*, *Oudneya Africana* and *Savignya parviflora*.

With climatic alteration (iron – manganese) and relentless wind and sand abrasion polishing the stone's surface, black *reg* (Photo 5.20) would be also common on the Hamada.

Photo 5.20. Black *reg* on Hamada (N 31.591 E 14.688, wadi Nefad - Seddada).

ROCK OUTCROPS ON JEBEL. Exposed geological layers and large rocky slabs without soils or shallow soils in pockets (top of Jebel Nafusa, plateaux in Jebel Al-Akhdar and Marmarica).

Soils resulting from material accumulation

ERG – SAND SEA. Large sand sea dunes, *erg* (also named *zemla* or *remla*), in the North African Sahara, are most common deep south into the Libyan desert (with patches on Hamada (Photo 5.21)), and large extents in Ash-Shati, Edeyen Murzuk, Edeyen Ubari, Calanshio Serir/Jalo to Kufra region. A small *erg* is located in the western Jeffara between El-Watiya and Takut. Mostly devoid of vegetation, the *erg* are occasionally used as grazing lands exclusively during winter/ spring. The dunes are perpendicular or parallel to dominant wind and formed with mobile sand (dunes up to 50–100 m high, up to 200–300 m between horns). The term *edeyen* is used for round-shaped dunes, while *iguidi* (rarely used in southern Libya, from Zenaga Berber from Mauritania) is sometimes used for long dunes with sharp crests (Capot-Rey *et al.*, 1963).

Photo 5.21. Mobile sand dunes along the road from Sinaween to Derj on Hamada.

EXTINCT VOLCANOES, VOLCANIC FLOWS AND LAVA. These Pleistocene extinct volcanoes are located around Gharyan (Bir Tuta, Qawasim etc.) and dotting the northern Hamada between Mizda and Beni-Walid (Photo 5.22) and further south into the desert.

Paleo-soils

Paleo-soils are widely represented in northern Libya with calcareous and gypsum powerful crusts (0.5–2 m thick). They are often exposed along wadis and gullies on rangelands, terraces and glacis (see Chapter 4). They result from past quaternary eras (Villafranchian and Tensiftian glaciations) when humid periods alternated with dry stages over millions of years. During humid episodes, calcium carbonate and gypsum from upper layers were slowly percolating by dissolution into the soil's profiles and then re-mobilized upwards by evaporation, evapotranspiration during interglaciation long dry periods, slowly recrystallizing and solidifying into the soil's profile. Two types of hard paleo-soil crusts are common: calcareous crust and gypseous crust.

Calcareous crust

The calcareous crusts (or hardpan) are the most widespread. They appear as fractured or solid large white slabs, either bare and exposed, or as lamellar/ribbon-like (pinkish, rose-tinted) slabs buried a few centimetres to 0.5–2 m below the upper soil's layer (Photos 5.23, 5.24). A few plants, such as *Gymnocarpos decander*, *Anthyllis henoniana*, *Helianthemum kahiricum*, *Pituranthos scoparius* and *Atractylis serratuloides*, may colonize the fractured calcareous crust. These crusts are common along the foothills north and south of the Jebel Nafusa, in the northern part of Sirte, and south Jebel Al-Akhdar.

Overhanging shelters and small caves (Photo 5.25) dug under the calcareous crust were used as temporary shelters by farmers during harvesting time or shepherds during hot days. Local population have long learned to dig through this powerful crust to reach the lower, softer marly layers to excavate underground

Photo 5.22. Basalt volcanic flow with bunches of *Cenchrus ciliaris* (Hamada, N 31.496 E 13.259, east of Mizda).

Photo 5.23. Calcareous hardpan underlying stands of *Stipa tenacissima* (Gdama, south Gharyan; N 31.916 E 13.098).

Photo 5.24. Calcareous hardpan exposed along a wadi (wadi Nefad; N 31.522 E 14.700, south-east Seddada).

well-concealed dwellings, storage rooms for barley grain (Oglat Lambach on the Hamada, wadi Maghur – south El-Abyar) and cisterns (Photo 5.26) having narrow concealed shafts.

Gypseous crust

They are common on terraces around *sebkha*, appearing as nearly bare and dazzling white sites in the centre of the Jeffara plain (south Sorman,

Photo 5.25. Shelters dug under the calcareous hardpan around a small doline (at Bir Saïd; N 32.523 E22.570, south Jebel Al-Akhdar).

Photo 5.26. Shaft of an antique cistern (or grain storage *matmura*) hand-dug into the thick calcareous crust (wadi Maghur, south El-Abyar).

Al-Assah – Hebiliyah), Sirte (wadi Bey-el-Kebir region, Ajdabiya), to the south-east. These gypseous crusts (Photo 5.27) are neither saline nor calcareous and may contain up to 70–80% gypsum either as concretions or pulverulent. In the steppe, they are extremely dry/arid environments with little vegetation cover (*Erodium glaucum, Nitraria retusa, Lycium arabicum, Astragalus spinosus*).

Pure gypsum *fer de lance* (iron of the lance) is commonly found with these crusts. In oases

Photo 5.27. Gypsum crust and concretions exposed in a thalweg (wadi Bey-el-Kebir; N 31.099 E 15.786, Sirte).

(Ghadames and Fezzan), desert gypsum roses or rock roses concretions, multifaceted flat (5–8 cm diameter), pinkish-reddish petal-like of sand grains and barytes, are continuously extracted from high water tables in palm groves encroached by sand dunes.

These paleo-soils may not always be exposed. They are often hidden under a topsoil of sand or sandy loam of various thicknesses and are present all over the northern part of Libya from west to east. They are often present and covered with alluvium, colluvium, or with loam or sand veils or under brown sub-desertic or grey sub-desertic soils. This calcareous crust has been a constant obstacle to large rangeland fodder shrub plantations necessitating deep ripping using powerful D9 bulldozers to allow rooting systems to reach deeper and softer horizons.

Azonal soils

These soils are poorly developed due to high salt content, low biological activity and are not linked to specific climatic zones.

Halomorphic soils

These soils are specific to depressions found along the coast (behind the coastal dunes – around Abu Kamash, Zuara, Taworga, Sirte, Tokra, Daryanah) and inland along the Tunisian border (west of Al-Assah, from Ras-Ej-Jdir to Bir Ougradia, Bir Beshiul), Solul-el-Khoder (Seddada), Sirte (Nofilia, Ben Jawad, Sebkha Kabira and from Marsa Brega to Ajdabiya-Zuwetina – Qaminis).

*Saline soils (*solonchatck*)*

These saline soils have an undifferentiated deep homogenous loam/clay profile. They overlay a fluctuating water table that comes to the surface with winter rain accumulating in endoreic depressions (Photos 5.28 and 5.29). The sodium chloride content rises up to 12–15% during the dry season with deposit of white salt efflorescence in summer, and sometimes gypsum.

The vegetation ranges from the centre of the depression to the periphery, following a decreasing gradient of salinity:

- $NaCl$ = > 10% on the hypersaline sites, when water has subsided, nearly pure stands of

Photo 5.28. Saline depression surrounded with gypseous soils (*sebkha* El-Hamraia (N 32.775 E 11.517), south-west Al-Assah.

Photo 5.29. Saline depression and fluffy gypseous soils (Nefad – Seddada, N 31.578 E 14.697).

an annual, *Halopeplis amplexicaulis* to the exclusion of other vegetation at the centre of depression

- NaCl = 1–10% with mostly *Arthrocnemum indicum*, *Salicornia arabica* and *Limoniastrum guyonianum*
- NaCl = 0.1–1%, often with fine pulverulent gypsum, *Atriplex halimus*, *Atriplex glauca*, *Halocnemum strobilaceum*, *Frankenia thymifolia*, *Suaeda fruticosa*, *Salsola vermiculata*, *Limonium pruinosum*, *Aeluropus littoralis*, *Sphenopus divaricatus*, *Tamarix* sp., and from

above from saline sites, all on the periphery of the depressions.

Crusted gypseous soils

Often, a sublayer of crusted gypsum ($CaSO_4.2H_2O$) is present under fine and coarse sand accumulation at the periphery of the saline depression. The NaCl content of the upper layer ranges from 0.3% to 0.8% in summer. Here, *Nitraria retusa* and *Lycium* sp. occur due to the high water table (–1 m to –2 m) along with *Suaeda mollis* and other *Salsola* sp.; stands of *Lygeum* spartum are common on shallow sand deposits due to gypsum in the soils profile (Al-Assah, south Sorman, Sirte, Ajdabiya).

Hydromorphic soils developing over gypsum sublayers

Soils evolving from a gypseous matrix and permanent water table; usually on deep fixed sandy sites with profile showing signs of hydromorphy (red-brown/greenish filaments of oxidized iron in the profile). This is common around *sebkha* covered with sand deposits; it is occurring in the 100–150 mm MAR zones with a common vegetation of *Traganum nudatum, Suaeda mollis, Arthrophytum schmittianum, Lygeum spartum* (Al-Assah – Jbebina, Hebiliyah). Small mound formations on sand with high gypsum content occur on low *sebkha* flats with high water table (–1 m to –3 m deep) overlaid with fixed gypseous sand in the 100–150 mm MAR. These sites (Al-Assah – Magdar-es-Sbot, Hebiliyah, west of Bir Ougradia) display small (0.5–1 m diameter) hemispherical white fragile mounds developing around the rooting system of *Limoniastrum guyonianum*. These mounds could either slowly disintegrate when the shrub dies (or is collected for fuelwood) or evolve into larger *nebkha* due to sand accumulation against the growing shrubs.

NB: all the above azonal soils offer vegetation mostly grazed by dromedaries during winter when plants' salt content is at its lowest, reducing the need for daily flock watering. If the winter season brings abundant rain, a flush of annuals (annual *Medicago* sp. and other Leguminosae, Cruciferae, Poaceae etc.) covers the sand veil deposits inside and around the saline depressions. This is when flocks of sheep are brought to these salty areas for 1–3 months' grazing, fast retreating to non-salty rangelands as soon as the summer drought approaches.

Photo 5.30. Non-saline soils on gypsum substrate (N 31.234 E 15.391; wadi Bey-el-Kebir to wadi Wishkah, Sirte).

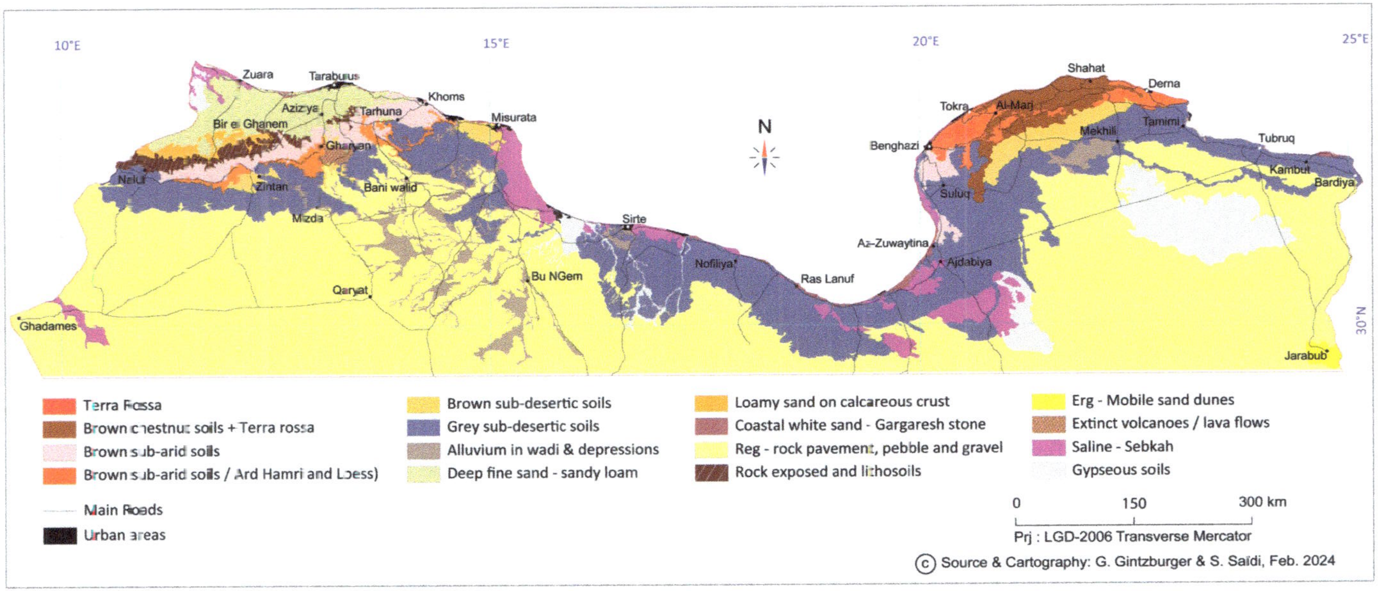

Map 5.1. Synoptic soil map of northern Libya.

Soils issued from compacted gypsum deposit or substrate

These are non-saline Saharan soil formations (in 50–75 mm MAR) and encountered:

- along the road from Nalut to Ghadames/Derj. They consist of hard pulverulent white gypsum often transported and deposited by wind and alluvium on nearby rangelands. The vegetation on these sites is mostly composed of *Oudneya africana* and *Anarrhinum* sp.;
- in Sirte region (wadi Bey-el Kebir, wadi Wishkah, Bu Ngem; Photo 5.30). Sites with pulverulent gypsum substrate exposing gypsum spars and crystals are common; they bear a vegetation of gypsum indicator plants such as *Erodium glaucophyllum*, *Astragalus spinosus* and *Anabasis aphylla*;
- at south Jebel Al-Akhdar, from South El-Abyar-Al-Jahishiyah – south Mekhili to Marmarica.

Soil Map of Northern Libya: Concept and Design

We present our soil map of northern Libya (Map 5.1) based on many years of fieldwork and rangeland surveys, our own description and soil classification following the one used by the Tunisian soil scientists. It is also an interpretation and update of the maps and work from Selkhozpromexport (1980a, b, c, d, e, f; 1984a, b).

We produced the soil map for northern Libya using the 617 soils polygons delimited by the ACSAD/Dutch teams, updating each of them according to our synthetic soil classification and specific legend, using our Libyan field experience and knowledge, completed with location from Google Earth. We also use the isohyets positions, as soil types in arid zones are closely related to rainfall zones. We then spatially aggregated each polygon using a GIS procedure. The limits between soil types on our soil map are not finite as there is often a continuum between soil types gradually shifting from one type to another according to the geological substrate, past and present climatic conditions (rainfall, temperature, wind direction), landforms and geomorphology, and macro- and micro-topography. This is usually poorly reflected on soil maps at different scales displaying *de facto* limits between soil types. The mapping scale is also important as mosaics of different soil types may appear at large scale but would disappear at small scale, the dominant soil type remaining the only one being represented on our soil map.

Conclusions

Arid soil descriptions and classifications are extremely complex and require specialized expertise not always available to rangeland surveyors. We presented the main Libyan soil types easily recognizable and accessible to rangeland ecologists and botanists who do not always have time to dig soil profiles. While our legend does not follow the US, the European or the Soviet soil classifications, it allows field scientists, botanists and environmentalists to easily identify each soil type described in simple terms (dominant colour, texture, structure, soil surface, associated native vegetation etc.). Arid soil types can often be simply determined from landscapes, vegetation types, ecological groups or plant association, or even by the presence of specific indicator plants we are going to explore.

Notes

[1] The *Terra rossa* are red heavy soils, also identified as Xeralfs in the American soil classification.

[2] The rendzina are rendolls in the American soil classification.

References

ACSAD-GWA (2005) SOTOR Soil map of Libya at scale 1.1M and SOTOR map of NW and NE of Libya at scale 1:250,000.

Belkhodja, K., Bortoli, L., Cointepas, J.P., Dimanche, P., Fournet, A., Jacquinet, J.C. and Mori, A. (1973) *Sols de Tunisie: les sols de la Tunisie septentrionale*. République Tunisienne, Ministère de l'Agriculture, Tunis.

Belobrov, V.P. (2015) Climatic features of desert soil formation in northern Libya. *Bulletin of V.V. Dokuchaev Soil Science Institute* 81, 631–634.

Capot-Rey, R., Cornet, A. and Blaudin-de-Thé, B. (1963) *Glossaire des Principaux Termes Géographiques et Hydrologiques Sahariens*. Université d'Alger, Institut de Recherches Sahariennes, Algiers.

Elaalem, M.M., Zurqani, H.A. and Ben Mahmoud, K.R. (2021) The history of soil mapping and research. In: Zurqani, H.A. (ed.) *The Soils of Libya*. Springer, Cham, Switzerland, pp. 13–31. DOI: 10.1007/978-3-030-66368-1_2

FAO (2006) Aggregated Land Cover Database for Libya – updated 2023. Available at: https://data.apps.fao.org/catalog/iso/ab3f6dca-e352-4b32-b4c1-255b4d3aa88f (accessed 10 August 2024).

FAO, ISRIC and ISSI (1998) World Reference Base for Soil Resources. Food and Agriculture Organization, International Soil Reference and Information Centre and International Society of Soil Science, Rome. Available at: https://www.fao.org/3/w8594e/w8594e00.htm (accessed 10 August 2024).

FAO Soils (2015). World reference base for soil resources. International soil classification system for naming soils and creating legends for soil maps. (No. 106, World Soil Resources Reports). Food and Agriculture Organization. Available at: https://www.fao.org/3/i3794en/I3794en.pdf (accessed 10 August 2024).

GEFLI (1972a) *Soil and Water Resources Survey for Hydro-agricultural Development, Central Zone (Sirte Area)*. Groupement D'Etude Française en Libye.

GEFLI (1972b) *Soil and Water Resources Survey for Hydro-agricultural Development, Eastern Region*, 2–2: Climatology. Groupement D'Etude Française en Libye.

GEFLI (1972c) *Soil and Water Resources Survey for Hydro-agricultural Development, Western Zone. Climatological Year Book*. GEFLI/Libyan Arab Republic.

GEFLI (1972d) *Soil and Water Resources Survey for Hydro-agricultural Development, Western Zone. Climatological Yearbook 3*. Groupement D'Etude Française en Libye.

GEFLI (1974a) Programme de mise en valeur de la zone pastorale développement de l'élevage ovin. Plaine de Gefara et djebel Nefusa. Rapport agrostologique. Groupement D'Etude Française en Libye.

GEFLI (1974b) *Survey for Agricultural Development (Wadi Hira, Bu Shaiba, and Gharyan Projects)*. Groupement D'Etude Française en Libye.

GEFLI (1977) *Survey for the Development of the Central Wadi Zone and Gulf of Sirte (Sirte, Wadi Zam-Zam Project)*. Groupement D'Etude Française en Libye.

Gintzburger, G. and Alidrissi, M. (1976) Range study on Jeffara Plain: Sheep Project at Bir-El-Ghanem (No. 83/76). FAO and Agricultural Research Center, Tripoli.

Gintzburger, G. and Skinner, P. (1985a) A simple single disc pitting and seeding machine for rangeland revegetation. *Australian Rangeland Journal* 7–1, 29–31.

Gintzburger, G. and Skinner, P. (1985b) Annual Medics species and pitting for rangeland regeneration. *Range Management Newsletter of the Australian Rangeland Society* 85(1), 13–15.

Gintzburger, G. and Skinner, P. (1985c) *Annual Medics Species for WA Rangelands: Selection and Establishment Techniques/Experimental Results 1984–85*. Western Australia Department of Agriculture Division of Plant Research, Perth, Australia.

Gintzburger, G., Skinner, P. and Liedel, L. (1986a) A simple single disc pitting and seeding machine for rangeland revegetation operation (Spain Patent No. 551912 1987). Available at: https://patentimages.storage.googleapis.com/55/f6/7c/a4a2828ef73870/WO1986004771A1.pdf (accessed 10 August 2024).

Gintzburger, G., Skinner, P. and Liedel, L. (1986b) A simple single disc pitting and seeding machine for rangeland revegetation operation (Australia Patent No. 577457 1989). Available at: https://patentimages.storage.googleapis.com/55/f6/7c/a4a2828ef73870/WO1986004771A1.pdf (accessed 10 August 2024).

Gintzburger, G., Toderich, K., Mardonov, B. and Mahmudov, M. (2003) *Rangelands of the Arid and Semi-arid Zones in Uzbekistan*. CIRAD-ICARDA, Montpellier, France.

Le Houérou, H.-N. (1959) Recherches écologiques et floristiques sur la végétation de la Tunisie méridionale. PhD dissertation, Université Alger Institut de Recherches Sahariennes, Algiers.

Le Houérou, H.-N. (1960) Contribution à la connaissance de la flore phanérogamique de la Libye. *Bulletin de la Société d'Histoire Naturelle de l'Afrique du Nord* 51, 175–200.

Le Houérou, H.-N. (1969) *La Végétation de la Tunisie Steppique (Avec Référence aux Végétations Analogues d'Algérie, de Libye et du Maroc)*. (Vol. 1–2). Annales de l'Institut National de la Recherche Agronomique de Tunisie, Tunis.

Novikoff, N. (1961) Contribution à l'étude des relations entre le sol et la végétation halophile de Tunisie. *Annales de l'Institut National de la Recherche Agronomique de Tunisie* 34, 339.

Omar, M.Y. (2023) Evaluation of soils of Qasr Libya Region (Jabal Al Akhdar, Libya) terms in their morphological and chemical characteristics. *Journal of Soil Sciences and Agricultural Engineering* 14(1), 11–15. DOI: 10.21608/jssae.2023.188175.1139

Orlovsky, L., Dourikov, M. and Babaev, A. (2004) Temporal dynamics and productivity of biogenic soil crusts in the central Karakum Desert, Turkmenistan. *Journal of Arid Environments* 56(4), 579–601. DOI: 10.1016/S0140-1963(03)00075-2

Rowland, J.W. (1945) *Survey of Land Resources in Tripolitania.* British Administration, Department of Agriculture.

Schulz, E., Hachicha, T., Marquer, L., Pomel, S., Salzmann, U. and Abichou, A. (2014) The distant chant: climate reconstruction and landscape history: the last two millennia in southeast Tunisia. *Zentralblatt Für Geologie Und Paläontologie* 1, 355–386.

Selkhozpromexport (1980a) *Soil and Geobotanical Studies in the Pastures of the Socialist Libyan Jamahariyah.* USSR.

Selkhozpromexport (1980b) *Soil map of the Central zone, Libya, Scale: 1/200 000, From Brega to Soluk.* Nauka, USSR.

Selkhozpromexport (1980c) *Soil map of the Eastern zone, Libya, Scale: 1/200 000, From Soluk to Derna and Northern Slope of Jebel Akhdar.* Nauka, USSR.

Selkhozpromexport (1980d) *Soil map of the Western zone, Libya, Scale: 1/200 000, From Zuara to Misurata and Part of Jebel Nefusa.* Nauka, USSR.

Selkhozpromexport (1980e) *Soil of the Socialist Libyan Jamahariya.* Nauka, USSR.

Selkhozpromexport (1980f) *Vegetation and Ecology of the Socialist Libyan Jamahariya* (Vol. 2). Nauka, USSR.

Selkhozpromexport (1984a) *Simplified soil map of the Eastern zone, Libya, Scale: 1/1 000 000, From Soluk to Derna and Northern Slope of Jebel Akhdar.* Nauka, USSR.

Selkhozpromexport (1984b) *Simplified soil map of the Western zone, Libya, Scale: 1/1 000 000, From Zuara to Misurata and Part of Jebel Nefusa.* Nauka, USSR.

Sishov, L.L. and Kharin, N.G. (1980) *Soil and Geobotanical Studies in the Pasture Zones of the Socialist People's Libyan Arab Jamahiriya* (Vol. 3). Selkhozpromexport, Nauka, USSR.

Stolbovoi, V. (2002) *Land Resources of Russia: Soils – Description – Version 1.1.* IIASA and Russian Academy of Sciences. Available at: http://webarchive.iiasa.ac.at/Research/FOR/russia_cd/soil_des.htm (accessed 10 August 2024).

USDA (1999) *Soil Taxonomy: A Basic System of Soil Classification for Making and Interpreting Soil Surveys* (2nd edn). Natural Resources Conservation Service, U.S. Department of Agriculture Handbook, Washington, DC.

Zurqani, H.A. (ed.) (2021) *The Soils of Libya.* Springer International Publishing, Cham, Switzerland. DOI: 10.1007/978-3-030-66368-1

6

The Vegetation Types

Abstract

Libyan landscapes are shaped by geology and geomorphology, resulting in various vegetation types, and ecological groups detailed region by region in this chapter. These groups are modelled by climate, soil types, and land use, forming specific plant associations. These associations are not stable and are altered over time by evolving land use and grazing pressure. Raunkiaer's biological spectrum helps assess their status and evolution but has its limitations. To define these groups, detailed phyto-sociological surveys are required, necessitating a deep understanding of environmental parameters and plant taxonomy. This process involves incessant field trips and observations, herbarium collection and accurate plant identification, complemented by local knowledge collected from shepherds and farmers.

Introduction

The vegetation is characterized by its physiognomy and structure, its floristic diversity and biological types distinguishing the plant associations composed by different ecological groups.

The Plant Biological Types

The biological spectrum of a vegetation, expressed as the percentage of plant biological types or life forms (Emberger and Godron, 1968; Raunkiaer, 1934), give an idea of the structure and physiognomy of a homogeneous vegetation on a specific site. The plant biological types are classified according to the position of the buds in relation to the soil surface. From a biological point of view, the position of the buds or reproductive organs allows plants to cope with stress and especially with the dry critical summer season in Libya. Most perennials survive the dry summer by losing their leaves, hence reducing their evapotranspiration rate (Orshan, 1972). The annual plants escape the summer drought by a winter/spring development ending with seed production before summer with seeds considered as buds. There is no relationship between the Raunkiaer's biological types and the plants' taxonomy, as any family, genus or species may belong to any of these biological types.

The plant biological types occurring in the Libyan vegetation are:

The phanerophytes

Large woody perennials, with buds at least 30 cm above-ground at different heights:

1. The megaphanerophytes are the tallest trees (> 30 m high): ex.: *Cupressus sempervirens*.

DOI: 10.1079/9781800627154.0006

2. The mesophanerophytes (8–30 m high): ex: *Pinus halepensis, Juniperus phoenicea, Pistacia atlantica, Quercus calliprinos, Tamarix* sp., etc.
3. The microphanerophytes (2–8 m high): ex.: *Retama raetam, Arbutus pavarii, Pistacia lentiscus, Calligonum* sp., *Genista saharae, Atriplex halimus, Rhus tripartita, Calotropis procera, Rhamnus* sp., *Ziziphus lotus*, etc.
4. The nanophanerophytes (< 2 m high): ex: *Thymelaea hirsuta, T. microphylla, Ephedra* sp., *Cistus* sp., *Genista* sp., *Atriplex halimus, Erica multiflora, Lycium* sp.
5. The phanerophytes-climbers, ex.: *Cocculus pendulus, Smilax aspera, Convolvulus althaeoides, Pergularia tomentosa, Tamus communis, Ephedra altissima.*

The chamaephytes

These medium-sized woody perennials have their buds within the 30–40 cm range above ground. They are the are most common on Libyan rangelands, ex.: *Artemisia herba-alba, Artemisia campestris, Argyrolobium uniflorum, Arthrophytum scoparium, Anthyllis henoniana, Gymnocarpos decander*, various *Suaeda* sp., *Salsola* sp., *Sarcopoterium spinosum, Helianthemum* sp., *Atractylis serratuloides, Rhanterium suaveolens, Polygonum equisetiforme, Pituranthos* sp., *Traganum nudatum, Noaea mucronata, Anabasis* sp., *Thymus* sp., *Euphorbia paralias, Ballota hirsuta, Marrubium deserti, Phlomis floccosa, Rosmarinus officinalis, Globularia* sp., *Limonium* sp., *Limoniastrum* sp., *Zygophyllum* sp.

The hemicryptophytes

Perennials, they have their buds (or rhizomes) very close to or in the topsoil, ex. stands of perennial Gramineae, *Stipa tenacissima, Lygeum spartum, Oryzopsis miliacea, Cenchrus ciliaris, Panicum turgidum, Stipagrostis pungens, Aeluropus lagopoides, Poa bulbosa, Cynodon dactylon*, etc. It also includes most herbaceous plants, ex.: *Sanguisorba minor, Atriplex mollis, Convolvulus dorycnium, Plantago albicans.*

The geophytes or cryptophytes

Still perennials, they have their vegetative reproductive bud, bulbs or corms or even rhizomes deep into the soils, ex.: *Asphodelus* sp., *Urginea maritima, Poa bulbosa, Allium* sp., *Scorzonera undulata, Cyclamen rohlfsianum, Erodium glaucophyllum, Dipcadi serotinum, Ophrys* sp., *Orchis* sp.

The therophytes or ephemerals

These are the annual plants with a very short cycle from germination to seed production with seeds considered as buds: ex: *Medicago truncatula, M. littoralis, M. laciniata, Anastatica hierunchuntica, Stipa capensis, Vicia* sp., *Vella annua, Sphenopus divaricatus, Atractylis cancellata* etc.

Lichens

Large populations of lichens (*Ramalina* sp., *Teloschistes villosus*) develop on *Juniperus* sp. in Jebel Al-Akhdar. On degraded rangelands, biocrust of lichens developing on topsoil are also worth recording.

All the above are autotrophic organisms.

The heterotrophs

Parasites developing on phanerophytes or chamaephytes, ex.: *Cistanche* sp., *Cuscuta* sp. We place desert truffles (*Terfezia* sp., *Tirmania* sp.) in this class as well.

The Raunkiaer's Biological Spectrum

The assortment of biological types (or life forms) represented by its biological spectrum, expressed in percentage of biological types, helping to characterize a specific site in a vegetation landscape.

El-Barasi and Barrani (2015) give an example of the Raunkiaer's biological spectrum in the Daphna region in Marmarica (Table 6.1).

Despite lacking information on the site studied, average rainfall and date of the list of species collected, we may visualize a degraded steppe vegetation landscape with a dominance of ephemerals and cryptophytes (65%) easily escaping grazing or unpalatable, some relic chamaephytes and few phanerophytes (35%) likely from pre-degradation stage and one hemicryptophytes.

At Bir-el-Ghanem, Gintzburger (1986) measured the spring standing biomass and other vegetation parameters on a well-protected sandy rangeland site (BGII, MAR ≈ 150 mm). When listing the species present in a homogeneous *Rhanterium suaveolens-Artemisia campestris* plant association (soil sub-association of *Retama raetam*), we noted that there is a dominance of therophytes due to absence of grazing, with still a clear presence of perennial microphanerophytes, chamaephytes and hemicryptophytes (Table 6.2).

This biological spectrum has little significance unless complemented with other ecological information and vegetation cover measurements (biomass and line intercept measurements per group of species) (Gintzburger and Saïdi, 2009). On the Bir-el-Ghanem vegetation, the perennials (microphanerophytes, chamaephytes and hemicryptophytes) contributed to the line intercept measurement 31.4% (SE = 3.7), the therophytes 21.8% (SE = 4.9), the dead wood 8.9% (SE = 2.0) and bare soil 35%. This rangeland has a well-balanced composition of biological types. It indicates that it is, at the time, in very good and healthy condition (Gintzburger, 1986).

The limit of the biological spectrum

Emberger and Godron (1968) state that:

- The Raunkiaer's biological spectrum must be established for a homogeneous identified vegetation association with ecological groups clearly defined. Many regional floristic papers present life forms spectrum (Abdalrhim, 2021; Abu-Aziza et al., 2017; Abusaief and Dakhil, 2013; Al-Khojah *et al.*, 2020; Al-Sodany *et al.*, 2003; Al-Traboulsi and Alaib, 2021; El-Barasi *et al.*, 2003; El-Mokasabi, 2017; Hegazy *et al.*, 2021; Mukassabi *et al.*, 2012; Sherif *et al.*, 2020). However, despite the many plants listed, the life forms spectrum presented has little or no ecological value when applied to a large region including unrelated site ecology, soil, climatic conditions and land use, or floristic diversity and various dates of surveys.
- The spectrum does not consider the taxonomic plant diversity and environment that will be explored when detailing the plant associations.
- The interpretation of the biological type of a plant may be subject to confusion in the field as, for example, *Stipagrostis pungens* or other rhizomatous grasses (*Panicum turgidum*, *Cynodon dactylon*, *Aeluropus littoralis*), in good condition, are typically classified as hemicryptophytes; however, when their aerial parts are heavily overgrazed and severely reduced or buried in mobile sand, as is the case on Libyan rangelands, they must be classified as geophytes. Therefore, the actual biological type recorded in the field may be different

Table 6.1. Raunkiaer's biological spectrum in the Daphna region in Marmarica. (Adapted from El-Barasi and Barrani, 2015)

Raunkiaer's biological types	No. of species recorded	% of biological spectrum
Therophytes	218	55.19%
Chamaephytes	112	28.35%
Cryptophytes	38	9.62%
Phanerophytes	26	6.58%
Hemicryptophytes	1	0.25%
Total no. of species	395	100%

Table 6.2. Raunkiaer's biological spectrum at Bir-el-Ghanem. (From Gintzburger, 1986)

Raunkiaer's biological types	No. of species recorded	% of biological spectrum
Microphanerophytes	2	4.4%
Chamaephytes	8	17.8%
Hemicryptophytes	5	11.1%
Therophytes	30	66.7%
Total no. of species	45	100%

from the standard published in the literature and florae (Nègre, 1961) or on websites.
- The survey date is important when establishing the life form spectrum. A late survey, early summer or early autumn, on Libyan vegetation would miss most annuals (therophytes) and present a bias towards perennials (phanerophytes, chamaephytes and hemicryptophytes).

In short, the Raunkiaer's biological spectrum is a simple and crude approach to describing a floristic diversity with little ecological value. It is of limited use unless generated from factual field observations, species by species, in a well-defined plant association at a specified date. It must be complemented with line intercept and biomass measurements.

Main Vegetation Landscapes: Physiognomy and Structure

Le Houérou (1959, 1969), working in southern Tunisia, identified different types of vegetation landscape and structure that fit Libya well. We follow le Houérou's classification that we have reorganized in order of importance in northern Libya.

The steppe

Located between the cultivated lands and the desert, the steppe is the domain of nomads and semi-nomads and their livestock. It covers most of the Libyan north-western part, Sirte to the southern Jebel Al-Akhdar (Cyrenaica and Marmarica). It consists of flat or hilly landscapes covered with a mostly treeless low vegetation of microphanerophytes, nanophanerophytes and hemicryptophytes, as in the Jeffara plain, Jebel Nafusa, northern Hamada, Sirte and southern Jebel Al-Akhdar.

We distinguish the following types:

The graminean steppe

It covers extensive areas in the west.

1. The *Stipagrostis pungens* steppe covers all the deep sand of the Jeffara plain, extending on sandy sites into the Sirte region and southern Jebel Al-Akhdar.
2. The *Stipa tenacissima* steppe on loamy soils on hills and northern slopes of the Jebel Nafusa from Yefren to Msallata-Homs-Zliten; *Stipa tenacissima* does not occur further east.
3. The *Lygeum spartum* steppe, of limited importance, is often associated with gypseous soils (Zuara coast, Al-Assah, Bir-el-Ghanem, small patches on loess in the Jebel Nafusa, to Taworga, Bu Gren-Sirte-Ajdabiya to South Benghazi).

The chamaephytic steppe

This is the most frequent vegetation landscape of low woody shrubs, mostly (< 2 m) occurring in northern Libya. It includes:

1. the *Rhanterium suaveolens* steppe, with only small areas left at Bir-el-Ghanem on shallow sand, near Cabao-Nalut, and some rare remnants at wadi Bey-el-Kebir to Sirte; it does not appear to occur further East;
2. the *Artemisia herba-alba* steppe on heavy soils, on the Jebel northern and southern piedmont and depressions, Sirte and Jebel Al-Akhdar;
3. the *Anthyllis henoniana* steppe of shallow rocky loamy soils, Gharyan, Yefren, Zintan, Harabah, Nalut, on northern Hamada;
4. the *Artemisia campestris* steppe on restricted areas as a post-cultural community and fallow land on sandy soils (Reqdalin, Al-Ajaylat, wadi Abu-Shayba, Daphniya);
5. the *Arthrophytum scoparium* steppe, colonizing large areas on heavy soils, often replacing the *Artemisia herba-alba* steppe with increasing aridity, mostly on the southern slope of Jebel Nafusa (Nalut, Harabah to Beni-Walid, northern Hamada), on the plateau of south Zliten, wadi Sassu and further south (Bu-Gren, Bu-N'Gem), inland Sirte, to southern slope of Jebel Al-Akhdar and Marmarica;
6. the *Salsola tetragona*, *S. tetrandra*, *S. vermiculata* steppe, common on rocky plateau (foothill Nalut, Hamada, Sirte, to south Jebel Al-Akhdar – Marmarica);
7. the *Anabasis articulata*/*A. oropediorum* steppe on sandy soils (south Al-Assah, Hamada, Sirte to south Jebel Al-Akhdar – Marmarica) at the limit of the desert;
8. the crassulescent azonal steppes of halophytic plants (*Atriplex* sp. *Salsola* sp., *Suaeda* sp.,

Arthrocnemum sp., *Halocnemum* sp., *Salicornia* sp., *Halopeplis* sp. etc.), found around *sebkha* on saline and gypseous soils, all northern coastal zone from inland along Tunisian border (Al-Assah) to Sirte, Ajdabiya, Benghazi coastal zone to Al-Sallum.

The pseudo-steppe

These are limited landscapes in areas dominated with microphanerophytes and nanophanerophytes in different ecological zones:

1. *Ziziphus lotus – Atriplex halimus* growing on *nebkha* in wadis and depressions (Solul-el-Khoder, wadi Sof-el-Jin) often with pure stand of large overgrown *Atriplex halimus* in wadi (south Gharyan, Gdama). They have often been cleared and levelled as target of rainfed barley cropping (wadi Abu-Shayba, wadi Hay, Sof-el-Jin) or for rangeland regeneration projects on the southern slope of the Jebel Nafusa.
2. *Retama raetam* pseudo-steppe is often associated with *Genista saharae* and *Stipagrostis pungens* (Bir-el-Ghanem, Sorman, Aziziya) on fixed sand dunes. It is replaced by sparse *Calligonum* sp. as aridity increases (foothill Nalut to north-western Hamada/Sinaween – Derj). Large stands of *Retama raetam* are also frequent on sandy areas of the southern slope of the Jebel Al-Akhdar.

The forests

In central and southern semi-arid and arid Tunisia, Le Houérou (1969) defined forest as a vegetation formation with trees at least 7–8 m high and density of about 100 trees/ha, a rarity in Libya.

Native forests in the Jebel Al-Akhdar

Remnants of native forests still occur these days in restricted highest rainfall zones of the Jebel Al-Akhdar, with mostly *Pinus halepensis* near Ras-Hilal, *Juniperus phoenicea* extending from the El-Marj escarpments to the central Jebel and southern slopes, and the remarkable large *Cupressus sempervirens* in wadi Kuf. These forests, though protected by law, are under immense pressure from poaching, fuelwood harvesting and charcoal production, land clearing for cropping, new city dwellings, road and track expansion, and grazing by goats and cattle. Accidental bush fires or voluntary arson contribute, year after year, to the destruction of native trees and precious forests.

Native forests and afforestation in the west

In the west, there are no native stands of forests left. Native forests may have existed in the eastern part of Jebel Nafusa with *Pinus halepensis* and *Juniperus phoenicea*. They disappeared long ago, likely since antiquity, due to intense fuelwood harvesting during the Roman-Byzantine colonization. All forested areas in western Libya are from modern afforestation projects. The case of isolated old *Pistacia atlantica* in the Jebel Nafusa (Gdama, south Gharyan to Mizda) raises the question of possibly denser stands of this remarkable drought-tolerant tree. There is no doubt that all these trees, native and planted, close to settlement, have been targeted by illegal fuelwood collectors and charcoal manufacturers.

Savanah-like clear tree formations

These are arid landscapes with *Acacia tortilis* ssp. *raddiana* occurring in the wadis of the Solul-el-Khoder region (several wadis from Beni Walid to Seddada, wadi Nefad, wadi Meimun, wadi Zorzor, Bir Dufan, Gasr Banat, wadi Ghirza). These native *Acacias* are associated with dense vegetation of Saharan large tufts of *Panicum turgidum*, *Pennisetum dichotomum* and other Saharan species. In the wadis of the Qasr Banat Roman site near Seddada, the *Acacia* density is about 1–2 trees/ha with a vegetation landscape that looks much like sub-Saharan savannah. Unfortunately, these wadis have long been targets for fuelwood harvesting/charcoal manufacturing and large irrigation projects (1980–2010), clearing all native trees and vegetation. A remarkable relict Savannah-like clear forest remains on a very limited area in wadis flowing down from the Jebel Nafusa into the Jeffara plain near Bir Ayyad-Wadi Ethel. In 2019, only 60–80 isolated large *Acacia tortilis* were still present on an area of about 9 km^2, i.e. a tree density of about one tree per 12–14 ha.

The matorral/garrigue

The matorral is a generic term usually describing a mosaic of degraded Mediterranean forests and dense scrublands with meso-, micro- and nanophanerophytes in southern Europe, from Portugal to Turkey extending into coastal Syria, Lebanon, Palestine and Jordan. This vegetation landscape occurs predominantly in Jebel and hilly country.

Two different matorral can be differentiated according to the soil type:

The maquis or macchia

This occurs on acid soils. As there are no proper acid soils in Libya, there is no proper maquis in Libya.

The garrigue

This covers calcareous soils and neutral to slightly acid *Terra rossa*, with patches of thick scrublands interspersed with bare soils and lithosols developing on karst landscape.

We prefer using the term garrigue to qualify the dominant vegetation landscape of eastern Libya. This is a mosaic of dense or sparse woody shrubs (< 7–8 m high) covering large areas of the Jebel Al-Akhdar from Benghazi plain to Derna above the 200–250 mm isohyets. The garrigue results from degradation of *Pinus halepensis* – *Quercus calliprinos* or *Juniperus phoenicea* forests through fire, fuelwood collection, land clearing and grazing. It results into a dense scrub of bushes of *Pistacia lentiscus*, *Arbutus pavarii* and degraded *Juniperus phoenicea*, with a cohort of low shrubs of *Cistus* sp., *Thymus* sp., *Rosmarinus* sp., *Globularia* sp., *Genista* sp., *Rhus tripartitum* and *Rhamnus lycioides* that colonize the landscape with increasing degradation (see Chapter 13). One should note that the latter low shrubs occur in the eastern part of the Jebel Nafusa, possibly indicating a relic vegetation landscape from past western forests and garrigues.

The erme

This is the utterly degraded rangeland vegetation from overgrazing hosting sparse unpalatable hemicryptophytes and geophytes (*Asphodelus* sp., *Urginea* sp., *Peganum harmala* etc.), and therophytes (*Stipa retorta*, *Neurada procumbens*), when all palatable plants have been eradicated and bare soil is dominant (Le Houérou, 1969). This is a common sight near settlements at many sites from Gharyan to Yefren on the top of the Jebel or on very degraded garrigue in the east, around wells and cisterns in Libya, and in North Africa, the Middle East and Central Asia (Gintzburger *et al.*, 2003).

Continuum or Mosaic of Vegetation Associations?

The different vegetation landscapes may occur in patches/mosaics or continuum according to climate, soil type and human and livestock activities. This could be quite confusing when studying and mapping the vegetation associations. Continuum or mosaic of vegetation associations would depend on the mapping scale when characterizing the vegetation associations and ecological groups (Gounot, 1969; Le Houérou, 1959, 1969).

A Brief Description of Main Vegetation Landscapes

Despite its aridity and often perceived monotony, the Libyan vegetation landscapes are diverse from west to east, and into the desert.

The western zone

Eight major western zones could be identified:

- Western Jeffara plain
- Jeffara plain centre and east
- Jebel Nafusa (west and central zones)
- Jebel Nafusa east
- Southern slope of the Jebel Nafusa
- Hamada
- South Zliten plateau
- Solul-el-Khoder

The western Jeffara plain

The vast saline, gypseous and arid *sebkha* (Photo 6.1), west of Al-Assah – Al-Watyah,

Photo 6.1. Saline *sebkha* along the Tunisian border (N 32.848 E 11.519, Al-Assah).

100–150 mm MAR[1]) limits the Jeffara plain along the Tunisian border. It is shaped into badlands by water and wind erosion and covered with sparse halophytes-gypsophytes-phreatophytes vegetation (*Salsola vermiculata*, *Suaeda* sp., *Lygeum spartum*, *Atriplex mollis*, *Atriplex dimorphostegia*, *Atriplex halimus*, *Tamarix* sp., *Lycium* sp., *Nitraria retusa*, *Reaumuria* sp., *Limoniastrum* sp., *Limonium* sp., *Aeluropus littoralis*, *Sphenopus divaricatus*, *Arthrocnemum glaucum*, *Halocnemum strobilaceum*). These are traditional winter season grazing territories when dromedaries come down from the Jebel through Nalut–Cabao and migrate back from the Hamada. Since the 1970s, due to availability of feed and water carted by trucks and tractors, sheep and goat breeding has become dominant from Assah to Al-Watyah. South-west of Al-Watyah, towards Bir Ougradia – Bir Shabunyah (100–125 mm MAR) to the foothill of Nalut, the Tunisian Saharan *erg*[2] expands into the Jeffara plain. This is where *Calligonum comosum* and *Calligonum arich* (both strictly western species) can be found on sand dunes roamed by gazelles and jackals.

Close to the coast, from Ras-el-Jdir to Tajura, *Artemisia campestris* and *Thymelaea microphylla*, *T. hirsuta* colonized abandoned barley fields on sandy soils (south of Az-Zawiyah, Al-Ajaylat and Al-Jmail-Zuara).

The Jeffara plain centre and east

The treeless Jeffara coastal plain (150–200 mm MAR) extends from the coast near Ra's-Ajdir to the foothill of Homs to the east and to the foothill of the Jebel Nafusa to the south. The plain is predominantly covered with *Stipagrostis pungens* and *Retama raetam* on deep sandy soils (north Bir-el-Ghanem, Photos 6.2 and 6.3). *Anthyllis henoniana*, *Gymnocarpos decander*, *Pituranthos* sp., colonize shallow soils of the most arid central zones near Bir Kuka–Bir-el-Ghanem foothills of the Jebel Nafusa as on the gypseous small spring mounds that pops out from the sand accumulation in central Jeffara (Zaharat-el-Mashad, Bir Masla, Sanyat-al-Malihah, Sanyat-el-Houd).

On wadi flats receiving run-off from the Jebel Nafusa (wadi Abu-Shayba, wadi El Hay), *Ziziphus lotus* and *Atriplex halimus* (Photo 6.4) accumulate sand and organic matter on *nebkha* now mostly levelled and cleared for rainfed barley and occasionally wheat cultivation.

On flat sandy-loam, rich *Rhanterium suaveolens* covers large areas (Bir-el-Ghanem (150–175 mm MAR), wadi Abu-Shayba). Loamy depressions were once covered with *Artemisia herba-alba* stands previously populated with bustards and devoted to sheep, goats and dromedary winter/spring grazing. These areas have been relentlessly turned into rainfed barley and large irrigation projects (1975–1990).

A few large, lonely and relic trees of *Acacia tortilis* ssp. *raddiana* are still present at Bir Ayyad (north of Yefren) and wadi Ethel (Photo 6.5). They produce valuable pods that used to be a favourite breeding and feeding zone for gazelles (*Gazella dorcas*) up to the mid-1960s.

Photo 6.2. Pristine *Stipagrostis pungens* landscape from central Jeffara (N 32.521 E 12.040).

Further west, deep wadis and gullies dissect the western (Nalut-Giado) and central region (Gharyan) of the northern slope of the Jebel Nafusa. *Cenchrus ciliaris* and *Pennisetum setaceum* are frequent in the gullies (south-east of Bir-el-Ghanem) lined with overgrazed *Periploca laevigata* from Qasr-el-Haj to Shakshuk, Kikla, Tiji, Nalut (Photo 6.6).

The Jebel Nafusa (west and central)

The mountainous arc of the Jebel Nafusa bordering the Jeffara plain to the south and the east is the continuation of the Tunisian Matmata. The Libyan Jebel displays massive and high alternating dolomitic limestone (Cenomanian and Turonian) and marls cliffs or *kreb* – cuestas culminating to 900 m near Gharyan (MAR = 300–350 mm/year), on which are perched modern and centuries-old villages and granaries (Photo 6.7).

Deep loess accumulations cover the synclinal and valleys on the top and the northern slope of the Jebel (Gharyan to Tarhuna). The local *damus* (*domus*, in Latin, 'home' from the Romans) are dug deep into these loess accumulations by the

Photo 6.3. Shepherd camp on *Aristida pungens* and *Retama raetam* landscape (BGII project, N 32.445 E 12.834, 1976).

Photo 6.4. *Ziziphus lotus* and *Atriplex halimus* on *nebkha* (wadi Abu-Shayba, 1976).

Photo 6.5. *Acacia tortilis* ssp. *raddiana* isolated trees at Bir Ayyad (N 32.131 E 12.433, 1976).

local population to create well-insulated underground houses (Despois, 1935).

This is *Stipa tenacissima* country with *Oryzopsis miliacea*, *Globularia alypum*, *Genista microcephala*, *Thymus* sp., *Ephedra* sp. (Photo 6.8); however, without any relic of antic clear forests (*Juniperus phoenicea*, *Pinus halepensis*, *Ceratonia siliqua*). Timber posts of these trees eradicated long ago are still found in old dwellings in Nalut-Cabao. Rainfed barley cropping is familiar on the top of the Jebel, with cultivated fig and almond trees and mostly olive (*Olea europea*) on terraces collecting rainwater. They have been grown there since antiquity as witnessed by local, Roman and Byzantine olive presses.

Stands of unpalatable *Asphodelus microcarpus* cover overgrazed shallow soils all over the Jebel around settlements and wells. South of Gharyan and the Jendouba plain, where large tracts of *Stipa tenacissima* rangelands start, centuries-old isolated *Pistacia atlantica* remains in the wadis near Gdama. These trees are fuelwood collectors' targets.

Alluvium and colluvium glacis on the northern piedmont create inter-wadi terraces (*foum*) that are traditional winter/spring grazing territories. These large glacis close to the Jebel are targeted for reforestation projects with *Pinus halepensis*, exotic *Acacia saligna* and *Parkinsonia aculeata* along the Al-Aziziya–Gharyan road.

Domes of black volcanic basalt dot the landscape around Gharyan on the Jebel and near Bir Rabta at the foot of the Jebel.

The Jebel Nafusa East

The eastern parts of the Jebel Nafusa culminate at 400–600 m and run through Tarhuna–Qasr-Ed-Daoun–Cussabat–Msallata (250–300 mm MAR). It ends up in the Mediterranean Sea at Naggaza – Ras-el-Msen west of

Photo 6.6. Overgrazed landscape of *Periploca laevigata* along gullies (northern foothill of the Jebel Nafusa, Shakshuk – Kikla, N 32.016 E 11.858, 1976).

Photo 6.7. A view of the northern cuesta of the Jebel Nafusa (Yefren, N 32.064 E 12.518, 1978).

Photo 6.8. *Stipa tenacissima* country with olive plantations on lower terraces in the Jebel Nafusa (north Tarhuna, N 32.501 E 13.619, 1978).

Homs. The gently undulating landscape is still covered with *Stipa tenacissima*. Large tracts of sandy-loam and loess inter-hills depressions have been cleared since antiquity and planted with olive trees that made the fortune of Leptis Magna. Many standing antique Roman and Byzantine olive presses can be seen in this region.

THE SOUTH-WESTERN SLOPE OF THE JEBEL NAFUSA. Close to the Jebel is the red sandy loam of the *dahar* speckled with olive and other fruit trees and rainfed barley. Further south start the large flattish areas (Zintan – Al-Harabah, 150–175 mm MAR) dissected by small wadis running southwards into the Hamada. These flat glacis lacking water resources are covered with chamaephytic *Arthrophytum scoparium*, *Anabasis articulata* (*A. oropediorum*), *Anthyllis henoniana* on loamy shallow soils on calcareous hardpan, with *Artemisia herba-alba* on heavy soils in depressions and *Arthrophytum schmittianum* on shallow sandy soils.

On the road to Mizda (75 mm MAR), a few large, isolated evergreen remarkable trees (*Pistacia atlantica*) border the wadis (Photo 6.9) colonized by *Zilla spinosa*, *Pergularia tomentosa* and *Moricandia arvensis*. Sparse Saharan vegetation with *Salsola tetragona* – *Salsola tetrandra* are common on bare flats and interfluves.

The Hamada-el-Hamra

Receiving less than 50–100 mm MAR, with 500–700 m elevation, it expands to 400–500 km from south Nalut – Zintan – Mizda, deep south towards Sebha – Brak (MAR < 50 mm). It is almost flat, bare, rocky-gravely-dusty plateau country sloping gently due south-south-east and intersected with several north-facing low cuestas, oriented south-west–north-east. Starting from the Jebel Nafusa, patches of *Stipagrostis pungens* hold on sand accumulation (Harabah – South Zintan) while *Anthyllis henoniana* and *Arthrophytum scoparium* cover the northern *reg*. The *reg* of the Hamada proper is mostly devoid of vegetation except sparse perennials (*Anvillea garcini*, *Farsetia aegyptiaca*, *Oudneya africana*, *Savignya aegyptiaca*, *Helianthemum* sp., *Fagonia* sp.) mostly on the Hamada *zrira* (Photo 6.10). South of the cuestas, with increasing aridity, the rare perennial vegetation (*Rhus tripartita*) retreats into rills and gullies. Occasional shallow depressions (*Hatiya*) capture scant run-off and may host a wealth of small annuals (*Aristida plumosa*,

Photo 6.9. *Pistacia atlantica* isolated tree (8 m high) in a wadi, along the road near Mizda (N 31.530 E 12.999, March 2009).

A. ciliata, Trigonella sp., *Erodium* sp., *Fagonia* sp., *Anastatica hierochuntica, Medicago laciniata* ssp. *brachyacantha*).

The Hamada is a winter/spring rangeland traditionally grazed by free dromedaries. Nowadays, due to new mechanized wells and water carts pulled by trucks all over the Hamada, these very poor rangelands remain open all-year-long to sheep and goats feeding on open-air feedlots. When spring rains are sufficient, white-brown truffles, *terfas* (*Terfezia boudieri* and *Tirmania ovalispora*), develop on *Helianthemum* sp. roots and the small Hamada becomes alive with avid collectors.

To the east of the Hamada, stands of tall woody *Tamarix passerinoides* can be found in remote wadis (wadi Harandal, wadi Ar-Rijl) adjoining the Solul-el-Khoder region.

The south Zliten plateau

East of the Msallata – Qasr Daoun region, limited by wadi Roman, wadi Yaoudi, wadi Jdari, wadi Majer and tributaries south of Zliten, the north-eastern Jebel Nafusa slopes gently towards the north-east to the Souk-el-Khamis – Zliten – Misurata coast into the vast olive/almond groves of Daphniya (Photo 6.11) and the coastal palm trees oasis (200–250 mm MAR).

All northern wadis bear well-maintained olive groves (wadi Sarata, Photo 6.12) with occasional *Oryzopsis miliacea.* On the northern part of the plateau near the village of Majer, the last remnants of *Stipa tenacissima* can be seen topping a few rocky hills. It is the most westerly extension of *Stipa tenacissima*. Antique terraces water-harvesting and -managing systems, still well maintained to date, display olive/fig tree plantations in wadi Roman, wadi Yaoudi etc.

Large modern quarries (Photo 6.13) are expanding over large areas, digging, and grinding limestones providing material for new roads and ever-expanding urbanization to Khoms – Zliten to Misurata. The white dust generated by this polluting industry, and associated traffic of trucks and heavy machinery, covers the surrounding landscape, olive groves, crops and large swaths of nearby rangelands. The damage is considerable and irreparable as at wadi Sassu, wadi Um-el-Jdari etc.

Further south-east, on the plateau (150–175 mm MAR) (Photo 6.14), and in large depressions (wadi Fayda, Um-el-Jdari), once excellent grazing lands of *Artemisia herba-alba* have

Photo 6.10. The Hamada *zrira* (about 105 km south Zintan, N 31.065 E 12.736, March 2010).

Photo 6.11. The large olive/almond plantations at Daphniya (N 32.360 E 14.975, January 2007).

Photo 6.12. Olive groves in wadi Sarata (25 km south Zliten, N 32.272 E 14.694, 2009).

Photo 6.13. Large modern quarries (south Zliten, south Misurata, N 32.235 E 14.859, January 2009).

been cleared and are relentlessly cropped with dicey rainfed barley.

The powerful wadi Sassu catchment reveals countless ruins of Roman and Byzantine fortified farms with large dams, proof of long-standing cropping on residual water in these wadis despite the low-rainfall environment. On the loamy-heavy soils, *Ziziphus lotus* and *Atriplex halimus* mounds speckle the wadi invaded by *Carrichtera annua* weeds and rare perennial grasses (*Hyparrhenia hirta*). Large stands of the best *Artemisia herba-alba* country speckle the loamy flats (Photo 6.15).

Further south (125–150 mm MAR), the rocky plateau (Photo 6.16) is covered with *Arthrophytum scoparium*, *Pituranthos tortuosus* (*P. scoparius*, *P. chloranthus*) progressively replaced, due to increasing aridity and shallow soils, with *Salsola tetragona-Salsola tetrandra* and patches of sand veils with *Anabasis articulata* on the way to Bir Dufan.

Wadi Um-el-Jdari and tributaries (referring to ancient abundant presence of *Rhus tripartita*), and wadi Al-Faid spread their water smoothly on broad-acre rainfed barley, flowing into wadi Sassu descending towards Kararim, Sebkha Taworga – El Hisha – Bu Gren and the western Gulf of Sirte. The Taworga region and other close-by *sebkha* (Photo 6.17) host mostly halophytic *nebkha* (*Atriplex halimus*,

Photo 6.14. Disc ploughing preparation for rainfed barley (30 km south Zliten, October 2007) on the south Zliten plateau (N 32.207 E 14.813) on the way to wadi Sassu.

Photo 6.15. *Artemisia herba-alba* on loamy flats at wadi Sassu (N 32.140 E 14.787, April 2008).

Photo 6.16. The rocky plateau towards Bir Dufan (N 32.021 E 14.557, January 2009).

Limoniastrum sp.), phreatophytes (*Lycium arabicum*, *Nitraria retusa*) ending with large ponds (Al-Hadariyah) colonized with reeds (*Phragmites communis*, *Juncus* sp. and *Scirpus* sp.) and attracting large flocks of resident and migrating water birds. The reeds are used by the local population for a most renowned traditional cottage industry of reeds woven mats and baskets.

The Solul-el-Khoder

South and southeast of the South Zliten plateau starts the Solul-el-Khoder region of the great wadis (Meimun, Merdum, Zorzor, Zamzam, Ghirza, Sof-el-Jin (Photo 6.18) (± 100 mm MAR) delimiting the western region from the Sirte region. Countless ruined Roman-Byzantine fortified farms dot the landscape along the wadis.

Photo 6.17. The Kararim-Taworga *sebkha* (Near Shaba, N 32.152 E 15.110, December 2008).

Photo 6.18. Wadis occasionally cropped with rainfed barley at wadi Sof-el-Jin and tributaries, with remaining *Ziziphus lotus* and *Atriplex halimus nebkha* (N 31.972 E 14.688, September 2008).

It links with the Hamada and extends from the Beni-Walid area to Nefad (Photo 6.19), El-Hisha – Bu-Gren. All these wadis flow to the north-east into the Taworga *sebkha*. This is a biodiversity-rich country with stands of large *Acacia tortilis* hosting exceptional and rare large climber *Cocculus pendulus* (near wadi Zorzor), *Zilla spinosa*, *Pergularia tomentosa* and a wealth of pseudo-tropical C4 perennial grasses (*Cenchrus ciliaris*, *Cymbopogon shoenanthus*, *Dichantium annulatum*, *Lasiurus hirsutus*, *Panicum turgidum*, *Pennisetum dichotomum*) in the wadis (wadi Nefad), and a rare but noticeable occurrence of *Globularia alypum* (or *G. arabica?*) north of Seddada.

These wadis (Nefad and others close-by) with a large and unique community of sparse *Acacia tortilis* should be protected as national flora and fauna reserves. Cutting any remaining *Acacia tortilis* should be strictly prohibited.

All over, sparse *Salsola tetragona* and *Salsola tetrandra* are prevalent on the stony plateaux (*daraâ* or *argoub*) turning into bare *reg*. Shallow depressions and wadis host stands of *Panicum turgidum*, *Pennisetum divisum*, *Zygophyllum album*, *Cornulaca monacantha*, *Traganum nudatum*, and *Bassia muricata* well appreciated by wandering dromedaries.

The El-Hisha – Bu Gren – Bu N'Gem line marks the most eastern limit of the *Limes Tripolitanus* with numerous ruins of Roman-Byzantine settlements, farms, ingenious water harvesting systems, mausoleum (wadi Merdum, Ghirza) and forts (Qasr-el-Banat at Seddada, Bu N'Gem) from the 3rd Roman legion (Barker, 1996; Barker *et al.*, 1996; Gilbertson, 1996; Gilbertson and Hunt, 1996; Goodchild, 1952a, 1952b, 1976; Mattingly, 1995).

The central Sirte region

This is where the Sahara desert meets the Mediterranean Sea separating the western North African and the eastern/Middle East biological region, i.e. the Ibero-Maghrebian and the Irano-Turanian florae (Le Houérou, 2005; Zohari, 1973). The central Sirte region is mostly lowland with vast desert gravely rocky flats patched with endoreic saline-gypseous depressions and coastal *sebkha*. The whole area is subject to strong wind erosion witnessed by flagging low shrubs with exposed collars and roots pointing to a 15–20 cm topsoil removal resulting from overgrazing and degradation.

We may recognize three main zones in this region (Gaston, 1977; GEFLI, 1977; Gintzburger *et al.*, 1983; Nègre, 1974).

The western Sirte zone

It covers from Qaryat Abu-Gren (El Hisha) – Buerat-el-Hsun to wadi Wishkah (Photo 6.20) – wadi Thamit – wadi Kharif). From a vegetation and landform point of view, the western part of Sirte (100–150 mm MAR) including the wadi Bey-el-Kebir – wadi Thamit is under the Saharan influence from the Hamada and the Solul-el-Khoder. This is due to precipitation coming usually from the north-west falling primarily on the eastern Jebel Nafusa (Msallata, Zliten, Homs,

Photo 6.19. Savannah-like tree formation of *Acacia tortilis* ssp. *raddiana* (wadi Nefad, N 31.498 E 14.754, 1979).

Photo 6.20. Remnant large *Rhus tripartita* at wadi Wishkah south (N 31.234 E 15.391, May 2009).

Misurata (250 mm MAR). The region has shallow soils/stony-gravely plateaux (*reg*) with large expanse of weathered black flints on calcareous crust with main wide wadis gently descending to the coast. The vegetation on the Sirte plateau, short of reaching the coast, is dominated by formations of *Gymnocarpos decander, Arthrophytum scoparium, Pituranthos tortuosus, Helianthemum lippii, Medicago laciniata, M. laciniata var. brachyacantha* with an increasing occurrence of *Anvillea radiata, Salsola tetragona, Pituranthos chloranthus, Helianthemum kahiricum* with escalating aridity to the south within 20–30 km from the coast. It extends to the east of *sebkha* Bayi deeper south and parallel to the coast. All over the western Sirte area, shallow mobile sands in shallow depressions are sparsely covered with *Zygophyllum album, Traganum nudatum* with *Stipagrostis pungens* and *Retama raetam*. Facies with relatively deeper (20–30 cm) loamy soils on the calcareous crust host *Salsola longifolia, Suaeda mollis* close to the coast gradually replaced by *Salsola sieberi, Zilla spinosa* further south. Small wadis and low non-saline flats display remnants of large live specimens of *Rhus tripartita*, overgrazed *Periploca laevigata*, lonely *Acacia tortilis* ssp. *raddiana* and occasional small bunches of perennial grasses (*Cymbopogon shoenanthus*). Along gullies running into gypseous terraces (east of El-Balik to El-Goshiyah), spiny *Astragalus spinosus* dominate. These gullies end up in saline depressions where *Atriplex halimus, Arthrocnemum* and *Halocnemum* are plentiful. The sandy-loam soils of wadi Zamzam (Al-Qaddahyiah) – Marah-Bey-el-Kebir – Thamit) are dotted with *nebkha* (*Ziziphus lotus, Retama raetam, Atriplex halimus* and *Tamarix* sp.). Large wadi flats close to the coast are still cropped or planted with olive and fig trees supported with harvested water or new irrigation projects.

The middle Sirte zone

The region between wadi Thamit and Sirte (wadi Tilal) – Sultan-Ben Jawad benefits from the exposure to the Gulf of Sirte, northern winds and marginally better rainfall. It still presents a steep gradient from lower arid Mediterranean on the coast to a strong Saharan influence inland. Like the vegetation found between south Surman-el-Ajaylat – Al-Jmail in western Jeffara, this is where the *Thymelaea microphylla, T. hirsuta, Pituranthos rholfsianus, Retama raetam, Calicotome intermedia, Helianthemum ruficomum* formation thrives on the coastal strip around Sirte (150–175 mm MAR) on 40–60 cm deep coarse to fine sand. *Artemisia campestris*, *Polygonum equisetiforme*, *Echiochilon fruticosum*, *Argyrolobium uniflorum*, *Launaea resedifolia*, and *Medicago littoralis* add to the formation close to the coast. All the area is under unsustainable rainfed cropping and grazing, except in the largely cultivated wadi Tilal ending near Sirte town. *Atractylis* sp. replaces most annuals on shallow sand deposits over the calcareous crust colonized with *Gymnocarpos decander*, *Thymelaea microphylla*, *Arthrophytum scoparium*. It ends in sparse stands of *Arthrophytum scoparium* with *Salsola tetragona* and *Salsola tetrandra* at the limit of the Sahara.

The eastern Sirte zone

Between Ben Jawad and Sabkha Al-Kabitah (ex-Marble Arch site/El-Gaus) the environment is desert-like with the 100–125 mm isohyet crossing the coast. Here, the Sahara dips into the Mediterranean Sea. This is the boundary between the west and the east of North Africa/ Middle East regions, also as historically defined by the Philaeni brothers' ordeal (350 BCE).

From Bisher-Marsa Brega (MAR < 100 mm) to Ajdabiya (± 130 mm MAR), the lowland is exposed to strong sea winds from the north and north-west and from the southern desert. This is a flat sandy country with low long crescent-shaped dunes continuously submitted to strong wind erosion and covered with sparse *Thymelaea microphylla*, *T. hirsuta*, *Stipagrostis pungens*, *Retama raetam*, *Traganum nudatum*, *Pituranthos tortuosus*, *Asphodelus* sp. The land degradation due to overgrazing compounded by utter wind erosion is appalling all over. The vegetation cover is nearly non-existent. However, one should notice the 1500 ha protected area established since 1998 at El-Bedan (27 km north-east of Ajdabiya on the Tobruk road, Photo 6.21). It shows the extraordinary vegetation regeneration potential despite the low rainfall (approx. 100 mm MAR) with the presence of *Stipagrostis pungens*, *Stipa lagascae*, *Argyrolobium uniflorum*, *Salsola vermiculata*, *Pituranthos* sp., *Arthrophytum schmitianum* on fixed sand and *Arthrophytum scoparium – Gymnocarpos decander* on nearby shallow crusted soils. Through community determination and nearly full protection and with the able support of the Range Development Project, many useful rangeland and desert plants resolutely colonized this previously ruined land, now amazingly hosting wildlife (rabbits, birds, reptiles) that totally deserted the surrounding rangeland for a generation.

North of Ajdabiya towards Az-Zuwaytinah, the environment becomes progressively less Saharan and more low-/middle-arid Mediterranean

Photo 6.21. Range regeneration at El-Bedan protected area (north-east Ajdabiya, N 30.859 E 20.506, 2009).

due to a better exposure to western winds and precipitation coming directly from the Gulf of Sirte. The dust bowl originating from the loamy soils of the Soluk region (160 mm MAR) gradually announces the proximity of the Jebel Al-Akhdar.

The Jebel Al-Akhdar

The Jebel Al-Akhdar embraces the antique western province of Cyrenaica (from Ajdabiya to Derna-Bumbah) and the eastern drier Marmarica extending into western Egypt (Al-Bukhari, 2019; Al-Habony, 1999; Al-Sodany *et al.*, 2003; Behnke, 1980; El-Barasi *et al.*, 2011, 2013; El-Barasi and Barrani, 2015; El-Barasi and Saaed, 2013; El-Mograby *et al.*, 2018; GEFLI, 1972; Gintzburger and Blesing, 1979; Le Houérou, 2004; Le Houérou and Barghati, 1982; MacPhie, 1980; Mahmoud *et al.*, 2016; Saaed, 2008; Saaed *et al.*, 2019 a, b, 2021, 2022; Vetter *et al.*, 2009).

The Benghazi Plain

From Daryanah-Tokra (200–300 mm MAR) to Tolmeita, it is covered with *Terra rossa* overlaying solid and karst thick limestone. A degraded matorral of *Pistacia lentiscus* bushes and small *Juniperus phoenicea* is dominant (Photo 6.22). It is replaced by wastelands of *Asphodelus microcarpus, Urginea maritima* due to local cattle, sheep and goats overgrazing when not cleared for occasional rainfed cropping or small orchards in dolines[3].

The El-Marj plain and surroundings

The road climbs east of the Benghazi airport (260 mm MAR) through the first limestone escarpments towards Ar-Rajmah – Bu-Maryam – El Abyar to the red cultivated El-Marj plain (350–400 mm MAR). These escarpments bear dense matorral of *Pistacia lentiscus* and *Juniperus phoenicea* on the hills. Here, *Sarcopoterium spinosum*, a northern Mediterranean, and Middle East plant species, appearing already in the Benghazi plain with *Phlomis floccosa*, expands in pure dense stands on hilly and rocky areas of rangelands cleared from *Pistacia* and *Juniperus*, a similar vegetation type found in nearby Crete island, and high-rainfall zones of the Middle East. The El-Marj plain benefits from deep red Mediterranean soils – clay almost exclusively allotted to modern broad-acre rainfed cereal cropping and

Photo 6.22. Degraded matorral of *Pistacia lentiscus* bushes and small *Juniperus phoenicea* (Benghazi plain, N 32.176 E 20.180, 2007).

occasional orchards. The cultivated plain extends north-east to Al Dirsiyah-Battah.

The Jebel Al-Akhdar northern slope and plateau

These are the areas receiving some of the highest precipitations in Libya with Marsa Susah (360 mm MAR) on the coast and Shahat-Al-Beida (580 mm MAR). The first plateau (200–250 m elevation – extending to the first escarpment east of Benghazi airport) below Shahat is a patchwork of small farms and garrigues (Photo 6.23).

This plateau, from Ad-Dirsiyah-al-Haniyah to Marsa Susah – Ras-Hilal-Derna, plunges abruptly into the sea to the north, faces a sharp escarpment (550–600 m elevation – the same escarpment that backs the El-Marj plain) to the south and ends up at Derna (250 mm MAR). All the area is solid karst limestone topped with variable depths of *Terra rossa* and brown soils covered with dense garrigues with thick 2–6 m bushes dominated by *Juniperus phoenicea, Pistacia lentiscus, Phlomis floccosa* with other trees (*Arbutus pavarii, Ceratonia siliqua, Quercus calliprinos, Olea europea sylvestris*), climbers (*Smilax aspera, Tamus communis, Asparagus* sp.), shrubs (*Calicotome villosa, Cistus villosus, Coronilla glauca, Erica multiflora, Erica sicula, Myrtus communis, Phillyrea media, Rhamnus alaternus, Rosmarinus officinalis, Ruscus aculeatus, Viburnum tinus*), and a wealth of others (*Cyclamen rholfsianum, Lonicera* sp., *Narcissus tazetta, Oryzopsis holciformis, Brachypodium ramosum, Urginea maritima, Asphodelus microcarpus, Ranunculus bullatus*). This is also where *Thapsia garganica*, possibly the famous *Silphium* of the Greeks and the Romans (Keith, 1965) and other tall Umbelliferae/Apiaceae (*Ferula* sp. or *Angelica* sp.) are common. Isolated trees or small patches of native *Cupressus sempervirens* var. *horizontalis* are still found near El-Beida. At Ras-Hilal, native degraded forests of *Pinus halepensis* are still present close to the sea. One should notice that large patches of *Juniperus phoenicea* matorral are affected by an ailment that appear to kill them (Photo 6.24). *Juniperus* stands host dense lichens (*Teloschistes villosus* and *Ramalina* sp., Saaed *et al.*, 2021), proof of fog and high air humidity in winter.

On the top of the Jebel Al-Akhdar, near Al-Beida airport, monospecific dense stands of *Pistacia lentiscus* cover large areas where *Juniperus* have been eradicated over centuries by fuelwood collectors. These garrigues are much-used grazing lands mostly for shepherded goats and free-ranging local cattle.

Photo 6.23. The matorral below Shahat towards Marsa-Suza (N 32.825 E 21.936, 2008).

Photo 6.24. Dead stand of *Juniperus phoenicea* (coastal plain matorral, 5 km south of El-Hamamah, N 32.884 E 21.651, 2009).

Most remarkable is the area of the deep wadi Al-Kuf (Photo 6.25) (17 km west-south-west of Al-Beida), rich in plant biodiversity and hosting *Cupressus sempervirens* ssp. *sempervirens* exceptional native trees reaching up to 30 m high.

All over, patches of afforestation projects with *Eucalyptus camaldulensis*, *E. gomphocephala*, and *Pinus halepensis* run along roads and tracks with moderate success due to shallow soils, low precipitations, and difficulties to protect against wandering livestock and illegal wood harvesting.

The Jebel Al-Akhdar southern slope

It starts some 25–30 km south of the 200 mm isohyet at the southern limit of the *Juniperus phoenicea* – *Pistacia lentiscus* vegetation type. Degraded facies are enclosed in the 200–300 mm isohyet that runs approximately from 25 km south of Benghazi – 10–20 km south of Al-Abyar (260 mm MAR) – Jardas-el-Abid – 20–25 km south of Taknis (275 mm MAR) and Marawah (260 mm MAR), 10 km north of Martuba and then to the coast at wadi Burur (25–30 km south-east from Derna). Quite interesting is the fact that tall and impenetrable *Juniperus phoenicea* bushes are common in the highest-rainfall zones garrigues while they turn into large, evenly distributed trees 6–10 m high; sometimes a lonely tree with one tree per 5–10 ha in the lowest-rainfall zones of their distribution (200–225 mm/year) such as at Al-Mzerah – Mafsel (15 km south of Gendoula). *Juniperus* have been systematically targeted for fuelwood over centuries and cleared for rainfed cropping in lowland (Photo 6.26). On now barren hills, *Juniperus* is replaced with thorny bushes of *Rhus oxycantha* and *Rhamnus alaternus* (Al-Guetara – wadi Iftilia) with a carpet of *Poa bulbosa*. Some large *Juniperus* retreats into deep rocky wadis (Abu-Erreba – 15 km east-south-east of El-Abyar), the latter being the most south-westerly

Photo 6.25. Wadi Kuf karst landscape (N 32.695 E 21.563, 2007).

Photo 6.26. Clear forest of *Juniperus phoenicea* on hilltop and rainfed barley in valley (10–12 km south Gendula, N 32.467 E 21.576, 2009).

extension of *Juniperus* in Jebel Al-Akhdar. From there, the *Juniperus* disappears. As aridity increases, it gives room to the chamaephyte steppe expanding on rocky flats and loamy depressions south of a line from Jahishiyah – Karuba – Marawa – Martuba to Bombah – Tobruk – Bardiya.

Further south of a line Kharuba – Derwa – Mekhili, below the 100 mm isohyet, the wadies descending from the southern slope of the Jebel are collecting large amount of water ending up in the *balat*, large baren and non-saline endorheic depressions opportunistically cropped with barley.

South Derna to east of Bumbah/Tamimi to Tobruk-Bardiya (Marmarica)

About 25–30 km south of Derna, the vegetation type changes abruptly to a steppic low cover chamaephytic shrubs (MAR < 150 mm, Photo 6.27) with *Arthrophytum scoparium, Anabasis articulata, Salsola tetragona, Salsola tetragona, Noaea mucronata* on calcareous hardpan and limestone shallow-soils plateaux. Rainfed barley is systematically cropped in the wadis. *Anabasis articulata, Arthrophytum schmitianum* dwells on shallow sandy soils while *Artemisia herba-alba, Atriplex glauca, Suaeda fruticosa* colonize depressions with loamy-clay soils. Further south, the large *balat* (± 50 mm MAR) are speculatively cropped on residual water after being inundated in winter (wadi Samalus, Balta Az-Zalak, 50 km south-east of Derwa) and south-west of Mekhili. Between Bumbah and Al-Qardabah, the 100 mm isohyet reaches the coast and then runs south of Tobruk (125 mm MAR) through El-Adem airport (100 mm MAR) to Bardia (130 mm MAR) – Sollum on the Egyptian border. This is *Periploca laevigata, Salsola vermiculata, Salsola tetragona, Salsola tetragona* country on shallow soils with *Thymelaea* sp. and others common on the most arid rocky environment and with *Salsola longifolia* along the coast (El-Barasi and Saaed, 2013; Le Houérou, 2004; Saaed, 2008; Saaed *et al*., 2021). Long depressions with deep heavy soils (Al-Qa'ararah, Kambut, Qasr-el-Jadi) parallel to the coast are rainfed cultivated and occasionally under irrigation.

Ecological Groups and Plant Associations

Ecological groups and plant associations present in arid North Africa were long-ago identified and

Photo 6.27. Bir Botna/wadi Deir (about 30 km SSW from Derna, N 32.536 E 22.409, 2007).

described especially in central Tunisia (Long, 1954) and southern Tunisia (Le Houérou, 1959, 1960, 1969; Novikoff, 1961). Le Houérou (1959, 1969) collected the plant lists of some 1600 southern Tunisian sites with detailed ecological information. He applied a manual phytosociological tabulation method (Braun-Blanquet, 1932; Gounot, 1969; Poore, 1955) about vegetation composition in relation to climate (rainfall and mean minimum of the coldest month), soil and land use. The manual or statistical treatment of these tables to order and structure the plant lists allows the identification of specific plant ecological groups with their relation to climate, soil, and human and livestock impact. One of these three factors could be dominant or could coexist. A set combination of one or several ecological groups defined by climate, soils or human/livestock factor constitutes a plant association. In short, we have a three-dimensional matrix of climate x soil x land use (human/livestock factors) defining the plant associations or sub-associations.

Ecological groups

Based on Le Houérou's work (1959, 1969), we list the main ecological groups identified in arid southern Tunisia, most of which are valid for Libya. We give some examples of common Libyan plants belonging to these ecological groups. These lists are neither exhaustive nor exclusive, i.e. a specific plant may belong to one, two or three ecological groups according to climate, soil type and land use.

Climatic ecological groups

These are defined by two major factors: the MAR and the mean minima of the coldest month (m), usually in January. Both are the major limiting factors for the growth of plants. Winter growth is further controlled by the mean MAXIMA of the coldest month (M') (see Chapter 3). These ecological groups are defined as bioclimatic variants.

Plants in ecological groups in relation to MAR:

- MAR ≥ 400 mm, upper semi-arid and above, *Cupressus sempervirens, Quercus calliprinos, Coronilla minima, Calicotome villosa, Sarcopoterium spinosum, Sanguisorba minor, Phalaris* sp., *Lolium perenne*
- MAR ≥ 300 mm, lower semi-arid and above, *Cistus salvifolius, Cistus villosus, Convolvulus dorycnium, Erica multiflora, Arbutus pavarii, Ceratonia siliqua, Ephedra altissima, Achillea santolina, Arum italicum, Eryngium* sp.
- MAR ≥ 200 mm, upper arid and above, *Pinus halepensis, Globularia alypum, Periploca laevigata, Rhus tripartitum, Oryzopsis miliacea, Thymus capitatus, Aegilops* sp., *Astragalus hamosus, Hyparrhenia hirta, Asphodelus microcarpus, Artemisia campestris, Urginea maritima*
- MAR ≥ 100 mm, lower arid and above, *Atriplex* sp., *Stipagrostis pungens, Hedysarum spinosissimum, Plantago albicans, Lygeum spartum, Artemisia herba-alba, Anthyllis henoniana, Arthrophytum* sp., *Anabasis* sp., *Salsola* sp., *Cenchrus ciliaris, Lasiurus hirsutus, Dichanthium annulatum, Stipa* sp., *Aristida* sp., *Periploca laevigata, Rhus tripartita*
- MAR ≥ 20 mm, Saharan, *Calligonum* sp., *Cornulaca monacantha, Bassia muricata, Oudneya africana, Savignya parviflora, Pennisetum dichotomum, Panicum turgidum, Ephedra alata, Zilla spinosa, Anvillea radiata, Tamarix* sp., *Zygophyllum* sp., *Pergularia tomentosa*

Plants in ecological groups in relation to the mean minima of the coldest month (m):

- m > 7°C, warm winter, *Calotropis procera, Solanum sodomaeum*
- m > 5.5°C, mild to warm winter, *Quercus calliprinos, Thymus capitatus, Phlomis floccosa, Convolvulus dorycnium, Lavandula multifida, Ononis* sp., *Arisarum vulgare, Arum cyrenaicum, Mesembryanthemum crystallinum, Juniperus macrocarpa*
- m > 3°C, temperate to warm winter, *Rhanterium suaveolens, Gymnocarpos decander, Salvia aegyptiaca, Cymbopogon shoenanthus, Lasiurus hirsutus, Dichanthium annulatum, Cenchrus ciliaris, Pituranthos* sp., *Zygophyllum* sp., *Carrichtera annua*
- m > 1°C, cool to warm winter, *Retama raetam, Genista saharae, Echiochilon fruticosum, Hyparrhenia hirta, Arthrophytum* sp., *Ziziphus lotus*
- m < 5.5°C, cool to temperate winter, *Thymelaea microphylla, Thymelaea hirsuta, Genista microcephala*
- m < 3°C, cool to cold winter, *Poa bulbosa, Sarcopoterium spinosum, Atractylis humilis*

Soil ecological groups

(See Chapter 5)

FORESTS AND GARRIGUE

- Organic soils (under shade): *Pistacia lentiscus, Quercus calliprinos, Smilax aspera, Tamus communis, Asparagus acutifolius*
- Alkaline soils (exposed to full sun): *Pinus halepensis, Stipa tenacissima, Juniperus phoenicea, Globularia alypum, Erica multiflora, Rosmarinus officinalis, Rhus tripartita, Olea europea*
- On rocky soils and cliffs: *Pistacia atlantica, Capparis spinosa, Phagnalon rupestre, Ephedra* sp., *Ballota hirsuta, Sedum* sp., *Thymus* sp., *Globularia arabica*

THE STEPPE

- Loamy to heavy soils: *Artemisia herba-alba, Arthrophytum scoparium, Carrichtera annua, Medicago laciniata, Asteriscus pigmaeus, Plantago ovata, Stipa retorta, Astragalus hamosus, Asparagus albus, Poa bulbosa, Achillea santolina, Psoralea bituminosa, Anastatica hierochuntica, Ajuga iva, Medicago truncatula, Medicago polymorpha, Medicago minima, Thymelaea hirsuta, Zilla spinosa, Noaea mucronata*
- Sandy soils: *Stipagrostis pungens, Rhanterium suaveolens, Calligonum* sp., *Retama raetam, Genista saharae, Artemisia campestris, Argyrolobium uniflorum, Citrullus colocynthis, Echiochilon fruticosum, Arthrophytum schmittianum, Ajuga iva, Cutandia* sp., *Teucrium polium, Salvia verbenaca, Hippocrepis bicontorta, Hedysarum spinosissimum, Onobrychis* sp., *Vicia* sp., *Helianthemum lippii, Koelpinia linearis, Lotus pusillus, Scabiosa arenaria, Plantago albicans, Schismus barbatus, Lupinus angustifolia, Scabiosa arenaria, Medicago littoralis, Stipa lagascae, Nolletia chrysocomoides, Cornulaca monacantha, Aristida obtusa, Neurada procumbens, Thymelaea microphylla, Cyperus conglomeratus, Fagonia glutinosa, Euphorbia retusa, Eremobium* sp., *Haplophyllum tuberculatum, Ephedra alata*
- Calcareous crust and shallow soils: *Gymnocarpos decander, Globularia alypum, Herniaria fontanesii, Pituranthos scoparius, Atractylis serratuloides, Salsola tetragona, Salsola tetrandra, Rhamnus* sp., *Calicotome* sp., *Rhus tripartita, Helianthemum kahiricum, Capparis spinosa*
- Gypseous soils: *Lygeum spartum, Anabasis* sp., *Arthrophytum schmittianum, Erodium arborescens, Erodium glaucophyllum, Astragalus spinosus, Moricandia arvensis, Frankenia thymifolia, Aeluropus littoralis, Pituranthos chloranthus, Gymnocarpos decander, Zygophyllum* sp., *Reaumuria vermiculata, Limonium pruinosum, Limoniastrum guyonianum, Traganum nudatum, Nitraria retusa, Salsola vermiculata, Spergularia* sp., *Herniaria* sp., *Paronychia* sp.
- Salty soils: *Halopeplis amplexicaulis, Salicornia* sp., *Suaeda mollis, Atriplex* sp., *Salsola* sp., *Lepturus incurvatus, Frankenia pulveranta, Aeluropus littoralis, Limoniastrum guyonianum, Limoniastrum monopetalum, Hordeum maritimum, Plantago coronopus, Tamarix* sp., *Aizoon* sp., *Frankenia* sp., *Spergularia* sp., *Spergula* sp., *Herniaria* sp.
- High water table and water logging: *Tamarix* sp., *Nerium oleander, Phragmites communis, Phoenix dactylifera, Juncus* sp., *Imperata cylindrica, Lycium* sp., *Nitraria retusa, Alhagi* sp.

Land-use ecological groups

- Post-cultural and fallow land: *Artemisia campestris, Thymelaea microphylla, Thymelaea hirsuta, Haplophyllum tuberculatum, Cynodon dactylon*
- Overgrazing and wasteland: *Asphodelus* sp., *Peganum harmala, Marrubium* sp., *Hordeum murinum, Mesembryanthemum* sp., *Cleome arabica, Pallenis spinosa, Thymelaea hirsuta, Hertia cheirifolia, Thapsia garganica, Stipa retorta, Nicotiana glauca, Malva aegyptiaca, Euphorbia* sp., *Scolymus* sp., *Centaurea* sp., *Emex spinosus, Hyoscyamus* sp., *Solanum nigrum, Notoceras bicorne, Malva sylvestris, Aizon* sp., *Tribulus terrestris, Neurada procumbens*

Most common Libyan plant associations

A matrix or complex combination of ecological groups characterize diverse plant associations studied with vegetation ecological surveys and mapping in southern Tunisia (Le Houérou,

1959, 1969) and in central and northern Tunisia (Gounot, 1969).

Some examples are:

- In the Jeffara plain (100–200 mm MAR), the nearly pure *Stipagrostis pungens* association covers aeolian and fixed deep sand (south Sorman, Saniet-el-Hod, Bir Shajarah). Fine sand and interdune depressions would host a *Retama raetam – Genista saharae* sub-association (north of the road Bir-el-Ghanem – Aziziya). With increasing aridity (100–125 mm MAR) on deeper sand, the *Stipagrostis pungens – Calligonum* sp. sub-association takes over (south-west El-Watiya to Bir Shabunyah-Takut).
- From Al-Jmail-Reqdalin to Al-Ajaylat (175–250 mm MAR), the pristine *Artemisia campestris* association colonizes stable sandy soils with companion species such as *Echiochilon fruticosum, Cutandia dichotoma, Scabiosa arenaria, Helianthemum lippii, Thymelaea hirsuta, Thymelaea microphylla* etc. Losing all these latter species with heavy grazing or cropping, it ends up with a pure dense stand of *A. campestris* on abandoned crops and fallow land.
- At Bir-el-Ghanem (150 mm MAR) (Gintzburger and Alidrissi, 1976), we find the original *Rhanterium suaveolens–Stipa lagascae* association with different soil facies. When in good condition, on shallow sand (up to 60–70 cm), the *Rhanterium suaveolens–Stipa lagascae* has companion species such as *Polygonum equisetiforme, Salsola vermiculata* and many annuals. With active wind erosion, the sand depth diminishes with the loamy subsoil exposed, annuals become rare and the soil sub-association of *Rhanterium suaveolens–Stipa lagascae* with *Artemisia herba-alba–Atractylis serratuloides* becomes dominant. On shallow soil with gypseous soil sublayers, the ecological group with *Lygeum spartum, Erodium glaucophyllum* appears. With the ultimate stage of wind erosion affecting the *Rhanterium suaveolens–Stipa lagascae* association still present on sparse low sand mounds, *Rhanterium* become rare, and the ecological group linked with the calcareous crust dominates with *Gymnocarpos decander–Atractylis serratuloides* sub-association.
- In the Jebel Nafusa (150–300 mm MAR), the *Stipa tenacissima–Genista microcephala* association with many degradation facies expands over hills and slopes. A sub-association with *Helianthemum kahiricum, Pituranthos tortuosus* and *Atractylis serratuloides* expands on shallow-eroded soils.
- On the southern and the northern slopes and foothills of the Jebel Nafusa, pure stands of *Artemisia herba-alba* or of *Arthrophytum scoparium* always occurs on loamy to heavy soils and therefore would be an indicator of this type of soils in the 100–300 mm MAR. However, *Artemisia herba-alba* would be replaced on this heavy soil type by pure stands of *Arthrophytum scoparium* association with increasing aridity.
- On the *reg* of the northern Hamada (Nalut, Al-Harabah, South Zintan), on the shallow soils of the rocky *reg* with 100–150 mm MAR, the *Anthyllis henoniana–Gymnocarpos decander* covers large areas. The same plant association on *reg* will display a *Farsetia aegyptiaca–Savignya parviflora* sub-association at the lower end of this rainfall zone. On sandy facies in the same area, the *Anabasis articulata–Plantago ciliata* sub-association prevails.
- Further east, the *Acacia tortilis–Panicum turgidum* association with rare *Cocculus pendulus* dots the Solul-el-Khoder zones (100–150 mm MAR) and the upper wadi Nefad, Zemzem, Zorzor, Merdum etc. Shallow sandy areas on gypsum host the *Traganum nudatum–Anabasis articulata* sub-association with *Cornulaca monacantha, Bassia muricata, Zygophyllum album, Salsola sieberi, Astragalus spinosus* etc. prevailing.
- In the Sirte region (Nègre, 1974), the *Gymnocarpos decander–Arthrophytum scoparium–Salsola tetragona* association is present on all calcareous crust covered or not with shallow soils associated with *Helianthemum lippii* on the coastal zone (150–175 mm MAR). The *Launaea nudicaulis–Picris coronopifolia* sub-association leads on loamy-heavy soils, or with abundant *Ifloga spicata–Medicago littoralis* on fixed sand. Inland, at the limit of the desert (100 mm MAR), the sub-association with *Anvillea radiata* takes over with abundance of *Medicago laciniata* on gravelly-sandy *reg*.

- East of Sirte to Ajdabiya (100–150 mm MAR), the *Thymelaea microphylla–Pituranthos rholfsianus* association colonizes fine sandy soils prone to wind erosion. Several sub-associations on different facies occur with *Stipagrostis pungens* on mobile sand, with *Thymelaea hirsuta* on compacted fine sand, with *Calicotome intermedia–Arthrophytum scoparium* on calcareous crusts. These extend to Zuetinah-Soluk.
- Reaching the Benghazi plain to the El-Marj low hills, we enter the dense garrigues with the nearly pure *Pistacia lentiscus* association with various levels of degradation due to overgrazing and fuelwood collection. It is often associated with the dense scrub sub-association of *Pistacia lentiscus–Juniperus phoenicea* that covers most of the top high-rainfall zones and karst of Cyrenaica.
- From below Shahat to Marsa Suza and further east along the coast to Derna, are 4–8 m high *Quercus calliprinos–Ceratonia siliqua–Arbutus pavarii* with remnants of *Pistacia lentiscus* sub-association degraded by repeated bushfires, fuelwood harvesting and browsing by wandering goats and cattle.
- On the coldest top of Jebel Al-Akhdar, the cushion-like low spinescent *Sarcopoterium spinosum–Marrubium vulgare* association covers barren hills with large rare and relict *Cupressus sempervirens* ssp. *horizontalis* dotting the landscape. When overgrazed, this association is slowly invaded with *Ballota* sp., *Phlomis floccosa* and geophytes.
- South Jebel Al-Akhdar, the nearly pure *Juniperus phoenicea* association with superb old trees (up to 8–10 m high) extends to the limit of the steppe (200 mm MAR) with a sub-association with *Poa bulbosa–Phlomis floccosa* on denuded hills where isolated *Juniperus* still survive. Below the 200 mm MAR, the steppic chamaephytic plant association of *Artemisia herba alba* and *Arthrophytum scoparium* common all over Libya takes over.
- In Marmarica (Saaed, 2008), receiving less than 150 mm MAR, the inland rocky-shallow soils plateau hosts an *Arthrophytum scoparium–Thymelaea hirsuta* association with sub-associations common to the Sirte region.

Conclusions

The geology and geomorphology model the Libyan landscapes supporting many different vegetation physiognomy and types. These in turn are conditioned by climate and bioclimates, soil types and land use defining ecological groups. Diverse combinations of ecological groups define plant associations. These plant associations, though well-defined and identified in southern and central Tunisia are not stable in time and are affected by land use and grazing pressure. The Raunkiaer's biological spectrum of these plant associations is one way to assess their status and evolution. However, ecological groups and plant associations can only be defined with detailed phytosociological surveys on sites combined with accurate climate, soil and land-use description. It implies that the surveyor must have a deep knowledge and clear understanding of ecological parameters and plant taxonomy of the local flora. It starts with the proper identification and correctly naming the rangelands plants. It is a long process that can be only acquired by repeated field trips, on-site ecological observations, herbarium collection, accurate plant identifications, complemented with information gathered from shepherds and farmers.

Notes

1 MAR = mean annual rainfall
2 *Erg* = large area with deep sand dunes
3 A doline is a depression in a karst landscape, sometime plugged or not with a sinkhole.

References

Abdalrhim, M.A.G. (2021) Size structure of *Arbutus pavarii* and *Juniperus phoenicea* populations in Al-Marj Plain, Libya. *International Journal of Multidisciplinary Sciences and Advanced Technology,* Special Issue 1, 10–16.

Abu-Aziza, F.B., El-Barasi, Y.M. and Rahil, R.O. (2017) Flora, vegetation and human activities of Wadi Derna-El-Jabal. *ContROL* 1(6).

Abusaief, H.M.A. and Dakhil, A.H. (2013) The floristic composition of rocky habitat of Al Mansora in Al-Jabal Al-Akhdar, *Libya. New York Science Journal* 6(5), 34–45.

Al-Bukhari, A.A. (2019) The use of remote sensing and GIS for monitoring rangeland degradation in Libya. PhD dissertation, Cranfield University, Bedford, UK.

Al-Habony, M.E.E. (1999) Vegetation and flora of a sector along Mediterranean Coast of Libya from Tobruk to Egyptian Border. MSc thesis, Department of Botany, Faculty of Science, Benghazi University.

Al-Khojah, H.M., Al-Yafour, L.A. and Mahklouf, M. (2020) Floristic study of selected districts in Sabratha region Libya. *Journal of Advanced Botany and Zoology* 7(4), 1–3. DOI: 10.5281/zenodo.3830169

Al-Sodany, Y.M., Shehata, M.N. and Shaltout, K.H. (2003) Vegetation along an elevation gradient in 000Al-Jabal Al-Akhdar, *Libya. Ecologia Mediterranea* 29(2), 125–137. DOI: 10.3406/ecmed.2003.1547

Al-Traboulsi, M. and Alaib, M.A. (2021) A survey of medicinal plants of Wadi Al-Kouf in Al-Jabal Al-Akhdar, *Libya. Natura Croatica Zagreb* 30(2), 389–404. DOI: 10.20302/NC.2021.30.25

Barker, G. (1996) *Farming the Desert: The UNESCO Libyan Valleys Archaeological Survey.* (Vol. 1–2). UNESCO, Paris.

Barker, G., Gilbertson, D., Hunt, C.O. and Mattingly, D. (1996) Romano Libyan agriculture: integrated models. In: Barker, G., Gilbertson, D., Jones, B. and Mattingly, D. (eds) *Farming the Desert: The UNESCO Libyan Valleys Archaeological Survey* Vol. 1. UNESCO, Paris, pp. 265–290.

Behnke, R.H. (1980) *The Herders of Cyrenaica: Ecology, Economy and Kinship among the Bedouin of Eastern Libya.* University of Illinois Press, Urbana, Illinois.

Braun-Blanquet, J. (1932) *Plant Sociology: The Study of Plant Communities.* McGraw-Hill, Inc., New York and London. Available at: https://www.cabdirect.org/cabdirect/abstract/19331600801 (accessed 12 August 2024).

Despois, J. (1935) *Le Djebel Nefousa (Tripolitaine), Etude Géographique.* Larousse, Paris.

El-Barasi, Y.M. and Barrani, M.W.B. (2015) Land use and disturbance effects on the natural vegetation of Daphna range zone (northeastern part of Libyan coast). *International Journal of Environment and Water* 4(2), 89–103.

El-Barasi, Y.M. and Saaed, M.W.B. (2013) Threats to plant diversity in the northeastern part of Libya (El-Jabal El-Akhdar and Marmarica Plateau). *Journal of Environmental Science and Engineering* 2, 41–58.

El-Barasi, Y.M., El-Sherif, I.M. and Gawhari, A.M.H. (2003) Checklist and analysis of the flora and vegetation of Wadi Zaza at Al-Jabal Al Akhdar (Cyrenaica, Libya). *Bocconea* 16(2), 1091–1105.

El-Barasi, Y.M., Barrani, M.W., El-Amrouni, A.O. and Mohamad, N.F. (2011) Check list of flora and vegetation on south El-Marj Zone: South El-Jabal El-Akhdar – Libya. *International Journal of Engineering* 9(3), 141–146.

El-Barasi, Y.M., Saaed, M.W.B. and Al Tajoury, R.O. (2013) Land deterioration of a semidesert grazing area in the north-eastern zone of Libya (Cyrenaica). *Journal of Environmental Science and Engineering B* 2, 357–373.

El-Mograby, A.S., Saaed, M.W.B., El-Barasi, Y.M. and Rahil, R.O.M. (2018) Characteristics of vegetation and soil seedbank of a transect in the coastal strip of the arid rangelands In Solouq Plain. *Cyrenaica, Eastern Libya, The Fifth Scientific Conference of Environment and Sustainable Development in the Arid and Semi-Arid Regions (ICESD)* Ajdabiya, 23–25 July, pp. 132–143.

El-Mokasabi, F.M. (2017) Studies on the flora of Libya. *ContROL* 1(1), 1–10. DOI: 10.28915/control.0008.1

Emberger, L. and Godron, M. (1968) *Code pour le Relevé Méthodique de la Végétation et du Milieu: Principes et Transcription sur Cartes Perforées.* Editions du Centre National de la Recherche Scientifique, Paris.

Gaston, A. (1977) *Survey for the Development of the Central Wadi Zones and Gulf of Sirte – Grazing Project – 3rd Agrostological Report – Mimeo.* IEMVT-GEFLI.

GEFLI (1972) *Soil and Water Resources Survey for Hydro-agricultural Development, Eastern Region, Water Resources Survey, 2–2 Climatology.* Groupement D'Etude Française en Libye.

GEFLI. (1977) *Survey for the Development of the Central Wadi Zone and Gulf of Sirte (Sirte, Wadi Zam-Zam Project).* Groupement D'Etude Française en Libye.

Gilbertson, D. (1996) Explanations: environment as agency. In: Barker, S.G., Gilbertson, D., Jones, B. and Mattingly, D. (eds) *Farming the Desert: The UNESCO Libyan Valleys Archaeological Survey* Vol. 1. UNESCO, Paris, pp. 291–317.

Gilbertson, D. and Hunt, C.O. (1996) Romano-Libyan agriculture: walls and floodwater farming. In: Barker, G., Gilbertson, D., Jones, B. and Mattingly, D. (eds) *Farming the Desert: The UNESCO Libyan Valleys Archaeological Survey* Vol. 1. UNESCO, Paris, pp. 191–225.

Gintzburger, G. (1986) Seasonal variation in above-ground annual and perennial phytomass of an arid rangeland in Libya. *Journal of Range Management* 39(4), 348–353.

Gintzburger, G. and Alidrissi, M. (1976) *Range Study on Jeffara Plain: Sheep Project at Bir-El-Ghanem* (No. 83/76; p. 51). FAO and Agricultural Research Center, Tripoli.

Gintzburger, G. and Blesing, L. (1979) *Genetic Conservation in Libya – Indigenous Forage Legumes Collection in Northern Libya, Distribution and Ecology of Medicago sp.* Food and Agriculture Organization, Rome.

Gintzburger, G. and Saïdi, S. (2009) From inventory to monitoring in semi-arid and arid rangelands. In: Squires, V.R. (ed.) *Range and Animal Sciences and Resources Management. Vol. 2. Encyclopedia of Life Support Systems*. EOLSS.

Gintzburger, G., Sbeita, A., Rojbani, H., Fituri, M. and Francis, C.M. (1983) Annual *Medicago* sp. collected from the Sirte Region of Libya. *Plant Genetic Resources* 56, 23–29.

Gintzburger, G., Toderich, K., Mardonov, B. and Mahmudov, M. (2003) *Rangelands of the Arid and Semi-Arid Zones in Uzbekistan.* CIRAD-ICARDA, Montpellier, France.

Goodchild, R.G. (1952a) Farming in Roman Libya. *The Geographical Magazine* 25, 70–80.

Goodchild, R.G. (1952b) The decline of Libyan agriculture. *The Geographical Magazine* 25, 147–156.

Goodchild, R.G. (1976). *Selected Papers of the Late R.G. Goodchild, Libyan Studies.* (J. Reynolds, Ed.). Paul Elek, London, UK.

Gounot, M. (1969) *Méthodes D'étude Quantitative de la Végétation.* Editorial Masson, Paris.

Hegazy, A.K., Kabiel, H.F., Al-Rowaily, S.L., Lovett-Doust, L. and Al Borki, A.S. (2021) Plant communities and reproductive phenology in mountainous regions of northern Libya. *Journal of Forestry Research*, 1–21. DOI: 10.1007/s11676-016-03

Keith, H.G. (1965) *A Preliminary Check List of Libyan Flora, Volumes 1 and 2.* (Vol. 2). United Middle East Executives Ltd, London.

Le Houérou, H.-N. (1959) Recherches écologiques et floristiques sur la végétation de la Tunisie méridionale. PhD dissertation, Université Alger Institut de Recherches Sahariennes, Algiers.

Le Houérou, H.-N. (1960) Contribution à l'étude des sols du Sud-Tunisien. *Annales Agronomiques* 11(3), 241–309.

Le Houérou, H.-N. (1969) *La Végétation de la Tunisie Steppique (Avec Référence aux Végétations Analogues d'Algérie, de Libye et du Maroc).* (Vol. 1–2). Annales de l'Institut National de la Recherche Agronomique de Tunisie, Tunis.

Le Houérou, H.-N. (2004) Plant diversity in Egyptian Marmarica: catalogue of the vascular species with their bioclimatic distribution patterns, frequency, habitat and main ecological attributes. *Candollea* 59(2), 259–308.

Le Houérou, H.-N. (2005) *The Isoclimatic Mediterranean Biomes: Bioclimatology, Diversity and Phytogeography.* (Vol. 2). Copymania, Montpellier, France.

Le Houérou, H.-N. and Barghati, M.S. (1982) An evaluation of fodder shrubs plantations in the Benghazi Plains. (UNTF/Lib 018, Technical Paper No 45). FAO and Agricultural Research Center, Tripoli.

Long, G. (1954) Contribution à l'étude de la végétation de la Tunisie Centrale. *Annale Service Botanique Agronomique Tunisie* 27.

MacPhie, G.R. (1980) *Final Report of Activities and Recommendations of the South Australian Agricultural Team Based on the Demonstration Farm at El-Marj (Libya) 1974–1980.* Overseas Project Division, South Australian Department of Agriculture, Adelaide, Australia.

Mahmoud, A.M.A., Hasmadi, I.M., Alias, M.S. and Azani, A.M. (2016) Rangeland degradation assessment in the south slope of the Al-Jabal Al-Akhdar, northeast Libya using remote sensing technology. *Journal of Rangeland Science* 6(1), 73–80.

Mattingly, D.J. (1995) *Tripolitania.* Routledge, New York. DOI: 10.4324/9780203481011

Mukassabi, T.A., Ahmidat, G., Sherif, I.M., Elmogasapi, A. and Thomas, P.A. (2012) Checklist and life forms of plant species in contrasting climatic zones of Libya. *Biological Diversity and Conservation* 5(3), 1–12.

Nègre, R. (1961) *Petite Flore des Régions Arides du Maroc Occidental.* (Vol. 2). Editions du Centre National de la Recherche Scientifique, Paris.

Nègre, R. (1974) Les pâturages de la région de Sirte (Libye): projet de régénération. *Feddes Repertorium* l(85–3), 185–243.

Novikoff, N. (1961) Contribution à l'étude des relations entre le sol et la végétation halophile de Tunisie. *Annales de l'Institut National de la Recherche Agronomique de Tunisie* 34, 339.

Orshan, G. (1972) Morphological and physiological plasticity in relation to drought. In: McKell, C.M., Blasdell, J.P. and Goodin, J.R. (eds) *Wildland Shrubs, Their Biology and Utilization: An International Symposium,* Logan, Utah, July 1971, pp. 245–254.

Poore, M.E.D. (1955) The use of phytosociological methods in ecological investigations. I. The Braun-Blanquet system. *Journal of Ecology* 43(1), 226–244. DOI: 10.2307/2257132

Raunkiaer, C. (1934) *Life Forms and Statistical Plant Geography.* Clarendon Press, Oxford, UK.

Saaed, M.W.B. (2008) Study of vegetation and land degradation factors in Daphna zone (north-eastern part of Libyan coast-Tobruk). MSc thesis, Department of Environmental and Engineering Science, the Libyan Academy of Postgraduate Studies, Benghazi, Libya.

Saaed, M.W.B., El-Barasi, Y.M. and Elhashani, N.A. (2019a) Insight into the soil seedbank characteristics of the arid rangelands in Libya: a case study in Marmarica Plateau, Cyrenaica (north-eastern part of Libya). *Libyan Journal of Science and Technology* 9(2), 184–193.

Saaed, M.W.B., El-Barasi, Y.M. and Rahil, R.O. (2019b) Our present knowledge about the history and composition of the vegetation and flora of Libya. *Webbia* 74(2), 325–338. DOI: 10.1080/00837792.2019.1669361

Saaed, M.W.B., El-Barasi, Y.M. and Rahil, R.O. (2021) An updated checklist and quantitative analysis of the Marmarica Plateau flora, in the north-eastern part of Libya. *Phytotaxa* 509(1), 001–055. DOI: 10.11646/phytotaxa.509.1.1

Saaed, M.W.B., Ali, A.B., El-Barasi, Y.M. and Rahil, R.O. (2022) Potential and perspectives for establishing protected areas in El-Jabal El-Akhdar region, northeast Libya: an overview and example from 'El-Abyar area'. *Journal of Degraded and Mining Lands Management* 9(3), 3453–3464. DOI: 10.15243/jdmlm.2022.093.3453

Sherif, A.S., Mahklouf, M.H. and Betelmal, A.G. (2020) Floristic study and species diversity of Msallata-Garaboulli Province in Libya. *Journal of Advanced Botany and Zoology* 7–3, 1–18. DOI: 10.33258/bioex.v2i2.232

Vetter, T., Rieger, A.-K. and Nicolay, A. (2009) Ancient water harvesting system in the north Eastern Marmarica (North Western Egypt). *Libyan Studies* 40, 9–23. DOI: 10.1017/S0263718900004489

Zohari, M. (1973) *Geobotanical Foundations of the Middle East.* Gustav Fisher Verlag, Stuttgart, Germany.

7

The Most Common Libyan Rangeland Plants

Abstract

The *Flora of Libya* (El-Gadi and Jafri, 1976) comprises 153 families and about 1800 genus and species. In this chapter, we illustrate 262 plants from 51 families commonly found on Libyan rangelands with their scientific and vernacular names, life form, description and morphology, reproduction, pastoral importance, palatability and fodder value, toxicity, economic interest, habitat, distribution and endemism.

Introduction

In our book, we mostly used and referred to the *Flora of Libya* from El-Gadi and Jafri (1976) and the compendium from Keith (1965) for identification and scientific names. Although the *Flora of Libya* needs updating with revised genus/ species names, and plants recently discovered in Libya, this extensive flora remains easily accessible to most botanists and environmentalists.

A review and taxonomic update of the *Flora of Libya* is available from Essokne *et al.* (2024) and Essokne and Mahklouf (2023).

Other information from plant collections and local vernacular names in Libya, North Africa and including Egypt are available from Battandier (1902); Bonnet and Barratte (1886); Boulos (1972, 1999); Cuénod (1954); Dobignard and Chatelain (2010); Durand *et al.* (1910); Faruqi (1980); Forskål (1775); Gawhari *et al.* (2018); Heyn (1963); Le Floc'h (1983); Maire (1925); Nègre (1961); Ozenda (1977); Pampanini (1930); Quézel and Santa (1962); Saaed *et al.* (2019); Scholz (1974); Täckholm (1974) and Trabut (1935).

Further taxonomic information, synonymy, distribution and illustrations are available on-line from:

http://www.worldfloraonline.org/
https://efloramaghreb.org/
https://flora.org.il/
https://powo.science.kew.org/
https://www.alyasmina.org/
https://www.alyasmina.org/alyasmina-2020/web.html
https://www.northafricatrees.org/
https://www.teline.fr/

Our plant descriptions and illustrations are organized in alphabetical order of family, then by genus and species. It follows the classification of the *Flora of Libya* (El-Gadi and Jafri, 1976).

In the next chapter, we review how much the rangeland plants and vegetation formations contribute to the biomass and feed available to the rangeland livestock.

DOI: 10.1079/9781800627154.0007

AGAVACEAE

Agave americana L.

Synonym(s): http://www.theplantlist.org/tpl1.1/record/kew-293578, or *Aloe americana* (L.) Crantz, http://www.theplantlist.org/tpl1.1/record/kew-297041
Exotic from Central America
Vernacular name: ?
Raunkiaer's life-form: Hemicryptophyte
Description and morphology: Succulent, large perennial, green-grey leaves (up to 1.5 m) in rosette, leaves lanceolate within gutter with margins armed with strong black prickles and ending with a sharp vulnerant spike, lifespan reported to be up to 30–40 years, inflorescence in tall unique spike (5–10 m) bearing upright yellow flowers arranged in many branched clusters, fruits in long beaked trilocular capsules (3–5 cm) containing many brown-black shiny seeds (5–8 mm), main plant dying after flowering, abundant new plants through suckers.
Reproduction: Consistently vegetative by suckers, rare by seeds
Pastoral importance: None *per se* except as fences/hedges near settlements
Palatability and fodder value: Not touched by livestock
Toxicity: *Agave* sap reported to induce dermatitis and skin lesions
Economic interest: None in Libya, just used as fences due to its prolific vegetative propagation and spines on leaves
Habitat: Suckers transplanted in many places, not invasive, sandy soils, drought-resistant (down to 150 mm MAR), not frost- or salt-resistant.
Distribution:

- In Libya, around fields and gardens as hedges and fences, all over coastal zones
- In arid zones, exotic introduced all over southern Europe, North Africa, the Middle East, west Asia to Australia

Endemism: NA

Photo 7.1. *Agave americana*, exotic (Jbebina, Bir roman, Al-Assah, likely from an old settlement).

AIZOACEAE

Mesembryanthemum crystallinum L.

Synonym(s): http://www.theplantlist.org/tpl1.1/record/kew-2363096, or *Gasoul crystallinum* (L.) Rothm., http://www.theplantlist.org/tpl1.1/record/kew-2817324, or *Cryophytum crystallinum* (L.) N.E.Br., http://www.theplantlist.org/tpl1.1/record/kew-2745181
Native:
Vernacular name: *Rhassul* (?)
Raunkiaer's life-form: Mesotherophyte
Description and morphology: Prostrate creeping (5–10 cm high) succulent annual, much branched, green-reddish spathulate – ovoid leaves fleshy frosted with undulate margins and with minute white papillae all over, leaves can be up to 10 cm large, axillary or terminal sessile white flowers (2–3 cm diam.) with linear staminodes longer that sepals, fruits red turning black penta-stellate dehiscent, many small reniform adpressed seeds.
Reproduction: Flowering April–May, seeds
Pastoral importance: None
Palatability and fodder value: None, not grazed
Toxicity: None
Economic interest: None
Habitat: Azonal, seashore to oasis, on saline – gypseous soils inland; nitrophilous, often on degraded land around sheep pen with abundance of faeces-urine.
Distribution:

- In Libya, all coastal zones and inland
- In arid zones, southern Europe, North Africa, ubiquitous

Endemism: NA

Photo 7.2. *Mesembryanthemum crystallinum* (south Zliten project, N 32.178 E 14.602, road to wadi Sassu).

Mesembryanthemum forskahlii Hochst. ex Boiss

Synonym(s): http://www.theplantlist.org/tpl1.1/record/kew-2361839, or *Opophytum forskahlii* (Hochst. ex Boiss.) N.E.Br., http://www.theplantlist.org/tpl1.1/record/kew-2390918, or ? *Mesembryanthemum cryptanthum* Hook F., http://www.ville-ge.ch/musinfo/bd/cjb/africa/details.php?langue=an&id=131289 ?
Native:
Vernacular name: *Rhassul* (?)
Raunkiaer's life-form: Mesotherophyte
Description and morphology: Prostrate small (3–5 cm high) creeping succulent annual, young plants looks like reddish gravel, lower leaves puffy, older leaves longer and cylindrical – rounded apex, very small white terminal flowers, calyx lobes (2) like but larger than leaves, ovary swollen.
Reproduction: Flowering February–April, seeds
Pastoral importance: None
Palatability and fodder value: None, not grazed (?)
Toxicity: None
Economic interest: Fruits reported collected and consumed by shepherds
Habitat: Gravelly soils, possibly azonal on slightly saline – gypseous soils inland on Hamada.
Distribution:

- In Libya, Saharan, on Hamada
- In arid zones, South Africa, North Africa

Endemism: NA

Photo 7.3. *Mesembryanthemum forskahlii* (south Zintan, pre-Hamada, N 31.216 E 11.647, Ouassa).

ALLIACEAE

Allium ampeloprasum L.

Synonym(s): http://www.theplantlist.org/tpl1.1/record/kew-294972, or *Allium porrum* L., http://www.theplantlist.org/tpl1.1/record/kew-296308
Native: exotic (?) invasive (?)
Vernacular name: *Kurrat* (?)
Raunkiaer's life-form: Geophyte
Description and morphology: Bulbs covered with white tunics, leaves V-shaped canaliculate along half-flowering stem, onion smell, purple flowers in spherical terminal umbels (3–5 cm diam.) on unique long stem (up to 150–200 cm high).
Reproduction: Vegetative with bulbs, seeds
Pastoral importance: None
Palatability and fodder value: None
Toxicity: None
Economic interest: Wild leek, collected fresh as vegetable, cultivated (?)
Habitat: in wadi, heavy soils, protected sites, down to 200 mm MAR
Distribution:

- In Libya, most exclusively Jebel Al-Akhdar, rare in Jebel Nafusa, not in Sirte
- In arid zones, Europe, North Africa, west Asia to Iran

Endemism: NA

Photo 7.4. *Allium ampeloprasum* in clump of *Atriplex halimus* (wadi Maghur, in protected area).

AMARYLLIDACEAE

Narcissus elegans (Haw.) Spach

Synonym(s): http://www.theplantlist.org/tpl1.1/record/kew-282677, or *Hermione elegans* Haw., http://www.theplantlist.org/tpl1.1/record/kew-277791
Native:
Vernacular name: *Kalil* (?)
Raunkiaer's life-form: Geophyte
Description and morphology: Small (10–15 cm) bulbous herb, ovoid bulb (2–3 cm) covered with scarious dark layers, long scape bearing 2–4 flowers in umbels, spathe papery translucent persistent, corona yellow with white lanceolate reflexed sepals, 1–2 linear leaves (10–15 cm) appearing after flowering, fruits as clear brown dehiscent papery capsules containing small black angular seeds.
Reproduction: Flowering October–November, seed, and bulbs
Pastoral importance: None
Palatability and fodder value: None, occasionally grazed (?)
Toxicity: None
Economic interest: Ornamental
Habitat: *Terra rossa*, heavy soils, common in garrigue with *Arbutus* sp., *Quercus* sp., down to 300 mm MAR
Distribution:

- In Libya, Jebel Al-Akhdar (in highest-rainfall zones)
- In arid zones, all Mediterranean zones

Endemism: Jebel Al-Akhdar

Note: Could be easily misidentified with *Narcissus tazetta* L., http://www.theplantlist.org/tpl1.1/record/kew-282289, that has shorter scape bearing more flowers with shorter and rounded multi-nerved sepals; same occurrence in Jebel Al-Akhdar with similar ecology

Photo 7.5. *Narcissus elegans*, plant (Shahat).

Photo 7.6. *Narcissus elegans*, flowers (Shahat).

ANACARDIACEAE (ex-Terebenthaceae)

Pistacia atlantica Desf.

Synonym(s): http://www.theplantlist.org/tpl1.1/record/kew-2408140, or *Terebinthus atlanticus* (Desf.) Dum. Cours., http://www.theplantlist.org/tpl1.1/record/kew-2608985
Native:
Vernacular name: *Batum, b'tum*
Raunkiaer's life-form: Mesophanerophyte
Description and morphology: Large deciduous dioecious tree (up to 10–12 m high), bulky hemispherical crown (up to 8–10 m diam.), robust superficial rooting systems, reticulated grey bark on old trees, barely winged petiolate alternate leaves (3–5 cm long) glossy green oval mostly imparipinnate (5–7 leaflets) with round rachis, resin/turpentine scented leaves, leaves often bearing red-yellow galls from aphids, inflorescence in pyramidal panicle, flowers unisexual small (5–8 mm diam.), red drupes turning black when dry.
Reproduction: Flowering March–April, seeds (dispersed by birds)
Pastoral importance: Limited
Palatability and fodder value: Poor, however lower branches grazed by dromedaries and goats
Toxicity: None
Economic interest: Used to be harvested for fuelwood and providing excellent charcoal, hardwood used to make the traditional (for minimum tillage) plough (*maharat*), *batum* should be used in reforestation and soil conservation projects in Jebel Nafusa and Jebel Al-Akhdar, easy to produce seedlings in nursery; dry fruits sold on local markets as food condiment (similar to Syrian *sumac*, fruits from *Pistacia coriacea* (?)) and long ago, for oil extraction from seeds; dry sap used as incense; leaves carry reddish galls (from aphids *Pemphigus utricularis* (?), *Slavum* sp.) rich in tannins that were locally used for leather processing.
Habitat: Common in *Stipa tenacissima* formations in Jebel Nafusa, *batum* occurrence down to 100 mm MAR in wadis and depressions (*daya*).
Distribution:

- In Libya, common south Jebel Nafusa (Mizda, Beni Walid etc.), not in Sirte, infrequent on the northern slope of Jebel Al-Akhdar
- In arid zones, Mediterranean regions, North Africa, Middle East, west Asia to Pakistan

Endemism: NA

Note: This exceptional Libyan tree should be protected (?), old trees often with hollow trunks used as shelter by herdsmen; sometimes these old trees are considered holy (*marabout* shelter) with pieces of material (*shellig* (?), voto (?)) hanging from branches.

Photo 7.7. *Pistacia atlantica*, lone tree (*Stipa tenacissima* country, south Gdama, Gharyan).

Photo 7.8. *Pistacia atlantica*, view of the trunk (*Stipa tenacissima* country, N 31.928 E 13.167, east Gdama, Gharyan).

Photo 7.9. *Pistacia atlantica*, hollow trunk (*Stipa tenacissima* country, east Gdama, Gharyan).

Photo 7.10. *Pistacia atlantica*, lone tree (11 km north of Mizda, N 31.530 E 13.007, 150 mm MAR).

Photo 7.11. *Pistacia atlantica*, leaves imparipinnate and drupes (Gdama, south Gharyan).

Photo 7.12. Galls on leaves of *Pistacia atlantica* (Gdama, south Gharyan).

Note: *Pistacia vera* L. Pistacia vera L. — The Plant List, a dioicous small tree with large imparipinnate leaves (3–5 leaflets), producing commercial pistachio nuts (*fustuk*), a species close to *Pistacia atlantica*; varieties introduced from Sicily and Tunisia have been cultivated in western Libya at a few places (small old plantation near Gasr-Ben-Gashir in the Jeffara plain, at Gdama (south Gharyan) and Jendouba on soils reclamation projects). Not successful in Libya as it may require enough cold requirement (chilling requirement) to flower and fructify, hence a better chance for fruit production on the top of Jebel Nafusa than in the Jeffara plain. *Pistacia vera* plantations require 1 male tree to 5–10 female trees to fructify. Providing that this is satisfied, it would start fruiting at 5–8 years old. *Pistacia atlantica* is commonly used as a root-stock for budding and grafting.

Photo 7.13. *Pistacia vera*, male flowers (Gdama, Gharyan).

Photo 7.14. *Pistacia vera*, fruits – pistachio (Gdama, Gharyan).

Pistacia lentiscus L.

Synonym(s): http://www.theplantlist.org/tpl1.1/record/kew-2407564
Native:
Vernacular name: *Dorru* (?) *Al-Daroo* (?)
Raunkiaer's life-form: Nanophanerophyte
Description and morphology: Resinous turpentine scented large (2–5 m diam., 2–3 m high) evergreen shrub densely branched sometimes prostrate, grey bark, coriaceous shiny dark green mostly paripinnate leaves (2–4 leaflets) with winged rachis, leaflets variation frequent, ovoid small (4–8 mm diam.) fruits in panicles, showy red apiculate drupes turning black when dry.
Reproduction: Flowering March–April, seeds
Pastoral importance: Limited
Palatability and fodder value: Poor, hardly touched by all livestock
Toxicity: None
Economic interest: Fuelwood, locally used as chewing whitish resin exuding from branches
Habitat: On *Terra rossa*, heavy soils, above 300 mm MAR, on calcareous karstic country, a main component of the vegetation thickets in Jebel Al-Akhdar.
Distribution:

- In Libya, extremely frequent and confined to Jebel Al-Akhdar, now rare/eradicated in western Libya (possibly rare on hills near Nazzaga, Homs)
- In arid zones, southern Europe, North Africa, Middle East to Pakistan

Endemism: NA

Photo 7.15. *Pistacia lentiscus*, bush (El-Marj plain).

Photo 7.16. *Pistacia lentiscus*, paripinnate leaves (El-Marj plain).

Photo 7.17. *Pistacia lentiscus*, fruits and paripinnate leaves with winged rachis (El-Marj plain).

Rhus tripartita (Ucria) Grande

Synonym(s): http://www.theplantlist.org/tpl1.1/record/kew-2873538, or *Searsia tripartita* (Ucria) Moffett, http://www.theplantlist.org/tpl1.1/record/kew-2608689, or *Rhamnus tripartita* Ucria, http://www.theplantlist.org/tpl1.1/record/kew-2522920
Native:
Vernacular name: *Jdari*
Raunkiaer's life-form: Nanophanerophyte
Description and morphology: Medium size (up to 3 m high) shrub, rarely in tree form (up to 4–6 m high), intricate spiky old grey branches, young red branches, short-winged petioled leaves trilobate (young entire to slightly dentate), tiny green flowers (dioecious) in panicles, all over branches red flattish glossy drupes turning black when ripe.
Reproduction: Flowering in autumn, seeds (possible dispersion of drupes by birds (?))
Pastoral importance: Limited
Palatability and fodder value: High, grazed by all livestock – mostly by dromedaries and goats (often heavily overgrazed to Bonsai size and shape).
Toxicity: None
Economic interest: Drupes edible (sweet taste when dried), fuelwood, charcoal, bark, and roots used for tanning hides
Habitat: Shallow rocky soils (on calcareous crusts), in wadi, depressions receiving run-off, above 100 mm MAR, on cliffs on Hamada.
Distribution:

- In Libya, coastal zone and inland, most abundant in wadi along Jebel Nafusa foothills, Hamada, south Zliten (wadi Um-el-Jdari, wadi Jaoudi, wadi Roman), wadi Sof-el-Jin, Beni Walid, inland Taworga, Bu-Gren (El Goshiyah) to wadi Bey, Sirte, to Jebel Al-Akhdar
- In arid zones, southern Europe, North Africa to the Middle East (Syria, Palestine) to the Arabian Peninsula

Endemism: NA

Note: Wood from tree-like may have been used for temporary dwellings and huts south-east of Bu Gren – Antique sites (20 km south Al-Buirat),

Photo 7.18. *Rhus tripartita*, upper leaves (south-east Bu-Gren, N 31.316 E 15.377).

Photo 7.19. *Rhus tripartita*, branches, thorns, and fruits (south-east Bu-Gren).

Photo 7.20. *Rhus tripartita*, bushes (Hamada, on limestone cliffs, Mged Shati N 30.279 E 11.569).

Photo 7.21. *Rhus tripartita*, large bushes (south-east Bu-Gren, N 31.316 E 15.377).

Photo 7.22. *Rhus tripartita*, tree (south-east Bu-Gren, 30 km south Al-Buirat N 31.145 E 15.644).

APIACEAE (Umbelliferae)

Crithmum maritimum L.

Synonym(s): http://www.theplantlist.org/tpl1.1/record/tro-1702306, unresolved name (?)
Native:
Vernacular name: *Shageret-al-Bahr* (?)
Raunkiaer's life-form: Nanophanerophyte
Description and morphology: Perennial stout fleshy shrublet (30–80 cm high), creeping/decumbent, glabrous, blue-green all over, thick round furrowed stems, fleshy triangular 2–3 pinnatisect leaves (4–8 cm long) into lanceolate – spathulate leaflets, profuse umbel with cream-green dense umbels, ovoid (4–6 mm long) spongious fruits with 5 prominent ribs.
Reproduction: Flowering January–May, rhizomatous, seeds
Pastoral importance: None
Palatability and fodder value: None
Toxicity: None
Economic interest: Leaves eaten as salad or pickles
Habitat: Along coast – seashore, mostly within 500 m from sea, on white sand, on Gargaresh coastal limestone.
Distribution:

- In Libya, all coastal zones in western and eastern Libya, not in Sirte (?)
- In arid zones, southern Europe, North Africa, Middle East to north-west-India

Endemism: NA

Photo 7.23. *Crithmum maritimum*, plant (Leptis magna – Lebda)

Photo 7.24. *Crithmum maritimum*, plant (Leptis magna – Lebda)

Ferula communis L.

Synonym(s): http://www.theplantlist.org/tpl1.1/record/kew-2808370, or *Ferula glauca* L., http://www.theplantlist.org/tpl1.1/record/kew-2808435

Native:

Vernacular name: *Khalk* (?) (*Derrias* (?))

Raunkiaer's life-form: Cryptophyte – geophyte

Description and morphology: Robust perennial with strong stump, unique inflorescence stalk (1–2 m high, 2–4 cm diam.) grooved and filled with spongy pith, big (50–80 cm long) basal leaves (green glaucous) finely dissected, upper leaves reduced to sheath on inflorescence, inflorescences as terminal compound umbel, long peduncled smaller umbels, yellow flowers, fruits on pedicel as long as fruit, winged fruits (up to 2 cm long) obovate with one main central vein and two marginal thin veins, papery wings, stylopodia bifid arched and retroverted.

Reproduction: Flowering February–April, seeds

Pastoral importance: Limited to none, poisonous (?)

Palatability and fodder value: Usually not grazed by livestock mostly due to ferulism.

Toxicity: Ferulism induces lethal haemorrhage (blood loss) (?) basal leaves contain Coumarin with anticoagulant properties; ingested dry leaves expand like sponges and induce intestinal tract blockage and subsequent death of small ruminants.

Economic interest: Very young inflorescence consumed as vegetable; dry spongy pith from stalk used to start and carry slow burning embers and fire.

Habitat: Loamy soils, abandoned crops, bushland, above 200 mm MAR

Distribution:

- In Libya, mostly Jebel Al-Akhdar, rarely mentioned in Jebel Nafusa (?)
- In arid zones, Mediterranean, southern Europe, all North Africa but not in Egypt and further east

Endemism: NA

Notes:

- *Ferula* sp. ecotypes collected from southern Europe present different level of toxicity.
- *Ferula communis* L., *Ferula tingitana* L., http://www.theplantlist.org/tpl1.1/

record/kew-2808690, and *Ferula marmarica* Asch. & Taub. ex Asch. & Schweinf., http://www.theplantlist.org/tpl1.1/record/kew-2808530, and other *Ferula* sp. all from Jebel Al-Akhdar, have been reported as possibly being the antic and mysterious *Silphium* sp. from Cyrenaica; *Thapsia garganica* L., http://www.theplantlist.org/tpl1.1/record/kew-2435941, is another possible candidate.

- *Ferula* sp. could easily be confused with *Thapsia garganica*, only found in Jebel Al-Akhdar; *Thapsia* has a large inflated fibrous rooting system and a mostly hollow inflorescence stalk; flowers yellow, fruits have wavy papery wings margins and erect parallel stylopods; plant grazed by livestock but reported vesicant causing blistering (leaves also used as a vesicant cataplasm), reported as putting to temporary sleep small ruminants; crushed root paste used in north-east Algeria to benumb and collect fishes from ponds.

Photo 7.25. *Ferula communis*, early flowering (Mzerah, N 32.428 E 21.552, south Gendoula).

Photo 7.26. *Ferula communis*, umbel flowering (Mzerah, south Gendoula).

Photo 7.27. *Ferula communis*, fruiting – end of season (Mzerah, south Gendoula).

Photo 7.28. *Ferula communis*, fruits (Mzerah, south Gendoula).

Pituranthos scoparius (Coss. & Durieu) Schinz

Synonym(s): http://www.theplantlist.org/tpl1.1/record/kew-2405106, or *Deverra scoparia* Coss. & Durieu, http://www.theplantlist.org/tpl1.1/record/kew-2763399
Native:
Vernacular name: *Guezar* (?) *Guzar* (?)
Raunkiaer's life-form: Nanophanerophyte
Description and morphology: Broom-shaped/caespitose perennial herb (50–100 cm high), cumin scented, strong taproot, abundantly branched, straight erect leafless flexuous stems, small leaves reduced to membranous sheaths, terminal or lateral pedunculate umbels, white flowers.
Reproduction: Flowering January–May, seeds
Pastoral importance: Limited
Palatability and fodder value: Grazed by all livestock
Toxicity: None
Economic interest: None
Habitat: Rocky/loamy soils, down to 100 mm MAR
Distribution:

- In Libya, all coastal zones to inland
- In arid zones, all North Africa

Photo 7.29. *Pituranthos scoparius* (Gdama, Gharyan).

Endemism: NA

Note: Could be confused with *Pituranthos rholfsianus* (Asch.) E.A. Durand & Barratte, http://www.theplantlist.org/tpl1.1/record/kew-2405121, = *Deverra scoparia* ssp. *tripolitana* (Andr.) Pfisterer & Podlech, http://www.theplantlist.org/tpl1.1/record/kew-2763400, an endemic to North-western Libya; very similar to *Pituranthos scoparius*, but smaller plant (20–40 cm high) (?)

Pituranthos tortuosus (Desf.) Benth. & Hook.f. ex Asch. & Schweinf.

Synonym(s): http://www.theplantlist.org/tpl1.1/record/kew-2405103, or *Deverra tortuosa* (Desf.) DC., http://www.theplantlist.org/tpl1.1/record/kew-2763402
Native:
Vernacular name: *Guzar* (?)
Raunkiaer's life-form: Nanophanerophyte
Description and morphology: Robust perennial herb (30–80 cm high), strong cumin scent, robust taproot, tortuose/intricate blue-green thick/rigid branches, leafless divaricate striate stems turning yellowish-white when drying, small caducous leaves reduced to membranous sheaths, terminal or lateral pedunculate umbels, white-beige flowers.
Reproduction: Flowering January–May, seeds
Pastoral importance: Limited
Palatability and fodder value: Well grazed by all livestock
Toxicity: None
Economic interest: None
Habitat: Fixed deep sand/sandy-loam soils, down to 100 mm MAR
Distribution:

- In Libya, all coastal zones to inland
- In arid zones, all North Africa

Endemism: NA

Photo 7.30. *Pituranthos tortuosus*, plant (Bir Samman – As-Sahn N 32.909 E 11.597, Al-Assah).

Photo 7.31. *Pituranthos tortuosus*, inflorescence (Bir Samman, Al-Assah).

Scandix pecten-veneris L.

Synonym(s): http://www.theplantlist.org/tpl1.1/record/kew-2479025.
Native:
Vernacular name: *Mischt-er-Raa'I* (?)
Raunkiaer's life-form: Mesotherophyte
Description and morphology: Small (20–40 cm) pubescent annual herb, furrowed stems mostly prostrate and branched, three-pinnatisect triangular-obovate leaves with furrowed petiole, terminal multiple umbels, umbelulles with white flowers developing into (2–4) long accrescent fruits, rigid linear fruits spear-like with bi-mucronate long beak (4–8 cm long and 3–5 times longer than mericarp), rugose/ribbed mericarps (2 per fruit).
Reproduction: Flowering January–May, seeds
Pastoral importance: None
Palatability and fodder value: None
Toxicity: None
Economic interest: None
Habitat: In well drained soils, stony loamy soils, above 200 mm MAR, often present in crops.
Distribution:

- In Libya, all coastal zones, not in Sirte (?), most common in Jebel Al-Akhdar
- In arid zones, southern Europe, North Africa, Middle East to north-west India

Endemism: NA

Photo 7.32. *Scandix pecten-veneris*, plant (wadi Aleg, N 32.018 E 20.742, Maghur).

Photo 7.33. *Scandix pecten-veneris*, fruits (wadi Aleg).

ARACEAE

Arum cyrenaicum Hruby

Synonym(s): no synonym
Native:
Vernacular name: *Nerish* (?)
Raunkiaer's life-form: Geophyte
Description and morphology: Perennial herb (20–40 cm high), shiny green glabrous all over, ovoid tuberous corm (2–4 cm diam.), hastate leaves (10–25 cm long, 8–10 cm wide) appearing before flowering with pointed base lobes and on long petiole, spathe apiculate/caped/funnelled (10–15 cm long), spathe outside green strips with purple patches (dots on stem) and dark purple/brown inside, spathe constricted in the lower third, fruits as red/orange berries.
Reproduction: Flowering March–May, vegetative (corm) and seeds
Pastoral importance: None
Palatability and fodder value: None
Toxicity: All *Arum* sp. are toxic, possible due to calcium oxalate and cyanogenic glycosides, however, corms of *Arum cyrenaicum* are reported to be collected and consumed as vegetables. (?)
Economic interest: Corms used as vegetables (?) Medicinal potential (?) (Azab, 2017)
Habitat: Heavy soils, on *Terra rossa*, in shade in bushes, above 300 mm MAR
Distribution:

- In Libya, endemic only in Jebel Al-Akhdar (El-Marj, Shahat, Marsa Suza, Derna, wadi Kuf)
- In arid zones, NA, south-west Crete

Endemism: Jebel Al-Akhdar only

Photo 7.34. *Arum cyrenaicum*, spathe (Shahat).

Photo 7.35. *Arum cyrenaicum*, spathe and leave (Shahat).

ASCLEPIADACEAE (now under Apocynaceae)

Calotropis procera (Aiton) Dryand.

Synonym(s): http://www.theplantlist.org/tpl1.1/record/kew-2693668, or *Asclepias procera* Aiton, http://www.theplantlist.org/tpl1.1/record/kew-2655086
Native:
Vernacular name: *Brumbach*
Raunkiaer's life-form: Mesophanerophyte
Description and morphology: Small tree (1–5 m high) branching from base, abundant milky sticky sap, branches white tomentose, sometimes developing cork on branches, persistent subsessile large (up to 20 cm long) ovate entire fleshy leaves drying up on branches and stems, Inflorescence in axillary cymes with large flowers (2–3 cm diam.) with white petals with top ending purple, pair of big (8–12 cm long) ovoid fruits, papery spongy when dry, fruit dehiscent freeing abundant small pyriform black seeds (4–8 mm) capped with long (20–40 mm) silky hairs.
Reproduction: Flowering winter to May, seeds dispersed by anemochory
Pastoral importance: None
Palatability and fodder value: Hardly touched by livestock, except dry leaves consumed at risk by desperate dromedaries.
Toxicity: Likely very aggressive irritant sticky sap reported to cause blindness to unwise livestock, used by herders for removing hair from goat fleeces to make containers for carrying water or milk (*shakwa*), sap reported (*Zaman ya zaman*) to punish and blind thieves or delinquents.
Economic interest: Medicinal, reported purgative (?), dry wood used for making fine charcoal for gunpowder, abundant and fine silky fibres once considered (1976/77) for weaving material for *jerd* or *jarid* traditional clothes (project abandoned).
Habitat: Desertic, sandy soils, sometimes on small dunes, 150–300 mm MAR
Distribution:

- In Libya, all along coastal zone (Sabratha, Suani, Lebda) to deep into Sahara (Hamada – Fezzan – Kufra – Uweinat), Sirte to Jebel Al-Akhdar
- In arid zones, all Sahara and south, south Middle East – Arabian Peninsula to Pakistan-India

Endemism: NA

Photo 7.36. *Calotropis procera*, small tree (2 m) (wadi Nefad, Seddada).

Photo 7.37. *Calotropis procera* (wadi Nefad, Seddada).

Photo 7.38. Flowers and white sticky sap of *Calotropis procera* (wadi Nefad, Seddada).

Caralluma europaea (Guss.) N.E.Br.

Synonym(s): http://www.theplantlist.org/tpl1.1/record/kew-2699128, or *Apteranthes gussoneana* J.C. Mikan, http://www.theplantlist.org/tpl1.1/record/kew-220668, or *Stapelia europaea* Guss, http://www.theplantlist.org/tpl1.1/record/kew-2472733
Native:
Vernacular name: *Daqmus* (?)
Raunkiaer's life-form: Chamaephyte – succulent
Description and morphology: Rare, perennial small (10–25 cm) inconspicuous cactoid with abundant stolons, glabrous undulate quadrangular stems (10–15 mm width) erect or ascending with flat or concave faces, very small caducous ovate acuminate leaves (2–5 mm long), inflorescence in terminal cluster with multiple shortly pedicellate flowers, putrid smell, calyx inconspicuous, spread star-like fleshy corolla (15–20 mm diam.) with 5 violet/chocolate (with yellow or white circular stains) pointed lobes with ciliate margins, elongated fruits in divergent pairs, seed with egret (if you are lucky to find one).
Reproduction: Mostly vegetative with stolons, flowering winter to May, seeds (?)
Pastoral importance: None
Palatability and fodder value: Scarcely touched by livestock (?) though reported heavily grazed by goats in Morocco (Claude Lemmel, Tiznit, March 2022, personal communication)
Toxicity:
Economic interest: The succulent stems reported to be chewed by thirsty herdsman, and consumed, though bitter taste, in time of famine in central Tunisia (Le Floc'h, 1983). Also a beautiful ornamental in garden.
Habitat: Loamy soils, rocky places, often protected inside large bushes, above 200–250 mm MAR
Distribution:

- In Libya, *Caralluma europaea* was spotted by several botanists since the early 1900s near Tarhuna, Gharyan, Derna. Recent sighting of *Caralluma* sp. was by Dr L. Boulos in 1967 along the road Shahat to Ras-Hillal and wadi Kuf, and later by Sherif *et al.* (2020) in the Msallata national reserve. It leads me to think that *Caralluma europaea* is a now-rare species in Libya and doomed to disappear due to trampling and overgrazing.
- In arid zones, North Africa, the Middle East, west Asia, South Africa.

Endemism: threatened in Libya (?)

The photos of the *Caralluma* sp. (?) flowering specimens presented below for the purpose of illustration are from Morocco (*Caralluma europea*) and from my succulent garden in Western Australia (*Orbea variegata*). Determination of *Caralluma, Ceropegia, Stapelia, Apteranthes, Orbea* sp., etc. is complex and under constant taxonomic revision.

Photo 7.39. *Caralluma europaea* flower, from Tourza region Morocco, N 31.121 E –5.482. (Photo: Claude Lemmel, http://alyasmina.org)

Note: Over the many years spent roaming the Libyan rangelands, I never saw a *Caralluma* sp. However, I have seen them twice, without flowers, on rangelands in very good condition:

- on the high plateau of Algeria (1970/71) on a protected *Stipa tenacissima* steppe, on shallow loamy soils on calcareous crusts, near Tadmit (Titteri Wilaya – approx. 250 mm MAR); and
- on a pristine and diversity-rich *Artemisia herba-alba*/*Pistacia khinjuk* steppe, north of Jebel Sinjar (N 36.419 E 41.482, northern Iraq – loamy soils, 300 mm MAR (1994)).

Photo 7.40. *Orbea variegata* (?), flower and articulate succulent stems (from my garden, Western Australia, March 2020), exotic from South Africa.

Pergularia tomentosa L.

Synonym(s): http://www.theplantlist.org/tpl1.1/record/kew-2404074, or *Daemia cordata* (Forssk.) R.Br., http://www.theplantlist.org/tpl1.1/record/kew-2755294, or *Asclepias cordata* Forssk., http://www.theplantlist.org/tpl1.1/record/kew-2654691
Native:
Vernacular name: *Umm-el-Laban* (?)
Raunkiaer's life-form: Nanophanerophyte
Description and morphology: Perennial shrub (50–150 cm high), climber, twining young stems, barely tomentose, short-petioled cordate opposite leaves (2–4 cm diam.), sticky white sap, inflorescences in lax pendulous cymes, pedicelled hanging flowers with triangular brown-reddish bearded sepals and white petals, echinulate soft dehiscent fruits with recurved horn and in pair, black seeds plumose.
Reproduction: Flowering January–April, seeds feathery dispersed by anemochory
Pastoral importance: Limited
Palatability and fodder value: Poor, dromedary feed (?)
Toxicity: Possible, sticky irritant sap.
Economic interest: None
Habitat: Loamy soils, in wadis, in depressions, often associated with *Acacia tortilis* ssp. *raddiana*./*Zilla spinosa*, below 200 mm MAR deep into Sahara.
Distribution:

- In Libya, inland, foothill Jebel Nafusa in Jeffara plain, south Jebel Nafusa (Zamzam) into Hamada, south Sirte (wadi Wishkah, Bey-el-Kebir), apparently not south Jebel Al-Akhdar (?)
- In arid zones, Sahara, North Africa (Morocco (?)) to Pakistan

Endemism: NA

Photo 7.41. *Pergularia tomentosa*, climbing and twining stems (wadi Sof-el-Jin).

Photo 7.42. *Pergularia tomentosa*, flowers (wadi Sof-el-Jin).

Photo 7.43. *Pergularia tomentosa*, leaves and fruits (wadi Zorzor – Seddada).

Periploca angustifolia Labill.

Synonym(s): http://www.theplantlist.org/tpl1.1/record/kew-2404106
Native:
Vernacular name: *Halaba, Sellouf* (?)
Raunkiaer's life-form: Nanophanerophyte
Description and morphology: Glabrous woody shrub (up to 2 m), stiff intricate criss-crossing branches, sometimes twining young stems, leaves sessile lanceolate (1–4 cm long) opposite (isolate or in bunch of 2–6) fleshy shiny, inflorescence in axillary cymes, small pendulous flowers (10–15 mm diam.), calyx of 5 free deflexed sepals (reddish-brown with white-greenish margins), corolla white rotate petals and five long twisted red stamens, long (5–10 cm) horny pedicelled upward-arched fruits (5–10 cm long) in pairs, seeds (3–5 mm) with profuse egret 5–8 x seed length.
Reproduction: Flowering March–June, seeds feathery dispersed by anemochory
Pastoral importance: Medium
Palatability and fodder value: Good, dromedary and goat feed, very much grazed
Toxicity: Possible, irritant sticky sap
Economic interest: None, except occasional fuelwood
Habitat: Shallow loamy soils, often on exposed calcareous crust in wadis, in depressions, below 200 mm MAR to deep into Sahara

Distribution:

- In Libya, inland, common wadi foothill Jebel Nafusa in Jeffara plain (Bir Ayyad to Nalut), south Jebel Nafusa (Sof-el-Jin, Zamzam) into Hamada, south Sirte (wadi Wishkah, Bey-el-Kebir), south Jebel Al-Akhdar to Tobruk – Solum (?).
- In arid zones, Sahara, North Africa (Morocco (?)), Egypt, the Middle East (Syria) to Pakistan.

Endemism: NA

Note: *Periploca angustifolia* reported by a Syrian visiting scientist as poisonous in Syria; this does not appear to be the case in Libya.

Photo 7.44. *Periploca angustifolia* heavily grazed bush (wadi Wishkah, N 31.208 E 15.410).

Photo 7.45. *Periploca angustifolia*, flowers (Shakshuk N 32.047 E 11.987 – Jeffara plain).

Photo 7.46. *Periploca angustifolia*, fruits (wadi Sof-el-Jin, N 31.590 E 14.410, Beni Walid).

ASTERACEAE

Achillea santolina L.

Synonym(s): *Achillea tenuifolia* Lam., http://www.theplantlist.org/tpl1.1/record/gcc-93656
Native:
Vernacular name:
Raunkiaer's life-form: Nanophanerophyte
Description and morphology: Perennial (sometimes annual), ascending, branched (10–30 cm) from base, strongly scented, all plant covered whitish tomentose-villous, leaves minute dentate, yellow capitula in terminal radiating multiple corymbs, pappus absent on cypselae (fruits).
Reproduction: Flowering February–April, seeds
Pastoral importance: None
Palatability and fodder value: Poor
Toxicity: None
Economic interest: Like many other Asteraceae used as a medicinal plant (used in Syria), reported to be insect repellent (?)
Habitat: Loamy to clay soils, down to 200 mm MAR
Distribution:

- In Libya, mostly from the east, from Ajdabiya – Jebel Al-Akhdar, possible in Jebel Nafusa
- In arid zones, Middle East, west Asia

Endemism: NA

Photo 7.47. *Achillea santolina* (Mzerah, south Gendula, Jebel Al-Akhdar).

Photo 7.48. *Achillea santolina* yellow capitula in terminal radiating multiple corymbs (Al Hagifat N 31.911 E 20.584, Suluk).

Anacyclus monanthos (L.) Thell.

Synonym(s): *Anacyclus monanthos* subsp. *cyrtolepidioides* (Pomel) Humphries, http://www.theplantlist.org/tpl1.1/record/gcc-57903, *Anacyclus cyrtolepidioides* Pomel, http://www.theplantlist.org/tpl1.1/record/gcc-28957, *Anacyclus alexandrinus* Willd., http://www.theplantlist.org/tpl1.1/record/gcc-81381, Many variations (?)
Vernacular name: *Sarat al kabsh* (?), *Tagrefta* (?)
Raunkiae's life-form: Nanochamerophyte
Description and morphology: Annual, usually rosetted prostrate flattish hemispheric
Reproduction: Flowering February–May, seeds
Pastoral importance: None
Palatability and fodder value: Poor
Toxicity: None
Economic interest: None
Habitat: Loamy-clay soils, common in rainfed crops, roadside
Distribution:

- In Libya, all coastal zones and inland in wadis
- In arid zones, North Africa

Endemism: likely a west-Mediterranean species (?)

Photo 7.49. *Anacyclus monanthos*, south Zliten, in rainfed crops.

Photo 7.50. *Anacyclus monanthos*, Jendouba, Jebel Nafusa.

Anvillea garcinii (Burm. f.) DC.

Synonym(s): *Anvillea radiata* Coss. & Durieu, http://www.theplantlist.org/tpl1.1/record/gcc-79922
Native:
Vernacular name: *Nogud, Negut* (?)
Raunkiaer's life-form: nanophanerophyte
Description and morphology: Desertic low (10–40 cm) woody shrub, intricate rigid branches, leaves triangular-shaped narrower at base, yellow capitula, with scarious involucral bracts tightly embracing dry black capitula, lower capitula bracts long radiating spiny, cypselae (achene) ending with one single thin filament.
Reproduction: Flowering March–May, seeds
Pastoral importance: Limited
Palatability and fodder value: Dromedaries forage, goats (?)
Toxicity: None
Economic interest: None
Habitat: Stony plateau in desert, loamy soils (with sand veil (?)), isolated individual or grouped in contracted vegetation in depressions.
Distribution:

- In Libya, south Jebel Nafusa deep into the Hamada (Derj, wadi Sof-el-Jin, Ghirza), Fezzan (to Ghat), to Sirte (wadi Wishkah, wadi Bey-el-Kebir), does not seem to occur in the east past Ajdabiya (?)
- In arid zones, North Africa (?) to Iran (?)

Endemism: Saharan endemic

Note: Could be confused with *Bubonium graveolens* (Forssk.) Maire, http://www.theplantlist.org/tpl1.1/record/gcc-20904, looks as *Anvillea* with same desert ecology and habitat, however *Bubonium* has white squamous bark on lower branches, with clearly longer dentate villous-hairy leaves embracing capitulum, cypselae ending with four membranous scales.

Photo 7.51. *Anvillea garcinii*, dry flower with radiating scarious involucral bracts (Bu-Gren).

Photo 7.52. *Anvillea garcinii* (Hamada, Bir Al-Kabasha N 29.802 E 12.090).

Photo 7.53. *Anvillea garcinii* (Hamada, Bir Al-Kabasha).

Artemisia campestris L.

Synonym(s): http://www.theplantlist.org/tpl1.1/record/gcc-25450
Native:
Vernacular name: *Chahal, Teguft*
Raunkiaer's life-form: Chamaephyte
Description and morphology: Perennial, slightly scented, woody glabrous dark green stems recumbent-ascending (30–80 cm) from base, stems brownish-red, leaves all sessile, basal leaves pinnatisect, upper leaves simple thin linear, long ample flowering panicle with very small capitula (1–2 mm) and scarious involucres.
Reproduction: Flowering from autumn, very small seeds, dispersed by anemochory (?)
Pastoral importance: Limited, sand binder and colonizer after rainfed cropping
Palatability and fodder value: Limited, goats and dromedaries, grazed by sheep when nothing else is available.
Toxicity: None
Economic interest: None
Habitat: Exclusively on sandy soils, withstand wind erosion when cropping abandoned, not occurring on dunes or mobile sand.
Distribution:

- In Libya, Jeffara (Regdalin, Agheilat, wadi Abu-Shayba, Bir-el-Ghanem, Daphnia, etc.), Sirte along the coast., not in the east
- In arid zones, North Africa, not in Egypt (?)

Endemism: Possibly a western species

Note: Sometimes sticky/gluey plant (occurs in more arid areas) = *Artemisia campestris* subsp. *glutinosa* (Besser) Batt., http://www.theplantlist.org/tpl1.1/record/gcc-33033)

Photo 7.54. *Artemisia campestris* (south Agheilat, Jmeil).

Artemisia herba-alba Asso

Synonym(s): http://www.theplantlist.org/tpl1.1/record/gcc-23310 = *Seriphidium herba-alba* (Asso) Soják, http://www.theplantlist.org/tpl1.1/record/gcc-118715
Native:
Vernacular name: *Shih*
Raunkiaer's life-form: Chamaephyte (hemicryptophyte when overgrazed)
Description and morphology: Small woody bush, up to 60–70 cm high, stems erect radiating from collar, hemispheric shape when not grazed, deep rooted, foliage strongly aromatic, tomentose, silvery-grey, inflorescences arranged in terminal broad-long capitula, with membranous reddish corolla.
Reproduction: Flowering from November to January, very small seeds (6–8000/g), dispersed by anemochory
Pastoral importance: One of the best arid rangeland plants in Libya, outstanding use as end of summer/autumn grazing, when flocks would have wasted spring–desert annual pastures and moving to summer–autumn rangelands (up on Jebel Nafusa and south, south of Jebel Al-Akhdar).
Palatability and fodder value: Excellent, reported to give a special flavour to Barbary sheep meat.
Toxicity: When grazed green in spring, reported inducing ewes' abortion and death of young lambs, hence *Artemisia herba-alba* stands avoided by shepherds as a spring pasture.
Economic interest: Medicinal, used as helminthicide, abortive; malaria treatment (?) *Artemisia herba-alba* stands targeted and cleared for rainfed (or irrigated) cereal cropping in depressions receiving run-off; often eradicated as fuelwood.
Habitat: Exclusively on loamy-clay soils, usually stands covering large areas of flat country, on foothills of Jebel, in depressions down to 100 mm MAR
Distribution:

- In Libya, Jebel Nafusa (foothills, top to south of Jebel), Sirte, south Jebel Al-Akhdar
- In arid zones, Spain, North Africa, Middle East, west Asia, many subspecies and var. to Pakistan and Central Asia

Endemism: NA

Note:

- White parasitic woolly globose (1–2 cm) galls from a diptera (*Rhopalomyia* sp.(?)) are often found inside *Artemisia herba-alba* bushes; these cotton balls used to be collected by shepherds as kindling to light a fire using flints sparks or hand or bow drill before matches were commonly available.
- Allelopathic effects of dry leaves of *Artemisia herba-alba* reported by Omar *et al.* (2017).

Photo 7.55. *Artemisia herba-alba* (wadi Sassu) – note the typical loamy topsoil.

Photo 7.56. *Artemisia herba-alba* stand in good condition (wadi Sassu).

Atractylis cancellata L.

Synonym(s): http://www.theplantlist.org/tpl1.1/record/gcc-144927
Native:
Vernacular name: *Ser, Seur* (?)
Raunkiaer's life-form: Tachytherophyte
Description and morphology: Annual, prostrate (10–20 cm), spinescent to spiny, central taproot, reddish stems, lower leaves soft, upper leaves spinescent, flowers (all tubular) white with a tinge of blue purple, solitary capitula embraced with involucral spiny leaves and spiny bifurcate combed involucral bracts forming like a lantern, cypsela with white dense egret.
Reproduction: Flowering March–May, seeds, dispersed by anemochory
Pastoral importance: Limited
Palatability and fodder value: Mostly grazed by goats, donkeys, dromedaries
Toxicity: None
Economic interest: None
Habitat: On rocky soils with loamy-clay texture, down to 100 mm MAR
Distribution:

- In Libya, Jebel Nafusa, Sirte, southern slope of Jebel Al-Akhdar
- In arid zones, North Africa to west Asia

Endemism: NA

Photo 7.57. *Atractylis cancellata* (Al-Jahishiyah, South Jebel Akhdar).

Photo 7.58. *Atractylis cancellata*, whole plant (Al-Jahishiyah, South Jebel Akhdar).

Atractylis serratuloides (Cass.) DC.

Synonym(s): http://www.theplantlist.org/tpl1.1/record/gcc-156600
Native:
Vernacular name: *Ser, Seur, Shawk-es-Serah* (?)
Raunkiaer's life-form: Chamaephyte
Description and morphology: Perennial small dense shrub (20–40 cm), branches white strongly interlaced, terminal inflorescence tubuliflora only, large pubescent involucral bracts ending with a short brown spiny mucron, cypselae black with long plumose pappus.
Reproduction: Flowering March–May, seeds, dispersed by anemochory
Pastoral importance: Limited
Palatability and fodder value: Mostly grazed by goats, donkeys, dromedaries
Toxicity: None
Economic interest: None
Habitat: On calcareous crust, rocky soils with loamy-clay texture, down to 100 mm MAR
Distribution:

- In Libya, Jebel Nafusa, Hamada, Solul-el-Khoder, Sirte, most southern slope of Jebel Al-Akhdar
- In arid zones, all North Africa

Endemism:

Photo 7.59. *Atractylis serratuloides* (Gdama, Jebel Nafusa).

Photo 7.60. *Atractylis serratuloides* capitula (Gdama, Jebel Nafusa).

Centaurea calcitrapa L.

Synonym(s): http://www.theplantlist.org/tpl1.1/record/gcc-97901
Native:
Vernacular name:
Raunkiaer's life-form: Therophyte
Description and morphology: Annual or biannual herbs (20–50 cm high) with long spiny bracts on flowering heads, much branched grooved stems, dull green long pinnatilobate linear soft leaves with wavy margins, ovoid globose terminal capitulum armed with long (2–3 cm long) yellow rigid (canaliculate and recurved) spines with small straight spines (2–5 mm long) at the base, yellow-purple flowers, bare seeds without pappus.
Pastoral importance: Limited
Palatability and fodder value: Grazed by dromedaries and donkeys, by sheep before flowering.
Toxicity: None
Economic interest: None
Habitat: Heavy soils, depressions, abandoned stony sites, often invading, above 300 mm MAR
Distribution:

- In Libya, west and east Libya coastal zone, not in Sirte
- In arid zones, Europe, North Africa, ubiquitous

Endemism: NA

Photo 7.61. *Centaurea calcitrapa* (Maghur, Gendoula).

Centaurea glomerata Vahl

Synonym(s): http://www.theplantlist.org/tpl1.1/record/gcc-27382 = *Centaurea contracta* Viv.
Native:
Vernacular name: *Sooret-en-Nagga* (?)
Raunkiaer's life-form: Therophyte
Description and morphology: Annual acaule, leaves long petiolate tomentose deep dissected with rolled margins and all in rosettes, yellow flowers all sessile clustered in centre of plant, yellow spines on involucral bracts of capitulum, cypselae have crown of setae on pappus.
Reproduction: Flowering March–May, seeds dispersed by anemochory
Pastoral importance: Limited
Palatability and fodder value: Young leaves grazed by sheep, mature plants grazed by equidae and dromedaries
Toxicity: None
Economic interest: None
Habitat: On loamy soils, down to 150 mm (overgrazed sites and in rainfed cereal crops).

Photo 7.62. *Centaurea glomerata* (22 km south-south-east Zliten).

Distribution:

- In Libya, coastal zone to inland, from west to east
- In arid zones, North Africa

Endemism: NA

Note: Many other *Centaurea* sp. present in Libya; identification difficult without flowers, involucral bracts of mature flowers, cypsela, papus, etc.

Cynara cardunculus L.

Synonym(s): http://www.theplantlist.org/tpl1.1/record/gcc-112928, or *Carduus cardunculus* (L.) Baill., http://www.theplantlist.org/tpl1.1/record/gcc-81276
Native:
Vernacular name: *Shawk-el-bel* (?), *Khorshoofe* (?) *Qardoon* (?)
Raunkiaer's life-form: Chamaephyte – Nanophanerophyte
Description and morphology: Perennial stout spiky herb with basal leaves prostrate in circle (up to 1 m diam.), pinnatisect leaves with clustered yellow-cream spines at base and lobes sharp pointed dissected lobes, large (3–4 cm diam. to 8–10 cm diam. with radiating spiny bracts) spherical armed terminal capitulum with blue-violet flowers (tubuliflora) on round (not winged) scabrous scape.
Reproduction: Flowering March–May, seeds with a long (20–50 mm) silky pappus
Pastoral importance: Limited
Palatability and fodder value: Grazed by dromedaries and donkeys
Toxicity: None
Economic interest: Leaves and young stems collected as vegetables (*qardoon*), taste as artichoke, sometimes cultivated (?)
Habitat: Heavy soils, depressions, abandoned stony sites, often invading, above 300 mm MAR
Distribution:

- In Libya, west and east Libya coastal zones, not in Sirte
- In arid zones, Europe, North Africa, ubiquitous

Endemism: NA

Photo 7.63. *Cynara cardunculus*, capitulum (El Maghur).

Echinops spinosissimus Turra

Synonym(s): http://www.theplantlist.org/tpl1.1/record/gcc-77452
Native:
Vernacular name: *Schouk-el-J'mel* (?)
Raunkiaer's life-form: Chamaephyte
Description and morphology: Perennial, spinescent, rigid stems and leaves veins with white – hairy tomentum (spiderweb-like), large capitulum (30–60 mm diam.) spherical terminal with tubular disc florets pale blue to whitish, bracteal spines protruding from capitulum, long spiny involucral bracts protruding from capitulum.
Reproduction: Flowering end of spring, seeds
Pastoral importance: Limited
Palatability and fodder value: Feed for dromedaries, donkeys and goats
Toxicity: None
Economic Interest: None
Habitat: Ruderal, on heavy soils, in scrub land, in wadis, down to 200 mm MAR
Distribution:

- In Libya, pre-desertic, all coastal zones, but rare in Sirte (in Jebel Al-Akhdar: *Echinops galalensis* Schweinf., http://www.theplantlist.org/tpl1.1/record/gcc-119857 with red stems, and *Echinops cyrenaicus* E.A. Durand & Barratte, Endemic to north-east Libya, unresolved, http://www.theplantlist.org/tpl1.1/record/gcc-132862)
- In arid zones, southern Europe and North Africa (not in Morocco (?)), Egypt

Endemism: NA

Photo 7.64. *Echinops spinosissimus*, capitulum pre-flowering (Mzerah – south Gendoula).

Photo 7.65. *Echinops spinosissimus*, Capitulum flowering (Mafsel – south Gendoula).

Helichrysum stoechas (L.) Moench

Synonym(s): http://www.theplantlist.org/tpl1.1/record/gcc-100508 = *Gnaphalium stoechas* L.
Native:
Vernacular name: *Hechbet l'arneb* (?)
Raunkiaer's life-form: Hemicryptophyte
Description and morphology: Small (10–40 cm) perennial silvery-whitish scented herb, ascending stems with narrow linear pubescent/tomentose leaves with rolled margins, flowers in small capitulum of yellow all tubular florets, involucral gold-coloured bracts scarious and fluffy enveloping each capitulum, cypsela with small white tubercles, papus with finely dented seta.
Reproduction: Flowering all year long, seeds dispersed by anemochory
Pastoral importance: Limited
Palatability and fodder value: Avidly grazed by all livestock and wildlife
Toxicity: None
Economic interest: None
Habitat: Stony sites, ruderal, ruins, loamy soils, on marl, minimum 250–300 mm MAR
Distribution:

- In Libya, on hills Jebel Nafusa (Yeffren to Homs), coast (on Gargaresh quarry, Tajoura), due to overgrazing, distribution often limited to protected sites
- In arid zones, North Africa to Turkey (Mediterranean Europe)

Endemism: NA

Photo 7.66. *Helichrysum stoechas* (Yeffren).

Photo 7.67. *Helichrysum stoechas*, flower and fruits (Yeffren).

Koelpinia linearis Pall.

Synonym(s): http://www.theplantlist.org/tpl1.1/record/gcc-401
Native:
Vernacular name: *Guiz el Ghazel* (?)
Raunkiaer's life-form: Tachytherophyte
Description and morphology: Annual, slender stems, skinny glabrous leaves, yellow ligulate flower with woolly indumentum, few (8–12) long (15–25 mm) cypsela strongly inward arched furrowed with dorsal hooked dents and end grapple-shaped.
Reproduction: Flowering late spring, seeds dispersed by zoochory and anemochory.
Pastoral importance: Limited but a valuable spring forage
Palatability and fodder value: Excellent, grazed by all livestock
Toxicity: None
Economic interest: None, fresh leaves reported to be consumed by shepherds
Habitat: Sandy loam/gravelly soils, down to 100 mm MAR
Distribution:

- In Libya, inland, Jeffara plain, Jebel Nafusa to the Hamada
- In arid zones, Spain to Central Asia

Endemism: NA

Photo 7.68. *Koelpinia linearis*, dented cypsela (Gdama, Gharyan).

Photo 7.69. *Koelpinia linearis* (Gdama, Gharyan).

Launaea resedifolia (L.) Kuntze

Synonym(s): http://www.theplantlist.org/tpl1.1/record/gcc-42452 = *Podospermum laciniatum* ssp. *decumbens* (Guss.) Gemeinholzer & Greuter

Native:
Vernacular name: *Lahiet Jeddi* (?), *Adeedaa* (?), *Azim* (?)
Raunkiaer's life-form: Hemicryptophyte
Description and morphology: Biannual or sometimes perennial, taproot sometimes woody, divaricated/dichotomous stems glabrous procumbent and ascending, milky sap turning yellowish, most leaves pinnatisect dentate laciniate basal in rosette, terminal yellow capitulum (pre-flowering elongated-pointed with inflated base) with white margins on involucral bracts, ligulate florets abruptly truncated-dented (3–5 dents), cypsela with long white smooth pappus.
Reproduction: Flowering March–May, seeds dispersed by anemochory
Pastoral importance: Abundant in spring pastures in good condition
Palatability and fodder value: Excellent, well grazed by all livestock
Toxicity: None
Economic interest: None, when in dense stands, used to be a reliable indicator of land suited for rainfed cereal cropping in the Jeffara plain
Habitat: Sandy/sandy-loam soils, down to 100 mm MAR in sandy depressions
Distribution:

- In Libya, all over the coastal zones and inland
- In arid zones, southern Spain, North Africa to Iran

Endemism: NA

Photo 7.70. *Launaea resedifolia* (Gdama, Gharyan).

Photo 7.71. *Launaea resedifolia* (south Al-Haraba, Hamada).

Nolletia chrysocomoides (Desf.) Cass.

Synonym(s): http://www.theplantlist.org/tpl1.1/record/gcc-125575 = *Conyza chrysocomoides* Desf., http://www.theplantlist.org/tpl1.1/record/gcc-106922
Native:
Vernacular name: *Guiz el Ghazel* (?)
Raunkiaer's life-form: Hemicryptophyte
Description and morphology: Inconspicuous slender suffruticose plant with deep taproot, perennial with annual procumbent then erect stems ending with a small (10–15 mm diam.) isolated capitulum, leaves sessile shaggy-linear adpressed on stems, few capitula per plant with small yellowish tubular florets with scarious deflexed involucral bracts at maturity, cypsela (1–2 mm long) with long equal pappus.
Reproduction: Flowering in autumn/winter, seeds dispersed by anemochory
Pastoral importance: Limited
Palatability and fodder value: Well grazed by all livestock
Toxicity: None
Economic interest: None
Habitat: On deep sandy/sandy-loam soils, sometimes on fixed dunes, down to 100 mm MAR
Distribution:

- In Libya, Jeffara plain, Jebel Nafusa, to Sirte; absent from eastern Libya (?)
- In arid zones, Spain to North Africa

Endemism: Spain, North Africa to west Libya (and Sirte)

Photo 7.72. *Nolletia chrysocomoides* (north of Tiji).

Photo 7.73. *Nolletia chrysocomoides* (south Sorman).

Pallenis spinosa (L.) Cass.

Synonym(s): http://www.theplantlist.org/tpl1.1/record/gcc-34594
Native:
Vernacular name: *Chabrom* (?)
Raunkiaer's life-form: Therophyte
Description and morphology: Low perennial (sometimes biannual) erected pubescent plant (20–40 cm), strong taproot, stems (dichotomic end (?)) reddish, basal rosette of long leaves, leaves alternate deeply furrowed becoming spinescent close to capitulum, heterogamous yellow solitary capitulum (3–5 cm diam.) ending stem, capitulum ringed with radiating involucral spiny bracts, fertile cypsela crowned with short cup-like laciniate pappus.
Reproduction: Flowering March–April, seeds
Pastoral importance: Limited
Palatability and fodder value: Poor, hardly grazed
Toxicity: None
Economic interest: None
Habitat: Loamy soils, often on rocky sites, also on *Terra rossa* (*Pallenis cyrenaica*, endemic Jebel Al-Akhdar)
Distribution:

- In Libya, rare in Jeffara plain (on rocky hills), Jebel Nafusa and south on to Hamada, inland Sirte, Jebel Al-Akhdar (possibly *Pallenis cyrenaica* Alavi (?)), http://www.theplantlist.org/tpl1.1/record/gcc-115285, endemic eastern Libya (?))
- In arid zones, pre-desertic and desert, North Africa, to the Middle East (?)

Endemism:

Photo 7.74. *Pallenis spinosa* (wadi Maghur).

Rhanterium suaveolens Desf.

Synonym(s): http://www.theplantlist.org/tpl1.1/record/gcc-13303
Native:
Vernacular name: *Arfedj*
Raunkiaer's life-form: Nanophanerophyte
Description and morphology: Low woody bush (often hemispherical shape, up to 80 cm high), deep taproot system, intricate silvery-grey pubescent branches and foliage, sessile narrow leaves

(10–20 mm long) alternate and slightly dented, terminal solitary yellow heterogamous globose capitula (10–15 mm diam.), involucral bracts scarious recurved in small hook, mature capitula scarious blackened, cypsela pappose with end of seta somewhat plumose.
Reproduction: Flowering March–May, seeds, easy to collect
Pastoral importance: Very valuable feed resource all year long, *Rhanterium* sites have a lot of valuable spring annual (annual *Medicago*, *Lotus*, *Trigonella*, annual grasses, *Schismus*, *Cutandia* and many other palatable annuals), rich and diverse rangeland; a valuable genus to use for degraded rangeland regeneration.
Palatability and fodder value: *Rhanterium* bushes grazed by all livestock, but preferably summer to late autumn
Toxicity: None
Economic interest: *Rhanterium* stands targeted for rainfed cereal cropping; however, once *Rhanterium* cleared or cropping abandoned, sites heavily subjected to wind erosion and later colonised with *Artemisia campestris*; clearing *Rhanterium* sites for cereal cultivation largely contribute to desertification and removal of one of the most valuable Libyan rangeland feed resource.
Habitat: Deep fixed sand (to sandy loam) soils, often over fragmented calcareous nodulous horizon (old Bir-el-Ghanem sheep project), down to 50 mm MAR in depressions with contracted vegetation.
Distribution:

- In Libya, Al-Assah, Bir-el-Ghanem, Cabao, Takut, Hamada, Sirte, does not appear further east (?)
- In arid zones, apparently not recorded in Morocco, from Algeria to Libya, not in Egypt (?) replaced by *Rhanterium epapposum* Oliv. Rhanterium epapposum Oliv. — The Plant List in the Middle East, Arabian Peninsula and the Emirates

Photo 7.75. *Rhanterium suaveolens* flowers (south Sorman).

Photo 7.76. *Rhanterium suaveolens* bush (Jbebina, Al-Assah).

Endemism: NA

Scolymus hispanicus L.

Synonym(s): http://www.theplantlist.org/tpl1.1/record/gcc-121678
Native: Exotic
Vernacular name: *Guernina* (?)
Raunkiaer's life-form: Therophyte – hemicryptophyte
Description and morphology: Thistle annual or biannual (perennial (?)), strong taproot, erect white stems slightly winged, spiny all over, alternate leaves lobular spiny pubescent, sessile yellow flowers with 3–5 spiny yellowish bracts, cypsela with shedding crown-papus.
Reproduction: Flowering March–May, seeds dispersed by anemochory
Pastoral importance: Limited
Palatability and fodder value: Poor, usually grazed by Equidae and dromedaries
Toxicity: None
Economic interest: None, sometimes young roots collected for cooking
Habitat: Ruderal, ruins, roadside, important as a reference for local rainfall equivalent as it requires at least 300–400 mm MAR (lower limit for rainfed wheat production in North Africa).
Distribution:

- In Libya, highest-rainfall zones in Jebel Nafusa and Jebel Al-Akhdar
- In arid zones, Mediterranean, ubiquitous, southern Europe, North Africa

Endemism: NA

Photo 7.77. *Scolymus hispanicus* (Gharyan, Jendouba top of Jebel Nafusa).

Photo 7.78. Flower, *Scolymus hispanicus* (Jendouba).

Scorzonera undulata Vahl

Synonym(s): http://www.theplantlist.org/tpl1.1/record/gcc-2069 = *Scorzonera alexandrina* Boiss., http://www.theplantlist.org/tpl1.1/record/gcc-14497
Native:
Vernacular name: *Guiz* (?)
Raunkiaer's life-form: Geophyte
Description and morphology: Slender perennial, tuber long cylindrical black, narrow entire furrowed leaves undulate-wavy linear faintly pubescent, unique violet flower subsessile vanilla scented, ray flowers long ligulated and disc flowers tubular, seeds with plumose sessile papus.
Reproduction: Flowering March–May, seeds dispersed by anemochory
Pastoral importance: Limited
Palatability and fodder value: Well grazed by all livestock
Toxicity: None
Economic interest: Tuber edible fresh or cooked (sweet taste – close to common *salsify*) and eagerly collected by shepherds
Habitat: Stony loamy sites, down to 50 mm MAR
Distribution:

- In Libya, all over deep into desert
- In arid zones, North Africa, Egypt, Middle East

Endemism:

Photo 7.79. *Scorzonera undulata* (Al-Haraba, N 31.663 E 11.514, west Zintan).

BORAGINACEAE

Cerinthe major L.

Synonym(s): http://www.theplantlist.org/tpl1.1/record/kew-2712572, or *Cerinthe aspera* Roth., http://www.theplantlist.org/tpl1.1/record/kew-2712541
Native: Possible exotic (?)
Vernacular name: (?)
Raunkiaer's life-form: Mesotherophyte
Description and morphology: Small annual (20–40 cm), glabrous/glaucous all over, erect, alternate amplexicaul hastate leaves, short pedicelled pendulous yellow flowers (red spots at base of corolla) with bluish ovate bracts, small pyramidal fruits.
Reproduction: Flowering February–March, seeds
Pastoral importance: None
Palatability and fodder value: None, not touched by livestock
Toxicity: Possible due to Pyrrolizidine alkaloids (?) leading to liver failure
Economic interest: Reported used for external treatment of eczema
Habitat: Ruderal, wasteland, on heavy soil, *Terra rossa*, above 300 mm MAR in well-watered sites
Distribution:

- In Libya, mostly highest-rainfall zones in Jebel Al-Akhdar and around Tripoli
- In arid zones, southern Europe, Mediterranean, ubiquitous

Endemism: NA

Photo 7.80. *Cerinthe major*, flower (Shahat).

Photo 7.81. *Cerinthe major*, flower (Shahat).

Echiochilon fruticosum Desf.

Synonym(s): http://www.theplantlist.org/tpl1.1/record/kew-2783905
Native:
Vernacular name: *Lerga* (?), *Zrega* (?), *Yidma* (?)
Raunkiaer's life-form: Nanophanerophyte
Description and morphology: Perennial shrublet (30–50 cm high), dense intricate branches from base, hairy all over, white stems, alternate linear sessile obtuse leaves (10 mm × 2mm), inflorescences in lax reddish spikes, abundant pale blue (or whitish) sessile zygomorphic small (8–10 mm) flowers solitary and axillary, beaked smooth seeds in tuberculate nutlets (4 per flowers).
Reproduction: Flowering all year long, seeds
Pastoral importance: Limited
Palatability and fodder value: High, excellent, grazed by all livestock, possibly one of the best range species in terms of palatability
Toxicity: None
Economic interest: None, could be used for range reseeding
Habitat: Desertic, sandy/sandy-loam fixed soils, down to 100 mm MAR, not on dunes
Distribution:

- In Libya, Jeffara plain, Jebel Nafusa, Hamada, Sirte, south Jebel Al-Akhdar
- In arid zones, Saharan north Africa, southern Middle East to Arabian peninsula

Endemism: NA

Photo 7.82. *Echiochilon fruticosum* (Bir-el-Ghanem, BGII).

Photo 7.83. *Echiochilon fruticosum*, flower (Al-Assah).

Echium angustifolium Mill.

Synonym(s): http://www.theplantlist.org/tpl1.1/record/kew-2783987, or *Echium diffusum* Sm., http://www.theplantlist.org/tpl1.1/record/kew-2784080, or *Echium hispidum* Sm., http://www.theplantlist.org/tpl1.1/record/kew-2784145, or *Echium sericeum* ssp. *halacsyi* Holmboe, http://www.theplantlist.org/tpl1.1/record/kew-2784348
Native:
Vernacular name: *Henna-el-Agreb* (?)
Raunkiaer's life-form: Nanophanerophyte
Description and morphology: Perennial herb, hispid/villous all over, erect white stems (20–50 cm) with basal rosette, entire leaves sessile lanceolate narrow, unilateral terminal long (up to 10 cm) inflorescence in recurved hairy spikes, conical red flowers (bluish when drying) upward on spike, corolla long (up to 2 cm) with round lobes, pyramidal tuberculate one-nerved fruits.
Reproduction: Flowering February–April, seeds
Pastoral importance: None

Palatability and fodder value: None, not touched by livestock
Toxicity: Possible due to Pyrrolizidine alkaloids (?) leading to liver failure
Economic interest: None
Habitat: Ruderal, wasteland, loamy-heavy soils, above 200 mm MAR
Distribution:

- In Libya, all over coastal zones
- In arid zones, Mediterranean to the Middle East

Endemism: NA

Photo 7.84. *Echium angustifolium*, plant (Bir Shawab N 32.420 E 21.474, Marawa).

Photo 7.85. *Echium angustifolium*, spike and flower (Bir Shawab, Marawa).

Echium humile Desf.

Synonym(s): http://www.theplantlist.org/tpl1.1/record/kew-2784150, or *Echium pycnanthum* ssp. *humile* (Desf.) Jahand. & Maire, http://www.theplantlist.org/tpl1.1/record/kew-2784308, or *Echium trygorrhizum* Pomel, http://www.theplantlist.org/tpl1.1/record/kew-2784395
Native:
Vernacular name: *Henna-el-Agreb* (?)
Raunkiaer's life-form: Mesotherophyte
Description and morphology: Small annual (15–30 cm), hairy/hispid all over, stem erect, leaves oblong with wavy/undulate margins, basal leaves narrowed in middle, often unique inflorescence in cyme, blue-violet conical flowers, pyramidal beaked grey seeds.
Reproduction: Flowering February–May, seeds
Pastoral importance: None
Palatability and fodder value: Not touched by all livestock
Toxicity: Possible due to Pyrrolizidine alkaloids (?) leading to liver failure
Economic interest: None
Habitat: Desertic, on fixed sandy/gravelly/rocky soils, down to 100 mm MAR
Distribution:

- In Libya, most arid sites in Jeffara plain, northern foothills of Jebel Nafusa (Cabao, Nalut), deep into Hamada, inland Sirte to southern Jebel Al-Akhdar to Tobruk–Solum
- In arid zones, North Africa to Arabian Peninsula

Endemism: NA

Photo 7.86. *Echium humile*, plant (Bir Alaq N 31.103 E 11.939, Hamada).

Echium rubrum Forssk.

Synonym(s): http://www.theplantlist.org/tpl1.1/record/kew-2784329, or illegitimate, *Echium setosum* Vahl., http://www.theplantlist.org/tpl1.1/record/kew-2784349
Native:
Vernacular name:
Raunkiaer's life-form: Mesotherophyte
Description and morphology: Small prostrate annual, red water-soluble pigment from roots, short (10–30 cm) decumbent stems, white hairy/tomentose all over, alternate leaves linear/spathulate, terminal inflorescence pauciflora, zygomorphic red flower (turning blue-purple when drying), small (2–3 mm diam.) triangular tuberculate seeds.
Reproduction: Flowering March–May, seeds
Pastoral importance: None
Palatability and fodder value: None, not touched by livestock
Toxicity: Possible due to Pyrrolizidine alkaloids (?) leading to liver failure
Economic interest: None, red pigment from roots used as dye on small items, leather
Habitat: Sub-desertic rangelands, sandy-loam soils, above 100 mm MAR
Distribution:

- In Libya, Jeffara plain, Jebel Nafusa, Hamada, Sirte, Jebel Al-Akhdar
- In arid zones, southern Tunisia to Middle East

Endemism: NA

Photo 7.87. *Echium rubrum* (El-Hisha N 32.489 E 22.490, south Beida, Jebel Al-Akhdar).

Photo 7.88. *Echium rubrum*, flowers (El-Hisha, south Beida, Jebel Al-Akhdar).

BRASSICACEAE (Cruciferae)

Anastatica hierochuntica L.

Synonym(s): http://www.theplantlist.org/tpl1.1/record/kew-2636323, or *Myagrum hierochunticum* (L.) Crantz, http://www.theplantlist.org/tpl1.1/record/kew-2358894
Native:
Vernacular name: *Akaraba* (?)
Raunkiaer's life-form: Therophyte
Description and morphology: Small (5–10 cm) prostrate annual villose herb, decumbent stems with star-like indumentum, oval-oboval basal leaves in rosette, all flowers subsessile white contracted at the base of plant, leaves falling at maturity, stems becoming woody curling inward to form a tight ball entrapping all the dry siliques, beaked siliques ovoid hispid with two cochlear valves, apterous small ovoid dormant seeds surviving many years of drought in desert environment.
Reproduction: Flower February–March, seeds, whole plant (not the seeds) dispersed by wind, plants uprooted by wind roll over kilometres (like tumbleweed – *Salsola kali*), when rain occurs, increasing hygrometry allows woody curled stems to uncurl – open up exposing dehiscent siliques and freeing seeds; if rain not sufficient stems will again be repeating curling and uncurling until amount of rain allows for seeds germination.
Pastoral importance: Limited
Palatability and fodder value: Grazed by all livestock when fresh
Toxicity: None
Economic interest: None, botanic curiosity – often called *Rose of Jericho*
Habitat: On heavy soils in depressions, down into the desert, occurring in large numbers in contracted vegetation in desert
Distribution:

- In Libya, west Libya (wadi zones, Hamada)
- In arid zones, Saharan, North Africa to northern Pakistan

Endemism: NA

Photo 7.89. *Anastatica hierochuntica* (dry plant with curled stems in ball – wadi Sof-el-Jin, N 31.434 E 14.151).

Carrichtera annua (L.) DC.

Synonym(s): *Vella annua* L., http://www.theplantlist.org/tpl1.1/record/kew-2456025
Native:
Vernacular name: *Kheshhina* (?)
Raunkiaer's life-form: Chamaephyte
Description and morphology: Dwarf annual (5–20 cm), powerful smell and taste, hispid, dissected pinnatisect leaves, many cream-yellowish flowers violet-veined short petiolate along all stems, silique pendulous in two segments (6–8 × 3–4 mm) – lower one fertile dehiscent short hairs-spiky with 1–3 seeds, upper one smooth sterile indehiscent woody and spatulate, small 1 mm black seeds.
Reproduction: Flowering March–April, seeds, dispersion by possible zoochory
Pastoral importance: Good as large seed production, recruitment with dense spring stands in depressions
Palatability and fodder value: Well grazed by all livestock at all stages
Toxicity: None
Economic interest: None
Habitat: On heavy soils in depressions and wadis, down to 100 mm
Distribution:

- In Libya, wadi zones south of Jebel Nafusa (Sof-el-Jin, Sassu), northern Hamada, Jebel Al-Akhdar (Soluk to Tobruk)
- In arid zones, south European, North Africa to Iran

Endemism: NA

Photo 7.90. *Carrichtera annua* (wadi Sassu).

Photo 7.91. Siliques on dry *Carrichtera annua* (wadi Sassu).

Farsetia aegyptiaca Turra

Synonym(s): http://www.theplantlist.org/tpl1.1/record/kew-2807702
Native:
Vernacular name: *Shajara-Azhub* (?)
Raunkiaer's life-form: Chamaephyte
Description and morphology: Small (20–40 cm) grey – pubescent all-over shrub including fruits, branching from base, leaves linear, flowers pinkish-brown (darker veins) with curly petals margins, oblong (3 cm × 1 cm) flat obovate erect silique (on short peduncle) round ended with short beak, silique two valves/dehiscent containing (6–10) papery winged seeds.
Reproduction: Flowering March–April, seeds dispersed by anemochory
Pastoral importance: Limited
Palatability and fodder value: Dromedary and goat feed
Toxicity: None
Economic interest: None
Habitat: Desertic, Saharan, on *reg* and small depressions on Hamada
Distribution:

- In Libya, Hamada, inland Sirte to southern Jebel Al-Akhdar, Saharan
- In arid zones, Saharan, North Africa to Arabian Peninsula

Endemism: NA

Photo 7.92. *Farsetia aegyptiaca* (flowers), Bir Al-Morhan N 31.085 E 12.350, Hamada.

Photo 7.93. *Farsetia aegyptiaca* (Siliqua), Al-Morhan, Hamada.

Lobularia maritima (L.) Desv.

Synonym(s): http://www.theplantlist.org/tpl1.1/record/kew-2338457, or *Alyssum maritimum* (L.) Lam., http://www.theplantlist.org/tpl1.1/record/kew-2632265, or *Clypeola maritima* L., http://www.theplantlist.org/tpl1.1/record/kew-2728760, or *Koniga maritima* (L.) R.Br., http://www.theplantlist.org/tpl1.1/record/kew-2334101
Native:
Vernacular name:
Raunkiaer's life-form: Hemicryptophyte
Description and morphology: Prostrate creeping perennial, strong taproot, woody stems, leaves linear – lanceolate, sweet scented white flowers, pedicel 2–3 times longer than ovate flattish silique, one marginally winged seed in each locule.
Reproduction: Flowering in winter, seeds (possible anemochory (?))
Pastoral importance: Limited
Palatability and fodder value: All livestock, rats, rabbits, birds
Toxicity: None
Economic interest: Nectar for bees, ornamental
Habitat: Rocky slopes, heavy soils, 300–500 mm MAR
Distribution:

- In Libya, Jebel Nafusa (Gharyan, Yeffren), Jebel Al-Akhdar (Shahat), on Gargaresh stones along Tripoli coast
- In arid zones, North Africa , Middle East

Endemism: NA

Note: *Lobularia maritima* green plants without white flowers can be confused with the perennial *Alyssum montanum* L. Alyssum montanum L. — The Plant List that has yellow flowers and similar ecology. *Lobularia libyca,* (Viv.) Webb & Berthel, Lobularia libyca (Viv.) Webb & Berthel. — The Plant List a silvery annual with white flowers is common in Jeffara plain on sandy soils (also in crops or on abandoned lands) and looks very much like *L. maritima*. All these annuals Brassicaceae green or dry are commonly grazed by all livestock.

Photo 7.94. *Lobularia maritima* (Gharyan).

Matthiola glutinosa Jafri

Synonym(s): http://www.theplantlist.org/tpl1.1/record/tro-100357814
Native:
Vernacular name:
Raunkiaer's life-form: Chamaephyte
Description and morphology: Low perennial (10–40 cm), erect branching, slightly pubescent to glabrous yellowish stems, few thick glaucous short lanceolate-dentate sticky leaves along stems, glandulous sticky all over with sand particles adhering, few terminal whitish/yellowish flowers subsessile/sessile with flat petals, elongated erect cylindrical (tri-horned) silique (3–7 cm long, 2 mm diam.) ending with two lateral short stigmatic recurved horns.
Reproduction: Flowering March–April, seeds by anemochory
Pastoral importance: Limited
Palatability and fodder value:
Toxicity: None
Economic interest: None
Habitat: Loamy/heavy soils, down to 150 mm MAR
Distribution:

- In Libya (?) Jafri reports this *Matthiola*, a new species (El-Gadi and Jafri, 1976–1989, Vol. 23, p. 155), along Hun-Sebha highway in flats with loamy soils in very arid site; our specimen is from the southern slope of the Jebel Al-Akhdar (Al Jahishiyah, 25 km southeast El-Abyar)
- In arid zones (?)

Note: *Matthiola glutinosa* could be confused with *Matthiola tricuspidata* (L.) R. Br. http://www.theplantlist.org/tpl1.1/record/kew-2357993, usually a coastal plant, with similar siliques but much abundant dentate obovate leaves at the base, abundant flowers along long stem above leaves.

Photo 7.95. *Matthiola glutinosa* (Al Jahishiyah, south Jebel Al-Akhdar).

Matthiola longipetala (Vent.) DC.

Synonym(s): Matthiola longipetala (Vent.) DC. — The Plant List
Native:
Vernacular name:
Raunkiaer's life-form: Therophyte
Description and morphology: Very variable plant morphology – low annual (10–30 cm) or biannual, erect branching, slightly glandular pubescent to glabrous, leaves linear – lanceolate covered with stellar indumentum, many terminal flowers white to purplish subsessile/sessile with undulate-margin petals, elongated cylindrical silique (3–5 cm long, 2 mm diam.) ending with two lateral short stigmatic erect horns.
Reproduction: flowering March–April, seeds by anemochory
Pastoral importance: Limited
Palatability and fodder value: Excellent
Toxicity: None
Economic interest: None
Habitat: Sandy loam soils, down to 150 mm MAR
Distribution:

- In Libya, all over coastal zones to the margin of desert
- In arid zones, southern Europe, North Africa to the Middle East

Endemism: NA

Note: *Matthiola* sp. could be confused with:

- *Lonchophora* sp. has a similar horned long tricuspid siliqua, but the silique has its base shortly two-pointed or horned
- *Malcolmia* sp. has an even larger horned tricuspid silique, but the silique has its base strongly horned, anchor-like

Photo 7.96. *Matthiola longipetala* ssp. *longipetala* (wadi Sassu).

Photo 7.97. *Matthiola longipetala* ssp. *pumilio* (?) (15 km north of Gumerah, Jebel Al-Akhdar).

Moricandia arvensis (L.) DC.

Synonym(s): http://www.theplantlist.org/tpl1.1/record/kew-2360753
Native:
Vernacular name: *Crumb-el-jmel* (?), *Bjik* (?)
Raunkiaer's life-form: Chamaephyte – therophyte
Description and morphology: Perennial shrub (up to 1 m), woody base, stem white, old leafless stems protruding from shrub, fleshy entire amplexicaule leaves round/oboval glaucous thick caulinar and embracing stem, showy violet flowers with dressed petals dark violet veined, dehiscent long thin compressed tetragon erect two-valves siliqua slightly beaked, seeds in two rows.
Reproduction: Flowering February–May, seeds
Pastoral importance: High but limited distribution
Palatability and fodder value: Avidly grazed by all livestock, fleshy leaves with high water content, reported by shepherds as sheep fattener.
Toxicity: None
Economic interest: Could be used for reseeding in wadis, on hill slopes
Habitat: Possibly gypsophyte on loamy soils down to 100 mm MAR, due to its high appetability usually found in protected areas (rocky sites/slopes, graveyards, hedge of irrigated garden etc.)
Distribution:

- In Libya, Jebel Nafusa (from Nalut to Homs/Misurata), does not appear to occur east of Misurata/Kararim (?)
- In arid zones, south-western Europe, North Africa to west Libya

Endemism: Likely a west Mediterranean genus and species

Note: Far south (Fezzan, Ghat), occurrence of *Schouwia purpurea* (Forssk.) Schweinf., a strictly Saharan plant (http://www.theplantlist.org/tpl1.1/record/kew 2475784) that could be confused with *Moricandia arvensis* because of the amplexicaule leaves. The *Schouwia purpurea* silique are large, flat and biwinged.

Photo 7.98. *Moricandia arvensis* bush (along Mizda road).

Photo 7.99. *Moricandia arvensis* – amplexicaul leaves (along Mizda road, N 31.547 E 12.992).

Photo 7.100. *Moricandia arvensis* flowers and siliquae (in wadi, along Mizda road).

Moricandia nitens (Viv.) E.A. Durand & Barratte

Synonym(s): *Hesperis nitens* Viv., http://www.theplantlist.org/tpl1.1/record/kew-2847589, or *Moricandia dumosa* Boiss., http://www.theplantlist.org/tpl1.1/record/kew-2360793
Native:
Vernacular name: *Crumb-el-jmel* (?), *Bjik* (?)
Raunkiaer's life-form: Chamaephyte – therophyte
Description and morphology: Very similar to *Moricandia arvensis*, perennial shrub (20–80 cm), woody base, stem white, however, fleshy obovate longer entire leaves (short petioled – subsessile) glaucous thick not embracing the stem, showy violet flowers with dressed petals veined violet, dehiscent long thin compressed tetragon erect two-valves siliqua slightly beaked, seed in one row.
Reproduction: Flowering February–May, seeds
Pastoral importance: High but limited distribution
Palatability and fodder value: Avidly grazed by all livestock, fleshy leaves with high water content, reported by shepherds as sheep fattener.
Toxicity: None
Economic interest: None
Habitat: Rocky outcrop in protected habitat from livestock, possibly gypsophyte
Distribution:

- In Libya, Jebel Nafusa (Gharyan Yeffren, Ain Zerga), Jebel Al-Akhdar
- In arid zones, North Africa to Egypt (cliffs from Solum to Marsa-Matruh), Palestine to Arabian Peninsula

Note: Could be confused with *Moricandia arvensis* that has amplexicaule leaves embracing stems, while *Moricandia nitens* has short petiolate leaves.

Photo 7.101. *Moricandia nitens* (Yeffren, Ain Zerga).

Oudneya africana R. Br.

Synonym(s): http://www.theplantlist.org/tpl1.1/record/tro-100357896, or *Henophyton deserti* (Coss. & Durieu) Coss. & Durieu, http://www.theplantlist.org/tpl1.1/record/tro-100358581, also reported as *Moricandia nitens* (Viv.) E. A. Durand & Barratte https://efloramaghreb.org/specie/142080
Native:
Vernacular name: *Eulga* (?) (Algeria)
Raunkiaer's life-form: Nanophanerophyte
Description and morphology: Medium-size bush (80–120 cm), strong woody trunk, intricate white branches and stems, sessile alternate narrow leaves entire fleshy-spathulate, flowers (in raceme) pinkish with whitish margin, long (20–30 × 3–5 mm) flat dehiscent silique bilocular compressed with septum, short beak, small seeds winged (wings twice larger than seed).
Reproduction: Flowering January–March, winged seeds dispersed by anemochory
Pastoral importance: Good but limited occurrence
Palatability and fodder value: Dromedary feed
Toxicity: None
Economic interest: Smashed leaves reported to be used in traditional medicine to prevent wound infection (?)
Habitat: Shallow sandy soils/rocky *reg* with sand veil, Saharan, below 100 mm MAR
Distribution:

- In Libya, Hamada (Ghadames, Sinaween), never seen in Sirte or further east
- In arid zones

Endemic: Southern Algeria, Tunisia, and Libya

Photo 7.102. *Oudneya africana* bush (Derj N 30.143 E 10.651).

Savignya parviflora (Delile) Webb

Synonym(s): http://www.theplantlist.org/tpl1.1/record/kew-2475906
Native:
Vernacular name: *Gulglan* (?)
Raunkiaer's life-form: Therophyte
Description and morphology: Small glabrous annual (5–15 cm), fleshy basal leaves in rosette with round-dented margins, upper leaves fleshy cylindrical, flowers clear violet, long thin pedicelled (20–30 mm) bearing flat pendulous obovate beaked siliques, small round flat black seeds surrounded with marginal papery wings.
Reproduction: Flowering February–April, seeds winged dispersed by anemochory
Pastoral importance: Good, large flat areas on Hamada covered with after rain
Palatability and fodder value: Grazed by all livestock in the Sahara
Toxicity: None
Economic interest: None
Habitat: Sandy-gravelly flats in most arid – *reg* desert sites, MAR < 150 mm
Distribution:

- In Libya, Hamada, south Sirte, south Jebel Al-Akhdar
- In arid zones, Saharan, North Africa to Pakistan

Endemism: NA

Photo 7.103. *Savignya parviflora* (Sinaween, N 31.009 E 10.620, Hamada).

Photo 7.104. Flower and young silique, *Savignya parviflora* (Sinaween, Hamada).

Photo 7.105. *Sinapis alba* with flower and siliquae (Bir-el-Ghanem).

Sinapis alba L.

Synonym(s): *Brassica alba* (L.) Rabenh, Brassica alba (L.) Rabenh. —The Plant List
Native:
Vernacular name: *Khardal* (?)
Raunkiaer's life-form: Therophyte
Description and morphology: Polymorphic hairy annual, pivotal root, erect single stems up to 1 m high, basal leaves petiolate lyrate with terminal lob larger, numerous yellow/whitish flowers on long floral spike, hairy siliquae (20–40 mm) with sabre-like reddish curved beak, beak equal or longer than siliquae, 1–3 pale brown seeds (2 mm diam.) in siliquae.
Reproduction: Seeds, flowering February–April
Pastoral importance: Important green and dry; when dry complement crop residues
Palatability and fodder value: Like all Brassicaceae, well grazed by all livestock
Toxicity: None
Economic interest: None, sometimes eaten as salad (?)
Habitat: Ruderal, common weed in rainfed crops on loamy-sand soil types, wasteland and rocky wadis
Distribution:

- In Libya, all over northern Libya
- In arid zones, all over North Africa

Endemism: NA

Note: Could be confused with *Sinapis arvensis* L. Sinapis arvensis L. —The Plant List, *Harra*, same ecology as *Sinapis alba*, collected as condiment (mustard) and salad, with yellow flowers, but short hairs on long straight cylindrical siliquae with three protruding veins and containing 2–7 dark seeds each.

Photo 7.106. *Sinapis alba* with hairy siliquae with sabre-like reddish curved beak (Bir-el-Ghanem).

Zilla spinosa (L.) Prantl

Synonym(s): http://www.theplantlist.org/tpl1.1/record/kew-2470433
Native:
Vernacular name: *Shebrum* (?)
Raunkiaer's life-form: Chamaephyte – nano-phanerophyte
Description and morphology: Open lax spiny hemi-spherical bush (up to 1 m diam.), intricate glabrous branches and stems with sharp thorns (sometimes bifurcated from base), basal leaves large lyrate, cauline leaves sessile elongated dentate, flowers pale lilac (or whitish veined) with darker veins, suberose tetra or hexagonal silique (5–10 mm diam.) capped with the old style as a long yellow pointed beak, silique spherical pointed (6–8 mm diam.) bilocular indehiscent with 1–2 seeds per valve, without wings.
Reproduction: Flowering November–March, seeds
Pastoral importance: Limited
Palatability and fodder value: Poor, dromedary feed
Toxicity: None
Economic interest: None
Habitat: Mostly in wadis, loamy to heavy soils (often covered with sand veil), in depressions to central Sahara, also on fractured rocky sites
Distribution:

- In Libya, wadi zones south-east Jebel Nafusa, Hamada, inland Sirte to southern Jebel Al-Akhdar, Sahara, MAR < 150 mm
- In arid zones, Sahara North Africa to Arabian Peninsula

Endemism: NA

Other main *Zilla* species in Libya (both Saharan and same ecology):

- *Zilla spinosa* ssp. *spinosa*, siliqua wingless, most common
- *Zilla spinosa* ssp. *macroptera*, large siliqua with four large wings, flowers violet, reported from Mizda (?) and Marmarica

Photo 7.107. *Zilla spinosa ssp. spinosa* (with flower), wadi Nefad – Seddada.

Photo 7.108. *Zilla spinosa* ssp. *spinosa* (siliquae and thorns), wadi Zamzam N 31.254 E 15.159.

CAPPARACEAE

Capparis ovata Desf.

Synonym(s): http://www.theplantlist.org/tpl1.1/record/kew-2697425 = *Capparis spinosa* var. *ovata* (Desf.) Sm., or *Capparis spinosa* subsp. *orientalis* (Duh.) Jafri
Native:
Vernacular name: *Kabbar*, *Khabbar* (?)
Raunkiaer's life-form: Chamaephyte
Description and morphology: Shrub (1–3 m) hanging on rocks with pendulous woody/reddish young branches or decumbent prostrate on rocky ground, entire oval-elliptic glaucous alternate short-petiolate fleshy leaves (shedding sometimes at end of summer), spiny curved retroverse short pair of stipules at nodes, solitary long-pedicellate erect oval flowers buds (6–10 mm diam.), flowers large (3–5 cm diam.) mostly white (tinged with pink – lilac), multiple stamens (white or lilac) longer than petals, fruit turning red when ripe, many small (2–3 mm) black seeds
Reproduction: Flowering April–June, seeds
Pastoral importance: Limited
Palatability and fodder value: Grazed by goats, cattle, dromedaries and barely by sheep
Toxicity: None
Economic interest: Flower buds commonly collected and pickled in vinegar. Entire or mashed leaves applied on head forefront to lessen heatstroke or insolation, reported anti-scorbutic, and many other medicinal properties reported in North Africa.
Habitat: On rocks, walls (on ruins), on limestone, on loamy/marly soils down to 100 mm MAR, not on sand, *Capparis ovata* large bushes often refuge for partridges, hence hunting dogs are prevented entering these bushes as may be caught by recurved spines.
Distribution:

- In Libya, coastal zones (on walls of antique ruins, Sabratha, Leptis Magna), Jebel Nafusa, Hamada, not in Sirte (?), Jebel Al-Akhdar
- In arid zones, southern Europe, North Africa

Endemism: NA

Note: A taxonomic revision of the *Capparis* sp. is available from (Inocencio *et al.*, 2006).

Photo 7.109. *Capparis ovata*, prostrate creeping bush with oval fleshy leaves (Toknis, Gendula).

Photo 7.110. *Capparis ovata*, flower buds and leaves (wadi Maghur) – note recurved short spiny stipules at nodes.

Photo 7.111. *Capparis spicula*, erect bush, buds and open flower (from Uzbekistan, Gintzburger *et al.*, 2003).

CHENOPODIACEAE (Amaranthaceae)

Atriplex halimus L. var. *halimus*

Synonym(s): http://www.theplantlist.org/tpl1.1/record/kew-2665336
Native:
Vernacular name: *G'taf, Kataf* (?)
Raunkiaer's life-form: Nanophanerophyte (sometimes chamaephyte)
Description and morphology: Large hemispherical bush (up to 3 m high and diameter), much branching from base, soft flexible stems, elliptical leaves (salty taste) silvery-white with salt vesicle, leafless inflorescence in lax panicle with male flowers at top and female flowers at base, fruits hard presses and congregated along stems, fruit with two ovate/round accrescent membranous perianth enclosing single seed.
Reproduction: Flowering April–November, abundant seed production; fruits must be soaked in running water to remove salt to allow faster germination. Higher germination rate is obtained with dehulled seeds.
Pastoral importance: High
Palatability and fodder value: Usually grazed by all livestock, however, due to rather pointed end of branches, sheep tends to avoid grazing, while dromedaries and goats make good use of it. One may notice that in a population of native *A. halimus*, some shrubs are not touched, indicating possible large variability in terms of palatability.
Toxicity: None
Economic interest: Source of (poor) fuelwood; mostly used for drought fodder reserves and steppic shrubs plantations from nursery seedlings; due to its strong growth, sometimes used as sand fixator on coastal dunes. *A. halimus* nursery seedlings have been used on very large-scale plantations (1970–1980) for rangelands improvement all over coastal Libya to the edge of the desert, and also in alley farming with double rows of *Atriplex* (as summer or drought standing fodder reserves) alternating with strips (10–15 m) of rainfed barley crops.
Habitat: Deep sandy-loam soils, halophytic, gypsum tolerant, highly drought-tolerant, 50–400 mm MAR, native stands covering large areas in wadis south Jebel Nafusa, Sirte, often forming *nebkha*.
Distribution:

- In Libya, all over coastal Libya, on sand close to seashore, from Zuara, Taworga, Sirte, coast of Jebel Al-Akhdar except in high elevation and on rocky sites, to Marmarica
- In arid zones, North Africa, Middle East

Endemism: NA

Other *Atriplex halimus* subspecies in Libya:

- *Atriplex halimus* ssp. *schweinfurthii*, present deep inland, http://www.theplantlist.org/tpl1.1/record/kew-2878282, more drought-tolerant, very woody, intricate branches ending with woody pointed stems, often mixed population or replacing *A. halimus* var. *halimus* with increasing aridity

Exotic *Atriplex* introduced in Libya:

- *Atriplex nummularia* Lindl., http://www.theplantlist.org/tpl1.1/record/kew-2665554, from Australia, large erect bush (up to 5 m in Benghazi plain project) widely used in large-scale plantations from nursery seedlings for rangeland improvement projects all over Libya, large salty leaves with salt crystal deposit, stems are very fragile and tends to break up when grazed by livestock, plantations plagued with exploding population of desert rodents (*Psammomys obesus* Cretzschmar, a fat-tailed rat reported to host *leishmania* parasites) and destroying nearby rainfed barley crops.
- *Atriplex semibaccata* R.Br., http://www.theplantlist.org/tpl1.1/record/kew-2665777, from Australia, small prostrate bush, failed introduction.
- *Atriplex canescens* (Pursh) Nutt., http://www.theplantlist.org/tpl1.1/record/kew-2665120, from north-west America, tested in fodder shrubs plantations at Wishtata (south Tarhuna), successful establishment, good growth but poorly grazed.

Photo 7.112. *Atriplex halimus nebkha* (in a wadi, south Gharyan, GN 1–2, 1979) with rainfed barley cropping.

Photo 7.113. Natural stand of *Atriplex halimus* ssp. *halimus* (in well protected rangeland at wadi Maghur, N 32.00 E 20.71, April 2010).

Photo 7.114. Thirty-year-old *Atriplex halimus* plantation (Gdama, south Gharyan, 2009).

Photo 7.115. Australian *Atriplex nummularia* new plantations (east Gdama, N 31.91 E 13.10, Gharyan, 1978).

Atriplex mollis Desf.

Synonym(s): http://www.theplantlist.org/tpl1.1/record/kew-2665510, or *Atriplex lunulata* Delile ex Moq. http://www.theplantlist.org/tpl1.1/record/kew-2665475
Native:
Vernacular name: *K'taf* (?)
Raunkiaer's life-form: Nanophanerophyte
Description and morphology: Small greyish shrub (30–50 cm high) with stiff erect branches, small alternate elongated oboval leaves, abundant red flowers all over the bush, flat fruits with round papery reticular perianth (8–12 mm diam., like a full moon) with a unique black seed.
Reproduction: Flowering April–November, abundant seed production; fruits must be soaked in running water to remove salt to allow faster germination. Higher germination rate is obtained with dehulled seeds.
Pastoral importance: High
Palatability and fodder value: Avidly grazed by dromedaries in winter
Toxicity: None
Economic interest: Poor fuelwood
Habitat: Found on waterlogged *sebkha* and muddy saline flats at Al-Assah along Tunisian border, with other halophytes.
Distribution:

- In Libya: azonal, all over coastal Libya, on saline mud flats, from Zuara, Taworga, Sirte, to Marmarica
- In arid zones, only in North Africa

Endemism: Endemic from south-east Algeria, southern Tunisia and Libya, apparently neither in Morocco, nor in Egypt and further east.

Photo 7.116. *Atriplex mollis* (Al-Hebiliyah, N 32.4 E 11.76, south Al-Assah, 1978).

Photo 7.117. *Atriplex mollis* flowering (N 32.84 E 11.54, Al-Assah west, November 1978).

Anabasis articulata (Forssk.) Moq.

Synonym(s): http://www.theplantlist.org/tpl1.1/record/kew-2635616
Native:
Vernacular name: *Ageram* (?) (*Bagel* (?), *Belbel* (?))
Raunkiaer's life-form: Chamaephyte (Hemicryptophyte)
Description and morphology: Low woody shrub with strong taproot, rather slender-thin articulated segments on intricate branches dark green turning white when drying, leafless, reddish fruit perianth with five winged tepals, seed lens-shaped coiled vertical.
Reproduction: Flowering January–May, seeds dispersed by anemochory
Pastoral importance: Limited, camel's pasture
Palatability and fodder value: Avidly grazed by dromedaries
Toxicity: Some toxicity reported (due to Anabasine and lupulin content)
Economic interest: None, fuelwood (ash rich in potassium)
Habitat: Steppic sandy/loamy brown soils over calcareous crust, often on shallow soils, down to 50 mm MAR
Distribution:

- In Libya, Al-Assah, Sirte, Ajdabiya to Tobruk – Sollum
- In arid zones, North Africa Middle East and Arabian Peninsula

Endemism: NA

Photo 7.118. *Anabasis articulata* var. *articulata*, heavily grazed by dromedaries (Touil-es-Souf, N 32.690 E 11.668, Al-Assah).

Anabasis oropediorum Maire

Synonym(s): http://www.theplantlist.org/tpl1.1/record/kew-2635672
Native:
Vernacular name: *Belbel* (?), *Bagel* (?) (*Ageram* (?))
Raunkiaer's life-form: Chamaephyte (Hemicryptophyte)
Description and morphology: Like *A. articulata*, low woody shrub with strong taproot, articulated stout mucronate segments on intricate erect branches remaining green when drying, leafless, end of stems mucronate retrorse, flower sessile, reddish (sometimes white (?)) fruit perianth with five winged tepals, seed lens-shaped coiled vertical.
Reproduction: Flowering January–May, seeds dispersed by anemochory
Pastoral importance: Valuable dromedary rangelands
Palatability and fodder value: Avidly grazed by dromedaries
Toxicity: Some toxicity reported (due to Anabasine and lupulin content)
Economic interest: None, fuelwood (ash rich in potassium)
Habitat: Sandy (gravelly) soils but not on dunes, down to 50 mm MAR, on shallow calcareous or gypseous crust covered with sand

Distribution:

- In Libya, inland Jeffara close to Tunisian border, Hamada, Derj, to Sirte, Ajdabiya to Tobruk, south Jebel Al-Akhdar
- In arid zones, North Africa

Endemism: Western Mediterranean, North African steppe and desert

Note: Often confused with *Arthrophytum schmitianum*

Photo 7.119. *Anabasis oropediorum* flowers (Jbebina, Al-Assah).

Photo 7.120. *Anabasis oropediorum* (El-Hisha, N 31.43 E 15.20, Bu-Gren).

Arthrophytum scoparium (Pomel) Iljin

Synonym(s): http://www.theplantlist.org/tpl1.1/record/kew-2653855 = *Hamada scoparia* (Pomel) Iljin, http://www.theplantlist.org/tpl1.1/record/kew-2839034 = *Haloxylon scoparium* Pomel, http://www.theplantlist.org/tpl1.1/record/kew-2838925, doubtful: *Haloxylon articulatum* Cav. Bung (Keith, 1965, p. 530)
Native:
Vernacular name: *Remth* (all over North Africa to the Middle East and Arabian Peninsula)
Raunkiaer's life-form: Chamaephyte (Hemicryptophyte)
Description and morphology: Low woody shrub (20–60 cm) with strong taproot, articulated slender dark green fresh segments on intricate branches turning black when drying, leafless or minute triangular scales, secondary branches ascending bearing terminal short group of flowers, fruit perianth pinkish to red, horizontal coiled seed.
Reproduction: Seeds dispersed by anemochory

Pastoral importance:
Palatability and fodder value: Goat and dromedary feed, not touched by sheep
Toxicity: None
Economic interest: Valuable fuelwood, target sites (as for *Artemisia herba-alba* sites) used for opportunistic rainfed cereal cropping above 150–175 mm MAR
Habitat: Large areas of heavy brown steppic soils, same ecology as and often associated with *Artemisia herba-alba*, large stands of *Arthrophytum scoparium* survive when *Artemisia herba-alba* uprooted for rainfed barley cropping is later abandoned, still found in depressions down to 50 mm MAR.
Distribution:

- In Libya, from south Al-Assah, south of Jebel Nafusa (Nalut, Al-Haraba, south Zintan), inland Sirte (Bu-Gren, Wishkah), Ajdabiya, south Jebel Al-Akhdar
- In arid zones, North Africa, to the Middle East

Endemism: NA

Photo 7.121. *Arthrophytum scoparium* (El-Hisha, Bu-Gren).

Photo 7.122. *Arthrophytum scoparium* (El-Hisha, Bu-Gren).

Photo 7.123. *Arthrophytum scoparium* (El-Hisha, Bu-Gren).

Other main *Arthrophytum*/*Haloxylon* species in Libya:

- *Haloxylon aphyllum* (Minkw.) Iljin, http://www.theplantlist.org/tpl1.1/record/tro-7201217, tree-like (up to 10 m), exotic introduced from rangelands from Uzbekistan (Gintzburger *et al.*, 2003, pp. 146–148) to Haschan forestry station early 1960 and successfully established with a few remaining still at Haschan forestry station old site in 1975; a remarkable source of fuelwood, tolerates hight saline water table.

Photo 7.124. *Haloxylon aphyllum* from Uzbekistan (plantation at Maragha, north-west Syria, with Dr H.N. Le Houérou, 1997.

Arthrophytum schmittianum (Pomel) Maire & Weiller

Synonym(s): http://www.theplantlist.org/tpl1.1/record/kew-2653854 = *Haloxylon schmittianum* Pomel, http://www.theplantlist.org/tpl1.1/record/kew-2838923, not as *Haloxylon salicornicum* (Moq.) Bunge ex Boiss., in Keith, 1965, p. 530), http://www.theplantlist.org/tpl1.1/record/kew-2838922
Native:
Vernacular name: *Bageul* (?)
Raunkiaer's life-form: Chamaephyte (Hemicryptophyte)
Description and morphology: Low woody shrub (20–60 cm) with strong taproot, articulated fatty glaucous green fresh segments on ascending branches turning white to yellow, leafless, flowers all along white branches, fruit with membranous white-winged perianth, seed flat and horizontal.

Note: Possible confusion with *Anabasis* sp. that has vertical seeds, similar ecology.

Reproduction: Flowering January–March, seeds dispersed by anemochory
Pastoral importance: Important for dromedaries, root used for fuelwood
Palatability and fodder value: Exclusively dromedary feed
Toxicity: None (reported toxic for dromedaries, doubtful)
Economic interest: None
Habitat: Inland far away from coast, mostly desert/Saharan plant, on deep fixed sandy/ gravelly soils, not on dunes though withstanding being partially buried in aeolian sand.
Distribution:

- In Libya, desertic, Tiji-Giado, Nalut, south Jebel Nafusa to Hamada, Sirte
- In arid zones, common on Algerian high plateau (Djelfa, Messaad, Laghouat)

Endemism: North Africa, possibly in Egypt as *Hammada elegans* (Bunge) Botsch (Täckholm, 1974, p. 127), http://www.theplantlist.org/tpl1.1/record/kew-2839024

Photo 7.125. *Arthrophytum schmittianum* (east El-Hisha, Bu-Gren).

Arthrocnemum/Salicornia sp.

Two halophytic genera looking very much alike and difficult to distinguish without the inflorescence and close examination.

Native:
Vernacular name: *Balbal* (?), *R'dem* (?), *Hamd* (?)
Raunkiaer's life-form: Chamaephyte – Hyperhalophyte
Description and morphology: Annual or perennial, fleshy, succulent, ascending articulate leafless stems (sometimes prostrate then ascending), inflorescences with three flowers

- *Arthrocnemum* sp. (*Arthrocnemum indicum* (Willd.) Moq.), http://www.theplantlist.org/tpl1.1/record/kew-2653800, generally perennial, light green turning yellowish at flowering
- *Salicornia* sp. (*Salicornia arabica* L. (?)), http://www.theplantlist.org/tpl1.1/record/kew-2480738, perennial (rooting at stems' nodes) or annuals, often reddish stems, seeds vertical

Reproduction: Flowering March–September, seeds
Pastoral importance: Limited, however a need for dromedaries during winter when they come back from limestone desert pasture
Palatability and fodder value: Camel feed
Toxicity: Camels grazing on both genera (rich in sodium chloride and sulphate) must be provided with good-quality water; the milk from dromedaries grazing these pastures is salty and induces diarrhoea when consumed by humans.
Economic interest: Source of potash in ashes
Habitat: Same ecology for both, in *sebkha*, on heavy muddy waterlogged saline soils sometimes covered with sand veil, high saline water table, azonal
Distribution:

- In Libya, coastal zones and inland on salty flats (Al-Assah, Zuara, Taworga, wadi Nefad, Sirte, Ajdabiya, Daryanah, Tobruk)
- In arid zones, all saline flats all over North Africa, the Middle East to Pakistan

Endemism: NA

Photo 7.126. *Arthrocnemum* or *Salicornia* sp. (Jbebina, Al-Assah).

Photo 7.127. *Arthrocnemum* or *Salicornia* sp., (Jbebina, Al-Assah).

Photo 7.128. Stand of *Arthrocnemum* sp. and *Salicornia* sp. (west wadi Nefad, Seddada).

Bassia muricata (L.) Asch.

Synonym(s): *Salsola muricata* L. (Salsola muricata L. — The Plant List), or *Kochia muricata* (L.) Schrad. (Kochia muricata (L.) Schrad. — The Plant List), or *Chenolea muricata* (L.) Hutch. & Dalziel Chenolea muricata (L.) Hutch. & Dalziel — The Plant List
Native:
Vernacular name: *Wass* (?)
Raunkiaer's life-form: Chamaephyte or therophyte
Description and morphology: Herbaceous, annual, or biannual, hemispherical shrub-like (30–100 cm), intricate canescent/woolly – grey ascending stems, young stems reddish, alternate linear lanceolate turgid hairy leaves, shy pentamerous woolly yellow flowers evolving in coriaceous fruits with a star-like perianth of five long yellow spines, discoid horizontal seed (1 mm diam.), dry bush rolling with the wind.
Reproduction: Flowering, fruiting February–October
Pastoral importance: Avidly grazed green or dry by dromedaries, hardly touched by small ruminants
Palatability and fodder value: Limited
Toxicity: None
Economic interest: None
Habitat: Mostly on shallow sandy soils and saline inland/coastal zones and *sebkha*
Distribution:

- In Libya, coastal Libya, Jeffara plain, Al-Assah to Derj, Sirte coast to Ajdabiya
- In arid zones, from Morocco to Arabia

Endemism: NA

Photo 7.129. Young *Bassia muricata* plant, south Erfoud, east Morocco (N 31.296654 W 04.147894). (Photo: C. Lemmel, Al Yasmina - revue botanique, *Bassia muricata* (L.) Asch. - eflora Maghreb)

Photo 7.130. Dry *Bassia muricata* bush, southern Tunisia (N 33.144 E 9.086). (Photo: C. Lemmel, Al Yasmina - revue botanique, *Bassia muricata* (L.) Asch. - eflora Maghreb).

Photo 7.131. *Bassia muricata* fruits with a star-like perianth of five long yellow spines, southern Tunisia Douz (N 33.3925 E 8.7427). (Photo: C. Lemmel, Al Yasmina - revue botanique, *Bassia muricata* (L.) Asch. - eflora Maghreb).

Cornulaca monacantha Del.

Synonym(s): http://www.theplantlist.org/tpl1.1/record/kew-2737867 = ? *Cornulaca arabica* Botsch. Cornulaca arabica Botsch. —The Plant List
Native:
Vernacular name: *L'had* (?)
Raunkiaer's life-form: Chamaephyte
Description and morphology: Small shrub (40–60 cm) deep-rooted in sand, with erect coriaceous branches, mature stems whitish with thin reticulated brown bark, short alternate leaves (5–10 mm) on short internodes, yellow spiny recurved trigone leaves woolly at base and with leathery white edges, shy flowers with two short-pointed bracteoles similar to leaves, ovoid utricle compressed containing vertical seed.
Reproduction: Flowering/fruiting September–November
Pastoral importance: Valuable dromedary feed
Palatability and fodder value: Green and dry when nothing else available, hardly touched by small ruminants
Toxicity: None
Economic interest: None
Habitat: On sandy Saharan areas, below 100–125 mm MAR
Distribution:

- In Libya, south Al-Assah, Hamada – south Ouassa, Derj – Sinaween, south Sirte/Bu N'gem – Bey-el-Kebir – Fezzan, deep south Jebel Al-Akhdar
- In arid zones, in desert, Mauretania to Iran

Endemism: NA

Photo 7.132. *Cornulaca monacantha* small bush, north Tchad (N 19.366 E 16.083). (Photo: C. Lemmel, Al Yasmina - revue botanique).

Photo 7.133. *Cornulaca monacantha* details of leaves and flowers, north Tchad (N 20.089 E 16.163). (Photo: C. Lemmel, Al Yasmina - revue botanique).

Halopeplis amplexicaulis (Vahl) Ung.-Sternb. ex Ces., Pass. & Gibelli

Synonym(s): http://www.theplantlist.org/tpl1.1/record/kew-2838568, or *Salicornia amplexicaulis* Vahl. http://www.theplantlist.org/tpl1.1/record/kew-2480741
Native:
Vernacular name:
Raunkiaer's life-form: Therophyte
Description and morphology: Small annual (5–10 cm), prostate, pivotal root, bluish all over, slender stems with succulent globulous amplexicaul glabrous alternate leaf rudiments not showing internode, very small whitish flowers hidden in elongated spikes, seeds wigless.
Reproduction: Flowering/fruiting May–September
Pastoral importance: Valuable dromedary feed
Palatability and fodder value: Green and dry when nothing else available, hardly touched by small ruminants
Toxicity: None
Economic interest: None
Habitat: Azonal, in *sebkha*, on very saline mud in the most saline environment in dense populations

Distribution:

- In Libya, Al-Assah, all coastal *sebkha* from west to east
- In arid zones, Spain, North Africa to the Middle East – Turkey

Endemism: NA

Photo 7.134. *Halopeplis amplexicaulis*, on highly saline muds, Jbebina Al-Assah, 1978.

Noaea mucronata (Forssk.) Asch. & Schweinf.

Synonym(s): Noaea mucronata (Forssk.) Asch. & Schweinf. — The Plant List
Native:
Vernacular name:
Raunkiaer's life-form: Chamaephyte
Description and morphology: Small woody shrub (30–50 cm) spiny-thorny with angular-intricate grey branches, alternate linear mucronate leaves, long upper leaves, lower leaves imbricated, red to pink flowers, fruit perianth with wings arranged in crown, vertical lens-shaped 1.5 mm seed.
Reproduction: Flowering/fruiting August–September
Pastoral importance: Valuable in spring, highly resistant to overgrazing
Palatability and fodder value: In spring, young twigs are eagerly browsed by all livestock, poor in summer due to sharp thorns.
Toxicity: None
Economic interest: Collected for fuelwood
Habitat: On loamy-clay soils, 200–400 mm MAR, withstands cold winter
Distribution:

- In Libya, plateau Jebel Al-Akhdar (Taknis – Gendoula) – Marmarica, rare occurrence in Wishtata (south Tarhuna) (?)
- In arid zones, from Morocco to the Middle East – Iran

Endemism: NA

Photo 7.135. *Noaea mucronata* with red-winged flowers (Oushtata rangeland project, south Tarhuna, 1984).

Note: Doubtfully reported near Zuara on the coast, possible confusion with *Alhagi maurorum* Medik. Alhagi maurorum Medik. — The Plant List.

Nucularia perrinii Batt.

Synonym(s): http://www.theplantlist.org/tpl1.1/record/kew-2378242
Native:
Vernacular name: *S'kaf* (?)
Raunkiaer's life-form: Chamaephyte
Description and morphology: Small woody shrub (30–50 cm), skinny branches (greyish-white) and leaves opposite, fleshy trigone leaves oblong lanceolate and apiculate, fasciculate flowers by two with scarious bracts and bracteoles and small leaves, false-fruit accrescent/hardened fused with bracts and bracteoles with curved end, vertical seed spiralled.

Reproduction: Fast-flowering after rare rain
Pastoral importance: Resistant to overgrazing
Palatability and fodder value: Mostly dromedary feed, reported stimulating lactation, gives a specific taste to dromedary milk (Volpato and Nardo, 2017)
Toxicity: None
Economic interest: Fuelwood
Habitat: On fixed sand, rocky sites, sometimes on slightly saline soils, strictly Saharan, MAR 25-100 mm
Distribution:

- In Libya, south – Fezzan to Akakus, north-west Tibesti
- In arid zones, from Mauritania to south-western Libya, apparently not present in south-eastern Libya

Endemism: Central-western Sahara only

Note: Often in association and confused with *Traganum nudatum*.

Photo 7.136. Stem of *Nucularia perrini* – opposite leaves and fruits, north Tchad (N 19.9735 E 16.2203). (Photos: C. Lemmel, Al Yasmina - revue botanique)

Photo 7.137. *Nucularia perrini* – fruits, north-west Ubari, Libya (N 26.6885 E 11.9668). (Photos: C. Lemmel, Al Yasmina - revue botanique)

Kochia indica Wight

Synonym(s): http://www.theplantlist.org/tpl1.1/record/kew-2334144 = *Bassia indica* (Wight) A.J. Scott, http://www.theplantlist.org/tpl1.1/record/kew-2670806
Native: Exotic (?)
Vernacular name:
Raunkiaer's life-form: Tachytherophyte
Description and morphology: Large annual woolly/hairy bush (up to 1 m diam.), stem brownish-red, leaves short sessile and covered with long white hairs, idem for yellow flowers with hairy perianth, fruits hairy (with variation of short horizontal rounded wing).
Reproduction: Flowering during summer months, seeds
Pastoral importance: Limited
Palatability and fodder value: Well grazed by all livestock
Toxicity: None
Economic interest: Could be used for reseeding in *sebkha*, thrive when supplemented with irrigation even with salty water
Habitat: around *sebkha*, on saline (temporary waterlogged) sites, on sand veil and muddy soils
Distribution:

- In Libya, azonal, all coastal zones (also at Al-Assah – never seen inland (?))
- In arid zones, not listed in Morocco, Algeria and Tunisia (?) from Libya to Pakistan – India

Endemism: NA

Note: *Kochia* sp. could easily be confused with *Bassia muricata*; the latter always has the perianth of the fruits with long thin yellow spines. Need fruits to make the difference.

Photo 7.138. *Kochia indica* (Jbebina, Surbia, Al-Assah).

Salsola longifolia Forssk.

Synonym(s): http://www.theplantlist.org/tpl1.1/record/kew-2481358, synonym doubtful (?), *Salsola oppositifolia* Desf., http://www.theplantlist.org/tpl1.1/record/kew-2481286, douteux (?) *Salsola sieberi* J.Presl, http://www.theplantlist.org/tpl1.1/record/tro-50001780.
Native:
Vernacular name: *S'med* (?)
Raunkiaer's life-form: Chamaephyte
Description and morphology: Small shrub (30–60 cm) with densely entangled opposite branches, old greyish branches, large rooting systems, pubescent/villose stems and acuminate opposite sessile deflexed fleshy and furrowed leaves, stems growth reddish soft and pubescent, dense spikes of flowers with five accrescent tepals/perianth cream, shiny yellow membranous wings, seeds horizontal and coiled.
Reproduction: Flowering all year round, seeds dispersed by anemochory, germination from fresh seeds after autumn rains
Pastoral importance: High, browsed by all livestock all year long, highly grazing-resistant.
Palatability and fodder value: Excellent, withstands heavy grazing and will regrow easily.
Toxicity: None
Economic interest: Fuelwood
Habitat: Azonal (?), sandy/sandy-loam soils over saline-gypsum crust, close to *sebkha*, slightly saline sites
Distribution:

- In Libya, all coastal zones and inland (Al-Assah sebkha, coastal Sirte, to Tobruk –Solum)
- In arid zones, southern Spain, North Africa to the Middle East

Endemism: NA

Note: Could be confused with *Nucularia perrini* that has similar opposite leaves but different ecology as *Nucularia* is a true Saharan occurring on sandy soils, never on saline site.

Photo 7.139. *Salsola longifolia* (opposite leaves) with taproot exposed, *sebkha*, west of Bir Ougradia – Al-Assah).

Photo 7.140. *Salsola longifolia* in flower (Al-Assah, Jbebina).

Salsola tetragona Delile

Synonym(s): http://www.theplantlist.org/tpl1.1/record/kew-2484186
Native:
Vernacular name: *Fersh* (?), *Belbal* (?)
Raunkiaer's life-form: Chamaephyte
Description and morphology: Small shrub (20–50 cm), strong woody taproot, stem fully covered with small silvery turgescent sessile leaves, internode not visible, opposed leaves firmly imbricated compressed giving the stem a four-angled shape, margin of leaves with long multicellular straight hair, flowers with five membranous reddish-winged perianth.
Reproduction: Flowering in autumn, seeds dispersed by anemochory
Pastoral importance: Limited
Palatability and fodder value: Poor, dromedary feed
Toxicity: None
Economic interest: Fuelwood for desert dwellers
Habitat: Desertic, Saharan, loamy soils (rocky/shallow) down to 50 mm MAR and less in depressions deep into Sahara
Distribution:

- In Libya, most arid part of Jeffara plain, south Jebel Nafusa, Sirte, south Jebel Al-Akhdar to Tobruk – Solum
- In arid zones, Sahara North Africa to the Middle East

Endemism: NA

Photo 7.141. *Salsola tetragona*, east El-Hisha Bu-Gren.

Photo 7.142. *Salsola tetragona* heavily grazed, wadi Nefad, north Seddada.

Photo 7.143. *Salsola tetragona* in flower, Seddada – Nefad.

Salsola tetrandra Forssk.

Synonym(s): Salsola tetrandra Forssk. — The Plant List
Native:
Vernacular name: *Fers* (?)
Raunkiaer's life-form: Chamaephyte (Nanophanerophyte (?))
Description and morphology: Small shrub (20–50 cm), superficial root system, opposite yellowish branches, main branches with bare internode (1–3 cm) exposed, lateral stem fully covered with small globular turgescent sessile opposed leaves in glomerular clumps, margin of leaves with long multicellular undulated hairs, flowers tetramerous (rarely seen) with membranous winged cream-yellow perianth.
Reproduction: Flowering in autumn, winged seeds dispersed by anemochory
Pastoral importance: Limited
Palatability and fodder value: Poor, dromedary feed
Toxicity: None
Economic interest: None
Habitat: Similar distribution as *Salsola tetragona*, desertic, sandy-loam gravelly deep soils, down to 50 mm MAR and less in depressions deep into Sahara, sometimes on slightly saline soils (?)

Distribution:

- In Libya, inland south Misurata, wadi Sassu, Bu-Gren to Ajdabiya
- In arid zones, Sahara, North Africa to Arabian Peninsula

Endemism: NA

Other *Salsola* species in Libya:

Salsola kali L., http://www.theplantlist.org/tpl1.1/record/kew-2481347, strong annual halophyte weed developing into a spherical bush (up to 1 m diam.), leaves sessile linear succulent spinescent when fresh ending with sharp yellow spine when dry, reddish flowers and fruits winged, individual bush easily uprooted by wind by the end of season turning into a brown-black globe rolling long distances with wind, in saline sites and *sebkha* all along the coast and inland (often a large weed in oasis).

Photo 7.144. *Salsola tetrandra*, east El-Hisha, Bu-Gren.

Salsola vermiculata L.

Synonym(s): http://www.theplantlist.org/tpl1.1/record/kew-2484195 = *Salsola microphylla* Cav.
Native:
Vernacular name: *Rhadam, Gaddam* (?)
Raunkiaer's life-form: Chamaephyte – Nanophanerophyte

Description and morphology: Small shrub (30–100 cm) with densely entangled woody branches, large superficial rooting systems, pubescent/villose stems and leaves, leaves short sessile conglomerate/densely imbricated/granulose along whitish stems, new stems growth reddish soft and pubescent, dense spikes of flowers with five accrescent tepals/perianth cream, shiny yellow (also pinkish-reddish) membranous wings, seeds horizontal and coiled.
Reproduction: Flowering all year round, seeds dispersed by anemochory, short-lived seed (2–3 months), germination from fresh seeds after autumn rains.
Pastoral importance: High, browsed by all livestock all year long, highly grazing-resistant.
Palatability and fodder value: Excellent, withstands very heavy grazing and will regrow easily.
Toxicity: None
Economic interest: Possibly one of the best rangeland plants, widely used for fodder shrubs plantations in northern Syria either by reseeding in autumn with fresh collected seeds or transplanted from nursery-grown seedlings; when established, self-reseeding in plantations; *Salsola vermiculata* definitely merits better use in rangeland regeneration projects on appropriate sites.
Habitat: Sandy/sandy-loam soils over gypsum crust, or sometimes close to *sebkha*, slightly saline sites
Distribution:

- In Libya, all coastal zone and inland (Al-Assah to Bir-el-Ghanem, Sirte, to Tobruk–Solum), also in Jebel Nafusa, in Sirte

Photo 7.145. *Salsola vermiculata*, Jbebina, Al-Assah.

Photo 7.146. *Salsola vermiculata* flowers, Al-Assah, 1981.

- In arid zones, southern Spain, North Africa, Middle East (in Syria – Jordan = well-named *Malikat-el-Badia*)

Endemism: NA

Note: Very variable, but appears to have two common varieties with similar ecology:

- *Salsola vermiculata* var. *vermiculata*, most common along coast
- *Salsola vermiculata* var. *spinescens* (*Salsola spinescens Moq.* (?)), stems ending with spiny leafless tips, possibly in more arid sites (?)

Suaeda vermiculata Forssk. ex J.F. Gmel.

Synonym(s): *Suaeda mollis* (Desf.) Delile, http://www.theplantlist.org/tpl1.1/record/kew-2484011, also *Salsola mollis* Desf., http://www.theplantlist.org/tpl1.1/record/kew-2481262
Native:
Vernacular name: *Swida* (?)
Raunkiaer's life-form: Chamaephyte – halophyte
Description and morphology: Small bush (40–80 cm), woody taproot, stems glabrous and red, leaves alternate subsessile (or slightly petiolate) glaucous/succulent/fleshy/oblong – cylindri-

cal/round-ended turning red in very saline sites, leaves often covered with white speckles of salt, leaves turning black when dry, inconspicuous flowers developing into wingless fruits.
Reproduction: Flowering November–April, seeds
Pastoral importance: High
Palatability and fodder value: Well grazed by all livestock, but especially by dromedaries
Toxicity: None
Economic interest: None, used as fuelwood
Habitat: Loamy saline soils on margin of *sebkha*
Distribution:

- In Libya, all saline sites, azonal, along coast and inland
- In arid zones, North Africa, Middle East, Arabian Peninsula to Pakistan

Endemism: NA

Other Libyan *Suaeda* sp.:

- *Suaeda* sp. often confused with *Salsola* sp.; *Salsola* sp. has apparently an accrescent calyx with winged fruits contrary to *Suaeda* sp.; leaves of *Suaeda* are glaucous /succulent/fleshy, oblong and branches are often turning black when dry, while leaves of *Salsola* are long, furrowed often hairy and recurved. Expect a lot of confusion between *Salsola* and *Suaeda* when consulting the flora of the region.
- *Suaeda fruticosa* Forssk. ex-J.F.Gmel., http://www.theplantlist.org/tpl1.1/record/kew-2483984, alternate leaves or *Suaeda longifolia* K. Koch, http://www.theplantlist.org/tpl1.1/record/kew-2484025, which have elongated succulent curved/reflexed/deflexed leaves. Same ecology as *Suaeda vermiculata.*

Photo 7.147. *Suaeda vermiculata* (alternate leaves), at Jbebina (Al-Assah).

Traganum nudatum Delile

Synonym(s): http://www.theplantlist.org/tpl1.1/record/kew-2431778, *Traganum acuminatum* Maire & Weiller, http://www.theplantlist.org/tpl1.1/record/kew-2431603
Native:
Vernacular name: *Damran*
Raunkiaer's life-form: Chamaephyte – nanophanerophyte
Description and morphology: Loose woody woolly shrub (30–100 cm), strong taproot, intricate white-grey branches longitudinally cracked brown, stems woolly, internodes can be very short figuring conglomerate leaves and fruits, micro-alternate ovoid trigone sessile leaves succulent and apiculate (yellow spines recurved downward), small cotton ball at base of leaves embracing single sessile flower with five scarious tepals and two bracts, thick and hardened wingless fruits, horizontal seed.
Reproduction: Flowering January–April, seeds
Pastoral importance: High in most arid sites and Sahara
Palatability and fodder value: Camel and goat feed
Toxicity: None
Economic interest: Fuelwood.
Habitat: Saharan, rocky/gravelly-sandy loam soils, below 150 mm MAR, on *reg* and rocky flats, also in depressions, would withstand salinity and possibly a gypsum crust indicator.
Distribution:

- In Libya, desertic, south Al-Assah to Tiji-Nalut foothills, south Jebel Nafusa – Seddada, wadi Nefad, south Ben-Walid, on Hamada, Sirte inland, to south Jebel Al-Akhdar to Tobruk – Sollum
- In arid zones, Sahara, North Africa to Arabian Peninsula

Endemism: NA

Note: Could be confused with *Cornulaca monacantha* (perennial, vertical seed) or even *Bassia muricata* (annual, horizontal seed), both with yellowish spines and occurring more on deep sandy soils, or with *Nucularia perrini* (opposite leaves and occurs on rocky soils deep into the southern desert – Fezzan to Ghat).

Photo 7.148. *Traganum nudatum*, bush, on plateau near wadi Zorzor (N 31.59 E 14.86), wadi Zemzem.

Photo 7.149. *Traganum nudatum*, stems and leaves, on sandy plateau near wadi Zorzor, wadi Zemzem.

Photo 7.150. *Traganum nudatum* woolly stem with alternate ovoid trigone sessile leaves succulent and apiculate, south Morocco (N 21.4544 W 16.9506). (Photo: C. Lemmel, Al Yasmina - revue botanique, *Traganum nudatum Delile* - eflora Maghreb)

CISTACEAE

Cistus salviifolius L.

Synonym(s): http://www.theplantlist.org/tpl1.1/record/kew-2723633
Native:
Vernacular name: *Berbesh* (?), *Tarrash* (?)
Raunkiaer's life-form: Chamaephyte
Description and morphology: Dense, leafy shrub (up to 1–1.5 m), much branched, elliptical undulate/wrinkled/embossed round base opposite short-petiolate leaves upper tomentose with lower face whitish covered with tiny stellate hairs, inflorescence in terminal solitary or in cyme (2–3 pedicellate flowers), flowers (2–3 cm) white with pink dot at base of fresh petal, long-petiolate fruits – pentagonal capsules woody persistent solid adpressed villous to hairy surrounded by mucronate woody sepals, small black polygonal triangular seeds.
Reproduction: Flowering February–May, seeds
Pastoral importance: None
Palatability and fodder value: None
Toxicity: None
Economic interest: None, fuel wood
Habitat: Common in matorral, rocky sites, limestone, karstic sites, on *Terra rossa*, heavy soils, above 300–400 mm MAR
Distribution:

- In Libya, mostly in Jebel Al-Akhdar, also occurrences in the west (Lebda, Khoms, Msallata, Tarhuna)
- In arid zones, south-eastern Europe, North Africa, west Asia, Crete, Cyprus, Greece, southern Turkey, Syria

Endemism: NA

Other *Cistus* sp., large variation in leaves shape, flower colour on all *Cistus* sp.:

- *Cistus parviflorus* Lam., http://www.theplantlist.org/tpl1.1/record/kew-2723561, common, small bush, leaves sub-sessile to sessile opposite covered with sparse straight hairs, small pink (or whitish with pink tinges) flowers (15–20 mm), shedding petals fast, on matorral/limestone/karst country in Jebel Al-Akhdar, also on hills near Khoms/Lebda.
- *Cistus incanus* Auct., http://www.theplantlist.org/tpl1.1/record/tro-7600221, large purple-reddish flowers (3–5 cm diam.), mostly in high-rainfall zones Jebel Al-Akhdar (Shahat, Marsa Suza), less common.

Photo 7.151. *Cistus salviifolius*, plant, Shahat, 2009.

Photo 7.152. *Cistus salviifolius*, dry persistent hairy capsules, Shahat, Marsa Susa.

Photo 7.153. *Cistus parviflorus*, pinkish flowers, Marsa Suza.

Helianthemum cinereum (Cav.) Pers.

Synonym(s): http://www.theplantlist.org/tpl1.1/record/kew-2842348
Native:
Vernacular name: *Jef-Jef* (?)
Raunkiaer's life-form: Chamaephyte
Description and morphology: Hemispherical woody shrublet (15–30 cm) abundantly branching from base, dark green, reddish stems, dense erect intricate leafy branches, leaves stipulates subsessile obovate–orbicular upper face glabrous and lower face tomentose, inflorescences in terminal branched spikes above leafy branches, flower buds woolly with reddish calyx, bright yellow pendulous flowers (8–12 mm diam.), small brown capsule shorter than calyx.
Reproduction: Flowering February–May, seeds
Pastoral importance: Medium
Palatability and fodder value: Grazed by all livestock
Toxicity: None
Economic Interest: None
Habitat: Stony/gravelly – sandy-loam soils, associated with degraded *Stipa tenacissima* association, down to 200 mm MAR

Distribution:

- In Libya: rare occurrence, on rocky hills, in Jebel Nafusa, in protected wadi rocky sites near Beni-Walid, also reported in Jebel Al-Akhdar (?)
- In arid zones, Spain, North Africa, not in Egypt and further east (?)

Endemism: NA

Photo 7.154. *Helianthemum cinereum*, Beni-Walid, 1978.

Helianthemum hirtum (L.) Mill.

Synonym(s): http://www.theplantlist.org/tpl1.1/record/kew-2842494, or *Cistus hirtus* L., http://www.theplantlist.org/tpl1.1/record/kew-2723434
Native:
Vernacular name: *Zef-Zef* (?)
Raunkiaer's life-form: Chamaephyte
Description and morphology: Hemispherical shrublet (15–30 cm), yellowish, pubescent-pilose to hairy-sticky, dense erect leafy branches from base, stipulate leaves subsessile oblong–orbicular, decumbent inflorescences in terminal spikes above leafy branches, flower buds with calyx densely woolly-hispid, bright yellow pedicellate (pendulous) flowers (10–15 mm diam.), small brown capsule not longer than yellowish calyx.
Reproduction: Flowering February–May, seeds
Pastoral importance: Medium
Palatability and fodder value: Grazed by all livestock

Toxicity: None
Economic interest: None
Habitat: Stony/gravelly – sandy-loam soils, often associated with degraded *Stipa tenacissima* association, down to 150 mm MAR
Distribution:

- In Libya, rocky hills of the Jeffara plain, Jebel Nafusa, northern Hamada, Sirte to Solum (?)
- In arid zones, south-west Europe, North Africa, Egypt (?)

Endemism: NA

Photo 7.155. *Helianthemum hirtum*, plant (Gdama, Gharyan).

Photo 7.156. *Helianthemum hirtum*, flowers (Gdama, Gharyan).

Helianthemum kahiricum Delile

Synonym(s): http://www.theplantlist.org/tpl1.1/record/kew-2842522, or *Cistus kahiricus* Steud. Cistus kahiricus Steud. — The Plant List
Native:
Vernacular name: *Zef-Zef* (?)
Raunkiaer's life-form: Chamaephyte
Description and morphology: Robust woody shrublet (20–30 cm) often on a solid woody stump, very similar to *Helianthemum hirtum*, but inflorescence spikes straight, leaves linear-oblong with revolute margin, pedicellate flower buds reddish or with reddish-purple ribs on helicoidal (more or less villose) calyx, smaller yellow flowers (8–10 mm diam.).
Reproduction: Flowering February–May, seeds
Pastoral importance: Medium
Palatability and fodder value: Grazed by all livestock
Toxicity: None
Economic interest: None, often collected as fuelwood due to its strong woody stump.
Habitat: In more arid sites, stony soils, often in crevices on calcareous crust, sandy-loam soils, often associated with degraded *Stipa tenacissima*, down to 100 mm MAR.
Distribution:

- In Libya, foothills of Jeffara plain, Jebel Nafusa, northern Hamada, Sirte to Solum
- In arid zones, south-west Europe, North Africa, Egypt, Middle East, Arabian Peninsula

Endemism: NA

Photo 7.157. *Helianthemum kahiricum*, plant, foothills Cabao.

Photo 7.158. *Helianthemum kahiricum*, flower and red-veined buds, foothills Cabao.

Helianthemum lippii (L.) Dum. Cours.

Synonym(s): http://www.theplantlist.org/tpl1.1/record/kew-2842562, or *Cistus lippii* L., http://www.theplantlist.org/tpl1.1/record/kew-2723495
Native:
Vernacular name: *Erga* (?), *R'gig* (?)
Raunkiaer's life-form: Chamaephyte
Description and morphology: Gracile shrublet (20–40 cm high), fine rooting system, slender lax stems, elliptic to linear subsessile stipulate leaves, lax terminal inflorescence with a string (3–5) of small yellow sessile flowers widely space on spike, fruits villose longer that calyx.
Reproduction: Flowering February–May, seeds
Pastoral importance: Medium
Palatability and fodder value: Grazed by all livestock
Toxicity: None
Economic interest: Rooting system often parasitized with edible underground tuberous fungi *Terfes* or *Trefez* (*Terfezia* sp. and *Tirmania* sp.)
Habitat: Deep fixed sand/sandy-loam soils, down to 100 mm MAR
Distribution:

- In Libya, Jeffara plain, Hamada, Sirte to Jebel Al-Akhdar – Marmarica
- In arid zones, south Europe, North Africa, Arabian Peninsula to Pakistan

Endemism: NA

Notes:
Two common var. often both present on sites:

- *Helianthemum lippii* var. *lippii*, erect, end of branches often spinescent
- *Helianthemum lippii* var. *sessiliflorum*, procumbent, end of branches not spinescent

Helianthemum sp., especially *H. lippii*, have a mycorrhizal association (Sheibani and Jamali, 2023) with *Terfes*, an hypogeous edible fungi commonly collected on sandy/sandy-loam rangelands on the Hamada, Jeffara plain and Sirte.

Photo 7.159. *Helianthemum lippii* var. *lippii*, south Al-Assah, Touil-er-Reschiada (N 32.712 E 11.652).

Photo 7.160. *Helianthemum lippii* var. *sessiliflorum*, Girzah, Hamada.

CONVOLVULACEAE

Convolvulus althaeoides L.

Synonym(s): http://www.theplantlist.org/tpl1.1/record/tro-8501085, no synonyms
Native:
Vernacular name: *Tuila* (?)
Raunkiaer's life-form: Hemicryptophyte
Description and morphology: Perennial rhizomatous, long (up to 2 m), voluble climber or prostrate, thin pubescent stems, pubescent petioled alternate leaves deeply dissected with linear lobes, flower buds with helicoidal corolla, flowers (1–2 per inflorescence) long pedicelled pink campanulate with centrifugal white strips, ovoid capsule in persistent calyx, oval-trigone dark brown seeds.
Reproduction: Flowering February–July, seeds and rhizomes
Pastoral importance: Limited
Palatability and fodder value: Grazed by all livestock
Toxicity: None
Economic interest: None
Habitat: Deep sandy-loam soils, climbing inside bushes or creeping on bare soils, often a weed in rainfed crops or fallows, down to 150 mm MAR.
Distribution:

- In Libya, all coastal zones from west to east
- In arid zones, Mediterranean, southern Europe, North Africa to Middle East, west Asia

Endemism: NA

Photo 7.161. *Convolvulus althaeoides*, climber in spiny *Ziziphus lotus* (wadi Sassu N 32.149 E 14.960).

Photo 7.162. *Convolvulus althaeoides*, leaves, details (wadi Sassu).

Convolvulus dorycnium L.

Synonym(s): Unresolved, http://www.theplantlist.org/tpl1.1/record/tro-8502187, however accepted in https://www.ville-ge.ch/musinfo/bd/cjb/africa/details.php?langue=an&id=139794
Native:
Vernacular name:
Raunkiaer's life-form: Chamaephyte
Description and morphology: Perennial (30–100 cm) hemispherical shrub, pubescent all over, intricately divaricated thin annual branches with few linear – spathulate sessile leaves, abundantly flowering, funnel-like corolla white-pinkish with internal pink rays, globose capsule protruding half-length from adpressed calyx and containing only one small (5–6 mm diam.) seed.
Reproduction: Flowering February–July, seeds
Pastoral importance: Limited
Palatability and fodder value: Poorly grazed (?)
Toxicity: None
Economic interest: Fuelwood, charcoal, garden
Habitat: Rocky slopes and sites, loamy to heavy soils, above 300 mm MAR

Distribution:

- In Libya, Jebel Nafusa (top and foothills – rare in Jeffara plain), not in Sirte (?), northern slope to top of Jebel Al-Akhdar
- In arid zones, south-west Europe, Tunisia to Egypt

Endemism: NA

Photo 7.163. *Convolvulus dorycnium*, bush (Gharyan).

Photo 7.164. *Convolvulus dorycnium*, flowers and stems (Gharyan).

Convolvulus supinus Coss. & Kralik

Synonym(s): http://www.theplantlist.org/tpl1.1/record/ifn-61357
Native:
Vernacular name: *Um Magrun* (?), *Ullak* (?)
Raunkiaer's life-form: Hemicryptophyte
Description and morphology: Small (10–40 cm spreading) perennial villous/hairy (all over) herb, sand sticking all over, creeping/decumbent, petioled alternate entire elliptical leaves, flowers buds villous, white campanulate corolla, long pedicelled axillary flower and erect above leaves and at end of creeping stems, flowers (1–2 per inflorescence), ovoid capsule (containing 3–4 seeds) protruding from persistent calyx with horizontal pointed sepals, oval-trigone dark tuberculate black ripe seeds (2–3 mm).
Reproduction: Flowering February–July, seeds
Pastoral importance: Limited
Palatability and fodder value: Grazed by all livestock
Toxicity: None
Economic interest: None
Habitat: Saharan, on deep soils (fixed sand, sandy-loam soils to heavy soils, also on rocky sites), down to 50 mm MAR
Distribution:

- In Libya, all coastal zones and inland from west (Jebel Nafusa, Jeffara plain, Hamada), Sirte to east
- In arid zones, Mediterranean, southern Europe, North Africa to Middle East, west Asia

Endemism: NA

Photo 7.165. *Convolvulus supinus*, plant (Al-Assah, Jbebina).

Photo 7.166. *Convolvulus supinus*, flowers, (Al-Assah, Jbebina).

CUCURBITACEAE

Citrullus colocynthis (L.) Schrad.

Synonym(s): http://www.theplantlist.org/tpl1.1/record/kew-2723897, or *Cucumis colocynthis* L., http://www.theplantlist.org/tpl1.1/record/kew-2746924, or *Colocynthis vulgaris* Schrad., http://www.theplantlist.org/tpl1.1/record/kew-2731336
Native:
Vernacular name: *Handal* (?), *Hanzal* (?)
Raunkiaer's life-form: Stenotherophyte
Description and morphology: Fleshy hairy annual, strong taproot, long leafy stems radiating and prostrate – several metres long, rough hairy alternate leaves embossed triangular dissected (with distinct ribs) barely dented with rippled margins, short-whirled tendrils, yellow flowers (1–2 cm diam.), yellow (green-veined) mature fruit smooth spherical (40–70 mm diam.) on short pedicels, dry fruit shrivelling becoming papery freeing seeds when crushed.
Reproduction: Flowering all year long, seeds
Pastoral importance: Dry papery fruits eaten by all livestock and gazelles, dispersing seeds
Palatability and fodder value: Poor, green plant not grazed
Toxicity: Strong purgative, fruits having very bitter taste
Economic interest: Medicinal, reported to treat pulmonary ailments by rubbing repeatedly fresh fruits cut in half on foot until the bitter taste reaches the mouth.
Habitat: On sandy soils (even on dunes), in wadis, weed, ruderal, ruined sites, all along coastal zone, fixed and mobile sand to loam, 50–500 mm MAR.

Photo 7.167. *Citrullus colocynthis*, flower and leaves (Al-Assah).

Distribution:

- In Libya, all over and deep into desert
- In arid zones, Mediterranean Europe to west Asia, North Africa, to India – Pakistan

Endemism: NA

Photo 7.168. Fruit of *Citrullus colocynthis*, south-west Algeria, courtesy of Claude Lemmel (http://alyasmina.org/).

Ecballium elaterium (L.) A.Rich.

Synonym(s): http://www.theplantlist.org/tpl1.1/record/kew-2780660, or *Momordica elaterium* L., http://www.theplantlist.org/tpl1.1/record/kew-2373004, or *Ecballium purgans* Schrad., http://www.theplantlist.org/tpl1.1/record/kew-2780666
Native:
Vernacular name: *Beit-el-Rhoul* (?)
Raunkiaer's life-form: Hemicryptophyte
Description and morphology: Fleshy hirsute monoecious perennial with long leafy stems radiating and prostrate, leaves (erect long petioled) embossed cordate – trilobe dented with rippled margins, underneath leaves tomentose – white, yellow flowers (2–3 cm diam.), fruit ovoid (3–4 cm long) hanging on long stalks at maturity, when touching mature fruit detaching from stalk and violently expelling gluey juice and sticky seeds from base, 20–30 brown seeds per fruit.
Reproduction: Flowering all year long, seeds violently dispersed within 50–100 cm.
Pastoral importance: None, fruits squirting usually frighten livestock
Palatability and fodder value: None
Toxicity: Likely, fruit juice irritating, may cause dermatitis (?) inducing abortion and depression (?)

Economic interest: Medicinal, fruit juice reported purgative
Habitat: Weed, ruderal, ruined sites, all along coastal zone, fixed sand to loam, 300–500 mm MAR

Distribution:

- In Libya, all over coastal zones
- In arid zones, Mediterranean Europe to west Asia, North Africa

Endemism: NA

Photo 7.169. *Ecballium elaterium*, flower with young fruits (Gharyan).

Photo 7.170. *Ecballium elaterium*, mature fruit on long stalk (Gharyan).

CUPRESSACEAE

Cupressus sempervirens L. var. *sempervirens*

Synonym(s): http://www.theplantlist.org/tpl1.1/record/kew-2748781
Native:
Vernacular name: *Arz* (?)
Raunkiaer's life-form: Mesophanerophyte
Description and morphology: A Libyan emblematic tall evergreen tree (up to 25–30 m high in wadi El-Kuf), solid base and trunk (up to 1.30–1.50 m diam. at 1.5 m high), wood scented, bark laciniate, short branches giving a nearly cylindrical overall tree structure, dark green leaves as small adpressed decussate scales on four ranks, cones (ripening over two years) sessile globose shiny/waxy clear brown with 6–12 woody scales adpressed in a polygonal fruit (15–20 mm diam.), angular brown shiny seeds barely winged.
Reproduction: Flowering, seeds (anemochory, bird dispersal (?))
Pastoral importance: High in Jebel Akhdar
Palatability and fodder value: None
Toxicity: None
Economic interest: Afforestation, ornamental, sometimes used in rare old forestry plantations and hedges (Beida, Shahat, Derna (?)); long ago – fuelwood, building material, medicinal: ground dry seeds in powder used to heal circumcision (?)
Habitat: Natural stands strictly in deep gorges in Jebel Al-Akhdar, well-watered site with deep soils

Distribution:

- In Libya, native specimens nearly exclusively in wadi El-Kuf (Dec. 2010)
- In arid zones, North Africa, Middle East, Türkiye, Iran

Endemism: Only in Jebel Al-Akhdar
Status: Threatened, natural regeneration problematic as seed germination challenging, all Libyan *Cupressus sempervirens* must be strictly protected.

Notes:

- *Cupressus sempervirens* L. var. *horizontalis* (Mill.) Loudon, http://www.theplantlist.org/tpl1.1/record/kew-2748784, close to *Cupressus sempervirens* L. var. *sempervirens*, however growing with longer horizontal branches displaying a shorter tree; much used as hedge trees at Beida planted 1930–1950.
- Another *Cupressus* sp. worth mentioning is *Cupressus dupreziana* A. Camus, http://www.theplantlist.org/tpl1.1/record/kew-2748641, a tall tree (up to 10–15 m), very long-lived (over 1000 years), a Saharan relict tree occurring only in deep south in central Sahara (Tassili n'Addjer/Tamrit – between Djanet (southern Algeria) and Ghat (southern Libya) on high altitude (1000–1800 m). Most of the few remaining trees have been decimated by fuelwood collectors and used as building material in Djanet and Ghat. Maybe still some specimens remain around Ghat in the Akakus ridge on the Libyan side.

Photo 7.171. *Cupressus sempervirens* var. *sempervirens*, about 25–30 m high (old road along wadi Kuf N).

Photo 7.172. *Cupressus sempervirens* var. *sempervirens*, fruits and leaves (wadi Kuf).

Photo 7.174. *Cupressus sempervirens* var. *sempervirens*, laciniate bark (wadi Kuf).

Photo 7.173. *Cupressus sempervirens* var. *sempervirens*, trunk is 1.35 m diam. at 1.50 m high (old road along wadi Kuf).

Photo 7.175. *Cupressus sempervirens* var. *horizontalis*, trees, Beida, likely planted.

Juniperus phoenicea L.

Synonym(s): http://www.theplantlist.org/tpl1.1/record/kew-2332707, or *Juniperus lycia* L., Juniperus lycia L. —The Plant List, or *Juniperus turbinata* Guss., https://powo.science.kew.org/taxon/urn:lsid:ipni.org:names:262382-1

> **Note:** A thorough review of the *Juniperus phoenicea* taxon has been published by Pavon *et al.* (2020). It appears that the name *Juniperus phoenicea* was wrongly attributed to this Libyan emblematic *Juniperus*. Now renamed *Juniperus phoenicea* L. subsp. *turbinata* (Guss.) Nyman [1881, Consp. Fl. Eur., 676] or *Juniperus turbinata* Guss. However, for simplicity, we keep naming it *Juniperus phoenicea* according to the *Flora of Libya* (El-Gadi and Jafri, 1976) available to Libyan scientists.

Native:
Vernacular name: *Ar'har* (?)
Raunkiaer's life-form: Micro-nanophanerophyte
Description and morphology: Bush or more often tree (up to 10–12 m high), long-lived (possibly several centuries for largest extremely rare specimen), irregular tormented foliage to hemispheric crown, mostly monoecious (but some are dioecious), strong trunk (unique or multiple) with laciniate grey-dark brown bark, wood fragrant, leaves are in two forms: (i) young leaves (on regrowth after fire or young seedling) are spiny – needle-like – with two white strips on upper surface; (ii) mature leaves are verticillate (3), small (1–3 mm), scale-like adpressed on six ranks, cones globose (8–15 mm diam.), somewhat polygonal wax-coated, turning brown when ripe, small wingless black seeds.
Reproduction: Flowering March–June, seeds (dispersed by birds)
Pastoral importance: High – iconic Libyan forest tree
Palatability and fodder value: Poor
Toxicity: None. However, ingested leave reported as abortive (?)
Economic interest: Fuelwood, used for making charcoal (poor quality – hardwood), resistant to insect borers, building material (I have seen pillars (25 cm diam.) and rafters in old houses in Jiado – Jebel Nafusa (1975), seasoned wood used for home manufacturing/carving small items (spoons, *khôl* applicator, toys, etc.); bark and leaves to tan leather. Medicinal: leaves reported to reduce headache, antidiarrheal, wood distillate use to be a source of tar (*gatran*), an antiseptic to treat water gourds (goats' skin *guerba*), to cure skin disease on dromedaries (*Jerbaa* due to *Sarcoptes scabiei* and *Trichophyton* sp.) and humans (ringworm = dermatophytosis), also used as antiseptic on wounds (domestic animal and human); tobacco substitute in Matmata and Jebel Nafusa dry leaves of *Juniperus* (from southern Tunisia, high plateaux in Algeria) mixed with leaves of *Arthrophytum scoparium* (*remeth*) used as snuff tobacco (*chemma*) or smoked alone or mixed with *hashish* (?) in small clay pipe (*sebsi*).
Habitat: Mostly inland, in garrigue/matorral (scrubland on limestone), on *Terra rossa*, on brown steppic soils, down to 175–200 mm MAR
Distribution:

- In Libya, Jebel Al-Akhdar only – extinct in Jebel Nafusa but possible rare specimen on old forestry plantations (around Shiarchara, Cussabat (?))
- In arid zones, southern Europe, North Africa, west Asia, Middle East

Endemism: Currently only in Jebel Al-Akhdar

Status: Abusively axed and collected in large quantities as fuelwood and charcoal manufacturing in Jebel Al-Akhdar

Notes:

- *Juniperus phoenicea* (inland *Juniperus*) must not be confused with *Juniperus oxycedrus* ssp. *macrocarpa* Sm., http://www.theplantlist.org/tpl1.1/record/kew-2332684, present on the coastal line in western Libya (Tajura, Sidi Ben Nur, prostrate (1–2 m high, several metres diam.) shrubs, exposed to sea spray on cliffs) and coast of eastern Libya, with always mature spiny leaves (by 3) spaced along branches, upper face of leaf has two white stomatic strips both sides of midrib vein, large cones (up to 20 mm diam.). This *Juniperus oxycedrus macrocarpa* Sm. is different from the *Juniperus communis* L. subsp. *communis*, http://www.theplantlist.org/tpl1.1/record/tro-9400052, a small tree or scrub also with spiny leaves (but with only one white strip on upper face of leaf – no visible midrib, and blue-black fruits) most common on Algerian mountains and high plateaux (Aures, Djurjura to Djelfa); does not appear to occur in Libya.

- *Juniperus phoenicea* dense scrublands on high land from El-Marj to Shahat and all top of Jebel Al-Akhdar are receiving the highest rainfall in Libya; fog is common occurrence during winter between the Shahat escarpment and Qaryat-el-Faidiyah, hence branches of *Juniperus phoenicea* are commonly heavily infested with two lichens (*Teloschistes villosus* and *Ramalina* sp.). These lichens do not develop on *Juniperus phoenicea* in dryer southern slope of Jebel Al-Akhdar towards Martuba-Gendula
- Prior to Libyan oil era, fuelwood from *Juniperus phoenicea* was extensively collected to supply families and bakeries; most was eradicated from low land and depressions turned into ever-expanding cropping land for rainfed barley.

Photo 7.176. *Juniperus phoenicea* (hills, El-Marj).

Photo 7.177. *Juniperus phoenicea*, old tree, hills, El-Marj, *c.*350 mm MAR.

Photo 7.178. *Juniperus phoenicea*, old tree, trunk 1.3 m diam., El-Marj hills, *c.*350 mm MAR).

Photo 7.179. Remnant *Juniperus phoenicea* open forest on hilltop, 16 km south-east Marawa, El-Mzerah, *c.*250 mm MAR and rainfed barley.

Photo 7.180. The last *Juniperus phoenicea* tree, south Mafsel, halfway between Gendula and Derwa, *c.*175–200 mm MAR, most southern limit of *Juniperus*'s extension (N 32.424 E 21.560).

Photo 7.181. *Juniperus phoenicea*, young leaves (*left*) and mature leaves (*right*) on same branch, Gumerah.

Photo 7.182. *Juniperus phoenicea*, mature leaves and ripening fruits, Gumerah.

Photo 7.183. *Juniperus phoenicea*, mature leaves and ripe fruits, Gumerah.

Photo 7.184. *Juniperus phoenicea* heavily infested with lichens, Shahat, *c*.550 mm MAR.

Photo 7.185. *Teloschistes villosus*, lichen on *Juniperus phoenicea*, Shahat, *c*.550 mm MAR.

Photo 7.186. *Ramalina* sp., lichen on *Juniperus phoenicea*, Shahat, *c*.550 MAR.

DIPSACEAE

Scabiosa arenaria Forssk.

Synonym(s): http://www.theplantlist.org/tpl1.1/record/kew-2593591, or *Sixalix arenaria* (Forssk.) Greuter & Burdet, https://powo.science.kew.org/taxon/urn:lsid:ipni.org:names:319708-1, or *Scabiosa fenestrata* Pomel, http://www.worldfloraonline.org/taxon/wfo-0000499996, or *Asterocephalus arenarius* (Forssk.) Vis., https://powo.science.kew.org/taxon/urn:lsid:ipni.org:names:318971-1
Native:
Vernacular name:
Raunkiaer's life-form: Mesotherophyte
Description and morphology: Annual (15–30 cm high), pivotal root, all basal deeply dissected linear leaves, terminal capitula (1.5–3 cm diam.) of numerous asymmetrical five-lip flowers on long peduncle with few dissected leaves, external flowers pale blue to pinkish, white central flowers, capitula has free pointed bracts, fruits remain on globulose capitula, each fruit with broad rotate papery cupula and five central fine star-like soft rays.
Reproduction: Flowering March–May, seeds, anemochory
Pastoral importance: High
Palatability and fodder value: Well grazed by all livestock
Toxicity: None
Economic interest: None
Habitat: Mostly on sandy soils
Distribution:

- In Libya, common all over northern Libya down to 100 mm MAR
- In arid zones, North Africa to the southern Middle East

Endemism: NA

Photo 7.187. *Scabiosa arenaria* (wadi Abu-Shayba, BGII, 1979).

EPHEDRACEAE

Ephedra alata Decne.

Synonym (s): http://www.theplantlist.org/tpl1.1/record/kew-332847
Native:
Vernacular name: *Alenda*
Raunkiaer's life-form: Chamaephyte
Description and morphology: Small (30–100 cm high) nearly leafless shrub, woody stump white bark, erect rigid pale green-yellowish branches, scabrous articulate stems, tiny leaves (deciduous), dioecious bush, flower cones subsessile axillary compacted along stems in glomerule with showy yellow-winged bracts, fruit as small red berries, very small seeds ovoid and pointed.
Reproduction: Flowering January–May, seeds (and stolons (?))
Pastoral importance: Medium
Palatability and fodder value: Grazed by all livestock, mostly dromedaries and gazelles
Toxicity: None
Economic interest: Fuelwood, medicinal: stems contain Ephedrine (vasoconstrictor) for the treatment of asthma and respiratory disorders; would raise blood pressure and cause insomnia, poisonous in high doses (?)
Habitat: Saharan, on gravel/sand, sand dunes, over calcareous crust, on rocky cliffs, down to 50 mm MAR
Distribution:

- In Libya, most arid parts of western Jeffara plain, Jebel Nafusa, Hamada (Derj, Sinaween)
- In arid zones, North Africa to Arabian Peninsula

Endemism: NA

Photo 7.188. *Ephedra alata* , bush, Sinaween, Derj.

Photo 7.189. *Ephedra alata*, grazed bush, Derj, south Nalut.

Photo 7.190. *Ephedra alata*: male flower (*left*); female flower (*right*). (Photo: Claude Lemmel, https://alyasmina.org/)

Photo 7.191. *Ephedra alata*, ungrazed bush in rocky site, Hamada – Girzah, south Beni-Walid.

Photo 7.192. *Ephedra alata*, grazed and regrowth, 30 km south of Taworga.

Ephedra altissima Desf.

Synonym(s): http://www.theplantlist.org/tpl1.1/record/kew-332855
Native:
Vernacular name: *Alenda*
Raunkiaer's life-form: Chamaephyte – climber
Description and morphology: Tall (2–3 m high), climber, dense growth with branches hanging inside other bushes, stems nearly leafless at swollen nodes, leaves as small scales opposite linear, diecious bush, pedunculate flowers and cones in terminal panicles, bracts not winged, berries reddish, seeds with micropillar appendices.
Reproduction: Flowering January–May, seeds
Pastoral importance: Medium
Palatability and fodder value: Low, hardly touched by livestock (?)
Toxicity: Poisonous (?)
Economic interest: Fuelwood, charcoal, medicinal: source of Ephedrine
Habitat: Matorral in Jebel Al-Akhdar, on *Terra rossa*, rocky sites in Jebel Nafusa, down to 200 mm MAR
Distribution:

- In Libya, mostly Jebel Al-Akhdar, some in Jebel Nafusa, not in Sirte (?)
- In arid zones, North Africa

Endemism: NA

Photo 7.193. *Ephedra altissima*, large, tall green bush in matorral, Toknis, Jebel Al-Akhdar.

ERICACEAE

Arbutus pavarii Pamp.

Synonym(s): *Arbutus unedo* L., Arbutus unedo L. — The Plant List (?)
Native:
Vernacular name: *Chmari*
Raunkiaer's life-form: Micro-phanerophyte
Description and morphology: Evergreen tree (2–6 m) or large shrub, uneven trunk and branches much distorted, bark reddish-brown peeling off, persistent shiny coriaceous petiolate blueish-green alternate leaves with white translucent veins and finely dented margin, pendulous inflorescence in raceme of white-cream flowers with urceolate corolla, pedicellate berries spherical (10–20 mm diam.) succulent overly verrucose, yellow when immature turning red mature.
Reproduction: Flowering January–March, fruiting ripe end-October–November, from seeds
Pastoral importance: Limited
Palatability and fodder value: Foliage not grazed, pods very palatable
Toxicity: None
Economic: Edible fruits when well ripened (otherwise astringent, granulose), very hard wood used for pickets (even pointed spears hardened in fire in old days for hunting), fuelwood, charcoal, root uses as a red dye and rich in tannins used for processing hides.
Habitat: On shallow calcareous soils with *Terra rossa*, karst sites, rocky slopes, 350–500 mm MAR
Distribution: (see De Santis *et al.*, 2023)

- In Libya, *Arbutus pavarii*, only on top of Jebel Al-Akhdar (Shahat, Marsa Suza)
- In arid zones, *Arbutus unedo*, recorded in Morocco, Algeria and Tunisia, southern Europe, Mediterranean predominantly on acidic soils

Endemism: Only in Jebel Al-Akhdar in Libya

Notes:

- The distinction between *Arbutus unedo* L. (exotic in Libya, in garden) and *Arbutus pavarii* Pamp. (endemic Jebel Al-Akhdar) is not clear. However, the bark on the trunk of *A. unedo* is dark grey, finely laciniate while it is reddish – smooth, desquamating in large pieces on *A. pavarii*. The leaves of *A. unedo* appear to be coarsely serrated while the leaves of *A. pavarii* are finely dented and with margins clearly translucent.
- *Arbutus pavarii* appears to be strictly east Mediterranean and looks like *Arbutus andrachne* L., https://flora.org.il/en/plants/arband/,http://www.lebanon-flora.org/species.php?id_pl=622 or http://www.theplantlist.org/tpl1.1/record/kew-2646651, from Palestine, Lebanon with identical desquamating red bark.

Photo 7.194. *Arbutus pavarii*, small tree with reddish-smooth bark, Shahat plateau, Jebel Al-Akhdar.

Photo 7.195. Reddish – smooth, desquamating bark on *Arbutus pavarii*, Shahat, Jebel Al-Akhdar.

Photo 7.196. *Arbutus pavarii*, translucent finely dented leaves and urceolate flowers, Shahat, Jebel Al-Akhdar.

Photo 7.197. *Arbutus pavarii*, fruits, Shahat, Jebel Al-Akhdar.

Photo 7.198. Grey laciniate bark and serrated leave of *Arbutus unedo*, author's garden in Western Australia.

Erica multiflora L.

Synonym(s): http://www.theplantlist.org/tpl1.1/record/kew-2793873 = *Erica peduncularis* C. Presl, Erica peduncularis C.Presl — The Plant List
Native:
Vernacular name: *Hamriyah* (?)
Raunkiaer's life-form: Chamaephyte
Description and morphology: Woody shrub with erect stiff white/grey long stems (up to 1–1.2 m), verticillate leaves in group of 2–5, leaves cylindrical revolute, inflorescences terminal verticillate with pendulous long-pedicelled full pink scented flowers with campanulate cylindrical corolla, violet anthers clearly protruding from corolla.
Reproduction: Flowering September–December, seeds
Pastoral importance: None
Palatability and fodder value: Barely touched, even by goats
Toxicity: None
Economic interest: Fuelwood, ornamental
Habitat: Karstic sites, limestone cliffs, on *Terra rossa*, above 300–350 mm MAR
Distribution:

- In Libya, Jebel Al-Akhdar only
- In arid zones, south Europe and Mediterranean

Endemism: Only in Jebel Al-Akhdar in Libya

Note: Doubtful presence in Jebel Al-Akhdar, also mentioned in *Flora of Libya* (El-Gadi and Jafri, 1976, p. 4), *Erica sicula* Guss., http://www.theplantlist.org/tpl1.1/record/kew-2794436, a dwarf hemispherical shrub (15–30 cm), pubescent, short-pedicelled campanulate pink flowers with stamens included, in terminal umbels. Present in Sicily, Libya, Cyprus, Anatolia and Lebanon (Pasta *et al.*, 2024).

Photo 7.199. *Erica multiflora*, bush on karst, between Shahat and Marsa Suza.

Photo 7.200. *Erica multiflora*, inflorescences, flowers with protruding stamens, between Shahat and Marsa Suza.

EUPHORBIACEAE

Euphorbia chamaepeplus Boiss. & Gaill.

Synonym: *Tithymalus chamaepeplus* (Boiss. & Gaill.) Soják, Tithymalus chamaepeplus (Boiss. & Gaill.) Soják — The Plant List
Native:
New species for Libya (?): Not in Keith (1965), not in *Flora of Libya: Euphorbiaceae*, Vol. 89, (El-Gadi and Jafri, 1976), but mentioned in *Student's Flora of Egypt* (Täckholm, 1974).
Vernacular name: *Leben zghir* (?)
Raunkiaer's life-form: Therophyte
Description and morphology: Small annual, slender taproot, mostly prostrate to semi-erect, zig-zagging stem, fleshy all over, glabrous-glaucous blue-green, nearly leafless small entire obovate alternate (triangular) leaves, some acuminate, cyathium with two large obovate cyathophyllum at node, long internodes, loose cyathium, umbels axillary or terminal with two rays, flowers sessile at node, milky sticky sap, tiny oblong seeds.
Reproduction: Flowering January–June, seeds
Pastoral importance: None
Palatability and fodder value: Not grazed by livestock
Toxicity: Poisonous (?), white sap irritant (?)
Economic interest: See note
Habitat: Common on fixed sandy flats, down to 150 mm MAR
Distribution:

- In Libya, from west to east
- In arid zones, southern Europe, North Africa, Middle East

Endemism: NA

Note:

- *Euphorbia* species are numerous in Libya from the coastal zone to deep inland into the desert, from very tiny ones (*Euphorbia chamaepeplus*) to larger ones (*Euphorbia dendroides, Euphorbia characias*); all usually not grazed by livestock.
- All *Euphorbia* sp. have white sticky sap with reported usage for early treatment of snake, scorpion or centipede bites.

Photo 7.201. *Euphorbia chamaepeplus*, Al-Assah, Magdar-es-Sbot.

Photo 7.202. *Euphorbia chamaepeplus*, flowers and stems, Al-Assah, Magdar-es-Sbot.

Euphorbia paralias L.

Synonym: http://www.theplantlist.org/tpl1.1/record/kew-81334
Native:
Vernacular name: *Leben* (?), *Guernit* (?)
Raunkiaer's life-form: Hemicryptophyte
Description and morphology: Perennial herb (20–60 cm high), strong taproot, many erect branches from woody base, glaucous glabrous plant, milky sticky sap, old stems turning red, small (10–25 mm long) fleshy elliptic entire leaves densely set along stems, compact umbel (Cyata) with 3–5 short rays and green flowers, capsule (5–10 mm) trilobed ovoid ribbed, small ovoid seeds.

Reproduction: Flowering January–June, seeds
Pastoral importance: None
Palatability and fodder value: Not grazed by livestock
Toxicity: Poisonous (?), white sap irritant (?)
Economic interest:
Habitat: Common on coastal sand, sandy flats and dunes, within short distance from seashore, down to 150 mm MAR

Photo 7.203. *Euphorbia paralias*, plant, coastal sand, near An-Nofilia, Sirte.

Photo 7.204. *Euphorbia paralias*, leaves and inflorescences, coastal sand, near An-Nofilia, Sirte.

Distribution:

- In Libya, coastal zones from west to east, rarely inland
- In arid zones, southern Europe, North Africa

Endemism: NA

Euphorbia retusa Forssk.

Synonym: Euphorbia retusa Forssk. — The Plant List, or *Euphorbia kahirensis* Raeusch., Euphorbia kahirensis Raeusch. —The Plant List, or *Euphorbia cornuta* Pers. Euphorbia cornuta Pers. —The Plant List
Native:
Vernacular name: *Lebena* (?)
Raunkiaer's life-form: Chamaephyte (?), Hemicryptophyte (?)
Description and morphology: Perennial glabrous herb (20–80 cm high), strong taproot, many erect intricate branches from woody base, fleshy lower leaves and bracts caulinar triangular deep furrowed with long-dented margins and with tip recurved, milky sticky sap, contracted inflorescences.
Reproduction: Flowering January–June, seeds
Pastoral importance: None
Palatability and fodder value: Not grazed by livestock
Toxicity: Likely poisonous (?), white sap irritant (?)
Economic interest:
Habitat: Saharan, on fixed sand, down to 100mm MAR
Distribution:

- In Libya, inland mostly, Nalut-Ghadames, Solul-el-Khoder, wadi Nefad, Zorzor, Seddada
- In arid zones, North Africa, Egypt to the Middle East

Endemism: NA

Note:

- Our *Euphorbia retusa* specimen from wadi Nefad is unusually solid and turgescent, with large, dented leaves and bracts, may be *species nova* (?) or a hybrid.
- Could be confused with *Euphorbia serrata* L., Euphorbia serrata L. — The Plant List, which is a smaller annual plant with narrow elongated serrated leaves, dichotomic terminal yellow-greenish inflorescences, common on the coastal zone in Libya, in crops, on road verges.

Photo 7.205. *Euphorbia retusa*, plant, wadi Nefad (N 31.591 E 14.716), Seddada.

Photo 7.206. *Euphorbia retusa* with protruding stamen and caulinar leaves and bracts, wadi Nefad – Seddada.

Ricinus communis L.

Synonym(s): http://www.theplantlist.org/tpl1.1/search?q=Ricinus+communis+
Exotic: Cosmopolitan
Vernacular name: *Kerouah* (?)
Raunkiaer's life-form: Nanophanerophyte
Description and morphology: Large shrub to small tree (up to 4–5 m high), extensive and aggressive root system, branches and stems glabrous (reddish-brown) pitted with small glands, petiolate palmate dark shiny green glabrous large leaves (20–30 cm diam.) with 6–9 pointed dentate lobes and often red-veined, inflorescences in long axillary and terminal pyramidal erect panicles with basal male flowers and upper female with red styles, fruits as spiny ovoid capsules (1.5–2 cm long) with three lodges containing each one carunculate smooth coated seed with black-and-white patched testa.
Reproduction: Flowering June–October, seeds, self-reseeding, invasive
Pastoral importance: None, considered as a hazard for human and livestock
Palatability and fodder value: Usually not touched by livestock.
Toxicity: Seeds ingested without testa are highly poisonous (even in small quantity, contain Ricin), however ingested full seeds with testa appears to be less toxic; sometimes used as purgative as crushed seed in hot water, all plant parts reported allergenic (may induce asthma), sap sticky irritant.
Economic interest: Oil extract from seed used as motor lubricant; plants once used as sand dunes binder along the western coast to Sirte area (not inland); many ornamental varieties with reddish foliage and fruits introduced in gardens (?).
Habitat: On wasteland, roadside, rubbish tips, garden hedges, wadi banks, down to 150 mm MAR, not on saline soils
Distribution:

- In Libya, ubiquitous mostly all over coastal zones
- In arid zones, southern Europe, North Africa, ubiquitous

Endemism: NA

Photo 7.207. *Ricinus communis*, inflorescence, Aradah, Tripoli.

Photo 7.208. Common *Ricinus communis*, leaves and fruits, Aradah, Tripoli.

Photo 7.209. *Ricinus communis*, spiny fruits, Aradah, Tripoli.

Photo 7.210. Ornamental *Ricinus communis* variety, red fruits, in garden, Sidi-Mesri, Tripoli.

FABACEAE

Anthyllis henoniana Coss.

Synonym(s): http://www.theplantlist.org/tpl1.1/record/ild-41159, or *Anthyllis subsimplex* Pomel http://www.worldfloraonline.org/taxon/wfo-0000194531
Native:
Vernacular name: *Rozdir, Hossir* (?)
Raunkiaer's life-form: Nanophanerophyte
Description and morphology: Small shrub (30–60 cm), strong woody taproot, dense branching, silky all over with adpressed short hair, leaves trifoliate (five (?)) with the terminal leaflet larger than the lateral, shy conspicuous axillary subsessile densely longhaired flowers (whitish-pinkish) in clumps of 3–8 all over the bush, pod with persistent hairy calyx with single seed.
Reproduction: Flowering March–May, seeds possibly by anemochory
Pastoral importance: Important
Palatability and fodder value: Well grazed by all rangeland livestock as summer–autumn feed
Toxicity: None
Economic interest: None, fuelwood
Habitat: Shallow loamy rocky soils, often on calcareous crust, down to 50 mm MAR, sub-Saharan
Distribution:

- In Libya, north of Nalut – Cabao (foothill Jebel), south Jebel Nafusa (Harabah) to Hamada, inland Sirte, Saharan
- In arid zones, southern Spain, North Africa to west Libya

Endemism: North Africa, Saharan

Other main native *Anthyllis* species in Libya:

- *Anthyllis vulneraria* L., http://www.theplantlist.org/tpl1.1/record/ild-7194, prostrate pubescent annual, leaves alternate imparipinnate with large terminal leaflet, flowers yellow-orange, red, turning brown, pods hairy and inflated, sandy to heavy soils, seed winged and beaked, all coastal zones (Regdalin, Zuara, Jebel Nafusa), rarely inland.
- *Anthyllis tetraphylla* L., http://www.theplantlist.org/tpl1.1/record/ild-5223, very similar and often confused with *Anthyllis vulneraria*, prostrate and densely woolly annual, leaves thick alternate, small yellow-reddish corolla barely protruding from large hairy and inflated calyx, pod (1–2 seeds), long-beaked remaining in calix and constricted between seeds, coastal zones on heavier soils, similar ecology to *A. vulneraria*.

Note: Both above 250–300 mm MAR.

Photo 7.211. *Anthyllis henoniana* bush (Cabao, Tiji).

Photo 7.212. Cluster of *Anthyllis henoniana* hairy flowers (Haraba, N 31.708 E 11.509), east Zintan.

Argyrolobium uniflorum (Decne.) Jaub. & Spach

Synonym(s): http://www.theplantlist.org/tpl1.1/record/ild-8309
Native:
Vernacular name: *Kheta* (?), *Ertha* (?)
Raunkiaer's life-form: Chamaephyte – xerophytic
Description and morphology: Small woody shrub (10–40 cm) sometimes prostrate, strong taprooting system, rigid centrifugal branches, leaves villous opposed petiolate oblong trifoliate (subequal) spaced on branches, flower solitary yellow, pods (20–30 mm long) linear flat hairy scarious twisted dehiscent with up to 10–12 ovoid yellowish small (2–3 mm) seeds.
Reproduction: Seeds
Pastoral importance: Forage all livestock, often overgrazed, extremely resistant to overgrazing and wind erosion
Palatability and fodder value: Excellent, very well grazed
Toxicity: None
Economic interest: Forage
Habitat: On loam/sandy-loam/coarse sand/gravelly-rocky, 300–100 mm MAR, not on saline soils
Distribution:

- In Libya, all coastal zones and inland
- In arid zones, North Africa to Middle East

Endemism: NA

Photo 7.213. *Argyrolobium uniflorum*, whole plant, Tiji, Giado, 1979.

Photo 7.214. *Argyrolobium uniflorum*, heavily grazed on eroded soil, Bir Ougradia – Al-Assah; note the yellow flower and pods.

Photo 7.215. *Argyrolobium uniflorum*, pods and leaves, Jbebina – Al-Assah.

Photo 7.216. Pod of *Argyrolobium uniflorum* and leaves.

Astragalus fruticosus Forssk.

Synonym(s): Astragalus fruticosus Forssk. —The Plant List or *Astragalus tomentosus* Lam., http://www.theplantlist.org/tpl1.1/record/ild-7577 or *Astragalus gombo* Bunge, http://www.theplantlist.org/tpl1.1/record/ild-7581
Native:
Vernacular name:
Raunkiaer's life-form: hemicryptophyte
Description and morphology: Perennial spineless herb, stems ascending pulverulent, long imparipinnate leaves with up to 15–25 pairs of leaflets folded obovate notched ended, axillary (sub)sessile yellow (up to 8) racemed flowers, ripe brown plump fluffy woody curved pendulous pods (30–40 x 10 mm) with prominent sutures and short strong beak, pods turning reddish when ripening, seeds yellow-brown reniform.
Reproduction: Flowering April–May, seeds
Pastoral importance: Limited
Palatability and fodder value: Poor
Toxicity: Possible as growing untouched on rangelands while other species around are overgrazed
Economic interest: None
Habitat: Sandy-gravelly soil, down to 100 mm MAR
Distribution:

- In Libya, most arid parts of the Jeffara plain, south Jebel Nafusa to northern Hamada
- In arid zones, Sahara, North Africa, Egypt, Arabian Peninsula (?)

Endemism: Saharan

Photo 7.217. *Astragalus fruticosus*, Cabao, May 1979.

Photo 7.218. *Astragalus fruticosus* with flowers and pods, Cabao, May 1979.

Photo 7.219. *Astragalus fruticosus* with ripening pods, Cabao, May 1979.

Astragalus spinosus (Forssk.) Muschl.

Synonym(s): http://www.theplantlist.org/tpl1.1/record/ild-7643
Native:
Vernacular name:
Raunkiaer's life-form: Chamaephyte
Description and morphology: Perennial, spiny low woody shrub, tomentose branches intricate, long recurved rather soft spines (as rachis of senescent leaves) arched and persistent, leaves with 5–6 pairs of oblong woolly leaflets, flower whitish-pink, pods sessile leathery short beaked embraced by inflated calyx post-flowering, pods dehiscent.

Reproduction: Seeds
Pastoral importance: Limited
Palatability and fodder value: Camel feed
Toxicity: None (?)
Economic interest: None
Habitat: Xerophyte, Gypsophyte, desertic, below 150 mm MAR
Distribution:

- In Libya, mostly occurring inland from Sirte (Wishkah, wadi Bey-el-Kebir, Bu-Ngem), Ajdabiya, and south Jebel Al-Akhdar (near Mekhili)
- In arid zones, Libya to Iran

Endemism: NA

Note: *A. spinosus* is different from *Astragalus armatus* Willd., Astragalus armatus Willd. — The Plant List, that is occurring on cold high plateaux in Tunisia and Algeria (Djelfa, Tadmit, Messad); *Astragalus armatus* has strong stems with rigid vulnerant long piercing white and straight spines (easily punching tyres from experience).

Photo 7.220. *Astragalus spinosus*, Bu-Ngem, Bey-el-Kebir, Wishkah – Bir-el-Gharouf (N 31.083 E 15.532).

Photo 7.221. *Astragalus spinosus* stems and spines, Bu-Ngem, Wishkah.

Astragalus tribuloides Delile

Synonym(s): http://www.theplantlist.org/tpl1.1/record/ild-7648
Native:
Vernacular name:
Raunkiaer's life-form: Therophyte
Description and morphology: Annual, mostly prostrate creeping, pubescent all over, leaves with short petiole and 11–15 leaflets with one terminal, flower whitish-bluish (yellowish (?)) in capitula with shorter peduncle than leaf or subsessile, cluster of pods arranged star-like, pods adpressed falcate pubescent with stiff valves with triangular section, dehiscent.
Reproduction: Seeds
Pastoral importance: Limited, sand binder
Palatability and fodder value: Hardly grazed by all livestock
Toxicity: Possible
Economic interest: None
Habitat: Sandy/sandy-loam soils, to 100 mm MAR
Distribution:

- In Libya, all coastal zones and inland, Hamada, Sirte and Jebel Al-Akhdar
- In arid zones, Mediterranean to Central Asia – Pakistan

Endemism: NA

Notes:

- Variable morphology, could be misidentified as *Astragalus asterias* Steven, http://www.theplantlist.org/tpl1.1/record/ild-7517, or *Astragalus sinaicus* Boiss., http://www.theplantlist.org/tpl1.1/record/ild-7641, with glabrous stem and pods.
- The genus *Astragalus* is widely distributed all over the Mediterranean zones. In Libya, 25 species are recorded with many morphological variations. These are mostly annuals with fresh pods necessary for proper determination. Two perennials (*A. spinosus*, *A. fruticosus*) are still commonly found. See *Flora of Libya* (El-Gadi and Jafri, 1976–1990), Fabaceae, Vol. 86.
- The *Astragalus*' wide occurrence is certainly due to their poor palatability and even possible toxicity.

Photo 7.222. *Astragalus tribuloides*, flower and leaves, wadi Sassu.

Photo 7.223. *Astragalus tribuloides*, prostrate stems with young pods, Bir Ougradia, Al-Assah.

Other main annual *Astragalus* species in Libya:

- *Astragalus annularis* Forssk., http://www.theplantlist.org/tpl1.1/record/ild-7510, herbaceous, erect, 2–4 leaflets and calyx hairy, flower racemed pinkish, pods red-spotted circular furrowed with short recurved stout beak, on sandy/sandy loam in Jeffara plain, Jebel Nafusa, Sirte to the Middle East.

Photo 7.224. *Astragalus annularis* with red spotted pods, Bir Ougradia, Al-Assah, 1979.

- *Astragalus hamosus* L., http://www.theplantlist.org/tpl1.1/record/ild-7589, herbaceous, ventral face of leaflets hairy, flower yellow with hairy calyx, green pods fish-hook-shaped turning red/papery/leathery, all coastal zones.
- *Astragalus corrugatus* Bertol., http://www.theplantlist.org/tpl1.1/record/ild-7554, = *Astragalus crenatus* Schult. (?), violet-purple flowers, pods on petioles usually longer that leaves, pods (furrowed section) arched/surface of valves reticulated with prominent veins, on peduncle longer or equal to leaf, on sandy/sandy-loam in Jeffara plain, Jebel Nafusa, Sirte to the Middle East.

Photo 7.225. *Astragalus corrugatus*, furrowed pods with reticulated valves, Magdar es Sbot, Al-Assah, 1978.

Calicotome spinosa (L.) Link

Synonym(s): http://www.theplantlist.org/tpl1.1/record/ild-8334
Native:
Vernacular name: *Gendul*
Raunkiaer's life-form: Nanophanerophyte
Description and morphology: Robust woody thorny perennial shrub (1–2 m), criss-crossing silky branches longitudinally grooved stems: branches – with strong linear vulnerant-glabrous thorns (20–40 mm long) with red or black tip, small trifoliate leaves, flowers yellow, pods thick papery flat – glabrous – smooth acuminate (20–40 mm long) with thick marginal (winged (?)) sutures with a couple of globular yellow seeds (2–3 mm) bulging through walls, pods dehiscent with twisted valves.
Reproduction: Flowering in January–March, seeds
Pastoral importance: Limited
Palatability and fodder value: Leaves barely grazed by goats and dromedaries, hardly touched by sheep due to sharp thorns.
Toxicity: None
Economic interest: None, sometimes used as fuelwood and *zeriba*
Habitat: On shallow soils on calcareous crust and rocky soils, in relict forest of *Pinus halepensis*, abandoned scrubland, invasive, down to 150 mm MAR.
Distribution:

- In Libya, all coastal zones (common Sabratha, Zawiya, Suani, Tajoura, Daphnia) and Jebel Nafusa (Gharyan), Jebel Al-Akhdar; rare in Sirte.
- In arid zones, southern Europe, North Africa, not in the Sahara, apparently absent from Egypt.

Endemism: NA

Notes:

- The spelling of this genus is sometimes *Calicotome* (accepted) or *Calycotome* (unresolved) in Keith (1965).
- Possible confusion with *Calicotome villosa* (Poiret), http://www.theplantlist.org/tpl1.1/record/ild-8336, that has woolly hairy pods, a shrub very similar aspect and ecology to *Calicotome spinosa*.
- Both *Calicotome villosa* and *Calicotome spinosa* are highly variable with many intermediate types, all present on sites.

Photo 7.226. *Calicotome spinosa*, spines and pods, Tajoura.

Photo 7.227. *Calicotome spinosa*, bush, Gharyan, near Tobi Nursery.

Photo 7.228. *Calicotome villosa*, bush, wadi Siouta, south Micurata.

Photo 7.229. *Calicotome villosa*, flower and leaves, Mzerah, south Gendoula.

Ceratonia siliqua L.

Synonym(s): http://www.theplantlist.org/tpl1.1/record/ild-1158
Native:
Vernacular name: *Kharub*
Raunkiaer's life-form: Mesophanerophyte
Description and morphology: Tree – coppice (native 2–8 m high) in Jebel Al-Akhdar, will develop into a tree with favourable conditions, generally dioceous, trunk twisted and knotted branches on female tree, splintered – grey bark, young stems bark reddish, young leaf pubescent, mature leaves paripinnate upper face shiny dark green rigid glabrous with 4–8 orbicular folioles, male (or hermaphrodite) flowers in axillary racemes, dark brown pendulous rigid pods (or beans – up to 20 cm × 3 cm) compressed (sometimes arched) indehiscent with yellow-brown sugary pulp, 10–16 brown ovoid shiny seeds per pod, seeds hard-coated, veined hard wood pink to reddish.
Reproduction: Flowering September–December, seeds
Pastoral importance: Medium, resistant to fire
Palatability and fodder value: Leaves not touched; pods – beans shed on ground grazed by wandering cattle (and wild pigs), pods used to feed all livestock and horses, donkeys and mules.
Toxicity: None
Economic interest: Building material, quality wood used in carpentry and to locally manufacture traditional (minimum tillage) mouldboard plough (*ma'harat*) or parts of olive or wine press since antiquity (*prelum*, in Latin), ground pods collected to feed cattle (and occasionally other livestock), flour from finely ground pod pulp used to treat babies' diarrhoea and generally gastroenteritis, resin used as mastic, tree commonly planted on farms (shade tree and source of pods) and on forestry projects, often grafted to obtain female trees abundantly producing pods, seeds are extremely hard-coated and must be scarified (or acid treated) to germinate.
Habitat: Native in eastern matorral only, on *Terra rossa* and brown forestry soils, 300–500 mm MAR, commonly planted on farms/gardens/public parks along coastal zone and Jebel Nafusa (not in Sirte (?)). When planted on farm, *Ceratonia* trees develop well with 10–15 m high and dense large flat crown (up to 10 m diam., Garabulli).
Distribution:

- In Libya, native in Jebel Al-Akhdar only
- In arid zones, southern Europe, Mediterranean – all over to the Middle East

Endemism: NA

Photo 7.230. *Ceratonia siliqua*, coppice regrowth after fire, Toknis.

Photo 7.231. *Ceratonia siliqua*, male inflorescence, Toknis.

Photo 7.232. *Ceratonia siliqua* female flowers, Toknis.

Coronilla scorpioides (L.) Koch

Synonym(s): http://www.theplantlist.org/tpl1.1/record/ild-5156, or *Ornithopus scorpioides* L., http://www.theplantlist.org/tpl1.1/record/ild-37755
Native:
Vernacular name: *Rhia-el-akreb* (?)
Raunkiaer's life-form: Stenotherophyte
Description and morphology: Small (10–30 cm) gracile annual with long glabrous stems with lax spaced leaves on stems, trifoliate ovate sessile glaucous alternate leaves with medium lobe larger that lateral (also only with one lobe on lower leaves), axillary inflorescence with 1–2 yellow (yellow-orange) flowers, quadrangular arched (3–6 cm long) pods (like scorpion tail) acuminate in indehiscent constricted corrugated segments, pod breaking in separate segments each containing one seed oval yellow to brown.
Reproduction: Flowering February–May, seeds
Pastoral importance: High
Palatability and fodder value: Well grazed by all livestock
Toxicity: None, however, seeds reported cardiotonic and may induce accidents when ingested in large quantities (?)
Economic interest: Seed production easy, could be used for rangeland reseeding.
Habitat: Sandy/sandy-loam soils, often in rainfed crops, down to 150 mm MAR
Distribution:

- In Libya, all western Libya to eastern Libya including Sirte
- In arid zones, southern Europe, North Africa, west Asia, Middle East to Iran-Pakistan

Endemism: NA

Photo 7.233. *Coronilla scorpioides*, pods, segments and seeds, Nalut.

Other *Coronilla* sp. in Libya:

- *Coronilla repanda* (Poir.) Guss., Coronilla repanda (Poir.) Guss. — The Plant List, = *Ornithopus repandus* Poir., Ornithopus repandus Poir. — The Plant List, annual, lower leaves trifoliate, upper leaves imparipinnate multifoliate (7–10), flower yellow, pods flat segmented and longitudinally reticulated, pod breaking in indehiscent segments, on sandy soils along western coast, in rainfed crops, waste land, MAR > 200 mm.
- Could be confused with *Ornithopus* sp., that does not appear to be present in Libya.

Photo 7.234. *Coronilla repanda*, full plant, Garabulli, Khoms.

Photo 7.235. *Coronilla repanda*, flower and pods, Garabulli, Khoms.

Genista saharae Coss. & Durieu

Synonym(s): *Spartidium saharae* (Coss. & Durieu) Pomel., http://www.theplantlist.org/tpl1.1/record/ild-6226, or *Calobota saharae* (Coss. & Dur.) Boatwr. & B.-E. van Wyk, Calobota saharae (Coss. & Durieu) Boatwr. & B.-E.van Wyk (worldfloraonline.org)
Native:
Vernacular name:
Raunkiaer's life-form: Mesophanerophyte
Description and morphology: Tall lax bush (2 m), erect bare/leafless glabrous stems branching from base, very few small leaves along branches, (short) pedicellate unique bright yellow flowers widely spaced at end of stems, smooth flat pods (30–80 mm long, 10 mm wide with prominent sutures) when ripe slightly arched flat-papery, with 3–5 bulging seeds.
Reproduction: Seeds
Pastoral importance: Limited
Palatability and fodder value: Hardly touched by livestock
Toxicity:
Economic interest: None
Habitat: Sandy sites, sand dunes (same ecology as *Stipagrostis pungens* and *Retama raetam*)
Distribution:

- In Libya, a rare occurrence, in the Jeffara plain near Bir-el-Ghanem (near entrance of old BGI grazing project, opposite entrance to wadi Abu-Shayba cereal project – never personally seen on other sites), reported south Jebel Nafusa (?)
- In arid zones, southern Algeria and Tunisia

Endemism: Saharan

Photo 7.236. *Genista saharae*, young pods and flowers, BGI along road to Bir-el-Ghanem.

Photo 7.237. *Genista saharae*, bush (1.5 m) at Bir-el-Ghanem (BGII).

Photo 7.238. *Genista saharae*, flower and pod, BGI along road to Bir-el-Ghanem.

Note: *Genista saharae* bush without flowers could be confused with *Retama raetam*, however *Retama raetam* has much more intricate branches; *Genista saharae* also could be confused with *Spartium junceum*, the latter being a south European species often escaped from garden and present in the highest-rainfall zone of Jebel Al-Akhdar (wadi Kuf to Derna); *Spartium junceum* is an unlikely occurrence in arid western Libya. *Spartium junceum* has hairy-woolly ripe (6–8 seeds) long slightly arched pods while *Genista saharae* has smooth flat glabrous pods with 3–5 seeds.

Genista acanthoclada DC.

Synonym(s): http://www.theplantlist.org/tpl1.1/record/ild-8377
Native:
Vernacular name:
Raunkiaer's life-form: Nanophanerophyte
Description and morphology: Low (up 70 cm high) hemispherical bush with stiff rigid-linear-cylindrical prickly twigs ending with black point, bearing few small trifoliate leaves, yellow flowers with trilobe lower lip in terminal raceme turning papery-white when faded and remaining attached to stem, small bulbous hairy pods containing one single seed.
Reproduction: Seed
Pastoral importance: Limited
Palatability and fodder value: Few small leaves, not sure if hemispherical bush shape is the result of goat's grazing or simply growth habit.
Toxicity: None (?)
Economic interest: None
Habitat: Rocky slopes, loamy soils, 200–400 mm MAR
Distribution:

- In Libya, Jebel Nafusa, on hills near Khoms, Gharyan, Yeffren, common in Jebel Al-Akhdar, not in Sirte
- In arid zones, southern Greece, Crete

Endemism: NA

Photo 7.239. *Genista acanthoclada* flowers and pointed twigs, near Tobi nursery, Gharyan.

Photo 7.240. *Genista acanthoclada* bush, near Tobi nursery, Gharyan.

Notes:

- More common in the western hills is *Genista microcephala* Coss. et Dur., Genista microcephala Coss. & Durieu — The Plant List, spineless, hemispherical low bush, leaves alternate unifoliate with two small spinescent stipules, flowers yellow with pubescent calyx in terminal capitula.
- Not listed in *Flora of Libya* (El-Gadi and Jafri, 1976–1989, Fabaceae, Vol. 86), possible occurrence of *Genista quadriflora* Munby, Genista quadriflora Munby — The Plant List, ephedroid aphyllous villous/pubescent shrub with stiff rigid blunt-ending.
- Twigs longitudinally ribbed, flowers in fascicules of four, sub-globular pods bearing persistent calyx and papery corolla, occurrence on rocky-loamy soils near Gharyan in *Stipa tenacissima* formation.

Photo 7.241. *Genista quadriflora* (?) low bush, near Gharyan in *Stipa tenacissima* formation.

Photo 7.242. Ephedroid *Genista quadriflora* (?) with blunt articles and drying yellow flowers, near Gharyan.

Hedysarum spinosissimum L.

Synonym(s): http://www.theplantlist.org/tpl1.1/record/ild-5108 or *Hedysarum palleus* (Moris) Ital., Hedysarum palleus (Moris) Ital. —The Plant List
Native:
Vernacular name:
Raunkiaer's life-form: Tachytherophyte
Description and morphology: Small annual, prostrate, centrifugal long pubescent reddish branches with long internodes, leaves alternate imparipinnate with erect oblong leaflets, raceme with 2–4 pink flowers at end of long petiole, pods constricted prickly with 2–5 flat-discoid segments (dissociating when dry) containing each a single seed, pod valves remaining attached to taproot after germination.
Reproduction: Flowering February–March, seeds

Pastoral importance: Forage
Palatability and fodder value: Well grazed by all livestock
Toxicity: None
Economic interest: Seeds easy to harvest could be used for rangeland reseeding
Habitat: Sandy-loam fine gravel soils, never on saline soils
Distribution:

- In Libya, Jeffara plain, Jebel Nafusa, Sirte, south Jebel Al-Akhdar
- In arid zones, North Africa to the Middle East

Endemism: NA

Note: Cropped and never on rangelands, at Tajoura, in oasis, sometimes *Hedysarum coronarium* L. (http://www.theplantlist.org/tpl1.1/record/ild-5100) may be cropped under irrigation (seeds from Malta) – bi-annual (perennial (?)) large glabrous plant, strong succulent hollow stems, panicle of red flowers (sometimes white (?)), pods papery and segmented also and possibly, *Hedysarum carnosum* Desf. (likely seeds from Tunisia), http://www.theplantlist.org/tpl1.1/record/ild-5099, similar to and possible confusion with *Hedysarum coronarium*, flower purple in dense elongated panicle, pods with square segments; both excellent forage used to feed lactating cows (and dromedaries) – cut-and-carry, do not tolerate heavy grazing or trampling. Both are cattle feed.

Photo 7.243. *Hedysarum spinosissimum*, flower, leaves and pods, Gdama, Gharyan.

Photo 7.244. *Hedysarum spinosissimum*, whole plant, Gdama, Gharyan.

Photo 7.245. *Hedysarum spinosissimum*, pods, Gdama, Gharyan.

Hippocrepis bicontorta Loisel.

Synonym(s): *Hippocrepis areolata* Desv., http://www.theplantlist.org/tpl1.1/record/ild-5168
Native:
Vernacular name: *Kheta* (?), *Nefel* (?)
Raunkiaer's life-form: Tachytherophyte
Description and morphology: Annual, radiating prostrate with zig-zagging stems, leaves imparipinnate with 3–5 pairs of obovate leaflets, yellow flowers on peduncle longer that leaf, horseshoe coiled compressed segmented pods (1–5 rings) slightly pubescent and horned opening outside, seeds thin yellowish-brown fragile and lunular difficult to extract from pods.
Reproduction: Flowering February–May, seeds
Pastoral importance: Part of the valuable spring annuals
Palatability and fodder value: Excellent forage
Toxicity: None
Economic interest: None
Habitat: Sandy/sandy-loam soils, 100–300 mm MAR, xerothermic
Distribution:

- In Libya, Jeffara plain, south Jebel Nafusa, Sirte, Ajdabiya, south Jebel Al-Akhdar
- In arid zones, North Africa, Middle East to Iran

Endemism: NA

Note: *Hippocrepis bicontorta* (one of the 6 Libyan *Hippocrepis* sp.) is widely distributed on sandy soils of Libya to the northern margin of the Sahara; all with imparipinnate leaves, flowers yellow and pods flat with horned segments.
Other main *Hippocrepis* species in Libya Annuals and looking very much alike, identification with pods:

- *Hippocrepis unisiliquosa* L., flowers sessile, one sessile pod, http://www.theplantlist.org/tpl1.1/record/ild-5195
- *Hippocrepis cyclocarpa* Murb., http://www.theplantlist.org/tpl1.1/record/ild-5177, reported in eastern Libya

Photo 7.246. *Hippocrepis bicontorta*, whole plant, Bir Ougradia.

Photo 7.247. *Hippocrepis bicontorta*, pods, Jbebina, Al-Assah.

- *Hippocrepis ciliata* Willd., pods arched with opening inside arch, Jebel Nafusa, http://www.theplantlist.org/tpl1.1/record/ild-5172
- *Hippocrepis multisiliquosa* L., pods arched with opening outside arch, South Nalut, http://www.theplantlist.org/tpl1.1/record/ild-5188

Hippocrepis scabra DC.

Synonym(s): http://www.theplantlist.org/tpl1.1/record/ild-10192
Native:
Vernacular name: *Nefel* (?)
Raunkiaer's life-form: Nanophanerophyte
Description and morphology: The only perennial *Hippocrepis* sp. in Libya, small shrub (20–30 cm), many woody branches at base, alternate imparipinnate leaves with 5–9 entire oboval turgescent leaflets, group of 6–8 yellow flowers turning orange-veined, on peduncle 3–5 longer than leaves, scabrous flat pods barely arched nearly linear with 6–8 horseshoe fragile yellow seeds with flat papillose seed interspace.
Reproduction: Flowering February–May, seeds
Pastoral importance: Valuable spring forage
Palatability and fodder value: Excellent forage
Toxicity: None
Economic interest: None
Habitat: Loamy/rocky soils, 300–350 mm MAR
Distribution:

- In Libya, only in Jebel Nafusa, Msalata to Yefren
- In arid zones, southern Spain – North Africa (apparently only Ibero-Maurusian)

Endemism: NA

Photo 7.248. *Hippocrepis scabra*, Jebel Nafusa, Jendouba, 1978.

Photo 7.249. *Hippocrepis scabra*, yellow flowers orange-veined, Jendouba, 1978.

Lotus corniculatus L.

Synonym(s): http://www.theplantlist.org/tpl1.1/record/ild-7260
Native:
Vernacular name: *Kheta*
Raunkiaer's life-form: Nanophanerophyte – hemicryptophyte
Description and morphology: Perennial with woody stout central rooting system, prostrate, creeping glabrous stems not stoloniferous, leaf imparipinnate with obovate- lanceolate leaflets, flowers yellow (sometimes yellow-veined with red or orange), pods black straight and cylindrical, multiseeded and pod valves dehiscent black and twisted remaining on stems.
Reproduction: Flowering February–April, abundant seeds production
Pastoral importance: Forage, sand binder
Palatability and fodder value: Well grazed by all livestock
Toxicity: None
Economic interest: Easy seed production, could be used for range regeneration on fixed sandy soils.
Habitat: Common on fixed sand/sandy-loam soils, 100–500 mm MAR
Distribution:

- In Libya, all Jeffara plain, Jebel Nafusa, Sirte, south Jebel Al-Akhdar (Soluk)
- In arid zones, North Africa to Central Asia

Endemism: NA

Note: *Lotus* sp. widely distributed in Mediterranean zones, with about 15 species in Libya in all rainfall zones and soil types, even on slightly saline soils. Other main *Lotus* species in Libya:

Photo 7.250. *Lotus corniculatus* grazed by a camel, Magdar-es-Sbot, Al-Assah.

Photo 7.251. *Lotus corniculatus*, Bir-el-Ghanem, BGII, 1978.

- *Lotus edulis* L., annual, stems pubescent, flower yellow, large fleshy pods, appear to sustain some mild salinity, http://www.theplantlist.org/tpl1.1/record/ild-7271.
- *Lotus halophylus* Boiss. & Spruner, gracile annual, prostrate, hairy stems, flower pale yellow, thin cylindrical beaked pods, occurs on sand along the coast and on saline substrate, http://www.theplantlist.org/tpl1.1/record/ild-7284.
- *Lotus polyphyllos* Clarke, strong perennial with creeping or erect silvery – woolly stems and leaves, small sessile leaves jam-packed along stems, wilted flowers papery turning dark brown, small obovate pod (containing 2–3 seeds) at end of culms, common on sand along the coast close to seashore, http://www.theplantlist.org/tpl1.1/record/ild-7313

Lotus halophilus Boiss. & Spruner

Synonym(s): http://www.theplantlist.org/tpl1.1/record/ild-7284, *Lotus pusillus* Viv., http://www.theplantlist.org/tpl1.1/record/ild-7285
Native:
Vernacular name: *Kheit* (?)
Raunkiaer's life-form: Nanophanerophyte
Description and morphology: Gracile annual, prostrate creeping, silvery stems and leaves, pedunculate pair (3 (?)) of sessile flowers pale yellow, thin long (20–25 mm × 2–3 mm diam.) cylindrical beaked black flattish pods, seeds protruding on pods, pods dehiscent with twisted valves, spherical yellow glossy seeds (1–2 mm diam.).
Reproduction: Flowering February–May, seeds
Pastoral importance: Valuable spring forage
Palatability and fodder value: Excellent forage
Toxicity: None
Economic interest: None
Habitat: Sandy soils and on dunes, often on sand veil on saline soils, 100–350 mm MAR
Distribution:

- In Libya, from west to east
- In arid zones, southern Spain – North Africa, the Middle East to Iran

Endemism: NA

Photo 7.252. *Lotus halophilus*, Magdar-es-Sbot, Al-Assah, 1978.

Annual *Medicago* sp.

Annual *Medicago* sp. is a very important and diverse genus present in semi-arid and arid zones of Libya from west to east, a major genus contributing to feeding small domesticated and wild ruminants, larger livestock and rangeland rodents. Annual *Medicago* (or medic) pastures are widely used in Western and Southern Australia in rotation with cereals under the rainfed ley farming system (Puckridge and French, 1983) introduced in Libya (Allen and Ewing, 1978; Gintzburger and Sbeita, 1995; Halse, 1978, 1982) and in Morocco, Algeria, Tunisia, Jordan, Syria, Iraq, Türkiye and Iran.

A large collection of native forage seeds was organized by the Agricultural Research Centre (Sidi-el-Mesri, 1976–79) and the FAO-TF-LIB 10 project (Gintzburger, 1978a, b, 1980; Gintzburger *et al.*, 1983; Gintzburger *et al.*, 1984a, b; Gintzburger and Blesing, 1979; Gintzburger and Kernick, 1979). Some 2400 seed samples of native forage ecotypes including 1800 annual *Medicago* ecotypes were collected in northern Libya from west to east above the 50 mm isohyet. The annual *Medicago* were sorted, identified and field-evaluated in Western Australia and in Libya under the joint Libyan Medic Evaluation Project (LMEP 1980–84) in Libya and Western Australia (the Western Australian Department of Agriculture, Perth, Australia). The joint Libyan-West Australian project was supervised by the late Dr C. Francis (Francis, 1978, 1982) and G. Gintzburger (Gintzburger *et al.*, 1984b) with the assistance of three Libyan agricultural engineers (Mr Adna. Sbeita, Mr Mohamed Fituri and Mr Abdelhafiz Rojbani) and the technical support of Mr Peter Skinner and Mr Kevin Foster.

All annual *Medicago* sp., locally named *Nefel*, look similar in the field; however, we can easily differentiate the different annual *Medicago* species using the specific morphological characteristic of the coiled pods, with or without spines, smooth tuberculate or not, with clockwise or anticlockwise coils, with coil surface with different vein patterns; for this, during the LMEP, we extensively used the publication from Heyn (1963) complemented with Lesins and Lesins (1979), Nègre (1956, 1959) and Gachet and Elmir (1972) to determine the Libyan annual *Medicago* sp. These publications and annual *Medicago* identification appear more reliable that the one from *Flora of Libya* (El-Gadi and Jafri, 1976, Vol. 86).

Green annual *Medicago* sp. without flowers and pods are difficult to determine and can be confused with *Trifolium* sp. (mostly present in high-rainfall zones in Libya). However, the difference on green plants is that the trifoliate leaves of annual *Medicago* have the middle foliole with a long rachis, while all folioles on *Trifolium* are sessile and not petiolate. Green annual *Medicago* sp. and *Trigonella* sp. have both the middle foliole with a long rachis adding to possible confusion. Leaflets often have coloured (black to copper-reddish) dots, marks or tiny white flecks. None of these markers appears to be specific. The pods remain the only way to correctly determine these plants. For identification purposes, we illustrate the different annual *Medicago* pods on specific photos on a millimetric scale or grid.

Below are the main annual *Medicago* (in alphabetical order), which were collected in Libya from west to east during the 1976–79 field collections or supposed to occur in Libya according to El-Gadi and Jafri (1976), Vol. 86, pp. 183–213). The identification details of individual annual medics ecotypes and collection sites (Ref. ecotypes: 101631–103437) of the whole LMEP collection were published in CSIRO (1985) with the Algerian annual *Medicago* collection (Ref. ecotypes: 103438–103677 – June 1982).

Medicago arabica (L.) Huds.

http://www.theplantlist.org/tpl1.1/record/ild-8482, or *Medicago maculata* Sibth., all leaflets with central black spots (not to be confused with the Australian cultivar *Medicago truncatula* cv. *Jemalong)*, pods truncated discoid, 3–5 anticlockwise tight coils containing 5–8 seeds, dorsal suture in double row bearing soft grooved spines, little interest for arid rangeland, common well watered or waterlogged sites, on road sides, in coastal oases (Tajoura), on rubbish sites, a common weed all along the coastal Libyan zone (pods not to be confused with pods of *M. polymorpha* or with *M. minima*).

Photo 7.253. Black-spotted leaflets of *Medicago arabica*, Tripoli, Sidi-el-Mesri.

Photo 7.254. Dry pods of *Medicago arabica*. (Pod sample courtesy of the Australian Pastures Genebank, SARDI)

Medicago constricta Durieu

http://www.theplantlist.org/tpl1.1/record/ild-37599, a strictly eastern Mediterranean species (Greece to Palestine) rare but present in Jebel Al-Akhdar, on heavy cracking soils, glabrous pods anticlockwise woody when mature with 6–8 adpressed coils, spines furrowed short at 180° to coils surface with apex hooked; confusion possible with *M. murex* (large pods, spine not furrowed, a western Mediterranean species, doubtful in Libya) and more likely confused with *M. rigidula* (from cold environment) or even *M. aculeata* (pubescent pods), both absent in Libya.

Photo 7.255. *Medicago constricta* (type), from Mzerah (south Gendoula, Jebel Al-Akhdar).

Medicago coronata (L.) Bartal.

http://www.theplantlist.org/tpl1.1/record/ild-8486, the smallest pods of all the annual *Medicago*, an eastern Mediterranean species, several crown-like pods at end of floral peduncle, very small cylindrical pods (1–4 mm diameter) with anticlockwise spaced coils (2–3 loose coils) with inwards-curved short spines inserted at 90° to coils' surface when dry, only present in eastern Libya, Benghazi plain, El-Marj, Derna, in matorral.

Photo 7.256. Pods of *Medicago coronata* (2–4 mm diam.). (Photo: Dr Ori Fragman-Sapir, https://flora.org.il/en/dr-ori-fragman-sapir/)

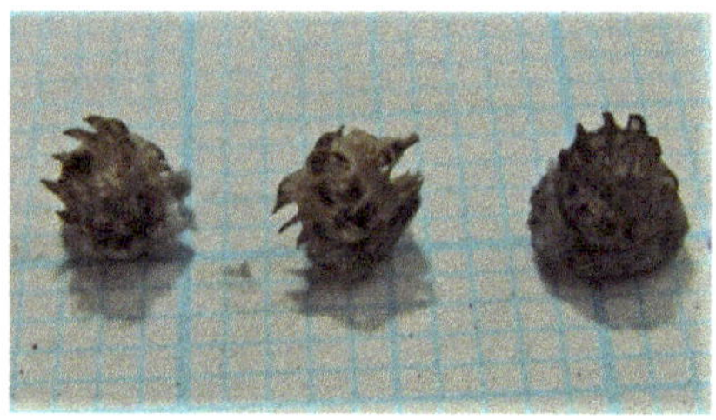

Photo 7.257. Dry pods of *Medicago coronata*. (Pod sample courtesy of the Australian Pastures Genebank, SARDI).

Medicago laciniata (L.) Mill.

http://www.theplantlist.org/tpl1.1/record/ild-8512, the most drought-resistant Libyan *Medicago* common down to 50 mm MAR zones, prostrate plants, small lax globose pods, with anticlockwise coils and radiating straight spines, prolific seed production (up to 10–12 seeds/pod), very common occurrence on heavier soils/rocky sites with thin sand veil in arid Jeffara plain, Jebel Nafusa, south of Beni-Walid, on Hamada, inland Sirte, south of Jebel Al-Akhdar to Solum; a strong contender for arid rangeland regeneration but not favourably considered by Australian commercial seed producers as potential wool contaminant that is irrelevant to Libyan rangeland systems. High value for arid rangeland rehabilitation.

Photo 7.258. *Medicago laciniata* pods (type), wadi Sassu (N 32.137 E 14.790).

Photo 7.259. *Medicago laciniata* plant, wadi Merdum (N 31.824 E 14.277), Hamada.

Medicago laciniata var. *brachycantha* Boiss.

Or *Medicago aschersoniana* Urb., http://www.theplantlist.org/tpl1.1/record/ild-49672, present in more arid sites than the *Medicago laciniata* (L.) Mill. type, much smaller globose pods (3–5 seeds) with very short criss-crossing spines (perpendicular to coil surface), with long radiating stems (up to 1 m long), often carpets of dry pods common on rocky patches with thin sand veils covering heavy soils in desert environment (Seddada, wadi Nefad, wadi Zorzor, Ghirza, Sof-el-Jin, inland Sirte, south Jebel Al-Akhdar), high value for arid rangeland rehabilitation.

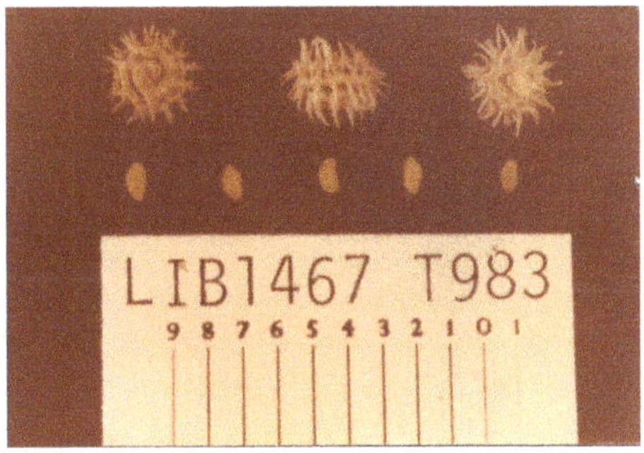

Photo 7.260. *Medicago laciniata* var. *brachycantha* pods, wadi Zamzam (N 31.422 E 15.250).

Photo 7.261. *Medicago laciniata* var. *brachycantha*, long radiating stems (up to 1 m long), wadi Ghirza (N 30.944 E 14.560, 1978.

Medicago littoralis Loisel.

http://www.theplantlist.org/tpl1.1/record/ild-8515, widespread on sandy/sandy-loam soils (even near the seashore, not on dunes) along coastal zone (Tajoura, Sabratha, Lebda) and best rainfall zones of Jeffara plain, silky-velvety leaves, pods whitish cylindrical truncated (3–6 clockwise or anticlockwise coils) with short spines on dorsal suture at close to 180° to coils' surface; misidentification is possible with *Medicago truncatula* var. *tricycla* (pods brownish), however, most spines of *M. littoralis* are irregular in length on the same pod and spreading fairly flat in the surface of the coils while the spines of the *M. truncatula* type (and *M. truncatula* var. *tricycla*) are mostly at 90° with the surface of the coil. High value for arid rangeland rehabilitation above 200 mm MAR.

Possibly similar to one of the most successful Australian medics (*Medicago littoralis* cv. *Harbinger* with short spines) was extensively sown on the sandy cereal project sites at wadi Abu-Shayba, El-Hira and Agheilat (1974–1985).

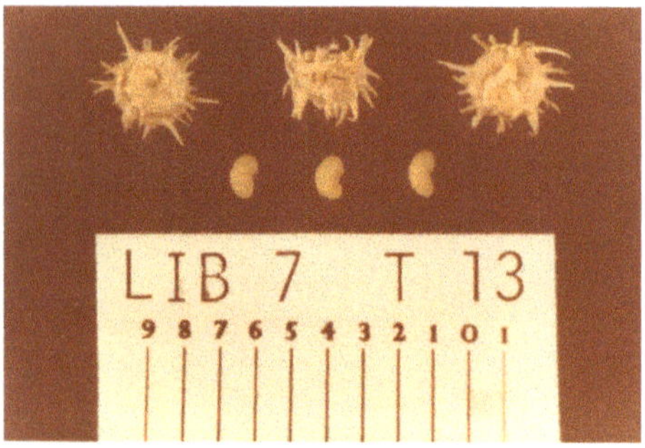

Photo 7.262. *Medicago littoralis* pods (type), Al-Jmeil – Zuara (N 32.882 E 12.075).

Photo 7.263. *Medicago littoralis* plant and pods (type), at Kararim – Misurata.

Notes:

- One native spineless *Medicago littoralis* var. *brevispina* with small nearly spineless (short tubercle on dorsal suture) pods is sometimes found on sites mixed with *Medicago littoralis* var. *littoralis* (type) with the same ecology.
- Many possible hybrids between *Medicago littoralis* and *Medicago truncatula*.

Medicago minima L.

http://www.theplantlist.org/tpl1.1/record/ild-31590, small ovoid pods with anticlockwise coils, centrifugal hooked spines (not to be confused with *M. laciniata* that has a smaller pod and straight spines), spines as long as diameter coil and insertion slightly separated from dorsal suture and not on dorsal suture, common all over coastal Libya on heavy soils and waterlogged sites, in irrigated crops, along roadsides, strong wool contaminant. Poor value for arid rangeland rehabilitation.

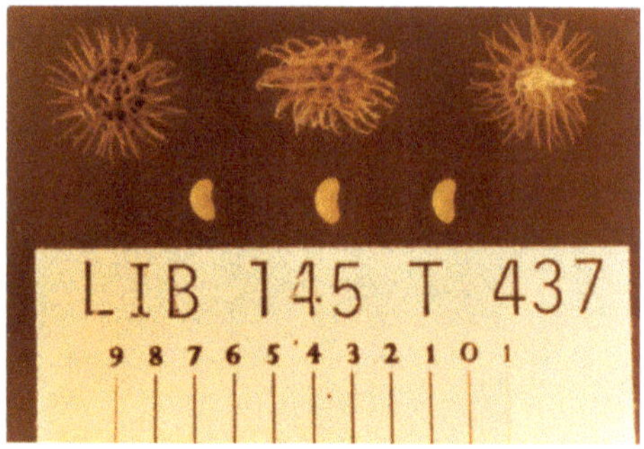

Photo 7.264. *Medicago minima* pods (type, Assaba), Gharyan (N 32.500 E 13.166).

Medicago minima var. *brevispina* Benth.

http://www.theplantlist.org/tpl1.1/record/tro-13076223, a small spineless pod form of *Medicago minima* is also commonly found in northern Libya.

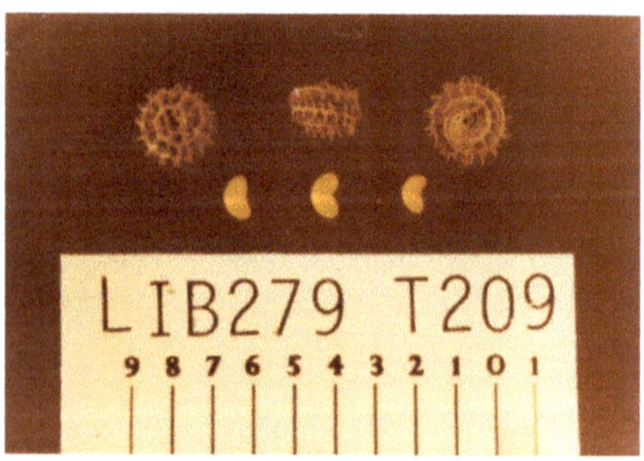

Photo 7.265. *Medicago minima* var. *brevispina* pods (type, Awiniya) (N 31.980 E 12.469).

Medicago orbicularis (L.) Bartal.

http://www.theplantlist.org/tpl1.1/record/ild-8527, one of the larger medic plants in Libya, mostly erected stems (sometimes prostrate), pods anticlockwise (12–15 mm diam.) flat discoid coils open, smooth papery spineless with 8–12 large triangular seeds/pods, in Jebel Nafusa, Jebel Al-Akhdar, in scrubland and crops in highest rainfall zones (> 300 mm) of Libya, poor value for arid rangeland rehabilitation but may be used in hay mixture in high-rainfall zones.

Photo 7.266. *Medicago orbicularis* pods (type).

Photo 7.267. *Medicago orbicularis*, Ain Rumia (N 31.850 E 11.810), 1978.

Medicago polymorpha var. *polymorpha* L.

http://www.theplantlist.org/tpl1.1/record/ild-8528, creeping plant, commonly flattish spiny pods with anticlockwise loose coils with many curly/spiral radial veins joining dorsal suture, long straight spines hardening at maturity in surface of anticlockwise coils (3–5 coils), dark brown or black mature pods, ubiquitous in Libya, in crops, on roadsides, mostly loamy to heavy soils even temporarily waterlogged, MAR > 150 mm. Some LMEP ecotypes of *Medicago polymorpha* are remarkably early flowering but are ecologically very site-specific, abundant in depressions on heavy soils.

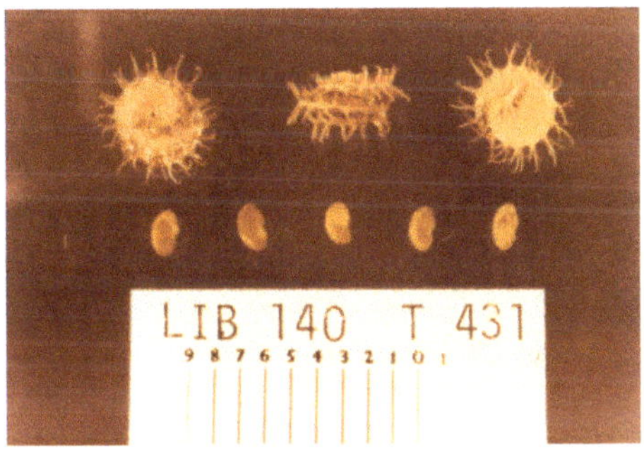

Photo 7.268. *Medicago polymorpha* var. *polymorpha* (type) from Tegrena (N 32.118 E 12.992).

Photo 7.269. Young pod of *Medicago polymorpha* var. *polymorpha*.

Photo 7.270. *Medicago polymorpha* var. *polymorpha* (type) from wadi Maghur, 20 km south-east El-Abyar.

Medicago polymorpha var. *brevispina* (Benth) Heyn

Medicago polymorpha var. brevispina (Benth.) ("Medicago polymorpha var. brevispina (Benth.) Heyn") Heyn (worldfloraonline.org), common spineless (tuberculate) pods found mixed in population of *Medicago polymorpha* var. *polymorpha*; *Medicago polymorpha* var. *brevispina* occurs in depressions temporarily waterlogged 8–10 km south-west of Aziziya (north of Ras-el-Loefa) along road to Bir-el-Ghanem; an early flowering spineless cultivar (*Medicago polymorpha* cv. *Serena* (cross from *M. polymorpha* from Chili and Coolgardie-WA) and *Medicago polymorpha* cv. *Circle Valley*) were introduced from Western Australia to Jendouba (?). Cv. *Serena* is the earliest flowering commercial annual *Medicago* available for heavy soils. These spineless *M. polymorpha* cultivars are favoured as not spoiling wool. Spineless ecotypes could be oversown in waterlogged depressions.

Photo 7.271. *Medicago polymorpha* var. *brevispina* pods, Jendouba – Gharyan.

Medicago scutellata (L.) Mill.

http://www.theplantlist.org/tpl1.1/record/ild-8541, one of the largest medic plants and pods in Libya, prostrate or erect plant when in crops, large pod (12–15 mm diam.), ovoid/cups stacked-like/papery smooth (4–7 lax coils), coils anticlockwise, rare due to overgrazing of large pods, occurrence only and most likely from commercial seeds introduced from southern Australia on El-Marj plain projects, used in irrigated or rainfed pasture mixtures; most used Australian cultivars tested with little success were *Medicago scutellata* cv. *Robinson, cv. Sava* and cv. *Sapa* from South Australia; rare native occurrence in eastern Libya, avidly grazed by all livestock, poorly resistant to grazing, possibly threatened, poor value for arid rangeland rehabilitation but may be used in hay mixture in high-rainfall zones.

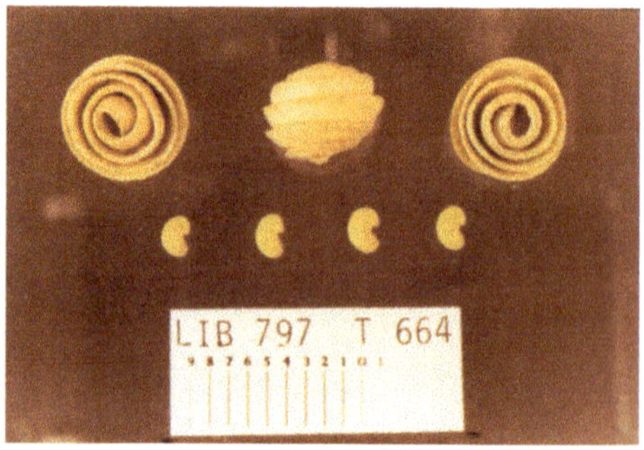

Photo 7.272. *Medicago scutellata* pods (type), Jendouba Project, Gharyan.

Photo 7.273. *Medicago scutellata* pods and seedlings, El-Marj plain – likely introduced by the South Australian Project).

Medicago tornata (L.) Mill.

http://www.theplantlist.org/tpl1.1/record/ild-8510, very flat smooth discoid pods with clockwise and anticlockwise (1–3 to 6) coils (5–8 mm diam.), short spaced slender spines on coil surface or rugulose-spineless, exclusively on sandy soils, limited distribution mostly western coastal zones around Suani (to Bir-el-Ghanem), to Garabulli – Lebda, a strict north-western Libyan *Medicago* sp. producing pods in abundance with only 2–3 seeds/pod; a west Australian spineless cultivar pods originating from Suani-Beni-Adem (*Medicago tornata* cv. *Swani* – as *M. tornata* var. *rugulosa*) was selected and is commercially available along with cv. *Tornafield.*, the latter from South Australia; never been successfully used in Libya as too late flowering.

Photo 7.274. *Medicago tornata* var. *aculeata* pods (type) from south Garabulli (N 32.640 E 13.712).

Photo 7.275. Smaller pods of *Medicago tornata* var. *rugulosa* (from Suani).

Medicago tornata subsp. *striata* (Bastard) Kerguelen

http://www.theplantlist.org/tpl1.1/record/ild-37630, present on several sites Garabulli, north of Suani and between Suani and Aziziya, an even smaller lenticular pod (1–2 coils) from *M. tornata* were collected, possibly a strict west Mediterranean species; there could also be hybrids (?) of the complex *Medicago tornata x M. littoralis* (?) types with small pods looking like *Medicago tornata* subsp. *striata*.

Photo 7.276. *Medicago tornata* subsp. *striata* pods, Garabulli, Suani – Aziziya.

Medicago truncatula Gaertn.

http://www.theplantlist.org/tpl1.1/record/ild-8545, vigorous growth prostrate or slightly erect, pods cylindrical barrel-like (both clockwise and anticlockwise types, 4–7 adpressed coils – 5–7 mm diam.) with prominent dorsal suture bearing strong stout criss-crossing spines (various lengths and curvature or even very short spines) at angle to coil surface or adpressed on the barrel, occurs on loamy-heavy soils, usually in higher-rainfall zones of Libya (above 200–300 mm), along the coast, in Jebel Nafusa, Sirte and Jebel Al-Akhdar (on *Terra rossa* soils). Several cultivars are commercially available from Australia and were used in the ley farming system in western Libya (cv. *Jemalong*, with a black spot on leaflets at Jendouba high-rainfall site doing well there and in Jeffara plain on heavy soils (however, not flowering early enough and not completing cycle to escape dry season), and cv. *Cyprus*, more early flowering hence quite drought-resistant and doing well on heavier soils in Jeffara plain (wadi Ethel, lower land of wadi Abu-Shayba, wadi Hira...); other Australian *Medicago truncatula* commercial cultivars (cv. *Borung*, *Ghor*, *Parabinga* and *Paraggio*) were tested in western and eastern Libya without success as too late flowering. The *Medicago truncatula* ecotypes from Sirte are extremely early flowering, likely escaping early spring grazing when flocks are moving into the desert (Gintzburger, 1978a; Gintzburger *et al.*, 1983; Gintzburger *et al.*, 1984b). High value for potential rangeland reseeding on heavy soils all over northern Libya, providing they are very early flowering.

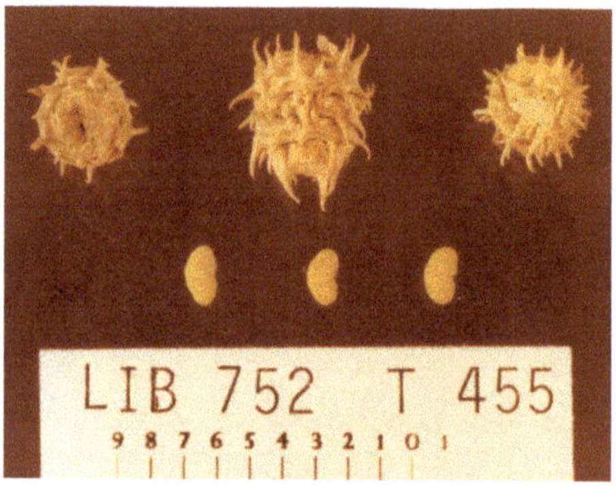

Photo 7.277. *Medicago truncatula* pods (type), south Aziziya (N 32.428 E 13.041).

Medicago truncatula var. *tricycla* (Gaertn.) Nègre

Described by Nègre (1956), an arid form/variety of *M. truncatula* (https://www.worldfloraonline.org/taxon/wfo-0001447224), which is in our opinion different from *Medicago truncatula* Gaertn. This arid form is common inland down to 100 mm, on heavy/loamy soils, inland from east to west (with smaller pod that *Medicago truncatula* but only 2–3 lax separate coils with short stout spines at 90° to coils surface – pods 3–5 mm diam.) that could be easily confused with *M. littoralis* (and especially the Australian cv. *Harbinger*, spines at to 180° to coil surface); from an ecological point of view *M. littoralis* occurs on sandy soils rarely below 200 mm predominantly on the coastal zone, while *Medicago truncatula* var. *tricycla* is found on loamy/heavy soils, rocky sites, even in arid wadis and plateau leading to the Hamada, inland from Sirte and south Jebel Al-Akhdar with MAR = 100–150 mm. Possibly the best candidate for all arid rangeland regeneration reseeding.

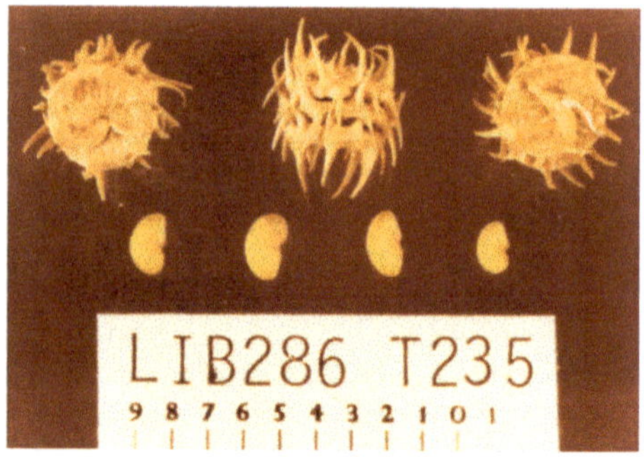

Photo 7.278. *Medicago truncatula* var. *tricycla* pods, from Giado – south Ain Zerga (N 31.848 E 12.067).

Photo 7.279. *Medicago truncatula* var. *tricycla* pods, from wadi Sassu, south Misurata.

Medicago turbinata (L.) All.

Medicago turbinata (L.) All. — The Plant List or *Medicago tuberculata* (Retz.) Willd., Medicago tuberculata (Retz.) Willd. — The Plant List, large ovoid – olive-shaped tuberculated pods (10–12 mm diam.) with tightly compressed coils, 6–8 large (2–4 mm long) reniform seeds, an eastern Mediterranean medic (Palestine), rare occurrence in eastern Libya only on *Terra rossa* in high-rainfall zone matorral of Jebel-Al-Akhdar, could be easily confused with another eastern Mediterranean medic *Medicago aculeata* Willd. = *Medicago doliata* Carmign., Medicago doliata Carmign. — The Plant List, that has tuberculated usually spiny pods with short stout recurved spines (also spineless, smooth forms); despite being reported in Jebel Al-Akhdar *Medicago turbinata* and *Medicago aculeata* are very doubtful in Libya and could be escaped from south Australian El-Marj project. The pod featured in the *Flora of Libya* (El-Gadi and Jafri, 1976–1989, Vol. 86, p. 202) is a *M. truncatula* – pod cylindrical truncated, spines at 90° to coils and not a *M. turbinata*.

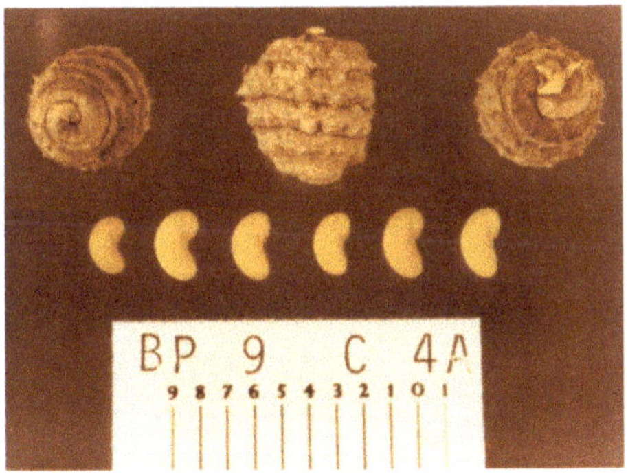

Photo 7.280. *Medicago turbinata* pods, near Shahat (N 32.832 E 21.879).

Other annual *Medicago* described by El-Gadi and Jafri (1976), in Vol. 86, were not found in Libya during our LMEP 1976–79 native forage field collections.

Medicago cyrenaica Maire & Weiller, http://www.theplantlist.org/tpl1.1/record/kew-2368329, also spelt *Medicago cyrenaea*, https://powo.science.kew.org/taxon/urn:lsid:ipni.org:names:506144-1, unresolved name (?) doubtful identification, reported in Jebel Al-Akhdar (Cyrene/Shahat) could have been confused with *M. minima* or *M. polymorpha* – see at http://www.worldfloraonline.org/taxon/wfo-0000374236, *Medicago murex* Willd., http://www.theplantlist.org/tpl1.1/record/ild-8523, large ovoid pods with short triangular spines (or rarely spineless) on dorsal suture, anticlockwise coils, pods black when ripe with 5–7 yellow seeds, few radial veins not reaching the dorsal suture, doubtfully reported in Gharyan, Cussabat and Benghazi (El-Gadi and Jafri, 1976, Vol. 86; Keith, 1965); possible around *sebkha* (Zuara, Sorman) as salt-tolerant, very limited use as ecologically limited and azonal.

Photo 7.281. Dry pods of *Medicago murex*. (Pod sample courtesy of the Australian Pastures Genebank, SARDI)

Medicago rigidula (L.) All., http://www.theplantlist.org/tpl1.1/record/ild-8533, pods ovoid – globose hardening/woody at maturity, coils (4–7) strongly adpressed and anticlockwise, spines as saw teeth at 90–180° (also tuberculate or spineless pods occur), a medic from cold Mediterranean regions, southern Europe, northern Syria, Turkey, very doubtful occurrence in Libya.

Medicago rugosa Desr., http://www.theplantlist.org/tpl1.1/record/ild-8535, pods discoid rugose spineless or slightly tuberculate with connecting radial veins and strong prominent dorsal veins, 3–5 anticlockwise coils with 2–3 seeds. Strictly on heavy red Mediterranean soils, possibly introduced from South Australia (*Medicago rugosa* cv. *Paragosa*), only rare occurrence in Jebel Al-Akhdar, on Benghazi plain/El-Marj on land developed by the South Australia ley farming project, a rare Mediterranean species, native doubtful in Libya (?). Very limited use.

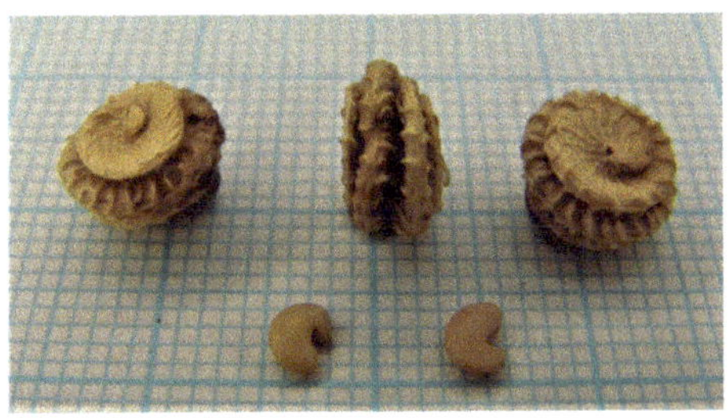

Photo 7.282. Dry pods of *Medicago rugosa*. (Pod sample courtesy of the Australian Pastures Genebank, SARDI)

Medicago disciformis DC, http://www.theplantlist.org/tpl1.1/record/ild-8488, pods flat discoid (4–5 coils with 6–10 small seeds) with long flexuous radiating spines equal or longer that the coil diameter and in the surface of the coil, a south European species from France to Cyprus and Turkey, reported from Derna region possible but extremely rare, very limited distribution and potential use in Libya.

Photo 7.283. Dry pods of *Medicago disciformis.* (Pod sample courtesy of the Australian Pastures Genebank, SARDI)

Perennial *Medicago* sp.

In Libya, mostly weeds and plants of limited importance with exception of *Medicago sativa* L. (http://www.theplantlist.org/tpl1.1/record/ild-8536), the common perennial Alfalfa or Lucerne (*Safsafa*) of which many local varieties are cropped under irrigation in Libyan oases. Alfalfa remains the foremost cultivated forage legume cropped under irrigation (with some salt tolerance) in Libya since antiquity with many old local cultivars (mostly from Tajura to Zliten) collected from Libya by FAO for the Bari Genebank.

Medicago falcata L., http://www.theplantlist.org/tpl1.1/record/ild-8492, creeping, rhizomatous, yellow flowers with brown veins, occurring in Spain, Morocco to eastern Algeria on marls – heavy soils, extremely doubtful in Libya.

Medicago marina L., http://www.theplantlist.org/tpl1.1/record/ild-8520, woollen whitish stems and leaves, pods villose with fine arachnean indumentum cylindrical truncated with short spines in surface of anticlockwise (2–3) coils with 2–4 seeds, plant commonly found on sand dunes along seashore, could be confused with *Lotus halophylus* or *Lotus polyphyllos.* Pods coiled/discoid on *M. marina* while straight on the *Lotus* sp., very limited use as ecologically narrow.

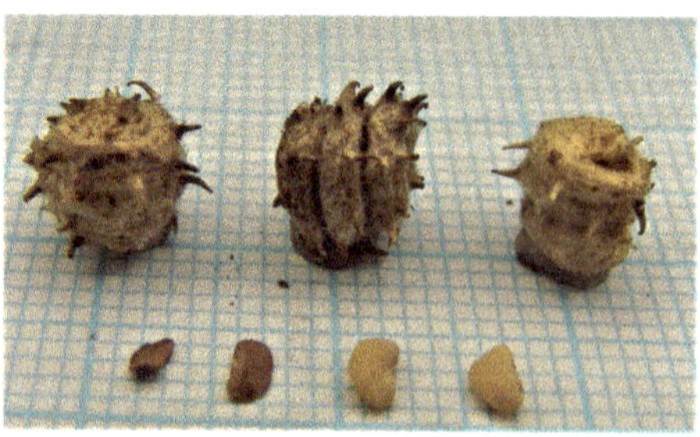

Photo 7.284. Dry pods of *Medicago marina.* (Pod sample courtesy of the Australian Pastures Genebank, SARDI)

Medicago arborea L., http://www.theplantlist.org/tpl1.1/record/ild-8484, a perennial bush (1–1.5 m), abundant green foliage, white stems and bark, flat papery pods one coil only, turning black mature; exotic introduced from Spain (Robledo *et al.*, 1993) and Tunisia on western rangeland projects. Successful establishments on heavy soils with 250 mm MAR or in depressions collecting run-off, tested at Wishtata (Le Houérou *et al.*, 1982) and Bir Ayyad, but avidly grazed by all livestock; does not withstand overgrazing and needs very strict controlled grazing management; often seen as an ornamental in coastal gardens.

Photo 7.285. Bushes of *Medicago arborea* plantations tested at Bir Ayyad, 1979.

Medicago lupulina L., http://www.theplantlist.org/tpl1.1/record/ild-8519, perennial or annual, long yellow inflorescence on long peduncle, very small pod reniform or lenticular with only one seed, mature pod and seed black. Very similar to *M. lupulina* is *Medicago secundiflora* Durieu, http://www.theplantlist.org/tpl1.1/record/ild-8542, inflorescence few yellow flowers on short peduncle, mature pods brown, both ruderal and a common weed on disturbed sites along the coast, in the Jebels, not in arid inland, sometimes cultivated under irrigation from exotic seeds.

Notes:

- *Medicago lupulina* and *Medicago secundiflora* could be confused with *Melilotus* sp. A very similar annual plant with the same ecology;

however, *Melilotus* small pods have strong concentric veining and striations with 2–3 (tuberculate – rugose) small seeds. *Melilotus alba* has white flower pendulous racemes while other *Melilotus* have yellow flowers; sometimes but rarely cultivated under irrigation in Libya from imported seeds.

- Similar to an annual *Medicago* sp. is the small inconspicuous hypogeous *Factorovskya aschersoniana* (Urb.) Eig, Factorovskya aschersoniana (Urb.) Eig — The Plant List, an amphicarpic legume, looks like a very small annual *Medicago* or a *Trigonella* and is only found in the Jebel Al-Akhdar, on *Terra rossa* around El-Marj and Benghazi plains and further east in high-rainfall zones. This *Factorovskya*, like *Trifolium subterraneum*, Trifolium subterraneum L. — The Plant List (El-Marj) and *Vicia amphicarpa*, Vicia amphicarpa L. — The Plant List, when fruiting, have their flower white pedicels elongating and burying their pods into soils cracks or under stones.

Onobrychis caput-galli (L.) Lam.

Synonym(s): *Hedysarum caput-gallii* L., Hedysarum caput-gallii L. — The Plant List
Native:
Vernacular name: *Haddida* (?)
Raunkiaer's life-form: Therophyte
Description and morphology: Annual, prostrate/decumbent, 10–30 cm, pod valves always remaining attached to base of plant, alternate imparipinnate leaves 5–10 cm long on long peduncle with 4–6 pairs of small oblong obovate hairy folioles, terminal solitary flower or raceme of 3–6 pink-violet flowers with hairy calyx on long peduncles longer than leaves, pod (5–8 mm) as unique article ovoid laterally compressed foveolate-pitted finely hairy with short thin spines (1–2 mm) on crest, unique seed per pod, dark brown to black seed (3–5 mm) reniform smooth.
Reproduction: Flowering January–April, seeds
Pastoral importance: High, spring forage
Palatability and fodder value: High, as green or dry forage
Toxicity: None
Economic interest: Forage, could be used for rangeland reseeding
Habitat: Sandy/sandy-loam soils, MAR > 100–150 mm, more common in higher rainfall zones
Distribution:

- In Libya, Al-Assah, Jebel Nafusa, Gharyan to Nalut, northern Hamada, Sirte, south Jebel Al-Akhdar to Marmarica
- In arid zones, North Africa, Middle East

Endemism: NA

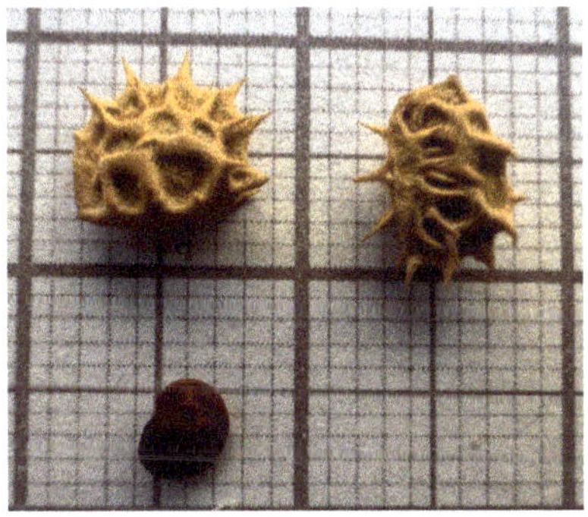

Photo 7.286. *Onobrychis caput-galli*, pods and seed. (Pod samples from South Australian Legumes collection PIRSA- SARDI)

Onobrychis crista-galli (L.) Lam.

Synonym(s): *Hedysarum crista-galli* L., Onobrychis crista-galli (L.) Lam. — The Plant List
Native:
Vernacular name: *Haddida* (?), *Senn el jej* (?)
Raunkiaer's life-form: Therophyte
Description and morphology: Similar to *Onobrychis caput-galli*, annual, prostrate/decumbent, 10–30 cm, pod always remaining attached to base of plant, alternate imparipinnate leaves 5–15 cm long on long peduncle with 8–12 pairs of oblong or narrow barely hairy folioles, mostly solitary (1 or 2) pink-violet flowers with hairy calyx on peduncles equal or shorter than leaves, pod (5–12 mm) as unique article glabrous/glossy ovoid laterally compressed deep foveolate-pitted with flat protuberant dentate crest with strong spines (2–4 mm), 1–2 seeds per pod, dark brown to black seed (2–5 mm) reniform angular smooth.
Reproduction: Flowering January–April, seeds
Pastoral importance: High, spring forage

Palatability and fodder value: High, as green or dry forage
Toxicity: None
Economic interest: Forage, could be used for rangeland reseeding
Habitat: Sandy/sandy-loam soils, gravelly, MAR > 100 mm, more common in lower-rainfall zones
Distribution:

- In Libya, Jebel Nafusa, Gharyan to Nalut, northern Hamada, Sirte, Jebel Al-Akhdar to Marmarica
- In arid zones, North Africa, Middle East

Endemism: NA

Photo 7.287. *Onobrychis crista-galli*, pods and seed. (Pod samples from South Australian Legumes collection PIRSA- SARDI (South Australia))

Ononis natrix L.

Synonym(s): Ononis natrix L. — The Plant List
Native:
Vernacular name: *Shedida*
Raunkiaer's life-form: Nanophanerophyte
Description and morphology: Small (perennial) shrub (20–50 cm high), very variable, woody base, plant sometimes hairy/pilose/sticky/glandular, erect stems, all leaves subsessile trifoliate (except floral leaves unifoliate) obovate/lanceolate serrate/dented, recurved, flower on awned pedicel with yellow corolla (veined red) and longer than calyx, pendulous cylindrical mucronate pods (10–20 mm long × 2–4 mm diam.) twice longer than calyx, seeds tuberculate.
Reproduction: Flowering December–May, seeds
Pastoral importance: Medium, colonizer on stable sand
Palatability and fodder value: Good, grazed mostly by dromedaries
Toxicity: None
Economic interest: None
Habitat: On deep fixed sand/sandy-loam soils (sometimes over saline soils – close to *sebkha*), above 150 mm MAR
Distribution:

- In Libya, all coastal zones, Jeffara plain, Sirte to Jebel Al-Akhdar
- In arid zones, southern Europe, North Africa to Middle East

Endemism: NA

Note: *Ononis natrix* is very variable; *Ononis* sp. are numerous in Libya, could be annuals or perennials, with violet, pink or yellow flowers, sometimes with viscous stems and leaves, occasionally spiny (*Ononis spinosa* subsp. *antiquorum* (L.) Briq., Ononis spinosa subsp. antiquorum (L.) Briq. — The Plant List, perennial, only on coastal Jebel Al-Akhdar.

Photo 7.288. *Ononis natrix*, Magdar-es-Sbot, Al-Assah.

Psoralea bituminosa L.

Synonym(s): http://www.theplantlist.org/tpl1.1/record/ild-7357, or *Bituminaria bituminosa* (L.) C.H. Stirt., http://www.theplantlist.org/tpl1.1/record/ild-7357
Native:
Vernacular name:
Raunkiaer's life-form: Nanophanerophyte
Description and morphology: Perennial lax herb with central rooting system and often remaining dry stems from previous season, strong tar scent, large alternate trifoliate leaves with entire leaflets with white margins, lower leaves obovate, upper leaves lanceolate, dense head with 10–30 small flowers, blue/violet (sometimes white) small flowers on long peduncle, hairy pods with glabrous elongated slightly curved – sword-like beak.
Reproduction: Flowering March–June, seeds, poor seed production, seed shedding over a long period not remaining on capitula, easy to germinate and propagate
Pastoral importance: Limited
Palatability and fodder value: Limited, poorly grazed by livestock (?), appear grazed by native rodents
Toxicity: Whole plant with a strong asphalt smell
Economic interest: None
Habitat: Heavy soils in wadis, down to 150 mm MAR
Distribution:

- In Libya, Lebda ruins, in wadis, wadi Sassu, wadi Sof-el-Jin, coastal Jebel Al-Akhdar to inland in matorral, Marmarica, not seen in Sirte
- In arid zones, Canary Islands, North Africa, to Turkey, northern Syria, Palestine

Photo 7.289. *Psoralea bituminosa* with seedlings, wadi Sassu.

Endemism: NA

Note: Currently evaluated as a potential perennial rainfed forage legume in Western Australia (Foster, 2015) from white flower plant material from the Canary Islands.

Photo 7.290. *Psoralea bituminosa* – in wadi Sassu, with a patch of *Medicago truncatula* var. *tricycle*.

Retama raetam (Forssk.) Webb

Synonym(s): *Genista raetam* Forssk., http://www.theplantlist.org/tpl1.1/record/ild-8466, or *Lygos raetam* (Forssk.) Heyw., http://www.theplantlist.org/tpl1.1/record/tro-13062624
Native:
Vernacular name: *R'tem, Er'tem*
Raunkiaer's life-form: Mesophanerophyte
Description and morphology: Tall silvery shrub (up to 2.5–3 m), branching from base, erect and then pendulous branches when loaded with pods, branches (longitudinally ribbed/grooved), nearly leafless, small spaced deciduous leaves along flexuous branches, abundant lax pendulous axillary raceme of abundant small white sweet-scented (1–5) flowers with reddish-purple calyx, pods indehiscent (shrivelled when mature) globous-spherical (one (2 (?)) seeded) with short, curved beak, yellow seeds extremely hard.
Reproduction: Seeds, seeds need scarifying (or through livestock intestinal tractus) for germination, seeds dispersion facilitated by zoochory
Pastoral importance: Feed as flowers and pods, high as role in wind erosion control
Palatability and fodder value: Barely grazed by dromedaries and goats, pods avidly ingested by

sheep (Bir-el-Ghanem at BGI project) at end of day, stem usually not or rarely grazed by livestock.
Toxicity: Alkaloids (Genistein etc.) in pods inducing lambs' sleepiness sometimes collapsing and recovering after a short time
Economic interest: Melliferous, fuelwood, charcoal, branches sometimes used as covering for temporary sheds; medicinal use/antiseptic ('Powdered dry leaves are used to heal circumcision wounds and used as an antiseptic for wounds, skin rash and pruritus' (Awen *et al.*, 2011)).
Habitat: Wide ecology, mostly on sand but extends to heavy soils, even on calcareous crust, down to 50 mm MAR on Hamada to Sirte and south Jebel Al-Akhdar

Distribution:

- In Libya, all over coastal and inland, from west to east
- In arid zones, North Africa to the Middle East/Arabian Peninsula

Endemism: NA

Note: Many variations appear in the shape of the bush, flowers and pods. *Retama raetam* bushes in Jeffara plain and along the coast have many low centrifugal branches then ascending, while *Retama raetam* on Hamada and south of Jebel Al-Akhdar have straight ascending branches.

Photo 7.291. Flowers of *Retama raetam*, Bir-el-Ghanem, Ras-el-Loefa (N 32.469 E 12.907).

Photo 7.292. Globular pods of *Retama raetam* var. *Bovei* (?), endemic Jebel Al-Akhdar – Marmarica (?), wadi Maghur, south Jebel Al-Akhdar.

Photo 7.293. Sheep feeding on pods of *Retama raetam* at dusk, from Bir-el-Ghanem BGI grazing project (N 32.400 E 12.743).

Scorpiurus muricatus L.

Synonym(s): http://www.theplantlist.org/tpl1.1/record/ild-5211
Native:
Vernacular name: *Loska* (?)
Raunkiaer's life-form: Mesotherophyte
Description and morphology: Annual herb with long (30 cm–1 m) stem on ground, stems prostrate to decumbent, leaves entire spathulate villous short-petiolate with stipules at base, yellow (sometimes orange) flowers on long peduncle (accrescent during fruiting), pods with 8 ribs coiled (½–2 coils) sometimes falcate (with longitudinal ribs either spiny/tuberculate/muricate or smooth), spiny most common, pods with 3–6 segments each containing one yellow seed, ripe pods disperse indehiscent segment.
Reproduction: Flowering March–May, seeds
Pastoral importance: High
Palatability and fodder value: Well grazed by all livestock
Toxicity: None
Economic interest: Valuable forage species, could be used for temporary well-managed pastures
Habitat: Along coastal zone, usually on loamy/heavy soils/*Terra rossa*, 300–500 mm MAR, often a ruderal plant in rainfed crops in Jebel Al-Akhdar
Distribution:

- In Libya, most common in Jebel Al-Akhdar, also present in Jebel Nafusa (Gharyan – Qusbat to Homs), rare around Tripoli, not in Sirte

Photo 7.294. *Scorpiurus muricatus*, plant, El-Marj.

Photo 7.295. *Scorpiurus muricatus* subsp. *sulcatus*, pods, Tokra.

Photo 7.296. *Scorpiurus muricatus* subsp. *sulcatus*, coils in one plan, 8–12 mm diam., nearly flat, spines only on external ribs.

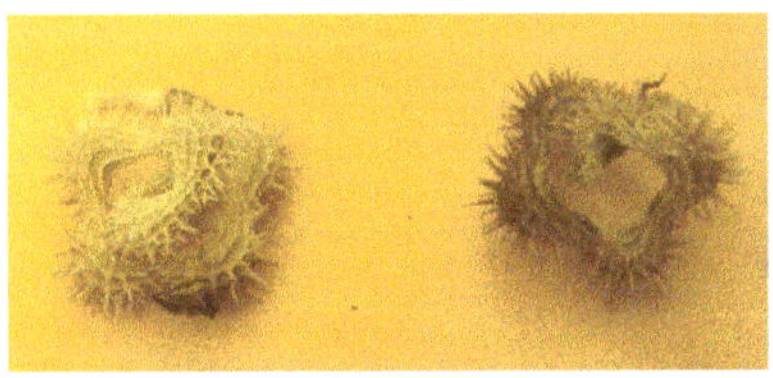

Photo 7.297. *Scorpiurus muricatus* subsp. *sub-villosus*, coils in corkscrew, 6–8 mm diam., spiny all over.

Photo 7.298. *Scorpiurus muricatus* subsp. *Inermis*, spineless pods 6–10 mm diam.

- In arid zones, southern Europe, North Africa, west Asia

Endemism: NA

Notes:
Identification of *Scorpiurus* subspecies requires pods. Pods coiled (½–2 coils) sometimes falcate with longitudinal ribs either:

- spiny all over: *Scorpiurus muricatus* subsp. *sub-villosus* (L.) Thell., http://www.theplantlist.org/tpl1.1/record/ild-37820, most common; or
- spiny on external ribs only (tuberculate, muricate): *Scorpiurus muricatus* subsp. *sulcatus* (L.) Thell., http://www.theplantlist.org/tpl1.1/record/ild-37821, all coils flat; or
- smooth: *Scorpiurus vermiculatus* L. var. *inermis*, a spineless form of *Scorpiurus muricatus*, with similar ecology.

Trigonella stellata Forssk.

Synonym(s): http://www.theplantlist.org/tpl1.1/record/ild-8035
Native:
Vernacular name:
Raunkiaer's life-form: Tachytherophyte
Description and morphology: Annual, prostrate, aromatic plant (sweet vanilla – coumarin fragrance even on dry plants), long petioled trifoliate leaves slightly dentate or serrate green emerald – glabrous sometimes thick and turgescent, many yellow and sessile flowers at nodes, numerous sessile hook-shaped pods (conglomerated star-shaped – compact at nodes) arched (with transverse prominent veins) and not beaked, small seeds ovoid-globose, smooth.
Reproduction: Flowering January–May, seeds, high seed production
Pastoral importance: Spring pasture
Palatability and fodder value: Excellent
Toxicity: None
Economic interest: None
Habitat: Sandy-loam – fixed sand, down to 50 mm MAR
Distribution:

- In Libya, from coastal zone to deep inland, from west to east
- In arid zones, from Canary Islands to Iran – north-west Pakistan

Endemism: NA

Other main *Trigonella* species in Libya (flowers yellow, rarely white or bluish):

- *Trigonella anguina* Delile, http://www.theplantlist.org/tpl1.1/record/ild-8013, prostrate to erect, leaves villose, flowers yellow, conglomerate pods sessile villose wavy-undulate, coastal zones.
- *Trigonella coerulescens* (M. Bieb.) Halacsy, http://www.theplantlist.org/tpl1.1/record/ild-8017, blue flowers, all plants densely

Photo 7.299. *Trigonella stellata*, pods, Kararim, June 1979.

hairy, pods flat erect pubescent to villous with curved beaked, only occurring in Jebel Al-Akhdar.

- *Trigonella maritima* Poir., http://www.theplantlist.org/tpl1.1/record/ild-8025, glabrous erect plant, glaucous green, conglomerate of pendulous pods on peduncle, weed, near seashore on sand, from west to east, possibly salt-tolerant.
- *Trigonella foenum-graecum* L., http://www.theplantlist.org/tpl1.1/record/ild-8021, possibly exotic, erect, strongly coumarin-scented, flowers white/cream, long (50–150 mm) unique subsessile long pods slightly arched and beaked, seeds (known as *helba*) reniform, cropped for forage and seeds used for cooking (North Africa to Indian subcontinent), a component of Libyan *bsissa* and *zumeita* mixed with barley flour; when forage fed to animal or seeds used in meals it induces strong scent in milk (lactating animal and humans), also reported as medicinal plant to control diabetes.

Notes:

- *Trigonella* sp. without flowers or pods can be confused with Annual *Medicago* or *Trifolium* sp.
- Both *Trigonella* and annual *Medicago* have the central leaflet longer and rachis contrary to *Trifolium* sp.
- *Trigonella* loose mature pods could also be confused with the ones from an *Astragalus* sp.

Vicia sativa L.

Synonym (s): Vicia sativa L. — The Plant List
Native and exotic:
Vernacular name: *Jilban, Jilbana*
Raunkiaer's life-form: Tachytherophyte
Description and morphology: Annual herbs, prostate to erect, climbing, paripinnate (sometimes barely villous) leaves, 3–8 oblong/triangular with top truncate – mucronate veined folioles, leaves ending with fine branched tendrils, flowers subsessile blue/violet, dehiscent glabrous cylindrical pods (5–12 mm long) with ripe black leathery valves remaining on plant after shedding seeds, seeds brown to black, smooth and globular, very polymorphic species sometimes amphicarpic variety rare and only in eastern Libya on *Terra rossa*.
Reproduction: Seeds, easy seed production
Pastoral importance: High
Palatability and fodder value: Excellent
Toxicity: None
Economic interest: Cultivated, seeds imported from other Mediterranean countries; used to be cropped mixed with oats and harvested as high-quality hay if harvested early (green) before *Vicia* sheds its seeds.
Habitat: Often mixed in rainfed crops, in gardens along the coast
Distribution:

- In Libya, all coastal zones, down to 150–200 mm MAR
- In arid zones, Europe, Middle East to Pakistan

Endemism: NA
Other main native *Vicia* sp. in Libya:

- *Vicia villosa* Roth, Vicia villosa Roth — The Plant List, a gracile climbing annual, leaves with 4–8 thin leaflets, tendrils at ends of leaves, flower blue/violet (sometimes with yellow wings), pods obovate, common in rainfed cereals and rangelands, at Tajura, Bir-el-Ghanem, in Jeffara plain, Dahar, on sandy/sandy loam soils.
- *Vicia lutea* L., Vicia lutea L. — The Plant List, flower yellow tinged with orange, all of plant hairy including pods, Gharyan – Yeffren, not in Sirte, to Beida – Derna, in crops, on rocky grounds and loamy-clay soils.

Photo 7.300. *Vicia sativa* (type – Aziziya).

FAGACEAE

Quercus coccifera L.

Synonym(s): http://www.theplantlist.org/tpl1.1/record/kew-172199, or *Quercus calliprinos* Webb, http://www.theplantlist.org/tpl1.1/record/kew-171917
Native:
Vernacular name: *Ballout* (?), *Karouch* (?)
Raunkiaer's life-form: Mesophanerophyte
Description and morphology: Robust evergreen monoecious tree (5–10 m high), many contorted limbs, coarse dark reticulated bark, whitish/pubescent young stems, leaves ovate dentate (spinescent to nearly entire) shiny green glabrous (both side) alternate on short petiole, male flowers cream in long spikes, female flowers (1–4) with spinescent cupola at base, cupola 1/3–1/2 acorn length, acorn (1–4 cm long, 10–12 mm diam.) containing one seed in woody mericarp, seeds usually with bitter taste, possible ecotypes with sweet taste seeds.
Reproduction: Flowering March–May, seeds
Pastoral importance: Medium
Palatability and fodder value: None, except acorns eaten by all livestock
Toxicity: None
Economic interest: Afforestation, fuelwood, (poor) building material, source of tannin from bark, acorns used as livestock feed, sweet acorns from particular trees once collected and consumed by locals in times of food shortage.
Habitat: In matorral, on *Terra rossa*, above 400 mm MAR
Distribution:

- in Libya, Jebel Al-Akhdar (El-Marj, Shahat-Marsa Suza, Tolmeita, wadi Kuf, Derna, etc.);
- in arid zones; southern Europe, North Africa, Aegean islands, west Asia

Endemism: Jebel Al-Akhdar only

Note: No *Quercus* sp. historically reported from highest-rainfall zones in western Libya (Gharyan, Jendouba – Yeffren); possibly eradicated long ago.
Other *Quercus* sp. in Libya

- *Quercus ilex* L, http://www.theplantlist.org/tpl1.1/record/kew-173162, could be confused with *Quercus coccifera. Quercus ilex* has coriaceous spiny leaves with ventral face white – pubescent, usually small dense low bushes (up to 1 m high (?) in matorral), strong invasive rooting system with root suckers aggressively coppicing. Doubtful in Libya (?). If currently present, only in Jebel Al-Akhdar in high-rainfall zones (?). Common in garrigue in southern Europe.
- *Quercus suber* L., http://www.theplantlist.org/tpl1.1/record/kew-175312, primary source of cork, possibly introduced from northern Tunisia – Algeria, or Sardinia, usually on soils of low pH, hence not native in Libya, poor specimens at Sidi-Mesri garden.

Photo 7.301. *Quercus coccifera*, Tree, (between Shahat and Marsa Suza)

Photo 7.302. *Quercus coccifera*, trunk and bark, between Shahat and Marsa Suza.

Photo 7.303. *Quercus coccifera*, male flowers, between Shahat and Marsa Suza.

Photo 7.304. *Quercus coccifera*, acorns, between Shahat and Marsa Suza.

Photo 7.305. *Quercus coccifera*, leaves polymorphism and ripe acorns, between Shahat and Marsa Suza.

FRANKENIACEAE

Frankenia laevis L.

Synonym(s): http://www.theplantlist.org/tpl1.1/record/kew-2814219
Native:
Vernacular name: *Sfishifa* (?)
Raunkiaer's life-form: Chamaephyte
Description and morphology: Hemispherical perennial shrublet (10–30 cm), all branches woody, flowering branches erect, non-flowering branches prostrate-decumbent, small verticillate leaves (3–6 mm) slightly fleshy with revolute margin and dotted with white crystal specks, discrete small flowers solitary along stems, corolla pink-violet with entire lobes largely protruding from calyx, small capsule hidden in calyx.
Reproduction: Flowering February–June, seeds
Pastoral importance: Medium
Palatability and fodder value: Grazed by dromedaries mostly
Toxicity: None
Economic interest: None
Habitat: Azonal (?), from coastal zone to inland, halophyte, in *sebkha*, saline sandy/loamy soils, down to 50 mm MAR
Distribution:

- In Libya, Al-Assah, Derj, Taworga, Sirte coastal, south Jebel Al-Akhdar – Marmarica
- In arid zones, southern Europe, North Africa

Endemism: NA

Notes:

- All Libyan *Frankenia* sp. in *sebkha*, saline sandy/loamy soils, gypsophyte (?), down to 50 mm MAR.
- Could be confused with *Frankenia hirsuta* L., Frankenia hirsuta L. — The Plant List, chamaephyte that also has solitary white to pink-lilac flowers along stems but leaves narrower (acuminate), tightly nested and not fleshy. Same ecology as *Frankenia laevis.*
- Common also found on saline (possibly temporary waterlogged) loamy soils is *Frankenia pulverulenta* L., http://www.theplantlist.org/tpl1.1/record/kew-2814273, a small annual, many dichotomous branches, small (2–5 mm long) sessile spathulate leaves in clusters of four, leaves pulverulent – white dots of crystals, solitary minute pink flowers along reddish stems.

Photo 7.306. *Frankenia laevis*, shrublet, Al-Assah, Jbebina.

Photo 7.307. *Frankenia laevis*, flowers along stems, Al-Assah, Jbebina.

GERANIACEAE

Erodium arborescens (Desf.) Willd.

Synonym(s): http://www.theplantlist.org/tpl1.1/record/kew-2798114, or *Erodium hussoni* Boiss., http://www.theplantlist.org/tpl1.1/record/kew-2798282, or *Geranium arborescens* Desf., http://www.theplantlist.org/tpl1.1/record/kew-2822837
Native:
Vernacular name: *Merjed* (?)
Raunkiaer's life-form: Hemicryptophyte
Description and morphology: Perennial shrublet (10–40 cm high), taproot with dark brown rounded tubers (2–3 cm long), triangular entire fleshy glabrous (dark green) cordate leaves with dentate/denticulate lobes, terminal inflorescences on peduncle longer than leaves, with 2–4 large (20–40 mm diam.) pink showy flowers with black centre, long (5–8 cm) fruits pendulous dehiscent, freeing beaked pointed seed with long straight plumose awn (10–15 times longer that seed) cork-screwed above seed.
Reproduction: Flowering February–May, seeds (anemochory), tubers
Pastoral importance: Medium
Palatability and fodder value: High, mostly small ruminants
Toxicity: None
Economic interest: Small tubers consumed fresh, sweet taste (?)
Habitat: Limestone foothills, loamy soils/colluvium, down to 150 mm MAR, also in wadis
Distribution:

- In Libya, northern slopes and foothills of Jebel Nafusa
- In arid zones, North Africa, Middle East (?)

Endemism: NA

Notes:

- *Erodium* sp. are numerous in Libya and problematic to determinate without flower, fruits and plumose awn.
- *Erodium* sp. have a remarkable seeding/anchoring mechanism facilitating germination. The plumose seeds are carried away by the wind and land away from the mother plant. The plumose awn curls when drying while the achene sits on the soil. When it rains, the arched plume anchors itself on the soil surface while the corkscrew part above the achene becomes hygroscopically active, unwinds itself and drills the achene into soil cracks or into the soft ground, hence allowing the seed to germinate into a favourable media.

Photo 7.308. *Erodium arborescens*, flowering, Tiji, Giado.

Photo 7.309. *Erodium arborescens*, plant, Beni Walid.

Photo 7.310. *Erodium arborescens*, flower and leaves, Beni Walid.

Erodium glaucophyllum (L.) L'Hér.

Synonym(s): http://www.theplantlist.org/tpl1.1/record/kew-2798248
Native:
Vernacular name: *Regma* (?)
Raunkiaer's life-form: Hemicryptophyte – gypsophyte
Description and morphology: Perennial herb (10–30 cm high), taproot with dark brown rounded tubers (2–3 cm long), elongated triangular fleshy glaucous (emerald) cordate basal leaves with dentate/denticulate lobes, smaller opposite leaves at swollen nodes, upright terminal inflorescences on long peduncle with 2–4 large purple flowers (15–20 mm diam.), long (5–8 cm) fruits erect dehiscent, freeing acute-beaked seed with plumose awn (6–10 times longer that seed), corkscrewed and at angle to short-haired seed.
Reproduction: Flowering December–April, seeds (anemochory), tubers
Pastoral importance: Medium
Palatability and fodder value: High, mostly small ruminants
Toxicity: None
Economic interest: Small tubers consumed fresh, sweet taste (?)
Habitat: Loamy soils, on gypsum crust, down to 50 mm MAR
Distribution:

- In Libya, all over coastal zone and inland, on gypsum soil and crust
- In arid zones, North Africa to Iran

Endemism: NA

Photo 7.311. *Erodium glaucophyllum*, plant, Gharyan.

Photo 7.312. *Erodium glaucophyllum*, flower and fruit, Gharyan.

Photo 7.313. *Erodium glaucophyllum*, flower and fruit with awns, Gharyan.

Erodium crassifolium subsp. *hirtum* (Forssk.) Guitt.

Synonym(s): https://www.ville-ge.ch/musinfo/bd/cjb/africa/details.php?langue=an&id=226129, or *Erodium hirtum* Willd., http://www.theplantlist.org/tpl1.1/record/kew-2798274
Native:
Vernacular name: *Kelkhel-el-rhoul* (?)
Raunkiaer's life-form: Hemicryptophyte
Description and morphology: Small (20–40 cm) perennial, lateral root with dark brown rounded tubers (10–20 mm long), mostly prostrate with decumbent stems with petiolate silvery white to glabrescens triangular deeply pinnatisect leaves (3–8 cm long), inflorescence 3–6 pink flowers long pedicellate above leaves, long dehiscent fruits with straight plume topping acute achene.
Reproduction: Flowering February–May, seeds (anemochory), tubers
Pastoral importance: Medium
Palatability and fodder value: Grazed by all livestock
Toxicity: None
Economic interest: None
Habitat: Sandy loam/loam soils, down to 150 mm MAR

Distribution:

- In Libya, Jeffara plain, Jebel Nafusa, Sirte, south Jebel Al-Akhdar
- In arid zones, North Africa, Middle East to Iraq (?)

Endemism: NA

Note: Usually known as *Erodium hirtum.*

Photo 7.314. *Erodium hirtum*, plant, Tarhuna, Wishtata.

Photo 7.315. *Erodium hirtum*, leaves and flower, Tarhuna, Wishtata.

CARYOPHYLLACEAE (including ILLECEBRACEAE)

Note: The Caryophyllaceae family includes the Illecebraceae family. The main difference is the Caryophyllaceae have hypogynous (tubular) flowers with proper petals and dehiscent fruits while the Illecebraceae have perigynous flowers with petaloid staminodes and indehiscent fruits.

Gymnocarpos decander Forssk.

Synonym(s): http://www.theplantlist.org/tpl1.1/record/tro-50211412
Native:
Vernacular name: *Guejroud* (?), *Jarrad* (?)
Raunkiaer's life-form: Chamaephyte
Description and morphology: Compact woody shrub (10–50 cm high), often hemispherical due to overgrazing, strong taproot, much divaricate intricate dichotomous branches and stems, grey/whitish bark, small opposite fleshy leaves (5–20 mm long) narrow cylindrical mucronate on swollen nodes, inflorescence in small (8–12 mm diam.) globose scarious clusters of flowers, petaloid green-white, sepals reddish-purple, minute oblong seed.
Reproduction: Flowering January–May, seeds (seed production is highly erratic)
Pastoral importance: High
Palatability and fodder value: Heavily grazed by all livestock
Toxicity: None
Economic interest: Fuelwood, one of the best Libyan rangelands plants in terms of drought and grazing resistance
Habitat: On calcareous crust, shallow rocky soils, on gypsum crust, down to 50 mm MAR
Distribution:

- In Libya, all Jeffara plain, Jebel Nafusa, Hamada, Sirte, south Jebel Al-Akhdar
- In arid zones, North Africa, Arabian Peninsula to Afghanistan

Endemism: NA

Note: Appears in Forskål (1775) as *Gymnocarpos decandrum* (CENTURIA III, PENTANDR. – DODECANDR, p. 65)

Photo 7.316. *Gymnocarpos decander*, ungrazed, wadi Nefad, Seddada, on *reg*.

Photo 7.317. *Gymnocarpos decander*, plant grazed, Al-Assah, Jbebina, Bir Saman, on gypsum crust.

Photo 7.318. *Gymnocarpos decander*, heavily grazed, El-Bedan, Ajdablya, on calcareous crust.

Photo 7.319. *Gymnocarpos decander*, inflorescences and leaves, Al-Assah, Jbebina, Bir Saman.

Photo 7.320. *Gymnocarpos decander*, end of flowering, fruiting and leaves, Gharyan northern foothill.

Herniaria hemistemon J. Gay

Synonym(s): *Herniaria hemistemon* J. Gay — The Plant List
Native:
Vernacular name:
Raunkiaer's life-form: Hemicryptophyte
Description and morphology: Small (10–30 cm diam.) prostrate/creeping herb (perennial (?), woody base (?), small (3–5 mm long) opposite elliptical acuminate sessile slightly hairy stipulate leaves, inflorescences minute in compact cluster of 4–6 tetramerous yellow-greenish flowers with four unequal sepals.
Reproduction: Flowering February–May, seeds

Pastoral importance: Low
Palatability and fodder value: Barely grazed
Toxicity: None
Economic interest: None
Habitat: Sandy loam, on gypseous soils (sometimes slightly saline), around *sebkha*, around overflowing artesian wells, down to 50 mm MAR
Distribution:

- In Libya, Jeffara plain, south Jebel Nafusa, Hamada, Sirte south Jebel Al-Akhdar
- In arid zones, southern Europe, North Africa, Middle East to Iran/Pakistan (?)

Endemism: NA

Photo 7.321. *Herniaria hemistemon*, plant,Al-Assah, Jbebina.

Photo 7.322. *Herniaria hemistemon*, flower and leaves, (Al-Assah, Jbebina).

Note:

- *Herniaria hemistemon* could be confused with *Herniaria fontanesii* J. Gay., http://www.theplantlist.org/tpl1.1/record/kew-2846920, a small yellowish-green shrub, very similar ecology and distribution as *H. hemistemon* but, clearly perennial, with erect branches (more like a small compact bush, with longer leaves and internodes; a western species extending to Marmarica, and apparently not extending further into the Middle East (?).
- Other Illecebraceae, such as *Herniaria* sp., are *Paronychia* sp. and *Sclerocephalus* sp., usually small annual or perennial prostrate plants.
- Classified as Caryophyllaceae are *Agrostemma* sp., *Arenaria* sp., *Cerastium* sp., *Gypsophylla* sp., *Minurtia* sp., *Spergula* sp., *Spergularia* sp., *Stellaria* sp., *Telephium* sp., *Vaccaria* sp. with rare *Dianthus* sp. and *Silene* sp.

Most are inconspicuous small annual or perennial plants contributing to feeding livestock in winter-spring.

Paronychia argentea Lam.

Synonym(s): *Chaetonychia paronychia* (L.) Samp., http://www.worldfloraonline.org/taxon/wfo-0000466806
Native:
Vernacular name: *Chai el Arab* (?)
Raunkiaer's life-form: Hemicryptophyte

Photo 7.323. Dry *Paronychia argentea*, Hamada, Bir-el-Kelb (N 31.093 E 12.285).

Description and morphology: Small prostrate creeping herb (10–30 cm diam.), annual with early stage stems reddish or perennial with long woody stems, small glabrous opposite mucronate flat leaves, small green-whitish flowers hidden in silvery bracts, globulous papery-like inflorescences when dry on long pedicels, calyx hairy, indehiscent fruits, remaining dry long during summer.
Reproduction: Flowering January–May, seeds April–June
Pastoral importance: Low
Palatability and fodder value: Grazed by all livestock
Toxicity: None
Economic interest: None, reported as substitute for tea (?)
Habitat: common on sandy soils all over Libya, down to deep into desert 50 mm MAR, also on heavy soils in Jebel Al-Akhdar (?)
Distribution:

- In Libya, Jeffara plain, Jebel Nafusa, Sirte, south Jebel Al-Akhdar
- In arid zones, southern Europe, North Africa, Middle East to Pakistan

Endemism: NA
Note: *Paronychia arabica* (L.) DC., http://www.theplantlist.org/tpl1.1/record/kew-2533099, very similar to *Paronychia argentea*, but an annual with red dichotomous prostrate branches with papery bracts on flowers at nodes along stem, both with similar ecology and distribution.

Vaccaria pyramidata Medik.

Synonym(s): http://www.theplantlist.org/tpl1.1/record/kew-2456758, or *Saponaria vaccaria* L., http://www.theplantlist.org/tpl1.1/record/kew-2485057
Native:
Vernacular name: *Suboun-el-Fahr* (?)
Raunkiaer's life-form: Mesotherophyte
Description and morphology: A gracile annual (20–40 cm high), erect, glabrous, thin dichotomous stems, leaves sessile cordate perfoliate (amplexicaule) elongated triangular (4–8 cm long × 1–2 cm wide at base), Inflorescences lax with terminal pink pedicelled flowers, calyx with inflated base and green nerves, dehiscent denticulate capsule ovoid concealed in calyx with four openings at apex, verrucose spherical seeds.

Reproduction: Flowering April–May, seeds
Pastoral importance: Low
Palatability and fodder value: Grazed by all livestock
Toxicity: None
Economic interest: Whole plant contains Saponin, reported used as substitute for soap – soap-like foam when dried roots crushed in water; seeds reported lactogenic (?)
Habitat: Common on sandy soils, often a common weed in Jeffara rainfed crops and nearby rangelands, down to 150 mm MAR, also on heavy soils in Jebel Al-Akhdar (?)
Distribution:

- In Libya, Jeffara plain, Jebel Nafusa, Sirte (?), Jebel Al-Akhdar
- In arid zones, southern Europe, North Africa, Middle East to Pakistan

Endemism: NA

Photo 7.324. *Vaccaria pyramidata*, plant and flowers, Aziziya.

IRIDACEAE

Crocus boulosii Greuter

Synonym(s): http://www.theplantlist.org/tpl1.1/record/kew-327190
Native:
Vernacular name:
Raunkiaer's life-form: Geophyte
Description and morphology: Bulb/corm ovoid (1–2 cm diam.) covered with older fibrous leaf base, acaulescent, basal new leaves erect linear narrow (1–3 mm) appear with flowers (often after a bushfire), short (3–6 cm (?) erect) scape white papery bearing unique white flower (also reported brown-yellowish (?)), end of style and yellow stigma divided, capsule plurilocular cylindrical (1.5–2.5 cm) trilocular dehiscent, seeds membranous.
Reproduction: Flowering October–December, seeds and corms
Pastoral importance: None
Palatability and fodder value: None
Toxicity: None
Economic interest: Ornamental
Habitat: On *Terra rossa*, heavy soils, down to 300 mm MAR in garrigue
Distribution:

- In Libya, eastern zones (endemic to Libya)
- In arid zones (?)

Endemism: Jebel Al-Akhdar only

Note: A close relationship with common cultivated *Crocus sativus* L., http://www.theplantlist.org/tpl1.1/record/kew-327454, larger leaves (3–5 mm wide), flowers whitish-violet, styles barely divided/entire, long stigma orange/red; usually cultivated (or escaped from garden), stigma collected source of commercial saffron spice (dried filiform style and stigma) used as condiment.

Photo 7.325. *Crocus boulosii*, plant, Toknis.

Photo 7.326. *Crocus boulosii*, flower, Toknis.

Iris sisyrinchium L.

Synonym(s): http://www.theplantlist.org/tpl1.1/record/kew-322387, or *Moraea sisyrinchium* (L.) Ker Gawl., http://www.theplantlist.org/tpl1.1/record/kew-323573
Native:
Vernacular name:
Raunkiaer's life-form: Geophyte
Description and morphology: Fibrous bulb/corm ovoid (1–2 cm diam.), short slender stem (6–10 cm) enveloped with lanceolate scape, long (10–40 cm) unique basal canaliculate leaf, fascicule of 2–4 blue zygomorphic flowers with white-yellow on label, fruits included in papery spathe as long capsule ending with rostrum, small rugose black seeds.
Reproduction: Flowering in spring, seeds and corms
Pastoral importance: None
Palatability and fodder value: None
Toxicity: None
Economic interest: Ornamental (?)
Habitat: Sandy soils
Distribution:

- In Libya, in Jeffara, Sirte to Ajdabiya, south Jebel Al-Akhdar
- In arid zones (?)

Endemism: NA

Photo 7.327. *Iris sisyrinchium*, BGII, Bir-el-Ghanem, April 1978.

LAMIACEAE

Ballota hirsuta Benth.

Synonym(s): http://www.theplantlist.org/tpl1.1/record/kew-19602, or *Pseudodictamus hirsutus*, https://powo.science.kew.org/taxon/urn:lsid:ipni.org:names:77167653-1
Native:
Vernacular name: *Rubia* (?)
Raunkiaer's life-form: Chamaephyte
Description and morphology: Perennial, large dense hemispherical bush (50–200 cm diam.), rhizomatous, woolly lanate – whitish shrub all over, all stems quadrangular, lower leaves petiolate, upper leaves subsessile, opposite leaves cordate denticulate embossed/goffered with upper surface green and lower face tomentose white, top of branch with interrupted conglomerate lax inflorescences (10–15 mm diam.) in verticils with 5–8 pinkish-white small flowers in glomerulus clusters at upper axis, villous stellar campanulate calyx with ten unequal pointed soft accrescent teeth, much polymorphic.
Reproduction: Flowering March–May, seeds
Pastoral importance: Medium
Palatability and fodder value: Not touched by small ruminants, dromedary feed
Toxicity: None
Economic interest: Potential soil fixing plants on bad loamy lands and wadis; medicinal (treatment of respiratory disorder), reported nectariferous for bees
Habitat: Deep loamy soils, 150–300 MAR, often ruderal on overgrazed sites near dwellings
Distribution:

- In Libya, Jebel Nafusa and along coast to Taworga/Bu-Gren and Jebel Al-Akhdar, not in Sirte (?)
- In arid zones, southern Europe, North Africa to northern Pakistan

Endemism: NA

Note: When not flowering, *Ballota* sp. is often confused with *Marrubium vulgare* or *Phlomis* sp., all with similar ecology.

Photo 7.328. *Ballota hirsuta*, bush, Mafsel, south Gendoula.

Photo 7.329. *Ballota hirsuta*, floral stem, Mafsel, south Gendoula.

Photo 7.330. *Ballota hirsuta*, flowers and calyx, Mafsel, south Gendoula.

Lavandula coronopifolia Poir.

Synonym(s): http://www.theplantlist.org/tpl1.1/record/kew-108999, Syn. *Lavandula stricta* Delile, http://www.theplantlist.org/tpl1.1/record/kew-109107

Native:
Vernacular name: *Kamoun-el-Jmel* (?)
Raunkiaer's life-form: Chamaephyte
Description and morphology: Small perennial (20–40 cm), no scent, woody taproot, smooth stems, lower leaves deeply dissected with narrow linear lobes, few or no upper leaves, single long stiff peduncle (15–20 cm) bearing short (2–4 cm) inflorescence in lax terminal spike with 2–3 pale blue flowers widely spaced along stem, glabrous calyx (half to quarter-length of flower).
Reproduction: Flowering March–May, seeds
Pastoral importance: Limited
Palatability and fodder value: Good, grazed by all livestock
Toxicity: None
Economic interest: None, no fragrance like *Lavandula multifida*
Habitat: Mostly Saharan, **w**adis, sandy=gravelly soils, often protected under larger plants, 100–200 MAR or in rocky sites not accessible to livestock.
Distribution:

- In Libya, quite rare, Jebel Nafusa (Gharyan, southern slope from Nalut to Mizda), not in Sirte or Jebel Al-Akhdar (?) could be confused with *Lavandula multifida* L., occurring in Jebel Nafusa in higher-rainfall zones
- In arid zones, Sahara, North Africa to Iran

Endemism: NA – new species for Libya

Photo 7.331. *Lavandula coronopifolia*, Mizda in rocky-sandy wadi (N 31.53 E 13.00).

Marrubium vulgare L.

Synonym(s): http://www.theplantlist.org/tpl1.1/record/kew-120952
Native:
Vernacular name: *Rubia* (?)
Raunkiaer's life-form: Chamaephyte
Description and morphology: Perennial, hemispherical woolly – whitish hirsute shrub (50–200 cm diam.), erect straight branches, old stems woody round, young stems quadrangular, petiolate lower leaves opposite, upper leaves subsessile, and deeply veined leaves cordiform crenate with upper face green lower face white, top of branch with conglomerate inflorescences in verticils with 8–15 whitish small white flowers in glomerulus clusters at upper axis, foul smell, scarious campanulate hairy calyx with five main pointed soft teeth and smaller teeth in-between.
Reproduction: Flowering March–May, seeds
Pastoral importance: Poor
Palatability and fodder value: Poor
Toxicity: None
Economic interest: Medicinal (fresh mashed leaves used as cataplasm to heal wounds, cuts), potential soil fixing plants on badlands and wadis

Photo 7.333. *Marrubium vulgare*, Inflorescence, wadi Maghur, Al-Jahishiyah.

Photo 7.332. *Marrubium vulgare*, bush, Mafsel, south Gendula.

Photo 7.334. *Marrubium vulgare*, flowers and calyx, (Wadi Maghur).

Habitat: Deep loamy soils, sometimes on rocky sites, 200–500 MAR

Distribution:

- In Libya, Jebel Nafusa and Jebel Al-Akhdar, not in Sirte
- In arid zones, southern Europe, North Africa

Endemism: NA

Note: When not flowering, *Marrubium vulgare* is often confused with *Ballota* sp. or *Phlomis* sp., all with similar ecology.

Phlomis floccosa D. Don

Synonym(s): http://www.theplantlist.org/tpl1.1/record/kew-152232, or *Phlomis bicolor* (Viv.) Benth, http://www.theplantlist.org/tpl1.1/record/kew-152139

Native:

Vernacular name: *Shaina* (?)

Raunkiaer's life-form: Chamaephyte

Description and morphology: Perennial, hemispheric woody shrub (50–150 cm), lanate all over, light green-yellowish, long petiole cordate large lower deeply veined – embossed opposite leaves (3–6 cm), upper leaves subsessile, globulose inflorescence large (15–20 mm) with 4–8 yellow flowers turning brown when dry, stellate densely woollen calyx with five radiating hooked teeth, large black trigone seed (4–6 mm).

Photo 7.335. *Phlomis floccosa*, bush, Mafsel, south Gendoula.

Photo 7.336. *Phlomis floccosa*, flowers, Mafsel, south Gendoula.

Reproduction: Flowering February–June, seeds

Pastoral importance: Limited, soil-fixing ability

Palatability and fodder value: Ungrazed, often colonizing overgrazed sites

Toxicity: None

Economic interest: Possibly medicinal, ornamental

Habitat: Heavy soils, *Terra rossa*, Mediterranean soils, common in Juniper forest, 200–500 mm MAR

Distribution:

- In Libya, Jebel Nafusa (Gharyan, Msallata, Cussabat – Homs), Jebel Al-Akhdar, not in Sirte (?)

Photo 7.337. *Phlomis floccosa*, Inflorescence and calyx, Mafsel, south Gendoula.

- In arid zones, southern Europe, Mediterranean, Greek islands, Turkey, Middle East (Syria)

Endemism: NA

Rosmarinus officinalis L.

Synonym(s): http://www.theplantlist.org/tpl1.1/record/kew-179873
Native:
Vernacular name: *Klil*
Raunkiaer's life-form: Chamaephyte
Description and morphology: Perennial aromatic shrublet (30–100 cm), long erect straight branches, old branches brown woody, young branches soft green; linear leathery leaves shiny green above and tomentose underneath with marked midrib, margins revolute; inflorescences in short spikes, purple or white flowers bilabiate with upper lip bifid and lower convex tri-lobed, accrescent tomentose/adpressed hairy calyx with two teeth, seeds oval dark brown.

Photo 7.338. *Rosmarinus officinalis*, wadi Turgut, Tarhuna.

Photo 7.339. *Rosmarinus officinalis*, flowers, wadi Turgut.

Reproduction: Flowering all year round, seeds
Pastoral importance: Limited
Palatability and fodder value: Not touched by livestock
Toxicity: None
Economic interest: Medicinal, antiseptic, aromatic, nectariferous for bees, collected and sold on local market, essential oil used for perfume, dry leaves widely used for seasoning food.

Habitat: Loamy soils, in Jebel, 300–500 mm MAR, forest relict
Distribution:

- In Libya, Jebel Nafusa (Yeffren to Homs), all over Jebel Al-Akhdar, not in Sirte
- In arid zones, southern Europe, Mediterranean regions, from North Africa to the Middle East

Endemism: NA

Salvia lanigera Poir.

Synonym(s): http://www.theplantlist.org/tpl1.1/record/kew-182986, or *Salvia verbenaca* ssp. *lanigera* (Poir.) O. Bolòs & Vigo, http://www.theplantlist.org/tpl1.1/record/kew-184125
Native:
Vernacular name: *Shejerat-el-Jmel* (?)
Raunkiaer's life-form: Hemicryptophyte
Description and morphology: Perennial woolly-hairy herbs or small shrub (10–40 cm), woody taproot, mostly erect stems, leaves petiolate opposite deeply dissected pinnatifid embossed, lobes narrow linear, lax interrupted verticillate inflorescences with 4–6 deep blue to violet robust flowers, calyx hairy accrescent, many pyriform black seeds.
Reproduction: Flowering February–June, seeds
Pastoral importance: High
Palatability and fodder value: Well gazed by all livestock
Toxicity: None
Economic interest: Possibly medicinal (?), mildly aromatic, strong soil colonizer
Habitat: Fixed sand to loamy soils, 150–400 mm MAR, not on dunes or on saline sites
Distribution:

- In Libya, Jeffara plain, Jebel Nafusa, Sirte, Jebel Al-Akhdar
- In arid zones, southern Europe, Mediterranean, North Africa

Endemism: NA

Notes:

Very close to:

- *Salvia verbenaca* L., http://www.theplantlist.org/tpl1.1/record/kew-184123, smaller than *Salvia lanigera*, glabrous plant with all basal oblong – triangular – cordate long petioled leaves with irregular round dents, flowers from white to pink – blue.
- Also could be confused with *Salvia aegyptiaca* L., http://www.theplantlist.org/tpl1.1/record/kew-181899, woody gracile glabrous small open bush with rigid stems, small opposite leaves narrow elliptical crenate-denticulated and short petioled, long internodes with pair of small pale blue flowers.

All three have similar ecology.

Photo 7.340. *Salvia lanigera*, BGII, Bir-el-Ghanem.

Teucrium pilosum (Decne.) Asch. & Schweinf.

Synonym(s): http://www.theplantlist.org/tpl1.1/record/kew-203213, or *Teucrium decaisnei* C. Presl., http://www.theplantlist.org/tpl1.1/record/kew-202863
Native:
Vernacular name: *Jaada soufi* (?)

Raunkiaer's life-form: Hemicryptophyte
Description and morphology: Perennial small, horizontal/prostrate sticky densely woolly shrublet (10–30 cm diam.), barely aromatic, leaves linear sessile with convolute-crenate margins on creeping stems, ending with densely hairy-lanate-furry spherical verticils, very small yellowish flowers with prominent concave shield and two pointed lateral lobes.
Reproduction: Flowering February–June, seeds
Pastoral importance: Poor
Palatability and fodder value: Appear not touched by small ruminants
Toxicity: None

Photo 7.341. *Teucrium pilosum*, wadi Maghur, Jebel Al-Akhdar.

Photo 7.342. *Teucrium pilosum*, lanate verticils and flowers, wadi Maghur, Jebel Al-Akhdar.

Economic interest: Possibly medicinal, barely aromatic
Habitat: Loamy soils, 150–300 mm MAR
Distribution:

- In Libya, south Jebel Al-Akhdar
- In arid zones, southern Europe, Mediterranean, North Africa

Endemism: Possibly only in Jebel Al-Akhdar

Teucrium polium L.

Synonym(s): http://www.theplantlist.org/tpl1.1/record/kew-203217
Native:
Vernacular name: *Jaada*
Raunkiaer's life-form: Hemicryptophyte
Description and morphology: Very polymorphic and most common, perennial aromatic spherical tomentose shrublet (10–30 cm), woody taproot, erect ascending branches, linear crenate sessile leaves with revolute margins and deep midrib, leaves green-grey both sides, globose verticils of white (pale yellow) flowers, oval seeds reticulated on surface.
Reproduction: Flowering February–June, seeds
Pastoral importance: High
Palatability and fodder value: Well grazed by all livestock
Toxicity: None
Economic interest: Possibly medicinal, frequently used as aromatic in tea by shepherds
Habitat: Sandy to loamy soils, 100–400 mm MAR

Photo 7.343. *Teucrium polium*, Al-Assah, Jbebina.

Photo 7.344. *Teucrium polium*, taproot exposed by wind erosion, south Sorman.

Photo 7.345. *Teucrium polium*, verticils and flowers, Al-Assah, Jbebina.

Distribution:

- In Libya, Jeffara plain, Jebel Nafusa, Sirte, Jebel Al-Akhdar
- In arid zones, southern Europe, Mediterranean, North Africa, Middle East to Iran

Endemism: NA

Thymus algeriensis Boiss. & Reut.

Synonym(s): http://www.theplantlist.org/tpl1.1/record/kew-204397, or *Thymus zattarellus* Pomel, http://www.theplantlist.org/tpl1.1/record/kew-205665
Native:
Vernacular name: *Zaatar*
Raunkiaer's life-form: Hemicryptophyte
Description and morphology: Polymorphic as other *Thymus* sp., perennial woody aromatic shrublet with erect dry bare pointed branches (10–40 cm high), rigid linear leaves narrow opposite with short beard on margin base, verticil flowers in long uninterrupted inflorescences along stem with some small terminal heads, corolla pink barely bilabiate, calyx ovoid with five crowned-exerted teeth and inner ciliate ring.
Reproduction: Flowering March–June, seeds
Pastoral importance: Limited
Palatability and fodder value: Well grazed by all livestock and wildlife (rabbits)
Toxicity: None
Economic interest: Medicinal, aromatic, nectariferous for bees, collected and sold on local market, commonly used for flavouring cooking
Habitat: Same ecology as *Thymus capitatus*, rocky sites (Karst), loamy soils, 200–500 mm MAR, relict of *Pinus halepensis* forests

Photo 7.346. *Thymus algeriensis*, bush, Shahat.

Photo 7.347. *Thymus algeriensis*, stem with inflorescence, Shahat.

Photo 7.348. *Thymus algeriensis*, flowers, Shahat.

Distribution:

- In Libya, Jebel Nafusa, Jebel Al-Akhdar, not in Sirte
- In arid zones, southern Europe, Mediterranean, North Africa to the Middle East and west Asia

Endemism: NA

Thymus capitatus (L.) Hoffmanns. & Link

Synonym(s): http://www.theplantlist.org/tpl1.1/search?q=Thymus+capitatus+, or *Thymbra capitata* (L.) Cav., http://www.theplantlist.org/tpl1.1/record/kew-204339, or *Coridothymus capitatus* (L.) Rchb.f., http://www.theplantlist.org/tpl1.1/record/kew-47242
Native:
Vernacular name: *Zaater* (?), *Zaatar* (?)
Raunkiaer's life-form: Chamaephyte
Description and morphology: Highly polymorphic, perennial woody aromatic shrublet often hemispherical (20–50 cm diam.) with inflorescences/flowers in terminal heads, basal branches round getting quadrangular on young stems, rigid navicular leaves narrow opposite decussate with stellar gland underneath, purple flowers upper lip corolla clearly bilabiate with trilobate lower shield, calyx long ovoid narrow at apex with teeth compressing corolla base.
Reproduction: Flowering March–June, seeds
Pastoral importance: Limited
Palatability and fodder value: Well grazed by all livestock and wildlife (rabbits)
Toxicity: None
Economic interest: Medicinal, aromatic, nectariferous for bees, collected and sold on local market, commonly used for flavouring cooking
Habitat: Rocky sites, loamy soils, 200–500 mm MAR, in relict forest of *Pinus halepensis*
Distribution:

- In Libya, Jebel Nafusa, Jebel Al-Akhdar, not in Sirte
- In arid zones, southern Europe, Mediterranean, North Africa to the Middle East and west Asia

Endemism: NA

Note: White flowers could be from *Thymus decussatus* Benth. (?), (http://www.theplantlist.org/tpl1.1/record/kew-204674).
Note: *Thymus* sp. is often parasitized with long red filaments from *Cuscuta epithymum* var. *rubella* (Engelm.) Trab. (http://www.theplantlist.org/tpl1.1/record/tro-50189627).

Photo 7.349. *Thymus capitatus*, Gharyan, Tobi nursery.

Photo 7.350. *Thymus capitatus*, purple flowers, Gharyan.

Photo 7.351. *Thymus decussatus* (?), bush, white flowers, Gharyan.

Photo 7.352. *Thymus decussatus* (?), white flowers, Gharyan.

Photo 7.353. *Cuscuta epithymum* var. *rubella*, Cuscutaceae or Convolvulaceae, red haustoria parasite on *Thymus* sp., Gharyan.

Ziziphora hispanica L.

Synonym(s): http://www.theplantlist.org/tpl1.1/record/kew-215866, or *Ziziphora spicata* Cav., http://www.theplantlist.org/tpl1.1/record/kew-215897
Native:
Vernacular name: *Chengoura* (?), *Chendgura* (?)
Raunkiaer's life-form: Therophyte
Description and morphology: Small (5–10 cm) inconspicuous pubescent annual, erect barely branched, strongly aromatic menthol scent, all leaves equal lanceolate ciliate at margins, inflorescence in spike verticillate along stem with few small flowers, long cylindrical calyx with pink corolla, corolla with entire upper lip and lower lip trilobed.
Reproduction: Flowering February–May, seeds
Pastoral importance: Limited
Palatability and fodder value: Not touched by livestock
Toxicity: None
Economic interest: Strongly aromatic, medicinal (?), commonly used by shepherds and to flavour black tea
Habitat: Fixed sand to sandy-loamy soils, 150–300 mm MAR
Distribution:

- In Libya, rare, mostly in Jeffara plain (Bir-el-Ghanem on *Rhanterium suaveolens steppe*), to Hisha-Bir Waddan (north Bu-Gren), Sirte (?)
- In arid zones, southern Europe, Mediterranean (Algeria, Tunisia, Libya)

Endemism: NA

Notes:

- *Ziziphora hispanica* not recorded in *Flora of Libya* (Lamiaceae, Vol. 118, Siddiqi, 1984) despite being recorded in North Africa (?), https://efloramaghreb.org/specie/145470, nearby Tunisia, https://powo.science.kew.org/taxon/urn:lsid:ipni.org:names:461993-1, and in Egypt (cf. *Ziziphora capitata* L., Täckholm, 1974).

Photo 7.354. *Ziziphora hispanica*, whole plant, Bir-el-Ghanem.

Photo 7.355. *Ziziphora tenuior*, flowers along stem, from Uzbekistan.

- *Ziziphora hispanica* L. from Libya looks very similar to *Ziziphora tenuior* L., http://www.theplantlist.org/tpl1.1/record/kew-215905, from Palestine – Syria, https://flora.org.il/en/plants/zizten/, and from Central Asia, https://powo.science.kew.org/taxon/urn:lsid:ipni.org:names:462033-1/images, and Gintzburger *et al.* (2003).
- In Libya, *Ziziphora* sp. could be confused with the common *Ajuga iva* (L.) Schreb., also locally named *Chengoura* (http://www.theplantlist.org/tpl1.1/record/kew-5293), a small perennial plant profusely hairy/lanate, branching from base, leaves linear denticulate, small yellow flowers, barely scented. Similar ecology.

LILIACEAE

Asparagus aphyllus L.

Synonym(s): Asparagus aphyllus L. — The Plant List, or *Asparagus erinaceus Borzì & Mattei* Asparagus erinaceus Borzì & Mattei — The Plant List
Native:
Vernacular name: *Jafaraz, Sakkoum* (?)
Raunkiaer's life-form: Chamaephyte – Geophyte
Description and morphology: Small (up to 1 m high) flexuous perennial rhizomatous bush, often a climber, green grooved stems and cladodes glabrous, tiny leaves as small axillary grey scales, bunch of 2–6 linear thin cylindrical cladophylls ending with sharp short spine, small pedunculate/pedicelled yellowish-greenish pedicellate flowers at base of cladodes, isolated fruit along stem as pedicellate/subsessile black spherical berries (4–8 mm diam.).
Reproduction: Vegetative through rhizomes, flowering August–October, seeds
Pastoral importance: Limited
Palatability and fodder value: Limited to young shoots – mostly by goats
Toxicity: None
Economic interest: Young spineless stems sprouting from rhizomes avidly collected as vegetables by locals
Habitat: In matorral, on Mediterranean *Terra rossa*, heavy soils, often on rocky sites/karst, above 300 mm MAR, common in *Pistacia lentiscus/Ceratonia siliqua* formations in Jebel Al-Akhdar.

Photo 7.356. *Asparagus aphyllus*, stem and leaves, Toknis, Jebel Al-Akhdar.

Distribution:

- In Libya, mostly in Jebel Al-Akhdar (El-Marj, Shahat, Derna), not in Sirte, in Jebel Nafusa (rare) (?)
- In arid zones, Mediterranean to Middle East

Endemism: NA

Asparagus stipularis Forssk.

Synonym(s): Asparagus stipularis Forssk. — The Plant List, or *Asparagus horridus* L, Asparagus horridus L. — The Plant List
Native:
Vernacular name: *Sakkoum*
Raunkiaer's life-form: Chamaephyte – Geophyte
Description and morphology: Small (up to 1 m high) perennial thorny rhizomatous bush, intricately rigidly branched, green stems and cladodes glabrous and grooved, alternate leaves as small axillary membranous grey scales, bunch of 2–6 stout linear cylindrical cladophylls ending

Photo 7.357. *Asparagus stipularis*, whole plant, regrowth after fire, Toknis, Jebel Al-Akhdar.

Photo 7.358. *Asparagus stipularis*, whole plant with berries.

Photo 7.359. *Asparagus stipularis*, conglomerate berries, Leptis Magna.

with sharp rigid short spine, small pedunculate/ pedicelled whitish flowers at base of cladodes along stems, numerous fruits as sessile bluish/ black spherical berries (4–8 mm diam.) could conglomerate mostly near the end of stems.
Reproduction: Vegetative through rhizomes, flowering August–October, seeds
Pastoral importance: Limited
Palatability and fodder value: Limited to young shoots – mostly by goats
Toxicity: None
Economic interest: Young spineless stems sprouting from rhizomes collected as vegetables
Habitat: On Mediterranean *Terra rossa*, heavy soils, rocky sites/karst, above 300 mm MAR, common in *Pistacia lentiscus*/*Ceratonia siliqua* formations in Jebel Al-Akhdar, recover fast after bush fires.
Distribution:

- In Libya, mostly in Jebel Al-Akhdar, not in Sirte, in the west, Leptis Magna, Jebel Nafusa (Msalata, rare (?))
- In arid zones, Mediterranean to the Middle East

Endemism: NA

Asphodelus tenuifolius Cav.

Synonym(s): http://www.theplantlist.org/tpl1.1/record/kew-275622, or *Asphodelus maroccanus* Gand., http://www.theplantlist.org/tpl1.1/record/kew-275555
Native:
Vernacular name: *Laïet, Laïet-el-Atrous* (?)
Raunkiaer's life-form: Tachytherophyte – geophyte
Description and morphology: Small annual or biannual (10–40 cm), fibrous roots, tubular scabrous erect (non-sticky) leaves ending flat in basal rosette, leafless lax inflorescence in straight erect stem 3–5 times longer than basal leaves, small white short pedicellate flowers with green to brown external midrib on tepals, spherical/ ovoid capsule on erect short articulated pedicel, black trigonal seeds.
Reproduction: Flowering winter–early spring, seeds (rhizomatous (?))
Pastoral importance: Limited
Palatability and fodder value: New shoots grazed, mature flowering plants ignored
Toxicity: None
Economic interest: Reported to be collected for silage in Morocco
Habitat: On sand/sandy-loam, rocky sites, down to 100 mm MAR
Distribution:

- In Libya, Jebel Nafusa, Hamada, Sirte, Jebel Al-Akhdar
- In arid zones, North Africa to India

Endemism: NA

Other main *Asphodelus* species in Libya:

- *Asphodelus microcarpus* Salzm. & Viv., http://www.theplantlist.org/tpl1.1/record/tro-18400528, or *Asphodelus ramosus* L., Asphodelus ramosus L. — The Plant List, (*Ansal, Ballouz*), a robust ubiquitous geophyte most common on all overgrazed sites from the Jebel Nafusa to Marmarica, the most common tall (inflorescence up to 120–150 cm) Libyan *Asphodelus* sp. with black tuberous oblong/ovoid roots, long flat entire leaves fibrous at base and bent 1/3 / ½ of length, strong multibranched inflorescence with many large flowers, on sandy-loamy/shallow rocky soils, above 150 mm MAR, *Asphodelus microcarpus* not touched by livestock except when dry, hence transplanted in bunches often with *Urginea maritima* to identify remote landmarks (tombs, field boundaries etc.) on rangelands, seeds of *Asphodelus microcarpus* used to be consumed by locals during drought and dearth years.
- *Asphodelus refractus* Boiss., http://www.theplantlist.org/tpl1.1/record/kew-275599, or *Asphodelus pendulinus* Coss. & Durieu, http://www.theplantlist.org/tpl1.1/record/kew-275580, looks like *A. tenuifolius* with basal rosette but with divergent viscous/sticky leaves dotted with sand particles, capsules pendulous retroflexed, desertic, down to 50 mm MAR, on sandy soils.

Photo 7.360. *Asphodelus tenuifolius*, plant, Girzah, Hamada.

Photo 7.361. Overgrazed site with dominant *Asphodelus microcarpus*, Jinawin, Gharyan.

Photo 7.362. *Asphodelus microcarpus* marking a solitary tomb, south Mafsel, south Gendula.

Dipcadi serotinum (L.) Medik.

Synonym(s): http://www.theplantlist.org/tpl1.1/record/kew-304368, or *Hyacinthus serotinus* L., http://www.theplantlist.org/tpl1.1/record/kew-278693, or *Hyacinthus lividus* Pers., http://www.theplantlist.org/tpl1.1/record/kew-278635
Native:
Vernacular name:
Raunkiaer's life-form: Geophyte
Description and morphology: Small (10–20 cm high) bulbous herb, small (2–4 cm diam.) bulb wrapped in grey papery membranes, few (1–4) long narrow arched fleshy entire leaves all basal, unique unilateral inflorescence with 2–10 orange (brown) pendulous campanulate flowers (10–20 mm long) with tepals in tubes on short pedicel with membranous triangular bract at base, three retroflexed sepals, capsule trigonal and trilocular, abundant small discoid black seeds.
Reproduction: Vegetative from bulbs, flowering January–May, seeds
Pastoral importance: Limited
Palatability and fodder value: Not grazed by livestock
Toxicity:
Economic interest: None
Habitat: Mostly on sandy soils, deep into desert, down to 50 mm MAR
Distribution:

- In Libya, Jeffara, Jebel Nafusa, Hamada, Sirte, Ajdabiya, south Jebel Al-Akhdar (?)
- In arid zones, south Europe, North Africa to Arabian Peninsula

Endemism: NA

Photo 7.363. *Dipcadi serotinum*, Girzah, Hamada.

Photo 7.364. *Dipcadi serotinum*, flowers, Girzah, Hamada.

Photo 7.365. Trigonal capsule of *Dipcadi serotinum*, Magdar es Sbot, Al-Assah.

Gagea reticulata (Pall.) Schult. & Schult. f.

Synonym(s): http://www.theplantlist.org/tpl1.1/record/kew-307396, or *Ornithogalum reticulatum* Pall., http://www.theplantlist.org/tpl1.1/record/kew-283418

Native:
Vernacular name:
Raunkiaer's life-form: Geophyte
Description and morphology: Small (10–25 cm high) bulbous herb growing in clump, 1–3 leaves filiform (1–2 mm diam.) basal erect (end arched) longer that flowers in false verticillate (caulinar) leaves under flower, flower pedicel ciliate-pubescent longer than flowers, flowers (1–3) with showy yellow-citrina lanceolate – narrow pointed tepals (15–25 mm long) veined with green, long trigonal style longer than six stamens, fruit trilocular oblong, seeds brownish.
Reproduction: Vegetative from bulbs, flowering December–March, seeds
Pastoral importance: None
Palatability and fodder value: None for livestock, appears to be grazed by rabbits and picked by partridges
Toxicity:
Economic interest: None, could be used as ornamental
Habitat: Coarse to fine sand (sandy-loam), gravelly, down to 200 mm MAR
Distribution:

- In Libya, only Jebel Al-Akhdar (?), eastern Mediterranean (?)
- In arid zones, south-eastern Europe, North Africa

Endemism: NA

Photo 7.366. *Gagea reticulata*, Mafsel, south Gendoula, March 2009.

Photo 7.367. *Gagea reticulata*, flowers, Mafsel, south Gendoula, March 2009.

Muscari comosum (L.) Mill.

Synonym(s): Muscari comosum (L.) Mill. — The Plant List, or *Leopoldia comosa (*L.) Parl. Leopoldia comosa (L.) Parl. — The Plant List
Native:
Vernacular name: *Ketout* (?)
Raunkiaer's life-form: Geophyte
Description and morphology: Small (10–20 cm high) bulbous glabrous herb, small (2–4 cm diam.) bulb wrapped in grey papery membranes, few short (1–6) lanceolate canaliculate fleshy entire curled leaves all basal, unique inflorescence (10–30 cm) with 10–30 cylindrical urceolate horizontal flowers on lax erect leafless raceme longer that leaves, yellow-brown fertile horizontal campanulate flowers (8–10 mm long) on short pedicel with small membranous triangular bract at base, tassel of few sterile violet erect flowers on top of inflorescence, hard capsule trigonal and trilocular, 1–2 black seeds per locule.
Reproduction: Vegetative from bulbs, flowering in spring, seeds
Pastoral importance: None
Palatability and fodder value: Not grazed
Toxicity:
Economic interest: None
Habitat: Sandy-gravelly well-drained soils, often weeds in rainfed crops, down to 150 mm MAR

Distribution:

- In Libya, western zone, foothills and south Jebel Nafusa, South Zliten, does not appear to occur in Sirte and over east (?)
- In arid zones, southern Europe, North Africa to the Middle East

Endemism: NA

Photo 7.368. *Muscari comosum*, south Zintan – Al-Haraba.

Urginea maritima (L.) Baker

Synonym(s): http://www.theplantlist.org/tpl1.1/record/kew-290809, or *Drimia maritima* (L.) Stearn, http://www.theplantlist.org/tpl1.1/record/kew-305015
Native:
Vernacular name: *Pharaoun, Faraoun*
Raunkiaer's life-form: Geophyte
Description and morphology: Robust plant with basal leaves on a group of large bulbs (5–15 cm diam. each) covered with membranous brown-whitish wrappers and white fleshy roots, leaves large (20–80 cm long, 10–20 cm wide) glabrous lanceolate usually erect entire dying (yellowing and prostrate on ground) before flowering, long pyramidal inflorescence on long stalk up to 1.5 m long (8–10 mm diam.), old dry stalks remain on bulb, large white flowers star-like with (green to reddish) nerved tepals, fruit as elongated trilocular capsule each containing 1–5 black seeds.
Reproduction: Mostly vegetative from bulbs, leaves appear in September/October developing before flowering April–June, seeds
Pastoral importance: Limited
Palatability and fodder value: Ungrazed/untouched by all livestock when green, dry leaves rarely grazed (?)
Toxicity: Bulbs highly poisonous containing Scilliroside, emetic, crushed dry bulbs added to bait used as rat poison, leaves with strong bitter taste
Economic interest: Medicinal (?), often in man-made large clumps commonly used as landmarks/boundary between *sikka* (crops, fields, pastures, garden, orchards etc.), more and more often nowadays associated or replaced with old tyres.
Habitat: Sandy loam soils to *Terra rossa*, down to 100–150 mm MAR
Distribution:

- In Libya, all over coastal zones, in steppe, to the upper limit of the desert
- In arid zones, North Africa

Endemism: NA

Photo 7.369. *Urginea maritima*, young leaves and bulb exposed, Toknis, October 2009.

Photo 7.370. *Urginea maritima*, flower, Jbebina, Al-Assah.

Photo 7.371. *Urginea maritima* clump, man-made *sikka* boundary on rangelands between Bir Saman and Jbebina, Al-Assah, March 2009.

Photo 7.372. *Urginea maritima* clump and tyre, man-made *sikka* boundary on sandy soils, Kararim.

MENISPERMACEAE

Cocculus pendulus (J.R.Forst. & G.Forst.) Diels

Synonym(s): http://www.theplantlist.org/tpl1.1/record/kew-2729715, or *Cocculus leaeba* (Delile) DC., http://www.theplantlist.org/tpl1.1/record/kew-2729675, or *Epibaterium pendulum* J.R. Forst. & G.Forst, http://www.theplantlist.org/tpl1.1/record/kew-2789903
Native:
Vernacular name: *Olleika* (?)
Raunkiaer's life-form: Chamaephyte – climber – creeper, thermophylous
Description and morphology: Perennial dioceous, robust woody climber or creeping on rocks, twisted white trunk with smooth bark, pendulous thin whitish branches with alternate leathery glabrous entire obovate-mucronate pedicelled leaves (20–40 mm × 10–15 mm), small male sessile green flowers in cyme, female flowers long-pedicelled, fruit as small (5–8 mm diam.) red drupe shrivelling black with one seed only.
Reproduction: flowering March–June, seeds
Pastoral importance: Limited
Palatability and fodder value: Avidly grazed by dromedaries
Toxicity: None
Economic interest: Medicinal, leaves used as diuretic, roots used against fever (?)
Habitat: Desertic, reported rare in southern Libya (El-Gadi and Jafri, 1976, Vol. 102, p. 2), however, we found few very large specimens (trunk 35–40 cm diam. with many younger specimens in *Acacia* population) way up north (N 31.689 E 14.958) climbing into *Acacia tortilis* subsp. *raddiana* in depressions/wadis; 50–150 mm MAR, deep loamy soils; this is the first report (2010) of *Cocculus pendulus* as north as the ones we recorded. Herdsmen report frequent presence of *Cocculus* in Bu N'Gem region.
Distribution:

- In Libya, only officially recorded in southern Libya, our record is exceptionally septentrional from a site between wadi Zorzor/Nefad and wadi Bey-el-Kebir.
- In arid zones, central Sahara, sub-Saharan Africa, Mauritania to southern Middle East, India, Pakistan.

Endemism: NA – rare, may be threatened in north-western Libya

Photo 7.373. *Cocculus pendulus*, see between the two persons the large liana climbing and taking over a large *Acacia tortilis* subsp. *raddiana* (N 31.689 E 14.958, Tarik-en-Nahar, along western man-made river track north of wadi Sof-el-Jin)

Photo 7.374. *Cocculus pendulus*, large liana (*right*) climbing an *Acacia tortilis* subsp. *raddiana* (left); Tarik-en-Nahar, along western man-made river track north of wadi Sof-el-Jin.

Photo 7.375. *Cocculus pendulus*, pendulous branches, leaves and red drupes, Tarik-en-Nahar, along western man-made river track north of wadi Sof-el-Jin.

Photo 7.376. *Cocculus pendulus*, leaves and red drupes, Tarik-en-Nahar, along western man-made river track north of wadi Sof-el-Jin.

MIMOSACEAE

Acacia saligna (Labil L.) H.L. Wend L.

Synonym(s): http://www.theplantlist.org/tpl1.1/record/ild-591, or *Acacia cyanophylla* Lindl., http://www.theplantlist.org/tpl1.1/record/ild-386
Exotic: Originating from south-western Australia having similar sandy soil and climatic conditions to coastal Libyan zones. Introduced from seeds to Haschan forestry station by Turkish foresters in the early 20th century from Australia.
Vernacular name: *Sunt* (?), *Kashia* (?)
Raunkiaer's life-form: Phanerophyte (sometimes Chamaephyte)
Description and morphology: Evergreen shrub to small tree, short-lived (15–20 years), or trees up to 6–8 m high; extremely polymorph (erect, dropping branches, prostrate) – possibly many subspecies, large superficial rooting systems, strong suckers, stoloniferous, N fixation nodules on roots, no taproot, withstand well wind erosion; bark even and smooth grey; blue-dark green long (up to 20 cm) lanceolate/falcate and undulate pedunculate pendulous phyllodes with prominent midrib, young shoots flattish soft reddish and turning green and round when older; long raceme of abundant globular golden yellow flowers, pods linear and flat, constricted between seeds, vigorous growth, dead trees appear twisted – flame-like shape.
Reproduction: Flowers (yellow, whitish, March–April) and abundant seeds production (black shiny seeds, May) in flat pods; and vegetative by developing stoloniferous root system.
Pastoral importance: High as used for afforestation projects and sand dune fixation all over coastal Libya but short lived (15–20 years)
Palatability and fodder value: Drought fodder, on standing trees, green/fresh foliage has poor intake and palatability except for hungry livestock. Young shoots and foliage chopped and usually used as cut-and-carry to let dry for 1–2 days in pens/paddocks; then reasonably consumed by all livestock. Protein 8–9%. Low overall digestibility.
Toxicity: Seeds contain protease inhibitors that are deactivated by heating or cooking.
Economic interest: Afforestation (easy seedlings mass production from nurseries), used for large-scale sand dune fixation; drought fodder, source of fuelwood (coppicing when harvested) and charcoal (24% yield, low tar); seeds may be used as poultry feed; leaves rich in tannins. Bark used for tanning, leaves used to dye wool yellow; seeds ground into flour and consumed by Aboriginal people in Australia.
Habitat: Minimum annual rainfall required 150 mm; in frost-free areas; deep sandy soils and fixed sand dunes (Jeffara plain), only coastal inland dunes; does not tolerate salt spray and salty soils but will thrive on deep sand over salty flats (from Al-Assah, Zuara to Ajdabiya) if roots not dipping into salt. When roots reach the salty water table, this *Acacia* will die. Not suitable on heavy soils (Benghazi plain – plateau, or on calcareous crust as in Gharyan area, top of Jebel Nafusa).
Distribution:

- In Libya, coastal zones only, mostly western Libya, Sirte to Ajdabiya, then some poor plantations on *Terra rossa* on Benghazi plain afforestation projects.
- In arid zones, all mild winter coastal zones of North Africa, Egypt, Lebanon, Palestine (rare in Syria – not surviving north of Hama).

Endemism: NA

Photo 7.377. Four-year-old *Acacia saligna* (= *Acacia cyanophylla*) plantation, south Sorman (N 32.38 E 12.16), Jeffara plain.

Photo 7.378. *Acacia saligna* flowers in terminal raceme with large leaves (Magdar-es-Sbot, Al-Assah, 1976).

Note: By 2009, the large *Acacia saligna* plantations at Magdar-es-Sbot (established 1970–75) have been totally eradicated for fuelwood and overgrazed by wandering dromedaries.

Other main exotic evergreen *Acacia* species introduced in Libya:

Photo 7.379. Fifteen-year-old *Acacia cyanophylla* plantation in flower, Sirte, 1976.

- *Acacia aneura* Benth., http://www.theplantlist.org/tpl1.1/record/ild-33174, originating from arid central Australia (commonly named there as *Mulga*), shrub type, linear cylindrical long (8–12 cm) but very narrow (1.5–3 mm) erect dull grey-green phyllodes with short pointed end, new shoots resinous, flowers in cylindrical spikes (2–4 cm long), small flat papery grey-brown pods (margin slightly winged) with 2–4 black seeds; tested at Bir-el-Ghanem and Wishtata rangeland projects. Good establishment and growth but poor palatability in Libya despite reported as drought feed in arid Australia.
- *Acacia cyclops* G.Don , http://www.theplantlist.org/tpl1.1/record/ild-387, originating from coastal south-western Australia, shrub to small tree type, thick-nerved (3–4) short round-ended obovate phyllodes, flower vivid yellow in long raceme with two pedunculate globular (1 cm diam.) flowers, ripe pods curved swollen at seed position, woody-leathery pods remain attached to stem after dehiscence, seed circled an orange (or yellow) aril; poor palatability, planted along the seashore, cliffs and littoral dunes fixation receiving salt spray (Tajoura, 15 km main highway east of Al-Hamidiya along cliff on seashore (N 32.810°, E 13.551°), Zuara, on Farwa Island), not to be planted on saline soils.

Photo 7.380. *Acacia cyclops*, ripe pods and seeds with yellow aril.

- *Acacia karroo* Hayne, http://www.theplantlist.org/tpl1.1/record/ild-470, or *Vachellia karroo* (Hayne) Banfi & Galasso; small bush with pair of white thorns at nodes, fine leaves paripinnate petiolate, globulous yellow long-pedicelled unique globular flower, pods arched nearly circular, originating from South Africa, very early introduction (1915) from Sicily, used for efficient fences on farmland along the western coastal zone and on some agricultural projects (Bir-el-Ghanem, wadi Megenin etc.), long (30–60 mm) white thorn (locally used to pick cooked snails (*babalouchi*) from shells), pods on ground consumed by livestock. Good fuelwood for charcoal manufacture.
- *Acacia ligulata* Benth., http://www.theplantlist.org/tpl1.1/record/ild-47889; small hemispherical bush, thick narrow linear/rectangular (6–10 cm long × 3–6 mm wide) flat phyllodes with gland at curved apex, globular yellow flowers on

foliate raceme, pods woody constricted around seeds turning black with red aril when ripe. Originating from northern Australia, introduced to Haschan (1960?), later tested on deep sandy soils at several Jeffara and Jebel Nafusa rangeland projects, unpalatable.

Photo 7.381. *Acacia ligulata* pods.

Further reading on revision of Australian *Acacia* at: http://worldwidewattle.com/infogallery/projects/acacia-saligna.pdf (retrieved 2 January 2020):

- Maslin (1997) Australia's golden future. *Landscope* 12(3), 16–22.
- Maslin and McDonald (2004) *Evaluation of Acacia as a Woody Crop Option for Southern Australia.* CSIRO, Rural Industries Research and Development Corporation, Canberra.

Acacia tortilis subsp. *raddiana* (Savi) Brenan

Synonym(s): http://www.theplantlist.org/tpl1.1/record/ild-637, or *Acacia raddiana* Savi, http://www.theplantlist.org/tpl1.1/record/ild-638, or *Acacia tortilis* (Forssk.) Hayne, http://www.theplantlist.org/tpl1.1/record/ild-632 or *Vachellia tortilis* (Forssk.) Galasso & Banfi —The Plant List
Native:
Vernacular name: *Talha*
Raunkiaer's life-form: Phanerophyte (sometimes Chamaephyte), long-lived (likely over 100 years)
Description and morphology: A large flat-top or umbrella-shaped tree (up to 10–15 m high and crown diameter), sometimes multibranched coppice, black bark, bipinnate leaves, white pairs of stipulate thorns, whitish-cream globular flowers on long peduncles; flowering in November; pod pubescent, spiralled, twisted contorted and constricted between seeds.
Reproduction: Flowering in November, propagation from seeds
Pastoral importance: Pods as a fodder source for gazelles, small ruminants and camels
Palatability and fodder value: Lower shoots, leaves and pods browsed by gazelles, small ruminants and dromedaries as well as dry pods fallen on ground
Toxicity: None reported
Economic interest: A valuable and highly drought-resistant tree that must be protected and replanted (from nurseries) on all projects and irrigated farms in the wadi areas (Solul-el-Khoder), used to be a source of fuelwood and charcoal, wood sometimes used as building material, pods used as fodder.
Habitat: Usually in Saharan conditions to 50 mm/year; in wadis and depressions with heavy soils; not occurring on sand dunes. The last northernmost relict open forest of *Acacia tortilis* subsp. *raddiana* in Libya is at Bir Ayyad (in the Jeffara plain, 20 km north-west of Yefren); pre-1970, a well-known hunting area, where shy Rim gazelles (*Gazella leptoceros*, coming from the Oriental *Erg* through Takut – Al-Assah – Hebiliyah) would be seen grazing *Acacia* pods and lower branches.
Distribution:

- In Libya, unique stand in Jeffara plain (relict open 'forest' (N 32.121° E 12.335°) Bir Ayyad), still common south-west of Jebel Nafusa, Beni-Walid, Nefad, Zamzam, Megenin, Zorzor, Seddada, Ghirza, Al-Hamada-al-Hamra, etc.).
- In arid zones, North Africa, Sahara, sub-Saharan Africa (Somalia to Senegal).

Endemism: NA
Status in Libya: Threatened as often uprooted to make room for new irrigation projects in the west (Solul-el-Khoder, Bir Dufan, wadi Merdum, Banai Walid, Sedadda etc.), fuelwood and charcoal.

Other native *Acacia* species in Libya:

- *Acacia seyal* Delile, http://www.theplantlist.org/tpl1.1/record/ild-609, red bark, doubtful, reported in Tibesti (?), present in most southern parts of North Africa Sahara, east Africa, Arabia to south Jordan, Oman, Emirates.

Photo 7.382. *Acacia tortilis* subsp. *raddiana* (N 31.690413, E 14.962749), south Misurata – MAR = 150 mm.

Photo 7.383. *Acacia tortilis* subsp. *raddiana*, relict open forest, Bir Ayyad (N 32.138 E 12.410), Jeffara plain.

Photo 7.384. *Acacia tortilis* subsp. *raddiana*, white pairs of stipulate thorns, wadi Zorzor (N 31.570 E 14.853).

Photo 7.385. *Acacia tortilis* subsp. *raddiana*, whitish-cream globular flowers on long peduncle, wadi Zorzor.

Photo 7.386. *Acacia tortilis* subsp. *raddiana* pods, Bir Ayyad, February 1978.

MYRTACEAE

Note: Despite not being native to Libya, two *Eucalyptus* sp. (*Eucalyptus camaldulensis* and *Eucalyptus gomphocephala*) and their hybrids are part of the northern Libyan rangeland landscape. They are profusely used in afforestation projects as windbreaks, source of fuelwood and charcoal industry, and largely contributing to honey production. A review of the Western Australian *Eucalyptus*, most relevant to Libyan conditions, is available with Gardner (1979).

Eucalyptus camaldulensis Dehnh.

Synonym(s): Eucalyptus camaldulensis Dehnh. —The Plant List, or *Eucalyptus acuminata* Hook, Eucalyptus acuminata Hook. —The Plant List, or *Eucalyptus longirostris* F.Muell. ex Miq., Eucalyptus longirostris F.Muell. ex Miq. —The Plant List
Exotic: Endemic and wide distribution all over arid Australia – common along watercourse that can be totally dry like wadis/introduced in Libya early 1910 by Turkish foresters and later by Italian foresters.
Vernacular name: *Kina* (?)
Raunkiaer's life-form: Macrophanerophyte
Description and morphology: A large tall fast-growing tree (up to 15–20 m high), could reach 4–5 m at girth in coastal Libya, deep (over 10 m) and large (up to 20 m horizontally) rooting systems in search of water, smooth white-cream bark longitudinally marbled, mature leaves are green to blue-green elongated coriaceous with white ventral midrib, flowers in umbels, each flower as a cupula with a ring of numerous white long stamens, fruits as a woody hard capsule (hypanthium) domed with a hooded pointed or round cap (= calyptra or operculum), many operculum variations from round-blunt to conical and beaked, at maturity the calyptra falls freeing very small seeds and chaff through four valves arranged in pointed cross. Wood is reddish with straight grain.
Reproduction: Flowering in spring/summer, seeds (anemochory)
Pastoral importance: Limited but much used in Libya in afforestation projects, windbreaks
Palatability and fodder value: None
Toxicity: None
Economic interest: Afforestation, source of fuelwood, charcoal, valuable for apiarists (flowers = rich source of nectar), source of medicinal Eucalyptus oil.
Habitat: All over northern Libya from west to east, on all soils (from deep fixed sand to loam (old Haschan forestry station) and moderate clay content, except on mobile sand dunes and saline-gypsum soils, poor salinity and salt-spray tolerance, also reported becoming chlorotic (yellowing leaves, stunt growth) on calcareous soils containing calcium nodules (Bir-el-Ghanem).
Distribution:

- In Libya, all over northern Libya, down to 150–200 mm MAR
- In arid zones, Australia, North Africa, southern Europe, California, and Ethiopia

Endemism: NA

Note: Keith (1965) reports that there are very few pure *Eucalyptus camaldulensis* in Libya; they are mostly hybrids *Eucalyptus camaldulensis* Dehnh. x *Eucalyptus rudis* Endl. = *Eucalyptus algeriensis* Trabut, the latter not being recognized as a *Eucalyptus* species (name unchecked, http://www.worldfloraonline.org/taxon/wfo-0000954377). This hybrid gathers the characteristics of the two species. *Eucalyptus camaldulensis* and *Eucalyptus rudis* have the same ecology in Western Australia. *Eucalyptus rudis* looks very much like *Eucalyptus camaldulensis* but has a whitish bark with patches of peeling rough fibrous-flaky brown bark. The capsules are red with white flowers clearly pedicellate with pyramidal pointed operculum.

Photo 7.387. Lonely hybrid *Eucalyptus camaldulensis* Dehnh. x *Eucalyptus rudis* Endl. (along south Sorman road (N 32.435 E 12.342), MAR = 150–175 mm, in *Aristida pungens* country).

Photo 7.388. Bark from hybrid *Eucalyptus camaldulensis* Dehnh. x *Eucalyptus rudis* Endl.

Photo 7.389. Fruits and leaves from hybrid *Eucalyptus camaldulensis* Dehnh. x *Eucalyptus rudis* Endl.

Photo 7.390. Pedunculate fruits and leaves from hybrid *Eucalyptus camaldulensis* Dehnh. x *Eucalyptus rudis* Endl.

Photo 7.391. Plantations of Eucalyptus (*E. camaldulensis* and *E. gomphocephala*, four years old), Jenoub Sorman rangeland project (MAR 175 mm, N 32.388 E12.168, 2008).

Eucalyptus gomphocephala DC.

Synonym(s): Florabase Quick Name Search (dpaw.wa.gov.au)
Exotic: Endemic to coastal limestone plain of western Australia – open forest on alkaline coastal sand
Vernacular name: *Kina* (?)
Raunkiaer's life-form: Macrophanerophyte
Description and morphology: A large tall fast-growing tree (up to 10–15 m high), 1–3 m at girth in coastal Libya, deep (over 10 m) rooting systems (not extending horizontally), solid persistent fibrous rough dark bark extending to young branches, pendulous alternate mature leaves light green elongated and curved like a sickle with prominent ventral reddish midrib,

Photo 7.392. *Eucalyptus gomphocephala*, near Sorman (N 32.598 E 12.457), MAR = 250 mm.

flowers in umbels with winged-thick stalks, flower buds (6–7) are sessile and mushroom-shaped with swollen caps, each flower as a cupula with a ring of numerous white-cream long stamens, woody hard capsule (hypanthium) with broad rim and domed with swollen cap, at maturity the calyptra falls freeing very small seeds and chaff through four valves arranged in flat cross. Solid and high-density wood is light yellow.
Reproduction: Flowering in spring/summer, seeds (anemochory)
Pastoral importance: Limited but much used in Libya in afforestation projects, windbreaks, withstands calcareous soils better than *Eucalyptus camaldulensis*
Palatability and fodder value: None
Toxicity: None
Economic interest: Afforestation, source of fuelwood, charcoal, valuable for apiarists (flowers = abundant source of nectar), honey produced is fluid and light
Habitat: All over northern Libya from west to east, on all soils (from deep fixed sand to loam and moderate clay content), except on mobile sand dunes and saline-gypsum soils, poor salinity and salt spray tolerance, less drought-resistant than *Eucalyptus camaldulensis*.
Eucalyptus gomphocephala (and other *Eucalyptus* sp.) is often parasited with larvae of *Phoracantha recurva* (Cerambycidae) drilling stunning contorted canaliculi and grooves while feeding on the wood under the bark. *Eucalyptus gomphocephala* is not affected by chlorosis and is better suited to sandy alkaline/calcareous soils than *Eucalyptus camaldulensis*.

Distribution:

- In Libya, all over northern Libya, down to 200–250 mm MAR
- In arid zones, Australia, North Africa, southern Europe

Endemism: NA

Note: Reported to hybridize with *Eucalyptus cornuta* Labill., Eucalyptus cornuta Labill. — The Plant List, that has small hypanthium and extremely long operculum.

Photo 7.393. *Eucalyptus gomphocephala* green fruits (Hypanthium capped with operculum).

Photo 7.394. *Eucalyptus gomphocephala* ripening fruits (without operculum) and leaf (note the winged pedicel under the sessile fruits).

Photo 7.395. Canaliculi of *Phoracantha recurva* larvae on Eucalyptus wood.

NEURADACEAE

Neurada procumbens L.

Synonym(s): http://www.theplantlist.org/tpl1.1/record/kew-2384148, or *Figaraea aegyptiaca* Viv., http://www.theplantlist.org/tpl1.1/record/kew-2812895
Native:
Vernacular name: *Saadan* (?)
Raunkiaer's life-form: Stenotherophyte
Description and morphology: Prostrate tomentose small annual (10–30 cm diam.), creeping yellowish stems, spathulate lobed alternate leaves, small white (yellow?) flowers, indehiscent sharp spiny discoid fruit with flat base and armed with strong erect wounding spines when dry, seed always germinates in fruit remaining attached to taproot.
Reproduction: Flowering March–June, seeds (frequently dispersed by zoochory)
Pastoral importance: Poor, sharp spiky fruit often lodged into dromedaries and small ruminants' soles and hooves or hooked to fleece, must be hand-removed by herdsmen noting limping animal.
Palatability and fodder value: Poor, camel feed
Toxicity: None
Economic interest: None
Habitat: Sandy soils, deep into Sahara on *reg*, most common on degraded rangelands and around wells and watering points.
Distribution:

- In Libya, desertic, Jeffara plain, south Jebel Nafusa Sirte, south Jebel Al-Akhdar
- In arid zones, North Africa, the Middle East, Arabian Peninsula to Pakistan

Endemism: NA

Photo 7.396. *Neurada procumbens*, plant, west Kararim.

Photo 7.397. *Neurada procumbens*, plant, west Kararim.

Photo 7.398. *Neurada procumbens*, spiky discoid fruit, west Kararim.

ORCHIDACEAE

Ophrys lutea Cav.

Synonym(s): http://www.theplantlist.org/tpl1.1/record/kew-141394
Native:
Vernacular name: *Narla* (?)
Raunkiaer's life-form: Geophyte
Description and morphology: Perennial, glabrous, tuber roots, erect floriferous single stem (10–30 cm), short basal leaves embracing stem, lax glabrous inflorescence with 1–3 lateral flowers (each 10–20 mm long), flower bract light green erect lanceolate acute, hooded yellow-green erect sepals, pendulous trilobed large convex round-ended glabrous spurred labellum with centre shield reddish to dark purple-blueish and attractive large yellow margin.
Reproduction: Flowering February–April, seeds
Pastoral importance: None
Palatability and fodder value: Flowering stem occasionally grazed by desperate livestock; tough leaves appear to remain untouched
Toxicity: None
Economic interest: None
Habitat: Shallow sandy-loam soils on Gargaresh limestone, 300–500 mm MAR
Distribution:

- In Libya, a rare occurrence, found only at one site at Tajoura ARC farm near Tripoli; possible in Jebel Al-Akhdar
- In arid zones, southern Europe

Endemism: NA

See Ashley *et al.* (1982).

Photo 7.399. *Ophrys lutea*, Tajoura, February 1980.

Photo 7.400. *Ophrys lutea*, ARC Tajoura farm, February 1980.

Ophrys speculum Link

Synonym(s): http://www.theplantlist.org/tpl1.1/record/kew-141747, or *Arachnites speculum* (Link) Tod, http://www.theplantlist.org/tpl1.1/record/kew-13723
Native:
Vernacular name: *Narla* (?)
Raunkiaer's life-form: Geophyte
Description and morphology: Perennial, glabrous, tuber roots, erect floriferous single stem (10–20 cm), basal obovate flat leaves in rosette, upper leaves wrapping floral stem, lax inflorescence with 1–4 lateral flowers (each 10–20 mm diam.), flower bract green erect lanceolate acute hooded, spreading lateral sepals, rear sepal hooded, sepals three-nerved red tinged, minute lateral petals, pendulous large gibbous velvet spurred entire labellum with centre shield shiny blue-purple and reddish-brown large profusely bearded margin.
Reproduction: Flowering February–April, seeds
Pastoral importance: None
Palatability and fodder value: Flowering stem occasionally grazed by desperate livestock; tough leaves appear to remain untouched
Toxicity: None

Economic interest: None
Habitat: Shallow sandy-loam soils (on Gargaresh limestone), 300–500 mm MAR
Distribution:

- In Libya, Tajoura ARC farm, Tarhuna (Shiarchara), Jebel Al-Akhdar (Tokra, wadi Kuf, Shahat, Marsa Suza), 300–500 mm MAR, not in Sirte
- In arid zones, Mediterranean, southern Europe, North Africa, west Asia

Endemism: NA

Photo 7.401. *Ophrys speculum*, flowers, Tajoura farm, March 1980.

Photo 7.402. *Ophrys speculum*, plant, Tajoura farm – March 1980.

Ophrys tenthredinifera Willd.

Synonym(s): http://www.theplantlist.org/tpl1.1/record/kew-141835, or *Ophrys rosea* (Desf.) Duf., http://www.theplantlist.org/tpl1.1/record/kew-141669
Native:
Vernacular name: *Narla* (?)
Raunkiaer's life-form: Geophyte
Description and morphology: Perennial, glabrous, tuber roots, erect floriferous single stem (10–30 cm), large basal recurved leaves in rosette, upper leaves wrapping floral stem, lax inflorescence with 1–2 lateral flowers (each 10–20 mm diam.), flower bract light green erect lanceolate acute. Pink spreading sepals and medium green nerved, pendulous large gibbous velvet spurred sub-trilobate labellum with centre shield dark purple (notched at apex with protruding central globule) and light green-yellowish bearded margin.
Reproduction: Flowering February–April, seeds
Pastoral importance: None
Palatability and fodder value: Flowering stem occasionally grazed by desperate livestock; tough leaves appear to remain untouched.
Toxicity: None
Economic interest: None
Habitat: Shallow sandy-loam soils (on Gargaresh limestone), 300–500 mm MAR
Distribution:

- In Libya, Tajoura ARC farm near Tripoli, Tarhuna – Shiarchara, Jebel Al-Akhdar (Tokra, wadi Kuf, Shahat, Marsa Suza), 300–500 mm MAR, not in Sirte
- In arid zones, Mediterranean, southern Europe, North Africa

Endemism: NA

Photo 7.403. *Ophrys tenthredinifera*, whole plant, Tajoura farm, March 1980.

Photo 7.404. *Ophrys tenthredinifera*, flower, Tajoura farm, March 1980.

Orchis collina Banks & Sol. ex Russell

Synonym(s): http://www.theplantlist.org/tpl1.1/record/kew-142363, or *Anacamptis collina* (Banks & Sol. ex Russell) R.M.Bateman, Pridgeon & M.W. Chase, http://www.theplantlist.org/tpl1.1/record/kew-8570, *Orchis saccata* Ten., http://www.theplantlist.org/tpl1.1/record/kew-143192
Native:
Vernacular name: *Debberb* (?)
Raunkiaer's life-form: Geophyte
Description and morphology: Perennial, glabrous, tuber roots, erect floriferous single stem (10–30 cm), basal oblong lanceolate canaliculate leaves in erect rosette, upper leaves wrapping brown floral stem, lax inflorescence with 5–10 lateral flowers (each 10–15 mm long), flower bract brown erect lanceolate acute short and hooded, spreading (45°) lateral sepals light green obtuse, rear sepal hooded, all sepals three-nerved, minute lateral petals, pendulous large pink to violet trigonate/cuneate labellum with long spur, labellum margin crenulate.
Reproduction: Flowering February–April, seeds
Pastoral importance: None
Palatability and fodder value: Flowering stem occasionally grazed by desperate livestock; tough leaves appear to remain untouched
Toxicity: None
Economic interest: None
Habitat: Shallow/sandy-loam/loamy soils, in damper sites and shallow depressions than previous *Ophrys* sp., 300–500 mm MAR
Distribution:

- In Libya, Tajoura ARC farm, Tarhuna (Shiarchara), Jebel Al-Akhdar (Benghazi, Tokra, wadi Kuf, Shahat, Marsa Suza), 300–500 mm MAR, not in Sirte
- In arid zones, Mediterranean, southern Europe, North Africa, Middle East, west Asia, Caspian region

Endemism: NA

Photo 7.405. *Orchis collina*, whole plant, Tajoura farm, January 1980.

Photo 7.406. *Orchis collina*, flowers, Tajoura farm, January 1980.

OROBANCHACEAE

Cistanche phelypaea (L.) Cout.

Synonym (s): http://www.theplantlist.org/tpl1.1/record/kew-2723172, or *Phelypaea lutea* Desf., http://www.theplantlist.org/tpl1.1/record/kew-2550690, or Cistanche tubulosa
Native:
Vernacular name: *Zeb-el-Ard* (?), *Danun* (?)
Raunkiaer's life-form: Geophyte (parasitic)
Description and morphology: Leafless parasitic plant, solid stem at base, underground suckers system hooked to Chenopodiaceae host plants, small scale (1–3 cm long) triangular acute short leaves along scape, scape solid (not hollow), unique pyramidal inflorescence (20–50 cm high) abundantly covered with bright yellow tubular recurved flowers, fruit beaked containing numerous minute (0.20–030 mm diam.) polygonal seeds.
Reproduction: Flowering March–May, seeds
Pastoral importance: None
Palatability and fodder value: None
Toxicity: None
Economic interest: Possible medicinal; stems reported consumed as vegetables (?)
Habitat: Sandy/sandy loam soils, parasitic of Chenopodiaceae (*Atriplex*, *Haloxylon*, *Salsola*, etc.), down to 100 mm MAR, sometimes on slightly saline soils.
Distribution:

- In Libya, all coastal zones to inland (Jeffara plain, Hamada, Sirte) from east to west
- In arid zones, southern Europe, North Africa to the Middle East

Endemism: NA

Other Libyan *Cistanche* sp.:

- *Cistanche tubulosa* (Schenk) Wight, http://www.theplantlist.org/tpl1.1/record/kew-2723186, desertic, similar to *Cistanche phelypaea*, yellow flowers with violet lobes, mostly parasitic on *Tamarix* sp.
- *Cistanche violacea* (Desf.) Hoffmanns. & Link, http://www.theplantlist.org/tpl1.1/record/kew-2723190, similar to *Cistanche phelypaea*, but pale violet-purple (with a tinge of yellow) flowers, mostly parasitic on Solanaceae, also on Chenopodiaceae, Plumbaginaceae and Zygophylaceae.

Note: *Orobranche* sp., also parasitic plants on Fabaceae (and many other families) are common on Libyan rangelands and crops, leafless scape, inflorescence (simple or branched) nearly cylindrical, much smaller scape (10–30 cm long) and flowers than *Cistanche* sp., tubular blue to violet/yellow/brownish campanulate flowers.

See Attioui (2022).

Photo 7.407. *Cistanche phelypaea*, parasitic plant, Nofilia, Sirte.

PAPAVERACEAE

Glaucium flavum Crantz

Synonym(s): http://www.theplantlist.org/tpl1.1/record/kew-2826358, or *Chelidonium glaucium* L., http://www.theplantlist.org/tpl1.1/record/kew-2716685, or *Glaucium luteum* Scop., http://www.theplantlist.org/tpl1.1/record/tro-24000591
Native:
Vernacular name:
Raunkiaer's life-form: Therophyte
Description and morphology: Perennial herb (20–40 cm), deep green, glaucous, yellow latex, pinnatisect amplexicaule wavy (near fleshy) alternate leaves with round lobes, flower buds twisted with few long hairs or tuberculate or soft-spiny, large (3–4 cm diam.) yellow flowers, long (10–20 cm) pedicelled fruits siliqua-like (bi-valve and dehiscent) ending with triangular stigma, small seeds reniform and reticulated.
Reproduction: Flowering February–May, seeds
Pastoral importance: None
Palatability and fodder value: None
Toxicity: None
Economic interest: None
Habitat: Often on seashore on Gargaresh limestone and coastal sand, also in northern slope of Jebel Al-Akhdar on stony heavy brown soils, above 200–300 mm MAR
Distribution:

- In Libya, all coastal zones (Sabratha, Tajura, Lebda, Taworga), Sirte (?), Jebel Al-Akhdar (Benghazi, Daryanah, Derna) and Marmarica
- In arid zones, Europe, North Africa, Middle East to Iraq

Endemism: NA

Note: Without flowers, could be confused with *Glaucium corniculatum* (L.) Curtis, http://www.theplantlist.org/tpl1.1/record/kew-2826321, an annual herb with orange flowers and subsessile leaves.

Photo 7.408. *Glaucium flavum*, plant, wadi Qattara dam (N 32.029 E 20.404).

Photo 7.409. *Glaucium flavum*, flower and fruits, wadi Qattara dam.

PINACEAE

Pinus halepensis Mill.

Synonym(s): http://www.theplantlist.org/tpl1.1/record/kew-2561998
Native:
Vernacular name: *Snouber*
Raunkiaer's life-form: Mesophanerophyte
Description and morphology: Medium-sized evergreen tree (up to 10–12 m high), crown irregular with mostly horizontal branches, yellow resin, silver-reddish reticulated bark, branches silver black-punctuated, pale-green flexuous leaves needle-like (50–80 mm long) in pairs, cones solitary or by 2–3 ovoid (7–10 cm long) clear brown (waxy-shiny, with lozenge scale dorsally pointed and silvery) pendulous persistent attached many years to branches, seeds (3–5 mm long) with membranous wings (1–5 times longer that seed).
Reproduction: Flowering, seeds (dispersed by anemochory)
Pastoral importance: Forestry
Palatability and fodder value: None
Toxicity: None
Economic interest: Mostly afforestation, fuelwood, poor-quality softwood, used as building material, crates, furniture; bark to tan leather, good-quality charcoal (also producing tar for marine use in antiquity), resin used as mastic, seeds are collected and chewed.
Habitat: All soils except sand, on limestone, on rocky slopes, on *Terra rossa* country, down to 200 mm MAR
Distribution:

- In Libya, these days, native stands exclusively in Jebel Al-Akhdar (Tolmeita – Shahat – Ras Hillal – Marsa Suza), present in forestry plantations all over coastal zone, Jebel Nafusa (from Yeffren to Homs), rarely planted in Sirte
- In arid zones, southern Europe, North Africa, west Asia, Middle East

Endemism: Currently, native only in Jebel Al-Akhdar

Notes:

- Fossil cones of *Pinus halepensis* have been found near Gharyan, native stand possibly extinct in western Libya.
- Monospecific *Pinus halepensis* forestry plantations (in western and eastern Libya) are highly susceptible to and regularly infested with large population of processionary caterpillars moving in single file (*Thaumetopoea pityocampa*), destroying leaves over large plantation areas; these caterpillars are nesting in *Pinus* branches during winter and are active in spring; hand removal and burning of nests were once common practices in North Africa and Libya: a painful task for a labourer as caterpillars' hairs are highly irritating to skin and eyes.

Other *Pinus* sp. planted in Libya:

- *Pinus pinea* L., http://www.theplantlist.org/tpl1.1/record/kew-2562556, a tall umbrella pine (exotic) tree that produces edible pine nuts used in confectionary and commonly served in green tea in Libya and all over Mediterranean regions; commonly planted as ornamental tree (Sabratha, Zawiya, Gargaresh, Tripoli, Tajura, El-Marj, Shahat, Derna).

Photo 7.410. *Pinus halepensis*, trees, forestry plantation, Gharyan.

Photo 7.411. *Pinus halepensis*, leaves, forestry plantation, Gharyan.

Photo 7.412. *Pinus halepensis*, cones, forestry plantation, Gharyan.

PLANTAGINACEAE (Globulariaceae)

Globularia alypum L.

Synonym(s): http://www.theplantlist.org/tpl1.1/record/kew-2826658
Native:
Vernacular name: *Zriga* (?)
Raunkiaer's life-form: Chamaephyte
Description and morphology: Small hemispherical bush (30–80 cm diam.), supple-lax green stems, white bark on older branches, leaves glabrous thick subsessile alternate (or in fascicule) spathulate with end lobe tri-pointed on lower leaves, young leaves entire spathulate, inflorescences terminal and axillary capitulum with silvery-blue (pale blue to rarely whitish) flowers (10–15 mm diam.) surrounded by involucral scarious bracts hairy on margins, achene enclosed in accrescent pubescent calyx.
Reproduction: Flowering November–April, seeds
Pastoral importance: Low
Palatability and fodder value: Occasionally young stems grazed by all livestock and wildlife
Toxicity: None
Economic interest: Leaves used as purgative, diuretic (?)
Habitat: Gravelly shallow soils, down to 150 mm MAR, a companion species of *Pinus halepensis* (present and relict) forests in matorral in the Jebel areas (west and east), not in Sirte (?)
Distribution:

- In Libya, Jebel Nafusa (Yeffren, Gharyan, Msallata etc.), not in Sirte (?), Jebel Al-Akhdar (El-Marj, Shahat, Gendula etc.)
- In arid zones, southern Europe, North Africa, west Asia

Endemism: NA

Photo 7.413. *Globularia alypum*, ungrazed plant in rocky wadi (wadi Nefad, Seddada).

Photo 7.414. *Globularia alypum*, inflorescence and leaves, wadi Nefad, Seddada.

Globularia arabica Jaub. & Spach

Synonym(s): http://www.theplantlist.org/tpl1.1/record/kew-2826662, or *Globularia eriocephala* Pomel, http://www.theplantlist.org/tpl1.1/record/kew-2826684
Native:
Vernacular name: *Zriga* (?), *Zerga* (?)
Raunkiaer's life-form: Chamaephyte
Description and morphology: Small woody bush (30–80 cm diam.), often Bonsai-shaped from overgrazing, stout pointed woody criss-crossed branches, white bark, leaves glabrous thick subsessile alternate spathulate or entire (with minute white dots) with end round lobe mucronate, inflorescences as terminal capitulum (above plant crown) with silvery-blue (pale blue) flowers (10–15 mm diam.) surrounded by involucral scarious bracts, hairy all over, achene enclosed in accrescent pubescent calyx.
Reproduction: Flowering November–April, seeds
Pastoral importance: Low
Palatability and fodder value: Can be heavily grazed (mostly goats, cattle)
Toxicity: None
Economic interest: Leaves used as purgative, diuretic (?)
Habitat: Rocky places, down to 150 mm MAR

Distribution:

- In Libya, mostly Jebel Al-Akhdar (Shahat, Derna, Tobruk, Bardiya, cliffs on the other side of the border in Egyptian Marmarica, south of As-Sallum, Sidi-Barani, Marsa-Matruh), not in Sirte (?), doubtful in the west
- In arid zones, North Africa, Middle East

Endemism: NA

Note: Possible confusion between *Globularia alypum* (rather herbaceous on gravelly soils) and *Globularia arabica* (woody on rocky sites).

Photo 7.415. *Globularia arabica*, ungrazed plant, on karst, Shahat.

Photo 7.416. *Globularia arabica*, capitulum and tripoint leaves, on karst, Shahat.

Photo 7.417. *Globularia arabica*, dry capitulum (with involucral scarious bracts hairy on margins) and tripoint leaves, on karst, Shahat.

Photo 7.418. *Globularia arabica*, woody, heavily grazed plant, Tobruk.

Plantago albicans L.

Synonym(s): http://www.theplantlist.org/tpl1.1/record/kew-2556637, or *Lagopus albicans* (L.) Fourr., http://www.theplantlist.org/tpl1.1/record/kew-2494920
Native:
Vernacular name: *Inem*
Raunkiaer's life-form: Hemicryptophyte
Description and morphology: Perennial densely villous/hairy/silvery herb (5–30cm), acaule, strong woody taproot with basal silvery narrow lanceolate or linear entire leaves in rosette on very short (black – woody) stems at soil surface level, abundantly branching from base, several inflorescences in long cylindrical dense spikes at end of erect straight peduncles (twice as long as leaves), small flowers with distinct white-greenish petals, small pyriform fruits as indehiscent capsules (pyxides) with an upper or lateral aperture, abundant small (1–2 mm) brown flattish oval seeds with narrow translucent winged (mucilaginous) margins.
Reproduction: Flowering all year long, seeds (possible anemochory and transported by water)
Pastoral importance: High, easy seed collection, could be used for rangeland reseeding
Palatability and fodder value: One of the most important Libyan rangeland species, due to its high palatability and high amount of seed production
Toxicity: None
Economic interest: Seed production and harvesting is easy, should be used for range regeneration on sandy/sandy-loam soils down to 150 mm MAR; seeds are used in local pharmacopoeia against constipation (mild laxative) due to its mucilaginous seed's coating, wet seeds becoming slimy-sticky.
Habitat: Pre-Saharan, on fixed sandy/sandy-loam soils, even on shallow soils, down to 50–100 mm MAR, even on small shallow depressions northern Hamada.
Distribution:

- In Libya, all over from western to eastern Libya
- In arid zones, southern Spain, North Africa (apparently not in Morocco (?)) to Iran

Endemism: NA

Note: *Plantago albicans* is common all over rangelands in Libya; often occurring in the same ecology is found (Jeffara Plain – Al-Assah, Zuara, south Sorman, Bir-el-Ghanem, Bir Touta, Sirte, Jebel Al-Akhdar), *Plantago psyllium* L. (http://www.theplantlist.org/tpl1.1/record/kew-2570621, or *Plantago arenaria* L (http://www.theplantlist.org/tpl1.1/record/kew-2557206), a small (10–30 cm high) annual psammophyte – tachytherophyte, villous, easily identified and distinct from all other *Plantago* sp. with an erect simple main stem (sometimes ramified) with long internodes, each internode bears opposite narrow lanceolate pointed leaves with long pedunculated inflorescences arranged lax pseudo-corymbs (3–5 cm diam.) and terminated with short dense globose inflorescence (10–20 flowers), navicular brown shiny seeds not winged; plant usually not grazed by livestock, local name in southern Tunisia = *Aslouj* (?). Medicinal plant used against dysentery. (?)

Photo 7.419. Well-developed *Plantago albicans*, protected area at El-Hisha – Bu-Gren, March 1979.

Photo 7.420. *Plantago albicans*, Al-Assah – Gazuliah, December 2009. Note: common aspect of several shoots from buried taproot.

Plantago amplexicaulis Cav.

Synonym(s): http://www.theplantlist.org/tpl1.1/record/kew-2571401
Native:
Vernacular name: *Inem* (?)
Raunkiaer's life-form: Tachytherophyte
Description and morphology: Annual villous herb (5–30 cm), short basal internodes, lanceolate entire silvery amplexicaul leaves in helicoidal rosette at soil-surface level, few inflorescences in short cylinder or obovate spikes at end of erect slightly arched peduncles, small pyriform fruits as indehiscent capsules (pyxides) with an upper or lateral aperture, many small navicular seeds with very narrow clear-winged (mucilaginous) margins.
Reproduction: Flowering all year long, seeds (possible anemochory and transported by water)
Pastoral importance: High
Palatability and fodder value: High palatability, all livestock
Toxicity: None
Economic interest: As *Plantago albicans*
Habitat: Sandy/sandy-loam soils, down to 150 mm MAR

Distribution:

- In Libya, Jeffara plain, Jebel Nafusa, Sirte, Jebel Al-Akhdar
- In arid zones, North Africa

Endemism: NA

Note: *Plantago amplexicaulis* (annual) may be confused with *Plantago albicans* (perennial).

Photo 7.421. *Plantago amplexicaulis*, Tobi nursery, Gharyan.

Plantago coronopus L.

Synonym(s): http://www.theplantlist.org/tpl1.1/record/kew-2571444
Native:
Vernacular name:
Raunkiaer's life-form: Tachytherophyte
Description and morphology: Annual herb, villous with adpressed hair, prostrate circular (5–20 cm diam.) flat on ground, leaves fleshy pubescent, leaves pinnatifid deeply dissected (2–6 pairs of acute linear lobes sometimes with denticulate margins), spikes of inflorescences longer than leaves, thin cylindrical straight or contorted creeping or partly ascending solid inflorescences, seeds ovate navicular upper part winged.
Reproduction: Flowering February–July, seeds (possible anemochory and transported by water)
Pastoral importance: Limited
Palatability and fodder value: Grazed by all livestock grazing around settlements
Toxicity: None
Economic interest: As other *Plantago*
Habitat: Ruderal, on compacted soils, on ruined sites, rare in arid zones, need to be well supplied with water to germinate, possibly halophytic
Distribution:

- In Libya, ubiquitous all over coastal zones
- In arid zones, Europe, North Africa to Pakistan

Endemism: NA

Photo 7.422. *Plantago coronopus*, Aradah, Tripoli, April 2010.

Plantago crypsoides Boiss.

Synonym(s): http://www.theplantlist.org/tpl1.1/record/kew-2571485
Native:
Vernacular name:
Raunkiaer's life-form: Tachytherophyte
Description and morphology: Annual herb prostrate, totally flat on ground (5–20 cm diam. including leaves), long dissected/pinnatisect hairy leaves, leaves twice as long as the inflorescences, enlarged inflorescences peduncles, inflorescences peduncles and short obovate retroflexed spikes enlarged at base and turning woody when dry, navicular seeds not marginally winged like most other *Plantago* sp.
Reproduction: Flowering in winter, seeds (possible anemochory and transported by water)
Pastoral importance: Poor
Palatability and fodder value: Grazed by all livestock
Toxicity: None
Economic interest: None
Habitat: Loamy-sand, pre-desertic, down to 100 mm MAR
Distribution:

- In Libya, mostly south Jebel Al-Akhdar (near Martuba, on nummulitic limestone (*Camelina* sp., a lenticular foraminifera Nummulites – tertiary geological deposits as on photo), possible in western Libya and Sirte
- In arid zones, Libya and Egypt

Endemism: Eastern Libya, coastal western Egypt

Photo 7.423. *Plantago crypsoides*, Martuba, Mekhili, south Jebel Al-Akhdar, April 2010, with nummulites.

PLUMBAGINACEAE

Limoniastrum monopetalum (L.) Boiss.

Synonym(s): http://www.theplantlist.org/tpl1.1/record/kew-2495505, or *Statice monopetala* L., Statice monopetala L. —The Plant List
Native:
Vernacular name: *Zita, Zeita*
Raunkiaer's life-form: Nanophanerophyte
Description and morphology: Perennial hemispherical bush (50–200 cm diam.), whitish-grey features, strong rooting system, ascending intricate rigid black stems, glabrous narrow lanceolate spathulate fleshy leaves often covered with small calcareous/salt nodules, petioles amplexicaule, abundant attractive terminal inflorescences, pink flowers (10–20 mm) with red calyx.
Reproduction: Flowering May–November, seeds
Pastoral importance: Limited
Palatability and fodder value: Winter camel feed
Toxicity: None
Economic interest: Fuelwood; bush parasitized with brown-reddish woody galls from a Lepidoptera (*Oecocecis guyonella*); these galls are collected and used in southern Tunisia for tanning leather. Also, infusion of galls and leaves reported to be used as treatment against dysentery.
Habitat: Azonal, in *sebkha*, loamy-sand on saline soils, bush forming *nebkha*, gypso-halophyte
Distribution:

- In Libya, Al-Assah, coastal salt marshes (Zuara to Kararim–Taworga), Sirte (Bu-Gren, Hisha to Ajdabiya), Jebel Al-Akhdar (Tokra to Tobruk)
- In arid zones, southern Spain – Portugal, North Africa to the Middle East

Endemism: NA

Note: In *sebkha* west of Al-Assah, along Tunisian border to Bu-Green, occurrence of *Limoniastrum guyonianum* Durieu ex Boiss. (unresolved name, http://www.theplantlist.org/tpl1.1/record/kew-2495501, accepted as *Limoniastrum guyonianum* Boiss., *Limoniastrum guyonianum* Boiss. - eflora Maghreb), *Zeita*, a North African west Saharan endemic, not recorded over eastern Libya/Egypt., with strong woody dark/black open slender branches and much narrower long hemicylindrical whitish (salt excretion (?)) leaves, divaricate spikes, purple-violet small showy terminal flowers; used to be an excellent source of slow-burning fuelwood when camping at Magdar-es-Sbot and west of Hebiliyah – Bir Ougradia in the 1970s, appeared to have all disappeared (uprooted for fuelwood) by 2007–2010; possibly threatened in Libya.

Photo 7.424. *Limoniastrum monopetalum*, Kararim *sebkha*, May 2008.

Photo 7.425. *Limoniastrum monopetalum*, Kararim *sebkha*, May 2008.

Limonium pruinosum (L.) O. Ktze.

Synonym(s): Unresolved name, http://www.theplantlist.org/tpl1.1/record/kew-2494912, or *Statice pruinosa* L., unresolved name, http://www.theplantlist.org/tpl1.1/record/kew-2598928, or *Limonium alleizettei* (Pau) Brullo & Erben, unresolved name, http://www.theplantlist.org/tpl1.1/record/kew-2495452
Native:
Vernacular name: *Rhadam-er-ghazel* (?)
Raunkiaer's life-form: Chamaephyte

Description and morphology: Small (20–50 cm) perennial, glabrous, erect intricate rigid grey zig-zagging stems covered with small calcareous nodules, basal leaves missing at flowering, inflorescences in open scorpioid cymes more or less dichotomic, flowers pink, unifloral spikes arranged in short dense glomerules of flowers.
Reproduction: Flowering May–November, seeds
Pastoral importance: Limited
Palatability and fodder value: Camel feed
Toxicity: None
Economic interest: None
Habitat: Azonal, in *sebkha*, loamy saline soils, temporary waterlogged sites
Distribution:

- In Libya, Al-Assah, coastal salt marshes (Zuara to Kararim – Taworga), Sirte (Bu-Gren, Hisha to Ajdabiya), Jebel Al-Akhdar (Tokra to Tobruk), Marmarica
- In arid zones, southern Europe, North Africa to the Middle East

Endemism: NA

Photo 7.426. *Limonium pruinosum*, El-Hisha, Nofilia, May 2007.

Photo 7.427. *Limonium pruinosum*, El-Hisha, Bu-Gren, May 2009.

Limonium thouinii (Viv.) Kuntze

Synonym(s): http://www.theplantlist.org/tpl1.1/record/kew-2494760, or *Limonium lobatum* (L. f.) Kuntze, http://www.theplantlist.org/tpl1.1/record/kew-2495823, or *Statice thouinii* Viv., http://www.theplantlist.org/tpl1.1/record/kew-2598785
Native:
Vernacular name: *Oudin-el-Jmel*
Raunkiaer's life-form: Stenotherophyte
Description and morphology: Small annual, glabrous, erect, stems not winged, pinnatifid basal elongated leaves (4–7 large lobes) in rosette, leaves alternate, top of leaves larger that base, flowers white (sometimes calyx reported in North Africa as bluish or yellow (?)) on bi- or tri-winged scape, flower panicle unilateral.
Reproduction: Flowering, June–November, seeds
Pastoral importance: Limited
Palatability and fodder value: Camel feed when green, not touched when dry
Toxicity: None
Economic interest: None
Habitat: Rocky – shallow loamy soils on calcareous crust, possibly gypsophyte and halophyte, may indicate soil profile containing phosphate, down to 100 mm MAR.
Distribution:

- In Libya, south Jebel Nafusa, north-eastern Hamada, Sirte (Bu-Gren), Jebel Al-Akhdar (wadi Maghur)
- In arid zones, southern Europe, North Africa to Iran

Endemism: NA

Photo 7.428. *Limonium thouinii*, plateau wadi Zorzor (N 31.545 E 14.814).

Photo 7.429. *Limonium thouinii*, Al-Jahishiyah (N 32.060 E 20.816).

Photo 7.430. *Limonium thouinii*, wadi Maghur, Al-Jahishiyah.

POACEAE

Aegilops triuncialis L.

Synonym(s): Aegilops triuncialis L. — The Plant List
Native:
Vernacular name: *Shahir-el-far* (?)
Raunkiaer's life-form: Therophyte
Description and morphology: A short annual grass, few basal leaves, single inflorescence on long stem, 2–3 ventricose spikelets with long straight awn ending glumes
Reproduction: Seeds
Pastoral importance: Grazed when green, tends to be ignored by small ruminants
Palatability and fodder value: Medium
Toxicity: None
Economic interest: An ancestor of *Triticum* sp.
Habitat: Ruderal, ruins, roadsides, loamy soils
Distribution:

- In Libya, all northern zones, rare in Sirte, down to 150 mm MAR
- In arid zones, Mediterranean to south-west Asia

Endemism: NA

Other common *Aegilops* species in Libya:

- *Aegilops ventricosa* Tausch; Aegilops ventricosa Tausch — The Plant List, long narrow moniliform spikes, short erect adpressed short awns on spikelet

Photo 7.431. *Aegilops triuncialis*, at Tobi nursery, Gharyan.

Aeluropus lagopoides (L.) Thwaites

Synonym(s): http://www.theplantlist.org/tpl1.1/record/kew-388021
Native:
Vernacular name: *Akresh*
Raunkiaer's life-form: Hemicryptophyte – Geophyte
Description and morphology: Perennial, creeping stems, rhizomatous, short globular inflorescence on long peduncle from stems nodes, leaves alternate convolute, pointed, coriaceous green-greyish often covered with salt efflorescence.
Reproduction: Mostly vegetative, rhizomatous, flowering March–June, seeds (?)
Pastoral importance: Limited, winter-spring growth
Palatability and fodder value: Winter-spring livestock feed, mostly dromedaries and cattle
Toxicity: None
Economic interest: None
Habitat: Inland salt marshes, *sebkha*, on shallow sand veils over salty flats
Distribution: All coastal zone, azonal, on salty flats

- In Libya: Al-Assah, Zuara, Taworga, Sirte, Ajdabiya to Tobruk
- In arid zones, North Africa, the Middle East

Endemism: NA

Other *Aeluropus* sp. in Libya:

- *Aeluropus littoralis* (Gouan) Parl., Aeluropus littoralis (Gouan) Parl. — The Plant List, elongated inflorescence, on long stems, North Africa to Central Asia (Uzbekistan)

Photo 7.432. *Aeluropus lagopoides*, globulous spikes, salty flats at Jbebina (Al-Assah).

Photo 7.433. *Aeluropus lagopoides*, general view, Jbebina (N 32.910 E 11.455), Al-Assah.

Photo 7.434. *Aeluropus lagopoides*, creeping stems with pointed, coriaceous green-greyish leaves.

Bromus scoparius L.

Synonym(s): http://www.theplantlist.org/tpl1.1/record/kew-401843
Native:
Vernacular name: *Sabal-el-far, Sboul-el-far* (?)
Raunkiaer's life-form: Tachytherophyte
Description and morphology: Annual grass, short leaves hairy on upper face, short glabrous culms (10–15 cm), inflorescences dense-contracted, spikelets globose, lemma with apex bifid and awned (all *Bromus* sp.)
Reproduction: Flowering March–June, seeds
Pastoral importance: Poor, sporadic-sparse stands, considered as weed
Palatability and fodder value: Limited to poor when inflorescences dry
Toxicity: None
Economic interest: None, considered a weed
Habitat: Ruderal, ruins, roadside, occurs down to 150 mm MAR
Distribution:

- In Libya, ubiquitous along the coast in dry areas, in wadis, in crops
- In arid zones, Mediterranean to Central Asia and north-west Pakistan

Endemism: NA

Photo 7.435. *Bromus scoparius*, wadi Maghur (N 32.039 E. 20.738), Al-Jahishiyah.

Other common *Bromus* sp. (all considered as weeds) in Libya:

- *Bromus rubens* L., small annual, inflorescence erected reddish-purplish compact, all coastal Libya, common weed in barley rainfed crops, synonyms = http://www.theplantlist.org/tpl1.1/record/kew-401803
- *Bromus madritensis* L., annual, tall with culms up to 50 cm, inflorescence loose spikelets lax pedicellate with linear awn, Jeffara, Shahat, wadi Kuf, synonyms = http://www.theplantlist.org/tpl1.1/record/kew-401383
- *Bromus alopecuros* Poir., annual, culms (20–30cm), inflorescence erect oblong flattish, lemma with reflexed awned, Jebel Al-Akhdar, Synonyms = http://www.theplantlist.org/tpl1.1/record/kew-400611
- *Bromus rigidus* Roth., annual, overlapping culms (20–60 cm) with long and large flexuous villous and spaced leaves, inflorescence contracted and erect, Zanzur, Tajoura, Khoms, synonyms = http://www.theplantlist.org/tpl1.1/record/kew-401784

Cenchrus ciliaris L.

Synonym(s): Cenchrus ciliaris L. — The Plant List, or *Pennisetum ciliare* (L.) Link, Pennisetum ciliare (L.) Link — The Plant List
Native:
Vernacular name: *Regma, Bu-regma, Lebed* (?)
Raunkiaer's life-form: Cespitose hemicryptophyte
Description and morphology: Perennial grass, rhizomatous, lax tuft, up to 50 cm in diameter, ascending stems and inflorescences, raceme/inflorescence plumose – silky purplish, glums plumose with long bare wavy awn, stem nodes with small ligulae lined with hairs, *Cenchrus ciliaris* often confused with *Pennisetum* sp.
Reproduction: Flowering from early spring into summer, seeds fluffy dispersal by anemochory, fresh seeds dormant, dormancy removed by exposure to sunlight and high soil temperature.
Pastoral importance: C4 grass, drought-resistant, end of spring to summer growing with leaves remaining green long into summer, withstands heavy grazing, easy establishment for rangeland restoration.
Palatability and fodder value: Excellent, end of spring/summer forage when late rains occur
Toxicity: None
Economic interest: Forage, high potential for revegetation of degraded rangelands
Habitat: Sandy/sandy-loam soils/in rocky wadis, 100–300 mm MAR
Distribution:

- In Libya, Jeffara plain (south Al-Assah, Bir Ougradia, El-Watiya), Bir-el-Ghanem, Bir Ayyad to North Nalut (along wadi Jebel Nafusa foothills), Misurata – Kararim (east of petrol station), Seddada (wadi Nefad), wadi Sof-el-Jin, Hamada (Girzah, on rocky/stony basalt slopes), Ajdabiya (El-Beda mahmiyat), south of Jebel Al-Akhdar to Tobruk
- In arid zones, Canary Islands, from Morocco, south Sicily to Iran, Afghanistan, South India – Pakistan, sub-Saharan Africa, South Africa

Endemism: NA
Other *Cenchrus* species in Libya:

- *Cenchrus biflorus* Roxb., Cenchrus biflorus Roxb. — The Plant List, annual grass, not present in northern Libya, strictly occurring southern Sahara (Fezzan, Tanezrouft), presence strictly defines southern limit of Sahara, on sandy soils, spikelets spiny/denticulate/vulnerant spiny burrs-like (*Kram-Kram*), burrs sticking to shoes, clothes, livestock fleece (in *Libyan Flora* = *Cenchrus incertus* M.A. Curtis, or *Cenchrus pauciflorus* Benth., unresolved)

Photo 7.436. *Cenchrus ciliaris*, near Kararim (N 32.164 E 15.103), Misurata.

Photo 7.437. *Cenchrus ciliaris*, inflorescence, Kararim.

Photo 7.438. *Cenchrus ciliaris*, axis of dry inflorescence scabrous, bare and undulating, wadi Nefad (N 31.527 E 14.692).

Photo 7.439. *Cenchrus ciliaris*, on basalt rocky slopes on Hamada, south-east Mizda (N 31.467 E 13.128), Hamada.

Cutandia dichotoma (Forssk.) Trab.

Synonym(s): Cutandia dichotoma (Forssk.) Trab. —The Plant List
Native:
Vernacular name: *Zawan* (?), *Ziwan* (?)
Raunkiaer's life-form: Tachytherophyte
Description and morphology: Small annual grass with ascending-erect stems and culms, culms geniculated (black or purplish) at base, partly enclosed in long swollen sheath, inflorescence dichotomous with lax spikelets, main axis of inflorescence with internodes shorter than spikelets.
Reproduction: Seeds
Pastoral importance: Sand binder, annual winter forage, tends to disintegrate when dry
Palatability and fodder value: Excellent winter feed, also straw well ingested when dry (*teben*)
Toxicity: None
Economic interest: None
Habitat: Down to 100 mm mean annual rainfall, on sandy soils (fixed, shifting and dunes)
Distribution:

- In Libya, all over sandy sites, Jeffara plain, Sirte, south Jebel Al-Akhdar
- In arid zones, Mediterranean to Iran

Endemism: NA

Other most common *Cutandia* species in Libya:

- *Cutandia memphitica* (Spreng.) Benth., very similar to *C. dichotoma*, reported difference is that *C. memphitica*' s main axis of inflorescence has internodes longer than spikelets. Synonym = http://www.theplantlist.org/tpl1.1/record/kew-406073

Photo 7.440. *Cutandia dichotoma*, Bir-el-Ghanem, wadi Abu-Shayba.

Photo 7.441. *Cutandia dichotoma*, Bir Ougradia (N 32.631 E 11.609), Al-Assah.

Cymbopogon schoenanthus (L.) Spreng.

Synonym(s): Cymbopogon schoenanthus (L.) Spreng. —The Plant List
Native:
Vernacular name: *Lemmad* (?)
Raunkiaer's life-form: Hemicryptophyte – Thermoxerophyte
Description and morphology: Perennial aromatic densely tufted grass, erect stems and culms above leaves, reddish inflorescence with long spatheoleate panicles, raceme in short pairs.
Reproduction: Flowering March–May, seeds
Pastoral importance: Saharo-Sudanian, C4 grass, extremely drought-resistant
Palatability and fodder value: Mostly dromedaries' and goats' forage
Toxicity: None
Economic interest: Aromatic grass, leaves sometimes added to tea by shepherds for lemony flavour, insect repellent, oil extract (close to tropical lemongrass *Cymbopogon citratus* oil used for flavouring cooking or commercial dish-washing liquid)
Habitat: Sandy loam/rocky soils, on gypsum crust in Sirte, 100–200 mm MAR
Distribution:

- In Libya, Nalut-Tiji, then south Sirte-wadi Bey-el-Kebir, wadi Wishkah
- In arid zones, from North Africa, Arabia to Pakistan-India.

Endemism: NA

Photo 7.442. *Cymbopogon schoenanthus*, wadi Nefad, Seddada.

Photo 7.443. *Cymbopogon schoenanthus*, inflorescence, wadi Nefad, Seddada.

Photo 7.444. *Cymbopogon schoenanthus* stands, overgrazed and protected, Ziarat – Quetta, north-western Pakistan.

Cynodon dactylon (L.) Pers.

Synonym(s): http://www.theplantlist.org/tpl1.1/record/kew-406339
Native:
Vernacular name: *Nedjem, Nejem*
Raunkiaer's life-form: Hemicryptophyte – geophyte
Description and morphology: Perennial creeping grass, rhizomatous, culms rooting at the lower nodes, digitate thin purplish awnless stellar spikes on elongated peduncles, tiny spikelets
Reproduction: Vigorous, rooting system and runners, considered as an aggressive weed in oases and gardens, withstands heavy grazing

and drought with rhizomes deep in soil, growing and flowering all year round on wet soils, remains dormant on dry soils.

Pastoral importance: Valuable green forage for all livestock and wildlife

Palatability and fodder value: Excellent as green biomass

Toxicity: Reported to be toxic when grazed dry and in abundance by livestock (?)

Economic interest: Forage, available all year round, extract reported to be anti-helminthic (?)

Habitat: Wet sandy/sandy-loam soils, when extra run-off (wadi)/irrigation available, withstands low salinity, never occurs on arid shallow soils and calcareous crusts

Distribution: Ubiquitous in temperate and Mediterranean zones:

- In Libya, all coastal zone, oases
- In arid zones, restricted to wet soils, in oases

Endemism: NA

Photo 7.445. *Cynodon dactylon*, Jbebina – Al-Assah, near artesian well.

Photo 7.446. *Cynodon dactylon* dense natural pasture in wadi with background of *Atriplex halimus* large bushes (in wadi Sassu – south Misurata).

Dactylis glomerata L.

Synonym(s): Dactylis glomerata L. —The Plant List

Native: Possibly *Dactylis glomerata* subsp. *hispanica* (Roth) Nyman in Libya

Vernacular name: *Nishia* (in Jebel Al-Akhdar)

Raunkiaer's life-form: Hemicryptophyte

Description and morphology: Perennial grass with deep rooting system. Most common in Libya is *D. glomerata* subsp. *Hispanica* – small plant (20–30cm), with only a few leaves, few and short tillers, often only one compacted inflorescence, with densely clustered pedicelled spikelets.

Reproduction: From seeds, flowering March–June

Pastoral importance: Limited in arid Libya, winter growing

Palatability and fodder value: Excellent forage but poor biomass production under dry conditions, heavily grazed when available to all livestock

Toxicity: None

Economic interest: Wide genetic variability, collected as phytogenetic resource of many highly productive selected forage cultivars for temperate agriculture (Europe, USA, Australia, etc.)

Habitat: Loamy to heavy soils, does not stand salinity, common in small water catchments around olive trees in Jebel Nafusa to Nalut, in water catchment on soil conservation projects, requires at least 300–400 mm MAR equivalent

Distribution:

- In Libya, coastal zone, Jebel Nafusa, in Jebel Al-Akhdar scrubland and forested areas
- In arid zones, from Morocco to Libya, in Saharan mountains, limited distribution to Mediterranean higher-rainfall zones to Afghanistan, type from temperate zones

Endemism: NA

Photo 7.447. *Dactylis glomerata* (possibly subsp. *hispanica* (Roth) Nyman), Gharyan – Jebel Nafusa.

Dichanthium annulatum (Forssk.) Stapf

Synonym(s): http://www.theplantlist.org/tpl1.1/record/kew-408597, or *Andropogon annulatus* Forssk., http://www.theplantlist.org/tpl1.1/record/tro-25504772
Native:
Vernacular name:
Raunkiaer's life-form: Thermophylous hemicryptophyte
Description and morphology: Rare perennial, cespitose, rhizomatous, erected geniculated culms (50–125 cm) with abundant basal leaves, circle of hairs at node, large lax stellar panicle high above leafy parts, narrow pedicellate 5–10 spikelets flexuous with reddish-violet spathe, kneed awns hairy
Reproduction: Flowering March–May, seeds
Pastoral importance: Limited except in wadis in Solul-el-Khoder
Palatability and fodder value: Poor appetability, moderately grazed when green, coarse and hardly grazed when dry
Toxicity: None
Economic interest: None
Habitat: Shallow rocky soils, rocky-stony-gravelly wadis to 50 mm MAR, never on saline soils, C4 grass likely originating from sub-Saharan regions (warm seasons and summer rainfall)
Distribution:

- In Libya, Solul-el Khoder, wadis between Girzah and Seddada
- In arid zones, Saharan (sub-Saharan), North Africa, Middle East to Afghanistan – India

Endemism: NA

Photo 7.448. *Dichanthium annulatum*, plant, Girzah, Beni-Walid.

Photo 7.449. *Dichanthium annulatum*, panicles, Girzah, Beni-Walid.

Note: *Dichanthium annulatum* (without inflorescence) could be confused with *Hyparrhenia hirta* and *Cymbopogon schoenanthus*. *Dichanthium annulatum* is found in rocky-gravelly wadis (wadi Ghirza and other wadis to Beni Walid) along with *Lasiurus hirsutus* (Forssk.) Boiss. = *Rottboellia hirsuta* (Forssk.) Vahl, = *Lasiurus scindicus* Henrard — The Plant List, *G'doum* local name, a thermophylous – desert C4 perennial rhizomatous and cespitose grass (50–100 cm high) woody at base, rachis and stems all erect silky scabrous with adpressed hard flakes, terminal solitary white silky spikes with sessile spikelets, rare occurrence only in southern Tripolitania and deep southern Libya, possible to southern Sirte, excellent camel forage, present in North African Sahara to Arabian Peninsula, southern Iran, India.

Photo 7.450. Rhizomatous tuft of *Lasiurus hirsutus*, near Erfoud, eastern Morocco (N 31.308 E -4.198). (Photo: Claude Lemmel, http://alyasmina.org)

Photo 7.451. *Lasiurus hirsutus* inflorescence, eastern Morocco (N 31.308 E -4.198) (Photo: Claude Lemmel, http://alyasmina.org)

Hordeum murinum L.

Synonym(s): http://www.theplantlist.org/tpl1.1/search?q=Hordeum+murinum+
Vernacular name: *Sboul-el-far* (?)
Raunkiaer's life-form: Stenotherophyte
Description and morphology: Annual grass, prostrate or erected culms, leaves hairy, inflorescence cylindrical, spikelets with long denticulate awn criss-crossing
Reproduction: From seeds, very fast growth, short cycle, flowering December–May
Pastoral importance: Limited, resistant to trampling
Palatability and fodder value: Well grazed by all livestock
Toxicity: None
Economic interest: Common weed
Habitat: Ruderal, nitrophilous, rich in organic matter, all abandoned and waste land following soil disturbance, not on saline soils
Distribution:

- In Libya, all over coastal zones, in irrigated crops, rare on arid land
- Ubiquitous in all temperate and Mediterranean zones, to Afghanistan

Endemism: NA

Photo 7.452. *Hordeum murinum*, Aradah, Tripoli.

Other main *Hordeum* sp.in Libya:

- *Hordeum maritimum* With. (?) = *Hordeum marinum* Huds. (?), Hordeum marinum Huds. The Plant List, annual grass, geniculate erect culms with flat inflorescence with long rectilinear and parallel awns at 90° to axis, on temporary flooded saline soils, around *sebkha*.
- *Hordeum bulbosum* L., Hordeum bulbosum L. —The Plant List, perennial grass, base of stem bulbous, long slender inflorescence, on heavy red soils in Jebel Al-Akhdar

(Tokra, El-Marj, Derna), used in interspecific hybridization of cultivated barley (*Hordeum vulgare* L.) with *H. bulbosum* L.

- *Hordeum vulgare* L., (the common cultivated *Shair*), Hordeum vulgare L. — The Plant List, annual, common cultivated barley, used to make local bread, dual-purpose barley (green feed and grain production), early grazing stimulating tillering, green barley (standing or cut) used to feed lactating ewes, grain commonly used as livestock feed, barley straw is a valuable livestock feed (Goodchild *et al.*, 2022) in summer until the first autumn rains, some old local varieties (*Tsaini* = 90-day cycle – Jebel Nafusa) have a very short cycle from seeding to harvesting; see large collection of local Libyan barley genetic resources done by Al-Jibouri (1966) and Hammer *et al.* (1988).

Photo 7.453. *Hordeum murinum*, spike, Aradah, Tripoli.

Photo 7.454. Local barley – *Hordeum vulgare*, rainfed, wadi Sassu, south Misurata.

Hyparrhenia hirta (L.) Stapf

Synonym(s): Hyparrhenia hirta (L.) Stapf — The Plant List = *Andropogon hirtus* L., Andropogon hirtus L. — The Plant List
Native:
Vernacular name: *Hamra* (?)
Raunkiaer's life-form: Hemicryptophyte, cespitose
Description and morphology: Perennial – tussock grass, erected culms (40–150 cm), large lax panicle inflorescence, narrow spikelets in pairs with reddish spathe, awns hairy
Reproduction: From seeds, flowering all year round
Pastoral importance: Limited, soil binder in wadis
Palatability and fodder value: Poor appetability, coarse and hardly touched when dry, moderately grazed when green
Toxicity: None
Economic interest: None
Habitat: Shallow rocky soils, on calcareous crust, roadside, rocky wadis to 150 mm MAR, never on saline soils
Distribution:

- In Libya, coastal zone, from Zuara to Tobruk, in wadis (Bir Ayyad to Nalut) – wadi Megenin, foothill of Jebel Nafusa
- In arid zones, Canarias to Iran, to Ethiopia

Endemism: NA

Photo 7.455. *Hyparrhenia hirta*, wadi Sassu (N 32.123 E 14.826), south Misurata.

Photo 7.456. *Hyparrhenia hirta* inflorescence, wadi Sassu, south Misurata.

Note: Without inflorescences, often confused with *Cymbopogon schoenanthus* or *Dichanthium annulatum*.

Lygeum spartum Loefl. ex L.

Synonym(s): http://www.theplantlist.org/tpl1.1/record/kew-423191
Native:
Vernacular name: *Sennak, Halfa Mahboula*
Raunkiaer's life-form: Cespitose hemicryptophyte, gypsophyte
Description and morphology: Perennial grass with centrifugal strong scarious coriaceous linear yellowish rhizomes bearing erect pointy shoots, long (30–50 cm) linear rolled glabrous yellowish leaves, spikelets single em braced in papery-scarious whitish spathe
Reproduction: Rhizomatous, flowering from March, seeds
Pastoral importance: Soil binder, very valuable
Palatability and fodder value: Poor, grazed by dromedaries along the coast
Toxicity: None
Economic interest: Widely used in North Africa for plaiting, basket and vegetal carpet work, sandals, stuffing mattresses, etc.; in time of drought and poor pasture, collected by farmers to feed their cows, leaves were used in Spain to make the famous English boat hat (or boater), sometimes leaves cut and used for sand dune fixation.
Habitat: Fixed sand/sandy-loam soils on gypsum crust, around *sebkha*, sometimes on slightly saline soils
Distribution:

- In Libya, limited to the coastal zones, from Al-Assah to coastal Jebel Al-Akhdar
- In arid zones, southern Spain, North Africa, southern Italy, Crete

Photo 7.457. *Lygeum spartum* tuft, (near artesian well at Jbebina, Al-Assah) (Author's own photo)

Photo 7.458. *Lygeum spartum*, lone spikelet embraced in wide, long awned papery spathe, Magdar-es-Sbot, Al-Assah.

Photo 7.459. *Lygeum spartum* woody rhizomes, Magdar-es-Sbot (N 32.824 E 11.728), Al-Assah.

Endemism: Western Mediterranean

Note: Often confused with *Stipa tenacissima*, differs by their very different ecology and the fact that the leaves of *Lygeum* cannot be hand-pulled like *Stipa tenacissima*.

Panicum turgidum Forssk.

Synonym(s): http://www.theplantlist.org/tpl1.1/record/kew-430705
Native:
Vernacular name: *Bu-rokba*
Raunkiaer's life-form:
Description and morphology: Perennial desertic grass, large tuft (1–1.5 m diam.), culms up to 150 cm high woody geniculate (hence the local name 'kneed') and strongly branched from base, long linear narrow and acuminate leaf-sheaths, reduced leaves at nodes, inflorescence in lax panicle, rooting at lower nodes, roots enclosed in tubular sand, small seeds easy to collect.
Reproduction: Rhizomatous, flowering February–April, seeds
Pastoral importance: Camel feed, sand binder
Palatability and fodder value: Young and dry leaves well grazed, high cellulose content
Toxicity: None
Economic interest: Easy to propagate by cuttings, rooting from nodes, seeds used to be easily collected during drought and famine by desert dwellers and nomads to substitute for flat barley bread.
Habitat: Sandy soils, shifting sand, withstand sand dune encroachment, in Sahara often present in association with *Acacia tortilis*
Distribution:

- In Libya, wadi Nefad-Seddada, Beni-Walid, Sirte, Ajdabiya, south Jebel Al-Akhdar
- In arid zones, Algeria to Libya, Egypt to further east

Endemism: NA

Photo 7.460. *Panicum turgidum*, tussock heavily grazed by camels, wadi Nefad, Seddada.

Photo 7.461. Branching geniculated node on *Panicum turgidum*, wadi Nefad, Seddada.

Photo 7.462. *Panicum turgidum*, panicle with spikelets, wadi Nefad, Seddada.

Pennisetum divisum (Forssk. ex J. F. Gmel.) Henrard

Synonym(s): http://www.theplantlist.org/tpl1.1/record/kew-432681, *Panicum divisum* Forssk.. ex J.F.Gmel, Panicum divisum Forssk.. ex J.F.Gmel. — The Plant List, or *Pennisetum dichotomum* Delile, Pennisetum dichotomum Delile — The Plant List
Native:
Vernacular name: *Bu-Rokba* (?)
Raunkiaer's life-form: Hemicryptophyte
Description and morphology: Desert C4 perennial grass, large superficial rooting system, woody stock, dense stiff stems straight-erect with swollen nods rarely branched, large junciformis clump-tussocks (1–2 m in diameter and 100–150 cm high), tall woody glabrous culms with sharp stiff leaves, false spikes, cylindrical spikelets with long scabrous bristles and white awns
Reproduction: Rhizomatous, and seeds
Pastoral importance: Used to be cultivated near Misurata, camel and horse feed
Palatability and fodder value: Moderate, mostly camel feed
Toxicity: None
Economic interest: Harvested and used to fix moving sand on afforestation projects on dunes
Habitat: Mostly Saharan, 50–200 mm MAR on deep sandy soils and dunes, never on saline soils
Distribution:

- In Libya, some at wadi Kam, wadi Nefad-Seddada, Beni-Walid, south Sirte, Ajdabiya, south Jebel Al-Akhdar
- In arid zones, Algeria to Libya, Egypt to further east

Endemism: NA

Note: Often confused with *Panicum turgidum* as it grows in the same ecology on sandy soils.

Photo 7.463. *Pennisetum divisum*, wadi Nefad, Seddada.

Photo 7.464. *Pennisetum divisum* inflorescence, wadi Nefad, Seddada.

Pennisetum setaceum (Forssk.) Chiov.

Synonym(s): Pennisetum setaceum (Forssk.) Chiov. — The Plant List
Native:
Vernacular name: *Harsha* (?)
Raunkiaer's life-form: Hemicryptophyte
Description and morphology: C4 perennial grass, strong erect tuft, stem straight unbranched emerging from base, stiff scabrous leaves longer that internodes fine linear with short hairs on margins, inflorescence in cylindrical long (8–10 cm) plumose, white (sometimes turning pinkish to violet) panicle bending at extremity, bristles longer than spikelets.
Reproduction: Seeds
Pastoral importance: Poor, often outstanding presence in landscapes as not grazed by small ruminants
Palatability and fodder value: Moderate to none
Toxicity: None
Economic interest: In previous times, harvested as winter feed for horses
Habitat: 150–400 mm along wadi banks on colluviums and rocky places
Distribution:

- In Libya, wadi Kam, along wadis Bir-Touta, Bir-el-Ghanem, Bir Ayyad, also near Gharyan
- In arid zones, Algeria to Libya, Egypt to further east

Endemism: NA

Note: Often confused with *Cenchrus ciliaris* (much smaller tufts than *P. setaceum*) as also present along wadi banks.

Photo 7.465. *Pennisetum setaceum*, near Tobi nursery, Gharyan.

Photo 7.466. Panicle of *Pennisetum setaceum*, near Tobi nursery, Gharyan.

Phalaris minor Retz.

Synonym(s): http://www.theplantlist.org/tpl1.1/record/kew-433505
Native:
Vernacular name: *Ziwan* (?)
Raunkiaer's life-form: Tachytherophyte
Description and morphology: Gracile annual grass (10–40 cm on dry sites, up to 60–80 cm in irrigated crops), culms geniculate at base then erected, leaves linear-lanceolate to tip of culm, obovate solitary contracted panicle, short peduncled and tight spikelets (compressed and flattish), glume not pointed and with short dorsal wing.
Reproduction: Flowering December–April, seeds
Pastoral importance: Limited, low biomass production
Palatability and fodder value:
Toxicity: Sometimes reported toxic to livestock
Economic interest: Forage
Habitat: Variable but mostly temporary wet areas, wadis, loamy soils, in crops, near leaking wells, withstand short waterlogging, not on saline soils
Distribution: Ubiquitous:

- in Libya, all coastal zones and inland
- in arid zones, worldwide

Endemism: NA

Other main *Phalaris* species in Libya:

- *Phalaris paradoxa* L, Phalaris paradoxa L. — The Plant List, annual, sometimes in large and dense stand, erect culms, common on red soils in Jebel Al-Akhdar, in crops, along roads, in temporary waterlogged areas, culms foliate largely embracing the capitula, glume pointed.

Photo 7.467. *Phalaris minor*, wadi Maghur, Jebel Al-Akhdar.

Photo 7.468. *Phalaris minor*, whole plant, wadi Maghur, Jebel Al-Akhdar.

Pholiurus incurvus (L.) Schinz & Thell.

Synonym(s): http://www.theplantlist.org/tpl1.1/record/kew-433911, or *Parapholis incurva* (L.) C.E.Hubb., http://www.theplantlist.org/tpl1.1/record/kew-431053, or *Aegilops incurva* L., http://www.theplantlist.org/tpl1.1/record/kew-387696
Native:
Vernacular name:
Raunkiaer's life-form: Stenotherophyte
Description and morphology: Small (10–20 cm) annual grass, fibrous roots, ascending leafy stems, leaves flat long lanceolate curved inwards, inflorescence narrow arched with alternating spikes adpressed/deeply inserted in spike rachis, glumes longer than spike and internode.
Reproduction: Flowering March–June, seeds
Pastoral importance: Limited
Palatability and fodder value: Grazed by all livestock
Toxicity: None
Economic interest: None
Habitat: Azonal, halophyte, on saline mud flats

Distribution:

- In Libya, coastal zone, on all mud flats/saline soils, *sebkha*
- In arid zones, southern Europe, North Africa, Middle East to Central Asia

Endemism: NA

Photo 7.469. *Pholiurus incurvus* (from Uzbekistan).

Piptatherum miliaceum (L.) Coss.

Synonym(s): http://www.theplantlist.org/tpl1.1/record/kew-434465, or *Oryzopsis miliacea* (L.) Asch. & Schweinf., http://www.theplantlist.org/tpl1.1/record/kew-426815
Native:
Vernacular name:
Raunkiaer's life-form:
Description and morphology: Large bunch grass, perennial, cespitose, rhizomatous, culms up to 120–150 cm, long leaves soft flexuous flat and large (up to 10 mm wide), long (20–40 cm) flexuous panicle lax and many branched.
Reproduction: Flowering December–February, high seed production, fine small seeds (shiny cream) easy to collect

Pastoral importance: Very valuable but limited extent due to its palatability, excellent regrow after grazing providing water in soil
Palatability and fodder value: Excellent, soft leaves even dry
Toxicity: None
Economic interest: Excellent forage, easy to reseed, should be extensively used in soil conservation and afforestation projects in Jebel Nafusa and Jebel Al-Akhdar
Habitat: Mesophytic, at least 250–300 mm mean annual rainfall, loamy to heavy soils, not on saline soils, in protected areas (sometimes in old graveyards, orchards) as avidly grazed by all livestock, often growing protected in bushes (*Atriplex halimus, Ziziphus lotus, Lycium* sp., *Calicotome* sp.).
Distribution:

- In Libya, Jebel Nafusa foothills in wadis (wadi Ayyad), Tarhuna (Shiarchara), Gharyan, Yeffren (Ain Zerga), Khoms, Misurata (Daphnia, wadi Saratan), to Jebel Al-Akhdar
- In arid zones, North Africa, Middle East to Iran

Endemism: NA

Note: *Piptatherum coerulescens* (Desf.) P.Beauv., Piptatherum coerulescens (Desf.) P.Beauv. — The Plant List, reported from Lebda and Marsa Suza, panicles lax, spikelets pedicelled, glumes bluish.

Photo 7.470. *Piptatherum miliaceum*, whole plant, wadi Saratan (N 32.303 E 14.732), south Daphniya – Misurata.

Photo 7.471. *Piptatherum miliaceum*, leaves, wadi Saratan, south Daphnia.

Photo 7.472. *Piptatherum miliaceum*, panicle with seeds, wadi Saratan, south Daphnia.

Poa bulbosa L.

Synonym(s): http://www.theplantlist.org/tpl1.1/record/kew-435458
Native:
Vernacular name:
Raunkiaer's life-form: Hemicryptophyte
Description and morphology: Short gracile perennial grass, carpet grass with membranous-bulbous base, short soft narrow linear leaves (up to 5–10 cm long), panicle purplish and dense, glume acuminate and woolly.
Reproduction: Vegetative from bulbils detached from mother plants, viviparous when flowering allowed
Pastoral importance: High, very well grazed by sheep in winter and spring, withstands heavy grazing and trampling, usually heavily grazed to a short green carpet, birds and wildlife also feed on bulbils.

Palatability and fodder value: Excellent winter-spring forage, usually heavily grazed
Toxicity: None
Economic interest: Soil binder; when overgrazed induces water erosion on site
Habitat: On loamy soils, on red soils, never on sand or saline soils
Distribution:

- In Libya, on mountains and southern plateau in Jebel Al-Akhdar (Gendoula, Martuba, Shahat), also Jebel Nafusa (Tarhuna, Gharyan, Ksar-el-Daoun, Msallata), absent in Jeffara plain and Sirte
- In arid zones, cold sites, North Africa, Middle East to Central Asia (Uzbekistan)

Endemism: NA

Photo 7.473. *Poa bulbosa*, south Gendula, Mafsel – El-Mzerah.

Photo 7.474. Panicle of *Poa bulbosa*. (Photo: Gintzburger *et al.*, 2003)

Schismus barbatus (Loefl. ex L.) Thell.

Synonym(s): Schismus barbatus (L.) Thell. — The Plant List, or *Festuca barbata* L., Festuca barbata L. — The Plant List
Native:
Vernacular name: *Khafoor* (?), *Zabal* (?)
Raunkiaer's life-form: Tachytherophyte
Description and morphology: Small (5–20 cm) tufted gracile annual, filamentous roots system, stems all ascending and geniculate, leaves filiform mostly erect and more or less pubescent, swollen brown nodes with ciliate ligule, glossy spikelets compressed, glumes subequal, lemma bifid.
Reproduction: Flowering in spring, seeds
Pastoral importance: High
Palatability and fodder value: Excellent annual for all livestock
Toxicity: None
Economic interest: None
Habitat: Fixed sandy/sandy-loam soils, dry plant easily uprooted by wind facilitating propagation
Distribution:

- In Libya, all over northern Libya into Hamada, Sirte, south Jebel Al-Akhdar, down into Sahara
- In arid zones, southern Spain, North Africa to the Middle East, South Africa

Endemism: NA

Note: *Schismus arabicus* Nees, Schismus arabicus Nees — The Plant List, is reported in Libya but difficult to distinguish from *Schismus barbatus* in the field. Hybrids between the two species are related and possible.

Photo 7.475. *Shismus barbatus*, Jbebina, Al-Assah.

Photo 7.476. *Schismus barbatus* plant (Al-Assah).

Photo 7.477. Dry panicle of *Schismus barbatus*.

Sphenopus divaricatus (Gouan) Rchb.

Synonym(s): http://www.theplantlist.org/tpl1.1/record/kew-443897, or *Poa divaricata* Gouan., http://www.theplantlist.org/tpl1.1/record/kew-435779
Native:
Vernacular name:
Raunkiaer's life-form: Stenotherophyte
Description and morphology: Minute (5–15 cm) annual grass, green turning violet all over leaves and inflorescences, stems ascending and barely geniculate, inflorescences branched in divaricate gracile lax panicles, pauciflora, tiny spikelets also divaricate.
Reproduction: Flowering March–April, seeds
Pastoral importance: Limited
Palatability and fodder value: Poor to medium, does not appear to be much grazed
Toxicity: None
Economic interest: None
Habitat: Halophytic on fixed sand/sandy-loam soils on periphery of *sebkha* down to 150 mm MAR
Distribution:

- In Libya, Jeffara plain, Hamada, Sirte, south Jebel Al-Akhdar to Solum
- In arid zones, southern Europe, North Africa to Iran

Endemism: NA

Photo 7.478. Divaricate panicle of *Sphenopus divaricatus*, protected area Magdar-es-Sbot, Al-Assah.

Stipa capensis Thunb.

Synonym(s): *Stipa retorta* Cav., Stipa retorta Cav. — The Plant List, *Stipa tortilis* Desf., Stipa tortilis Desf. — The Plant List

Native:
Vernacular name: *Behma, Ziwan* (?)
Raunkiaer's life-form: Tachytherophyte, thermophile
Description and morphology: Annual with short cycle, tufted 5–40 cm high, often in dense stands, culms branching from base, upper leaves pubescent, compact silvery tail-like inflorescence with long (8–10 cm) vulnerant twice-geniculated twisted awns, base of awn sharply pointed.
Reproduction: Flowering February–April, seeds, anemochory
Pastoral importance: Limited despite large biomass produced
Palatability and fodder value: Well grazed before flowering; at flowering and after, not touched by livestock as vulnerant awns accumulate and stab mouth, eyes, hide and legs (legpit) creating ulcerous wounds; when awns ingested induce intestinal ulcers; shepherds evade stands of flowering or dry *Behma*.
Toxicity: None, but vulnerant after flowering
Economic interest: None, considered a weed
Habitat: Warm sites, does not stand cold or frost, all soil types on overgrazed sites except on waterlogged or saline areas, fast colonizer of bare soils, a clear indicator of degradation or overgrazing.
Distribution:

- In Libya, all over, dry overgrazed areas
- In arid zones, Canarias, Mediterranean to India

Endemism: NA

Photo 7.479. *Stipa capensis*, wadi Sassu, south Misurata.

Photo 7.480. Dense stand of *Stipa capensis*, wadi Sassu, south Misurata.

Stipa lagascae Roem. & Schult.

Synonym(s): http://www.theplantlist.org/tpl1.1/record/kew-445249
Native:
Vernacular name: *Adem*
Raunkiaer's life-form: Hemicryptophyte
Description and morphology: Cespitose perennial tufted, deep rooted, 60–80 cm tall, long firm pointy leaves but not sharp, gracile lax inflorescence well above basal leaves, very long (15–20 cm) kneed awn above seed, awn flexible naked yellowish translucent.
Reproduction: Flowering February–June, seeds
Pastoral importance: A very valuable pastoral plant indicating a rangeland in good condition.
Palatability and fodder value: Excellent all year round
Toxicity: None
Economic interest: None
Habitat: 150-300 mm mean annual rainfall, on sandy-loam, rocky–gravelly soils, on shallow soils on calcareous crust.
Distribution:

- In Libya, Jeffara plain, Jebel Nafusa, wadi Sof-el-Jin, Sirte, Jebel Al-Akhdar
- In arid zones, southern Europe, Mediterranean to Afghanistan

Endemism: NA

Note: *Stipa parviflora* Desf., Stipa parviflora Desf. —The Plant List, perennial, looks like *S. lagascae* but smaller, desertic, very loose inflorescence, long plumose awn slightly kneed and corkscrewed below knee, foothill Jebel Nafusa (Giado, Nalut, Cabao), south Jebel Nafusa (Zintan), reported in Tibesti and Fezzan. (?)

Photo 7.481. *Stipa lagascae*, near Gdama, south Gharyan.

Stipa tenacissima L.

Synonym(s): http://www.theplantlist.org/tpl1.1/record/kew-445814
Native:
Vernacular name: *Halfa, Halfa Gdima*
Raunkiaer's life-form: Hemicryptophyte
Description and morphology: Large perennial grass (1–1.5 m high), rhizomatous, large round compact tussock evolving into a circle of younger plants while centre is progressively dying off, long and strong culms (up to 130–150 cm), leaves rolled-dry pointed-spiny, lax inflorescence with spikelets with long awn hairy – twisted on first half and last half geniculate – bent.
Reproduction: Flowering May–June, seeds, vegetative (?)
Pastoral importance: When in good condition, valuable role of soil protection on soil conservation projects
Palatability and fodder value: Leaves have poor palatability, high in celluloses, mostly for dromedaries and goats when desperate, sheep will only take inflorescences (= *Bouss*, Nalut, Tunisia, Algeria).
Toxicity: None
Economic interest: 1900–1911 and later 1950–1960, young leaves used to be extensively harvested/hand-pulled (labour-intensive work) and exported to Europe as a source of high-quality paper paste (for bank notes), locally used in plaiting, basket and vegetal carpet work, sandals, stuffing mattresses etc.
Habitat: 150–500 mm MAR, used to occur on flat loamy to heavy soils (now all cultivated), mostly remaining on hills and mountain slopes, on shallow rocky soils and calcareous crust.
Distribution:

- In Libya, all Jebel Nafusa, most eastern extent is on top of hills, along the road (near wadi Saratan) from 20–30 km south Misurata – road from Daphnia to wadi Sassu, last *Stipa tenacissima* remnants on rocky hills now destroyed by large quarries.
- In arid zones, western Mediterranean only

Endemism: Southern Spain, North Africa from Morocco (?) to western Libya to Zliten – south Misurata, not occurring Jebel Al-Akhdar and further east

Note: Sometimes confused on the coast with *Ampelodesmos mauritanicus* (Poir.) T. Durand & Schinz, Ampelodesmos mauritanicus (Poir.) T. Durand & Schinz — The Plant List, common on coastal zones in Algeria and Tunisia, but not in Libya, a tall robust perennial tufted rhizomatous, long flexuous leaves (1 m), *Ampelodesmos* sp., doubtful occurrence in Libya, looks very much like *Stipa tenacissima* but totally different ecology and distribution.

Photo 7.482. *Stipa tenacissima* in flower, Gdama, south Gharyan.

Photo 7.483. A remaining press (*c.*1960) to prepare bails of *Halfa* (*Stipa tenacissima*) young leaves for export (near Gdama – south Gharyan, 2009).

Stipagrostis obtusa (Delile) Nees

Synonym(s): *Aristida obtusa* Delile, http://www.theplantlist.org/tpl1.1/record/kew-445979
Native:
Vernacular name: *Nsiah* (?)
Raunkiaer's life-form: Hemicryptophyte
Description and morphology: Small circular clump of perennial grass (10–20 cm), leaves fine scarious scythe-like, tiller only one geniculate, lax panicle above leaves, yellow spikelet with black dot at base, with three awns, only central awn long and plumose.
Reproduction: Winter flowering, seeds, anemochory
Pastoral importance: Forage, limited, low biomass production
Palatability and fodder value: Well grazed (green and dry) by all livestock and gazelles
Toxicity: None
Economic interest: Forage
Habitat: Saharan, down to 50–100 mm MAR, in small depressions on Hamada, coarse sandy soils, not on shifting sand or dunes (often confused with *Stipagrostis plumosa*).
Distribution:

- In Libya, Bir-el-Ghanem, Bir-Tuta, Nalut, Mizda, wadi Sof-el-Jin, Seddada, Hamada, inland Sirte to south Jebel Al-Akhdar
- In arid zones, North Africa to Pakistan, Mauretania, Chad to Ethiopia, southern Africa

Endemism: Saharan endemic (?)

Photo 7.484. *Stipagrostis obtusa,* curly leaves, Bir Saoul Lian (N 30.070 E 12.097), on Hamada.

Photo 7.485. *Stipagrostis obtusa*, Bir Saoul Lian, on Hamada.

Photo 7.486. *Stipagrostis obtusa* and *S. plumosa* in small shallow sandy depression on Hamada (Contracted vegetation, at Saoul Lian),

Stipagrostis plumosa Munro ex T. Anderson

Synonym(s): Stipagrostis plumosa Munro ex T. Anderson —The Plant List = *Aristida plumosa* L., Aristida plumosa L. —The Plant List
Native:
Vernacular name: *N'see* (?), *Naussiyah* (?)
Raunkiaer's life-form: Hemicryptophyte
Description and morphology: Small desert cespitose annual sometimes perennial, erect stems (10–40 cm) geniculated twice or more, leaves fine scarious scythe-like and pointed, ligulae on culm node are bare, lax panicle, silver-whitish spikelet with three awns, only central awns long plumose, feather-like at angle with pointed tip , glumes 1–3 nerved.
Reproduction: Seeds, anemochory
Pastoral importance: Well grazed (green and dry) by all livestock and gazelles
Palatability and fodder value: Excellent fresh and dry
Toxicity: None
Economic interest: Forage
Habitat: To 50–100 mm MAR, sandy-gravelly soils
Distribution:

- In Libya, Jeffara plain, Hamada, Sirte to south Jebel Al-Akhdar
- In arid zones, North Africa to northern Pakistan

Endemism: NA

Other main *Stipagrostis* species with central awns plumose – feathery in Libya:

- *Stipagrostis ciliata* (Desf.) De Winter, Stipagrostis ciliata (Desf.) De Winter —The Plant List, very similar to *S. plumosa*, difference is that the culms nodes of *Stipagrostis ciliata* are bearded-hairy, same distribution and habitat.

Photo 7.487. *Stipagrostis plumosa,* at El-Hisha (N 31.514 E 15.252), Bu Gren, with sand 'tubes' protecting roots against desiccation.

Photo 7.488. *Stipagrostis plumosa,* Jbebina, Al-Assah.

Stipagrostis pungens (Desf.) De Winter

Synonym(s): *Aristida pungens* Desf., http://www.theplantlist.org/tpl1.1/record/kew-446000
Native:
Vernacular name: *Sbot* (Libya), *Drinn* (Algeria, Tunisia)
Raunkiaer's life-form: Hemicryptophyte
Description and morphology: Desert perennial grass, rhizomatous with large rooting system, straggling bushes, large clump-tussocks (several m in diameter and 150–180 m high) and accumulating sand (*nebkha*), arched woody stems, tall woody glabrous culms with rolled sharp pointed curved rigid leaves, spikelets in lax-open panicle, three awns sub-equal and all plumose.
Reproduction: Rhizomatous, and seeds
Pastoral importance: Sand binding plant, camel feed
Palatability and fodder value: Moderate for the leaves, good for inflorescences
Toxicity: None
Economic interest: Leaves harvested and used to fix moving sand on afforestation projects on dunes
Habitat: 50–400 mm MAR on deep sandy soils and dunes, never on saline soils
Distribution:

- In Libya, dominant in all Jeffara plain, in Sirte to Ajdabiya, south Jebel Al-Akhdar, possible in Marmarica
- In arid zones, Morocco, Algeria, Tunisia to Libya, does not appear to occur in Egyptian Marmarica (not mentioned in Täckholm, 1974)

Endemism: NA

Photo 7.489. *Stipagrostis pungens* on sand dunes, central Jeffara plain (N 32.525 E 12.029), south Sorman, 45 km south-west Sabratha.

Photo 7.490. Exposed roots of *Stipagrostis pungens* with sand 'tubes' protecting roots against desiccation.

POLYGONACEAE

Calligonum comosum L'Hér.

Synonym(s): http://www.theplantlist.org/tpl1.1/record/kew-2692499, or *Pallasia comosa* (L'Hér.) Raeusch., http://www.theplantlist.org/tpl1.1/record/kew-2571836, or *Calligonum polygonoides* ssp. *comosum* (L'Hér.) Soskov, *Calligonum polygonoides* subsp. *comosum* (L'Hér.) Soskov - eflora Maghreb
Native:
Vernacular name: *Harta* (?), *Ressu* (?)
Raunkiaer's life-form: Nanophanerophyte
Description and morphology: Small desert shrub (up to 1.5 m high), strong woody stump, intricate old stiff branches with white bark breaking easily, erect slender articulated leafless young stems geniculate swollen at nodes, young branches flexuous green, stipules fused in bi-lobed papery ochrea, small (5–10 mm long) linear caducous alternate scale-like leaves, white-greenish small flowers along branches, globose-ovoid fruits as indehiscent achenes (8–15 mm long) with quadrilateral section bearing long dichotomous soft bristles on ribs.
Reproduction: Flowering in February–June, seeds (fruits easily dispersed by anemochory).
Pastoral importance: High
Palatability and fodder value: Camel's forage all year long
Toxicity: None
Economic interest: Used to be collected for fuelwood – charcoal, nowadays plantations for sand dunes fixation, commonly used for range rehabilitations and transplanted from nursery seedlings at BGI–BGII (Bir-el-Ghanem to wadi Nefad – along Seddada road), fruits (achenes) are easy to collect and reseed, easy seedlings production in nursery for transplanting on sandy country; never seen self-regeneration from achenes in the field, roots used as anthelmintic and bark for tanning goat skins for *guerba*.
Habitat: Lower-arid – Saharan, desertic, sandy soils, even on sand dunes (would accumulate and withstand being buried in flowing sand – initiating *nebkha*), MAR 150 mm down to 50 mm.
Distribution:

- In Libya, west Jeffara plain (between south Al-Assah and foothills Nalut, Hamada), not in Sirte (?) or further east Libya (?)
- In arid zones, North Africa to north-west Pakistan

Endemism: NA

Note: Considering the successful plantations and establishment of *Calligonum comosum* in the late 1970s on sand dunes colonized by *Stipagrostis pungens* in the Bir-el-Ghanem region, it is quite possible it would have been present all over the sandy Jeffara plain from Al-Assah-Takut to Aziziya long ago before it was systematically eradicated for fuelwood and by relentless grazing.

Photo 7.491. *Calligonum comosum*, bush, south Al-Assah, El-Ajemiya – Takut, north-east Nalut (N 32.047 E 11.133).

Photo 7.492. *Calligonum comosum*, with achenes, south Al-Assah, El-Ajemiya, north Nalut.

Photo 7.493. *Calligonum comosum*, young achene, south Al-Assah, El-Ajemiya, north Nalut.

Photo 7.494. Seeds collection of *Calligonum comosum*, bush, 1975 plantation at BGII, near police station (N 32.559 E 12.424).

Other *Calligonum* sp. in Libya:

- *Calligonum azel* Maire, http://www.theplantlist.org/tpl1.1/record/kew-2692479, Saharan, endemic, small tree up to 4–5 m high, white bark peeling off, on sand dunes in lowest-rainfall zones, in Sahara (North Africa), doubtfully mentioned near Tarhuna.

Photo 7.495. *Calligonum comosum*, branches and achenes, plantation at BGII, near police station.

Photo 7.496. *Calligonum comosum*, ripe achenes, plantation at BGII, near police station.

Photo 7.497. *Calligonum comosum*, ripe achenes under shrubs, plantation at BGII, near police station.

Photo 7.498. *Calligonum comosum*, branches, plantation at BGII, near police station.

Photo 7.499. *Calligonum arich* tree in the great *erg* in Tunisian Sahara (N 32.859 E 9.563). (Photo: Claude Lemmel, http://alyasmina.org/)

Photo 7.500. Pendulous branches from *Calligonum arich* in the great *erg* in Tunisian Sahara (N 32.859 E 9.563). Photo: Claude Lemmel, http://alyasmina.org/)

- *Calligonum arich* Le Houér., http://www.theplantlist.org/tpl1.1/record/kew-2692477, or *Calligonum calvescens* Maire (?), http://www.theplantlist.org/tpl1.1/record/kew-2692491, remarkable rare desert tree (5–10 m high) on high sand dunes (50 mm MAR), trunk reaching 30–50 cm diam. at girth on large specimen, long erect branches with pendulous flexible young stems, yellow bark desquamating, reddish wood, oblong achenes with large ribs twisted at end, rare tetragonal achenes with few bristles exposing bare surface, fruit capped with a tuft of bristle.

Both endemic to Algerian and Tunisian Sahara (great eastern sand sea and western Libya). Reported by Keith (1965) in the *erg* south-west of Ghadames in 1958, by Le Houérou (1959, 2002) near Bir Aouine (southern Tunisia (N 32.417 E 9.300). Both *Calligonum* sp. were still present in the mid-1950s to early 1960s, nearly extinct from south Tunisia and Libya (south-west Ghadames) in the oriental *erg*, due to relentless fuelwood collection.

Status: Likely extinct in Libya

These *Calligonum* sp. were avidly collected for fuelwood and making charcoal by desert nomads, caravans and dwellers, hence its eradication from the Libyan desert.

These three *Calligonum* sp. should be actively used for sand dunes fixation in Libya. Seedlings can easily be nursery-grown from achenes, and seeds obtained from southern Tunisia.

Other *Calligonum* sp. are abundant in Central Asian and northern Chinese arid rangelands (Gintzburger *et al.*, 2003) and could be beneficially tested in Libya.

Photo 7.501. *Calligonum azel* in the great *erg* in Tunisian Sahara (N 32.859 E 9.563). (Photo: Claude Lemmel, http://alyasmina.org/)

Photo 7.502. Stems of *Calligonum azel* in the great *erg* in Tunisian Sahara (N 32.859 E 9.563). (Photo: Claude Lemmel, http://alyasmina.org/)

Polygonum equisetiforme Sibth. & Sm.

Synonym(s): http://www.theplantlist.org/tpl1L..1/record/kew-2574556
Native:
Vernacular name. *Gurdab*
Raunkiaer's life-form: Hemicryptophyte
Description and morphology: Perennial herb with stout woody stump and deep pivotal root, sometimes creeping, rhizomatous, slender erect straight parallel stems (sometimes decumbent on large specimens), small (1–3 cm long) lanceolate alternate acute leaves, leaves caducous in summer, thin translucent ochrea (brown base and bifid) shorter that internode, rich inflorescence in long raceme with (1–2) axillary white (pinkish) small (3–5 mm diam.) flowers, trigonal brown minute achene (1–3 mm long)

Photo 7.503. *Polygonum equisetiforme*, plant buried in sand, Al-Assah, Jbebina.

Photo 7.504. *Polygonum equisetiforme*, stems and flowers, Al-Assah, Jbebina.

Photo 7.505. *Polygonum equisetiforme*, heavily grazed plant, Al-Assah, Jbebina.

Photo 7.506. *Polygonum equisetiforme*, roots exposed by wind erosion, Al-Assah, Jbebina.

Reproduction: Flowering February–May, seeds, possibly rhizomatous
Pastoral importance: High, likely one of the best and most resistant plants to grazing from Libyan rangeland (with *Salsola vermiculata*)
Palatability and fodder value: Well grazed by all livestock

Toxicity: None
Economic interest: None
Habitat: Sandy soils (fixed or slightly mobile), withstands wind erosion and being buried in flowing sand, down to 50 mm MAR
Distribution:

- In Libya, all coastal zones and inland, from east to west
- In arid zones, North Africa to Afghanistan

Endemism: NA

Other main *Polygonum* sp. in Libya:
- *Polygonum aviculare* L., Polygonum aviculare L. —The Plant List, small annual, prostrate with persistent leaves on stems, ubiquitous along the coast in high-rainfall zones, on wasteland.

Rumex vesicarius L.

Photo 7.507. *Rumex vesicarius*, plant, Gharyan, near Tobi nursery.

Synonym(s): http://www.theplantlist.org/tpl1.1/record/kew-2424478, or *Acetosa vesicaria* (L.) Á.Löve, http://www.theplantlist.org/tpl1.1/record/kew-2616829
Native:
Vernacular name: *Umadet Hammam* (?)
Raunkiaer's life-form: Mesotherophyte – thermophylous
Description and morphology: Annual glaucous herb (20–50 cm high), taproot, multicaule, erect furrowed stems branching from thick collar, entire petiolate triangular – spearheaded fleshy alternate leaves with round base lobes, petiole longer than leaf, ochrea brown or white, inflorescence in top panicles, abundant and showy trigone fruits with large membranous entire wings (20–30 mm diam.) veined pink to red, woody achene dark brown, seeds pear-shaped and acuminate.
Reproduction: Flowering January–March, seeds
Pastoral importance: Limited, can stand heavy grazing and will reshoot during the same season
Palatability and fodder value: Well grazed by all livestock
Toxicity: None, however high content in Oxalates (acid taste)
Economic interest: Leaves collected and consumed as vegetables (?)
Habitat: Saharan, gravelly – rocky sites, sandy loam to clay soils, in wadis, rocky slopes, down to 100 mm MAR, usually on sites protected from heavy grazing.
Distribution:

Photo 7.508. *Rumex vesicarius*, winged fruits, Gharyan, near Tobi nursery.

- In Libya, Jebel Nafusa (Msallata to Nalut), Hamada, Sirte (Bu-Green, wadi Wishkah), Jebel Al-Akhdar (Derna to Solum)
- In arid zones, southern Europe, North Africa to south-west Asia, Iran

Endemism: NA

PRIMULACEAE

Anagallis monelli L.

Synonym(s): http://www.theplantlist.org/tpl1.1/record/kew-2636090, or *Anagallis linifolia* L., http://www.theplantlist.org/tpl1.1/record/kew-2636075
Native:
Vernacular name: *Zerega* (?)
Raunkiaer's life-form: Hemicryptophyte
Description and morphology: Biannual (perennial – woody base (?)), small prostrate herb, many branches flat – decumbent on ground, stems four-ribbed, small (15–20 mm long, 3–5 mm wide) linear – lanceolate sessile opposite leaves, round-lobed blue (sometimes orange to reddish (?)) flowers on long pedicels, capsule spherical (capped with persistent filiform style before dehiscence) shorter that stellar calyx, ripe fruit pedicels erect – arched upward, capsule dehiscent from top lid, many small (0.5–1 mm) black conical seeds.
Reproduction: Flowering February–June, seeds
Pastoral importance: None
Palatability and fodder value: None (?), not touched (?)
Toxicity: Reported poisonous to all livestock
Economic interest: None, common weed
Habitat: Ruderal, wasteland, in crops, ubiquitous, mostly on fixed sandy/sandy-loam, down to 200 mm MAR in rainfed crops
Distribution:

- In Libya, all coastal Libya
- In arid zones, south-west Europe, North Africa

Endemism: NA

Note: Could be confused with very common ruderal *Anagallis arvensis* L., http://www.theplantlist.org/tpl1.1/record/kew-2636008, with a similar ecology but tends to occur in higher-rainfall zones in gardens, leaves are much larger (heart-shaped, as long as large), corolla with lobes finely denticulate, ripe fruit pedicels retroflexed – pendulous.

Photo 7.509. *Anagallis monelli*, flower and leaves, Aziziya, Bir-el-Ghanem.

Cyclamen rohlfsianum Asher.

Synonym(s): http://www.theplantlist.org/tpl1.1/record/kew-2749845
Native:
Vernacular name:
Raunkiaer's life-form: Geophyte
Description and morphology: Small (8–12 cm high) herbaceous perennial, flat black tubers (to 6–8 cm diam.), flowering in autumn, flowers slightly scented appearing before leaves, leaves in basal rosette (with long reddish soft peduncles) dark-green leaves with centre speckled with white marks, leaves variable entire/ovate/dentate/denticulate with base cordate, leaves lower face reddish with prominent veins, twisted flower buds, pink-whitish reflexed flowers (pendulous, 2–3 cm long) with violet ring at base of corolla, corolla lobes pointed reflexed free except

at base, flowers on long reddish petioles (8–12 cm), stamens triangular protruding from corolla with white style longer than stamens, flower petiole corkscrewed and bent (amphicarpic (?)) when fruiting, fruits globose (8–10 mm diam.) with style persisting.
Reproduction: Flowering October–December, seeds (amphicarpic) and vegetative (tubers)
Pastoral importance: None
Palatability and fodder value: None
Toxicity: None
Economic interest: Valuable ornamental, commonly transplanted in gardens
Habitat: In matorral, karstic rocky – limestone sites, *Terra rossa* or brown forestry soils, 300–500 mm MAR
Distribution:

- In Libya, only in Jebel Al-Akhdar, in matorral
- In arid zones, others *Cyclamen* sp. (southern Europe, North Africa, Middle East, west Asia)

Endemism: Jebel Al-Akhdar only

Notes:

- Leaves are very variable in shape and marks.
- Young plants have a brown-black flattish tuber (2–5cm diam.) with lateral roots.

Photo 7.510. *Cyclamen rohlfsianum*, plant flowering, Shahat.

Photo 7.511. *Cyclamen rohlfsianum*, flowers with stamens and style prominent, wadi El-Kuf.

Photo 7.512. *Cyclamen rohlfsianum*, tuber, wadi El-Kuf.

Photo 7.513. *Cyclamen rohlfsianum*, on karst, leaves, El-Beyda.

RANUNCULACEAE

Adonis dentata Delile

Synonym(s): http://www.theplantlist.org/tpl1.1/record/kew-2623116
Native:
Vernacular name:
Raunkiaer's life-form: Therophyte
Description and morphology: Small (8–20 cm high) glabrous erect annual, often growing in bunches, thin leaves bi- or tri-pinnatisect linear with curved margins, unique apical yellow flower with black spot at base of petal, short dense black/violet stamens, 20–30 black ripe beaked rugose achenes borne on elongated accrescent conical rostrum.
Reproduction: Flowering January–May, abundant seeds
Pastoral importance: Poor
Palatability and fodder value: Barely touched when green
Toxicity: Possible glucoside and saponin (?)
Economic interest: None
Habitat: Sandy soils
Distribution:

- In Libya, Jeffara plain, Sirte, south Jebel Al-Akhdar
- In arid zones, North Africa, Middle East to Pakistan

Endemism: NA

Photo 7.514. *Adonis dentata*, Magdar-es-Sbot, Al-Assah, 1978.

Ranunculus trilobus Desf.

Synonym(s): http://www.theplantlist.org/tpl1.1/record/kew-2524543
Native:
Vernacular name:
Raunkiaer's life-form: Therophyte
Description and morphology: Slender annual (8–15 cm high), fibrous root system, plant light shiny green all over, basal leaves shortly pedunculate palmate serrated/dented with minute white bristles, bristle also on flowering stem, apical yellow flower with five petals and five sepals as long as petals, dense yellow stamens, at fruiting thalamus bearing many beaked tuberculate carpels (achenes) arranged in spirals.
Reproduction: Flowering October–December, abundant seeds
Pastoral importance: Poor
Palatability and fodder value:
Toxicity:
Economic interest: None
Habitat: on *Terra rossa*, on heavy waterlogged soils
Distribution:

- In Libya, Jebel Al-Akhdar only (?)
- In arid zones, North Africa to the Middle East

Endemism: NA

Photo 7.515. *Ranunculus trilobus*, El-Marj plain on *Terra rossa*, 2010.

RHAMNACEAE

Rhamnus lycioides L.

Synonym(s): http://www.theplantlist.org/tpl1.1/record/kew-2523132.
Native:
Vernacular name: *S'luf* (?), *Salluf* (?), *Slufa* (?)
Raunkiaer's life-form: Nanophanerophyte
Description and morphology: Robust shrub (up to 2 m high), much intricate thorny branches (due to continuous overgrazing), whitish-grey glabrous bark, branches ending in long (2–5 cm) sharp thorns, short petiolate elliptical alternate glabrous round-ended leaves with prominent unique midrib on grey scabrous lower face, small quadri- or pentamerous flowers with green-yellow calyx and without petals, yellow ovoid (5–10 mm diam.) drupes turning red-violet when ripening.
Reproduction: Flowering March–June, seeds (dispersed by possible zoochory)
Pastoral importance: High
Palatability and fodder value: Good, heavily grazed by all livestock (mostly goats), seeds consumed by birds
Toxicity: None
Economic interest: Fuelwood, thorny branches used as *zeriba*.
Habitat: Heavy soils, rocky site (karst in Jebel Al-Akhdar), above 200 mm MAR
Distribution:

- In Libya, mostly Jebel Al-Akhdar (Tokra, El Marj, Shahat, to Marmarica); a few occurrences in the west, Nazzaga, south Zliten, south Daphnia on rocky hills
- In arid zones, southern Europe, North Africa to coastal western Egypt (Solum – Marsa – Matruh)

Endemism: NA

Notes:

Other *Rhamnus* in Libya:

- *Rhamnus oleoides* L., http://www.theplantlist.org/tpl1.1/record/kew-2523110, similar to *R. lycioides*, but with yellowish ripe drupes, reported from Toknis to Marmarica.
- *Rhamnus alaternus* L., Rhamnus alaternus L. — The Plant List, thornless woody contorted shrub (1–1.5 m), bark pubescent, ovate leathery veined leaves with finely serrated; denticulate translucent margin, shy quadri- or pentamerous flowers with small or without petals, fruit globular turning red to black when ripe, in rocky places in Jebel Nafusa and Jebel Al-Akhdar, MAR > 250 mm, not in Sirte.

Photo 7.516. *Rhamnus lycioides*, leaves and sharp thorns, Toknis, El-Marj.

Photo 7.517. *Rhamnus lycioides*, leaves and thorn, Toknis, El-Marj.

Ziziphus lotus (L.) Lam.

Synonym(s): http://www.theplantlist.org/tpl1.1/record/kew-2470687, or *Rhamnus lotus* L. Rhamnus lotus L. — The Plant List
Native:
Vernacular name: *Sedder*, *Sedra*
Raunkiaer's life-form: Phanerophyte
Description and morphology: Medium- to large-sized spinescent bush (in *nebkha* up to 10–12 m diam., 2–4 m high above mound), sometimes as tree-like form (with 2–3 m truck) when

nebkha removed, thickly intricate zig-zagging white spiny branches, backward arched pair of stipular spines (5–10 mm long) at node, short petiolate alternate elliptical entire glabrous shiny green trinerved leaves, yellow small pentamerous flowers, small globular pea-size, yellow turning brown drupes containing one hard seed.
Reproduction: Vegetative by developing suckers, flowering March–April, seeds dispersed by zoochory; propagation possible by cutting
Pastoral importance: High; *Ziziphus nebkha* is a small ecosystem by itself, contributing to soils fixation (with organic matter and nitrogen accumulation) against water and wind erosion; it is a refuge for many plant species that would be otherwise overgrazed; it is rich in wildlife (rabbits, rats, jerboa, snakes (vipers, cobra, *Coluber* sp., lizards etc.), foxes, jacals, small birds, partridges, insects, etc.).
Palatability and fodder value: Young stems and fruits grazed by all livestock (mostly goats and camels)
Toxicity: None
Economic interest: High value to apiarists for specific honey production, used to be collected for fuelwood, fruits reported collected and consumed long ago, *Ziziphus* spiny branches commonly used as *zeriba* to corral young livestock at night near countryside temporary or permanent settlements.
Habitat: In area receiving run-off, wadis, fixed sand to heavy soils, forming large mounds (*nebkha*) of accumulated sand and organic matter, a clear indicator of potential cropping land; removal of *nebkha* for cultivation contributes to land degradation and considerable loss of biodiversity.
Distribution:

- In Libya, all over coastal zones to inland, from east to west and deep into the desert in wadis and depressions
- In arid zones, southern Europe, North Africa to Arabian Peninsula

Endemism: NA

Other *Ziziphus* species in Libya:

- *Ziziphus spina-christi* (L.) Desf., http://www.theplantlist.org/tpl1.1/record/tro-27501041, occasionally cultivated in gardens in Libya (Tajoura, Nalut, Sebha, Fezzan, etc.), southern Europe, North Africa, Middle East, slow-growing developing into a large tree (up to 8–10 m under favourable conditions), spineless cultivars available, producing edible berries (10–20 mm diam.) yellow turning dark brown when dry, known as *nabbak* or *nabek* often sold on local markets, reported rich in vitamin C, possibly introduced to Libya from sub-Saharan Africa or the Middle East.

Photo 7.518. *Ziziphus lotus*, *nebkha*, Bir Saman, Al-Assah, grazed by dromedaries.

Photo 7.519. *Ziziphus lotus*, *nebkha* in wadi Sassu, occasionally cropped with rainfed barley.

Photo 7.520. *Ziziphus lotus* branches and flowers, wadi Turgut, Tarhuna.

Photo 7.521. *Ziziphus lotus*, branches and pair of stipular spines, Bir Saman, Al-Assah.

Photo 7.522. Rare *Ziziphus lotus* (or *Ziziphus spina-christi* (?)) tree on old cropping zone, track from south Kararim to wadi Sassu.

ROSACEAE

Sanguisorba minor Scop.

Synonym(s): http://www.theplantlist.org/tpl1.1/record/rjp-643, or *Poterium sanguisorba* L., http://www.theplantlist.org/tpl1.1/record/rjp-1040, or *Sanguisorba pimpinella* Spenn., http://www.theplantlist.org/tpl1.1/record/rjp-26855
Native:
Vernacular name: *Sh'breg* (?)
Raunkiaer's life-form: Hemicryptophyte
Description and morphology: Small (10–40 cm diam.) perennial herb (annual in most arid sites), strong taproot, stems green or reddish, basal leaves in rosettes prostrate/ascending, all leaves imparipinnate deeply serrated/dented with 10–20 lateral folioles, drying leaves turning red, few smaller similar leaves on erect flowering stems, inflorescence (green to red) in globules (1–3 cm diam.) with male flowers at base and female flowers at top, four green sepals, petals absent, style plumose reddish, calyx containing inferior ovary with 1–3 tuberculated/reticulated achenes.
Reproduction: Flowering April–June, seeds
Pastoral importance: High, easy to propagate
Palatability and fodder value: Excellent forage for all livestock
Toxicity: None
Economic interest: Reported to be collected by Turkish troops to feed horses, European cultivars introduced during Italian colonization and later, many cultivars selected and commercialized, easy commercial seed production, could be certainly used for rangelands reseeding on highest-rainfall zones (MAR 350–500 mm) in Libya.
Habitat: Loamy-clay soils, above 200 mm MAR
Distribution:

- In Libya, top ridge of Jebel Nafusa, common all over Jebel Al-Akhdar to edge of southern slope
- In arid zones, Europe, North Africa, west Asia, Middle East to Iran, Caucasus, Central Asia

Endemism: NA

Photo 7.523. *Sanguisorba minor*, plant, El-Mzerah, south Gendoula.

Photo 7.524. *Sanguisorba minor*, plant, El-Mzerah, south Gendoula.

Photo 7.525. *Sanguisorba minor*, fruits, El-Mzerah, south Gendoula.

Sarcopoterium spinosum (L.) Spach

Synonym(s): http://www.theplantlist.org/tpl1.1/record/rjp-1340, or *Poterium spinosum* L., http://www.theplantlist.org/tpl1.1/record/rjp-2049, or *Sanguisorba spinosa* (L.) Bertol., http://www.theplantlist.org/tpl1.1/record/rjp-20557
Native:
Vernacular name: *Shebreg* (?), *Kebrek* (?)
Raunkiaer's life-form: Chamaephyte – hemicryptophyte
Description and morphology: Small hemispherical rigid spiky shrub (up to 70–100 cm diam.), end of branches with sharp spine, branches dichotomous and densely intricate, small oblong imparipinnate (11–25 lobes) leaves with dented leaflets, small green flowers with inferior ovary, small (2–5 mm diam.) globose fleshy yellowish to reddish berries capped with calyx ending with four retroflexed adpressed sepals, two achenes per berry.
Reproduction: Flowering April–July, seeds
Pastoral importance: Limited
Palatability and fodder value: Poor due to its spikiness
Toxicity: None
Economic interest: Used to be mostly collected for fuelwood
Habitat: On heavy soils, on limestone slopes on karst, on red Mediterranean soils, above 200 mm MAR, relict of *Juniperus phoenicea* forests, covering large degraded areas in the coldest part of the Jebel Al-Akhdar.
Distribution:

- In Libya, occurs only in Jebel Al-Akhdar, marks the extreme western limit of its distribution
- In arid zones, a strict east Mediterranean genus (Turkey, Syria, Palestine, Lebanon, Jordan)

Endemism: NA

Photo 7.526. *Sarcopoterium spinosum*, spiny branches and fruits, wadi Haleg, Maghur.

Photo 7.527. *Sarcopoterium spinosum*, Spiny branches and fruits, wadi Haleg, wadi Maghur.

RUTACEAE

Haplophyllum tuberculatum (Forssk.) A. Juss.

Synonym(s): http://www.worldfloraonline.org/taxon/wfo-0000715624, or *Haplophyllum vermiculare* Hand.-Mazz., http://www.theplantlist.org/tpl1.1/record/kew-2839460
Native:
Vernacular name: *Shejeret er rih*, (*Fedjel*) (?)
Raunkiaer's life-form: Chamaephyte
Description and morphology: Perennial (30–50 cm high), glaucous greenish-yellow, multi-ascending branches woody at base, strong foetid smell all over, glandulous and sticky all over, small entire lanceolate cuneate leaves often with crenulated margins and recurved, all stems with open corymbose inflorescences, yellow flowers with five entire petals, globose smooth fruits in five dehiscent papery locules.
Reproduction: Flowering October–April, seeds
Pastoral importance: None
Palatability and fodder value: None
Toxicity: Abortive with high dose
Economic interest: Medicinal, used as herbal tea against abdominal pain (colic, flatulence, nausea, constipation), also reported against malaria, aphrodisiac (Keith, 1965; Le Floc'h, 1983).
Habitat: On fixed sandy soils, often in *Artemisia campestris* formation (150–300 mm MAR)
Distribution:

- In Libya, from west to east
- In arid zones, North Africa to Irano-Turanian

Endemism: NA

Photo 7.528. *Haplophyllum tuberculatum*, Jbebina, Al-Assah, April 2010.

Photo 7.529. Flowers *Haplophyllum tuberculatum*, Jbebina, Al-Assah, April 2010.

Ruta chalepensis L.

Synonym(s): http://www.theplantlist.org/tpl1.1/record/kew-2527496, or *Ruta bracteosa* DC., http://www.theplantlist.org/tpl1.1/record/kew-2527524
Native:
Vernacular name: *Fejel*
Raunkiaer's life-form: Chamaephyte
Description and morphology: Perennial herbaceous multi-stems, dark dense green-yellowish foliage with bad smell, bi- or tri-pinnatisect leaves with oval leaflets, corymbose apical inflorescences above foliage, yellow flowers with four ciliate petals, dehiscent globous fruit with four rugose locules, black seeds.
Reproduction: Flowering January–May, seeds
Pastoral importance: None
Palatability and fodder value: None
Toxicity: Aphrodisiac (?)
Economic interest: Medicinal, reported keeping scorpions away (Keith, 1965; Le Floc'h, 1983)
Habitat: Loamy soils, protected on slopes, in rocky places, 200–300 mm MAR
Distribution:

- In Libya, only in Jebel Nafusa, apparently not in Jebel Al-Akhdar
- In arid zones, North Africa to Middle East

Endemism: NA

Photo 7.530. *Ruta chalepensis*, protected in rocky slopes, Tarhuna to Yeffren, 1977.

SCROPHULARIACEAE

Linaria haelava Delile

Synonym(s): Unresolved name, http://www.theplantlist.org/tpl1.1/record/kew-2499452, or *Antirrhinum haelava* Forssk., Antirrhinum haelava Forssk. — The Plant List
Native:
Vernacular name: *Ayn-el-Gat* (?)
Raunkiaer's life-form:
Description and morphology: Annual glabrous herb (10–30 cm), arced ascending stems, leaves fleshy opposite with hairy margins often in spiral of 3–4, glandular and hairy racemed inflorescence, short pedicellate flowers purple-white (also yellow) with stronger yellow blotch on throat lower lip, long narrow conical curved-ended spur (10–25 mm long) 3 times longer than flower, ovoid/spherical capsule concealed into glandular long hairy calyx lobes, verrucose seeds round/kidney-shaped.
Reproduction: Flowering winter to May, seeds
Pastoral importance: Poor
Palatability and fodder value: Hardly touched by small ruminants
Toxicity: None
Economic interest: None
Habitat: Loamy soils, 150–300 mm MAR.
Distribution:

- In Libya, Jeffara plain, Tarhuna, south Zliten and Misurata, wadi Sassu, Sirte to Jebel Al-Akhdar
- In arid zones, North Africa, Middle East – Arabian Peninsula

Endemism: NA

Note: The determination of *Linaria* sp. is difficult due to the many revisions of the genus. See also *Antirrhinum* sp.

Photo 7.531. *Linaria haelava*, plant, south Zliten, wadi Sassu.

Photo 7.532. *Linaria haelava*, glandular – hairy inflorescences with capsules, south Zliten, wadi Sassu.

Kickxia aegyptiaca (L.) Nábělek

Synonym(s): http://www.theplantlist.org/tpl1.1/record/kew-2335408, or *Linaria aegyptiaca* (L.) Dum.Cours., http://www.theplantlist.org/tpl1.1/record/kew-2498138, or *Antirrhinum aegyptiacum* L., http://www.theplantlist.org/tpl1.1/record/kew-2642461
Native:
Vernacular name: *Eshb-ed-dib* (?)
Raunkiaer's life-form: Hemicryptophyte
Description and morphology: Perennial, small dense shrub (20–50 cm), glandulous hairy – white, intricate rigid woody branches ending spinescent, young stems pale yellowish, glandulous leaves alternate subsessile hastate apiculate denticulate or serrate, upward pedicellate isolated yellow spurred flowers all along young stems.
Reproduction: Flowering winter to May, seeds
Pastoral importance: Poor
Palatability and fodder value: Hardly touched by small ruminants

Photo 7.533. *Kickxia aegyptiaca*, Plant, south Zliten.

Toxicity: None
Economic interest: None
Habitat: Loamy soils, 100–300 mm MAR
Distribution:

- In Libya, slopes Jebel Nafusa, south Zliten and Misurata, Sirte to Jebel Al-Akhdar
- In arid zones, North Africa, Middle East – Arabian Peninsula

Endemism: NA

Photo 7.534. *Kickxia aegyptiaca*, flower, south Zliten.

SOLANACEAE

Lycium europaeum L.

Synonym(s): Lycium europaeum L. — The Plant List
Native:
Vernacular name: *Ausej, Sekkoum*
Raunkiaer's life-form: Nanophanerophyte – Phreatophyte
Description and morphology: Robust spiny shrub (up to 2–3 m high), branches erect intricate, top branches arched, with (15–25 mm long) rigid acute spines, young spine bearing lone leaves, subsessile leaves thin spathulate fleshy in bunches (4–6), erect flowers with very short calyx barely dented, flowers long (10–15 mm) tubuliform whitish/lilac corolla with short lobes, fruits spherical reddish to black.
Reproduction: Flowering March–May, seeds
Pastoral importance: Poor
Palatability and fodder value: Grazed by small ruminants, mostly goats, more by camels
Toxicity: None
Economic interest: Fuelwood, berries, dense bush offering refuge for other plants and fauna
Habitat: Phreatophyte, halophyte (?), mostly on deep sandy-loamy soils, sometimes in cracks in calcareous crust, down to 100 mm MAR, indicate presence of (high) water table or wet sub-horizons (?)
Distribution:

- In Libya, most common in *sebkha*, along the coast and inland
- In arid zones, southern Europe, North Africa

Endemism: NA

Notes: Very similar bush to *Lycium europaeum* is *Lycium shawii* Roem. & Schult.. http://www.theplantlist.org/tpl1.1/record/kew-2496628, but with pendulous flowers and orange berries, in Jebel Nafusa near springs, also phreatophyte.

Photo 7.535. *Lycium europaeum*, dishevelled ungrazed 2 m high bush, Kararim *sebkha*.

Photo 7.536. *Lycium europaeum*, south Taworga.

Photo 7.537. *Lycium europaeum*, erect flower and lone leaves on spines, south Taworga.

Lycium schweinfurthii Dammer

Synonym(s): http://www.theplantlist.org/tpl1.1/record/kew-2496523, no synonym recorded
Native:
Vernacular name: *Ausej* (?), *Sekkoum* (?)
Raunkiaer's life-form: Nanophanerophyte – phreatophyte
Description and morphology: Robust spiny shrub (up to 1.5–2 m high), branches compactly intricate, bark grey-white, with short (10–20 mm long) rigid acute leafless spines, subsessile thin spathulate fleshy leaves in bunches (4–6), flower with small calyx barely dented, corolla long (10–15 mm) tubuliform white with short lobes and violet throat, fruits spherical bluish-black when ripe.
Reproduction: Flowering March–May, seeds
Pastoral importance: Medium
Palatability and fodder value: Grazed by small ruminants, mostly goats, more by camels
Toxicity: None
Economic interest: Fuelwood
Habitat: Phreatophyte, sandy soils, down to 100 mm MAR, indicate presence of (high) water table
Distribution:

- In Libya, common along coast and inland, in gardens, oases
- In arid zones, southern Europe, North Africa

Endemism: NA

Photo 7.538. *Lycium schweinfurthii*, heavily grazed bush, Jbebina, Al -Assah.

Photo 7.539. *Lycium schweinfurthii*, leaves, flowers and thorns, Jbebina, Al- Assah.

Photo 7.540. *Lycium schweinfurthii*, leaves and flowers, Al-Assah.

Nicotiana glauca Graham

Synonym(s): http://www.theplantlist.org/tpl1.1/record/kew-2382892
Exotic from South America – naturalized
Vernacular name: *Akus Musa* (?)
Raunkiaer's life-form: Mesophanerophyte
Description and morphology: Glabrous dull green tall perennial shrub (up to 3–5 m high),

erect long flexuous leafy green-blueish branches covered with a veil of whitish wax, alternate long-petiolate large oval entire glabrous leaves (5–10 cm long) spearhead-shape, pendulous lax inflorescences in terminal racemes of many yellow flowers, calyx tubular persistent with acute lobes, yellow long (20–40 mm) corolla with minute dentate lobes, hanging capsules biloculate dehiscent remaining in calyx, opening with two valves releasing many tiny (0.5 mm) rectangular black reticulated seeds.

Reproduction: Flowering all year round, seeds

Pastoral importance: None

Palatability and fodder value: Not touched by all livestock

Toxicity: All plant poisonous (pyridine alkaloids, anabasine and nicotine), desperate and hungry livestock may ingest leaves leading to acute respiratory failure, coma and sudden death.

Economic interest: None, very poor fuelwood

Habitat: Ruderal, naturalized, invasive (?), common on wasteland, ruins, along roads, along wadis (?)), rare on rangelands except near dwellings (possibly nitrophilous > 200–250 mm MAR.

Distribution:

- Common all over Libya
- In arid zones, southern Europe, North Africa, Middle East

Endemism: NA

Photo 7.541. *Nicotiana glauca*, plant, Aradah, Tripoli.

Photo 7.542. *Nicotiana glauca*, leaves and flowers, Aradah, Tripoli.

TAMARICACEAE

Reaumuria vermiculata L.

Synonym(s): http://www.theplantlist.org/tpl1.1/record/kew-2527785, or *Reaumuria mucronata* Jaub. & Spach, http://www.theplantlist.org/tpl1.1/record/kew-2527795
Native:
Vernacular name: *Um neda* (?)
Raunkiaer's life-form: Chamaephyte
Description and morphology: Shrublet (30–60 cm high), intricate erect branches, lanceolate fleshy acuminate short (8–12 mm) glaucous glabrous sessile leaves (possible confusion with *Salsola* sp. or *Suaeda* sp. or even *Frankenia*), leaves often speckled with salt crystals and sand, white flower (8–12 mm diam.) terminal or axillary with membranous sharp-pointed bracts as long as calyx, small (8–10 mm) capsule containing minute (2–4 mm long) ovoid hairy seeds.
Reproduction: Flowering May–October, seeds (possible anemochory)
Pastoral importance: Medium
Palatability and fodder value: Grazed by all livestock
Toxicity: None
Economic interest: None
Habitat: Gypsophyte/halophile, periphery of *sebkha*, from coastal zone to deep inland, on sandy soils, down to 100 mm MAR
Distribution:

- In Libya, Jeffara plain, south Jebel Nafusa, Hamada, Sirte, Jebel Al-Akhdar
- In arid zones, southern Europe, North Africa.

Endemism: NA

Photo 7.543. *Reaumuria vermiculata*, bush, Haraba, Zintan.

Photo 7.544. *Reaumuria vermiculata*, leaves and flowers, Al-Assah, Jbebina.

Tamarix aphylla (L.) H. Karst.

Synonym(s): http://www.theplantlist.org/tpl1.1/search?q=Tamarix+aphylla, or *Tamarix orientalis* Forssk., http://www.theplantlist.org/tpl1.1/record/kew-2520138, or *Tamarix articulata* Vahl., http://www.theplantlist.org/tpl1.1/record/kew-2520077
Native:
Vernacular name: *Athel, Ethel*
Raunkiaer's life-form: Mesophanerophyte
Description and morphology: Tree (up to 8–10 m high) or shrub, dense flexuous articulated branches with young red bark, branches look articulate, minute articles (2–4 mm long) acuminate alternate leaves (scales) rolled around twigs or leaves absent, inflorescences on top young branches in long (30–60 cm long) panicles with red-pinkish small pentamerous sessile flowers inserted at base of each article, small fruits (3–6 mm long) in brown-reddish ovoid capsules containing small (2–3 mm) seeds capped with an apical aigrette.
Reproduction: Flowering July–November, seeds (possible anemochory), also vegetative with suckers
Pastoral importance: Medium
Palatability and fodder value: Mostly grazed by camels, goats and gazelles, desperate sheep will occasionally graze it too
Toxicity: None
Economic interest: Excellent fuelwood, building material, commonly used in forestry projects, windbreak, edges, easy propagation from cuttings and fast growing, branches on grounds would take root, highly melliferous.

Habitat: Deep sandy soils, phreatophyte – halophyte, withstands waterlogging, down deep into desert in wadis, withstands dunes encroachment, would form *nebkha*, all over to 50 mm country, and in gardens, in oases.

Distribution:

- In Libya, from coastal zones to deep inland, in the desert, on sandy soils
- In arid zones, southern Europe, North Africa to Pakistan

Endemism: NA

Notes:

- Determination of *Tamarix* is complex and requires inflorescences and flowers.
- *Tamarix aphylla* is flowering late summer/ early autumn contrary to other Libyan *Tamarix* flowering early spring.
- Most remarkable is *Tamarix nilotica* (Ehrenb.) Bunge, Tamarix nilotica (Ehrenb.) Bunge — The Plant List, with long erect branches and stems, on saline sandy depressions in south-east Hamada, in desert; their long flexible, light and straight branches may have been a source for manufacturing spears and javelins in antiquity. Also possibly a source of wood for the horse-drawn chariots used by the Garamantes and Egyptians.

Photo 7.545. *Tamarix aphylla*, tree with lower branches grazed by dromedaries, south-east Al-Assah, Hebiliyah.

Photo 7.546. *Tamarix aphylla*, tree, wadi Girzah, Hamada.

Photo 7.547. *Tamarix nilotica* (Ehrenb.) Bunge, possibly, shrubs 2–3 m high in coppice, wadi Arandel, south-east Hamada (N 29.970 E 12.263), MAR ± 50 mm.

Photo 7.548. Collecting *Tamarix aphylla* seedlings, Al-Assah, for transplanting at Jbebina artesian pond.

THYMELAEACEAE

Daphne jasminea Sibth. et Sm.

Synonym(s): http://www.theplantlist.org/tpl1.1/record/kew-2757020 or *Daphne oleoides* Schreb. (?), https://powo.science.kew.org/taxon/urn:lsid:ipni.org:names:831316-1
Native:
Vernacular name:
Raunkiaer's life-form: Chamaephyte
Description and morphology: Small evergreen woody shrub (30–80 cm), erect; scarious glossy alternate leaves entire subsessile oboval narrowing towards base, folded along midrib, apex apiculate; abundant flowers in terminal clusters of 2–4 sweet-scented white (yellowish) flowers with four recurved petaloid sepals (with tinge of purple outside), calyx pointed and scarious at maturity, fruits as orange ovoid berries.
Reproduction: Flowering January–May, seeds
Pastoral importance: None
Palatability and fodder value: None
Toxicity: All plant parts and berries are poisonous
Economic interest: Ornamental, locally used as scented decoration on head cover
Habitat: Rare, occurs on karstic limestone, on heavy Mediterranean red soils
Distribution:

- In Libya, only in Jebel Al-Akhdar (along the coast Benghazi – Derna and plateau), 300–500 mm MAR, rare occurrence
- In arid zones, southern Europe, North Africa

Endemism: NA

Photo 7.549. *Daphne jasminea*, leaves and mature scarious calyx, Al Haniyah – Marsa-Suza, Shahat.

Thymelaea hirsuta (L.) Endl.

Synonym(s): http://www.theplantlist.org/tpl1.1/record/kew-2518409, or *Passerina hirsuta* L., http://www.theplantlist.org/tpl1.1/record/kew-2551074
Native:
Vernacular name: *Methnan*
Raunkiaer's life-form: Chamaephyte
Description and morphology: Dense hemispherical bush (up to 1.5 m diam. and height) with whitish-cream pendulous branches and stems, densely leafy with very short internodes, small triangular alternate interlocked adpressed dark-green scale-like leaves pubescent on upper face and margins upward-recurved, yellow-whitish minute short-lived pubescent flowers, small scarious fruits enclosed in tiny scarious calyx.
Reproduction: Flowering October–June, seeds
Pastoral importance: Medium, sand binder
Palatability and fodder value: Not touched by livestock
Toxicity: Reported poisonous
Economic interest: Collected as (poor) fuelwood, sometimes locally used as substitute for small rope (long solid strips of stem skin/bark easily dissociated from broken stems) hence considered as a potential source of fibre; reported to use as a potential soil/sand binder due to its compact growth habit (?), not practical or even efficient.
Habitat: Sandy/overgrazed/wind-eroded sites, down to 100 mm MAR, often the only green present large bushes on highly overgrazed sites.
Distribution:

- In Libya, mostly Sirte area and eastwards (Nofilia, Brega, Ajdabiya, Al-Bedan, to Tobruk plateau to Solum), some in south Jebel Nafusa (Giado, Tiji, on highly overgrazed sites)
- In arid zones, North Africa to Egypt

Endemism: NA

Note: Another common *Thymelaea* sp. occurs mostly along the coast (rarely inland except in Sirte and Ajdabiya), *Thymelaea microphylla* Coss. & Durieu ex Meisn., http://www.theplantlist.org/tpl1.1/record/tro-50321391 (*Methnan-el-Abiod*), densely pubescent, leaves alternate sessile entire widely spaced and adpressed on whitish-woolly stem, flower cream, on overgrazed and wind-eroded sandy sites (Al-Assah north, Magdar-es-Sbot, Regdalin, Jamil, El-Watiya), often intermixed on flat sandy sites rich with *Artemisia campestris* (to Taworga – Bu-Gren), erect white pubescent branches and stems, small oblong/spatulate leaves hairy on both sides, leaves spaced on long internodes; appears to be grazed by dromedaries at Al-Assah.

Photo 7.550. *Thymelaea hirsuta*, plant, Al-Jahishiyah, Jebel al-Akhdar.

Photo 7.551. *Thymelaea hirsuta*, pendulous stem and leaves, Al-Jahishiyah.

Photo 7.552. *Thymelaea hirsuta*, leaves and capsules, Al-Jahishiyah.

ZYGOPHYLLACEAE

Fagonia cretica L.

Synonym(s): http://www.theplantlist.org/tpl1.1/record/kew-2807156, or *Zygophyllum creticum* (L.) Christenh. & Byng, Zygophyllum creticum (L.) Christenh. & Byng (worldfloraonline.org)
Native:
Vernacular name: *Shoukiyah Zerga* (?)
Raunkiaer's life-form: Chamaephyte
Description and morphology: Perennial erect shrublet (20–60 cm) dark green all over, glabrous, often profusely climbing in *nebkha/ Ziziphus* bush, much divaricated glabrous ribbed stems, leaves all trifoliate with spiny retro-arched stipules (shorter that petiole or leaves), solitary pale blue flowers (10–15 mm diam.) along stems, fruit capsule on glandulous pedicel, dark brown pendulous top-pointed with five heart-shaped dehiscent dark lobes, pitted dark reniform seeds (3–4 mm long).
Reproduction: Flowering November–May, seeds
Pastoral importance: High
Palatability and fodder value: Grazed when accessible
Toxicity: None

Photo 7.554. *Fagonia cretica*, Flower and spiny stipules, Wadi Siouta, South Misurata.

Economic interest: Reported used as soap (?)
Habitat: From coastal zone to inland on rocky sites, sandy loam soil, down to 50 mm MAR
Distribution:

- In Libya, Jebel Nafusa, Hamada
- In arid zones, southern Europe, North Africa

Endemism: NA

Note: Proper determination of *Fagonia* sp. is difficult and requiring flowers and fruits.

Photo 7.553. *Fagonia cretica*, flower and fruits, Girzah.

Fagonia glutinosa Delile

Synonym(s): http://www.theplantlist.org/tpl1.1/record/kew-2807179
Native:
Vernacular name:
Raunkiaer's life-form: Hemicryptophyte
Description and morphology: Small shy perennial prostrate herb (10–30 cm) dark green all over, all plant exceptionally glandulous with adhering sand particles, woody base, dichotomous glabrous ribbed quadrangular purplish stems with long internodes, tiny opposite trifoliate leaves (all leaflets equal) on short petioles, short (2–3 mm) spiny stipules on swollen node, solitary axillary purple-lilac long pedicellate flowers (10–15 mm diam.) along stems, fruit capsule (few hairs) pendulous yellowish on glandulous pedicel, pitted dark reniform seeds (3–4 mm long).
Reproduction: Flowering February–May, seeds
Pastoral importance: Low
Palatability and fodder value: Grazed when accessible

Toxicity: None
Economic interest: None
Habitat: Inland on sandy-gravelly sites, sandy loam soil, down to 50 mm MAR, often accumulating small sand mounds
Distribution:

- In Libya, Jebel Nafusa foothills, deep into Hamada, Sirte
- In arid zones, southern Europe, North Africa, Middle East, Arabian Peninsula

Endemism: NA

Notes:

- Similar ecology as *Fagonia tenuifolia* Steud. & Houhst. ex Boiss., http://www.theplantlist.org/tpl1.1/record/kew-2807279, or *Fagonia orientalis* C.Presl, http://www.theplantlist.org/tpl1.1/record/kew-2807248
- Determination of *Fagonia* sp. is difficult and requires whole plant, flowers and fruits.

Photo 7.555. *Fagonia glutinosa*, plant, Hamada, Girzah.

Peganum harmala L.

Synonym(s): http://www.theplantlist.org/tpl1.1/record/kew-2548698
Native:
Vernacular name: *Harmel*
Raunkiaer's life-form: Chamaephyte
Description and morphology: Small (30–80 cm) perennial hemispherical dark-green bush, unpleasant scent, woody base, densely branched, glabrous, sessile leaves finely dissected in narrow strips, apical white-cream pedicellate flowers (25 mm diam.) with long elliptic petals, golden fruit as globose dehiscent capsule with persistent calyx, opening

Photo 7.556. *Peganum harmala* bush, El-Hisha, Benghazi.

Photo 7.557. *Peganum harmala*, flowers, Al-Assah.

Photo 7.558. *Peganum harmala*, dry fruit, Al-Assah.

from 3–4 apertures, small (2–3 mm) black pyramidal seeds.

Reproduction: Flowering February–October, seeds

Pastoral importance: Low, strong indicator of overgrazing, degradation

Palatability and fodder value: Not grazed when green, desperate livestock would consume when dry

Toxicity: Green plants reported poisonous

Economic interest: Much used as medicinal (leaves – anthelmintic, healing, etc.), as drug (seeds – soporific, euphoria, well-being, sedation), seeds ingested in large quantity may lead to death (cardiotonic), reported to keep away snakes and scorpions, associated with magic practices (fumigation as protection from *jinn* and bad spirits).

Habitat: Wasteland near dwellings, overgrazed sites, sandy/sandy-loam soils around watering places, down deep into Sahara around wells and watering places, often the only plant remaining on heavily overgrazed and stomped sites near encampments.

Distribution:

- In Libya, all coastal zones and inland on overgrazed sites
- In arid zones, southern Europe, North Africa, Middle East, west Asia to Central Asia, Pakistan

Endemism: NA

Zygophyllum album L. f.

Synonym(s): http://www.theplantlist.org/tpl1.1/record/kew-2469665, or *Tetraena alba* (L.f.) Beier & Thulin, http://www.theplantlist.org/tpl1.1/record/kew-2522149

Native:

Vernacular name: *Bougreibah* (?)

Raunkiaer's life-form: Chamaephyte

Photo 7.559. *Zygophyllum album*, shrublet, (Wadi Nefad, Seddada).

Photo 7.560. *Zygophyllum album*, fruits and leaves, south-east El-Hisha, Bu-Gren.

Description and morphology: Perennial small (20–80 cm high) shrublet, pubescent erect stems, white bark cracked on old plant, leaves (5–8 mm × 2–4 mm) fleshy opposite ovoid, flowers cream-white on short pedicel, short pedunculi shorter than fruit, fruit tetragonal reddish – five dilated lobes from base to top, lower half lobes coalescent.
Reproduction: Flowering October–May, seeds
Pastoral importance: Low
Palatability and fodder value: Not touched by all livestock
Toxicity: Reported poisonous
Economic interest: None, reported as medicinal tea to treat toothache (?)
Habitat: Saharan, sand (on gypseous subsoils (?)), down to 50 mm MAR, withstands being buried in mobile sand
Distribution:

- In Libya, most arid part of western Jeffara plain, Hamada, Sirte to Ajdabiya, south Jebel Al-Akhdar (Tobruk)
- In arid zones, North Africa, Middle East, Arabian Peninsula to South Africa

Endemism: NA
Notes:
Fruits are necessary to distinguish different *Zygophyllum* sp. Apparently many hybrids may exist rendering identification debatable:

- *Zygophyllum simplex* L., http://www.theplantlist.org/tpl1.1/record/kew-2469722, the only Annual *Zygophyllum*, prostrate herb, very small yellow flowers, fruits pendulous elongated with five round lobes (wadi Wishkah – Sirte to Ajdabiya).
- *Zygophyllum geslinii* Coss., http://www.theplantlist.org/tpl1.1/record/kew-2469587, perennial, Saharan, erect bush, fruit peduncle as long or longer than fruit, fruit lobes round and coalescent, on coastal sand.
- *Zygophyllum gaetulum* Emb. & Maire, perennial, erect bush, elongated thin fruit with tubular coalescent lobes, upper part of lobes round dilated, Hamada, Saharan.

References

Al-Jibouri, H.A. (1966) *Description and Distribution of Wheat and Barley Varieties Grown in Libya*. Food and Agriculture Organization of the United Nations, Rome.

Allen, J. and Ewing, M.A. (1978) *Report on 4 Years' Experience (1974–78) Growing Annual Medics in the Jeffara Region*. Western Australian Advisory Group (Cereal Project – Jeffara Plain Authority), Libya.

Ashley, J., Gintzburger, G. and Hossain, E. (1982) *Ophris lutea* (Gouan) Cav. var. *minor* and *Ophris fusca* Link. subsp. *fusca* Nels: new records for Libya. *American Orchid Society* 51(1), 15–20.

Attioui, Z. (2022) Les cistanches du Maghreb. *Al-Yasmina* 3(2), 1–101.

Awen, B.Z.S., Unnithan, C.R., Ravi, S., Kermagy, A., Sasikumar, J.M., Khrbash, A.S. and Ekreem, W.L. (2011) Essential oils of *Rotama raetam* from Libya: chemical composition and antimicrobial activity. *Natural Product Research* 25(9), 927–933. DOI: 10.1080/14786419.2010.503612

Azab, A. (2017) Arum: a plant genus with great medicinal potential. *European Chemical Bulletin* 6(2), 59. DOI: 10.17628/ecb.2017.6.59-68

Battandier, J.A. (1902) *Flore analytique et synoptique de l'Algérie et de la Tunisie*. Vve Giralt, Imprimeur-Éditeur, Paris.

Bonnet, E. and Barratte, G. (eds) (1886) *Catalogue Raisonné des Plantes Vasculaires de la Tunisie*. Imprimerie Nationale, Paris.

Boulos, L. (1972) Our present knowledge on the flora and vegetation of Libya bibliography. *Webbia* 26(2), 365–400. DOI: 10.1080/00837792.1972.10669962

Boulos, L. (1999) *Flora of Egypt* (Vols 1–4). Al Hadara Publishing, Cairo.

CSIRO (1985) Australian plant introduction review. *Australian Plant Introduction Review* 18(1), 182.

Cuénod, A. (1954) *Flore analytique et synoptique de la Tunisie*. S.E.F.A.N, Tunis.

De Santis, S., Spada, F. and Magri, D. (2023) Geographic range vs. occurrence records in plant distribution mapping: the case of Arbutus in the Old World. *Forests* 14(5), 1010. DOI: 10.3390/f14051010

Dobignard, A. and Chatelain, C. (2010) Index synonymique et bibliographique de la flore d'Afrique du Nord. In: *Conservatoire et Jardin Botaniques de la Ville de Genève (Suisse)*.

Durand, E., Barratte, G., Ascherson, P., Barbey, W., Muschler, R. and Meunier, S. (1910) *Florae Libycae Prodromus, ou Catalogue Raisonné des Plantes de Tripolitaine*. Imprimerie Romet, Geneva, Switzerland.

El-Gadi, A.A. and Jafri, S.M.H. (eds) (1976) *Flora of Libya*. El Fateh University of Tripoli/Arab Development Institute, Tripoli.

Essokne, R. and Mahklouf, M. (2023) Index to new updated names of the *Flora of Libya* Volume I. *Journal of Science* 16, 41–52.

Essokne, R., Mahklouf, M. and Isweiri, H. (2024) Index to new updated names of the *Flora of Libya* Volume II. *Journal of Science* 17, 70–78.

Faruqi, S.A. (1980) Studies on Libyan grasses: an annotated catalogue and key in the species. *Willdenowia* 10, 173–225.

Forskål, P. (1775) *Flora Aegyptiaco-Arabica: sive descriptiones plantarum quas per Aegyptum inferiorem et Arabiam felicem*. Hauniae, Ex officina Mlleri.

Foster, K.J. (2015) Drought resistance and recovery mechanisms of Tedera (*Bituminaria bituminosa* var. *albomarginata*). PhD dissertation, University of Western Australia, Perth, Australia.

Francis, C.M. (1978) *Distribution and Ecology of Annual Medicago sp. in Northwest Libya – A Report Based on a Plant Collection Tour 1978*. Western Australia Department of Agriculture, Perth, Australia.

Francis, C.M. (1982) *Libyan Medic Evaluation Project*. Western Australian Overseas Project Authority, Perth, mission report.

Gachet, J.P. and Elmir, A. (1972) *Etude monographique des* Medicago annuelles. *Annales de l'Institut National Agronomique de Tunisie* (Vols 45–1). INRAT, Tunisia.

Gardner, C.A. (1979) Eucalyptus of Western Australia. *Western Australian Deptartment of Agriculture Bulletin*, 4013.

Gawhari, A.M.H., Jury, S.L. and Culham, A. (2018) Towards an updated checklist of the Libyan flora. *Phytotaxa* 338(1), 1–16. DOI: 10.11646/phytotaxa.338.1.1

Gintzburger, G. (1978a) *A Prospection and Collecting Tour in the Middle Zone and the Eastern Part of Zone* 7 (No. 231/78). FAO Agricultural Research Centre, Tripoli, pp. 118–151).

Gintzburger, G. (1978b) *Collection of Seeds and Shrubs (Bir-el-Ghanem, Bir Ayyat and Nalut Area)* (No. 183/78). FAO Agricultural Research Centre, Tripoli.

Gintzburger, G. (1980) Forage legumes in Libya. *Plant Genetic Resources* 43, 7–11.

Gintzburger, G. and Blesing, L. (1979) *Genetic Conservation in Libya – Indigenous Forage Legumes Collection in Northern Libya, Distribution and Ecology of Medicago sp*. Food and Agriculture Organization, Rome.

Gintzburger, G. and Kernick, M.D. (1979) *Plant Genetic Conservation, Forage Legume Collection in Libya, Part II: Minutes of Meetings, Field Trip Reports and Proposals for Agronomic Evaluation* (No. 231-233/79). FAO Agricultural Research Centre, Tripoli.

Gintzburger, G. and Sbeita, A. (1995) The ley-farming system in Libya: a case study. Regional Symposium on Integrated Crop-Livestock Systems in the Dry Areas of West Asia and North Africa.

Gintzburger, G., Sbeita, A., Rojbani, H., Fituri, M. and Francis, C.M. (1983) Annual *Medicago* sp. collected from the Sirte region of Libya. *Plant Genetic Resources* 56, 23–29.

Gintzburger, G., Francis, C.H., Sbeita, A., Rojbani, H.R., Fituri, M. and Skinner, P.W. (1984a) *LMEP – Catalogue of the Annual Medicago sp. from the Western and Central Zones of Northern Libya* (Vol. 2). Western Australian Overseas Project Authority, Perth, Australia.

Gintzburger, G., Sbeita, A. and Francis, C.M. (1984b) *Libyan Medic Evaluation Project – Final Report*. WAOPA-Western Australian Department of Agriculture, Perth, Australia.

Gintzburger, G., Toderich, K., Mardonov, B. and Mahmudov, M. (2003) *Rangelands of the Arid and Semi-arid Zones in Uzbekistan*. CIRAD-ICARDA, Montpellier, France.

Goodchild, A.V., Grando, S., Thomson, E.F., Haylani, M. and Ceccarelli, S. (2022) Heritability and genetic gain of digestible organic matter intake of barley straw genotypes by sheep, using repeatable in situ and laboratory-based indicator traits. *The Journal of Agricultural Science*, 1–11. DOI: 10.1017/S0021859622000399

Halse, N.J. (1978) *Land Use Recommendations for the Jeffara Plain, Tripoli, Libya*. Western Australia Department of Agriculture, Perth, Australia.

Halse, N.J. (1982) *Annual Inspection Visits to Cereal Project – Libya*. Western Australian Advisory Group (Cereal Project – Jeffara Plain Authority).

Hammer, K., Lehmann, C. and Perrino, P. (1988) A checklist of the Libyan cultivated plants including an inventory of the germplasm collected in the years 1981, 1982 and 1983. *Kulturpflanze* 36, 475–527.

Heyn, C.C. (1963) The annual species of *Medicago*. *Scripta Hierosolymitana* 12(1), 1–154.

Inocencio, C., Rivera, D., Concepción Obón, M.C., Alcaraz, F. and Barreña, J.-A. (2006) A systematic revision of *Capparis* section *Capparis* (Capparaceae) 1, 2. *Annals of the Missouri Botanical Garden* 93(1), 122–149. DOI: 10.3417/0026-6493(2006)93

Keith, H.G. (1965) *A Preliminary Check List of Libyan Flora, Volumes 1 and 2* (Vol. 2). United Middle East Executives Ltd, London.

Le Floc'h, E. (1983) *Contribution à une étude ethnobotanique de la flore tunisienne*. Imprimerie Officielle de la République Tunisienne, Tunis.

Le Houerou, H.-N. (1959) *Recherches écologiques et floristiques sur la végétation de la Tunisie méridionale*. PhD dissertation, Universite Alger Institut de Recherches Sahariennes, Algiers.

Le Houérou, H.-N. (2002) *Multipurpose Germplasm of Fodder Shrubs and Trees for the Rehabilitation of Arid and Semi-arid Land in the Mediterranean Isoclimatic Zone: A Photographic Catalogue*. CIHEAM, Zaragoza, Spain. Available at: https://om.ciheam.org/om/pdf/b37/02600003.pdf (accessed 9 August 2024).

Le Houérou, H.-N., Dumancic, D. and Eskileh, M. (1982) *Study of Fodder Shrubs Consumption by Sheep in the Wishtata Range Development Project* (UNTF/LIB 018, Technical Paper no. 43, 11). FAO Agricultural Research Centre, Tripoli.

Lesins, K.A. and Lesins, I. (1979) *Genus Medicago (Leguminosae): A Taxogenetic Study*. Dr W. Junk Publishers, The Hague, Boston, London.

Maire, R. (1925) *Carte phytogéographiques de l'Algérie et de la Tunisie*. Dir. Agric. Commerce et Colonisation.

Maslin, B.R. (1997) Australia's golden future. *Landscope* 12(3), 16–22.

Maslin, B.R. and McDonald, M.W. (2004) *Evaluation of Acacia as a Woody Crop Option for Southern Australia*. CSIRO, Rural Industries Research and Development Corporation.

Nègre, R. (1956) *Les luzernes du Maroc* (No. 5, Série Botanique). Travaux de l'Institut Scientifique Chérifien, Rabat.

Nègre, R. (1959) *Révision de Medicago d'Afrique du Nord, Extrait de la Société d'histoire Naturelle de l'Afrique du Nord*. (pp. 50, 267–314).

Nègre, R. (1961) *Petite Flore des Régions Arides du Maroc Occidental* (Vol. 2). Editions du Centre National de la Recherche Scientifique, Paris.

Omar, M.K., Masoud, M. and Ali, H. (2017) The allelopathic effects of *Artemisia herba-alba* Asso: aqueous extracts on seed germination and seedling development of *Ceratonia siliqua* L. *Journal of Marine Sciences and Environmental Technologies* 3(2).

Ozenda, P. (1977) *Flore du Sahara*. Editions du Centre National de la Recherche Scientifique. Paris.

Pampanini, R. (1930) *Prodromo della flora Cirenaica*. Ministero delle colonie, Forli. (Vol. 577). Tipografia Valbonesi, Italy.

Pasta, S., El Zein, H., Şentürk, O., Gücel, S., De Simone, L. and De Montmollin, B. (2024) *Broad-ranged, highly disjunct, locally rare and severely endangered: the challenging risk assessment and a global conservation strategy for Erica sicula Guss. sensu lato (Ericaceae)*. DOI: 10.3897/arphapreprints.e124671

Pavon, D., Véla, E. and Médail, F. (2020) Are Mediterranean trees well known? *Juniperus turbinata* (Cupressaceae), a common but misunderstood taxon. *Ecologia Mediterranea* 46(2), 77–104. DOI: 10.3406/ecmed.2020.2110

Puckridge, D.W. and French, R.J. (1983) The annual legume pasture in cereal ley-farming system of Southern Australia: a review. *Agricultural Ecosystem and Environment* 91, 229–237.

Quézel, P. and Santa, S. (1962) *Nouvelle flore de l'Algérie et des régions désertiques méridionales* (Vol. 2). Editions du Centre National de la Recherche Scientifique, Paris.

Robledo, A., Rios, S. and Correal, E. (1993) The *Medicago arborea* group in the Mediterranean Basin: I. Origin, distribution and morphology. *Pastos* 23(2), 55–67.

Saaed, M.W.B., El-Barasi, Y.M. and Rahil, R.O. (2019) Our present knowledge about the history and composition of the vegetation and flora of Libya. *Webbia* 74(2), 325–338. DOI: 10.1080/00837792.2019.1669361

Scholz, H. (1974) Liste der Gräser Libyens. *Willdenowia* 7, 419–458.

Sheibani, M. and Jamali, S. (2023) Mycorrhizal association of *Terfezia claveryi* with several Cistaceae plant species. *Rostaniha* 24(1), 21–38. DOI: 10.22092/BOT.J.IRAN.2023.361421.1348

Sherif, A.S., Mahklouf, M.H. and Betelmal, A.G. (2020) Floristic study and species diversity of Msallata-Garaboulli Province in Libya. *Journal of Advanced Botany and Zoology* 7–3, 1–18. DOI: 10.33258/bioex.v2i2.232

Täckholm, V. (1974) *Students' Flora of Egypt,* 2nd edn. Cairo University.

Trabut, L. (1935) *Noms Indigènes des Plantes d'Afrique du Nord. Répertoire des Noms des Plantes Spontanées, Cultivées et Utilisées dans le Nord de l'Afrique*. Imprimerie La Litho-type et Jules Carbonel réunies, Algiers.

Volpato, G. and Nardo, A. (2017) The role of *Nucularia perrinii* Batt. (Chenopodiaceae) in the camel-based Sahrawi social-ecological system. *Journal of Ethnobiology and Ethnomedicine* 13(1), 12. DOI: 10.1186/s13002-017-0141-3

8

From Rangeland Vegetation Standing Biomass to Stocking Rates

Abstract

Sustainable rangeland management requires biomass field measurements or their evaluation using the rain-use efficiency coefficient. Biomass measurements are time-consuming and must be repeated on different vegetation types due to variable annual amounts and spatially uneven distribution of rainfall in arid zones. New technologies like satellite imagery and GIS allow assessment on large areas but require complex computer skills. With these assessments, management decisions can be made after calculating carrying capacity and stocking rates. Sustainable arid rangeland grazing remains with the rangeland users or the experienced managers evaluating the standing biomass and potential stocking rates. However, western grazing concepts appropriate for managing well-controlled grazing areas may not be suitable for shepherding in arid zones due to logistical constraints, water availability, flock management and land tenure issues. Overgrazing in Libya further limits management options, leading to reduced rangeland grazing and expensive imported complementary feed.

Introduction

Rangeland yearly vegetation production evaluation is the foundation for proper livestock rearing, especially in arid zones. It must be sustainable in the long term despite high intra- and inter-annual rainfall variability, spatial distribution, and differences in soil types. Maintaining a satisfactory vegetation cover is the best way to avoid rangeland desertification due to droughts (Saïdi *et al.*, 2016), climatic vagaries and grazing pressure. Measuring rangeland production and evaluating rangeland health (National Research Council, 1994) are paramount to rangeland sustainable management (Gintzburger *et al.*, 2009; Pellant *et al.*, 2005; Stoddart *et al.*, 1975).

It is essential to assess how much forage is available for grazing from different vegetation types at peak standing biomass for sustainable rangeland management. There is therefore a need to measure or, if possible, evaluate by different techniques the rangeland yearly production, then decide how much of the dry matter production should be available for sustainable livestock feeding, i.e. by determining what is the proper rangeland use factor, the potential carrying capacity and the feasible stocking rate that should be applied. Grazing more than the available yearly rangeland production will stimulate wind and water erosion, reduce the vegetation cover and lead to degradation and ultimate desertification.

Rangeland Proper Use Factor, Carrying Capacity and Stocking Rate

Heady (1975) and Stoddart *et al.* (1975) defined the rangeland proper use factor, carrying capacity

DOI: 10.1079/9781800627154.0008

and stocking rate supported by the climax theory (Clements, 1916).

Proper use factor

(Synonyms: proper utilization, proper grazing use, allowable use)

A percentage of the standing biomass of the annual plants or perennial seasonal growth allowed for grazing to maintain or improve the long-term production of the vegetation type. The proper use factor is different for annuals and for perennial plants and is a function of plant's palatability or toxicity. Deciding about a range use factor is a rangeland manager's decision that always carries a risk of overstocking or understocking and damaging the vegetation cover.

Carrying capacity

Carrying capacity of a specific vegetation type at a defined time is the ideal number of livestock units (sheep, goats, camels) per unit of area (ha) that the rangeland can sustain long term without being damaged, without altering its long-term production, and maintaining its good health. It is a function of the yearly biomass production in relation to rainfall, soil fertility, percentage vegetation cover, proper use factor, plant palatability, season of grazing, site grazing history etc. It is a technical element that, in some cases, has to consider the presence of herbivorous wildlife.

Stocking rate

The number of livestock (and wildlife) units per ha that the rangeland manager decides to apply to a vegetation type during a year, a season or a set time. The rangeland may be understocked; hence, grazing a rangeland with a number of livestock per ha below the carrying capacity. The range manager may decide to overstock the rangeland for various reasons; hence, risking overgrazing and long-term rangeland degradation.

Arid rangelands' carrying capacity has long been studied, measured and defined for many vegetation types in North Africa, the Middle East and Central Asia by scientists and consultant companies. However, stocking rates practised by local rangeland users in North Africa and the Middle East remained ignored for cultural and land tenure reasons. In the ex-Soviet republics of Central Asia, rangeland stocking rates determined by scientists after lengthy field studies were enforced on large Sovkhozes and Kolkhozes by the government (Gintzburger *et al.*, 2009).

Rangeland Standing Biomass and Rangeland Production

For any vegetation (trees, shrubs, annuals), survival is failsafe when plants complete their growth cycle, season after season, producing vegetative and reproductive elements, producing seeds replenishing the seed soil stock, and strengthening the rooting systems of perennials, all ensuring the perennation of the vegetation cover.

When continuously grazed, defoliation, browsing of young green twigs, and fruits ingested before ripening slowly weaken each individual plant. Some plants are more resistant to grazing (*Salsola vermiculata*, *Gymnocarpos decander*) or may survive longer and resuscitate thanks to high seed production on good years and seed dormancy mechanisms (Annual *Medicago* sp.). Other plants will resist grazing or browsing with spines, thorns (*Atractylis* sp., *Astragalus spinosus*, *Neurada procumbens*, *Ziziphus lotus*, *Lycium* sp., *Rhus tripartita*, *Callicotome* sp.) or sharp woody twigs (*Atriplex halimus*, *Anthyllis henoniana*, *Gymnocarpos decander*, etc.). Others are not grazed as poorly palatable (*Haloxylon scoparium*, *Artemisia campestris*), unpalatable (*Asphodelus microcarpus*, *Urginea maritima*, *Thymelaea* sp.) or poisonous (*Ferula communis*, *Ephedra altissima*, *Euphorbia* sp., *Calotropis procera*, *Haplophyllum tuberculatum*), and will mostly remain untouched and increase in number year after year on overgrazed and degraded rangelands.

All these rangeland plants, palatable or not, producing biomass, contribute to the health and stability of the ecosystem.

Measuring Standing Biomass and Production on Libyan Rangelands

Rangeland production or dry matter (DM) productivity is approached by measuring the seasonal evolution of the standing biomass of annuals and perennials on a site protected from grazing. The standing biomass (kg DM/ha) is measured by field clipping at a specific date on random quadrats of different sizes. Rangeland production evolves over the seasons according to rainfall, water stored in soil and temperatures on site (Floret and Pontanier, 1978).

The rangeland production or productivity is evaluated from the above-ground biomass (standing biomass) evolution on a site over the season or the year. It is expressed in kg DM/ha per unit of time, as kg DM/ha/year (Gintzburger, 1986) or during a season.

The rangeland standing biomass is composed of:

- annual plants (therophytes, geophytes), which have a short cycle during the wet season, as soon as the first autumn rainfall, during the winter and spring, and dry out at the beginning of the summer; and
- perennial plants (shrubs and small trees), which have mostly woody parts, grow during the wet season and may lose their leaves and yearly growth during the dry summer (Evenari *et al.*, 1976; Fahn, 1964; Fahn and Sarnat, 1963).

Quite reasonably and conservatively for sustainable rangeland management, livestock should only graze or browse part of the year's production (Stoddart *et al.*, 1975); i.e. only part of what was produced in excess of the summer standing biomass, usually measured when the Mediterranean vegetation is at rest.

In Libya, the biomass production of annual plants is measured at peak standing biomass, habitually by late April/early May before the hot days wilt them. Only one single biomass measurement of annual plants at time of peak standing biomass may be necessary to approach its productivity. We consider that the peak standing biomass of annual plants is equal to its productivity as most of it is palatable. However, when open to grazing, livestock should only take up a portion of the peak standing biomass of annual plants, conservatively with a range use factor of about 50–75% to allow for reseeding and leaving some organic matter (OM) on the soil surface to prevent soil erosion.

The rangeland productivity of perennial plants, when not grazed, is approximately equal to the peak perennial standing biomass minus the perennial biomass at rest, usually at the end of the summer. It is more complex to assess as livestock would not graze the woody parts. They would only nibble the green parts (new leaves, stems, flowers, fruits) when available. Therefore, when measuring the peak or at-rest standing biomass of perennials, it is necessary to separate the woody parts from the green parts (Gintzburger, 1986). This is a tedious process. For sustainable rangeland management, we would allow a range use factor of 30–50% of the perennial vegetation annual production.

Little work was done on Libyan rangeland measuring standing biomass production. Very few papers on the subject were published in refereed journals in the Sirte region (Nègre, 1974) and near Bir-el-Ghanem (Gintzburger, 1986) in the Jeffara plain. Other occasional measurements were published in restricted technical reports from GEFLI consultant firm (*Groupement d'Etudes Françaises en Libye*) and FAO experts (TF Lib06, Lib10, LIB11, LIB04, LIB18). Comparatively, many biomass measurements were performed on various arid steppic vegetation types in southern Tunisia (Floret *et al.*, 1981, 1982, 1983; Floret and Le Floc'h, 1973; Le Houérou, 1959, 1969, 1975; Telahigue *et al.*, 1987) and in the Middle East (Evenari *et al.*, 1976; Gutman, 1978; Thalen, 1979).

Examples of measured standing biomass and fodder value on Libyan rangelands

Biomass measurements and percentage vegetation cover are only valid at the date when they are measured. They represent a photograph of the rangeland condition at a set date. The information cannot be extrapolated or assumed to be identical for other years on the same site as the vegetation cover and composition may change due to rest or overgrazing, fuelwood harvesting, or inter- or intra-annual rainfall variability. A dry or a wet winter or spring hugely impact the amount of standing biomass, annual plants contribution and productivity on one site. It is important to list the vegetation composition by species and measure the percentage vegetation cover at the time of the biomass measurements (Gintzburger *et al.*, 1986, 2009).

In the Sirte region

Nègre (1974) and his team measured the standing biomass in 1972 and feed value (Table 8. 1) in the

Table 8.1. Standing biomass, vegetation cover and feed value of the Sirte region rangelands in 1972. (Adapted from Nègre, 1974)

Rainfall zone (mm MAR)	Vegetation and soil type	Location	Vegetation cover (%)	Perennial standing biomass (kg DM/ha)	Annuals standing biomass (kg DM/ha)	Max. total peak standing biomass (kg DM/ha)	Estimated FU/ha	Rangeland health < 1 (very poor) to 5 (excellent)	Grazing period	Used mostly by	Remarks
> 125–150 + run-off	*Arthrocnemum* sp., *Halocnemum* sp., *Salicornia* sp., in *sebkha*	*Sebkha* Coastal Sirte (Pre-desertic)	25–40% (20–50 cm high)	2500–3500	0	3000	300–600	3	Early winter to late summer	Camels	
125–150	*Salsola* sp. *Atriplex* sp. *Lycium* sp., *Nitraria* sp., *Limonium* sp. *Limoniastrum* sp., on saline loam (100–500 sodium mEq/g) in wadis and depressions	Coastal Sirte (Pre-desertic)	40–70%	130–150	5–10	150	50–80	2	Early winter to late summer	Small ruminants and camels	
125–150	*Thymelaea microphylla*, *Pituranthos rholfsianus*, *Retama raetam* on deep sand (> 50 cm approx.) with *Aristida pungens* and some *Panicum* sp.	Coastal Sirte (Pre-desertic)	Degraded mobile sand	50–100	20–50	100–150	20–40	1	March April May	Small ruminants and camels	Occasionally cropped (poor rainfed barley), wind erosion
100–125	*Gymnocarpos decander*, *Arthrophytum scoparium*, *Pituranthos tortuosus* on calcareous crust on shallow sand veil (10–20 cm)	Inland Sirte (Pre-desertic)	Best shrub's stand = 50%, 30–50 cm high, down to 1%	15–20	5–20	35–20	2–8	< 1	March April May	Small ruminants and camels	

Continued

Table 8.1. Continued.

Rainfall zone (mm MAR)	Vegetation and soil type	Location	Vegetation cover (%)	Perennial standing biomass (kg DM/ha)	Annuals standing biomass (kg DM/ha)	Max. total peak standing biomass (kg DM/ha)	Estimated FU/ha	Rangeland health < 1 (very poor) to 5 (excellent)	Grazing period	Used mostly by	Remarks
50–100	*Anabasis articulata*, *Traganum nudatum*, *Salsola tetragona*, *Pituranthos chloranthus*, *Aristida plumosa* on calcareous crust on shallow soils	Inland Sirte –(desertic)	5–20%	15–20	15–20	30–40	2–8	< 1	March April May	Camels	
50–100	*Zilla spinosa*, *Cleome arabica*, *Aristida plumosa*, *Acacia raddiana* in desert depressions (also *Panicum turgidum*) on loam/sandy loamy soils (colluvium) and sometimes gypsum (*Astragalus spinosus*)	Inland Sirte (desertic)	20–40%	10–23 (from *Panicum turgidum*)	10–20	20–50	10–20	1	March to June	Small ruminants and camels + gazelles and mouflons	

FU = Feed Unit

Sirte region, from Buirat-Al-Hasun to the west, and to As-Sultan and wadi Harawah to the east. All rangelands of these areas are extremely poor, receiving between 75 mm and 175 mm MAR with high intra- and inter-annual variability. Some years, pre-desert and desert areas may be totally dry, receiving rainfall far below the annual average. Furthermore, the soils are often shallow over calcareous or gypseous crust covered with sand veils on the arid plateau inland, while depressions and wadies close to the coast end up in large saline *sebkha*. At the time of the measurements, very low vegetation cover due to overgrazing by local sheep, goats and camels was rampant with ubiquitous damages due to wind erosion. Occasional rainfed barley cropping in sandy depressions close to the Sirte coast are rarely successful and leaves bare soil prone to wind erosion (Photo 8.1).

At Bir-el-Ghanem (Jeffara plain)

The above-ground biomass and productivity was measured (Gintzburger, 1986) on an *Aristida (Stipagrostis) pungens/Retama raetam* vegetation type covering some 100–200 000 ha of deep aeolian fixed sand in the Jeffara plain. The site was located between Aziziya and Bir-el-Ghanem (BG-II, N 32.44 E 12.83). The range was in very good condition as protected from trespassing for three years prior to measurements. During spring 1978, the peak total standing biomass climbed from 2 to 3 tonnes/ha between early February and April, with about 0.9 T DM/ha mostly from *Retama raetam* and about 0. 22 T DM/ha from annuals, and kept growing as the spring season progressed. In short, it produced a total of 2 T and 0.95 T DM/ha of net above-ground phytomass with 200 mm and 171 mm of rainfall during the 1977/78 and 1978/79 growing seasons, respectively. The annual plants contributed 460 and 300 kg DM/ha in the two seasons (Photo 8.2).

Photo 8.1. Failed rainfed barley crop on coastal Sirte rangelands (N 31.78 E 16.37, 1975) (From Gintzburger, 1986)

The perennial shrubs lost their 1978 winter-spring aerial production during the long 1978 summer and dry winter when grazeable forage was most needed. In contrast, spring production of annual plants on these protected rangelands remained mostly available for grazing as standing dry matter until late summer. During the 1977/78 good growing season, the total ungrazed rangeland biomass growth rate was about 21.7 kg DM/ha/day, with a contribution of 17.6 kg DM/ha/day from the perennials and 4.1 kg DM/ha/day from the annuals. During the poor 1978/79 growing season, it was down to 9.2 kg DM/ha/day (6.9 and 2.2. respectively, for perennials and annuals) comparable to similar arid zones in the Middle East and Tunisia (Floret *et al.*, 1983; Gutman, 1978; Tadmor *et al.*, 1974).

Most interesting is the relative growth rate (RGR) of perennials and annuals that allows comparing the speed at which each group or plant produces dry matter per kg of standing dry phytomass per unit of time (Hunt, 1978). On the BGII site, the RGR of the perennials and of the annuals, respectively, reached 0.057 kg DM/week and 0.085 kg DM/week during February and April 1978. It was 0.036 (RGR perennials) and 0.104 (RGR annuals) between December 1978 and April 1979. It illustrates the fast-growing capacity of the annuals compared to the perennials during the spring and also the difference in biomass production between years.

On this *Aristida pungens/Retama raetam* vegetation type, fuelwood availability, a valuable resource, was measured (Photo 8.3). However, settlement of shepherds' families on site is not advised as the dead standing wood from *Retama raetam* (about 270 kg dry wood/ha) may be rapidly

Photo 8.2. Sorting the annuals after clipping for biomass measurements (BG II site, Bir-el-Ghanem, N 32.44 E 12.83, April 1977)

Photo 8.3. Weighing wood standing biomass of *Retama raetam* (BG II site, Bir-el-Ghanem, April 1977)

exhausted leading to wind erosion problems on presently fixed sand dunes.

In the South Sorman Protected area

Early in 2007, it was decided to open for grazing the South Sorman protected area (4235 ha, 125–150 mm MAR) to registered shepherds and flocks from the surrounding areas. Grazing inside the protected area had been strictly banned for the previous four years. The landscape is:

- moderately mobile sand dune ridges vegetation (15% of the area) with unpalatable *sbot* (*Stipagrostis pungens*) vegetation type;
- fixed sandy soils inter-dunes flats (85% of the area) covered with a good stand of annuals.

We opted to have only the annuals grazed by the flocks by the end of the spring when the annuals were dry and available as *guech*. To determine the number of sheep and acceptable number of grazing days, we conducted the biomass measurements only of the annuals using the repeated random quadrats clipping technique (Gintzburger,

1986; Gintzburger *et al.*, 2009). The dry annual standing biomass measured amounted to about 260 kg DM/ha (Photo 8.4), approximately available on 3600 ha of inter-dunes flats, or a total of 936 tonnes DM of edible dry annuals. We effected a conservative 50% proper use factor of this total annual biomass to avoid denuding the sandy soil and stimulating soil erosion. Considering that one sheep would graze from rangeland about 1.5–2.0 kg DM/day and supplemented with about 100 g of barley grain/day/head, we calculated that the South Sorman protected area could conservatively support about 4000 head of sheep and 400 head of goats, or 1.2 sheep equivalent/ha for about 70 days from early May to mid-July 2008. It was implemented. Registered shepherds and their flocks were allowed to enter and disperse inside the protected area. The number of heads were counted, but the authorities present on site could not stop extra animals and many flocks entering, or disregarding the allowed stocking rate and number of grazing days, or forcing them to use the whole area. As the annuals were dry, water was made available by the project at set places, consequently creating rings of overgrazing. All the annuals and more were wiped out. Despite proper technical grazing planning, we did not anticipate that the shepherds would be allowed to camp and have water available inside the protected area, making it very difficult to force the flocks to exit and control them. We simply forgot to consider the human and cultural dimensions. Lessons were learned from this experience, and likely soon forgotten.

Evaluating rangeland biomass using satellite imagery: the case of the Ravnina sheep co-operative in Turkmenistan

The availability of satellite imagery offers new techniques for evaluating the rangeland standing biomass. This was done at Ravnina (N 37.93 E 62.66) in the Karakum desert in Turkmenistan (Gintzburger *et al.*, 2009). The vegetation type

Photo 8.4. Measurement of the standing biomass of the annuals on randomly repeated quadrats at South Sorman protected area (N 32.39 E 12.19, 24 April 2008).

and physiognomy on fixed sand dunes is very similar to the protected one from the BGII/Bir-el-Ghanem site and receives a similar average 140–175 mm of winter precipitation per year. No rain occurs during the summer months.

The sandy rangeland Ravnina sites are covered with abundant annual and ephemerals in spring. These rangelands are exclusively used by co-operative large flocks of sheep guarded by a team of shepherds. The Ravnina co-operative hosts an average of about 30,000 to 35,000 sheep units on 320,000 ha of grazing territories distributed around 47 wells. Each flock (1000–2500 SU) (SU = Sheep Unit) are fed on rangelands surrounding the assigned well during 7–8 months from mid-spring to early autumn. Each grazing territory per well covers 3000–6000 ha. The stocking rate during these grazing periods is from 3.5 to 5 sheep/ha. During the cold and snowy winters, the flock is kept in large sheep sheds close to the well and fed with cotton residues, barley grain, and locally made concentrate (*kombicorn*). The rangelands are at rest during the cold months.

We assessed the vegetation types, the standing biomass and monitored the degradation or restoration processes on a large area (110 km × 90 km, about 990,000 ha) of the Ravnina co-operative. The qualitative and quantitative appraisal and monitoring of the vegetation cover of the Ravnina territory were evaluated using a combination of three seasons of field surveys and biomass measurement (2001–2003), with matching satellite imagery (LANDSAT and SPOT) and up-to-date remote-sensing processing coupled with GIS techniques. We first identified the main vegetation types with field surveys in relation to image classification. We correlated the measured annual and perennial vegetation (annual plant and ephemeroids biomass (APEB)) and perennial vegetation line intercept measurements (LIM) with three different vegetation indices:

- Normalized Difference Vegetation Index – NDVI (Leblond, 2004; Rouse *et al.*, 1974)
- Transformed Soil Adjusted Vegetation Index – TSAVI (Baret *et al.*, 1989)
- Perpendicular Vegetation Index – PVI (Richardson and Wiegand, 1977).

We used radiometric values of a SPOT 4 image from the end of May 2002 and LANDSAT imagery from May 2003. Mapping the area using the three available vegetation indices showed that, during ground truthing, NDVI ignores low vegetation cover and classifies them as bare soil. NDVI is definitely not an appropriate vegetation index for low (< 20–30 %) vegetation cover, a common occurrence in arid zones, steppe or desert. Such problems were confirmed from our results in the Ravnina area. By contrast, all the areas covered with annual and ephemeroids vegetation during spring 2002 were accurately identified on the PVI and TSAVI maps while they were identified as bare soil using the NDVI. However, NDVI gave a better discrimination between the annual–ephemeroids vs. perennial vegetation on dense vegetation (reeds and irrigated crops) along the Karakum canal.

The PVI and TSAVI were highly discriminating for sparse vegetation cover and bare soil and were used to map vegetation degradation or regeneration on the Ravnina territory.

Relationship between APEB and vegetation index

All biomass measurement sites were GPS-located with a precision of 5–10 m. We used the average brightness of 10 pixels × 10 pixels (approximately 4 ha on SPOT and 9 ha on LANDSAT images) to compensate for GPS precision of the exact location of APEB measurement site (Map 8.1). We established a linear relationship between our field biomass measurements (spring 2002) and the PVI and TSAVI calculated from the spring SPOT image (31 May 2002). We could confirm the linear regression between the measured APEB and mean PVI or TSAVI values computed on each 4 ha area. We removed sites outside the satellite image or sites due to unexplained abnormality and stretched the pixel histogram with 0 minimum and 255 maximum values to improve the image contrast. The equations were quite similar for PVI and TSAVI, with correlation coefficients of 0.79 (n = 8 and 11). We used this APEB relation with PVI to elaborate the APEB map (Map 8.2) for the whole of the Ravnina territory.

The annual plants and ephemeroids spring biomass at Ravnina ranges from 150 to 650 kg DM/ha during spring 2002 and 2003. We measured an annual plant biomass of 300–400 kg DM/ha on a very similar protected vegetation site at Bir-el-Ghanem receiving 175–200 mm during the 1977/79 seasons (Gintzburger, 1986).

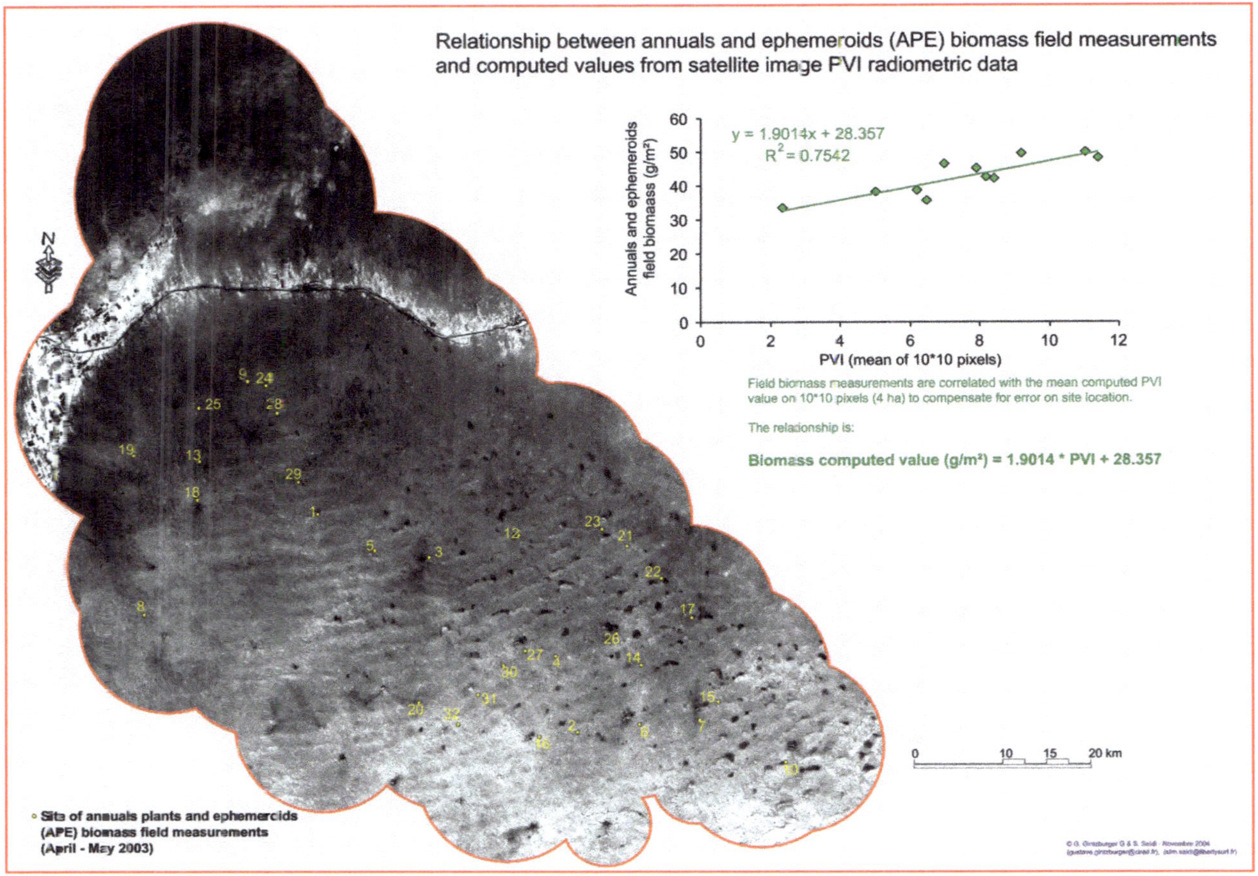

Map 8.1. Relation between PVI and annual plants' and ephemeroids' standing biomass (Ravnina sheep co-operative, Turkmenistan; spot image spring 2003). (From Gintzburger *et al.*, 2009)

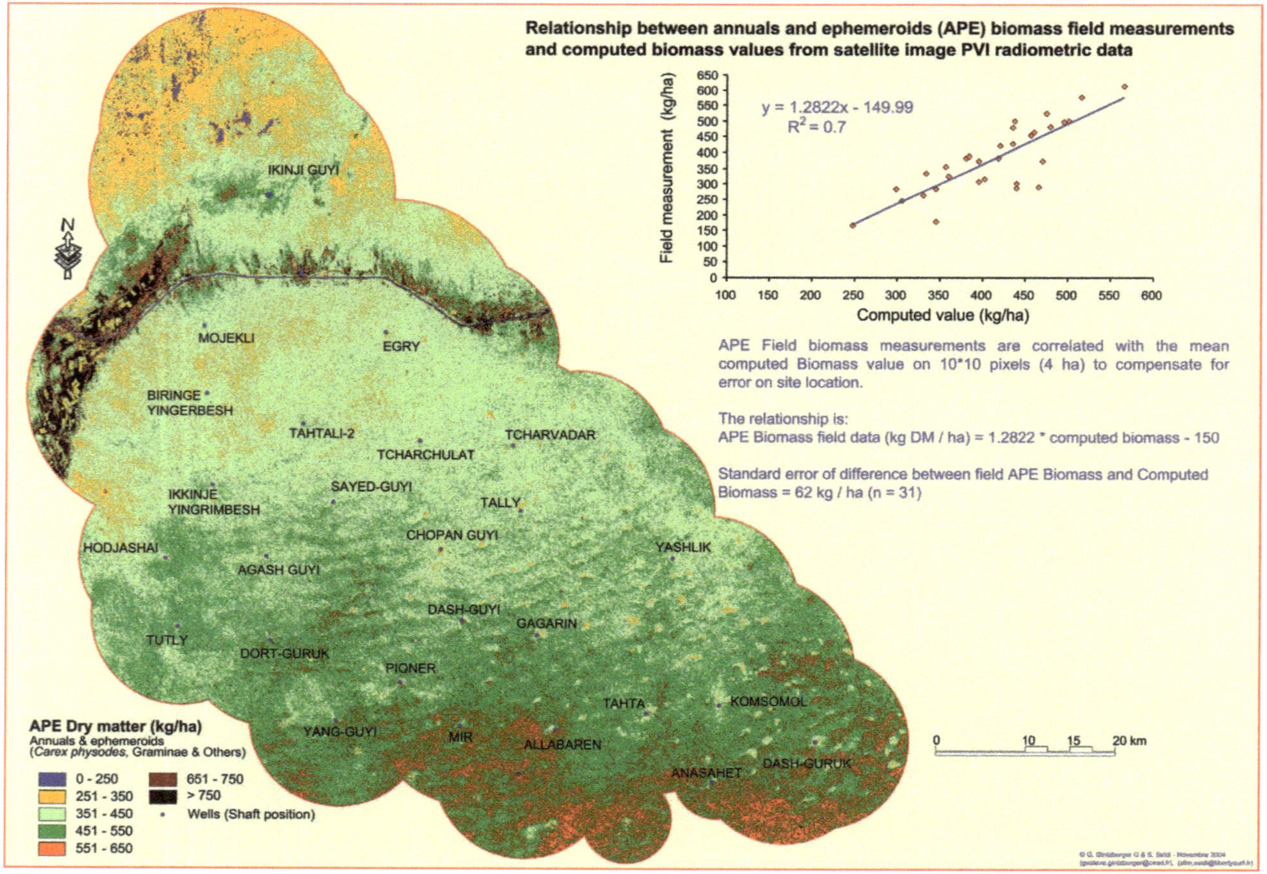

Map 8.2. Annual plants and ephemeroids' standing biomass map (Ravnina sheep co-operative, Turkmenistan, spring 2003). (From Gintzburger *et al.*, 2009)

We produced biomass maps (phase 2) by testing various vegetation indices and using field measurements. They were conducted during the springs of 2001, 2002 and 2003. In spite of constraints of image quality, limited dates available, and the difficulty in discriminating between vegetation and soil in this arid environment, the results were encouraging. High-resolution satellite images for this type of vegetation study are worthwhile, providing the appropriate vegetation index is used.

Satellite imagery application on Libyan rangelands

The techniques we used on the Karakum rangelands could be used on Libyan rangelands or any rangelands of the region to evaluate the standing annual rangeland biomass (or production of annual plant in spring). This is providing that:

a. plant biomass measurements are performed on many GPS located sites during the growing season and when the vegetation is at rest in summer. These biomass measurements should be done on the same sites over several successive years. This is to evaluate the annual rangeland production of the annuals and perennials in relation to the annual rainfall (October to next September);
b. the relationship is established between these biomass measurements with the PVI calculated on each site and from corresponding satellite imagery selected for the time of the biomass measurements;
c. NDVI is not used, as it is not suitable for sparse rangeland steppe and desert vegetation from arid Libya; NDVI will not discriminate between sparse arid vegetation and bare soil and will return a bare soil signal (Gintzburger *et al.*, 2009).

When no biomass measurements are available: apply the Rain Use Efficiency

When biomass measurements are not available, another way to approach the total rangeland production is to refer to the Mediterranean Rain Use Efficiency (RUE) defined by Le Houérou (1984). The RUE is based on the linear relationship between annual rainfall and measured production from rangeland sites in Morocco, Algeria, Tunisia and Libya (Le Houérou and Hoste, 1977) and Turkmenistan (Gintzburger *et al.*, 2009), and envisioned in Syria (Van der Veen, 1967) and Iraq (Thalen, 1979). The RUE is the predicted amount of dry matter produced annually from unfertilized unirrigated rangeland vegetation per mm of rainfall received on site. The RUE is expressed in kg DM/ha/mm rainfall/year.

Is the Mediterranean RUE a constant?

The RUE calculated is a function of the rangeland condition from fully protected with high perennial and annuals vegetation cover, to highly degraded and overgrazed with dominant bare soil, little perennials and occasional annuals (Fig. 8.1).

The calculated RUE values ranges are (Gintzburger and Saïdi, 2009; Le Houérou, 1984):

- 0–1.0: almost bare soil, highly degraded vegetation, very low annual production, no edible perennials left on site
- 1.0–2.0: very poor condition and low productivity, degraded vegetation around watering points with continuous long-term open grazing and fuelwood collection
- 2.0–3.0: poor to medium condition (frequent grazing, long-term open grazing)
- 3.0–4.0: medium to good condition (new grazing territories, occasional grazing due to remote limited access and poor watering access)
- 4.0–5.0: very good condition (remote rangelands, occasional seasonal grazing and little or no permanent water supply)
- > 5.0: very good condition (fully protected areas, rare grazing, territory rarely accessed)

Le Houérou and Hoste (1977) calculated that the average RUE for measured biomass production is about 4 kg DM/ha per mm of rainfall from 41 North African rangeland sites in the mid-1970s, in various conditions and located in rainfall zones between 50 and 800 mm/year.

In Libya, the protected Bir-el-Ghanem/BG II site, in excellent condition, had a rangeland production of 1150 kg DM/ha (annuals and perennials) after receiving about 188 mm of rainfall, hence an RUE of about 6.1 kg DM/ha/mm rainfall (Gintzburger, 1986). On a nearby Bir-el-Ghanem site, on highly degraded *Rhanterium suaveolens*

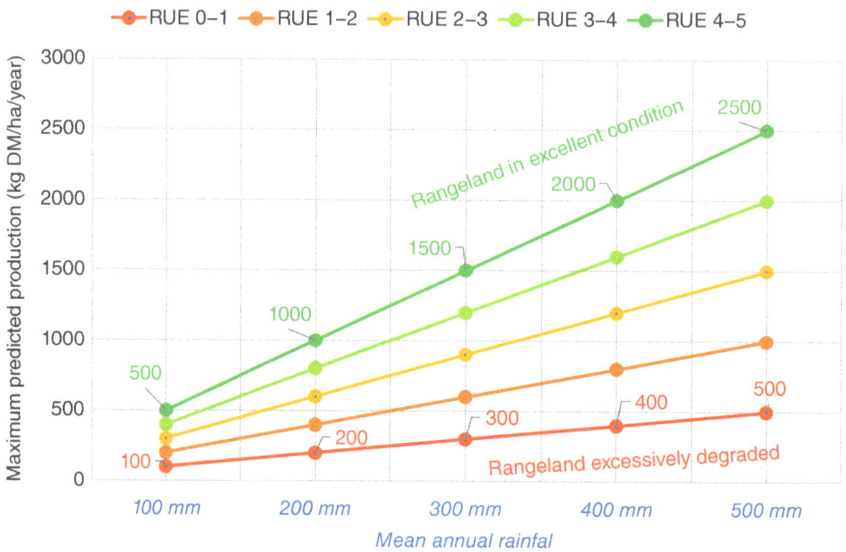

Fig. 8.1. Predicted rangeland production as a function of the RUE and the mean annual rainfall in Mediterranean zones. (After Gintzburger and Saïdi, 2009; Le Houérou, 1984)

receiving 150 mm, the RUE is as low as 0.9 (Telahique and Le Houérou, Tripoli, 1979, personal communication). In the Syrian steppe, in an area receiving an average of 150 mm/year (Van der Veen, 1967), an RUE of 3.4 on a six-year protected *Artemisia herba-alba* was calculated while the same but grazed vegetation outside the exclosures had a 1.8 RUE. In the Karakum desert of Turkmenistan (Gintzburger *et al.*, 2009), in a sandy region receiving an average of 150 mm, the RUE changes from about 2.4 close to the co-operative centre at Ravnina, where heavy grazing occurs, to about 3.6 on rangelands in good condition and difficult to access some 100 km from Ravnina.

Using this simple relationship, it is then possible to rapidly evaluate the rangeland production at the end of the spring growing season in a known rainfall zone considering the estimated RUE according to the rangeland condition from excellent to overgrazed.

As plant production is closely related to plant transpiration, the Mediterranean RUEs mentioned above are only valid on flat rangelands that are neither receiving run-on water (depressions or irrigation), nor located in depression or above high-water tables, as on *sebkha*. In these latter cases, the RUE is much higher. It is not valid in sub-tropical regions (sub-Sahara) zones receiving summer rainfall (Saïdi and Gintzburger, 2013).

Rangeland Proper Use Factor

Once the rangeland biomass production is assessed, either by field measurements or using the RUE, one must decide what is the sustainable amount that can be grazed without handicapping future vegetation growth. As a rule of thumb, it is advised to allow for about 30–50% of the biomass production (annual growth) of the palatable perennials only. In case of abundance on site of poorly palatable or unpalatable perennials such as *Thymelaea hirsuta*, *Thymelaea microphylla*, *Artemisia campestris*, *Retama raetam*, *Stipagrostis pungens*, *Arthrophytum scoparium*, *Peganum harmala*, etc., their biomass and production must be ignored.

As for production of annual plants that are almost edible at all stages, a range use factor of 50–75% may be applied only after flowering and fruiting at the peak of standing biomass at the end of spring or later early summer as dry matter (*guech*). This is to allow restitution of organic matter to the soil and to prevent soil erosion. The dry matter of annuals gets very brittle when dry, as summer progresses, and disintegrates quickly when trampled by livestock. The same use factor is commonly applied to cereal crop residues grazed by livestock after harvest.

Chemical Analysis and Nutritive Value of Some Libyan Rangeland Species and Fodder Shrubs

Fodder value of rangeland plants is determined by chemical and digestibility analysis followed by interpretation by livestock scientists. Chemical analysis and fodder value of rangeland plants in North African arid zones are available from Algeria (Chehma, 2003) and Libya (Le Houérou *et al.*, 1982; Shawesh and Abuzakhm, 1993).

As for Libya, 82 fresh samples of 45 rangeland common species were collected during spring 1978, 1979 and 1980 in western Libya (Le Houérou *et al.*, 1982). Fifteen samples are from annuals and the rest are from various perennials. The Laboratory of Animal Nutrition and the Laboratory of Soil Analysis from the Libyan Agricultural Research Centre (Sidi-Mesri, Tripoli, 1980) performed the following standard chemical analyses on samples oven-dried and ground through a 0.1 mm sieve, ashes in furnace at 550°C for 2 h, nitrogen (Kjeldahl method), crude fibre (Wende method), crude fibre (sulfuric acid extract method).

All data are expressed in % DM:

Crude protein (CP) = N (Kjeldahl) x 6.25
Digestible crude protein (DCP) according to Demarquilly and Weiss (1970) formula
Nitrogen free extract (NFE)
Energy value (NE) according to Breirem formula

Digestibility was assumed from literature and data from arid rangeland species. From Libya, we do have the data from Suleiman (1981) who worked mostly on shrubs from the Taworga region. Evaluating 43 range common species, he obtained an average digestibility of 53%. Soviet scientists (Sishov and Kharin, 1980), surveying the rangelands of the Ajdabiya region, performed *in vitro* OM digestibility and obtained the following results:

Shrubs (120 samples from 35 species): early growth = 58%, mature stage = 50%
Herbs and annuals (75 samples from 34 species): early growth = 61%, mature stage = 54%

It certainly varies according to the season; on the Algerian steppe, OM digestibility annuals' and perennials' early growth in spring may reach 60–80% while in dry years, or at mature stage, the OM digestibility would drop to 30–40%. It is obvious that this figure could be refined with *in vivo* measurements. From literature and these figures, an average 55% OM *in vitro* digestibility of edible parts was used for all calculations. This gross figure of 55% OM digestibility remains a yearly average for all rangeland species.

The following general results ensued:

- Ash content is often high, averaging 16% on 85 samples, but reached as high as 30–40% on Chenopodiaceae growing on saline soils of *sebkha*.
- Protein content is usually around 12%, leading to a DCP (digestible crude protein) of 7–8%, close to barley grain DCP, hence quite satisfactory.
- Crude fibre content (about 20%) is low in most plants except in grasses (25–35%), hence rangeland livestock would always be fibre-hungry.
- Calcium content (0.7%) is always very high, about three times the desirable content – not surprising as Libyan rangeland plants are all growing on limestone and soils with neutral to high pH.
- Phosphorus is deficient (0.06%) in all samples while proper level should reach about 0.12%. This result is in accordance with Libyan soils analysis all displaying a clear phosphorus deficiency. *Atriplex halimus* flowers and young twigs from Sabkha display a high-level P content of 0.11–0.12%; *Echiochilon fruticosum* and *Lotus creticus*, both well-grazed plants from Al-Assah, also reveal a P content of 0.14% and 0.10%, respectively. *Retama raetam* flowers and twigs, avidly browsed by small ruminants from Bir-el-Ghanem, have a high P content between 0.10% and 0.16%. *Thymelaea microphylla*, though mostly consumed by camels, reports a 0.11% P content. *Moricandia arvensis* leaves and flowers from Nalut (wadi Zigzaw) have a P content of 0.10%. All these plants,

when growing, flowering and fruiting, are well grazed and avidly selected and ingested by lactating ewes. It shows the importance of shrubs and perennial herbs for the livestock diet on Libyan rangelands, especially for lactating ewes in spring.

- Average energy expressed in FU (feed unit = energy equivalent to 1 kg barley grain) average for all analysed species of 0.368 FU/kg DM (= equivalent to 370 g of barley grain per kg of DM of rangeland plants). This average value may climb to 0.40–0.45 FU/kg DM with fresh annual or perennial edible plant material.

Converting biomass to feed units

The standing biomass is composed of different species at different growth stages, some being palatable, some not, with different chemical composition (mostly protein content) and digestibility. It is commonly accepted that North African rangeland managers express rangeland standing biomass or production in FU/ha. FU may stand for feed unit, forage unit or fodder unit, a common way to evaluate the feeding value of rangeland biomass available for grazing by livestock (Le Houérou, 1981; Le Houérou *et al.*, 1982, 1983). It is a convenient reference when it comes to feeding livestock as barley grain is the usual supplementary feed used in North Africa and in Mediterranean arid rangelands. Most annuals are palatable, easily ingested and digested, and have the highest fodder value, usually around 0.3–0.4 FU/kg DM when fresh at peak standing biomass, declining to 0.2 FU/kg DM when dry (*guech*) at the end of the growing season (Le Houérou *et al.*, 1982). From perennials, only fresh parts are usually nibbled by small ruminants while camels and equidae would chew even woody parts of rangeland plants.

Rangeland Sustainable Carrying Capacity and Stocking Rate

Libyan rangelands are supplying feed to different types of livestock throughout the year. The energy maintenance requirement for each of them is not related to body weight but to body surface. A small animal having a larger body surface per unit of weight, requires a larger energy need per body weight than a larger animal. The surface:-weight ratio or metabolic weight is the power 0.75 of the body weight.

Conversion factors between different types of livestock on Libyan rangelands

Average live body weight and metabolic weight of adult livestock on Libyan rangelands are shown in Table 8.2.

Therefore, when fed on rangeland, the accepted feeding conversion factors are: one goat ≈ 0.8 sheep; one cattle ≈ 4 sheep; one camel ≈ 5 sheep. Often, instead of using one adult sheep, you may also find in the Libyan reports reference to a sheep unit (SU), with one SU equal to one ewe 40 kg + one lamb (0–3 months) + 0.2 yearling ewe + 0.05 ram.

Daily roughage intake estimates for livestock maintenance on rangelands

The estimated daily roughage intake for animals on rangelands is shown in Table 8.3.

These intakes are average daily values that will drastically increase on poor overgrazed rangelands when animals have to walk long hours and long distances to satisfy their maintenance requirements. When feed requirements are not met, shepherds see their animals losing

Table 8.2. Average livestock body weights and metabolic weights. (Adapted from Le Houérou, 1981)

Libyan livestock on rangeland	Average body weight (kg)	Metabolic weight (kg) = body weight$^{0.75}$	Ratio metabolic weight compared to one sheep
Goat	30	12.8	0.81
Sheep	40	15.9	1.00
Cattle	250	62.8	3.95
Camel	350	80.9	5.09

Table 8.3. Daily roughage intake estimates for livestock maintenance on rangelands. (Adapted from Le Houérou, 1981)

Libyan livestock on rangeland	Average live body weight (kg)	≈ % live body weight roughage intake	≈ Roughage intake per adult animal (kg DM/day)	In FU/day (at 1 kg DM = 0.368 FU)
Goat	30	4	1.2	0.44
Sheep	40	4	1.6	0.59
Local cattle	250	2.5	6.25	2.3
Camel	350	2.3	8.05	2.96

weight and know well when there is no point leading their animals onto poor rangelands. They soon turn to supplementing their animals with barley grain, straw and chaff, now a common sight on degraded Libyan rangelands.

Calculating sustainable rangeland vegetation stocking rate

Feed requirement for one SU is about 400 FU/year (Le Houérou, 1981) equivalent to 640 kg DM/year. If we hypothesize that three-quarters of the FU (for 270 days grazing) would come from rangeland feed (approx. 300 FU/SU or 480 kg DM/SU), the rest issued from summer supplementation, a standard flock of 200 SU would need approximately 96 tonnes of DM feed. With a rangeland use factor of 50%, it means that the rangeland should supply to the flock about 190–200 T DM during the 270 days of the grazing season on about 400 ha to warrant rangeland sustainability and proper feeding of the flock. This would be a stocking rate of 2 ha/SU.

If the rangeland is in very good condition and produces about one tonne DM/ha, the flock would need access to about 400 ha of grazing land (cf. Bir-el-Ghanem 1977–79; about 175–200 mm MAR). It will graze within a radius of 3–4 km from the shepherd's base camp. A small ruminant flock is moving at an average 1.5–2 km/h and would reach the end of its grazing territory in about two hours in the morning and need another two hours at the end of the day to get back to base camp. This is still feasible and acceptable to the shepherd within a grazing day of ten hours.

If the rangeland is in poor condition, a common present situation, producing only 0.1 T DM/ha, the 200 SU flock would need to have access to about 4000 ha; i.e. exploring within a radius of about 11–12 km around the shepherd's camp. However, 5 km is the maximum daily walking distance on rather flat country from base camp for a small ruminant flock while grazing. Walking longer daily distances while grazing will lead to animals losing body weight and increase the need for, and costs of, supplementary feeding and watering.

In short, 200 SU on 400 ha of rangeland in very good condition is a stocking rate of 2 ha/SU, a long-past memory on Libyan rangelands. With the current poor and degraded Libyan rangeland conditions, with 200 SU needing 4000 ha or 1 SU/20 ha, we have already passed the breaking point and the shepherd has an impossible task to properly feed his small ruminants' flock. Hence, livestock owners and shepherds accessing poor overgrazed rangelands are relying more and more on necessary supplementary feeding with barley grain on stationary open-air feedlots now common on Libyan rangelands and a trend long ago identified by Boutonnet (1989) in Algeria, and in North Africa, the Middle East and Central Asia (Gintzburger *et al.*, 1997; Nordblom *et al.*, 1997; Nordblom and Shomo, 1993).

Summarizing the Process from Biomass Measurements to Evaluating Stocking Rate

The general process from biomass measurements to evaluating stocking rate is summarized in Fig. 8.2. It must be performed in a homogeneous vegetation association on a specific site at peak standing biomass. In case the available rangeland feed is insufficient, the livestock owner must resort to supplementary feeding to maintain his flock in acceptable condition.

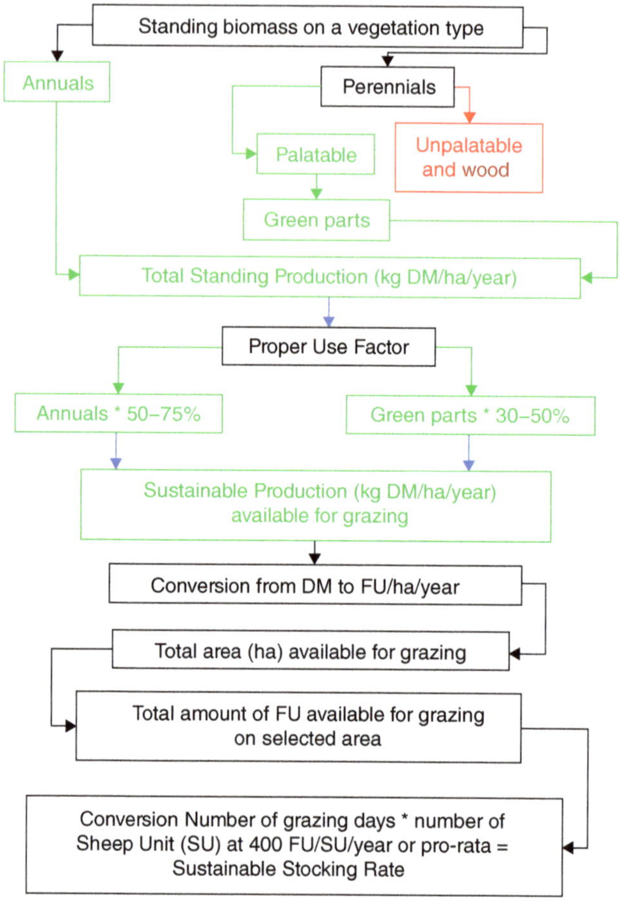

Fig. 8.2. Process from biomass measurement to stocking rate.

Are the standard carrying capacity, proper use factor and stocking rate concepts relevant to open grazing performed by Libyan shepherds?

The concepts of carrying capacity, stocking rate and proper use factors based on biomass measurements or evaluations commonly implemented in the USA (Heady, 1975, p. 19; Stoddart *et al.*, 1975), Europe and the Soviet Union may not be appropriate for Libyan conditions and Mediterranean zones where flocks are herded and wander over large open areas, not on set, well-defined areas with fenced perimeters. These concepts have been contested and reviewed by Behnke *et al.* (1993) with examples from sub-Saharan African flocks grazing open communal lands where rainfall and biomass are spatially and seasonally fluctuating from year to year. It applies to Libyan rangelands where the shepherd is constrained by the rainfall season, the seasonal rangeland condition on sites, the position of his base camp inside his vast allowed grazing territory (*watan*) and accessibility to water. The shepherd is the sole decision maker regarding where he leads his flock when evaluating the vegetation availability and quality. He does not perform biomass measurements, rely on chemical analysis or calculate stocking rates, but he is certainly guided by his knowledge of the local rangelands for the benefit of his flock and the return he is expecting. To our knowledge, there has been no long-term comparison between the performances of a flock led by a shepherd on open rangeland and a flock grazing on a set perimeter based on biomass measurement and subsequent evaluation of carrying capacity, stocking rate and proper use factor: a tremendously complex grazing experiment to set up!

Conclusions

Sustainable rangeland management by graziers necessitates field measurements or evaluation using the rain use efficiency coefficient of the standing biomass and production. These are tedious and lengthy field measurements and laboratory processes that have to be repeated year after year on different vegetation types as inter- and intra-rainfall is highly variable in amount and unevenly distributed in arid zones. Biomass measurements remain site- and date-specific and difficult to upscale to large areas. New technologies using satellite imagery and GIS allow assessment on very large areas. These technologies are progressing fast, requiring a high degree of computer skills, with complex satellite imagery preparation and processing. The proper vegetation index must be applied with timely field calibration, and essential final ground-truthing subsequently performed.

Without rangeland biomass and production measurements, proper and sustainable rangeland management remains hazardous, not counting land tenure problems. This information is necessary to evaluate theoretical carrying capacity, rangeland use factors, and practical stocking rates. Rangeland users and managers with long-practised field experience are able to assess standing biomass and possible stocking rate. These concepts developed by western rangelands experts may not be appropriate to shepherding in arid zones. Practical logistic position of base camp, water availability and the flock's ability to exploit finite rangeland areas within a few kilometres of base camp limit the shepherd's options. These options are even further limited by the current overall Libyan rangelands' poor and highly overgrazed conditions. They compel livestock owners to resort to increasing amounts of expensive imported complementary feed as less and less rangeland grazing is available.

References

Baret, F., Guyot, G. and Major, D.J. (1989) TSAVI: a vegetation index which minimizes soil brightness effects on LAI and APAR estimation. *12th Canadian Symposium on Remote Sensing Geoscience and Remote Sensing Symposium* 3, 1355–1358. DOI: 10.1109/IGARSS.1989.576128

Behnke, R.H., Scoones, I. and Kerven, C. (eds) (1993) *Range Ecology at Disequilibrium: New Models of Natural Variability and Pastoral Adaptation in African Savannas.* Overseas Development Institute, London.

Boutonnet, J.P. (1989) *La spéculation ovine en Algérie: un produit clé de la céréaliculture.* Institut National de la Recherche Agronomique, Montpellier, France.

Chehma, A. (2003) *Etude floristique et nutritive des parcours camelins du Sahara Septentrional Algérien: Cas des Régions de Ouargla et Ghardaïa.* Thesis, Université Badji Mokhtar.

Clements, F.E. (1916) *Plant Succession: An Analysis of the Development of Vegetation.* Carnegie Institution of Washington.

Demarquilly, C. and Weiss, P. (1970) *Tableaux de la valeur alimentaire des fourrages* (No. 42). INRA-SEI Série Etudes, Versailles, France.

Evenari, M., Schulze, E.-D., Lange, O.L., Kappen, L. and Buschbom, U. (1976) Plant production in arid and semi-arid areas. In: Lange, O.L., Kappen, L. and Schulze, E.-D. (eds) *Water and Plant Life: Problems and Modern Approaches*. Springer, Berlin, Heidelberg, pp 439–451. DOI: 10.1007/978-3-642-66429-8_27

Fahn, A. (1964) Some anatomical adaptation of desert plants. *Phytomorphology* 14, 93–102.

Fahn, A. and Sarnat, C. (1963) Xylem structure and annual rhythm of development in trees and shrubs of the desert. IV. Shrubs. *Bulletin of the Research Council of Israel*, 11D4, pp. 198–209.

Floret, C. and Le Floc'h, E. (1973) Production, sensibilité et évolution de la végétation et du milieu en Tunisie présaharienne, Projet – Parcours Sud, Doc. *CEPE*, 71.

Floret, C. and Pontanier, R. (1978) Études des Relations Climat-Sol-Végétation de Quelques Formations Végétales Naturelles du Sud Tunisien (Production, Bilan Hydrique Des Sols). IRA Médenine/CNRS Paris/CEPE Montpellier/ORSTOM Paris.

Floret, C., Le Floc'h, E. and Romane, F. (1981) Dynamique des systèmes écologiques de la zone aride: Application à l'aménagement sur des bases écologiques d'une zone de la Tunisie Présaharienne. *Acta Œcologica: Œcologia Applicata* 2(3), 195–214.

Floret, C., Pontanier, R. and Rambal, S. (1982) Measurement and modelling of primary production and water use in a South Tunisian Steppe. *Journal of Arid Environments* 5(1), 77–90. DOI: 10.1016/S0140-1963(18)31466-6

Floret, C., Le Floc'h, E. and Pontanier, R. (1983) Phytomasse et production végétale en Tunisie Pre-Saharienne. *Acta Oecologica: Oecologia Plantarum* 4, 133–152.

Gintzburger, G. (1986) Seasonal variation in above-ground annual and perennial phytomass of an arid rangeland in Libya. *Journal of Range Management* 39(4), 348–353.

Gintzburger, G. and Saïdi, S. (2009) From inventory to monitoring in semi-arid and arid rangelands. In: Squires, V.R. (ed.) *Range and Animal Sciences and Resources Management.* Vol. 2: *Encyclopedia of Life Support Systems*. EOLSS Publications, Oxford, UK.

Gintzburger, G., Nordblom, T. and Osman, A. (1997) Agro-pastoral systems, feed calendars and feed resources in the Arab countries of the Mediterranean region. In: Haddad, N., Tutwiller, R. and Thomson, E.F. (eds) *Improvement of Crop–Livestock Integration Systems in West Asia and North Africa*. International Center for Agricultural Research in the Dry Areas, Aleppo, Syria. Available at: https://agris.fao.org/search/en/providers/122511/records/6471fa9377fd37171a726820 (accessed 3 August 2024).

Gintzburger, G., Saïdi, S. and Soti, V. (2009) Rangelands of the Ravnina region in the Kara-Kum Desert of Turkmenistan: current condition and utilisation. In: *Sustainable Agriculture in Central Asia and the Caucasus Series, No. 2. CGIAR PFU*. Tashkent, Uzbekistan.

Gutman, M. (1978) Primary production of transitional Mediterranean steppe. *First International Rangeland Congress*, 225–228.

Heady, H.F. (1975) *Rangeland Management.* McGraw-Hill, New York.

Hunt, R. (1978) *Plant Growth Analysis.* Edward Arnold, London.

Le Houérou, H.-N. (1959) *Recherches écologiques et floristiques sur la végétation de la Tunisie méridionale.* PhD dissertation, Université Alger Institut de Recherches Sahariennes, Algiers.

Le Houérou, H.-N. (1969) *La Végétation de la Tunisie Steppique (Avec Référence aux Végétations Analogues d'Algérie, de Libye et du Maroc)*, Vols 1-2, N° 42/5, 622 p. + maps and annexes. Annales de l'Institut National de la Recherche Agronomique de Tunisie, Tunis, Tunisia.

Le Houérou, H.-N. (1975) Les pâturages naturels de l'Afrique du Nord: typologie, production, productivité et développement. *Proceedings International Symposium on the Evaluation and Mapping of Tropical African Rangelands*, 41–55.

Le Houérou, H.-N. (1981).*How to Calculate Livestock Conversion Factors on the Libyan Rangelands.* (UNTF/LIB 018; p. 15). FAO and Agricultural Research Center, Tripoli.

Le Houérou, H.-N. (1984) Rain use efficiency: a unifying concept in arid-land ecology. *Journal of Arid Environments* 7(3), 213–247. DOI: 10.1016/S0140-1963(18)31362-4

Le Houérou, H.-N. and Hoste, C.H. (1977) Rangeland production and annual rainfall relations in the Mediterranean Basin and in the African Sahelo–Sudanian zone. *Journal of Range Management* 30(3), 181–189. DOI: 10.2307/3897463

Le Houérou, H.-N., Gintzburger, G. and Khoja, N. (1982) *Chemical Composition and Nutritive Value of Some Range Plants and Fodder Shrubs of Libya.* (UNTF/LIB 018, Technical Paper No. 44). FAO and Agricultural Research Center, Tripoli.

Le Houérou, H.-N., Dumancic, D. and Eskileh, M. (1983) *Feeding Shrubs to Sheep. Intake, Feed Value and Performance.* (UNTF/Lib 018, Technical Paper No. 50, p. 60). FAO and Agricultural Research Center, Tripoli.

Leblond, B. (2004) Soil and vegetation optical properties. Available at: http://www.r-s-c-c.org/rscc/Volume4/Leblon/leblon.htm (accessed 3 August 2024).

National Research Council (1994) *Rangeland Health: New Methods to Classify, Inventory, and Monitor Rangelands.* National Academies Press, Washington, DC. DOI: 10.17226/2212

Nègre, R. (1974) Les pâturages de la région de Sirte (Libye): projet de régénération. *Feddes Repertorium* I(85–3), 185–243.

Nordblom, T.L. and Shomo, F. (1993) Livestock and feed trends in West Asia and North Africa: past, present and future. *CIHEAM – Options Méditerranéennes* 1(5), 15–30.

Nordblom, T.L., Goodchild, A., Shomo, F., Gintzburger, G. and Renard, C. (1997) Dynamics of feed resources in mixed farming systems of West/Central Asia-North Africa. In: Renard, C. (ed.) *Crop Residues in Sustainable Mixed Crops/Livestock Farming Systems.* CAB International, Wallingford, UK, pp. 131–148.

Pellant, M., Pyke, D.A., Herrick, J.E. and Shaver, P. (2005) *Interpreting Indicators of Range Health.* (Technical Reference 1734-6). U.S. Department of the Interior, BLM, National Science and Technology, Denver, Colorado.

Richardson, A.J. and Wiegand, C.L. (1977) Distinguishing vegetation from soil background information. *Photogrammetric Engineering and Remote Sensing* 43, 1541–1552.

Rouse, J.W., Haas, R.H., Schell, J.A. and Deering, D.W. (1974) *Monitoring Vegetation Systems in the Great Plains with ERTS.* Third Earth Resources Technology Satellite-1 Symposium. Volume 1: Technical Presentations, section A. Washington, DC. Available at: https://ntrs.nasa.gov/citations/19740022614 (accessed 3 August 2024).

Saïdi, S. and Gintzburger, G. (2013) A spatial desertification indicator for Mediterranean arid rangelands: a case study in Algeria. *The Rangeland Journal* 35(1), 47–62. DOI: 10.1071/RJ12021

Saïdi, S., Gintzburger, G., Bonnet, P., Daoud, I. and Alary, V. (2016) GIS-modelling of land-use trends: impact of drought in the Naghamish Basin (North-Western Egypt). *The Rangeland Journal* 38(6), 605–618. DOI: 10.1071/RJ16062

Shawesh, O.M. and Abuzakhm, A.T. (1993) Basic nutrient constituents for some range plants in Bir-al-Ghanam Range Project. *Libyan Journal of Agriculture* 4.

Sishov, L.L. and Kharin, N.G. (1980) *Soil and Geobotanical Studies in the Pasture Zones of the Socialist People's Libyan Arab Jamahiriya* (Vol. 3). Selkhozpromexport, Nauka, USSR.

Stoddart, L.A., Smith, A.D. and Box, T.D. (1975) *Range Management,* 3rd edn. McGraw-Hill College, Blacklick, Ohio.

Suleiman, S.A. (1981) *Phytoécologie des milieux salés de Taworga (Libye).* PhD dissertation, Université St-Jérôme, Marseille, France.

Tadmor, H.N., Eyal, E. and Benjamin, R.W. (1974) Plant and sheep production on semi-arid annual grasslands. *Journal of Range Management* 27, 427–433.

Telahigue, T., Floret, C. and Le Floc'h, E. (1987) Succession post-culturale en zone aride de Tunisie. *Acta Oecologica: Oecologia Plantarum* 8(22), 45–58.

Thalen, D.C.P. (1979) *Ecology and Utilization of Desert Shrub Rangelands in Iraq.* Springer, Dordrecht, The Netherlands. DOI: 10.1007/978-94-009-9622-9

Van der Veen, J.P.H. (1967) Preliminary results of a grazing trial in the Syrian Steppe. *Netherlands Journal of Agricultural Science* 15(3), 198–206. DOI: 10.18174/njas.v15i3.17438

9

The Rainfed Ley Farming System on Rangelands in Western Libya

Abstract

The Australian ley farming system, based on a cereal-annual *Medicago* sp. pasture rotation, was implemented on 50,000 ha of the best rangelands in western Libya from 1973 to 1984. The rotation improved rainfed cereal yield and sheep production compared to traditional cropping and sheep rearing. It was successful as long as annual rainfall was about 250–275 mm and that the West Australian technical support and commercial medic seed supply remained available. The system was difficult to manage due to low and erratic rainfall. It requires constant agricultural expertise and training of Libyan colleagues in modern machinery, land preparation, seeding techniques, fertilization, harvesting, post-harvesting, and sheep-grazing management. The ley farming system technology transfer to private farmers eventually failed due to lack of community participation, small farm sizes and the complexity of the input system. It was ultimately abandoned for more profitable agricultural options like irrigated vegetable and fruit tree production or simply reverting to customary free rangeland grazing.

Introduction

Early in 1970, Libyan oil money was intensively channelled towards agriculture to reach food self-sufficiency. Various projects, including rangeland development, afforestation, sand dune fixation, irrigation projects, poultry, dairy, cattle, sheep, dromedary production and, most important, rainfed cereal cultivation were established under the executive authority in the western and eastern regions of Libya. More specifically, a greater effort on cereal cultivation and meat production was launched to meet people's expectations for better living conditions. Despite approximately 115,000 ha of rainfed wheat, 58,000 ha under irrigation, twice this area under barley (280,000 ha rainfed, 140,000 ha irrigated (1974 season)) with an average production of 0.3 tons wheat/ha and 0.6 tons barley/ha, the Libyan agriculture was not able to satisfy the national demand (ARC, 1977). Furthermore, demand for local meat, especially lamb and mutton, was drastically snowballing with the increase in the Libyan population and its wealth. Quite sensibly, the Libyan government turned towards Australia, having a similar environment and wide experience in the integrated cereal-sheep system known as the ley farming system.

The Australian Ley Farming System

It was conceptualized and implemented in South Australia by 1930 and later in Western Australia (Puckridge and French, 1983). It is a rotation of self-regenerating annual legumes with cereals, minimum tillage and applications of superphosphate. Annual legumes, naturally fixing

DOI: 10.1079/9781800627154.0009

nitrogen, improved soil fertility leading to increased cereal yields and sheep production in an integrated crop-livestock system. The value of annual legume pastures with *Trifolium subterraneum* and various *Medicago* species was recognized (Byerlee, 2023). Both species were accidentally introduced by the early colonizers in Mediterranean southern Australia, naturally expanding as weeds and their agricultural value was soon accepted. Australian farmers from low-rainfall zones rapidly adopted the rotation of annual legumes-cereals and integrated sheep rearing with the minimum tillage technique. The ley farming system was born, exported and tested in Morocco, Algeria, Tunisia, Libya, Syria, Turkey, Jordan, Iraq, South America and Chile (Ates *et al.*, 2013; Cocks, 1988). Annual legumes seed genetic resources collections were subsequently conducted in the Mediterranean regions, especially in Libya, by Australian agronomists (Francis, 1978; Gintzburger, 1978, 1980; Gintzburger *et al.*, 1983; Gintzburger and Blesing, 1979; Gintzburger and Kernick, 1979) to find drought-resistant ecotypes suitable for Libyan arid conditions.

But was the Australian ley farming system really a new agricultural system?

Traditional Rainfed Cereal Cropping in Libya

From time immemorial, prior to launching the ley farming system, North African and Libyan farmers practised an integrated crop-livestock system, feeding sheep, goats and dromedaries from rangeland and fallows in winter-spring, and cereal crop grain and residues during the dry season. Long before ley farming was developed in Australia, a similar system was practised by Mediterranean farmers cropping rainfed cereals, then running a fallow year rich in annual legumes to feed their flocks. To till their land, they used a *mihrath* (Photo 9.1), a minimum-tillage wooden plough

Photo 9.1. The traditional Libyan plough (*mihrath*) pulled by a donkey, a horse or a dromedary. (Photo: Misurata-Zliten)

pulled by a donkey or a dromedary, just scratching the soil, avoiding burying and destroying annuals and small legume seeds. But the antique *mihrath* was progressively abandoned due to the introduction of modern farm machinery by the Italian colonizers. By the mid-1970s, traditional rainfed cereal cropping was a fairly simple process. Small tractors were largely available but modern seeders were rare. Land preparation before seeding was basic using a three-offset disc plough to remove the native vegetation, leaving the soil surface rough and uneven with uprooted bushes. Fertilization was rarely practised due to limited availability and because of the unsecure land tenure system. The cereal seeds were hand-broadcast at high seeding rate and covered with a simple harrow or by criss-crossing the field with the disc plough. The seeding depth was uneven and germination emergence was erratic. The yield remained low. As modern harvesters were few, the crop was mostly hand-harvested and threshed using stationary Marinoni threshers. It was time to test the move to the modern ley farming system.

Introducing the Ley Farming System to Western Libya

Within the Ministry of Agriculture and Land Reclamation, two Regional Executive Authorities respectively in charge of agricultural development of the Jeffara and the Jebel Al-Akhdar regions were to tackle rainfed cereal and sheep production under integrated rainfed projects. The Libyan government invited Western and Southern Australian expertise (Allen, 1979) to help develop the ley farming system in western and eastern Libya, respectively, where rainfed cereal cultivation could technically be expanded and improved. This new system for Libya, but practised on millions of hectares in Mediterranean Australia, had to be adjusted to lower-rainfall zones unlike that encountered in Australia. However, if the 350–450 mm zone and heavy soils in the Jebel Al-Akhdar region were not a new story for South Australian expertise, it was a challenging environment for the West Australians in the sandy Jeffara plain where the ley farming system had to be established in rainfall zones as low as 150–200 mm/year; a difficult task as these lower rainfall zones render rainfed cereal cropping highly unreliable, more so for wheat than for barley. Cool winters were not considered a problem as frost was rare both in the western and eastern zones, though not unknown.

This chapter deals only with the implementation of the ley farming system on the cereal projects on new sites suitable for large-scale modern agriculture practices and for machine seeding and harvesting in the Jeffara plain of Western Libya. A similar approach was developed at El-Marj, in the Jebel Al-Akhdar region of eastern Libya (MacPhie, 1980).

Introducing the ley farming system in the Jeffara plain was to be achieved through the Australian package technology and know-how. Therefore, from 1974 to 1984, an Australian group comprising one or two officers of the Western Australian Department of Agriculture, seven or eight farmers and a mechanical service team was present on site, on five-year contracts, to support and train Libyan counterparts in all stages of land development and ley farming implementation and management. By then, Libyan agriculturalists would have been confident to run all ley farming operations established in the Jeffara plain.

Land clearing and project establishment: 1973–1974

Preliminary surveys took place to locate potential sites and areas for establishing new agricultural projects. In the case of rainfed cereal-based projects, the obvious target sites were either wadi flats and depressions or large tracts of rangeland within adequate rainfall zones, i.e.:

- wadi beds and large depressions located in the 175–200 mm rainfall zones but benefiting from run-off from large water catchment areas onto fertile alluvial deposits. These areas were occasionally cropped by semi-nomadic populations. This was the case of the wadis Abu-Shayba, El-Hira, Diga, Ethel and Bir-el-Ghanem, all located in the Jeffara plain;
- flat or gently undulating sandy/sandy-loam rangeland close to the coast, receiving at least an average 200 mm but not developed

as lacking underground water for irrigation. Sites in the Jeffara plain such as Abu-Aisha and Agheilat, and in the Jebel Nafusa, at Jendouba and south Tarhuna (wadi Martha), were also identified.

Nine sites (Map 9.1; Table 9.1) covering about 50,000 ha were allocated to the Jeffara Plain Authority under the supervision of the Cereal Project. One should notice that these new cereal project sites were primarily aimed at reclaiming government agricultural land not otherwise permanently used by landowners. Prior to project establishment, these sites were mainly rangelands devoted to open grazing and occasional cropping of rainfed cereals by local communities.

In the Jeffara plain, natural vegetation consists mainly of gently undulating sandy/sandy loam country of low fixed dunes covered with perennial grass *Stipagrostis pungens* and bushes of *Retama raetam* (Gintzburger, 1986).These were not targeted as potential cereal land development and remained assigned to new rangeland projects. The Cereal Project instead targeted sandy rangelands at Agheilat in the inter-dunes flats, large

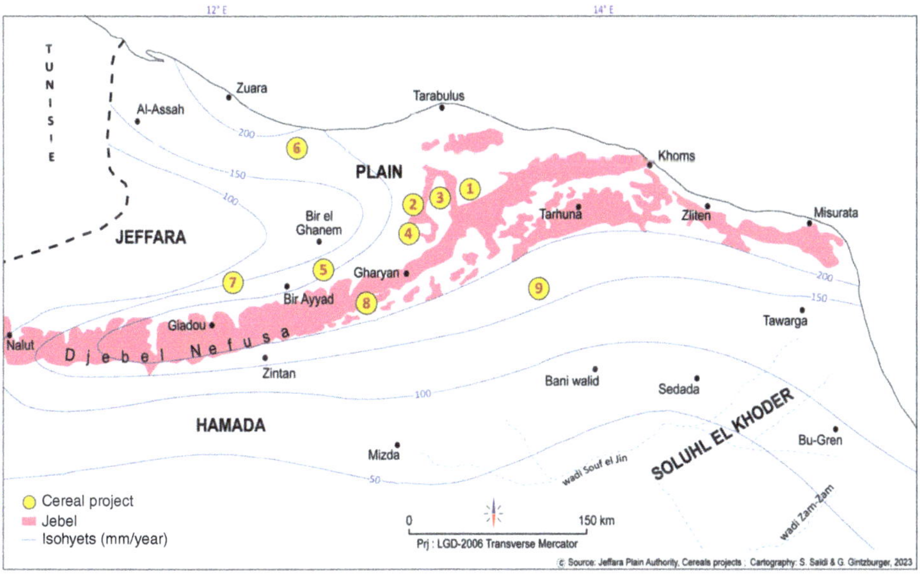

Map 9.1. Cereal Project locations in western Libya in 1978.

Table 9.1. Cereal project sites under the Jeffara Plain Authority (planned in 1974; see Map 9.1 for location).

Site 'Cereal Project' Western Libya	ha	Rainfall (mm/year)	Soil type
1) Abu-Aisha	3090	200–250	Sand
2) Wadi El-Hira	4700	175–200	Sandy loam
3) Wadi Diga	3000	175–200	Sandy loam
4) Wadi Abu-Shayba	9250	175–200	Sand
5) Bir-el-Ghanem	5000	150–175	Loam
6) Ageilat	5750	175–200	Sand
7) Wadi Ethel	9550	150–175	Loam/clay
8) Jendouba	4690	300–350	Loam
9) South Tarhuna	5275	150–175	Loamy sand
Total (ha)	**50 305**		

depressions receiving run-off usually covered with low bushes of *Rhanterium suaveolens* (wadis Abu-Shayba, Hira, Bir-el-Ghanem) on sandy-loam soils or with *Artemisia herba-alba* – *Arthrophytum scoparium* on loamy soils (wadi Ethel, South Tarhuna). Native vegetation was cleared, large sandy nebkha of spiny *Ziziphus lotus* were uprooted and levelled in the wadi beds to allow for operations from large modern farm machinery. A combination of deep-rooting implements, subsoiling, heavy offset disc plough and crawler tractors (Allen, 1979) were used for land preparation (Photos 9.2 and 9.3).

Photo 9.2. Deep-uprooting implements and subsoiling for rangeland clearing of *Retama raetam* dense shrubs, 1976, at Abu-Shayba.

Easier land clearing using heavy-disc ploughs (Photo 9.3) was possible on the flat sandy site (Agheilat) covered with sparse *Artemisia campestris*, a post-cultural low-shrub vegetation (Le Houérou, 1959, 1969). In most cases, organic matter was buried deep during land clearing. On sandy soils, wind erosion was immediately controlled with cereal seeding using cultitrash seeders and with minimum fertilization. Cereals were cropped in the first instance and medics sowing started soon afterward ensuring a stable soil and surface organic matter build-up. Sandy sites at Abu-Aisha, Diga and Al-Hira required reseeding most years due to strong wind erosion.

Medic and cereal establishment progress: 1974–1978

After a slow start due to ongoing training of Libyan agriculturalists and land-clearing operations (Table 9.2), the first 300 ha of medics were sown to start in the 1974/75 season. The

Photo 9.3. Heavy-disc ploughs used to level sandy soils after rangeland vegetation clearing, 1976, at Abu-Shayba.

Table 9.2. Medics sown (ha) and cereals under the supervision of the Cereal Project (Jeffara Plain Authority, Libya, 1973–1983). (After Allen and Ewing, 1978; Ewing, 1978; Richardson, 1983)

Winter season	1973/74	1974/75	1975/76	1976/77	1977/78	1982/83
New medic sown	0	300	5252	8500	7700	645
(Initially planned)						(3316)
Medic top-dressed or resown	0	0	300	5137	9185	18,840
Medic area cropped with cereals	0	0	0	30	5092	?
Total medics	**0**	**300**	**5552**	**13,667**	**21,977**	**19,485**
Wheat	14,358	30,984	44,560	29,783	23,926	23,578
Barley	7847	8184	8845	4766	5205	3227
Oats	0	266	2832	2735	1704	1095
Total cereals	**22,205**	**39,434**	**56,237**	**37,284**	**30,835**	**27,900**
Total area Cereal Project (ha)	**22,205**	**39,734**	**61,789**	**50,951**	**52,812**	**47,385**

following season, some 5250 ha of medics were established but no medic area was cropped with cereal. By 1976/77, an additional 8550 ha of medics were sown, the 1974/75 - 5250 ha of medic regeneration were top-dressed with super-phosphate, while only 30 ha were cropped with cereals. It was considered that two successive years of medics were necessary to acquire a large medics seed bank. It was only during the 1977/78 season that 5000 ha of medics, established during the two previous seasons, were cropped into cereals. That was the beginning of a large-scale ley farming operation in western Libya. Gradually, paddocks with Australian sheep-proof fences were installed on most sites to start grazing the medics pasture.

By 1978, the Cereal Project had about 22,000 ha of medics established, mostly with the Australian cultivar 'Harbinger' on sandy soils and 'Cyprus' on loamy soils. These 22,500 ha were divided into 5000 ha cropped with cereals, 4455 ha in medic regeneration (2nd year pasture) and 13,515 ha in 1st year medic pasture establishment (Allen and Ewing, 1978; Halse, 1978).

Distributing new land to private farmers: 1980–1984

By 1978, the Jeffara Plain Authority Cereal Project site was to be progressively subdivided into private farms of 100–300 ha depending on rainfall (Halse, 1982). An annual return of LYD 2200[1] per farm was projected based on average cereal yield and sheep carrying capacity on each site without the need for imported supplements (Watson, Perth, 2023, personal communication). Individual farmers were starting to test the ley farming system. By 1983, 43 farms were operating at Jendouba, 17 at wadi El-Hira, 16 at wadi Diga, and 25 at wadi Abu-Shayba. The Cereal Project had stopped all operations and farmland allocation by 1984.

Plant Material Used with the Ley Farming System in Western Libya

Cereals

Wheat and barley were sown on most sites despite the expected low rainfall more appropriate to rainfed barley. Superior grain yield of 'Gameyna' wheat was consistently outlined with three-year experiments on selected sites (1974–1978), followed by 'Gambee' yielding 7% less on average (Ewing, 1978). The most locally used varieties, 'Florence Aurora' and 'Sidi Mesri', yield 17% less than 'Gameyna'. Among barley, 'California Marriout' yields 10% more grain than the reference 'Athenais'. However, the six-row 'Beecher' barley was the best performing variety on the project when compared with the named varieties, as well as 'Mahmoudi', a black-awned local six row barley, while oats such as 'Swan' were planted for hay and seed production at most wadi sites (Watson, Perth, 2023, personal communication).

Best yield with 'Gameyna' was achieved with a seeding rate of 60–80 kg/ha sown at a depth of

3–5 cm by mid-November. Local farmers tend to use higher seeding rate up to 150 kg/ha to counter establishment problems due to poor soil preparation and seeding depth adjustment.

On clean and perfectly levelled project sites, best seeding operations are with tine seeders achieving accurate seed depth control. However, early seeding operation on land newly cleared always gave better yield with a disc seeder less susceptible to catch residual native vegetation (Ewing, 1978).

Weed competition in cereal crops was a recurrent problem with native annual grasses such as *Lolium rigidum*, *Cutandia* sp., and various Brassicaceae (*Brassica* sp., *Sinapis* sp., *Diplotaxis* sp., *Carrichtera annua*, etc.) and even medics. Weed control was achieved with various levels of success with cultivation after opening rain or with selective post-emergent spraying.

Large-scale yield (Table 9.3, Fig. 9.1) varies from site to site, soil texture and year to year and is indeed lower than achieved on experimental results. Ewing (1978) quotes potential figures of water-use efficiency of 8–9 kg/ha of grain production per mm of winter-spring rain on rainfed experiments in the Jeffara plain, and on large-scale operations, 0.3–2.8 kg/ha per mm at wadi Abu-Aysha for the 1977/78 season. By 1982/83, these latter figures improved slightly due to better management, improved soil fertility, adequate fertilization, and better growing seasons. Best-run farms reach experimental yield of 6–7 kg grain/mm of winter rain at wadi Abu-Shayba and at Diga. However, the average

Table 9.3. Large-scale wheat grain yield recorded on three Cereal Project sites in the Jeffara plain (1973–1983).

	Wadi El-Hira		Wadi Abu-Shayba		Wadi Ethel	
Season	kg/ha	mm rain	kg/ha	mm rain	kg/ha	mm rain
1973/74	224	?	37	?	10	?
1974/75	635	?	259	?	177	?
1975/76	240	?	219	?	281	?
1976/77	30	153	107	149	0	86
1977/78	341	262	625	225	62	144
NA	?	?	?	?	?	?
1982/83	846	227	703	228	422	210

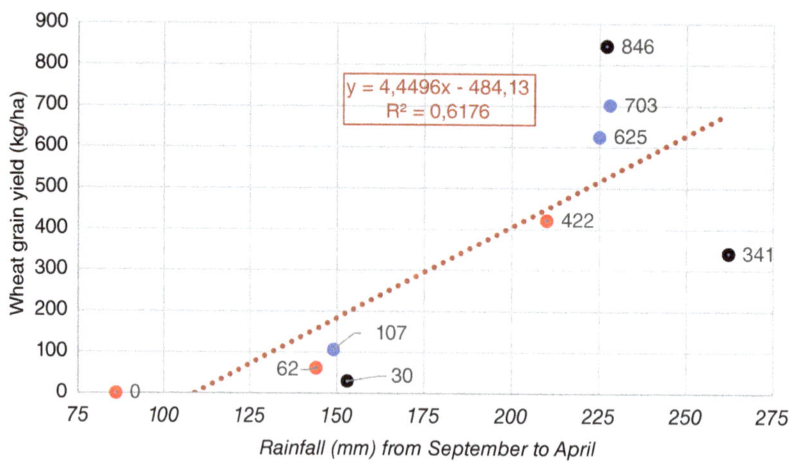

Fig. 9.1. Large-scale wheat grain yield (kg/ha) in relation to winter-spring rainfall (mm) at wadi Ethel, wadi Abu-Shayba and wadi El-Hira (1973–1983).

final cereal yield on all project sites and farms is 3.3 kg grain/mm (Richardson, 1983), mostly due to losses during post-harvest handling.

From Fig. 9.1, it is evident that wheat grain yields are close to or under 100 kg/ha, barely the seeding rate, with seasonal rainfall below or close to 100 mm/year at wadi Ethel and climbing to a poor 400–800 kg/ha with 225–275 mm/year at Abu-Shayba and El-Hira. One just wonders why wheat was grown on low-rainfall Cereal Project sites receiving less than 300 mm MAR, a long-known lower limit for sustainable but still risky rainfed wheat cropping in North African arid zones (Le Houérou, 1969). We would have thought that barley would have been a more rational choice considering it is regularly rainfed-cropped in Libya down to 175–200 mm MAR. Unfortunately, we do not have large-scale barley grain yields available from the Cereal Project records.

Wadi Ethel (150–175 mm MAR and silty soils) has the lowest wheat grain water-use efficiency (WGWUE) of 0.00–2.01 kg/ha per mm of rain, while Abu-Shayba and El-Hira sites (sandy soils), usually receive a better, but marginal, 175–200 mm MAR. Their WGWUE ranges from 0.72 to 3.73 kg/ha per mm of rain with an exceptional 0.20, not surprising with only 30 mm during the growing season. In low-rainfall environments below 300 mm MAR, the soil texture is critical as rainfed crops on sandy/sandy loam soils always perform better than on loamy-clay soils.

Cropping on the Cereal Project sites at wadi Ethel and South Tarhuna in the 150–175 MAR on heavy soils proved to be unreliable and were slowly abandoned and returned to rangeland projects. The Ageilat sandy Cereal Project site was also abandoned due to wind erosion and turned into small, irrigated farms. One should note that nowhere else in the world does large-scale rainfed cereal production successfully operate below the 250 mm isohyet. As such, the Libyan experience with the Cereal Project in the Jeffara plain is unique and definitely confirms that cereal cropping below the 200 mm MAR in North Africa is adventurous and remains a risky enterprise.

Annual medics

Australian agronomists early identified medics as the only annual legumes to be used for the ley farming system in the Jeffara plain due to alkaline soils and, above all, the very low rainfall conditions and short growing seasons encountered on project sites. This was confirmed with visits and preliminary seed collections on Libyan sites highlighting the overall occurrence of native annual medics in the northern coastal region of Libya (Francis, 1978; Gintzburger, 1980, 1983; Gintzburger and Blesing, 1979). Various subclovers (*Trifolium subterraneum*) are the other self-reseeding annual legume used in the Australian ley farming system on acid soils and rainfall above 300–350 mm/year. Subclovers would not survive on projects even in higher-rainfall zones of the Jebel AL-Akhdar region; even the cultivar 'Clare' that is able to thrive on neutral soils. Growing seasons were, even there, too short and rainfall too erratic to allow proper development, seed setting and regeneration of subclovers. Other subclovers, though early maturing, are not adapted to alkaline soils.

One prerequisite to using annual medics was readily available large commercial seed quantity for large-scale seeding operations. This was only possible with medics' seeds produced in Australia. A wide range of commercial registered annual medics from the Register of Australian Herbage Plant Cultivars (CSIRO) (https://research.csiro.au/cultivars/legumes/ [accessed 8/1/24]) was initially tested on several sites both in Jeffara plain and in the cooler environment of the Jebel Nafusa. The hunt for appropriate early flowering/maturing medics started.

Need for early flowering-maturing annual medics

Preliminary tests in the western Libyan region confirmed the need for very early flowering cultivars completing their cycle from seeding to pod maturity in around 120–130 days, i.e. flowering in 95–105 days (Gintzburger, 1983; Gintzburger *et al.*, 1984a). Both cultivars – 'Harbinger' on sandy soils and 'Cyprus' on loamy soils – performed reasonably well mostly in the Jeffara plain. One should notice that cv. 'Harbinger', identified as a *Medicago littoralis*, was also performing well in a similar environment on sandy soils of Western Australia. On heavier soils of the Jebel Nafusa, best vegetative and seed production was achieved with *Medicago truncatula* cv. 'Cyprus', still flowering in 95–100 days at Tarhuna.

Cultivar 'Jemalong' (*M. truncatula*), as well as other cultivars such as 'Tornafield' and 'Murrayland' (*Medicago tornata*) were flowering too late, in 125–130 days, and never had their place in western Libya. A new spineless *Medicago tornata* ssp. *rugulosa* cv. 'Swani', collected from Suani-Aziziya (Khalil *et al.*, 1977), adapted to sandy soils and flowering earlier than 'Harbinger', seems to have a limited use to the very sandy coastal strip of the Jeffara plain. *Medicago polymorpha* cv. 'Circle Valley', though early flowering, always faced nodulation problems on sandy soils and needed specific inoculation before seeding, thus was not in favour.

The Libyan Medic Collection and the Libyan Medic Evaluation Project

The overall presence of a wealth of native annual legumes and especially annual medics in Libya led the Agricultural Research Centre to organize an Indigenous Forage Legume Collection with the support of FAO and IBPGR (Gintzburger, 1980). This collection gathered a large genetic basis of native annual medics for further evaluation and integration in the Libyan farming systems. Collecting missions took place in 1976, 1977 and 1978 all over northern Libya from the coast into the desert. The agronomic evaluation of this large genetic pool of native medic ecotypes collected from 1800 Libyan sites commenced in 1980 under the Libyan Medic Evaluation Project (LMEP) operating both in Western Australia and Libya to take advantage of running two seasons per year, alternately in both hemispheres.

The objectives of the LMEP were:

- select an alternative to cv. 'Harbinger', having a higher seed set, better or equivalent vigour, more persistence under the soil and climatic conditions prevailing in western Libya;
- select medics for ultra-low, i.e. rainfall zones (approx. 150 mm/year), for rangeland regeneration/rehabilitation purposes; and
- train Libyan counterparts in plant evaluation and all stages of seed production of medics.

By 1984, the Libyan medic ecotypes were taxonomically identified, documented and evaluated (Gintzburger *et al.*, 1984a) at the WA Agricultural Medina research station South of Perth in Western Australia. This was done with the active laboratory and fieldwork of three Libyan agricultural engineers (A. Sbeita, M. Fituri and A. Rojbani) assigned to the project.

Passport ecologic and morpho-agronomic information were recorded and available with the Libyan Medic Database System (LMDBS), an interactive computer system (Webb *et al.*, 1984) facilitating the early selection process and agronomic ranking.

- The passport information (40 descriptors) containing the ecological description of the Libyan collection sites was published by CSIRO (1985).
- The morphologic information (25 descriptors) details, for each unique plant, accession identity and morphologic characters (florets per peduncles, leaf characteristic, pod characteristics, pod's coiling direction, pod's spininess, seed per pods, etc.).
- The agronomic data (23 descriptors) contain the summary of experimental results obtained annually over three years at different sites either in Western Australia (Medina, Merredin and Tenindewa) or in Libya (Tajura, Bir-el-Ghanem, Oushtata and South Zliten).

The most promising ecotypes were swiftly transferred and evaluated in the field in Libya (Photo 9.4), taking advantage of opposite growing seasons in Australia (southern hemisphere) and Libya (northern hemisphere), hence benefiting from five growing seasons during the three years of agronomic evaluation.

Among the 10% outstanding medics preselected and screened from ungrazed small plots valuation, 79 of them appear to be potential replacements for 'Harbinger' in terms of flowering earliness, winter vigour and seed setting. The outstanding contribution of ecotypes originating from low-rainfall zones from the Sirte and the Solul-El-Khoder regions were noticed (Gintzburger *et al.*, 1983). The best-bet selection LIB 1493.3/4 (later named 'El-Wassat') was from a native hybrid from *M. truncatula* ssp. *tricycla* x *M. littoralis*, and LIB 1484.1 (*M. tornata* ssp. *rugulosa*) went into large-scale seed increase under irrigation on 20 ha at wadi Diga and Bir-el-Ghanem sites in 1982/83. Later, Engineer Adnan Sbeita with colleagues from the Range Development

Photo 9.4. Testing the Libyan annual *Medicago* sp. collection at Abu-Shayba (1981–1982, with Engineer Adnan Sbeita and Dr Abubaker Maddur).

Project (Suani) kept producing medic seed of the best selected ecotypes at different irrigated sites in the Jeffara plain and at wadi Sassu. Unfortunately, due to lack of finance and logistical support, the pods and seeds harvested and stored were never used, either on large-scale rainfed cropping or rangeland projects.

Was there a need for medic inoculation prior to seeding in the Jeffara plain?

Despite the absence of published experimental results, the need for commercial seed inoculation was never demonstrated on standard Australian cultivars such as 'Harbinger' and 'Cyprus' established in Jeffara plain. This is certainly due to the wide distribution of native *Medicago littoralis* and *Medicago truncatula* in Libya and especially in the western and central zones (Francis, 1978; Gintzburger and Blesing, 1979).

However, Francis (1982) and Sbeita (Tripoli, 1981, personal communication) reported nitrogen deficiency and poor growth of non-inoculated *Medicago polymorpha* ssp. *inermis*, especially with LIB 793.4, an outstanding LMEP ecotype grown on sandy soil at the Tajura ARC research station, while all other medics tested in similar conditions nodulated naturally and grew well without displaying nitrogen deficiency. Contrary to *M. littoralis* and *M. truncatula* ubiquitous on sandy-loam soils in Libya, *M. polymorpha* occurrence in western Libya is restricted to loamy-clay soils often in winter water-logged depressions such as small depressions between Aziziya and Ras-el-Loefa, and are known to have very specific inoculum requirement.

During the LMEP in Western Australia, early field evaluation of Libyan medics inoculated with the standard commercial U45 inoculum performed satisfactorily. Later, the use of an allegedly better Australian commercial rhizobia strain identified as CC169 proved to be a poor choice for most of the Libyan *M. littoralis* but passably working with Libyan *M. truncatula*, *M. tornata* and on all Australian medic cultivars (Gintzburger *et al.*, 1984b). In short, there was no real need to inoculate the Libyan medics when used on soils of north-western Libya.

Medic vegetative production

All experimental work (Ewing *et al.*, 1979) stresses the need for phosphate fertilization (100–200 kg/ha 18–20% superphosphate) to stimulate medic growth and seed production on soils of the Jeffara region highly deficient in phosphorus. Seeding rate ranges from 10 to 14 kg/ha clean scarified inoculated medics seeds. Best seeding depth is achieved at 3–5 cm depth on light sandy soils to avoid seed exposure by wind erosion during the early germination stage. On heavier soils, best seeding depth is 1–3 cm. Ewing *et al.* (1979) report medics vegetative dry matter production ranging from 1500 to 2500 kg/ha with 150–175 mm or 175–200 mm of well distributed rainfall during the growing season on sandy soils or on heavier soils. Such highly nutritive dry matter medic production would allow 1–1.8 dry sheep equivalent /ha/ year with a 50% use factor of the standing biomass. In comparison, a nearby unfertilized and protected rangeland would produce 400–500 kg DM/ha/year of annual plants (Gintzburger, 1986) and proportionally allows far lower stocking rate.

Medic seed production and regeneration

Annual medic seed production and regeneration are crucial for the sustainability of the ley farming system based upon hard-seeded annual legumes. It is usually considered that a medic seed stock of about 200 kg/ha is necessary to have a good regeneration of the medic pasture in the ley farming system in Australia.

Allen and Ewing (1978) stated that 'Harbinger' sown at 10 kg clean scarified seed/ha produced 50 kg medics seed/ha at Agheilat with 135 mm rainfall during the 1977/78 season and 240 kg seed/ha at wadi Abu-Shayba with 225 mm during the same season. In their recommendations, they stress the point that there was little hope of getting an adequate medics seed bank (150–200 kg/ha) established on sandy soils receiving less than 150 mm rainfall/ growing season. They also indicated that a minimum rainfall of 175 mm/growing season was required on heavy soil sites such as at Bir-el-Ghanem and wadi Ethel. The best medics regeneration was obtained in the Jebel Nafusa at the Jendouba site (1480 ha), on the highest rainfall site, which received 266 mm and 241 mm for the 1976/77 and 1977/78 seasons, respectively (Allen and Ewing, 1978). Most other sites had barely adequate regeneration on about 400–500 ha following poor rain (80–170 mm) in 1976/77 and a better season in 1977/78 (225–300 mm). No, or poor, regeneration occurred at Agheilat, wadi Ethel and Bir-el-Ghanem with seasons receiving less than 150 mm. The 150 mm isohyet appears to be the lowest limit for any reliable and sustained cropping on land not receiving run-off in the Libyan Mediterranean environment as indicated earlier by Le Houérou (1969). However, after a few years' experience, Halse (1978, 1982) quoted good medics (likely cv. 'Cyprus') regeneration despite a poor 80 mm post-germination in 1981/82 at the Tarhuna site, which grew medics with only 130 mm in 1979/80. The site was abandoned and later cropped with barley in 1980/81. He made a similar observation at the Bir-el-Ghanem site on medics first sown in 1975/76. These sites may have benefited from significant water harvesting on low land.

Sheep Production: Animal Performance and Stocking Rate

Within the 150–200 mm rainfall isohyets, the best Libyan rangeland may carry about one dry sheep (fat-tailed Barbary breed) per 4–6 ha/year with 30–50% supplementation as barley grain by the end of summer/early autumn before lambing starts. Overgrazing and overuse of rangeland is frequent as settlement becomes the rule (Gintzburger, 1986; Gintzburger and Bayoumi, 1977), especially when newly cropping land further reduces rangeland availability.

Proper medic pasture stocking rate within the ley farming system established in Libya had to be ascertained as deficient medic pasture management may lead to overgrazing, inducing catastrophic wind erosion (Ewing *et al.*, 1979). Therefore, Barbary sheep performances were experimentally assessed (Lightfoot, 1988). An 80 ha established medic pasture (Ewing *et al.*, 1979) was divided into four paddocks (11.1, 14.3, 20 and 33.3 ha). Each paddock was grazed with 50 ewes, 15 months old, with average body weight about 34 kg/head; hence, stocking rates of 1.5, 2.5, 3.5 and 4.5 ewes/ha. The experiment started in early February 1978. Measurements on sheep and the pasture were carried out. All sheep started grazing their respective medic paddocks which averaged 3000 kg DM/ha (difference initially not significant between stocking rate) by May 1978 after the site had received a total of 225 mm. To compare this with a close-by site on the BGII Bir-el-Ghanem Grazing Project, native rangeland vegetation receiving a similar rainfall yielded about 2000 kg DM/ha of which 460 kg/ha were from various annuals (Gintzburger, 1986).

All sheep gained weight throughout summer and reached 46–49 kg by September when they started losing weight. Supplementary feeding was then allowed as residual medic phytomass was measured (October 78) indicating a sharp decline in available forage and medic burrs (Table 9.4), especially with the high stocking rate. Considering that medic burr of cv. 'Harbinger' contains approximately 30% seeds in weight, one could question the regeneration potential of the medic paddocks with a barely adequate seed stock of 140–150 kg/ha of seed left by the end of the first season on any of the tested stocking rates. No further indication is given of the pasture regeneration in the following year. By the end of August, the highest stocking rates (3.5 and 4.5 sheep/ha) were subject to wind erosion due to overgrazing (Photo 9.5).

Considering that Harbinger burrs contain about 30% of seed per burr weight, the medic seed stock on all stocking rates would barely reach 150 kg/ha to allow for a successful medic pasture regeneration the following season, even at the lower stocking rate.

Supplementary feeding was then given to all sheep. By February 1979, after one year of experiment, the animals had received a total of 3.5 kg and 5.5 kg barley grain/ewe for the 1.5 and 2.5 ewe/ha treatments, respectively; higher stocking rates required 9.3 kg and 27.8 kg barley grain/ewe to keep the animals in good condition

Table 9.4. Residual forage of medic pasture and seeds by the end of October 1978 at wadi Abu-Shayba grazing experiment (Ewing *et al.*, 1979).

	Stocking rate (ewes/ha)			
	1.5	2.5	3.5	4.5
Forage (kg DM/ha)	1420	1015	1055	615
Medic burrs (kg DM/ha)	440	340	210	320
Total phytomass (kg DM/ha)	1860	1355	1265	935
Residual Medic seeds (kg/ha)	132	102	63	96

Photo 9.5. Stocking rate experiment at wadi Abu-Shayba (end of August 1978) (*right:* high stocking rate 4.5 ewes/ha; *left:* low stocking rate 1.5 ewes/ha).

on their respective paddocks, certainly indicating a deterioration and a poor regeneration of the medic pasture during the second season that received only 200 mm rainfall.

No data are given on lamb growth, however, during the same season, the Cereal Project commercial flock grazing exclusively on medic pastures reported average growth rate of 160 g/lamb/day over a period of 100 days from birth to weaning in the lowest rainfall zones where medics were poor, to 317 g/lamb/day in the highest rainfall zones with good medic pastures (Ewing, 1979). Lightfoot (1988) mentions 220–280 g/head/day under average conditions with some astonishing 400 g/head/day for the elite rams of the nucleus flock weighing up to 50 kg at weaning.

Problems Encountered while Establishing the Ley Farming System in Western Libya

Desert rodents

Desert rodent populations are common in arid zones. They were thriving on newly developed land of the Cereal Project sites on crops and pastures especially on sandy soils at wadi Abu-Shayba and South Tarhuna. This is not a new problem as cereal crops are regularly pillaged by rodents before and after harvesting. This seems to be even worse during the medic phase. Most susceptible is cv. 'Harbinger' whose seeds are easily extracted from soft ripe pods. Pods and seeds are rapidly collected and stored in desert rodents' burrows. Thus, the medic seed bank is quickly depleted. 'Cyprus' medic pastures are less affected by desert rodents because they are usually grown on loamy/clay soils where these desert rats have difficulty digging their burrows and establishing their nests. 'Cyprus' also appears to be less susceptible to seed extraction by desert rodents due to its hard woody adpressed coils and spiny burrs. These two characteristics could be included in further selection of new medics for local farming systems and rangeland regeneration.

Desert rodents are identified as either *Meriones libycus*, *Gerbillus* sp., or on mobile sand, *Jaculus Jaculus*. They are locally known as '*Jird*, '*Gerbil*' and '*Jarboa*' (Gintzburger and Alidrissi, 1976; Hufnagl, 1972). *Meriones libycus*, a fat-tailed rat, is known to be very destructive to agriculture. It can breed at three to five months old, with three to four young/litter, a lactation period of 25–29 days and live for about two years (Hufnagl, 1972). They live in burrows with two to eight openings to the exterior where they store feed such as various seeds, cereal grain and medic seeds. This has been exacerbated by overall land clearing operations during the project's establishment. It destroyed or heavily disturbed the natural habitat (spiny *Ziziphus lotus* nebkha) of desert rodents' predators such as snakes (*Cerastes* sp., *Naja* sp., *Coluber* sp. etc.), foxes (*Vulpes* sp.), jackals (*Canis aureus*) or prey birds (*Falco* sp.), thus allowing a drastic increase in the rat population benefiting from abundant and rich feed from new crops and pasture on the Cereal Project. Problems with desert rodents (Ewing, 1978) became quite noticeable on medic pastures during the second year after establishment (Photo 9.6). Rats' holes (very likely *Meriones* sp.) at wadi Abu-Shayba could amount to 1/m^2 (i.e. 10,000 holes/ha), indicating average population of about 2000 families by the early summer, or about 80–100 kg live weight of desert rodents/ha, a stocking rate equivalent to 1.5–2 sheep/ha. No wonder that damage to medic pastures and surrounding cereal crops can be extensive. In captivity, these desert rats fed with medic burrs extracted seeds from an average 100 burrs/night/rat (Halse, 1982). Various control methods ranging from poisoning with zinc phosphide, anticoagulants (Cransberg, 1984) and cultivation destroying burrows were tested. In this regard, 10 kg of treated cereal seeds/ha with 3% zinc phosphide is most efficient; it appears to be an economic practice as seeds are dispersed with the farmer's fertilizer spreader, bearing in mind that it is highly poisonous to humans and sheep.

It was further noticed that second-year medic pasture regeneration had high rat infestation throughout summer causing poor regeneration and further damage to nearby cereal crops (Cransberg, 1984). On the contrary, cultivation after medics always reduces rat populations, likely by destroying burrows.

Photo 9.6. Rodents' infestation on second-year medic pastures (wadi Abu-Shayba, March 1978).

Photo 9.7. Sand dunes encroachment on Cereal Project site after land allocation to private farmers, 1984–1985 at Abu-Shayba.

Private farmers later settled on the Cereal Project land efficiently used a combination of chemical application and hand stopping-up of burrows which proved to be efficient enough (Halse, 1982) but required a lot of labour.

Nematodes

As early as the 1979/80 season, attention was drawn to bare patches of cereals and medics at wadi Aisha and other sites. Cyst nematode infestation was diagnosed and identified by the ARC. Trial testing of various nematicides (Richardson, 1983) associated with nematodes population counts (Cransberg, 1984) confirmed the effect of a single nematicide treatment with Nemacur® (8–12.5 g/ha), allowing a 50 % increase in wheat yield. However, experts noticed that proper fertilization response was greater than those achieved with nematicide treatments (Cransberg, 1984).

Land tenure problems and farm size

Throughout the establishment phase, medic pastures within the Cereal Project sites were regularly intruded on by private flocks. This was due to the very attractive look and forage quality offered by such medic pastures. As noticed by Halse (1982), '...there is nothing wrong with allowing grazing with private flocks [...], when the Cereal Project has more than enough feed available for their own sheep, [...] but it should be done in a controlled way...'. However, all project developers know how difficult it is to control encroaching flocks when feed is scarce. The underlying problem was also related to the traditional land tenure system and local farmers' and shepherds' grazing rights on land now occupied *de facto* by the Cereal Project, targeting and occupying the best rangelands, depressions and areas receiving regular run-off from wadi descending from the Jebel Nafusa. Local farmers not benefiting from irrigation facilities and livestock owners felt strongly deprived of the land they traditionally use for rainfed cropping or rangeland grazing. Community participation and governmental consultations remained non-existent (Sbeita, 2005). Unrestrained grazing by large private flocks led to systematic overgrazing. Wind erosion and deposits became a plague due to poor cultural practices and using inadequate farm machinery. Sand dunes slowly built up on project sites and new farms (Photo 9.7). The other problem was that the Libyan farmers were not convinced of the economic benefit of the ley farming system with investment in new farm machinery, especially when running small farms, a common problem facing the ley farming system adoption in the Mediterranean regions (Nordblom and Pannell, 1994).

Conclusions

The Australian ley farming system was implemented on 50,000 ha of mostly traditional rangelands in western Libya between 1973 and 1984. Based on a cereal-medic pasture rotation, it proved to be rather successful as long as the seasonal rainfall was above 250–275 mm MAR and the West Australian technical support and commercial medic seed supply remained available. The cereal-annual medic pastures rotation appears to improve rainfed cereal and sheep production compared to traditional cropping practices despite low and erratic rainfall. However, the local farmers and livestock owners, while interested, poorly accepted it due to social and economic reasons. Much was learned about the establishment and the management of medic pastures in the Jeffara plain and in the Jebel Nafusa. It certainly required agricultural expertise and training of Libyan agricultural engineers, technicians and labourers in mastering modern farm machinery, land preparation, seeding techniques, fertilization, harvesting, post-harvesting techniques, sheep management and adjusting proper stocking rates. This was certainly the most successful part of the project. However, the technology transfer to the local farmers failed due to poor community participation and acceptance of the rather complex and expensive input system. The size of the farms assigned to farmers to practise the ley farming was too small. It also faced continuous problems with trespassing private flocks from surrounding farmers and shepherds deprived of their good traditional cropping and grazing lands. Once the Cereal Project areas were subdivided and allocated to local farmers by 1978–1983, the creeping availability of modern irrigation from uncontrolled private wells and facilities on the Cereal Project areas led to the slow abandonment of the ley farming system for economically far more profitable agricultural options such as vegetable and fruit tree production, as long as irrigation water remains available in quality and quantity.

Note

[1] In 1980, 1 LYD approximately equalled 5 $US.

References

Allen, J.M. and Ewing, M.A. (1978). *Report on 4 years' experience (1974–78) growing annual medics in the Jeffara region.* Western Australian Advisory Group (Cereal Project – Jeffara Plain Authority), Libya.

Allen, J.M. (1979) Ley farming in Libya-North Africa. *Wool Technology and Sheep Breeding* 27, 5–9.

ARC (1977) Some background information on Libyan agriculture. Agricultural Research Centre, Tripoli.

Ates, S., Nefzaoui, A. and Keles, G. (2013) The role and management of pastures and fodder resources in integrated crop-livestock systems in West Asia-North Africa. In: *Proceedings of VII National Animal Feeding Congress, 25–27 September, Ankara*, pp. 26–27.

Byerlee, D. (2023) A global North–South Mediterranean Exchange 1891–1990: pastures for dryland crop-livestock systems. In: Broughton, O. (ed.) *Livestock as Global and Imperial Commodities: Economics, Ecologies and Knowledge Regimes*, Berlin, pp. 14–15.

Cocks, P.S. (1988) The role of pasture and forage legumes in livestock based farming systems. In: Beck, D.P. and Materon, L.A. (eds) *Nitrogen Fixation by Legumes in Mediterranean Agriculture: Proceedings of a Workshop on Biological Nitrogen Fixation on Mediterranean-type Agriculture, ICARDA, Syria, April 14–17, 1986*, pp. 3–10. Springer Dordrecht, The Netherlands. DOI: 10.1007/978-94-009-1387-5_1

Cransberg, L. (1984) *Experimental Summary – Cereal Project 1983–84.* Western Australian Advisory Group (Cereal Project – Jeffara Authority), Libya.

CSIRO (1985) *Australian Plant Introduction Review*, 18(1), 182.

Ewing, M.A. (1978) *Report on experience with cereals and medic pastures on the Cereals Project – Neffusa region.* Western Australian Advisory Group (Cereal Project – Jeffara Plain Authority), Libya.

Ewing, M.A. (1979) *Progress with Annual Medics (*Medicago *sp.) in the Jeffara region of the Socialist People's Libyan Arab Jamahiriya.*

Ewing, M.A., Standing, W., Fituri, M. and El-Taash, S. (1979) *Medic Experiments in the Cereal Project – Jeffara Plain.*

Francis, C.M. (1978) *Distribution and Ecology of Annual* Medicago *sp. in Northwest Libya – A Report Based on a Plant Collection Tour 1978.* Western Australia Department of Agriculture, Perth, Australia.

Francis, C.M. (1982) *Libyan Medic Evaluation Project.* Western Australian Overseas Project Authority, Perth, Australia.

Gintzburger, G. (1978) *Collection of Seeds and Shrubs (Bir el Ghanem, Bir Ayyat and Nalut area).* No. 183/78. FAO Agricultural Research Centre, Tripoli.

Gintzburger, G. (1980) Forage legumes in Libya. *Plant Genetic Resources* 43, 7–11.

Gintzburger, G. (1983) *Libyan Medic Evaluation Project.* Mission report. Western Australian Overseas Project Authority, Perth, Australia.

Gintzburger, G. (1986) Seasonal variation in above-ground annual and perennial phytomass of an arid rangeland in Libya. *Journal of Range Management* 39(4), 348–353.

Gintzburger, G. and Alidrissi, M. (1976). *Range Study on Jeffara Plain: Sheep Project at Bir-El-Ghanem.* No. 83/76. FAO and Agricultural Research Center, Tripoli.

Gintzburger, G. and Bayoumi, M. (1977) *Survey of the Present Situation and Production of the Libyan Rangelands.* No. 118/77. FAO and Agricultural Research Center, Tripoli.

Gintzburger, G. and Blesing, L. (1979) *Genetic Conservation in Libya – Indigenous Forage Legumes Collection in Northern Libya, Distribution and Ecology of* Medicago *sp.* Food and Agriculture Organization, Rome.

Gintzburger, G. and Kernick, M.D. (1979) *Plant Genetic Conservation, Forage Legume Collection in Libya, Part II. Minutes of meetings, field trip reports and proposals for agronomic evaluation.* No. 231–233/79. FAO and Agricultural Research Center, Tripoli.

Gintzburger, G., Francis, C.H., Sbeita, A., Rojbani, H.R., Fituri, M. and Skinner, P.W. (1984a) *LMEP – Catalogue of the Annual* Medicago *sp. from the Western and Central Zones of Northern Libya* Vol. 2. Western Australian Overseas Project Authority, Perth, Australia.

Gintzburger, G., Sbeita, A., Rojbani, H., Fituri, M. and Francis, C.M. (1983) Annual *Medicago* sp. collected from the Sirte Region of Libya. *Plant Genetic Resources* 56, 23–29.

Gintzburger, G., Sbeita, A. and Francis, C.M. (1984b) *Libyan Medic Evaluation Project – Final Report.* WAOPA-Western Australian Department of Agriculture, Perth, Australia.

Halse, N.J. (1978) *Land Use Recommendations for the Jeffara Plain, Tripoli, Libya.* Western Australia Department of Agriculture, Perth, Australia.

Halse, N.J. (1982) *Annual Inspection Visits to Cereal Project – Libya.* Western Australian Advisory Group Cereal Project – Jeffara Plain Authority.

Hufnagl, E. (1972) *Libyan Mammals.* The Oleander Press, Cambridge, UK.

Khalil, H., Francis, C.M. and Halse, N.J. (1977) *Annual* Medicago *sp. ecotypes from Northwest Libya. I – Preliminary Agronomic Evaluation.* Technical Bulletin no. 37). Western Australia Department of Agriculture.

Le Houérou, H.-N. (1959) *Recherches écologiques et floristiques sur la végétation de la Tunisie méridionale.* PhD dissertation, Université Alger Institut de Recherches Sahariennes.

Le Houérou, H.-N. (1969) *La Végétation de la Tunisie Steppique (Avec Référence aux Végétations Analogues d'Algérie, de Libye et du Maroc).* Vols 1–2. Annales de l'Institut National de la Recherche Agronomique de Tunisie.

Lightfoot, R.J. (1988) Selection for lamb growth in Libyan Barbary sheep. In: Thomson, E.F. and Thomson, F.S. (eds) *Increasing Small Ruminant Productivity in Semi-arid Areas: Proceedings of a Workshop Held at the International Center for Agricultural Research in the Dry Areas, Aleppo, Syria, 30 November to 3 December 1987*, pp. 145–154. Springer, Dordrecht, The Netherlands. DOI: 10.1007/978-94-009-1317-2_11

MacPhie, G.R. (1980) *Final Report of Activities and Recommendations of the South Australian Agricultural Team Based on the Demonstration Farm at El-Marj (Libya) 1974–1980.* Overseas Project Division, South Australian Department of Agriculture, Adelaide, Australia.

Nordblom, T.L. and Pannell, D.J. (1994) From weed to wealth? Prospects for medic pastures in the Mediterranean farming system of North-West Syria. *Agricultural Economics* 11, 29–42.

Puckridge, D.W. and French, R.J. (1983) The annual legume pasture in cereal ley-farming system of Southern Australia: a review. *Agricultural Ecosystem and Environment* 91, 229–237.

Richardson, J. (1983) *Annual Report, 1982–83 Season.* Western Australian Advisory Group (Cereal Project – Jeffara Plain Authority), Libya.

Sbeita, A.A. (2005) *Pasture/crop/livestock integration in Libya.* FAO-RNE Technical Report. FAO and Agricultural Research Center.

Webb, L.G., Gintzburger, G. and d'Antuono, M.F. (1984) *The Libyan Medic Data Base: A Computerised Database for Plant Genetic Resources Using INFO HENC01 Software.* Western Australia Department of Agriculture, Perth, Australia.

10

Livestock on Libyan Rangelands

Abstract
Libyan rangelands have been home to herded flocks of local breeds of sheep, goats and dromedaries fed on native vegetation for centuries. Feed and water availability limited roaming livestock. Droughts would occasionally reduce their numbers, taking years to restock. Today, Libyan rangelands are no longer producing the necessary feed to sustain rocketing livestock numbers. Overstocking is likely three to four times higher than the native vegetation, even in fair condition, could today sustain. The reason is that livestock are more and more artificially fed under open-air feedlots business, with large quantities of subsidized imported barley grain. Local feed production and barley grain are insufficient despite expanding rainfed cropping, land clearing and new irrigation projects targeting the best rangelands. Overstocking is not only causing desertification but also it stimulates the spread of infectious diseases barely controlled. The livestock situation is dire and cannot be solely attributed to global warming.

Introduction

The livestock roaming the Libyan rangelands are local breeds of sheep, goats and dromedaries, generally privately owned. Local breeds of cattle are mainly located in the Jebel Al-Akhdar, grazing the matorral in the highest rainfall zones. Dairy cattle, mostly imported breeds, do not enter the rangelands and are exclusively confined to irrigated farms and feedlots on private farms or government projects. In this chapter, we review the local livestock breeds' traits, the catastrophic livestock population evolution to the present day, and explain the reasons for this trend leading to unavoidable rangeland desertification.

Rangelands Livestock Systems

A general review of the Middle Eastern and North African livestock systems was published by Mohamed *et al.* (2019); however, the Libyan livestock systems present specificities.

The nomadic system

Pre-WWII and up to the 1960s, most rangeland livestock were following a nomadic system across borders, practised by the Arab tribes in the Jeffara plain into Tunisia (Louis, 1979) and Algeria, on the southern slope of the Jebel Nafusa and the Hamada. A similar nomadic system would be common in Sirte and the east, the southern slope of the Jebel Al-Akhdar and across into Egyptian Marmarica. Customary border crossing became more difficult for flocks and herders due to concomitant smuggling of goods increasingly controlled by border and military forces. Hence, seasonal nomadic wandering across borders into Tunisia, Algeria and Egypt soon ceased, and they strictly retreated to

DOI: 10.1079/9781800627154.0010

their customary territories (*watan*) within Libya proper.

The system was simple: flocks managed by mobile shepherds would graze within the tribal land covering large areas, moving during winter/spring deep into the desert in search of green pastures and watering places, then retreating to coastal rangeland zones or summer camps as the dry summer months would wilt the meagre desert pastures. Little grain supplementation was provided to nomadic livestock. Grazing and rainfed cropping rights and territories were not always clear between tribes, easily degenerating into enduring disputes. During poor seasons or drought, shepherds and their flocks would move to friendly tribal land with better pastures under mutual agreements (see Chapter 12). In the mid-1970s, due to a lasting regional drought in the east, a large number of small ruminants were temporarily displaced from the east to the Hamada according to arrangements between eastern and western tribes. Earlier, unrest during the early Italian colonization would see large numbers of dromedary flocks moving to neighbouring countries. These days, private flocks would still be managed according to semi-nomadic systems largely supported with modern trucking (Chatty, 2013) providing feed and water to remote locations in the west, in Sirte, in the east to deep into the southern desert.

The semi-nomadic system

The full nomadic system slowly faded away post-WWII while the semi-nomadic systems took over with social mutations and modern facilities becoming available. The semi-nomadic systems always existed besides the nomadic system, especially with the settled Jebel Nafusa Berber and Arab populations (Despois, 1935). Their flocks would descend into the Jeffara plain during winter/spring, grazing mostly annuals (Photo 10.1), then would graze cereal crop residues after harvest into summer, then slowly climb back onto the perennial pastures of the Jebel and southern slopes, ending up being fed with stored barley grain for a couple of months awaiting the first autumn rains triggering once more the vegetation growth. The semi-nomadic system only operates in interaction with rainfed cereal cropping by providing valuable grains and cereal crop residues. However, with private and governmental project rainfed cereal acreage expanding onto the best and most productive rangelands and producing more grain and crop residues, the livestock numbers expanded, too, and much faster while rangeland areas available for traditional grazing contracted to breaking point (see Chapter 9). The locally produced barley grain tonnage became insufficient to complement rangeland livestock feed during the summer months or even during the year, triggering the

Photo 10.1. A semi-nomadic sheep flock in the Jeffara plain.

necessary and costly importation of vast quantities of barley grain and other feed. At the same time, depleted and overgrazed rangeland vegetation could supply feed to less and less livestock. By 1960, Libyan rangelands would supply about 70–80% of the annual feed requirement to wandering livestock, while these days, with constant overstocking, available complementary feeding, more watering facilities in the most remote places and ensuing rangeland degradation, it dropped to less than 10% (Gintzburger and Bayoumi, 1977).

In short, the nomadic traditions with small flocks are hardly surviving and tend to slowly disappear. It is replaced with a semi-nomadic system with larger flocks and business feedlots increasingly depending on local or imported supplementary feed as rangelands are excessively overgrazed. This is a general trend observed all over arid North Africa and the Middle East (Boutonnet, 1989; Gintzburger *et al.*, 1997).

Government rangeland and rainfed cereal projects

By early-1970 (GEFLI, 1972a, 1972b, 1972c, 1972d, 1972e, 1974a, 1974b, 1977), the government started rangelands development planning in the west (Map 10.1), in Sirte and in the east, and rainfed cereal projects in the Jeffara plain and El-Marj with the commendable intentions to promote modern agriculture and land management. The western and Sirte rangeland projects were located between the 100 mm and 175 mm isohyets.

The rangelands projects in the Jeffara plain were gradually implemented with sometimes problematic land delineations due to community opposition. Large nurseries were established to raise fodder shrubs and trees for large-scale plantations. Tens of thousands of hectares were transplanted (1981–1985) by thousands of manual workers with access to unrestricted modern logistical support (land preparation, seedlings transport, manual plantations and watering). The Sirte rangeland projects sites (Buirat-Hisha, wadi Wishkah, Sidra), though identified and delimited, sometimes entirely fenced and with veterinarian buildings constructed, were lagging behind due to very poor rangeland conditions in the 100–150 mm MAR (Nègre, 1974) and abandoned. In the east, only the Benghazi plain project was implemented (1976) under Engineer Mohamed Salem Bargathi's wise guidance. It covered about 120,000 ha of matorral north of Benghazi in a triangle between Sidi Khalifah-el-Guetara to Tolmeita, benefiting from 200–300 mm MAR. Fodder shrubs, mainly *Atriplex nummularia*, were tested and planted but land management faced many problems with encroaching flocks of sheep, goats and local cattle. By 1985, the Benghazi rangeland project was progressively abandoned and open to private flocks.

The rangeland projects were totally fenced off with expensive sheep-proof fences and barbed wire. Grazing, banned for years, allowed remarkable range vegetation regeneration. Fodder shrubs plantations of *Atriplex* sp., spineless cactus, exotic *Acacia* sp., and few native species were established at high cost. Cereal projects were also established using the Australian ley farming system supported with modern cropping technologies (see Chapter 9). However, this was not well accepted by the local farming and shepherd communities, not consulted and losing, *de facto*, considerable large grazing and rainfed cropping areas they traditionally used. Local conflicts ensued, fences were pulled down, and local flocks regularly encroached protected rangeland and cereal projects. By 1980, some of the projects were subdivided into small farms allotted to individuals with the weak support of extension services. However, it mostly failed as farmers soon reverted to traditional land management, livestock rearing, overstocking and unabated overgrazing. New wells were dug, legally or illegally, and irrigated crops crept in, changing the initial purpose of the rangelands and rainfed cereal projects. Recommended stocking rates and fodder shrubs plantation management were ignored. Private livestock numbers on projects soared with disastrous consequences on the native vegetation and fodder shrubs plantations. It was a total failure.

Starting in 2005, the Suani Rangeland Development Project took over, gradually establishing 82 fully fenced protected rangeland areas (*mahmiyat*) in collaboration with the local communities, each covering 3,000–

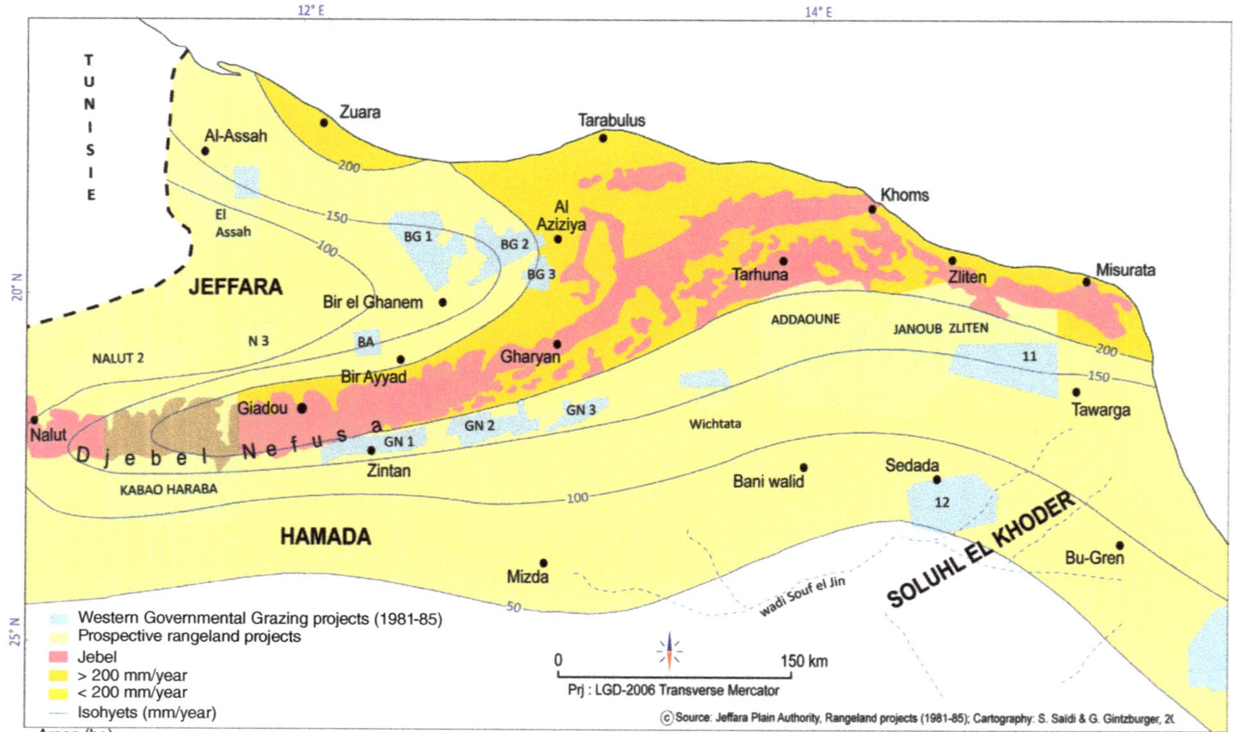

Areas (ha)

El-Assah	59 265
Nalut II	71 411
N 3	22 629
Kabao-Haraba	242 384

BA	8 830
BG 1	46 150
BG 2	31 043
BG 3	11 432

GN 1	25 006
GN 2	28 795
GN 3	13 426
Wishtata	13 386

Addaoune	132 585
Jenoub Zliten	232 971
11	60 874
12	58 583

Map 10.1. Government rangeland grazing projects implemented or planned (1981–1985) in the western zone.

8,000 ha in the west and in the east. Some regenerated well (see Chapter 16), but again, sustainable management with proper stocking rates were ignored. Quite interesting is the fact that these protected rangeland areas saw the return of wildlife, starting with birds such as bustards (*Chlamydotis undulata – Houbara*), hares, partridges, and even gazelles, all soon indiscriminately hunted despite being legally protected.

Open-air feedlots on rangelands

The semi-nomadic system, while persisting, slowly gave room to stationary flocks (Photo 10.2) mostly fed with locally produced and imported grain and cereal residues, and stale bread on rudimentary to sophisticated open-air feedlots (Boutonnet, 1989) due to rangeland overgrazing. Stale-bread sheep feeding is becoming common, the bread being collected and sold in marketplaces, attracting livestock businesses owners, as 5–6 kg of subsidized cheap stale bread produce 1 kg of meat (Le Houérou, Montpellier, 2004, personal communication) sold at a high price. These rangeland-damaging trends creep in, unabated, all over arid rangelands of North Africa and the Middle East.

The semi-nomadic system evolved. Attractive market meat price, escalating legal and illegal irrigation facilities, increasing number of settled farmers located on the coastal zone from west to east started basic feedlots rearing and fattening, but still allowing their flocks to overgraze surrounding rangelands to total destruction. One sarcastic shepherd told us that he daily keeps leading his flock for a leisurely stroll on rangelands just for *breathing fresh air* (.... شم الهواء ...). The rangelands destruction continues relentlessly.

Local Livestock Breeds on Rangelands

The Barbary sheep is present all over northern Libya and across into southern Tunisia. Sheep flocks are mixed with ever-present local goats. Dromedaries, once abundant, still present in small numbers, lost their role as draft animals and are mostly kept for meat and sometimes for milk production.

The Barbary sheep

The Barbary fat-tail sheep is related to the Anatolian fat-tail sheep and the Awassi fat-tail sheep

Photo 10.2. A rudimentary open-air feedlot (Bir Haraba, Hamada, N 31.708 E 11.509, March 2009).

from Syria, the Middle Eastern and Central Asian steppe (Gintzburger *et al.*, 2003). They may have been introduced by the Phoenicians (4th century BC) to Cartago in northern Tunisia, and then, later, into Libya with the Arab colonizers (Mason, 1967) coming from the east. They are also present in central and southern Tunisia, extending into eastern Algeria, mixing with thin-tail sheep where they are common (Sagne, 1950).

Sheep breeds encountered on northern Libyan rangelands are 95% Barbary sheep consistently present in the Jeffara, Jebel Nafusa, Hamada and Sirte regions. Another Libyan Barbary type, the *Barki* (or *Dernawi*) is mostly found in the Jebel Al-Akhdar and Marmarica, with flocks flowing over the Egyptian border to Sollum, Marsa Matruh–Sidi Barani. Besides the northern Barbary sheep, the *Ghimi* Barbary sheep, a possible cross of northern Barbary sheep and Algerian breed, is present in the Hon and Zella oasis (Mason, 1967). Sub-Saharan Sudanese *Dubasi* and *Watish* sheep are found in the south, mostly in the oasis of the Fezzan and more rarely in southern Hamada. These sheep are much larger and taller than the Barbary breed; they have a smooth piebald white, hairy, brown or black fleece and nearly no wool; they have a convex forehead and a thin, rather long tail. They are rarely seen in the northern part of Libya, appearing sometimes with dromedary flocks brought in from the south.

The Barbary sheep is a tough animal and a good walker, both attributes that allow him to survive and produce under rough arid rangeland conditions. In the old days, feeding came from native annual rich green vegetation during winter/spring and then from dry cereal stubbles and dry shrubs during summer and early autumn. Watering was not an issue in winter/spring as moist annual plants would amply supply water. As summer progresses and natural vegetation dries off, watering the sheep flock must be a daily necessity with three to five litres of fresh water/adult head, and much more with saline water.

The sheep flock usually move within 4–5 km around the shepherd camp on rangeland during the day and return to the camp at night. When the flock is mixed with goats, the whole flock may walk and graze on larger areas, 6–7 km from the shepherd's base camp, or even more when dromedaries lead the flock. However, shepherds tend to restrain their dromedaries as the small ruminants would waste their energy mostly walking without grazing. A peculiar overgrazing pattern appeared with the availability of mobile phones, with foreign shepherds tending to congregate with their flocks on better phone-reception node areas even in the most remote rangelands, hence remaining quasi sedentary and snowballing overgrazing.

Barbary traits

The main characteristic of the Barbary sheep is its fat-tail (Photo 10.3), a bilobed flat sack of fat on each side of the tail reaching the top of the hock. The related fat-tail sheep breed, Gissara or Hissar, are also present in Uzbekistan (Gintzburger *et al.*, 2003). The weight of the Barbary's fat tail is between 3 kg and 5 kg. Thanks to this adipose tail reserve, the Barbary sheep is very resilient and well adapted to extensive husbandry, adverse arid climatic conditions, and feeding variations common in pre-desertic environments. The body fat reserve is used during time of feed and water shortage, a common feature in animals from hot desert environments such as those found with the small sand rodent *Psammomys obesus* and dromedaries.

The Barbary fleece is usually white, much preferred by local livestock owners, though brown and black fleeces are current (Magid

Photo 10.3. A Barbary ewe (at Bir Ougradia, south Al-Assah) with its characteristic fat tail.

et al., 1987; Magid and Mursy, 1987; Magid and Zaied, 1987). By the end of spring (mid-April) when mild weather settles in, wool is shorn averaging 1.5–2 kg/head. The first shorn wool is the cleanest and is usually kept for the family. If the shearing drags on, as the hot months progress, heat flares and dust spoil the late-shorn wool, often full of vegetal impurities and sand.

Ahtash (2006) reports, from a mid-side wool sample of 391 adult Barbary sheep from different regions of Libya, an average 1.6 kg of greasy fleece yielding about 0.90 kg of clean fleece. Fibre diameter is coarse with 30.1 μ on average. The medullated fibre percentage is high (9.6%). Staple length and fibre length are 8.3 cm and 11.1 cm, respectively; fibre crimp (CR/2 cm) is high and reflects variation in climatic and feeding conditions. Finally, from the cleaned fleece, counting percentage for fine (F) and coarse (C) wool is 67% and 22%, respectively, the rest being kemp (K), corresponding by weight percentage (F) 46, (C) 43 and (K) 10. This is a typical coarse wool used for making bags (*molkha*), sacks for transporting grain (*ghrarah*), rugs, carpets (re-carpet industry from Misurata) and tent strips (*flij, hammel*). The finer wool quality from the plastron-breast is kept for weaving, as a cottage industry, on a home vertical loom (*masda*) to produce blankets or the fine material for the elegant white *holy*, or the brown winter heavy and warm *wazra* worn by shepherds. The *holy* or *jard* is exclusively made of white sheep wool; the material used for *wazra* is a woven mix of fine brown sheep wool, dromedaries' and goats' hairs, giving it good waterproof quality. *Flij* and *hammel* (see Chapter 12) are the traditional strong tent elements such as dividers, or narrow or wide roof strips; all are made with the coarse raw sheep wool woven with dromedaries' and goats' hairs to render them waterproof and woven on a rudimentary antique horizontal loom.

Barbary sheep's legs and faces are bare. The forehead is flat to slightly convex, with long pendulous ears, and generally white, but also could be part-coloured red-brownish or black. Horns are generally short or absent, but rams with imposing outwardly spiralled horns (Photo 10.4) are much valued. Adult body weight ranges from 40–70 kg for males and 25–60 kg for females.

Production

The Libyan Barbary sheep is mostly for meat production. Male lambs are slaughtered when reaching about 20–25 kg at four to six months depending on the season and delivering 40–50% carcase weight composed of about 60–65% muscle, 10–20% body fat and the rest bones. Lamb and fat mutton meat is preferred during the winter. The valuable fat from the tail averages 6–7% of carcase weight and is mostly used for cooking. It is locally named *leyyah* or *layeh*. Contrary to Middle Eastern *Awassi* sheep, Barbary's milk production under rangeland conditions is low (in the order of 25–35 kg/4–5 months/adult ewe), essentially allocated to lambs aged up to 4–6 weeks, though many shepherds keep their suckling lambs with the mother for 3–4 months. Milk is rarely used for human consumption and is mostly transformed into *smen*, a sour salted and fermented butter.

Usually just-born lambs (*alluch*) are weaned at four to five months (as *khruf*) and sold for fattening along with old infertile ewes. Shepherds tend to hold onto old ewes in their flocks if possible (5–6 years old) and until teeth are worn out by grinding sand. This is a simple way to capitalize wealth and often barter old ewes for barley grain, straw or water. Old ewes may represent 15–20% of the adult sheep in the flock. The

Photo 10.4. A Barbary ram (at Sriah N 32.71 E 11.64, Al-Assah).

standard flock (*rlem*) usually managed by one shepherd is about 200–250 adult animals of which 4–6 are rams and 10–15 are goats. Dogs are rarely seen with the flocks, and if they are present, this is mostly to keep jackals away. Sometimes, many more rams are kept with the flock as these do attract a high price for the Eid festival. Flock management follows ancestral rules. Rams are rarely castrated; they are kept with the Barbary females all year long, and as the ewes are non-seasonal, i.e. fertile nearly all year round, it results in a very long mating season, often extending from April to November (Magid *et al*., 1987).

Reproduction

Young males mature by their fourth year, start to mate ewes, but face contests from the master ram (*kebs'h*).

Rams and fertile ewes are often flushed with supplementary barley grain when available at a daily rate of 300–400 g/head and with barley straw. Hay supplement is not normally available on rangelands but may be available in small quantities on irrigated coastal farms.

The fat tail of ewes is an obstacle to mating, with shepherds assisting rams by holding sideways the tails of oestrous females. Lambing is spread between September and March, with two peaks in November and December (lambs, *kharf'i)* and late spring February and March (lambs, *reba'i*). Twinning rate is low. Usually, providing the winter is mild, early-winter lambing and lactating ewes benefit from nutritious annual vegetation growth. During this time, the ewes with their lambs are kept close to the camp and are well looked after by the shepherd.

Late lambing may encounter an early summer and poor dry pastures and therefore ensuing poor lambs' growth rate. Lambs born between April and July always have a lower vigour and scoring than the ones born during early autumn and winter. Under rangeland conditions, with winter lambing, lambs' (10–30 days old) growth rate ranges between 150 and 200 g/day according to feed availability for lactating ewes. During this period and when the season and rangelands are poor, livestock owners would sacrifice some of their rainfed or irrigated green barley and let their lactating ewes graze it. Then, as the season progresses, two options are available to the shepherd: the season is definitely worsening, and the lactating ewes and lambs are kept grazing on the green barley until no barley is left. In case the season is improving with late spring rain and milder weather, permitting rangeland vegetation growth, the shepherd will then move his lactating ewes and lambs to the rangeland with the rest of his flock, saving as much as possible from his barley to harvest some grain.

Mortality rate varies and is accentuated during cold winters, when late spring rains occur after shearing or during drought. During the memorable 1935/36 drought (Rowland, 1945), the number of sheep in Tripolitania dropped from about 750,000 (1935) to 148,000 (1936). It is quite possible that, at that time, part of the threatened flocks would have been moved across the border to nearby southern Tunisia.

Other issues – climatic, economic, banknote changeover (1985), unrest – would make the Libyan sheep flock fluctuate as it remains a capital that livestock owners would dearly control as the local banking system is seen as deficient and unreliable. In short, regardless of constant climatic and economic fluctuations, sheep rearing in Libya remains a secure investment and comfortable return for sheep owners.

Coming back to the non-seasonal ewes' breeding of Barbary ewes, some GEFLI and FAO livestock experts advocated increasing the number of lambings to three per two years. This has been definitely rejected by sheep owners as multiple lambing occurrences over two years would be subject to limited seasonal feed availability on depleted rangelands and on expensive feedlots.

Attempt to improve the Libyan Barbary sheep through crossbreeding

Selection programmes (Mason, 1967) were conducted to improve Barbary's wool quality during the Italian colonization. Crossing with the Central Asian Karakul sheep was attempted at Sidi-el-Mesri experimental station during 1934–1939 to boost milk production. These programmes were abandoned due to WWII. Later (1953/54), a selected Karaman flock of 117 rams and 204 ewes was imported from Turkey. These Karaman rams were used to cross with selected Barbary ewes at the Garabulli sheep experimental station. The Barbary half-bred offspring produced were much larger than the local Barbary, more productive, but appeared to be less resistant to drought. The project was abandoned.

The rangeland goats

Contribution to the flock

Goats, like sheep, are omnipresent on northern rangelands with about 45% in the west, 20% in the Sirte region and 35% in the east. They are from the *Mahali* breed (Photo 10.5), long black-hair fleece or piebald white, black and reddish-brown (Ahtash, 2006). The small face and legs are almost bare. The neck is long; the forehead is flattish to convex, with pendulous ears encircled with short forward-curved horns for the females. Males have longer twisted horizontal horns. The tail is thin, short and always curved upwards. Body weight is 30–40 kg for females and 40–50 kg for males.

Their role is important for the shepherd. Goats provide milk, hair, hide and usually lead the sheep flocks. Goats are lively, tireless walkers and very robust. The shepherds keep them out of the cold and wet weather up to mid-February (*Gerret el-Anz*). After this, they get back in the open and on the range.

Goats are mostly browsers, i.e. feeding habitually but not exclusively on shrubs. Therefore, a few goats with the sheep flock would consume twigs from woody shrubs not touched by sheep. Large flocks of sheep (200–300 heads) with a *ma'az* of 25–30 goats of exclusively *Mahali* breed are common on the southern slopes of the Jebel Nafusa (south Zintan, Haraba, Nalut), the northern Hamada and in Sirte, grazing desolated large areas of *remeth* (*Arthrophytum scoparium*), a rangeland vegetation type hardly touched by sheep. In this case, the flock would only be composed of goats as in the dense matorral in the east.

Reproduction and production

Mahali goats are non-seasonal, and as rams (*atrus*) are kept within the flock. Mating and kidding may occur anytime during the year (Ahtash, 2006). However, as for Barbary sheep, kids born in autumn/winter when native vegetation is growing are the best in terms of vigour and survival rate. At birth, kids are small

Photo 10.5. *Mahali* females and male goats (at Touil-er-Reschiada, south Al-Assah).

(2–3 kg). Daily gain before weaning is reported to be about 90–100 g/day. Weaning occurs at about three months with body weight reaching 9–10 kg. Twinning is common (10–40%).

Milk production ranges from 80 to 130 kg/year/head, containing little 2% fat and 3% protein. Carcase dressing yields 45–50% of body weight of which about 70% is meat and 30% is bone. Goats' meat is leaner than that of Barbary sheep, making goats' meat much preferred during summer months. Goats' hair yields about 0.5–1 kg/head/year. It is mostly used to make ropes and for weaving, mixed with sheep's wool, and with young dromedaries' hair for tent runners. The goatskin (*shakwa*) was a valuable shepherd's attire to carry water or milk before plastic containers appeared. The goatskin was waterproofed and preserved with tar (*gatrun*) coated on the internal side of the hide. The tar used to be extracted by distillation from *Juniperus* wood (*Ar'har*), a tree now extinct in the Nafusa.

Attempt to improve goats' production through crossbreeding

Experimental work was performed at the Bir-el-Ghanem Sheep Project crossing *Mahali* goats with imported Damascus and Maltese goats to improve meat and milk production (Ahtash, 2006). While experimental results were encouraging, it was not adopted by rangeland livestock owners for the simple reason that increasing animal performance requires proper management and increased quality and quantity feeding, both of which are often lacking on traditional rangeland husbandry systems.

Other breeds of goats (*Tebagi* and *Targi*), in small numbers (about 2–3% of Libyan goats), are found exclusively in the south and mostly fed with crop residues from the oasis.

The dromedaries

Dromedaries (*Camelus dromedarius*), the one-hump camels, are a common feature on most arid Libyan rangelands. While several races are found from Mauretania to Egypt, they all belong to the large group of *Maghrebi* dromedaries (Diop *et al.*, 2020; Faye *et al.*, 2014) related to the African races and very different from the larger and more productive Asian dromedaries from Saudi Arabia to India-Pakistan. Often neglected, the dromedary remains a mythical animal of the desert and kept in high esteem by local and marginalized populations of North Africa (Faye *et al.*, 2014, 2022). From antiquity, dromedaries were used in combat by the Libyan tribes (Roques, 2009), as pack animals across desert or as draft animals in agriculture, from the North African desert fringes deep into sub-Saharan regions. The availability of hauling trucks from the mid-1920s soon made dromedaries redundant.

In Libya, from west to east, despite fluctuation in numbers and redundancy, dromedaries are still omnipresent. Revered for their ability to survive in extreme desert conditions, dromedaries produce meat, milk, hide and hair, all of value where neither sheep nor goats, horses nor cattle could survive. Large private herds of dromedaries are still seen around Bu N'Gem, roaming in the Solul-el-Khoder and wadi Bey-el-Kebir areas, on the Hamada (Photo 10.6), south Sirte, deep into southern Jebel Al-Akhdar. These herds graze freely (*himliya*) without herdsmen and are regularly checked by their owners at watering points where they regularly gather.

In the west, in the Jeffara plain and Sirte, up to the 1970s, a few dromedaries would carry tents and equipment from the shepherd, but would also encourage small ruminants to graze larger areas. This is a sight from the past as livestock owners now supply their shepherds and flocks using trucks.

Dromedary traits

We will mostly consider the pack dromedary (Photo 10.7) common in the northern part of Libya. This is a bulky and stout animal. The fleece is often brown/dark and woolly/hairy. Their soles are bulky and hardened making them able to walk sandy and rocky rangelands. They are indefatigable slow walkers. In the old days, one of these dromedaries would carry loads of 200–250 kg as pack animals in large trans-Saharan caravans travelling some 20–40 km/day for several weeks along the main north-south caravan tracks: Tripoli – Gharyan – Rojban/Zintan to Ghadames, then to Fezzan and further south to Ghat, or Tripoli-Beni Walid – Sokhna – Murzuk, or in the

Photo 10.6. A large dromedary herd grazing freely (*himliya*) without herdsmen on Hamada Zrira.

Photo 10.7. The northern Libyan pack dromedary (here a gravid female, in the *sebkha* Jbebina, Al-Assah).

east: from Beida – Shahat – Barca – Benghazi, then to Awjilah-Jalo to Tazerbo-Kufra, and further into sub-Saharan Africa (Passon *et al.*, 2020). Caravans of 250–300 dromedaries were common up to the 1930s.

In the south, another type of dromedary is present and commonly named *Mehari* (Photo 10.8). They are rarely seen up north. Slim, lean with long, fine legs, the fleece is clear (beige to white) and nearly hairless. The soles are large to manage sand dune country. It was and still is used in the desert as a saddle/riding animal able to travel up to 100 km/day. These fast dromedaries were formerly, and regularly, used by Saharan raiders attacking and plundering isolated caravans. Today, *Mehari* are still used by scouts/long-range riders on the Hamada and in the desert (Photo 10.9).

Photo 10.8. A *Mehari* from the south of Libya (at Touil-es-Souf (N 32.68 E 11.67), south Al-Assah, 2007).

Photo 10.9. A lone, silent and suspicious *meharist* (a scout and long-range camel rider) appearing from nowhere on the empty Hamada (east of Bir Al-Kababsha, N 29.82 E 12.14, April 2010).

The Al-Assah Camel project

A herd of pack dromedaries was long kept at the Al-Assah (N 32.83 E 11.63) on a government camel project from which we collected information (Dumancic and Gintzburger, 1979).

In the old days, the Al-Assah region had been a long-standing rearing and roaming place for dromedaries belonging to the people from Regdalin and Jmeil, for the purpose of carting goods back and forth from the coast (Zuara to Ghadames) and further south. By 1974, a camel project was initiated by the government on 21,000 ha, first around the Al-Assah *sebkha*, then south to 87,000 ha in 1975 and, by 1978, expanding to a total of 350,000 ha even further south towards Ksar El-Watiya – Bir Shabunyah – Bir Hebiliyah to the foothills of Ksar El-Haj/ Shakshuk into sandy country. The average annual rainfall at Al-Assah ranges between 125 mm and 150 mm, dropping to about 100 mm south with very low productivity and fragile sandy rangelands. The project area was already heavily overgrazed, especially around watering places (Bir Ougradia, Bir As-Sahn), with many private small ruminant flocks and camel herds. While small ruminants are mostly resident the whole year round, the project and private dromedaries would move from the winter/spring grazing in the *sebkha* and slowly move to the most southern sandy part of Al-Assah and even to the Hamada (south Nalut to Derj) as spring ended and summer progressed, then return to the Assah area for the next wintering. This traditional alternating grazing pattern between saline and sandy–non-saline rangelands is considered necessary by the shepherds to maintain a proper health and mineral balance (sodium/calcium/phosphorus) of the dromedaries' feeding calendar.

The camel project ended these migrations and resulted in set-stocking the whole year on the project land. By 1979, the project hosted about 2900 *Maghrebi* pack dromedaries, coming from nearby Tunisia, from Al-Assah, Sorman and some from Sirte. They were divided into 40 herds of 40–80 animals (Photo 10.10). Each herd was looked after, days and nights, by two experienced shepherds, mostly southern Tunisians, often from the Touazine, Ouderna or Merazigue tribes from Medenine, Tataouine and Kebili (Moslam and Megdiche, 1989). According to tradition, besides their salary, the shepherd would receive one new *holy* and a *wazra* each year and sometimes be allowed to run their own small, unregistered number of dromedaries with the project's herd.

Herd management

Shepherds operate on a 15-day roster all year long. Living on the range 24/7 with their herd, rudimentary camping, the shepherds would be responsible for leading their herds wherever on the range they like, as far as 80–100 km south from Al-Assah. They would be met regularly at wells by project officials providing them with food and fresh water. Watering the herd during winter and grazing the halophytic *sebkha* vegetation occurs every two to four days, though more frequently when watered with saline water, such as the one at Bir Ougradia, or on fresh artesian water, as at Jbebina. In full spring with a proficient annual vegetation, intervals between watering can be much longer. When on non-saline vegetation, during summer, watering the herd can be spaced five to seven days, and still at a shorter interval when available water is saline.

The project or private animals are identified with the project's or owner's hot-iron brand

Photo 10.10. Part of the Al-Assah camel herd (watering at Touil-es-Souf, N 32.684 E 11.674, south Al-Assah, 2009).

(*sima* – Photo 10.11) on the rear or the neck. They are not individually tagged. Libyan researchers occasionally recorded individual project animals' performances.

Reproduction and production

Female dromedaries (*naga)* have a late sexual maturity (3–4 years old), a 13-month gestation and average calving every two years when in good condition. Twinning is exceptional. Males *(Jmel)* mature at three to four years but are usually mating when they reach five years old. Mating period stretches between November and the following February/March when the herds are on halophytic vegetation of the *sebkha*. The same selected male bull is kept for up to 15 years with the herd, hence common consanguinity problems. Young males two years old are separated from the herd to avoid fierce fights between males. Castration is sometimes practised to avoid disruption in the herd. The dominant bull is very protective of his females and may attack foreigners (human or animals) approaching them (Photo 10.12). At mating time, usually in autumn/winter, the bull would cover two or three females each day. Several servings per female are necessary for proper fecundation. Bulls are heavy and every care is taken by the shepherds to keep young fragile females far away during mating season.

Delivery, usually easy, does not require shepherds' assistance. At birth, the calves weigh between 15 kg and 30 kg in relation to feeding over the previous 12–24 months. Shepherds keep an eye on calves as mortality can be high and jackal attacks at night are not uncommon. The calf suckles all of its mother's milk during the first six months. Weaning is not regularly practised, and the calf may remain suckling until the next delivery. However, after six months after delivery of the calf, the shepherd handmilks daily half the production. For this he fixes a crude udder-bag (*chemla* or *chmal)* to cover two of the four udders or covering all the udders to deter the calf from suckling. The shepherd would expertly hand-spin the camel's hair and wool and hand-knit the *chemla* (Photo 10.13). When exclusively under the lactating females, calves are reported to show a daily growth rate of 350–700 g/day at 30 days and 120 days, respectively (Moslam and Megdiche, 1989).

Camels' milk production is essential for shepherds living on rangelands. Most of the camels' milk production collected is consumed by shepherds, as his daily meal, often mixed with barley flour (*zumeita*). Dromedaries' milk is also a valuable product, finding its way in minute quantities to the local market and fetching a higher price

Photo 10.11. A *sima* branded on a private dromedary at Bir Hebiliyah (south Al-Assah, 2009).

Photo 10.12. With his handler (at Sriah N 32.70 E 11.64, Al-Assah), a hefty bull pack camel in rutting season, menacing, frothing from mouth, projecting his soft palate (*hadhdara*) with deep gurgling sounds and teeth grinding.

Photo 10.13. The *chemla* covering the udders (Al-Assah).

Photo 10.14. A *h'war* just born (1976, Al-Assah).

than cows' and goats' milk. Lactation lasts about 12 months. Al-Assah cameleers report a daily milking of 2–4 litres/head according to rangeland vegetation conditions. In Tunisia, by comparison, dromedaries' milk production under intensive farming system, varies between 4.2 kg and 2.3 kg/machine milking (Atigui *et al.*, 2014; Kamoun and Jemali, 2012) and Hammadi *et al.* (2012) report up to 6 litres/day.

Supplementing feed rations to the Al-Assah camel herds is rare and erratic. It should be noted that when lactating females are grazing saline rangelands and watered with saline water, the milk produced is slightly salty, and when consumed by humans may induce strong diarrhoea.

Meat is the main product issued from young dromedaries *h'war* up to six months old (Photo 10.14) and then *gaoud*, up to four years old, fetching a prime price (Faye *et al.*, 2014). A two-year old calf would weigh about 120–200 kg (50–80 kg of meat). A four-year-old would weigh about 300–350 kg (120–160 kg of meat). As calves have difficulties following the herd, they tend to be sold after weaning, with the extra young males. In the mid-1970s, dromedaries, brought from the deep south, were sold and slaughtered at the Gasr er-khiah camel market. Adult camel's meat is rather chewy, and customers would much prefer *h'war* and *gaoud* meat.

Dromedaries' hair and wool production is variable according to the phenotype. Northern dromedaries have a good covering of wool and hair on the hump and the flanks. Wool and hair are either shorn or plucked in April from animals reaching their third year. Yield ranges from 0.5 kg to 1.5 kg/head. These are not collected by the Al-Assah project and remain with the shepherds to spin and hand-make crude ropes, *chemla*, or weave the material for making the *wazra* (Photo 10.15). The white *jard* is used as a blanket and light tent by the shepherd on the range in mid-season while the heftier, warmer and waterproof *wazra* is used during cold/wet weather in winter.

Livestock diseases and unidentified mortality on rangelands

Sheep and goats

Most common diseases are Sheeppox, Brucellosis and Foot-and-Mouth (FMD).

Photo 10.15. Draped in a *wazra* at Al-Assah (January 2007).

- Sheeppox symptoms are acute fever with dermatosis developing papules or nodules, necrotic vesicles on the face, rhinitis, conjunctivitis and internal lung lesions.
- Brucellosis symptoms in ewes are abortion at the end of pregnancy. It is propagated by infected rams with lesions and inflammation of the epididymis that transports semen from the testes. Young rams appear to be less affected.
- FMD symptoms are development of blisters in the mouth and on the feet. It has been reported in 2019 in western Libya and Fezzan.

These three diseases are all highly contagious between animals. They can be prevented with vaccination, which is generally not available or rarely practised on Libyan rangelands.

Infectious sites of *Peste des petits ruminants* (PPR – Sheep and Goats Plague) is an acute viral, highly infectious disease of goats and sheep characterized by fever, necrotic stomatitis, gastroenteritis, pneumonia and diarrhoea, often leading to death. It was reported in Jebel Al-Akhdar in 2019 (Bouguedour and Ripani, 2019).

Internal parasites (worms) are common, and treatments are occasionally performed.

Dromedaries

- **Osteophagy** (dromedaries chewing bones) has been observed at the Al-Assah Camel Project. It is a sign of mineral deficiencies. It may also affect small ruminants.
- **Kraft Disease** has been diagnosed in southern Tunisia (Mabrouk, 2007) and likely to be present across the border at Al-Assah. It affects sedentary dromedaries not any more transhuming between *sebkha*, calcareous vegetation and phosphate-rich vegetation. It appears to be the result of an imbalance between calcium and phosphorus in the diet.
- **Large ticks** (*garad or ougrad*), *Hyalomma dromedarii*, are common. Only massive infestations of calves are perceived as potential problems causing nervous symptoms and weakness (Antoine-Moussiaux *et al.*, 2005; Faye, Montpellier, 2023, personal communication).
- More worrying is a form of **'Mad Camel Disease'** identified by 2007 at the Ouargla slaughterhouse in south-east Algeria. Affected dromedaries display weight loss, tremors, aggressiveness, hyperreactivity, typical down-and-upward movements of the head, and other neurologic signs such as 'drunken' walk, loss of control of hind legs, unexpected falls and difficulty getting up. It appears that this could be related to a prion disease or Camel Prion Disease (CPD). It has been confirmed in 2018 by Algerian researchers in the region of Ouargla and found subsequently in 2019 in south-east Tunisia near Tataouine along the Libyan border (Babelhadj *et al.*, 2018).

- In western Sahara, Volpato *et al.* (2015) list the most common diseases (i.e. mange, dermatomycosis, respiratory infections and mastitis) affecting dromedaries and traditional treatments used by Sahrawi cameleers. Much could be learnt from them.

Poisoning or unidentified cause of mortality

Poisoning from native vegetation is an uncommon occurrence and often difficult to identify. Rare are the poisonous plants that livestock would voluntarily consume from Libyan rangelands. Local flocks are accustomed to local vegetation and apparently learn from an early age to avoid consuming poisonous plants. It is more often the result of hasty trucking of flocks from one type of rangeland to another and/or rapid change in diet. This was possibly the case in the mid-1970s when small ruminant flocks displayed an unusual level of high mortality after being transferred from drought-affected Jebel Al-Akhdar region (Ajdabiya) to the Hamada near Bir-el-Kelb (N 31.101 E 12.286). Veterinary *post mortem* examination of sheep on site found balls (1–3 cm diameter) of wool obstructing the intestinal tract. Sheep devouring their own wool could be the result of a high-protein diet deficiency due to drought (Faye, Montpellier, 2023, personal communication), like osteophagy.

Local cattle on eastern matorral

Local cattle breeds are issued from crossbred Italian cattle imported from Sardinia and Sicily during the Italian colonization. We mention them for memory as they are often only seen alone or in small herds in the Jebel Al-Akhdar matorral. They mostly roam freely on the Benghazi plain and the scrub on the escarpment, venturing not far from the farm where they belong. They are rather small, rugged animals, often underfed, with a beige to brownish/red coat. They are kept at little cost for providing milk (about 100–120 kg/year) for the family. When slaughtered, the carcase yields 30–40 kg of meat. None of them is seen in the steppe below the 300 mm isohyet or in the west and Sirte.

Local cattle breeds, as well as horses and donkeys, will be ignored in our work as they represent a small contribution to the livestock numbers on Libyan rangelands.

Livestock Numbers on Rangelands

Libyan livestock data are scarce and unreliable before 1961. By the early 1960s, the FAO of the UN started collecting these data from countries of the world using occasional censuses performed in each country. While livestock censuses for on-farm animals may be more dependable, data from rangelands animals remain subject to 'guesstimate' as herdsmen are difficult to reach in remote areas, and in any case would be suspicious and reluctant to disclose their actual animal numbers that could be used for taxation purposes. Furthermore, when consulting the FAO, sheep number data on FAOSTAT (2021) for Libya spanning 1961–2019, data for 13 years (22%) are FAO estimate, 4 years (7%) are unofficial figures, and 5 years (8%) are FAO data based on imputation methodology. Only data for 37 years are from official Libyan data provided to FAO. Most noticeable data gaps are for the periods 1991–1997, 2003–2005 and 2009–2019. Similar gaps are to be reported for the goats and dromedaries numbers. Even in the best case, it is very unlikely that a proper and full livestock census would have been conducted every year all over Libya. It also looks like the Libyan annual official data reported to FAO are simply computed. Ideally, and still even with much caution, only official census data from the Ministry of Agriculture should be considered. Unfortunately, these census data are not available to us. Therefore, our Libyan livestock data are from FAOSTAT (2021) published data and figures found in the literature.

The first Italian livestock census was performed in 1928 (Rowland, 1945) only in Tripolitania. It reports about 565,000 sheep, 350,000 goats and some 48,000 dromedaries. Then figures appeared sporadically, only showing a spectacular reduction in small ruminant numbers after the 1935/36 drought (Fig 10.1), slowly recuperating by 1944, while the dromedary number remains constant. It looks as if, during this period, climatic conditions and drought drastically impacted the animal numbers on Tripolitanian rangelands.

During this period, still in Tripolitania, the overall ratio of sheep to goat heads was 1.6 for 1928, 1.7 (1930), 1.4 (1935), 0.9 (1936), 1.1 (1937) and 0.75 (1944). This overall ratio of sheep to goats would be around 1.4 to 1.7 in normal years; it drastically dropped during 1935/36 exposing the drought impact on the sheep population (reduced to one fifth) while the goat population, also affected by the drought, was reduced to one third. Likewise, as WWII was ending, there were fewer sheep than goats in Tripolitania.

This overall ratio of sheep to goats does not reflect the shepherds' flock composition. On good rangelands in the Jeffara and Nafusa, the standard sheep flock usually comprises about 200 heads with 10–15 goats (5–7%) providing milk to the shepherd. South of the Nafusa and in the Jebel Al-Akhdar, due to poorly palatable shrubs (*Arthrophytum scoparium* and matorral respectively), the flocks are often exclusively goats.

Livestock data from eastern Libya – Cyrenaica (Allan *et al.*, 1973) show a ratio sheep/goat of 1.14 in 1925, 0.93 in 1945, 1.49 in 1950, 1.04 in 1960 and 1.25 in 1965 (Table 10.1). Here, this overall ratio is leaning towards a higher number of goats' flocks explained by the predominant bushy vegetation in the east more suited to browsers.

Distribution of small ruminants and dromedaries in northern Libya in 1975

A 1974/75 livestock census covering the whole country is reported in a rare document by Khoja *et al.* (1976). From these data (Gintzburger and Bayoumi, 1977), we have an indication of the

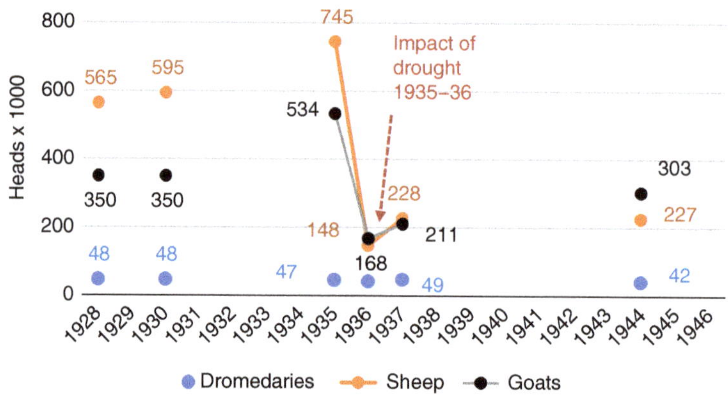

Fig. 10.1. Heads of sheep, goats and dromedaries (1928–1944) in Tripolitania. (After Rowland, 1945)

Table 10.1. Livestock data from eastern Libya – Cyrenaica (after Allan *et al.*, 1973, from British Military Admin.).

	Heads (in Cyrenaica)			
	Sheep	Goats	Dromedaries	Ratio sheep/goats
1925	800,000	700,000	75,000	1.14
1945	413,000	445,000	32,000	0.93
1955	1,032,000	691,000	76,000	1.49
1960	350,000	335,000	67,000	1.04
1965	578,510	464,606	58,163	1.25

small ruminants and dromedaries numbers distribution in northern Libya (Table 10.2) and the ratio of sheep to goats.

These data do not match the published official FAOSTAT (2021), reporting for 1975 about 4.2 million sheep, 1.4 million goats and some 353,000 dromedaries. It shows how careful we must be with the FAO published data officially released by each country. We, however, would anticipate that Khoja *et al.* (1976) had directly accessed the census data from the Ministry of Agriculture and Agrarian Reform. From their data, we do have a one-time indication of the relative rangelands' livestock distribution in the three northern provinces (Table 10.3).

The west, having open rangelands, has the largest population of sheep and goats while the east has, comparatively, a smaller number of sheep and a larger population of goats owing to dense scrublands and matorral rangelands. The central part of Sirte, oddly, displays a rather smaller number of dromedaries compared to the west and the east. We would have expected the Sirte region to be more suitable for dromedaries' grazing. These low numbers may simply reflect the difficulty to account for all animals belonging to semi-nomadic herders, including sheep and goats of the Sirte region. It may also reflect the abandonment of dromedaries as a means of transportation.

Inferring the number of small ruminants and dromedaries per province in 1928

Hypothesizing that the percentage of livestock per province would be similar between the 1975 census for all of Libya (Table 10.2) and the 1928 census for Tripolitania, we estimate the total number of small ruminants and dromedaries in 1928 for all Libya would be as shown in Table 10.4

While our speculation may be questionable, it is an indication that the Libyan sheep flock may have increased by 2.5 times in 50 years (1928–1975), by 1.5 times for the goats, while the dromedaries' number dropped by 60%. The ratio of sheep to goats' heads for all Libya would have been averaging 1.4, decreasing from 1.6 in the west to 1 in the east.

Evolution of the ratio sheep:goats heads

The ratio of sheep to goats in Libya evolves according to climatic and economic events (Fig 10.2). We have seen that this ratio fluctuates around 1.4 in normal years pre-WWII and post-WWII up to 1970 and then starts climbing up from 3.5 to 4 by 1984. This may be the result of an increasing demand for meat from lamb and

Table 10.2. Small ruminants and dromedaries numbers in northern Libya in 1975. (From Gintzburger and Bayoumi, 1977; Khoja *et al.*, 1976)

	Heads livestock 1974/75 census			
	West	Sirte	East	Total Libya
Sheep	1,477,000	432,000	703,000	**2,612,000**
Goats	518,000	162,000	410,000	**1,090,000**
Dromedaries	31,000	13,000	30,000	**74,000**
Ratio sheep/goats	**2.85**	**2.67**	**1.71**	**2.40**

Table 10.3. Small ruminants and dromedaries distribution in the three northern provinces in 1975. (After Khoja *et al.*, 1976)

	% heads livestock 1975 census		
	West %	Sirte %	East %
Sheep	56.5	16.5	26.9
Goats	47.5	14.9	37.6
Dromedaries	41.9	17.6	40.5

Table 10.4. Guesstimate of the number of small ruminants and dromedaries heads for all northern Libya in 1928.

	1928 heads estimates per province (x 1000)			
	Tripolitania 1928 census	Sirte estimate 1928	East estimate 1928	Estimate total Libya
Sheep	565	165	269	999
Goats	350	109	277	736
Dromedaries	48	20	46	115
Ratio sheep/goats	**1.61**	**1.51**	**0.97**	**1.36**

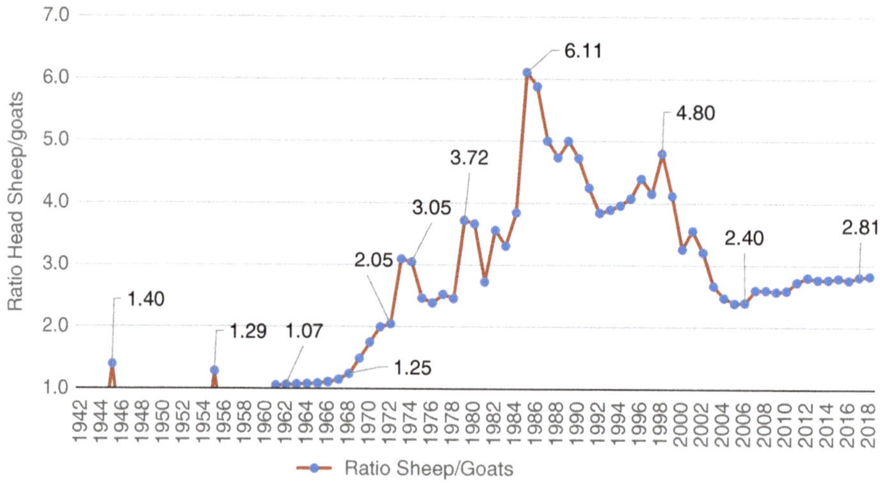

Fig. 10.2. Evolution of the ratio sheep:goats (1945–2018).

mutton with the human population getting wealthier. The 1985/86 peak rockets to an astonishing and unique ratio of 6.1 sheep for each goat and can only be explained by the startling and unexpected change of Libyan banknotes with ensuing banking restrictions that occurred by 1980. This would have seriously dented the population's trust in the local banking system, hence capitalizing by stealthily increasing their sheep head numbers, a long-standing capital refuge with livestock on rangelands not easily accessible to government officials in case of perceived economic instability. By 2003/4, the ratio dropped down to 2.5 to 2.8 depicting lamb and mutton meat remaining much in favour on the local market. It is however doubtful that this ratio remained stable around 2.8 past the 2011 events.

Livestock population evolution on Libyan rangelands (1928–2018)

Using the livestock data (Allan *et al.*, 1973; FAOSTAT, 2021; Rowland, 1945), despite their imperfection and approximation, it gives us a view of the evolution of the sheep, goats and dromedaries numbers (Fig. 10.3) from 1942 to 2017 on the whole of Libyan rangelands. We consider sheep, goats and dromedaries' evolution between 1928 and 2019 after converting each livestock category into sheep equivalent (SE[1]) according to Le Houérou (1981).

In 1928, our estimate is that the Libyan rangeland sustained about two million SE. It was still around 1.9 million by 1945, rising to 3.14 million in 1955. It then stabilized to around 3.5–4 million between 1961 and 1972 and rocketed to around 6–7 million from 1975 to 2000, again climbing steadily to 9–10 million SE by 2009 (Fig 10.4).

These are staggering figures past 2005–2007 with some 9 million SE compared to the 1928–1960 period with about 2 million SE when we would consider the Libyan rangelands in fair condition and when herds would live mostly out of rangeland vegetation occasionally complemented with post-harvest crop residues and local barley grain production. As early as 1977, Gintzburger and Bayoumi (1977) raised the alarm with a report to the Agricultural Steering Committee from the Ministry of Agriculture and Agrarian Reform. They calculated that the Libyan rangelands, from east to west, were already heavily overstocked by 1977, possibly by 200%, considering that only 50% of the range production should be available for grazing and 25% of the annual feed supply would come

Fig. 10.3. Evolution of small ruminants' and dromedaries' heads in Libya (1945–2018). (After Allan *et al.*, 1973; FAOSTAT, 2021)

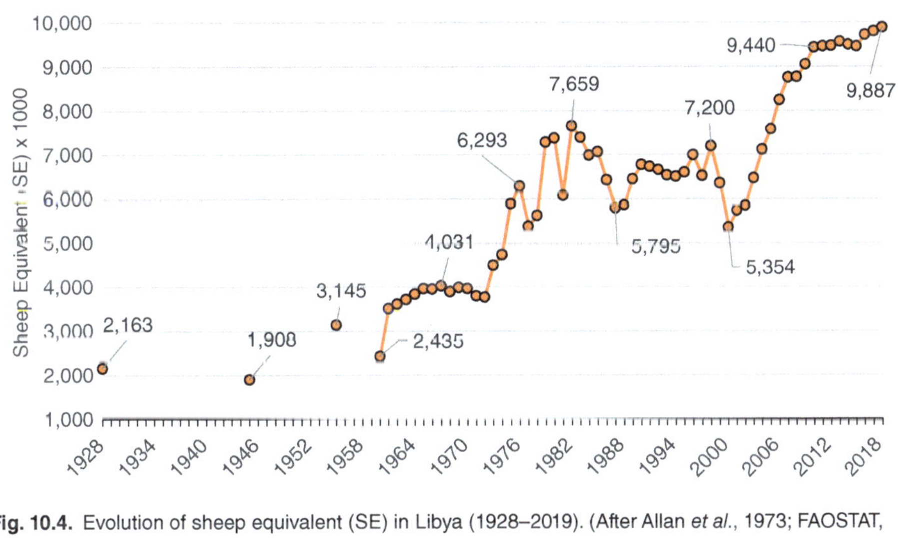

Fig. 10.4. Evolution of sheep equivalent (SE) in Libya (1928–2019). (After Allan *et al.*, 1973; FAOSTAT, 2021; Rowland, 1945)

from barley grain and crop residues during the three to four summer months.

How is it possible that the livestock numbers reached such an unsustainable level?

Importation of feed

FAOSTAT (2021) provides data about imported feed in Libya between 1961 and 2019 such as barley grain, alfalfa meal and pellets, wheat bran, oats and miscellaneous forage products.

- The import of alfalfa meal and pellets started only in 2001 with an average of about 17,000 tonnes/year (STDev. = 9300, n = 15) with a peak in 2010 with 40,000 tonnes. It dropped to a minimal amount, 150–400 tonnes/year, post-2015.
- Wheat bran was imported early in the 1960s for a few years and stopped. It resumed sporadically in early 2000s, dropping to a very low quantity by 2015, likely assigned to the dairy cattle and poultry industry.
- Oat was officially imported starting in mid-1980s in minimal quantities, most likely for seeds to produce forage on irrigated farms and as feed for horses.
- Forage products (not specified) entered Libya in the late 1970s and early 80s, up to 1998, at an average of 39,000 tonnes/year (STDev. = 22 000, n = 19) and dropped abruptly in 1999 to very low quantities (average = 330 tonnes/year, STDev = 406, n = 21) to 2000.

During our field surveys and visits, as early as 1975, we witnessed the impact of overstocking on many rangeland places easily accessible. Gradually, with new wells dug in remote low-rainfall areas, water carting, feed trucking availability, and new cereal and rangeland government projects on the best rangelands being fenced off, the majority of private livestock were slowly relocating onto the most marginal areas where rangeland production was already poor, hence accelerating desertification. This should have slowly strangled the livestock number expansion. In fact, it did not, for the simple reason that increasing feed import, especially barley grain (Fig. 10.5), slowly crept into the system, artificially sustaining more animals on rangelands despite ominous and pervasive overgrazing and environmental degradation.

During this 57-year period, the barley grain locally produced stagnated around 80–150,000 tonnes per year. As for imported barley grain, from mid-1980s, it boomed reaching close to 1 million tonnes by 1990–1993. It dropped

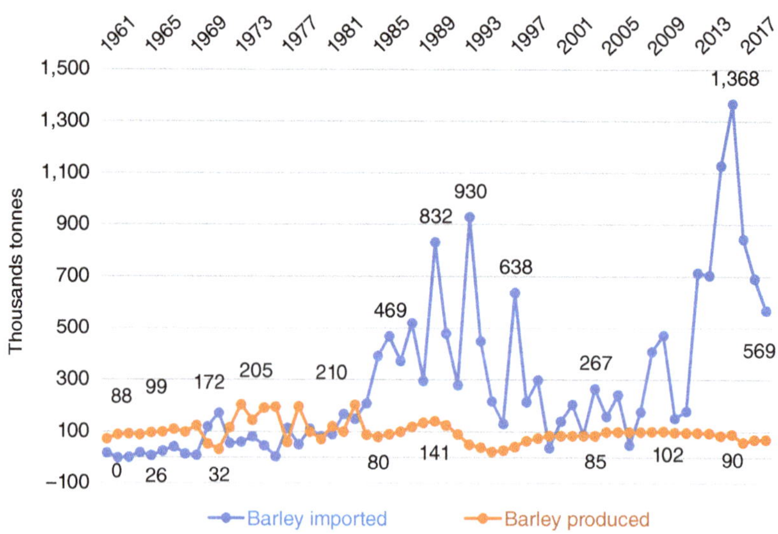

Fig. 10.5. Evolution of barley grain locally produced and imported in Libya (1961–2019). (After FAOSTAT, 2021).

to 150–250,000 tonnes up to 2006 and then to 1.4 million tonnes by 2016, dipping to 600,000 tonnes by 2019.

When looking at the evolution of sheep equivalent (Fig. 10.4), it was by early to mid-1970 that the livestock population exploded, likely exposing a swelling feed deficit. It triggered a massive barley grain import starting in the mid-1980s, filling the feed gap. The availability of imported barley grain largely contributed to artificially increasing the number of livestock on rangelands.

Local barley production

Barley grain is mostly used as supplementary or complementary livestock (sheep, goats, dromedaries) feed on rangelands. We removed the cattle numbers from our calculations as the local cattle breed rarely graze rangelands except in the Jebel Al-Akhdar in dense matorral vegetation, and dairy cattle are exclusively kept on irrigated farms and feedlots on coastal farms.

We hypothesize that possibly 80% is used as livestock feed; the 20% remaining would be for human consumption (Saïdi *et al.*, 2005).

Looking at barley locally produced (FAOSTAT, 2021), we calculate the theoretical daily amount of barley grain available per sheep equivalent from 1961 to 2019 (Fig 10.6). Even if only the barley locally produced was allotted during the four months (120 days) of summertime when rangelands are not providing any grazing, this would amount to between 150 and 200 g/SE/day (1961–1970), reaching a peak of 295 g (1973) and steadily declining to about 40–50 g/SE/day (2015–2018). While we would hypothesize that rangelands would still provide enough grazed feed during the rest of the year, a complementation of less than 250 g barley grain/SE/day would be barely enough for maintenance during the summer months, more so past 1980 when it steadily dropped from about 150–200 g in the 1980s to 50–100 g locally produced barley grain/SE/day past 2000. The barley grain locally produced is not sufficient to cover the ever-increasing livestock numbers with supplementary feed during the summer months. Hence, the increasing importation of barley grain (Fig. 10.5) starting in the early 1980s.

Looking at the total tonnage of barley grain (locally produced and imported) and still hypothesizing that 80% of the total barley production and import is used for sheep, goats and dromedaries feed, then the theoretical daily amount of barley grain available per SE (Fig 10.6) during the summer months is barely sufficient up to 1982/83 (150–400 g/SE/day). It jumped

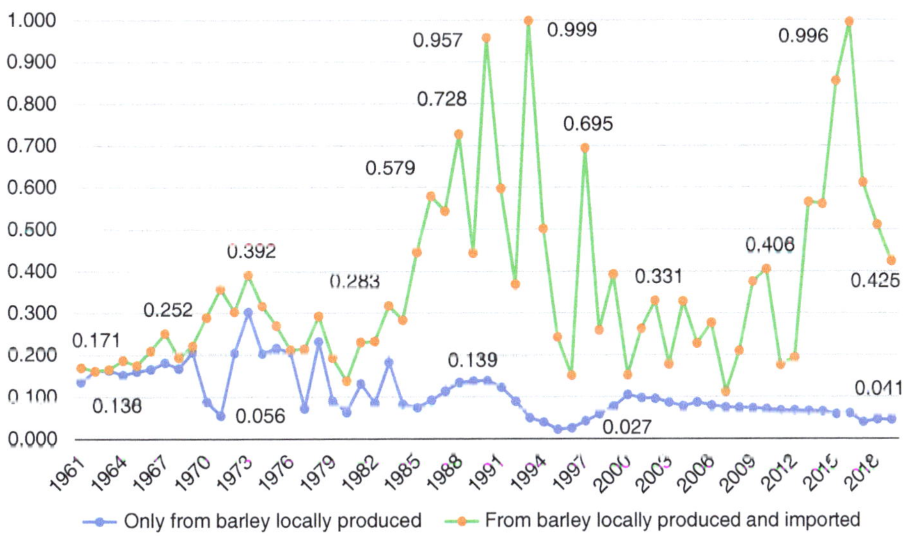

Fig. 10.6. Barley grain locally produced and imported available per sheep equivalent per day during summer months (1961–2019).

to high levels from 1986 to 1997 (500–900 g/SE/day), fell back from 1998 to 2012 (150–300 g/SE/day), and rose again to close to 1 kg/SE/day peak by 2016. These fluctuations of available barley grain feed are undoubtedly linked to climatic conditions affecting the local amount of barley harvested, economic decisions and unrest events. It remains that with the huge import of barley grain, rangeland livestock production has become a country-wide open-air unsustainable feedlot, a common trend in semi-arid regions already identified in Algeria (Boutonnet, 1989) and in North Africa, the Middle East and Central Asia (Gintzburger *et al.*, 1997; Nordblom *et al.*, 1997; Nordblom and Shomo, 1993). The rocketing livestock numbers on rangelands stimulated with barley grain import is relentlessly fuelling rangeland degradation and desertification. This is not a new story in Libya as back in the mid-1930s, Despois (1935), walking the western Jebel Nafusa, mentioned rangeland degradation by overgrazing. Later, Al-Idrissi *et al.* (1996), Gintzburger and Bayoumi (1977), Le Houérou (1968), Nègre (1974), Rowland (1945), Sbeita (2005) and many others repeatedly raised the alarm without being heard. It is therefore clear that without imported barley grain, the current number of SE (2019) in Libya is unsustainable and will lead to irreversible rangeland desertification unless Libyan agriculture can produce more barley grain if extra cropping lands remain available (see Chapter 17).

Conclusions

For centuries, Libyan rangelands were the domain of herded flocks of local breeds of sheep, goats and dromedaries mostly fed on native vegetation up to the mid-1960s with occasional seasonal complementation. Availability of locally produced feed and water limited the number of animals on rangelands. Occasional recurrent national or regional droughts would drastically reduce the livestock numbers. Nowadays, rangelands are not any more supplying the basic feed in quality and quantity to livestock. Large amounts of imported barley grain is a game-changer leading to more rangeland degradation. Rangeland livestock are now mostly artificially fed for months under open-air feedlots with copiously available subsidized imported barley grain and stale bread even into the most remote rangelands reachable by carting feed and water and availability of new wells and cisterns. Today, the Libyan rangelands are likely overstocked by four times more livestock than the native vegetation, even in fair condition, can sustain. The best rangeland gives way to land clearing and pervading rainfed cropping and irrigation projects. Desertification is widespread. High livestock concentration on these open-air feedlots stimulates the prevalence and propagation of infectious diseases not always under control. The situation is dire and simply cannot be purely attributed to climate change or global warming.

Note

[1] Sheep equivalent (SE): 1 adult sheep (40 kg) = 1 SE, 1 goat (30 kg) = 0.8 SE, 1 camel (350 kg) = 5 SE (after Le Houérou, 1981)

References

Ahtash, A. (2006) Goat production in Libya. *Proceedings of the Recent Advances in Goat Production Under Arid Conditions.* DRC Cairo, pp. 303–308.

Al-Idrissi, M., Sbeita, A., Jebriel, A., Zintani, A., Shreidi, A. and Ghawawi, H. (1996) *Libya Country Report.* FAO International Technical Conference on Plant Genetic Resources, 29.

Allan, J.A., McLachlan, K.S. and Penrose, E. (1973) *Libya: Agriculture and Economic Development.* Frank Cass and Company Ltd., London.

Antoine-Moussiaux, N., Faye, B. and Vias, G.F. (2005) Tuareg ethnodiagnostic skill of camel diseases in Agadez area (Niger). *Journal of Camel Practice and Research* 12, 85.

Atigui, M., Hammadi, M., Barmat, A., Farhat, M., Khorchani, T. and Marnet, P.G. (2014) First description of milk flow traits in Tunisian dairy dromedary camel under intensive farming system. *The Journal of Dairy Research* 81(2), 173–182. DOI: 10.1017/S0022029914000089

Babelhadj, B., Bari, M.A., Pirisinu, L., Chiappini, B., Gaouar, S.B.S. *et al.* (2018) Prion disease in dromedary camels, Algeria. *Emerging Infectious Diseases* 24(6), 1029–1036. DOI: 10.3201/eid2406.172007

Bouguedour, R. and Ripani, A. (2019) Overview on animal diseases situation in the Mediterranean basin. Available at: https://rr-africa.oie.int/wp-content/uploads/2019/12/overview-on-animal-diseases-situation-in-the-mediterranean-basin.pdf (accessed 4 August 2024).

Boutonnet, J.P. (1989) *La Spéculation Ovine en Algérie: Un Produit clé de la Céréaliculture.* Institut National de la Recherche Agronomique, Montpellier, France.

Chatty, D. (2013) *From Camel to Truck: The Bedouin in the Modern World.* White Horse Press, Winwick, UK. Available at: https://www.jstor.org/stable/j.ctv289dtw3 (accessed 4 August 2024).

Despois, J. (1935) *Le Djebel Nefousa (Tripolitaine), Etude Géographique.* Larousse, Paris.

Diop, A.K.M., Chrif Ahmed, M.S., Biya, M.B., Haki, M.L., Konuspayeva, G. and Faye, B. (2020) Comparaison des phénotypes camelins de Mauritanie aux écotypes d'Afrique et d'Asie. *Revue d'Élevage et de Médecine Vétérinaire des Pays Tropicaux* 73(4), 247–254. DOI: 10.19182/remvt.31948

Dumancic, D. and Gintzburger, G. (1979) *El-Assah Camel Project – General Information and Prospect for the Future.* Agricultural Research Centre, Tripoli.

FAOSTAT (2021) Livestock data Libya. Data set. Food and Agriculture Organization, Rome. Available at: https://www.fao.org/faostat/en/#data/EK (accessed 4 August 2024).

Faye, B., Jaouad, M., Bhrawi, K., Senoussi, A. and Bengoumi, M. (2014) Elevage camelin en Afrique du Nord: État des lieux et perspectives, Revue d'élevage et de médecine vétérinaire des pays tropicaux. *Revue d'Élevage et de Médecine Vétérinaire des Pays Tropicaux* 67(4), 213–221. DOI: 10.19182/remvt.20563

Faye, B., Konuspayeva, K. and Magnan, C. (2022) *L'Élevage des Grands Camélidés.* Editions Quae.

GEFLI (1972a) *Soil and Water Resources Survey for Hydro-agricultural Development, Central Zone (Sirte area).* Groupement D'Etude Française en Libye.

GEFLI (1972b) *Soil and Water Resources Survey for Hydro-agricultural Development, Eastern Region. Water Resources Survey, 2–2: Climatology*. Groupement D'Etude Française en Libye.

GEFLI (1972c) *Soil and Water Resources Survey for Hydro-agricultural Dévelopment, Western Zone. Water Resources Survey. Climatological Year Book.* GEFLI/Libyan Arab Republic.

GEFLI (1972d) *Soil and Water Resources Survey for Hydro-agricultural Development, Western Zone. Water Resources Survey, 2–2: Climatology.* Groupement D'Etude Française en Libye.

GEFLI (1972e) *Soil and Water Resources Survey for Hydro-agricultural Development, Western Zone. Water Resources Survey, Climatological Yearbook 3.* Groupement D'Etude Française en Libye.

GEFLI (1974a) *Programme de mise en valeur de la zone pastorale développement de l'élevage ovin. Plaine de Gefara et djebel Nefusa. Rapport Agrostologique*. Groupement D'Etude Française en Libye.

GEFLI (1974b) *Survey for Agricultural Development (Wadi Hira, Bu Shaiba, and Gharyan Projects).* Groupement D'Etude Française en Libye.

GEFLI (1977) *Survey for the Development of the Central Wadi Zone and Gulf of Sirte (Sirte, Wadi Zam-Zam Project).* Groupement D'Etude Française en Libye.

Gintzburger, G. and Bayoumi, M. (1977) *Survey of the Present Situation and Production of the Libyan Rangelands* (No. 118/77). FAO and Agricultural Research Center, Tripoli.

Gintzburger, G., Nordblom, T. and Osman, A. (1997) Agro-pastoral systems, feed calendars and feed resources in the Arab countries of the Mediterranean region. In: Haddad, N., Tutwiller, R., and Thomson, E.F. (eds) *Improvement of Crop–Livestock Integration Systems in West Asia and North Africa.* International Center for Agricultural Research in the Dry Areas, Aleppo.

Gintzburger, G., Toderich, K., Mardonov, B. and Mahmudov, M. (2003) *Rangelands of the Arid and Semi-Arid zones in Uzbekistan.* CIRAD-ICARDA, Montpellier, France.

Hammadi, M., Barmat, A., Khorchani, T., Fatnassi, B. and Ben Ahmed, B. (2012) Feed intake, digestibility and milk production in mid lactation of Tunisian *maghrebi* camels fed alfalfa based diet. 3rd Conference ISOCARD, pp. 251–252.

Kamoun, M. and Jemali, B. (2012) Milk yield and characteristics of Tunisian camel. *Journal of Animal Scientist* 1(1), 12–13.

Khoja, M.N., Alidricci, M.S., Al-Ansari, M.D. and Al-Modwak, A.M. (1976) *Study of the Animal Production Sector and the Effect of the Development Plan.* ARC, Tripoli. (In Arabic)

Le Houérou, H.-N. (1968) La désertisation du Sahara septentrional et des steppes limitrophes (Libye, Tunisie, Algérie). Comptes rendus du colloque de Hammamet. *Programme Biologique International, Section Conservation Terrestre et Annales Algériennes de Géographie* 6, 2–27.

Le Houérou, H.-N. (1981) *How to Calculate Livestock Conversion Factors on the Libyan Rangelands* (UNTF/LIB 018; p. 15). FAO and Agricultural Research Center, Tripoli.

Louis, A. (1979) *Nomades d'hier et d'aujoud'hui dans le sud Tunisien, Collection Mondes Méditerranéens.* EDISUD, Aix-en-Provence, France.

Mabrouk, S. (2007) Kraft disease in Tunisia: Dromadaires et chameaux. Available at: http://camelides.cirad.fr/fr/science/kraft_tunisie.html (accessed 4 August 2024).

Magid, A.F. and Mursy, A. (1987) Productivity of the Libyan Barbary sheep in relation to face colour, Libyan. *Journal of Agriculture* 13, 21–29.

Magid, A.F. and Zaied, A.A. (1987) The Libyan Barbary sheep. II: Growth and wool traits. *Libyan Journal of Agriculture* 13, 9–18.

Magid, A.F., Zaied, A.A. and Sharieha, A.M. (1987) The Libyan Barbary sheep. I: Reproductive traits. *Libyan Journal of Agriculture* 13, 1–8.

Mason, I.L. (1967) *Sheep Breeds of the Mediterranean.* Commonwealth Agricultural Bureaux, Farnham Royal, UK. Available at: https://www.cabdirect.org/cabdirect/abstract/19670104208 (accessed 4 August 2024).

Mohamed, A.H., El Hawy, A.S., Sawallah, M.N. and Squires, V. (2019) Middle East and North Africa livestock systems. In: Squires, V.R. and Bryden, W.L. (eds) *Livestock: Production, Management Strategies and Challenges, Agriculture Issues and Policies, Animal Science, Issues and Research.* Nova Science Publishers, Hauppauge, New York.

Moslam, M. and Megdiche, F. (1989) *L'élevage camelin en Tunisie*. Available at: chrome-extension://oemmndcbldboiebfnladdacbdfmadadm/https://www.doc-developpement-durable.org/file/Elevages/chameau-dromadaire/L-%C3%A9levage%20camelin%20en%20Tunisie.pdf (accessed 4 August 2024).

Nègre, R. (1974) Les pâturages de la région de Sirte (Libye): projet de régénération. *Feddes Repertorium* I(85–3), 185–243.

Nordblom, T.L. and Shomo, F. (1993) Livestock and feed trends in West Asia and North Africa: Past, present and future. *CIHEAM – Options Méditerranéennes* 1(5), 15–30.

Nordblom, T., Goodchild, A., Shomo, F., Gintzburger, G. and Renard, C. (1997) Dynamics of feed resources in mixed farming systems of West/Central Asia-North Africa. In: Renard, C. (ed.) *Crop Residues in Sustainable Mixed Crops/Livestock Farming Systems,* CAB International, Wallingford, UK, pp. 131–148.

Passon, J., Braun, K., Hamid, S., al-Mahdi Khalifa, S., as-Sadeq at-Tellisi, N., El Nayedh, M. and Metzger, J. (2020) Places of trade, communication and religion: important oases. In: Braun, K. and Passon, J. (eds) *Across the Sahara: Tracks, Trade and Cross-Cultural Exchange in Libya.* Springer International Publishing, Cham, Switzerland, pp. 165–247. DOI: 10.1007/978-3-030-00145-2_6

Roques, D. (2009) Procope de Césarée, La guerre contre les Vandales (Guerres de Justinien Livre III et IV). In: *Les Belles Lettres - La Roue à Livre*. Paris.

Rowland, J.W. (1945) *Survey of Land Resources in Tripolitania.* British Administration, Department of Agriculture.

Sagne, J. (1950) *L'Algérie pastorale, ses origines, sa formation, son passé, son présent, son avenir.* Imprimerie Fontana, Algiers.

Saïdi, S., Lemtouni, A., Amri, A. and Moudden, M. (2005) Use of barley grain for food in Morocco. In: Grando, S. and Macpherson, H.G. (eds) *Food Barley: Importance, Uses and Local Knowledge*. International Center for Agricultural Research in the Dry Areas. Available at: https://hdl.handle.net/20.500.11766/67677 (accessed 4 August 2024).

Sbeita, A.A. (2005) *Pasture/crop/livestock integration in Libya.* FAO-RNE Technical Report. FAO and Agricultural Research Center, Tripoli.

Volpato, G., Lamin Saleh, S.M. and Nardo, A. (2015) Ethnoveterinary of Sahrawi pastoralists of Western Sahara: camel diseases and remedies. *Journal of Ethnobiology and Ethnomedicine* 11, 54. DOI: 10.1186/s13002-015-0040-4

11

Water on Rangelands

Abstract

Water availability is vital for livestock owners and farmers on arid rangelands. Communities have, since antiquity, through hard labour, dug wells and cisterns, developed efficient water harvesting, flood farming and management techniques to harness every drop of water. Water accessibility from widely spaced traditional rangeland wells and watering points was the norm up to the end of WWII. It was soon perceived as a limitation to modern rangelands use and livestock production, hence uncontrolled implementation of legal and illegal wells and cisterns. Lack of control led to chaotic competition between government authorities, communities and private entities, resulting in accessing and overgrazing the most remote rangelands. Modern drilling and pumping stations accessing the deep aquifers also contributed, leading to rangeland destruction in Libya over the past 70 years. Without strict control and policing of water resources, Libyan rangelands will be totally degraded, leading to livestock being exclusively fed on open-air feedlots.

Introduction

In arid zones, seasonal winter rainfall is scarce and highly variable within and between years, randomly and spatially distributed, hence securing water and collecting every drop is vital to populations, to livestock and rainfed agriculture, even in desert areas.

With sparse vegetation (often less than 30% cover), very low organic matter in the soils (less than 0.5%) and extremely large, bare water catchments, run-off and run-on are important features, and violent precipitations, even occurring far away from the wadis, may cause havoc downstream with torrential water walls loaded with mud and rocks. These flash floods are extremely dangerous, cutting roads and tracks, devastating infrastructure and houses. Some wadis may flow once or twice a year, but most would remain dry over many years. During the rainy season, flooded wadis are difficult if not impossible to pass. Local populations have long developed parallel tracks along wadis, one for summer that is running into the wadi itself and the winter track that is on the dry terrace along the high wadi banks. In any case, herdsmen, steppe and desert dwellers will never set camp in a wadi or lowland susceptible to sudden flooding.

Water Resources

Wadis

There are very few permanent rivers in Libya. Most wadis (*wadi* = a valley) remain dry and may trickle or violently flow with the first autumn rains (October/November) or mid-spring

DOI: 10.1079/9781800627154.0011

(February/March). The exceptions are in eastern Libya with short permanent courses like wadi Derna, Al-Kuf and Al-Athrun, with variable amounts of water running most seasons. Extraordinary dramatic and destructive floods, such as with storm Daniel (10–11 September 2023) saw some 400 mm of rain fall during one night in the Derna region. Mention must be made of the mythical underground Lethe river flowing in the fractured karst near Benghazi.

In the west, the water collected on the Al-Assah plateau trickles slowly into the large saline and gypseous lowland *sebkha* Tadet along the Tunisian border. In the Jeffara plain, wadis (Abu-Shayba, Al-Hira, El-Hay etc.) descend from the northern slope of the Jebel Nafusa and end up lost under the sand of the Jeffara plain. At the eastern end of the Jebel Nafusa, wadi Lebda at Leptis Magna, and wadi Kham (also known as wadi Targhalat), 20 km east of Zliten, flow sporadically in winter, discharging into the sea. On the gentle south-western slope of the Jebel Nafusa and wadi Suzam, temporary large lakes (near Derj) may last several years, attracting scores of migratory birds. East of the last escarpment of the Jebel Nafusa and south-east of Homs-Misurata, in the Solul-el-Khoder region, wadis Sassu, Merdum, Zorzor, Sof-el-Jin and Zamzam roll towards the coast at Taworga, Al-Hisha, Al-Huwaynat, Marah and coastal *sebkha*.

In the Sirte region, all waters run to the sea (wadis Bey-el-Kebir, Thamit, Wishkah, Harawah, At-Hamar, Al-Aqr and Ad-Dawmaran) or in endoreic coastal or continental saline and gypseous depressions. In the Gulf of Sirte, from Ras-Lanuf-Ajdabiya, large saline shallow coastal *sebkha* (Al-Kabitah, Al-Hatirah, Ash-Shuawirh, Kurkura) collect the surface water from adjacent large flat desert areas. There are no significant wadis from Ajdabiya to Soluq, Benghazi,

From Benghazi to Tobruk, on the northern slope of the Jebel Al-Akhdar and Marmarica, a multitude of short and deep wadis descend towards the sea. On the southern slope, most wadis, like wadi Samalus, drain water southward, vanishing into large flat loamy-clay endoreic depressions such as Balta Az-Zalak and many others.

Springs

Springs with drinkable water are rare in Libya with few of them in the Jebel Nafusa. Despois (1935) visited and located some 80 of them between Nalut and Kikla (Map 11.1).

Twenty-five of the most noticeable ones are located between Giosh and Tarhuna (Allan *et al.*, 1973) but remain anecdotal in terms of water output. The best known are Ain Zerga (N 31.92 E 12.00) near Yeffren and Ain Shiarchara (N 32.46 E 13.62). The latter nearly dried-up by 2010 north of Tarhuna, close to a Neolithic site (Fathallah, 1970) with many petroglyphs witnessing an outstanding and rich prehistoric settlement.

The Sirte region does not have any springs.

On the Northern face of the Jebel Al-Akhdar, about 40 significant springs are known and located in the highest rainfall zone and along the coast (Allan *et al.*, 1973). Surveys along the coast took place by 2000 and later to locate and possibly tap into freshwater resurgences into the sea.

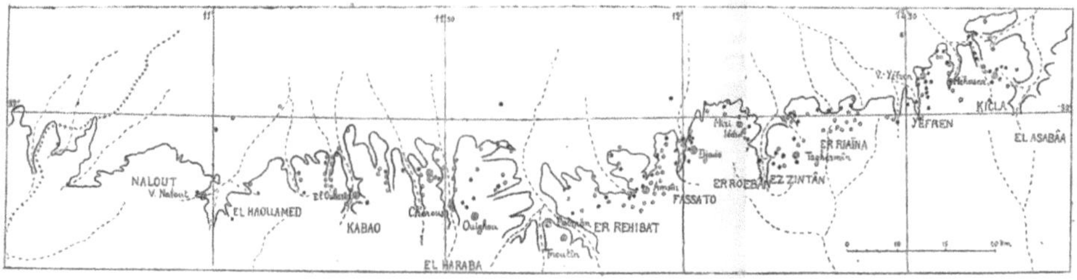

Map 11.1. Location of main springs (.) in the Jebel Nafusa. (From Despois, 1935)

All these springs, water quality and quantity permitting, were or are mostly used to irrigate small gardening areas or orchards mostly on the western and eastern Jebels. These are of little use to livestock owners and rainfed cropping on remote arid rangelands.

Wadis and surface water from run-off

Wadis are most common all over the country where torrential flows occur during spatially and quantitatively erratic winter rainfalls. Local populations and colonizers had to resort to water harvesting techniques to collect drinkable water for human or livestock consumption or to sustain rainfed cropping for either cereals or fruit trees. Libyan local communities have developed many ways to make the best use of scarce surface water with small-scale water conservation works like in the Jebel Nafusa, while Roman, Greek and Byzantine colonizers expanded dams, weirs and cisterns deep into the pre-desert zones up to the 5th century. Many ruins of these antique water conservation works are still to be seen in northern Libya. This has been practised for millenaries in many arid environments (Helms, 1981; Prinz, 1996; Vetter *et al.*, 2009; Vetter and Rieger, 2019a,b) as well as in Libya (Barker, 1996; Barker *et al.*, 1996; Despois, 1935; Gilbertson, 1996; Gilbertson and Hunt, 1996; Goodchild, 1952a, b, 1976). Cisterns were commonly dug and are well maintained to this day in mosques' yards for the benefit of worshippers and the population. Old and some modern private houses all over northern Libya still have their own cisterns collecting rainwater on terraces or earth-compacted surfaces.

Modern drilling techniques give access to water available from artesian aquifers like the one at Jbebina (Al-Assah) or at wadi Zamzam. Despites large flows, water from artesian wells is too irregular and of poor quality (saline, sulphurous) to be used for drinking water by locals or even for agricultural purposes, except the very deep underground water resources extracted at high cost from the Nubian Sandstone Aquifer System in the Kufra basin. This possibly non-renewable huge amount of fresh water has been used since the early 1970s on local pivot irrigation projects and then progressively diverted towards the coast with the Great Man-Made River Project (*Tariq al-Nahr*) providing fresh drinking water to the main Libyan cities. It is therefore of not much use on rangelands.

While many large-scale and small-scale water harvesting systems are commonly available in arid zones, a distinction is generally accepted between floodwater harvesting and rainwater harvesting (Vetter *et al.*, 2009).

Floodwater harvesting and flood farming

Floodwater harvesting consists in benefiting from floods and large amounts of water channelled in wadis, spilling, spreading, or accumulating in cropped depressions locally called *feidah* in the west (Photos 11.1 and 11.2), in flooded areas often speculatively cropped post-flood such as the *Balat*, south of Jebel Al-Akhdar, or in *gar'a*, sporadically inundated desert depressions and occasional partially cropped.

Floodwater comes from large bare watersheds from the Jebels. It escalates into sudden floods or even fast-moving water walls originating far away from the target depression. The wadi discharge may be bursting and violent, causing large amounts of water and alluviums to be transported into the depression. In this case, and over time, local farmers would build chequered dry boulders and stone embankments, weirs and low dams across the depression to abate the fierce flow and ensure that water is gently spread across the largest possible areas. It also encourages water infiltration, supplying water tables. Fine alluviums and organic matter would sediment and form low terraces against dams and embankments used for opportunistic cropping. These banks would be maintained and raised accordingly over years and centuries. It allows opportunistic rainfed cropping even in extremely low rainfall zones. It usually does not require any bulky earth or stonework, but just finding on the vast rangeland a wadi discharging remote run-off in a depression having suitable soils to crop. These soils are usually loam to clay without trace of salinity and have a higher field water capacity than

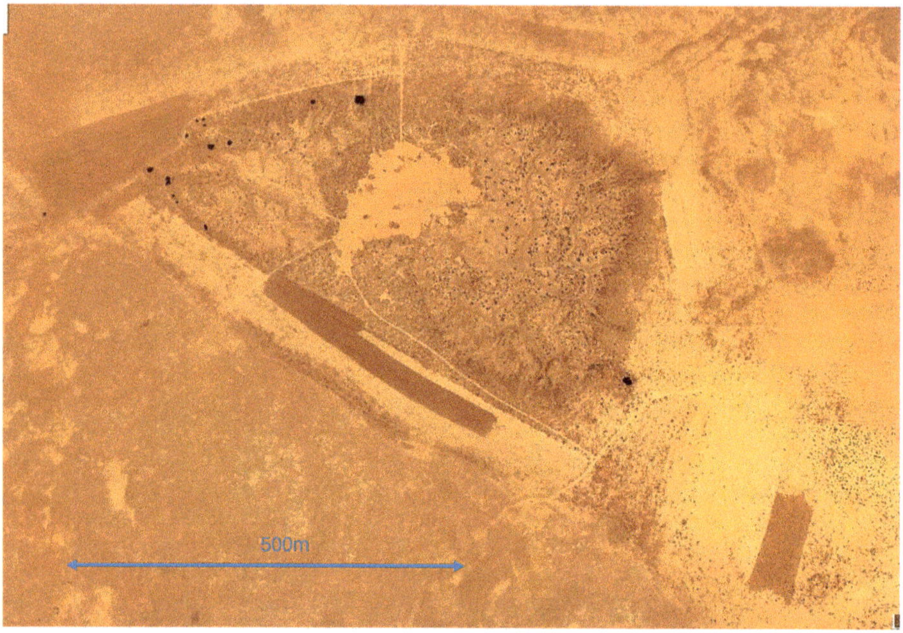

Photo 11.1. Balta with opportunistic rainfed cropping (N 31.864 E 21.600, head of wadi Samalus (MAR ± 100 mm), about 35 km south of Derwa). (From Google Earth)

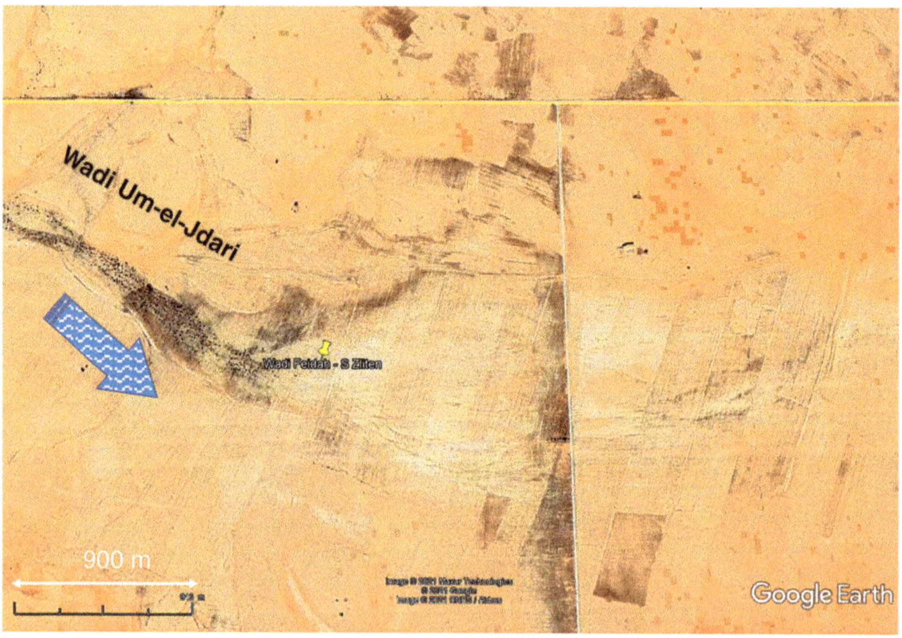

Photo 11.2. Wadi Um-el-Jdari / *feida* (N 32.085 E 14.385); MAR ≈ 150 mm; flood farming in large depression for rainfed barley ≈ 2700 ha. (From Google Earth)

sandy soils. These valuable depressions, from a few hectares to several thousand hectares, are divided into cropping sections (*sikka* or *sekkah*) allotted to members of a family or a clan, or a tribe having the grazing rights on the surrounding rangelands.

Further south from Derwa-Mekhili (south Jebel Al-Akhdar) extremely large depressions collect water flowing from the southern slope of the Jebel into the Balat region in the desert.

These depressions are often inundated, receive large run-off from wadis and are then rainfed-cropped exclusively with barley and sometimes, in limited and better condition, with wheat. Barley is preferred as it is more drought-resistant than wheat. Green barley is often grazed by lactating ewes in spring while the rest of the flock is grazing the rangeland vegetation away from the depression. Barley grain is commonly used to feed livestock, and post-harvest barley stubbles, either standing in the field or finely ground and fed as *tebben*, is a valuable feed supplement for sheep. Moreover, coarsely ground barley grain flour is commonly used to make local and much appreciated flat unleavened bread rich in fibre.

Rainwater harvesting – wadi cropping and fruit trees

Rainwater harvesting is habitually done on a smaller scale than flood farming, though both may overlap.

Rainwater harvesting and wadi farming is an antique practice implemented probably for millenaries in the Jebel Nafusa by local populations and later during the Roman-Byzantine era. It results from water collected from sheet flow from small watershed with gentle slopes (1–3%) and on short distance from the wadi. Sheet flow is a few-centimetres-thick slow water blade occurring on adjacent wadi banks. In low-rainfall zones (MAR = 75–150 mm), water harvesting of this sheet flow is collected close to the water course and gently directed through low stone levees (*jarraya*) (Despois, 1935), which can be several hundred metres long. The rainwater collected with the *jarraya* (Photos 11.3 and 11.4) enters the side of the walled wadi banks into the cultivated wadi sections through a moveable boulder sluice (Photo 11.5). It allows controlling the amount of water flowing into the cropped walled wadi section.

The stone *jarraya* are exclusively established on loamy/marly/heavy and crusted soils common on low-rainfall rangelands but also along the small wadis of the Jebel Nafusa providing that the slope is negligible (1–3%) to prevent starting water erosion.

On rangelands, the limestone in place or bits and pieces of the paleo hardpan are a readily accessible source of stones to construct the *jarraya*. These can be a simple single alignment of stones for the most recent ones (Photo 11.4) or a more sophisticated low structure about half a metre wide and high, constructed with two parallel walls of standing flat stones (Photo 11.3),

Photo 11.3. An antique double-wall *jarraya* near a Byzantine fortified farm at wadi Zorzor (N 31.568 E 14.840, MAR = 75 mm).

Photo 11.4. A recent simple stone *jarraya* alignment at Wishtata (MAR ≈ 150 mm), south Tarhuna.

Photo 11.5. Sluice (watergate) on side wall – wadi farming near Byzantine farms at wadi Zorzor (N 31.570 E 14.841); MAR = 75 mm.

filled in-between with smaller stones as occurring on Roman-Byzantine sites in the Solul-el-Khoder areas. These water harvesting structures common from central to southern Tunisia do not extend into the Sirte area outside the *Limes Tripolitanus* (Mattingly, 1995; Ruprechtsberger, 1993), but are occasionally encountered on the southern arid slopes of the Jebel Al-Akhdar.

Inside the wadi, the slow flow is further managed on successive terraces delimited with stone embankments or low dams (*tabia*) made of dry stones and boulders across each *sekkah*. Water collected is then distributed into the *sekkah* and gathered into several spaced moon-shaped low earth levees hosting either an olive tree, fig tree or grapes. The *tabia* is cut with a solid stoned spillway (*menfes*), allowing excess water to flow to the next terrace. These terraces are nearly exclusively allocated to olive groves and fruit trees (Photos 11.6, 11.7, 11.8, 11.9) and rarely cropped with barley.

In very large wadis (wadi Kaam, wadi Sassu, wadi Nefad), where *tabia* would not stand powerful water flush, long antique solid

Photo 11.6. Present water harvesting in Jebel Nafusa, 4 km south-west of Yeffren with olive and fig trees in gullies. (From Google Earth)

Photo 11.7. Olive trees with water harvesting on well-maintained terraces in Jebel Nafusa, 3 km east of Gharyan; MAR ≈ 250 mm.

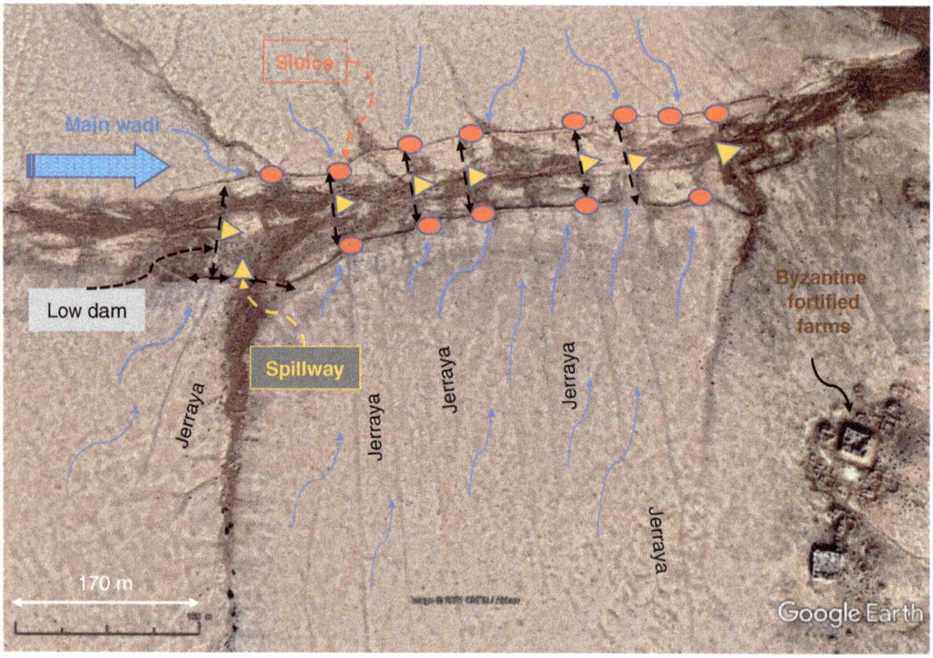

Photo 11.8. Fifth century AD wadi farming at wadi Zorzor near Byzantine fortified farms (N 31.570 E 14.843) ; MAR ≈ 75 mm. (From Google Earth)

Photo 11.9. Wadi farming in wadi Ghirza (N 30.940 E 14.550); rainfed barley + native trees (*Acacia raddiana*); MAR ≈ 50 mm. (From Google Earth)

dams were built across the water flow (Photos 11.10 and 11.11).

Modern water harvesting on mountain rangelands

Starting from the early 1970s, large rangeland rehabilitation and soil conservation projects were implemented in the west. The aim is to allow range regeneration with forestry plantations along embankments (1.5–2 m high, spaced 10–50 m) expected to capture run-off on slopes. These embankments are cut with bulldozers along contour lines on slopes from 2% to 6% (Photos 11.12 and 11.13). The area is protected, and grazing is not allowed once the embankments are constructed. As water and sediments accumulate upstream against the embankments, forestry nursery seedlings

Photo 11.10. Constructed Roman dam (about 310 m long x 3 m wide, stones and mortar) across wadi Sassu (MAR ≈ 125 mm); N 32.130 E 14.803.

Photo 11.11. Roman dam (3rd century AD?) across wadi Nefad (N 31.462 E 14.713) near Qasr-Banat (MAR ≈ 75 mm), with background of *Acacia tortilis* ssp. raddiana trees.

Photo 11.12. Modern water harvesting with rangeland embankments in Jebel Nafusa (Gdama, south Gharyan, N 31.915 E 13.108); MAR ≈ 200 mm; slope: 2–3%.

Photo 11.13. Modern soil conservation and water harvesting with forestry embankments in Jebel Nafusa (Jendouba – Al-Asabah, 26 km SSW Gharyan; N 31.987 E 12.826; MAR ≈ 250–300 mm. (From Google Earth)

of *Pinus halepensis* are transplanted in areas receiving at least 200–250 mm MAR. In the highest rainfall zones (MAR > 300 m), establishment of *Ceratonia siliqua* and sometimes olive trees is attempted. Success is slow and not guaranteed as often the soils removed to construct the embankments expose the underlying hard calcareous crust or the limestone geological layers. Embankments are often interrupted with constructed stone spillways to allow excess water running into the next embankment downstream.

Sometimes, the rangeland space between embankments is deep-ripped using a 1.5 m single-tooth pulled by a D9 bulldozer to allow for deep water infiltration. Unfortunately, deep-ripping tends to destroy the remaining native vegetation (*Stipa tenacissima*) in place, stimulates water erosion, and makes soil surface very rocky and uneven. Shepherds avoid these areas as sheep would easily break their legs.

Water Storage and Supply on Rangelands

Antique and modern cisterns

Water harvesting systems supply antique cisterns or *fasquiya* (Fig. 11.1) from Roman-Byzantine times to the most recent ones dispersed on rangelands on appropriate small water catchments or even in wadis (Photo 11.14). Their locations are well known to livestock owners.

When located in wadis or small water catchments, it catches the surface flow with a constructed stone parapet (head wall) 1–1.5 m high, protruding from the wadi floor. This is to avoid filling up the cisterns with heavy gravel and alluvium transported on the wadi floor with the early flash flood. Once the flow recedes and settles, only the relatively clean water blade enters the shaft and fills the cistern through a decanter/filter or sedimentation pit. These old cistern shafts are cut into the limestone paleo hardpan or geological limestone, marl sublayers or softer stones and excavated to 5–8 m deep in various forms, horizontally digited or ovoid to avoid collapsing. The cisterns have internal walls rendered with an antique Roman-Byzantine very long-lasting waterproof lime mortar (*mortier de tuilleau*) mixed with finely crushed bricks, tiles and charcoal (Büttner, 2020; Malinowski, 1979; Ortega, 2019) still waterproof these days. The mortar lime was produced often on steppic sites, from local limestone (calcium carbonate) crushed and calcined in lime kilns requiring heavy loads of fuelwood to reach temperatures above 900°C. No doubt that large amounts of fuelwood collected on nearby rangelands or forests would have contributed to deforestation, vegetation clearing and degradation – a small price to pay for cisterns lasting several centuries.

Nowadays, old cisterns on rangelands are maintained and regularly cleaned providing fresh water to locals and livestock. These are most common in the west and in the east but are rare in the Sirte region. Abandoned, collapsed or

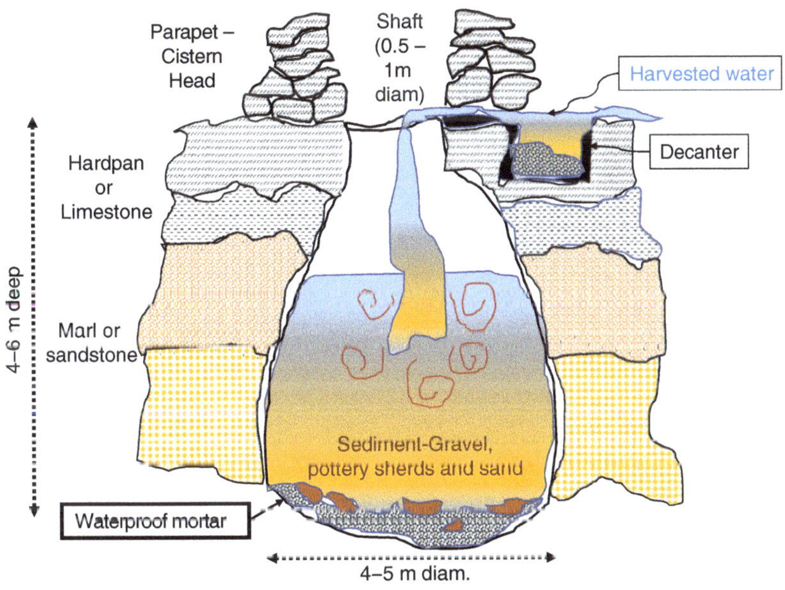

Fig. 11.1. Schematic of an antique Roman cistern at Bir Jubaybinah (N 32.900 E 11.561, 3 km SSE of Jbebina veterinarian buildings – Al-Assah) ; approx. 25 m³ ; MAR ≈ 125–150 mm.

Photo 11.14. Parapet and protruding shafts of Roman cisterns in wadi Turgut (east Jeffara plain, Oct. 1979).

filled-in cisterns can be often located near forgotten antique settlements with stones and marl mounds from excavation (e.g. small roman fort at N 32.635 E 11.679, 25 km south of Al-Assah, east of Bir Ougradia).

Modern cisterns (100–400 m^3) were built all over on recent rangeland development projects, even deep into the Hamada. They are standard reinforced cement-constructed water reservoirs with decanter and livestock watering troughs (e.g. N 32.402 E 12.160, south Sorman) with deep well and submersible electric pumps. While these watering places are welcomed by shepherds, the availability of these modern cisterns poses problems in terms of rangeland management as livestock remains grazing for longer periods on site, hence persistently exhausting rangeland vegetation.

By 2005, other mobile water storage using large water tank bladders (Photo 11.15), up to 20 m^3), made of industrial flexible-coated PVC, became available on rangeland facilitating even more camps and livestock mobility to remote dry sites.

Traditional wells

Traditional shallow wells (*bir* sing., *abyar* plur.) are common all over Libyan rangelands where a perched water table is close enough to the soil surface. Shepherds and farmers have long ago learned to read the landscape and the vegetation indicating the presence of these water tables. This is often confirmed with the presence of phreatophytes such as *Lycium* sp., *Tamarix* sp., *Atriplex* sp., and even reeds. Water level may fluctuate according to seasonal rainfalls. These water tables may withdraw or even disappear during drought. The depth of the shaft is a few metres and is usually not cased. It is reported that wells dug by Romans or traditional wells dug by the locals could reach a maximum depth of 50–60 m but would usually be very shallow wells due to difficult and dangerous digging operations and later lifting water by hand (Photo 11.16) using a rope and a goat's skin or a rubber bucket (*dellous*).

Shallow wells close to the northern or southern piedmont of the Jebel Nafusa provide water of acceptable quality but in limited low quantity. Traditional shallow wells far away from the slope of the Jebel, towards Al-Assah, are close to the *sebkha* and therefore supply mostly brackish water (Bir Ougradia, N 32.623 E 11.610) with 5000–8000 ppm just suitable for dromedaries.

With the availability of windmills as early as 1920, traditional good wells were equipped with this cheap means of drawing water as deep as 40–50 m. Close to the windmills, water storage tanks and troughs were constructed (Photo 11.17) to facilitate and speed up livestock watering. While this was clear progress from drawing water by hand, maintenance of wind-

Photo 11.15. Mobile water storage bladder (N 32.709 E 11.653, Touil-er-Reschiada, Al-Assah).

Photo 11.16. Hand watering flock from a traditional well on the Hamada (1979, Bir-el-Kelb, N 31.097 E 12.285), 75 km SW Mizda).

mills on remote rangelands was a persistent problem.

In the Tripoli region, between 1945 and 1950, about 350 traditional wells and 88 cisterns were repaired, and 60 new wells drilled by the British military administration. During that time, 16 old wells were cleaned close to Yeffren, and three deep wells were drilled for communal use between Yeffren and Gharyan. In the Misurata region, 75 wells were repaired, and 300 concrete troughs were constructed to facilitate livestock watering (Rowland, 1945).

Modern wells

Over time, the effort to supply water to remote areas quickly expanded. Drilling wells to 400 m deep and deeper, in search of more water,

Photo 11.17. Abandoned windmill with concrete water tank and watering trough at Bir Azeteria (N 32.52 E 12.04); 26 km N-NE of El-Watiya.)

became common practice by competing administrations pressed by local livestock owners. These wells are equipped with submersible electric pumps run with diesel generators or even, by 2005, solar panels (Photo 11.18). Near modern wells, water troughs and storage facilities are available to nearby flocks. Extracted water is also carted (Photo 11.19) to livestock owners anywhere, even on the most remote degraded rangelands, even deep onto the Hamada and the desert. The usual incentive to move and rotate their flocks to rangelands in better condition was soon lost as livestock owners turned to operating stationary open-air feedlots, having nothing left to graze or keep their flocks in place where there is very little to graze but with available water on site. It is now common to see resident flocks exclusively fed with about 0.5–1.0 kg barley grain/head/day and providing fresh water at a rate of 3–5 m^3/day for about 800 head of small ruminants. Carted water to refill cisterns or barrels at faraway encampment cost about 5–10 LD for 10 m^3 (2009/10). It would not deter livestock owners paying high prices for water considering the high local market meat price.

Artesian wells

Ancient artesian springs and wells characterized with white gypseous overflow mounds may have been common in the Jeffara plain. Water collected on the Jebel feeds the shallow aquifer through geological layers. Some relics' dry low mounds are still seen south from Sorman on gypseous flats in the middle of sandy rangelands (south Sorman, Saniet-el-Houd, N 32.393 E 12.276).

With modern drilling, artesian wells have become a common feature on Libyan rangelands (Photos 11.20 and 11.21, Al-Assah – Jbebina N 32.910 E 11.546, Darat-el-Mahala N 32.912 E 11.650, Becheria N 32.725 E 11.665 or Gazuliah N 32.950 E 11.588). At Gazuliah, the water surges from 320 m deep, with a pH of 7.5 and a conductivity of 670–690 mS/m (or about 3400 ppm_{500}, winter 2009); this water is suitable for dromedaries and would lead to problems when watering small ruminants for long periods. As a reference, the accepted upper conductivity limit for drinking water for human consumption is about 400 mS/m (or about 2000 ppm), compared

Photo 11.18. Solar panels supplying electricity to the Bir-el-Kababsha well (N 29.802 E 12.090) on Hamada; MAR ≈ 50 mm.

Photo 11.19. Water trucking from modern well at Bir-al-Habet on the Hamada (N 30.070 E 12.097); MAR ≈ 50 mm.

to seawater which has a conductivity of 5000 mS/m (or about 30,000–35,000 ppm).

The artesian water under pressure flows naturally from an aquifer without the need for pumping.

At Jbebina (Photo 11.21), a large pond (about 20 ha) issued from the artesian well expanded in depressions and flows into the nearby *sebkha* along the Tunisian border. The surging water, close to the empty veterinary building, is brackish (4000 5000 ppm in summer) and provides watering for the large resident flock of dromedaries from the Al-Assah project. The pound, largely colonized with reeds, offers refuge, resting and nesting places for a rich diversity of resident and migratory birds (herons, egrets, bitterns, white stork, ducks, grebes, cranes, avocets, kingfishers etc.) transiting from southern Europe to Africa. A bird sanctuary inside a 70 ha enclosure, where hunting was prohibited, was established (Photos 11.21, 11.22 and 11.23 – 20 April 2010) with local community approval and Al Assah project support. The idea was to offer visiting national and international birdwatchers accommodation facilities at the veterinarian building nearby, hence providing work and income to local community members.

Photo 11.20. Artesian well at Darat-el-Mahala – Al-Assah (N 32.912 E 11.650).

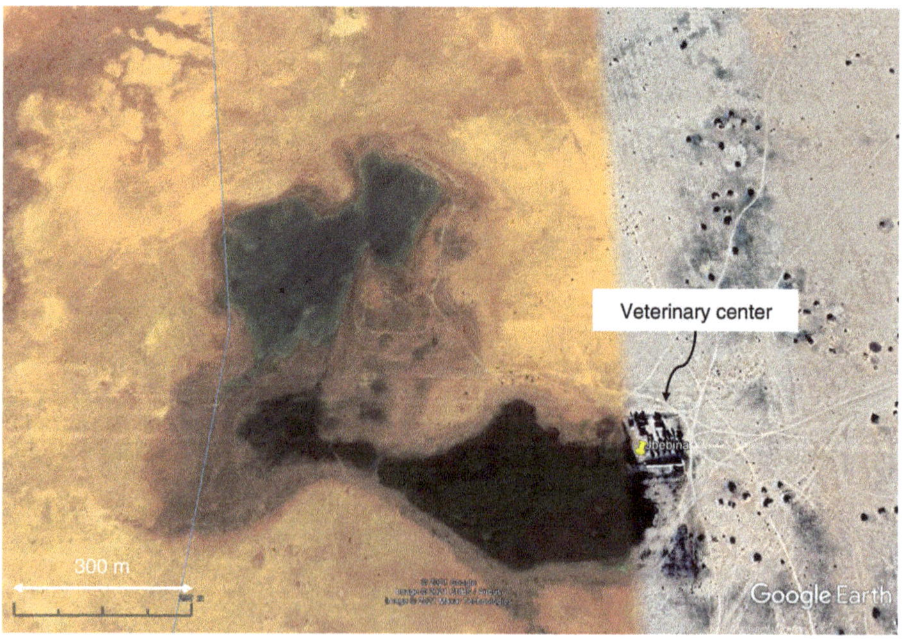

Photo 11.21. The Jbebina ponds (N 32.910 E 11.546, Al-Assah). (From Google Earth)

Photo 11.22. The Jbebina Bird Sanctuary established by Ali Nezdef Sultan and Gus Gintzburger at Jbebina artesian well (Al-Assah – 20 April 2010).

Photo 11.23. Overflow of the Jbebina artesian well (Al-Assah Camel Project, 5 December 2009) ; N 32.908 E 11.545; MAR ≈ 150mm.

Most artesian wells are either brackish, sulphurous (*moya quibrity*) or hot, or loaded with iron from iron bacteria (*Gallionella, Leptothrix* and *Crenothrix* sp.) giving an oily sheen on the water surface (wadi Zamzam). This water is unsuitable for irrigation and barely usable for watering livestock.

Conclusions

Water is critical to livestock owners on arid rangelands and to opportunistic farmers practising rainfed cropping/flood farming since time immemorial. Local dwellers from arid zones have long learned and developed astute water management by hand digging wells, creating cisterns, developing sophisticated water harvesting techniques, building dams, channelling surface water, and harnessing every possible drop of water to permit life even in the most desolate environment. Over centuries, these simple and efficient facilities brought benefits and easier life on rangelands. Water availability from widely spaced traditional wells and watering points on rangeland was the norm up to the end of WWII. It was, however, perceived as a limitation to modern rangeland use and livestock production, hence uncontrolled implementation of legal and illegal new wells and cisterns everywhere. Unfortunately, by lacking water distribution control, it led to chaotic competition between government authorities and private entities drilling legal and illegal new wells, tapping into soon overused water resources. To add to the competition, oil drilling facilities and powerful pumping stations progressively gave access to deep water and artesian water. Furthermore, ever-expanding trucking facilities and mobile water storage allowed not only moving entire flocks to faraway rangelands but also to cart water to these flocks to places where water was never available. This certainly contributed to accessing remote pastures that were once only seasonally accessible. What was, *de facto*, the traditional rotational grazing in space and time due to limited water resources has been disrupted and abandoned, now allowing unrestrained persistent overgrazing on most rangelands. This is certainly not the only reason leading to overgrazing, but it has contributed to rangeland destruction over the past 70 years in Libya.

Without strict controls and policing of water resources on rangelands, we will reach soon a breaking point where Libyan rangelands would be totally desertified and would become a large open-air feedlot for most livestock – just for the flocks to take a stroll and breathe some fresh air on rangeland, without anything to graze.

References

Allan, J.A., McLachlan, K.S. and Penrose, E. (1973) *Libya: Agriculture and Economic Development.* Frank Cass and Company Ltd., London.

Barker, G. (1996) *Farming the Desert: The UNESCO Libyan Valleys Archaeological Survey* (Vol. 1–2). UNESCO, Paris.

Barker, G., Gilbertson, D., Hunt, C.O. and Mattingly, D. (1996) Romano-Libyan agriculture: integrated models. In: Barker, G., Gilbertson, D., Jones, B., and Mattingly, D. (eds) *Farming the Desert: The UNESCO Libyan Valleys Archaeological Survey* Vol. 1, pp. 265–290. UNESCO, Paris.

Büttner, S. (2020) Le mortier de tuileau en France de l'Antiquité à l'époque moderne: Héritage ou réinvention permanente? In: Blary, F. and Gély, J. (eds) *Ressources et Construction: La Transmission des Savoirs sur les Chantiers*. Éditions du Comité des travaux historiques et scientifiques, Paris. DOI: 10.4000/books.cths.10932

Despois, J. (1935) *Le Djebel Nefousa (Tripolitaine), Etude Géographique.* Larousse, Paris.

Fathallah, E.E. (1970) Tarhuna Zone – prehistoric rupestral pictures. *Archaeological News*, 146–149.

Gilbertson, D. (1996) Explanations: environment as agency. In: Barker, S.G., Gilbertson, D., Jones, B. and Mattingly, D. (eds) *Farming the Desert: The UNESCO Libyan Valleys Archaeological Survey* Vol. 1, pp. 291–317. UNESCO, Paris.

Gilbertson, D. and Hunt, C.O. (1996) Romano-Libyan agriculture: walls and floodwater farming. In: Barker, G., Gilbertson, D., Jones, B., and Mattingly, D. (eds) *Farming the Desert: The UNESCO Libyan Valleys Archaeological Survey* Vol. 1, pp. 191–225. UNESCO, Paris.

Goodchild, R.G. (1952a) Farming in Roman Libya. *The Geographical Magazine* 25, 70–80.

Goodchild, R.G. (1952b) The decline of Libyan agriculture. *The Geographical Magazine* 25, 147–156.

Goodchild, R.G. (1976) *Selected Papers of the Late R.G. Goodchild, Libyan Studies.* (ed. J. Reynolds). Paul Elek, London.

Helms, S.W. (1981) *Jawa: Lost City of the Black Desert.* Cornell University Press, Ithica, New York.

Malinowski, R. (1979) Concretes and mortars in ancient aqueducts. *Concrete International,* 66–76.

Mattingly, D.J. (1995) *Tripolitania.* Routledge, New York. DOI: 10.4324/9780203481011

Ortega, I.F. (2019) L'apparition et la diffusion des citernes en méditerranée phénico-punique: un bilan. In: Bouffier, S., Belvedere, O. and Vassalo, O. (eds) *Gérer l'eau en Méditerranée au premier millénaire avant J.-C.* Presses universitaires de Provence, Aix-en-Provence, France.

Prinz, D. (1996) Water harvesting: past and future. In: Pereira, L.S., Feddes, R.A., Gilley, J.R. and Lesaffre, B. (eds) *Sustainability of Irrigated Agriculture*. Springer, Dordrecht, The Netherlands, pp. 137–168. DOI: 10.1007/978-94-015-8700-6_10

Rowland, J.W. (1945) *Survey of Land Resources in Tripolitania*. British Administration, Department of Agriculture.

Ruprechtsberger, E.M. (1993) Die Römische limes Zone in Tripolitanien und der Kyrenaika Tunesien – Libyen. *Schriften des Limesmuseums Aaalen* 47, 120.

Vetter, T. and Rieger, A.-K. (2019a) Ancient water harvesting in the Old-World dry belt – synopsis and outlook. *Journal of Arid Environments* 169, 42–53. DOI: 10.1016/j.jaridenv.2019.103996

Vetter, T. and Rieger, A.-K. (2019b) Water harvesting as a key for understanding adapted ancient livelihoods in an arid environment – approaches to dryland archaeology in the Eastern Marmarica (NW-Egypt). *Journal of Arid Environments* 171, 102940, 22. DOI: 10.1016/j.jaridenv.2019.02.009

Vetter, T., Rieger, A.-K. and Nicolay, A. (2009) Ancient water harvesting system in the north Eastern Marmarica (North Western Egypt). *Libyan Studies* 40, 9–23. DOI: 10.1017/S0263718900004489

12

Tribes and Territories

Abstract

Berber and Arab Libyan tribes have long managed their rangelands despite an endless history of battles, invasions and colonizations, fiercely defending and controlling their territories, often sharing their scant resources, their rainfed cropping and water in times of scarcity or drought. Each tribal territory covers a diversity of landscapes and rainfall zones allowing for a sustainable land management often hardly providing sufficient living conditions. Tribal rules about territory and shared resource management are reviewed with examples from the west, Sirte and the east extending deep into the southern desert. The weaving art, composing the customary bedouin tent, is portrayed as steadily disappearing. With the discovery of immense oil reserves by 1959, the Libyan tribal system and economy gradually evolved, leading impoverished nomads and semi-nomads to quit the rangelands life for financially more secure but humble employment. Yet the *esprit de corps*, traditions and rules within and between tribes remains strong to this day.

Introduction

It is impossible to write about Libyan rangelands without mentioning the complex ancient and recent tribal systems, networks, alliances, feuds and interaction over centuries, co-operating or fighting between themselves to secure grazing territories, watering places, cropping land, and uniting against encroaching invaders or colonizers.

The Ancient Berber Tribes in Libya

Ancient tribes and communities inhabited and used the Libyan steppe and the desert up to the 6th century CE[1] and possibly as late as the 11th–12th centuries, successively confronting or associating with the Carthaginians, Romans, Vandals, Byzantines, Arab Muslims, Turks and Italian conquerors. This is documented by Herodotus (484–425 BCE), Pliny the Elder (23–79 CE) and Ibn Khaldun (1332–1406 CE), Bates (1914), Brett and Fentress (1996), Desanges (1962), Ibn Khaldūn (1852a, 1852b, 1856a, 1856b), Mattingly (1995), Najem (2004), Smith (1854), and many other authors (see http://www.perseus.tufts.edu/hopper/searchresults?q=Libya [accessed 11/1/23]) for Libya.

The Indigenous Libyan tribes were called *Berber* by the ancient Greeks. They named 'uncivilized populations' as *βάρβαρος* (barbarous), all foreigners not belonging to their own civilization including the Romans. Later the Roman empire applied the same word to any local populations they subjugated. In fact, the Berbers call themselves *Imazighen* often translated as 'the freemen'. Native to North Africa and ancient Libya, they shared a common language (*Tamazight*) with many regional dialects. They lived on

DOI: 10.1079/9781800627154.0012

the coastal land, the steppe and desert territories from the Atlantic to the Nile, and specifically would cover more than the limits of the modern Libyan borders to the west, east and south.

Five leading ancient tribal Berber confederations are recognized in western Libya (Mattingly, 1995). The names and English spellings of these ancient tribes may vary according to the authors, from disambiguation, interpretation and transliteration from Berber, Punic, Greek or Latin script:

- The *Lotophages* – on the coast from the island of Djerba (Small Sirte) to Great Sirte and possibly to Cyrenaica, reported consumers of lotus fruits likely identified as *Ziziphus lotus* (*sedder*), hence associated to country where *sedder* usually grows.
- The *Libyphoenices* – mixed population of locals and Phoenicians on coastal settlements in Tripolitania.
- The *Gaetuli* – a large confederation expanding from actual Atlantic Morocco to the high plateaux of Algeria, to southern Tunisia (Byzacena), to Libya. They were pastoralists in the steppe and the desert, based in the Nefzaoua (south Tunisia) to Ghadames, western Jeffara to the Dahar, where they had agricultural settlements. They were often associated with or fighting against the Numidians controlling western Libya during the 2nd–1st century BCE.
- The *Nasamones* – from the Sirte coastal region to Awjilah in the desert; they were transhuman pastoralists and date cultivators in the oasis.
- The *Macae* – located west of the Nasamones to wadi Kaam – Leptis Magna into the Solul-el-Khoder (Bu Ngem).
- The *Garamantes* a powerful confederation from the actual Fezzan/Akakus into the Hamada, the Ubari and Murzuk Edeyen, practising irrigated oasis cropping, date palm cultivation and pastoralism, fiercely opposing Roman colonization.

However, many more Berber tribes are known, covering territories from west to east, deep into the southern desert and further. These locations may vary according to the period considered and expansion or retreat of tribes' confederation. Their territories are still fiercely discussed, confirmed or contested by ancient and modern authors (Mattingly, 1995). The main ancient native Libyan tribes mentioned up to the 3rd century CE with their probable original locations were as follows.

In the east:

- The *Marmaridae* (generic name for the tribes of Marmarica) but apparently extending their territory from actual Benghazi to Marsa-Matruh:
  - the *Adyrmachidae* from Marmarica (actual Sollum to Bombah?);
  - the *Giligamae* around Cyrene/actual Shahat/Derna and the Silphium country; they are reported to ride light chariots pulled by four horses;
  - the *Asbystae* (or *Bashitae*) further south from the *Giligamae*.
- The *Auschisae*, west of the *Asbystae* mostly in Cyrenaica, from Berenice (actual Benghazi) – south of Barca (actual Al-Marj) to the gulf of Sirte (actual Ajdabiya).
- The *Nasamones*, from the gulf of Sirte into the desert between actual Siwa and Awjilah oasis, often reported raiding the coastal Greek settlements of Cyrenaica.
- The *Mazices*, also known as Maxyes, Mazyes, Mazikes and Mazazaces, a large confederation of eastern Berber tribes that regularly plundered Cyrenaica up to the 4th century CE. Gsell (1927) reports that at one stage their territory was covering the land up to the Cynips river (wadi Kaam).

In the Sirte:

- The *Macae*, reported to carry shields made of ostrich skins, lived to the west of the *Nasamones*, possibly south of the actual Misurata to wadi Kaam.
- The *Lotophagi*, along the gulf of Sirte, eaters of fruits of *Ziziphus lotus*, extending to Djerba.
- The *Alachroe*, along the coast, who were seafarers.

In the west:

- The *Luwata* (wrongly named 'Mauri' by Procopius (Roques, 2009)), also spelt as Llwāta, Louata, Lauata, Leuathai, Laguatan, llaguas, Hilaguas, Leucada in antique literature.

They were from the east and extended to the Jebel Nafusa and the Jeffara. From the 3rd century CE, all native populations were named 'Mauri' as a generic name used by the Romans, the Vandals and the Byzantines for all North African tribes including the ones from Libya. One of the outstanding leaders of this western Libyan Berber-Botr tribe was Kabaon[2] who fought (536 CE) against the Vandals and the Byzantines in Tripolitania (Roques, 2009). The *Luwata* tribal territory and allied tribes extended from east Barca and Marmarica into southern Algeria/central Tunisia according to Ibn-Khaldun (Ibn Khaldūn, 1852a). Their associated tribes settled as far as the Nile valley as recorded by Ibn-Khaldun (Ibn Khaldūn, 1852a, 1852b, 1856a, 1856b) and by De Agostini (1917b, 1917a).

- The *Ausenses* (or 'Ausurii' or 'Auses'), nomads, inhabited the northern region of the Lake *Tritonis* (likely the Chott-el-Jerid to actual Gabes – southern Tunisia) while the *Machlyes* were on the southern side of the lake, towards Byzacena (all southern Tunisia extending into Jeffara).
- Their neighbours were the *Maxyes* or 'Maxues', settled farmers (possibly in the Matmata and Jebel Nafusa).
- The *Gaetulians*, nomads from the Sahara, reported to roam the steppe from Tripolitania to Cyrenaica. They were subjects of the King of Numidia, served as mercenaries with Hannibal (247–183 BCE), and specifically used javelins during battles.

In the south:

- The *Garamantes* (400 BCE–600 CE), located in actual Fezzan, raiding regularly Oea, Leptis Magna and the north-western coast, and extending to Misurata; appeared to have taken over the 'Psylli' tribe roaming the southern desert.
- The *Atarantes* in actual western Fezzan. The name *Adrar* may derive from the name of this tribe.

These tribes were mostly nomadic while others were semi-nomadic settled in west and east Libya. The ancient Egyptians (1000 BCE) called 'Lebou' (Bates, 1914) the populations living between the Nile and the gulf of Sirte. The Greek used to call all inhabitants of North Africa 'Libya'. The Punic Carthaginians also mention the name of 'Loubi' or 'Loubat' (Gsell, 1927).

According to alliance and enmities of the moment, they would launch expeditions far away from their original territories. While they often fought against each other for scant resources, they also relentlessly battled against Greek, Roman, Vandal, Byzantine and Arab intruders. They also served as slaves or enlisted as auxiliary forces and mercenaries according to reciprocal and temporary agreements with the invaders. Their territories were fluctuating, expanding or retracting, according to endless inter-tribal conflicts for resources and external pressures due to invaders.

Horses, dromedaries and warriors

All these ancient tribes were initially horses-mounted at war (Bates, 1914), for their displacements and trade and using horse-drawn light chariots (see Garamantes, rock engravings in the Akakus mountains). Using horses for travel and transport in the steppe and the desert was limited by water availability. The dromedaries, appearing in North Africa possibly about 2500 years ago from the east (Saber, 1998), changed the warfare landscape. In 331 BCE, dromedaries were used by Alexander's expedition to Siwa oasis. Later (46 BCE), the troops of Juba I, king of Berber Numidia, mounted on dromedaries were defeated by Julius Caesar's troops at the battle of Thapsus (30 km south of the actual Monastir in Tunisia). In Libya, dromedaries are depicted on sculptures at the Roman settlement of Ghirza (Photo 12.1). Dromedaries were efficiently used in battles by the famous *Kabaon*, leading the 'Luwata' western Berber tribes, in their battles (536 CE) in Byzacena (southern Tunisia adjacent to Tripolitania) against the horse-mounted Vandals and the Byzantines (Procopius of Caesarea, in Roques, 2009). Dromedaries not only allowed longer range of operations in arid zones compared to using horses, but war-horses would refuse charging challengers mounted on dromedaries due to the smell of the latter, as reported in the *Vandalic Wars* (533–535 CE) by Procopius (Roques, 2009).

At war, early Berber tribes used mostly javelins but also maces, boomerangs, slings, bows and poisoned arrows kept in leather quivers. Javelins were powerfully hurled using a throwing-thong similar to the spear thrower. They often used a small leather shield and always individually carried a leather water-bottle.

With the Arab conquest, dromedaries became the preferred means of displacement and warfare for all tribes in arid North Africa and desert of the region.

By the 5th–6th century CE, the Berber tribes with their approximate location are on Map 12.1 (Najem, 2004).

Photo 12.1. Dromedaries on sculptures at the Ghirza Roman settlement (wadi Zamzam) – possibly 1st–3rd century CE (from the Tripoli Museum).

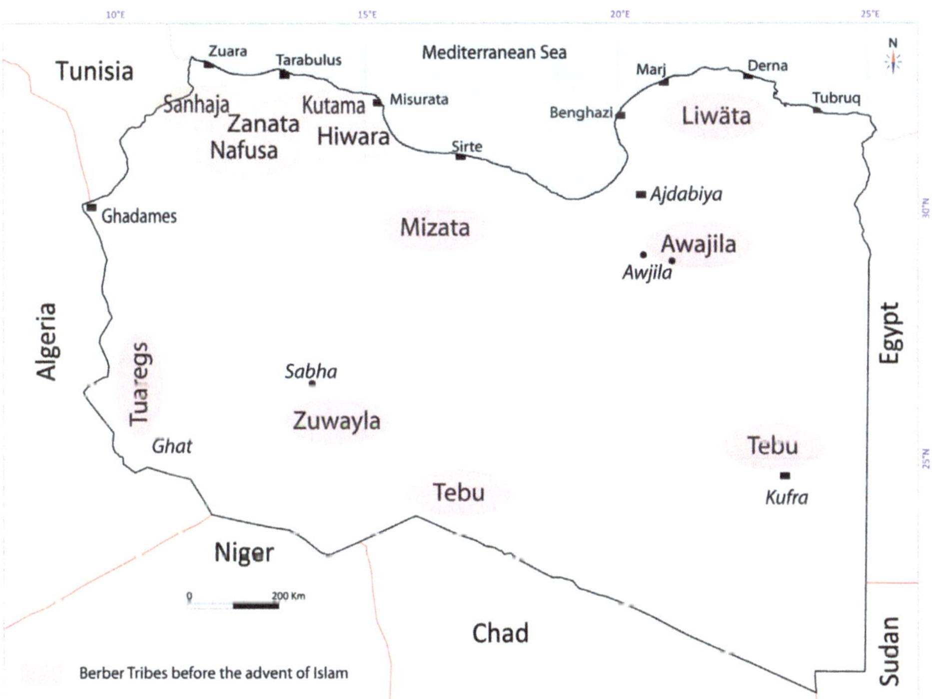

Map 12.1. Main ancient Berber tribes' location before the advent of Islam. (Adapted from Najem, 2004)

Berber and Arab Tribes after the Early Arab Conquest

While the Byzantines held most of western Libya by 536 CE, their rulers were in continuous internal religious and economic battles weakening their grip of the land and the local tribes. It opens the way for the Arab and the Muslim conquests of North Africa. The early Arab control of Byzantine Libya came by 642 CE for eastern Libya and by 643 CE for western Libya. Several stronger waves come one after the other. By 647 CE, the Arab tribes first gained control of northern Libya and southern Tunisia held by the Byzantines at the battle of Sufetula (actual Sbeitla in central Tunisia). The Arabs soon withdrew further east after the Byzantines agreed paying a tribute. But the Byzantines, still holding Barca-Cyrene, were trying to regain control of the Libyan coast. This is when a second Arab wave led by Uqba-ibn-Nafi destroyed the Byzantine army and their Berber allies. Progressing fast westward, Uqba-ibn-Nafi established his capital in central Tunisia at Kairouan (670 CE) and kept conquering North Africa, reaching as far as the Atlantic Ocean. Returning to Kairouan, he was killed (683 CE) in battle and his troops were defeated near Biskra (southern Algeria) by a Byzantine and Berber coalition, the latter led by the chief *Kusayla*. Intertribal wars (*fitna barbaryya*) erupted again for control of the land and resources. This was even further complicated with conflicts between religious factions and heretics within the local Christian communities.

The early Arab conquest re-established order and peace between local tribes for several years, regularly allowing the Berbers to manage their people and land, and using them as much-feared mercenaries to facilitate their North African conquests. Famous was the Aghlabid kingdom (800–900 CE), vassal of the Sunni Abbasids, developing trade and advanced irrigated agriculture around their capital at Al-Abbasiyah near Kairouan. Early 900 CE, the Ismalis, helped by the southern Berber Sinhajah tribes established the Fatimid-Shi'a Zirid kingdom in Kairouan (Idris, 1968b, 1968a), fighting against the Sunni Abbasid Caliphate (Bagdad), swiftly conquering Egypt (969 CE) and establishing their capital in Cairo. The Fatimid, while pressing their wars against the Abbasid to the east, neglected controlling the western part of their empire. Long internecine wars erupted between the Berber tribes supporting either the Fatimids from Egypt or the Umayyad Caliphate established in Spain. The Fatimids from Kairouan slowly collapsed (972 CE) due to the financial crisis, leading to social and regional unrest (Poncet, 1967). They were replaced (1018 CE) by the Hammadids, Sunni's rulers and now vassals of the Abbasid Caliphs of Baghdad (Idris, 1968b, 1968a). The Fatimid in Cairo, facing a prolonged famine and dreading an attack from the west, with the support of the Coptic Egyptian Church, were very happy to get rid of the unruly Arab Bedouin tribes of the Banu-Hilal and Bani-Sulaym (Idris, 1968a). These true Arab nomad tribes had migrated during the 8th century CE from the Hejaz and Najd and temporarily settled in upper Egypt. They were highly disruptive and an embarrassment to the Fatimid rulers who unleashed them towards the western provinces of *Ifriqiya*.

Starting from upper Egypt, with a troop of about 7000–8000 Banu-Hilal warriors, followed by families, tents, camels and small ruminants, they slowly moved along coastal Libya invading first the well-off Cyrenaica and Marmarica (1050 CE), then pushing further west, conquering southern and central Tunisia and Algeria. The Banu-Hilal tribes' confederation included the Djochem(?), the El-Athbedj, the Zoghba, the Rîah[3], the Rebia and the Corra or the Adi, after (Ibn Khaldūn, 1852a,b). If we count 7000–8000 Banu-Hilal mounted warriors supported by as many or more infantry, the whole group will number about 80,000–100,000 heads (with women, children and slaves), followed with likely 15,000–20,000 tents, with 2–3 horses per warrior (15,000–25,000 horses) and as many camels, and 10–20 small ruminants per tent (150,000–300,000 heads). No doubt it would be a heavy burden on the local agriculture and rangelands. Ibn-Khaldun (1852a,b) states that they were 'like a swarm of grasshoppers, destroying everything on their passage', a theory contested by Poncet (1967) in an argument with Idris (1968a,b). There is no doubt that such a large multitude of nomads would have to live on the land, steadily moving fast westward towards Tripolitania and southern Tunisia and crushing any opposition. The invaders spread over large land areas along the way and were rushed to move fast forward by rear guards burning tents if stragglers were not moving camps fast enough. The Banu-Hilal were soon followed by the Banu-Sulaym, a confederation comprising the Rouaha, the Nacera and the Oraeira, settling mostly around Barca in Cyrenaica and in Marmarica. Part of the Sulaym, the Dabbab, the Awf and the Zaghb followed the Banu-Hilal, settling along the way, up to present day in southern Tunisia and Algeria (Map 12.2).

All these Arabs Muslim nomad tribes supported one or several local tribes in their relentless fights against local leaders, changing allegiances according to potential benefits and threats. They slowly, over years, settled in *Ifriqiya*, controlling large areas of rangelands, coastal and inland towns deep into the desert. They blended with and integrated with the local population (Berbers, Christians and Jews), converting them to Islam sometimes with difficulty. Berber tribes retreated to Jebel Nafusa (*al-Jabaliya*) or on the coast (Zuara – *Zwariyya*) while Christian small communities slowly disappeared from western and eastern Libya pockets. A small population of Jews remained in Libya, mostly in Jebel Nafusa (Yeffren, Gharyan, Giado) for many years working in crafts or simply servicing the towns and the rangeland communities as peddlers (De Felice, 1985; Goldberg, 1990).

Féraud (1927) published a comprehensive history of western Libya. He was the French Consul in Tripoli (1878–1884). A French army officer, a diplomat and an Arabic and Turkish interpreter, he spent his life and career in Morocco, Algeria, Tunisia and Libya. He was well-informed about the people, the tribes, the religions and the cultures of North Africa. He compiled an exhaustive *History of Libya from 642 to 1912* covering the Arab period (642–1510 CE), the Spanish Conquest (1510–1551 CE),

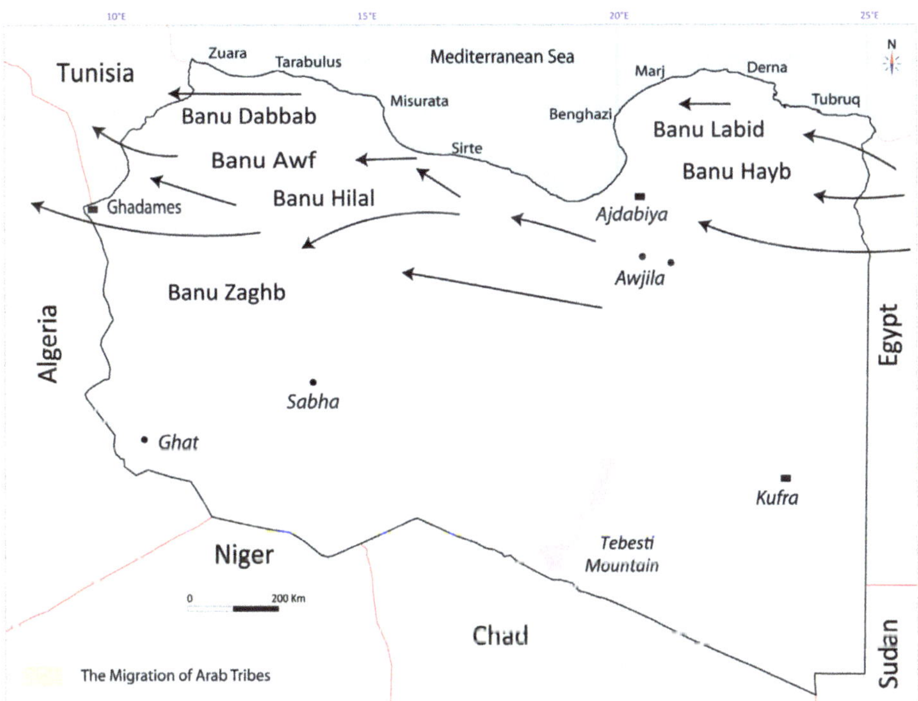

Map 12.2. The 11th century Arab tribes' migrations (1040 CE, 442 H) into Libya. (Adapted from Najem, 2004)

the Turkish period (1551–1711 CE), the Karamanli's Dynasty (1711–1835 CE), the Tripoli Ottoman Wilayat (1835–1879 CE) and up to the Italian colonization (1912 CE). He mostly focused on the Tripolitanian events. He discussed Tripoli leaders, regional and international interactions with neighbouring countries, and with their opportunistic alliances and conflicts with the Berbers population and Arab tribes from west to east Libya, and deep into the Fezzan.

Despite all this turbulent long history, the Libyan tribes had to make do with the scant resources offered by their dry land, steppe and desert. Some of these resilient ancient native Libyan tribes were mentioned up to the 12th century, strongly interacting, fighting against or siding with the Arab conquerors. Slowly and inexorably, they were assimilated into the tribes of the Arab Muslim conquerors.

How the Libyan Tribes Traditionally Used Their Territories and Resources

All over the world, pastoral communities and tribes' territories cover a diversity of ecologically complementary landscapes to provide for their members' and their flocks' sustenance throughout the year. This is not different in Libya where tribal territories are usually stretching north-south across rainfall gradients to benefit from seasonal grazing and wadis' flats for cropping. However, in the west, the semi-sedentary and nomadic pastoral communities have long organized a symbiotic system between the Jebel Nafusa producing barley and fruit trees (oil from olive trees) with the southern oases producing dates in the Chati/Fezzan they control. In the Sirte region, the mostly nomadic communities have also long ago similarly taken control of the date palms oases in Zillah and Jofra and further into northern Chad. In the east, in the Jebel Al-Akhdar, benefiting from high rainfall but devoid of significant palm groves in the southern desert, the pastoral system remains centred on the Jebel itself, providing sufficient grazing and farming resources. The tribes from Libyan Marmarica, having fewer farming resources, relied on exchanging farm products with the other tribes from the Jebel and nearby Egyptian Marmarica.

> The names and lineage of Libyan tribes with maps of the location of their territories are detailed in Behnke (1980); Cauneille (1954, 1957, 1958, 1963); De Agostini (1917a, 1917b; Despois (1935); Najem (2004).
>
> **DISCLAIMER: We do not intend to confirm or invalidate the territorial limits, watering places or cropping zones of the Libyan tribes, factions, clans or families presented on our historical maps and documents. Our aim is simply to illustrate and explain the geography and the many ways the tribes spatially used their rangeland territories and resources over centuries.**

As examples, we explore four main tribes or group of tribes which roamed the Libyan rangelands in the west (up to the mid-1960s) and in the east (up to 1975):

- the Zentan from the Jebel Nafusa to the Chati;
- the Magarha from Ash-Shwayrif (Chouiref) to Brak;
- the Gadadfa from inland Sirte; and
- the tribes from the Jebel Al-Akhdar.

The tribes from western Libya and Sirte

The main tribes with strong nomadic traditions and practising long-distance transhumance, from western Libya and Sirte, use the rangelands stretching from the Jebel Nafusa or the coast to the oases in the south (Map 12.3). We will consider neither the tribes nor clans traditionally grazing the Jeffara plain and the slopes of the Jebel Nafusa, such as the Sian, the Nouail, the Bu-Agheila, Belassa, Warshefena, Guiara, Hamadat, etc. (Map 12.4; Ricchieri, 1913), nor the factions from the Ouerghemma, the Ouderna, and the Mrazig usually roaming nearby southern Tunisia and often entering Tripolitania (Fozzard, 1987; Louis, 1979). In fact, all the above tribes did not recognize any border as the Jeffara and southern rangelands extend from modern Libya into Tunisia, and often reciprocated cross-border raiding up to the beginning of the 20th century.

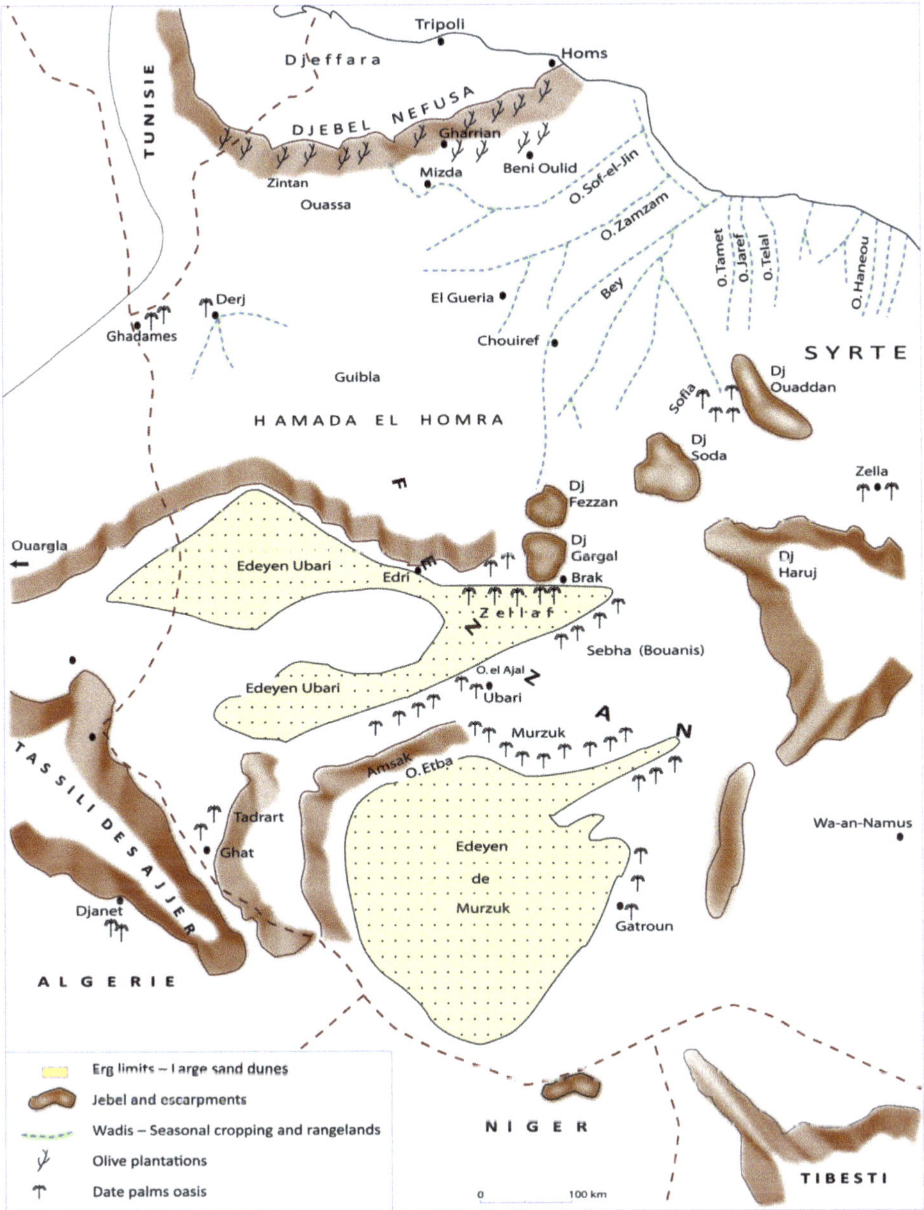

Map 12.3. Western Libya and Sirte from the coast to the deep south. (Adapted from Cauneille, 1963)

The Zentani

The Zentani appear to be originating from ancient Zanata (or possibly the Sanhaja) Berber tribe overpowered by the Banu-Sulaym associated with marabout tribes. They have their main settlement at Zentan (Cauneille, 1957; De Agostini, 1917a, b; Despois, 1935). They include two tribal divisions: the Ould-Belohl, mostly Arabs, and the Arabo-Berbers Ould-Douib. The Zentani

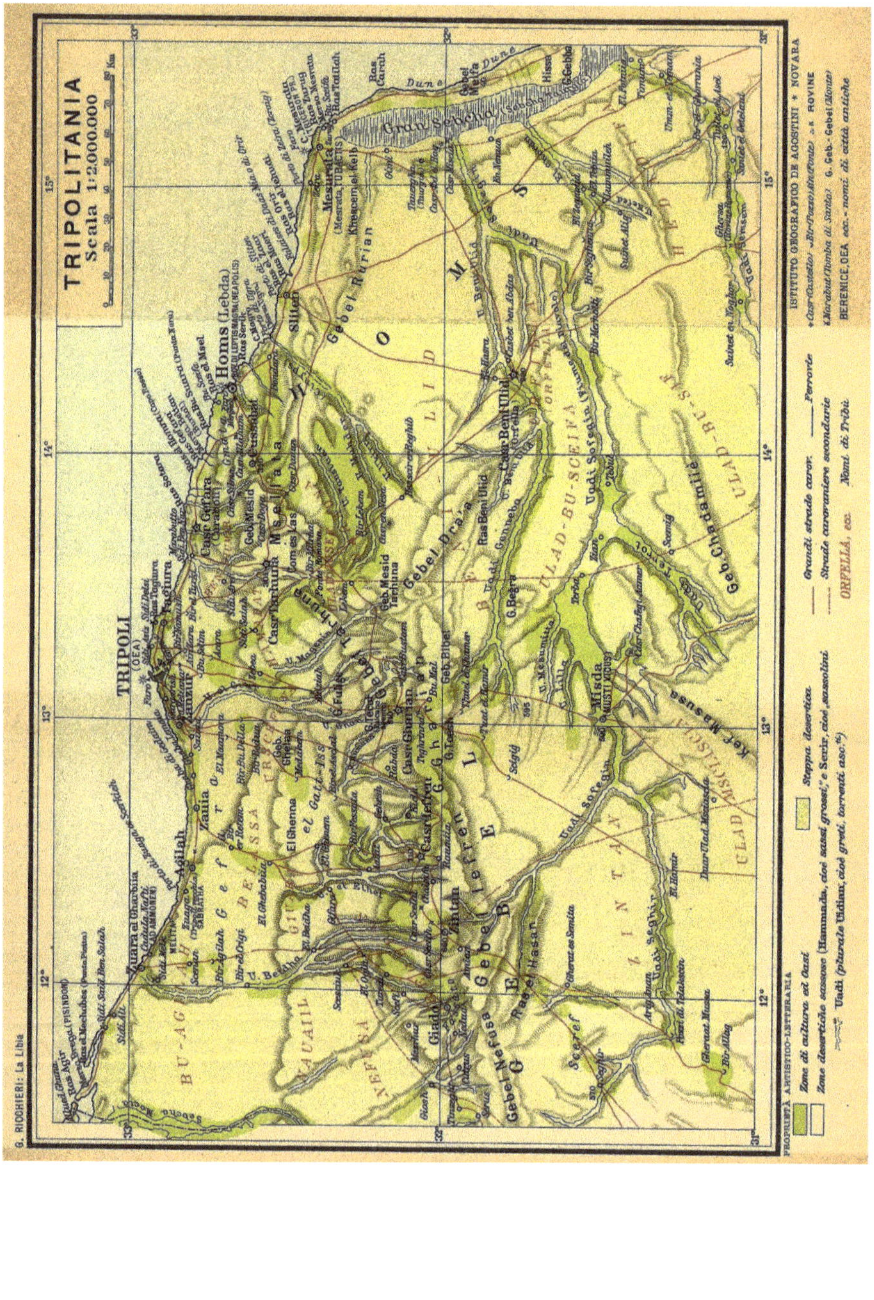

Map 12.4. Map of western Libya (Tripolitania) with main tribes' locations, early 20th century. (From Ricchieri, 1913)

have a strong status among the Libyan tribes. They have a war drum (*tobbel*) held by the Ould-Douib to call to arms and '*Ya Zina*' is their war-cry. In 1917, Agostini (De Agostini, 1917a, b) reported that the Zentani gathered 7300 heads of which 6000 were Arabs and 1300 were Arabo-Berbers.

The Zentani run a complex symbiotic system associating livestock on rangelands, rainfed cropping in wadis and depressions, and date palms and irrigated cereals (wheat, barley, sorghum and millet) in the southern oases in the Fezzan (Maps 12.3 and 12.5). This diversification compels them to allocate time and manpower to these complementary activities that make their strength and wealth. Their territory traditionally stretches from the town of Zentan (Jebel Nafusa) to Ash-Shati (Fezzan), to the west to Derj and south-east to Qariyah Al-Gharbiya, 130 km south of Mizda.

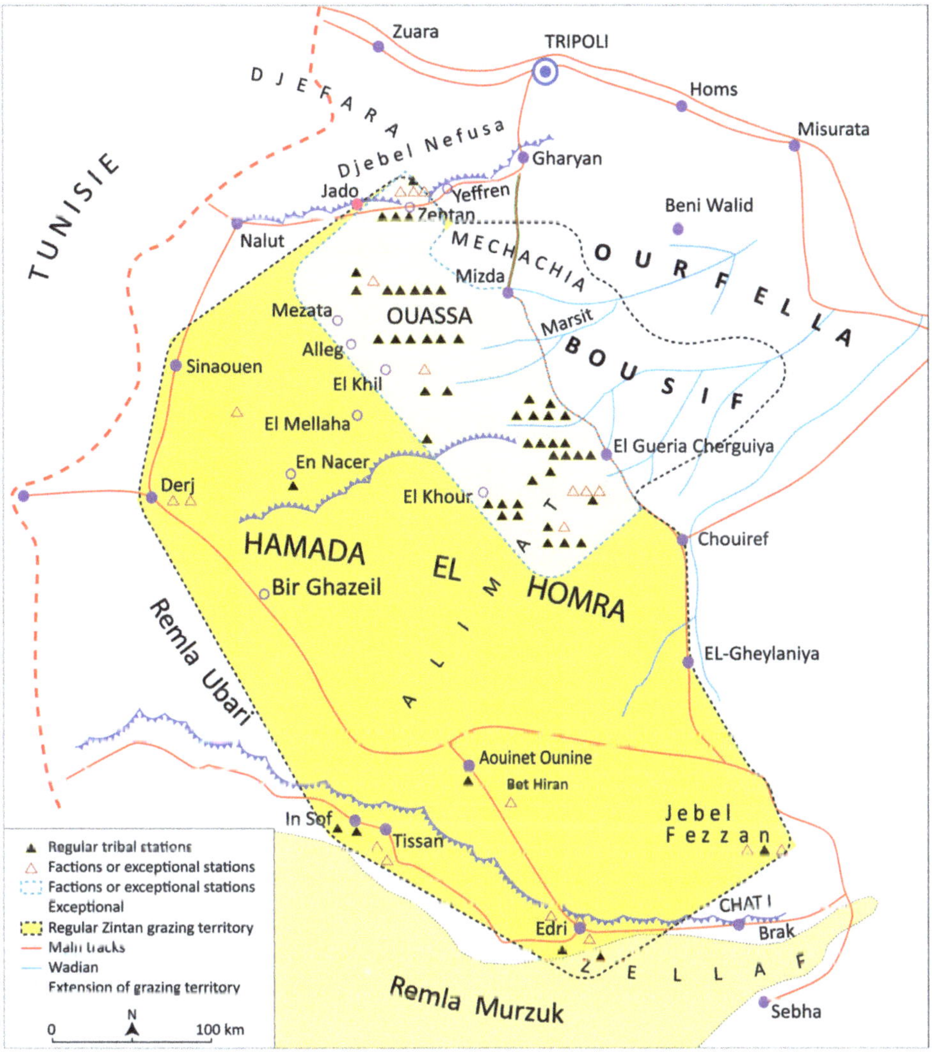

Map 12.5. Customary grazing territory of the Zentani. (Adapted from Cauneille, 1963)

They also occasionally extend their spring grazing lands in the Jeffara from Qasr-el-Hajj up to Zuara on the coast. The link between the people from Al-Assah region seems to hold as, in spring 2010, shepherds (Nouail [Nwaiyl] or Sian tribes?) from Al-Assah were grazing on the Zentani's territory on the Hamada near Bir-Tlakhsin (N 31.244 E 12.056). The Zentani's presence in the western part of the Jeffara plain may be relics of their control of the caravan tracks starting from Zuara – Al-Jmail – Regdalin/Sabratha to Ghadames and further south (Passon *et al.*, 2020). Several tribes such as the Rojban (west of Zentan), the Goualich (Kikla), the Gontrar (Mizda) and the Jaafra, are part of the Zentani. Occasional conflicts occur with the Riaynah as well as with the Ould-Bu-Sif neighbours about cropping land in wadis. To the east, Zentani and Ourfella had a reciprocal agreement to forbid access to wells to other tribes except at two watering stations in wadi Sof-el-Jin. To the west, the Rojban and Haraba would share their rangelands with the Zentani while the Mahmid (Sulaym?) would hold the Jebel from Nalut to Cabao. The Zentani would host flocks from other tribes from Derj and Ghadames and further west providing that they submit to Zentani's rules.

As for the vegetation available to the Zentani, we find from the north large areas of *Artemisia herba-alba* and *Anthyllis henoniana* in the Jebel suitable for sheep grazing in summer and autumn. As aridity increases, large expanses of *Arthrophytum scoparium* are mostly used by goats. On the rocky Hamada and further south, the vegetation gets thinner and very scarce (patches of *Aristida obtusa*, *Farsetia aegyptiaca*, *Savignya parviflora*), essentially contracted in very small depressions. Light sand dunes offer sparse cover of *Stipagrostis pungens*, *Retama raetam*, *Calligonum comosum*, *Oudneya africana*, *Cornulaca monacantha*, *Bassia muricata*, *Panicum turgidum* and *Nucularia perrini*, mostly suitable for dromedaries. The large expanse of bare shifting sand dunes (Ubari Edeyen) close to the Shati are exceptionally used and only during spring.

All over their territory and wadis, winter and spring annuals are available for a few months before the summer settles in. As the annuals wilt and temperatures rise, flocks retreat either to the northern pastures, to wadis and finally to encampments near wells and cisterns on the Hamada and rainfed crops ready for harvest.

The Zentani from the Jebel (Zentan-el-Outen/Thagermine) are semi-sedentary. They used to inhabit stone masonry and troglodyte dwellings common in the Jebel Nafusa. They own (Table 12.1) many olive trees around Zentan, fig trees, almonds and pomegranates close to their settlements in the Dahar south of and adjacent to the Jebel. They crop rainfed barley in the wadis, the Sof-el-Jin and tributaries and in the Ouassa. They also possess palms trees in the deep of the valleys in the Jebel and mostly palm groves in the Shati around Edri (or Adiri) in the Fezzan and near Derj, so diversifying their resources (olive oil, dry figs, grain, dates) that they sell or exchange with other tribes. They also used to trade salt from the Sebkha-el-Mellaha (100 km south-east of Sinaween).

Cauneille (1957) reports an average of 16 olive trees, 15 fig trees and one palm tree per Zentani family of the Jebel, while the Zentani of the *Guibla*[4] own, per family, 27 and 17 palms trees, respectively, in the Fezzan and at Derj.

During winter and spring, the Zentani-el-Outen used to roam the rangelands on the Hamada living under black tents along with their

Table 12.1. Fruits trees owned by the Zentani. (Adapted from Cauneille, 1957)

		South			
	Jebel and Dahar	El-Gueria-Tobga	Chati	Wadi-el-Ajel	Derj
Olive trees	20,000				
Fig trees	18,000				
Palm trees	1,800	6000	6300	40	700

fellow nomads, the Zentani of the *Guibla*. Winter and spring watering are not a problem as the annual vegetation provides enough water to wandering flocks. When summer comes, the Zentani from the Jebel return to the Dahar and their mountains for barley harvesting, fruit picking, olive pressing etc. The Zentani of the *Guibla* remain on the Hamada during the summer, retreating with their flocks near small vegetable gardens (Saniet Er-Ramah, Oglat Lambesh), wells and cisterns (Bir El-Hamra, Bir Tlakhsin, Bir Halaq, Bir El-Kelb, Bir El-Abot, Bir El-Kabash etc.) they control.

Despois (1935) gives an indication of the livestock owned by the Zentani in 1932 (Table 12.2)

Cauneille (1957) estimates that the Zentani from the Jebel run an average eight sheep and 1.2 dromedaries per family while the Zentani of the *Guibla* have an average ten sheep and three dromedaries per family. These figures have evolved over years.

In conclusion, the Zentani, mostly nomads, settled in the Jebel Nafusa due to the fruit trees they possess in the Dahar and in the Shati (Fezzan). Their shepherds lead their livestock deep south and west into the Hamada and the upper Sof-el-Jin, wadis and depressions they control west of the Mizda–Ash-Shwayrif road leading to Ash-Shati.

The Magarha

The territory of the Magarha tribe (Cauneille, 1954; De Agostini, 1917a) stretches on a narrow strip north-south from approximately wadi Zazamet to wadi Bey el-Kebir to Ash-Shwayrif, with some limited access to Buirat-el-Hsun in the north and to Brak/Guira in the Fezzan (Map 12.3). This is mostly in the very arid zones with poor desertic rangelands and without access to any fruit trees cultivation like the Zentani, but they used to crop many of the wadi flats of their territory and own palm groves and small settlements in Zellaf (Maps 12.6 and 12.7). In the old days, the Magarha claimed that they occasionally wandered with their flocks as far as Tarhuna, Sirte (Hamraya) to Ghadames and southern Tunisia (Cauneille, 1954). However, having powerful neighbours (Zentani, Ould Bou-Sif and Ourfella) and lesser ones to the east, south-east (Riah) and south-west (Hassouna), the Magarha's freedom on their rangelands and rainfed cropping (Map 12.7) was constantly challenged. Rainfed cropping in the north-eastern wadis (wadi Bey el-Kebir and wadi Rouaous) repeatedly triggered serious conflicts with the Riah tribe. During the Turkish time, the Magarha acquired a special status and were exempted from taxes provided that they would serve as police and military auxiliaries.

Cauneille (1954) reports his 1948 census of the Magarha identifying five main groups:

1. A nomadic/semi-nomadic group of 482 families moving to Sirte in winter/spring and owning an average of 5 dromedaries and 7 sheep and goats per tent and returning during summer to Ash-Shwayrif. They also own an average of 6 palm trees per tent in Zellaf. Despite being deeply nomadic, this group keeps acquiring palm trees in the south.

2. A group of wealthy livestock owners (Brakis and Mechelcha? Mechachiya?) a total of 97 families settled in the Chati but sent their shepherds and flocks to Sirte. They own an average nine dromedaries, ten sheep and goats, and five palm trees per family. Quite rich, this sedentary group keep buying palm trees, considering that this is a much more secure long-term investment than their livestock managed by shepherds.

3. A semi-nomadic group of 50 families lead their small flocks to the region of wadi Gheylan/ Bir el-Aghelanyah during winter/spring, 170 km north of Brak and retreat to the Chati and Zellaf in summer. Each family owns an average 112 palm trees and less than four dromedaries and five small ruminants. This group capitalizes on palm groves and tends to sedentarism.

4. The semi-nomadic group of 31 families still sent their meagre flocks to Sirte and estivates back in the Chati. There, each family owns about 125 palm trees but little livestock with about two dromedaries and four small ruminants. Having lost most of their livestock, they are slowly trending to sedentarism in the south.

Table 12.2. Zentani's livestock in the Jebel, Dahar and *Guibla*. (Adapted from Despois, 1935)

In 1932	Jebel and Dahar	*Guibla*
Sheep	4500	5000
Goats	4600	
Dromedaries	1500	1500

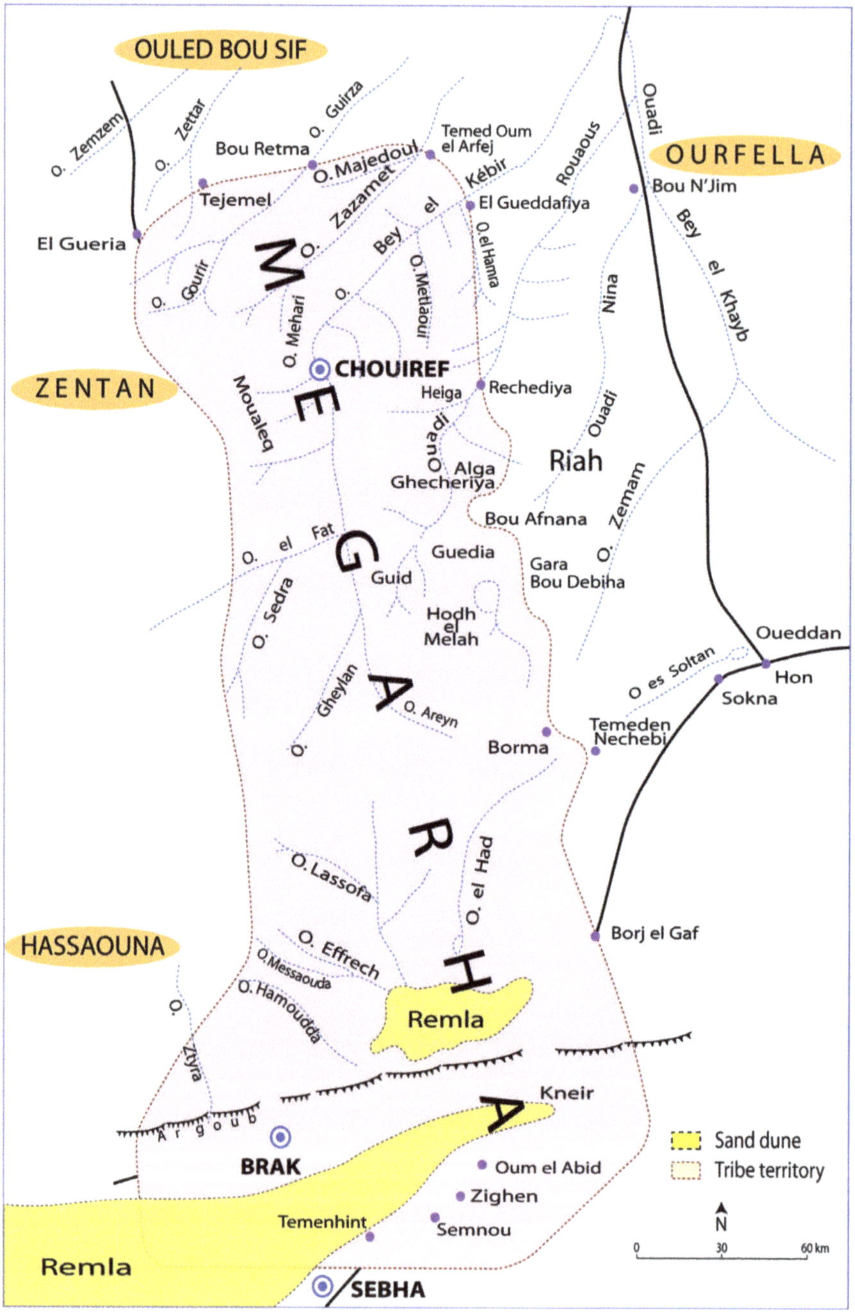

Map 12.6. Customary grazing territory of the Magarha. (Adapted from Cauneille, 1954)

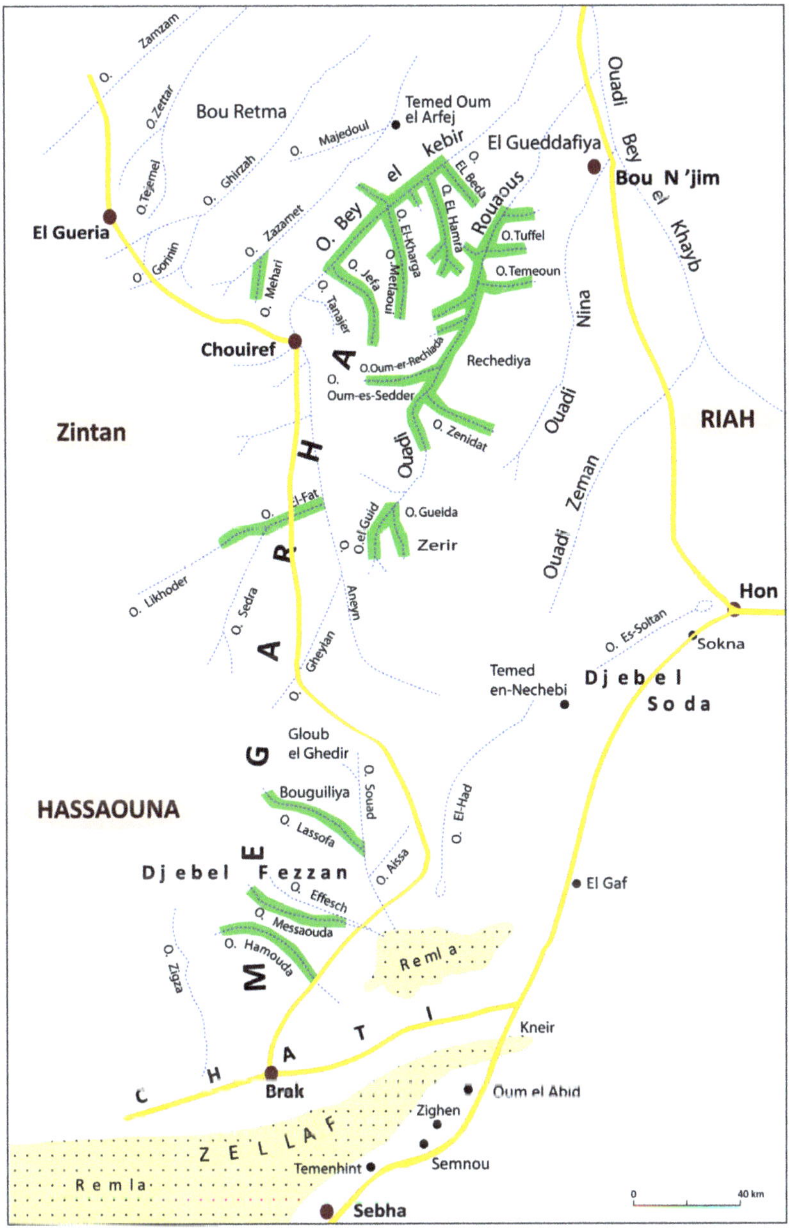

Map 12.7. Wadis occasionally cropped with rainfed barley by the Magarha tribe. (Adapted from Cauneille, 1954)

5. Lastly, the main group of some 344 families are almost sedentary owning an average 60 palm trees per family, but hardly any livestock; only a quarter still own a single dromedary, and a third have no small ruminants. At the time of the census (1948), 30 families of the El-Gueria clan owned about 40% of the Magarha's palm trees in the Chati and Zellaf.

By the late 1940s, Cauneille (1954) categorized the Magarha mostly as semi-nomads with a trend to sedentarization due to the many constraints they faced with their ecologically limited and poor grazing territory, with their rainfed cropping areas diminishing due to recurring clashes with neighbours. Losing access to wadis and depressions where they could crop barley essential for their livestock's survival accelerated their slow retreat to palm groves country in the south.

The Gadadfa

This Gadadfa tribe (Cauneille, 1958; De Agostini, 1917a, 1917b) was largely nomadic, raising small ruminants and many dromedaries. They traditionally used the rangelands some 40–50 km inland south of Sirte as part of the *Saff-el-Foghi* (inland tribes). Their territory is very arid with rainfall ranging from about 150 mm/year in the north and down to true desert in the south. In spring, the Gadadfa used to move south to the Hamraya, Argoub and the Bsat-Chorfa regions and even further to the Jebel Waddan, Al-DJoufra and Jebel Souda in good years. They own a few palm trees in Zella and Hon. They are also transhumed to northern Chad where they own date palms in the Kanem region between Bir-Ali and N'Djamena (Najem, 2004).

In the north, by the end of spring, they retreat to around wadis Jaref, Telal and Haneouah where they own wells and cisterns, harvest their occasional rainfed barley and graze crop residues. To the west, they extend their spring grazing to Bu N'gem, the lower wadi Bey-el-Kebir, to wadi Zamzam and even up to wadi Nefad in friendly Ourfela and Magarha territories with which they have a long-lasting and strong alliance. To the east, they would reach wadi El-Ajer in welcoming Ould-Sliman and even Margarba land (Ajdabiya) at the western limit of Cyrenaica. Over centuries, they had little access to the rich and permanent saline rangelands of the coast controlled by the *Saff-el-Bahar* (coastal tribes) comprising, from west to east, the Ferjan, the Hamamla and the Maadan (Map 12.8).

The rangelands available to the Gadadfa are mostly inland from the coast, arid and desert (Nègre, 1974). To the north, edging the coastal *sebkha* in areas receiving about 150 mm MAR, we find the best and most productive, but saline rangelands of the region (mostly *Atriplex halimus*, *Salsola* and *Suaeda* sp., *Limonium sp.*, *Limoniastrum* sp. *and Arthrocnemum*). These can be grazed all year long by dromedaries; however, due to the high sodium chloride content of these plants, livestock must have access to fresh, non-saline water especially during summer. The dromedaries and small ruminants must move part of the year to vegetation on limestone country to balance their sodium/calcium/phosphorus physiology. About 20–25 km south of Sirte, they find this on shallow soils covering the limestone crust where *Haloxylon scoparium*, *Gymnocarpos decander*, *Pituranthos tortuosus* and *Salsola vermiculata* are dominant with a carpet of annuals during spring. Remnants of bushes of *Rhus tripartita*, *Periploca laevigata*, a few *Acacia tortilis* trees dot thalweg and wadis where *Cymbopogon shoenanthus* and *Zilla spinosa* subsist. This is the domain of gazelles feeding on acacia pods. To the east, on fine and coarse fixed sand is the domain of poor rangelands with *Thymelaea microphylla* – *Pituranthos rholfsianus*, a land that is prone to wind erosion, displaying rare patches of spring annuals and offering some limited and risky rainfed cropping in shallow depressions. On low fixed sand dunes to the west, *Stipagrostis pungens* and sometimes *Panicum turgidum*, *Pennisetum dichotomum*, *Bassia muricata* and *Cornulaca monacantha* offer sparse dromedary pastures. Further west, south of wadi Bey-el-Kebir, spiny *Astragalus spinosus* appears on gypseous crusts. With aridity increasing in areas receiving less than 100 mm MAR (about 100–120 km south of Sirte), the vegetation definitely turns desertic getting even thinner with *Anabasis articulata*, *Traganum nudatum*, *Pituranthos chloranthus*, *Salsola tetrandra*, *Salsola tetragona*, *Cornulaca monacantha*, *Bassia muricata* and *Nucularia perrini* colonizing shallow soils on calcareous crust, often covered with a sand veil. Annuals are rare on this desert rangeland that remains the typical dromedary's territory.

In short, the Sirte inland rangelands are poor and desertic, subject to wind erosion and overgrazing, with increasing aridity. The better vegetation in the many inland small wadis and depressions allowed the Gadadfa to maintain a presence with short flock displacements.

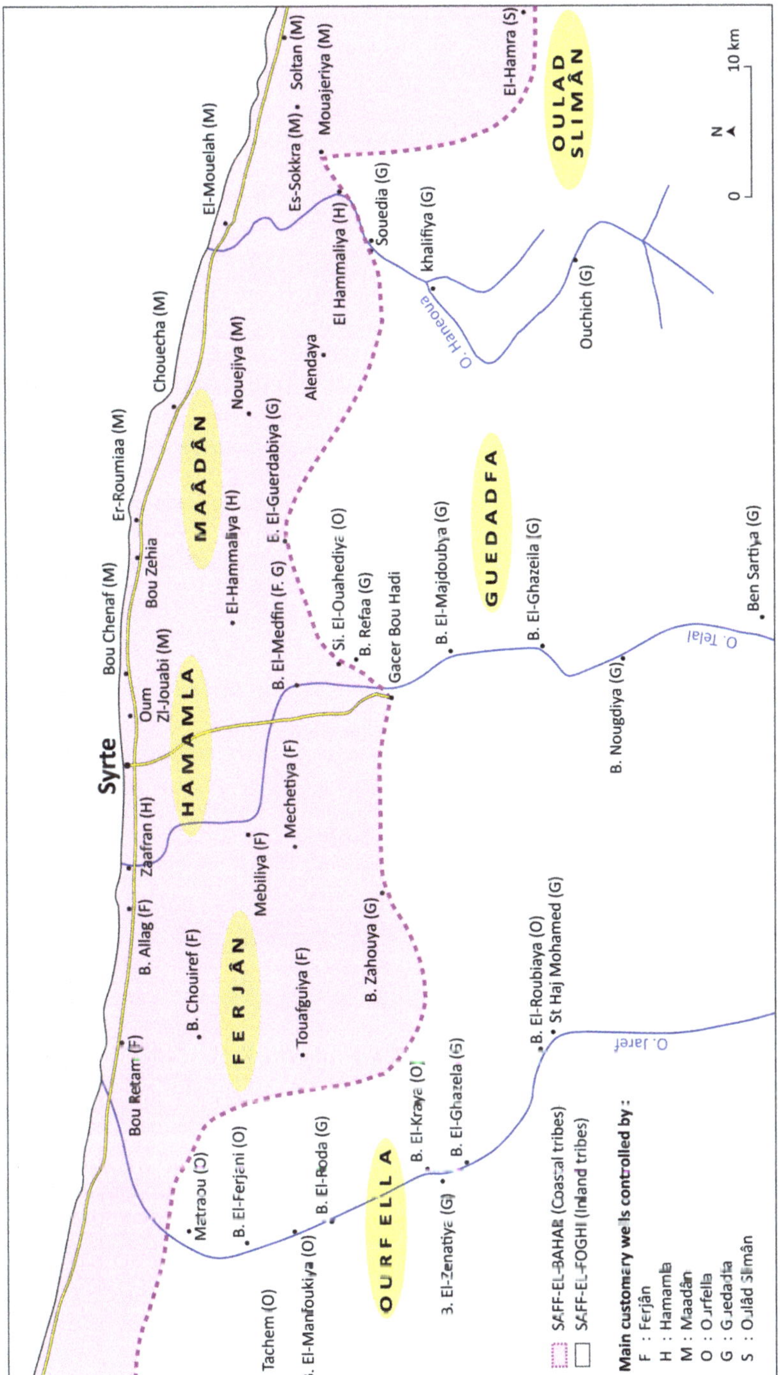

Map 12.8. The tribes from the Sirte region with their wells. (Adapted from Cauneille, 1958)

However, the Gadadfa always had a persistent motivation to gain access to the rich coastline *sebkha*, often sparking conflicts with or between the seaside Sirte tribes (Ferjan, Ould Sliman).

The tribes from the Jebel Al-Akhdar

The eastern region (Cyrenaica and Marmarica) includes eight main tribes belonging to the Saadi, a large tribe named after a common female ancestor *Saada*, according to a matrilineage. The legend reports (Najem, 2004) that Dhu'ayb (Dhuiyib) Abi al-Layl al-Sulaymi of the Hayb tribe and chief of both the Banu Sulaym and Banu Hilal, settling in the region, married Saada, the daughter of the Berber chief Khalifa al-Zanati (one of Ibn Badis's commanders – 11th century CE). Saada had three sons, Salam, Barguth and Aqar, from which originated the main Saadi eastern Arab tribes.

All of them are issued from the Banu-Sulaym confederation (Salem Essaker, El Marj, personal communication, 2020). These tribes are, from west to east (Map 12.9):

1. the Magharba, settled around Ajdabiya, roughly from Agheila to Sultan along the coast with many *sebkha*, and inland on a desertic region;
2. the Awqir (or Awaquir), around Benghazi, along the coast from Sultan-Magrun to Tokra and south inland;
3. the Arafa (or Urfa), centred on the rich Al-Marj plain with a small access window to the coast from Tokra to Tolmeita;
4. the Abeid (or Abid), south of the Arafa, distributed between Jabal Al-Abeid, south-east of Al-Marj to Al-Beyda on a territory that is land-locked but open to the south;
5. the Bra'asa, land-locked, too, and adjacent east of the Abeid land, covering from wadi Al-Kuf to Al-Beyda and south to the Marawa region;
6. the Addrasa (or Drusa), along the coastal plain north of the Jebel escarpment, from Tolmeita, north-east of the city of Al-Marj, to Al-Haniyah, north-west of the city of Al-Beyda;

Map 12.9. The main tribes of Cyrenaica and Marmarica (from an Italian military map of 1916, corrected and updated by Salem Essaker). (El-Marj, October 2021, personal communication)

7. the Hasa, who access the coast from El-Hamamah to Marsa Suza with land on the plateau around Shahat; and
8. the Abeidat (or Ibidat), which spreads its territory all over Marmarica from Labraq, east of Al-Beyda, Derna, to Tobruk and Sollum at the Egyptian border and further east.

The rainfall inter- and intra-annual variations, vegetation and water resources command the mode of life of these tribes and associates. The coastal tribes, the Arafa, the Addrasa and the Hasa benefit from the highest average annual rainfall and the best agricultural land, hence being mostly sedentary and farmers. The matorral bush vegetation on the coastal plain and lower escarpments is used by goats and local cattle browsing, and sparse cereal cropping and fruit trees on cleared matorral. The communities on the escarpments and the plateau are mostly sedentary where the rainfall is the highest. Cropping is difficult on this rugged and rocky bushland, though there are more irrigated gardens and fruit trees in smallholdings they own in full. Further south, with rainfall decreasing, sheep flocks from semi-sedentary communities are common. They practise a spring seasonal migration between the Jebel crest line, descending onto the southern slope of the Jebel covered with *Juniperus phoenicea* and to the more southern arid rangelands. Summer coming, they retreat to the Jebel where water, forage and cereal crop residues remain available late in the summer. They also cleverly and sporadically crop rainfed barley in wadis, depressions (*gar'â*) and large flood plains (*balat*) on south-eastern tribal lands.

The tribes from the crest line of the Jebel have their territories stretching from the top and have access deep into the southern arid rangelands.

Land and Resource Management in Eastern Libya

This part is largely inspired from the outstanding study *The Herders of Cyrenaica*, by Behnke (1980). While it was specifically written about Cyrenaica's herders, most of Behnke's observations about rangelands resources usage remains valid all over Libya.

Three main resources are of equal importance to herders:

- rangelands
- water resources
- cropping land (rainfed mostly).

Rangelands

Flocks are always privately owned. The *watan* territory of the tribe and associated sub-tribes, clans and families (lineage) is owned and controlled as common ownership. Individuals of the tribe do not own the *watan* land but inherit the right to access and use it for their livestock. Everybody from the tribe and their neighbours know where their *watan* boundaries are located; it could be a wadi course, a track, a remarkable feature of the landscape, an escarpment, the line dividing two watersheds, depressions for rainfed cropping, etc. While these boundaries are regularly accepted by consensus, fiery disputes may arise mostly about cropping land and use of cisterns or wells. Every member of the lineage is free to use the pastures and rangelands for their livestock wherever they want only within their *watan*. However, as in arid zones, rainfall may be spatially and randomly distributed and may affect one part of the territory and no other, hence temporary reciprocal agreements between tribes and associates permit grazing outside their *watan* when drought occurs.

Cropping land

In high-rainfall country in the north and on the Jebel top smallholding and gardens may be privately used by sedentary communities within the *watan* and own in full the production. The land cropped is *de facto* owned by the one who first cleared the land and cropped it. It cannot be sold to members of another tribe. If a piece of land is fallowed or temporarily abandoned for years, the usual user keeps the right to crop it.

In the Juniper forest zone, because of the relative abundance of vegetation and tall shrubs and trees, it is the domain of goat herders who tend to camp always at the same place and crop near their camps. In this case, they also own and control the land they plough year after year.

Further south, in the steppe, the cropping land is firmly collectively owned. But the land to crop does not remain the asset of one individual from the tribe: 'First come, first served' is here the essence in the steppe. If one fails to start cropping and another member of the tribe starts ploughing, the former loses his turn on this piece of land up until the following year. Once a male family member of the tribe cultivates a piece of land, he can crop as many successive years as he wishes, as long as he is the first to cultivate and crop, and the harvest (fruit, vegetables, grain) and crop residues belong to him. This is understandable by the fact that in lower-rainfall zones in the south, cropping land and water are scarce, rainfall is unpredictable in quantity and spatial distribution, and an individual may hesitate too long to crop, while another one would gamble, plough and seed without delay. This situation implies that no one would use any fertilizer on these transitory assets; hence, the usual poor grain yields due to soil fertility being slowly depleted.

In the west, the Turks implemented the *sekkah* (or *sikka*) system in flood plains (see wadi Al-Feida – south Zliten) by which field strips were allocated to different farmers buying the right to crop every year, in fact a cropping tax. Here again, farmers would not use fertilizers because of the compound risk, besides the variable rainfall, of not being able to secure, year after year, a specific piece of land.

In the flood plains in the south (*balat* region), cropping lands are generally recognized as belonging quasi-permanently to someone. Fields are delimited by stone cairn (*redjem*), old tyres, posts, unpalatable perennial plants (*pharaoun*), etc.

Water availability from north to south

Water is essential to households and livestock. Water is accessible from wells and cisterns, or from ponds on rangelands as well from vegetation. While it is not a problem in the north and high-rainfall zones, and available in quantity and quality, it decreases in available quantity and quality and increases in value when moving deeper in the steppe and further south.

Water in winter and early spring is available as livestock would get enough water from fresh and green pastures they graze or browse. Access to watering facilities for livestock is more challenging at the end of spring and summer when vegetation dries off. This is usually the time when tents retreat to the north or, in the south, close to wells and cisterns they own as members of their lineage and inside their own *watan*. When members of one tribe or associates graze rangelands that are outside their *watan*, and invited by their neighbours from another descent, the latter would have access to water for the family but may have to buy the water from wells or cisterns for watering their livestock. Water accumulated in ponds or in wadis remains free access all over rangelands.

Cisterns are fed by small water catchments either of compacted earth or from large limestone slabs. These cisterns could be either from antiquity or more recently, either dug by the community or wealthy private livestock owners, or recently constructed by the government. Rules of accessibility to tribal or private wells and cisterns change with increasing aridity. Well and cisterns, built long ago by religious legacy, the Italian colonizers (Photo 12.2), the British occupants, and lately by the Libyan government, are, in principle, accessible to all, wherever they are.

On the highest-rainfall zones on the Jebel top and karstic plateau, water from wells is shared without constraints by all, even if access in quantity and use in turn are commonly prearranged. However, private cisterns remain individual property and are rarely shared. In the Juniper zone, water available from lineage and private cisterns is often locked and more commonly sold. There the rules change as wells are less frequent and most water comes from antique, old or recent cisterns, collecting water from small catchments. Once the Juniper disappears, entering the steppe country, dug wells are rare. Water from cisterns strictly belongs to lineage groups. However, over years, privately owned cisterns with restricted access are becoming increasingly frequent.

When shepherds move from the mountain top further south to the Juniper forest, wells are uncommon, and cisterns and water resources are owned by lineage. Access is therefore restricted to the owners and sold to

Photo 12.2. The Italian memorial stone at Bir el-Ghilba (near Mafsel, 20 km south of Gendoula) with the late Engineer Mohamed Saleh Alshhibi (Benghazi).

other groups when requested and if available. In short, water in the mountain is usually simply shared, while it is sold for watering livestock in the south with increasing aridity. However, water for any household consumption always remains available without restriction for other groups not from the lineage and camping in the vicinity.

Moving to the steppe, cisterns are all owned by lineage and the above rules remain enforced also for the most southern flood plains. Exceptions are the 20 cisterns at Mekhili belonging to the religious Senussi *zawiya*. These cisterns hold a large amount of water that is freely available to all from Mekhili and all herdsmen grazing around either from lineage or not.

Several of these wells dug during the Italian colonization under Marshal Italo Balbo (Governor-General of Libya, 1933–1943) remain in good condition and well-maintained to date, south of Jebel Al-Akhdar (e.g. Bir el-Ghilba, Bir Bu Abirah etc.). On the headstone of the wells (erected *c.*1938), it reads (translation Dr M. Saaed):

> O son of the desert, I have ordered you and your animals to build this well in the heart of this barren dryland, so drink from its water and water your animals from it, and remember what was mentioned in the Holy Qur'an about water (and we made everything alive from water), and be keen to maintain and preserve this well, the most conservative, and woe to who dares to

> damage it or destroy it in any way, because the damage falls on the one who causes it and is held accountable before Allah Almighty for his bad deed.

In the most southern extension of flood plains into the fringe of the desert and in southern Marmarica, this is dromedaries' country with very sparse vegetation and little water resources; tribal boundaries and rangeland access are strictly respected.

The Bedouins' tent and the art of *kilim* in Libya

The essence of Bedouins' life is mobility, hence simplicity, light weight and portability of all essential daily items. All items are sourced out of rangelands and especially from livestock products providing wool and hairs for spinning, weaving and embroidering elements of tents, an essential craft passed from women to women and young girls.

The Bedouins are renowned for their unfailing hospitality under their stretch tents. Described in detail by Louis (1979) in southern Tunisia and Fahd (1976) for the Middle East and Arabia, the traditional Libyan Bedouin tents have not changed over centuries, like the ones we see all over the North African steppe. The tent (*bayt-es-sha'ar* or *khayma*; Photo 12.3) is rectangular and comprises the roof, the poles and the ropes. The roof is made of 10–12 strips or *flij* that are 8–12 m long and 60–100 cm wide, woven from sheep's wool, camels' and goats' hair on a simple horizontal lightweight backstrap loom (*masda* or *musada*), a simple and easily transportable weaving loom which evolved into the common vertical loom (*nool*) used at home or in shops. The strips are then assembled and strongly sewn together along the length of the fabric edge. It is heavy and waterproof once wetted. Undyed, monochrome, the colour remains natural, black or white, but with brown and white alternating strips, according to the tribes. In Libya, the western tribes in Jeffara and eastern Nafusa tend to have dark-coloured roofs, while the eastern and Sirte tribes to Solul-el-Khoder/Bani-Walid region have white roofs with narrow dark strips.

The tent is always erected with its entrance open to the east with an awning (*riwaq?*) to close the entrance. The back of the tent is shut and facing west to protect against dominant wind and

Photo 12.3. The Zentani black tent with fuelwood collected on the rangelands during winter (south Dahar/near Rojban, 1976).

rain. The roof is supported with wooden poles (*emidat alkhayma*), with a 2–4 m vertical one in the middle (*wassat't*) and two shorter ones both sides that are tilted outwards. The wood to make the long poles is rare in the steppe, hence precious. They are usually made from light wood of *Pinus halepensis* (*Snouber*), *Tamarix* (*Ethel*) or heavier wood from *Acacia tortilis (Talah)*, *Pistacia atlantica* (*Batum*) or olive trees. Inside the tent, the roof material is protected with either a small wooden disk or with decorated long narrower *flij* between the top of the poles and the fabric. The set of three poles is repeated two or three times according to the width of the tent with the higher one at the back of the tent. The tent structure is secured with ropes linking the roof and wooden (*Ziziphus* sp., *Rhus* sp. in west and Sirte, *Arbutus* sp. or Juniper in the east) or metal stakes pegged into the ground. In winter, the peripheral walls are secured against wind with stones or sand and can be rolled up during warm weather to facilitate ventilation.

The tent is divided into two compartments, one for the men and entertaining the male guests, and the other private side for the women, female guests and the children. The separation (*al-maksura*) between the two compartments is made of a long piece of fabric (*hamel*, *flij* or *margum*), as long as the depth of the tent and 2–2.5 m high, either a patchwork of cotton material sewn together or with woven long *flij* similar to the one forming the roof but often decorated with geometric coloured patterns (*raqma?*). The latter is a *kilim* weave. The Bedouin *kilim* are traditional flat rugs and decoration inside the tent. They are woven by women as several-metres-long runners on a rudimentary horizontal narrow ground loom (Photo 12.3), from carded and hand-spun sheep wool, camels' or goats' hair for the warp and the weft, both in natural colours or dyed. The traditional horizontal loom is still used by all nomadic, semi-sedentary communities all over the rangelands of North Africa, the Middle East, and middle and central Asia. It is light, easy to roll and transport. The *kilim* is rather rough, tightly woven, highly resistant, and woven without a pile and not knotted like the Turkish or Persian rugs (Liebetrau, 1963). The warp is coloured (Photo 12.4) while the weft, made of cotton or wool, is white. The pattern on the *kilim* is created by manually raising or lowering the coloured warp yarns in a similar technique known as inkle weaving, a type of warp-faced weaving. The women expertly move the shuttle left-right and then back right-left, above or below the warp to create geometric plain, stripped, chevron, or chequered

Photo 12.4. A traditional horizontal ground loom commonly used to weave *kilim* and runners (from Tym-Karnap region, Uzbekistan).

Photo 12.5. A decorated Libyan runner (45 cm wide x 370 long, from Sirte, 1976) placed between the roof material and the wooden poles of the tent.

patterns transitioning from one to another. The runners (Photo 12.5), once completed, are about a cubit wide (about 45–60 cm) and several metres long. They are then assembled together by overstitching, as larger strips, as tent partitions (*sqifa* or *teqaq*) or cut and sewn to provide different elements (Photos 12.6 and 12.7) for the tent, rugs/blankets (*hamel*), cushions, bags (*ghrara,* or decorated *ushada)*. The colours used to be vegetal dyes such as the red colour processed from madder's roots (*Rubia tinctorum, Fuwwa,* Keith, 1965) cultivated in Libyan gardens and oases. Other vegetal dyes (Cardon, 2003) were sometimes used such as the leaves of *Pistacia atlantica* and *Pistacia lentiscus* or the skin of fruits of *Punica granatum* or the fruits of *Rhamnus alaternus* for yellow, the leaves of *Quercus ilex* or *Quercus calliprinos* for light brown/grey to black. It was gradually replaced with chemical dyes processed in southern Tunisia (Ayed and Alatrache, 2001), and imported and marketed in Tripoli and Misurata.

Up to the early 1990s, a large diversity of original Libyan tribal *kilim* were mostly sold at the Misurata souk. Decorated *kilim* rugs and carpets (*sajada* or *b'sat*) have nearly disappeared and been replaced with reed material or even plastic mats. Nowadays, the traditional tent is often covered with lighter white or a light-green cotton tarpaulin. Dromedaries slowly carrying heaving tents, poles and common utensils have disappeared from Libyan rangelands having long been replaced with pickup trucks (Photo 12.8) fast covering long distances.

Conclusions

Berber and Arab Libyan tribes have managed their rangelands from time immemorial, fiercely defending and controlling their territories and often sharing with neighbours their scant resources, rainfed cropping and water in times of scarcity or local droughts. Each tribal territory covers a diversity of landscapes, ecology and rainfall zones, allowing for a sustainable land management and sometimes hardly sufficient living conditions for the community. Association, *Esprit de corps* (*asabiyyah*), reciprocal support and enmity remain strong and durable to this day. With the discovery of immense oil reserves by 1959, the Libyan tribal system and economy slowly evolved, leading impoverished nomads and semi-nomads to quit the rangelands

Photo 12.6. Various *kilim* and *margum* from Bu N'Gem: a decorated tent separation, a decorated cushion, brown grain bags (brown and white), and narrow roof protection runners (Nov. 2009).

Photo 12.7. An old Libyan *kilim* (150 cm wide x > 170 long) woven with ancestral 'evil eye' patterns.

life for financially more secure but humble employment. It strongly impacted the population of rangeland shepherds and *jebbad*, traditionally tending the palm groves in the south with strong migration to main settlements in the north. Oil wealth slowly percolated throughout Libya, stimulating population growth, new buildings and urbanization, road improvements,

Photo 12.8. Transporting tent and camping equipment during a trip on the Hamada with Zentani's friends (April 2010).

immigration and business. Population growth led to increased demand for meat and farm products, inducing livestock numbers escalating with more grazing and cropping pressure compounded with unremitting imported feed to compensate for unbridled rangeland degradation. All these trends have contributed to undoing the traditional rangeland systems and altering rangeland management. Nevertheless, the Libyan tribal social affiliations, traditions and rules are still very strong and clearly resilient when managing the rangeland resources despite the many changes brought by successive invasions and colonization periods, political adjustments, sporadic upheavals and unstoppable modernity.

Notes

[1] CE = Christian era; BCE = before Christian era

[2] Would the name of the famous 'Kabaon' still survive in Libya with the Berber town of Cabao or Kabaw (N 31.845 E 11.340), in the Western Nafusa, about 36 km East of Nalut?

[3] The Riah tribes? Are they the ones we still find adjacent east of the territory of the Megharha, and south-East of the Ourfella?

[4] The *Guibla* is all the rangelands south of the Jebel Nafusa covering the Hamada to the Edeyen Ubari and sand dunes country north of the Fezzan.

References

Ayed, N. and Alatrache, A. (2001) Traditional recipes for natural dyeing of wool in the south of Tunisia. *Dyes in History and Archaeology* 16(17), 43–53.

Bates, O. (1914) *The Eastern Libyans: An Essay.* Macmillan and Co. Limited, London.

Behnke, R.H. (1980) *The Herders of Cyrenaica: Ecology, Economy and Kinship Among the Bedouin of Eastern Libya.* University of Illinois Press, Urbana, Illinois.

Brett, M. and Fentress, E. (1996) *The Berbers: The People of Africa.* Wiley-Blackwell, Oxford, UK.

Cardon, D. (2003) *Le Monde des Teintures Naturelles.* Editions Belin, Paris.

Cauneille, A. (1954) Le nomadisme des Megarha (Fezzan). *Travaux de l'Institut de Recherche Saharienne (I.R.S),* XII, 41–47.

Cauneille, A. (1957) Le nomadisme des Zentan (Tripolitain et Fezzan). *Travaux de l'Institut de Recherche Saharienne (I.R.S.),* XXVI, 73–99.

Cauneille, A. (1958) Le nomadisme des Guedadfa, Tribu de Tripolitaine (Syrte). *Bulletin de Liaison Saharienne* 32, 338–353.

Cauneille, A. (1963) Le semi-nomadisme dans l'ouest Libyen. In: Dupire, M. (ed.) *Nomades et nomadisme au Sahara* pp. 101–112. UNESCO, Paris.

De Agostini, E. (1917a) *La popolazioni della Tripolitania, Governo della Tripolitania, Parte I.* Tipografia Pirrotta et Breschiano, Tripoli. Available at: http://archive.org/details/DeAgostiniPopolazioniDellaTripolitania-ParteUno (accessed 7 August 2-24).

De Agostini, E. (1917b) *La popolazioni della Tripolitania, Governo della Tripolitania, Parte II.* Tipografia Pirrotta et Breschiano, Tripoli.

De Felice, R. (1985) *Jews in an Arab Land: Libya 1835–1970.* University of Texas Press, Austin, Texas.

Desanges, J. (1962) *Catalogue des tribus africaines de l'antiquité classique à l'ouest du Nil.* Université de Dakar, Faculté des Lettres et Sciences Humaines, Dakar.

Despois, J. (1935) *Le Djebel Nefousa (Tripolitaine), Etude Géographique.* Larousse, Paris.

Fahd, T. (1976) La tente bédouine. *Ktéma* 1(1), 53–62. DOI: 10.3406/ktema.1976.1774

Féraud, L.-C. (1927) Tournier et Vuibert. *Annales Tripolitaines* 437.

Fozzard, A. (1987) Tribesmen and the colonial encounter: southern Tunisia during the French protectorate 1882 to 1940. Thesis, Durham University, UK.

Goldberg, H.E. (1990) *Jewish Life in Muslim Libya: Rivals and Relatives.* The University of Chicago Press, Chicago, Illinois.

Gsell, S. (1927) *Histoire ancienne de l'Afrique du nord. Les royaumes indigènes, organisation sociale, politique et économique.* Librairie Hachette, Paris.

Ibn Khaldūn (1852a) *Histoire des Berbères et des Dynasties Musulmanes de l'Afrique Septentrionale* (B. de Slane, Trans.; Vol. 1). Imprimerie du Gouvernement, Algiers.

Ibn Khaldūn (1852b) *Histoire des Berbères et des Dynasties Musulmanes de l'Afrique Septentrionale* (B. de Slane, Trans.; Vol. 2). Imprimerie du Gouvernement, Algiers.

Ibn Khaldūn (1856a) *Histoire des Berbères et des Dynasties Musulmanes de l'Afrique Septentrionale* (B. de Slane, Trans.; Vol. 3). Imprimerie du Gouvernement, Algiers.

Ibn Khaldūn (1856b) *Histoire des Berbères et des Dynasties Musulmanes de l'Afrique Septentrionale* (B. de Slane, Trans.; Vol. 4). Imprimerie du Gouvernement, Algiers.

Idris, H.R. (1968a) De la réalité de la catastrophe hilâlienne. *Annales Histoire, Sciences Sociales* 23(2), 390–396. DOI: 10.3406/ahess.1968.421920

Idris, H.R. (1968b) L'invasion hilâlienne et ses conséquences. *Cahiers de Civilisation Médiévale* 43(Juillet–Septembre), 353–369. DOI: 10.3406/ccmed.1968.1452

Keith, H.G. (1965) *A Preliminary Check List of Libyan Flora, Volumes 1 and 2.* United Middle East Executives Ltd, London.

Liebetrau, P. (1963) *Oriental Rugs in Colour.* Simon & Schuster, London.

Louis, A. (1979) *Nomades d'hier et d'aujourd'hui dans le sud tunisien, Collection Mondes Méditerranéens.* EDISUD, Aix-en-Provence, France.

Mattingly, D.J. (1995) *Tripolitania.* Routledge, New York. DOI: 10.4324/9780203481011

Najem, F.A. (2004) Tribes, Islam and state in Libya: analytical study of the roots of the Libyan tribal society and interaction up to the Qaramänli rule (1711–1835). PhD dissertation, University of Westminster, London.

Nègre, R. (1974) Les pâturages de la région de Sirte (Libye): projet de régénération. *Feddes Repertorium* I(85 3), 185 243.

Passon, J., Braun, K., Hamid, S., al-Mahdi Khalifa, S., as-Sadeq at-Tellisi, N., El Nayedh, M. and Metzger, J. (2020) Places of trade, communication and religion: important oases. In: Braun, K. and Passon, J. (eds) *Across the Sahara: Tracks, Trade and Cross-Cultural Exchange in Libya*. Springer International Publishing, Cham, Switzerland, pp. 165–247. DOI: 10.1007/978-3-030-00145-2_6

Poncet, J. (1967) Le mythe de la catastrophe hilalienne. *Année* 5, 1099–1120. DOI: 10.3406/ahess.1967.421605
Ricchieri, G. (1913) *La Libia.* Italian Federation of Popular Libraries.
Roques, D. (2009) Procope de Césarée, La guerre contre les Vandales (Guerres de Justinien Livre III et IV). In: *Les Belles Lettres – La Roue à Livre*. Paris.
Saber, A.A. (1998) The camel in ancient Egypt. *Proceedings of the Third Annual Meeting for the Animal Production Under Arid Conditions, United Arab Emirates* 1, 208–215.
Smith, W. (1854) *Dictionary of Greek and Roman Geography.* Little, Brown, and Co., Boston, Massachusetts.

13

Forests, Afforestation, Fuelwood and Charcoal

Abstract

Libya still has a wealth of native trees and shrubs in high- to low-rainfall zones. The use of fuelwood and charcoal is a part of Libyan culture. Native forests, indigenous trees, woody shrubs from the steppe, and long-established afforested areas are relentlessly threatened by illegal logging and charcoal manufacturing. Current forestry and rangelands legislation are inadequate and inefficient. Without policing, accurate mapping and monitoring, these fuelwood resources will continue to be plundered. A contractual system regulating these activities is needed. Alternatives to fuelwood harvesting and charcoal manufacturing are advised. Large private and governmental afforestation projects must be encouraged and supported for large plantations of native and recommended appropriate exotic species listed in this chapter. Without immediate and sustained actions to remedy illegal unsustainable logging and charcoal manufacturing, Libya will soon see its native forests, exceptionally drought-resistant trees, and forest plantations decline, increasing already rampant rangeland desertification.

Introduction

The Food and Agriculture Organization of the United Nations defines a forest as land covering more than 0.5 ha with trees higher than 5–8 m and with a canopy cover of more than 10%. Native forests are a common feature in the Jebel Al-Akhdar, providing feed resources and refuge to wildlife, roaming small ruminants and local cattle nearly exclusively. These eastern native forests are threatened. In the west, native forests from the Jebel Nafusa have long been eradicated since antiquity. From the beginning of the 20th century, the successive Turkish and Italian forestry departments supported government afforestation projects using *Eucalyptus* sp. all over northern Libya (Keith, 1965). These commendable projects, also supporting private afforestation initiatives, were upheld by the Libyan government from the 1950s to the mid-1990s, but slowly ground to a halt. The remaining Libyan forests and afforestation projects are being slowly plundered.

Ancient Libyan Trees and Forests

Petrified wood and large tree trunks have been reported from Libya in the desert between Ubari and El-Uweinat deep in south Libya. Less known are the fossilized large tree trunks of *Metapodocarpoxylon*(?) scattered along the track (N 31.902 E 11.324) from Kabao to Tiji on the northern slope of the Jebel Nafusa and at the base of red silt and yellow Cretaceous sandstone cliffs. Keith (1965) reports fossil cones' imprints of *Pinus* sp. found near Gharyan attesting to its presence in geological time. Carbonized wood of *Pinus* and

DOI: 10.1079/9781800627154.0013

other plants were found at the wadi Gahham – wadi Derna junction (N 32.711 E 22.610) in tuffaceous-Pleistocene deposits.

In antiquity, forested areas inhabited by wild beasts (see Chapter 14) were reported around the present Homs, Misurata, the Jebel Nafusa and in the Jebel Al-Akhdar. Common tree species would have been mostly clear forests of *Pinus halepensis* (*Snouber*) and *Juniperus phoenicea* (*Ar'har*).

Current Status of Libyan Forests

Forestry data from Libya are scarce, issued from experts' estimates and unreliable. The FAO's Global Forest Resources Assessment (FAO, 2005) credits Libya with 217,000 ha of forest and about an extra 330,000 ha of other wooded areas, without defining them. It further indicates that there has been no change in forested areas and other wooded areas between 1990 and 2005, a statement difficult to agree with considering the persistent and heavy pressure from fuelwood harvesting for charcoal manufacturing. In this report, it states that only 16% are privately owned and 83% of forests and 100% of other wooded areas are government domain. The FAO estimates growing stock (total aboveground and below-ground biomass) of Libyan forests in 2005 at about 36 m^3/ha while other wooded areas are credited with about 12 m^3/ha, without real measurement to confirm however. We anticipate that the aerial biomass would be about half to two-thirds of the total biomass.

Forests and forest-like formations

Forest and matorral in the east

In the east, dense stands of *Pinus halepensis* and *Juniperus phoenicea* once covered large areas. In fact, not much is left; after thousands of years of relentless destruction by fire, wood harvesting for cooking, or as building and manufacturing material, nearly all of what would have been called 'forest' has been destroyed. Remnants of original forest of *Pinus halepensis* would likely still be seen on cliffs facing the sea from Tolmeita to Ras-Al-Hilal. Today, native forests of *Pinus halepensis* do not exist any more at wadi Derna and further east. Mixed dense tree stands such as *Quercus calliprinos* (*ballut*), *Ceratonia siliqua* (*k'harub*), and smaller *Arbutus pavarii* (*sh'meri*) are more common on the flats between Shahat and Marsa Suza, receiving about 300–400 mm/year and replaced with low matorral formations of dense stands of scraggy senescent *Juniperus phoenicea*, *Pistacia lentiscus*, *Cistus* sp. etc., degraded by bushfires, fuelwood overharvesting and creeping urbanization (Photo 13.1).

Most remarkable are the relict tall endemic trees of *Cupressus sempervirens var sempervirens* (*Arz*) in wadi Kuf, possibly part of long-gone forests patches colonizing the deep and cool wadies in Cyrenaica. Often seen also are the straggly *Cupressus sempervirens var pyramidalis* (*Erez*(?)) on the Cyrenaican plateau, but most likely resulting from modern afforestation.

Some old and large specimens and dense stands of *Juniperus* are common on the plateau east of El-Marj towards Taknis – Marawa. Further east, 14 km south of Gendoula, large areas are still covered with stands of mature *Juniperus phoenicea* trees on the north-western steep slopes of the hills. Juniper density is about 100–125 trees/ha (N 32.472 E 21.573; Photo 13.2), i.e. with aggregated crowns of up to 55–100 m^2 or 8–15 m in diameter. These 'open forests' cover about 10% of the area considered. The *Juniperus* nearly all disappeared about 25 km south of Gendoula, near Mafsel (N 32.419 E 21.529) – Bir-el-Ghilba with some rare, isolated specimens still surviving in wadis and mostly on northern slopes in an area receiving about 175–200 mm of rain each year. This is the most southern extension of what we could consider the Juniper forests in the Jebel Al-Akhdar.

'Open forests', 'savannah-like' formations and isolated trees in the west

On the southern slope of the Jebel Nafusa, sparse or isolated *Pistacia atlantica* (*batum*) would be a common feature of the landscape. Further south in the wadi zones (Nefad, Zorzor) extending into the south Sirte region, *Acacia tortilis* ssp. *raddiana* (*talha*) trees and robust specimens of *Rhus tripartita* (*jdari*) and *Periploca laevigata* (*haleb*) would sparsely colonize the wadis and some of the northern Nafusa foothills and depressions from Bir-el-Ghanem – Bir-Ayyad to Takut. Most remarkable

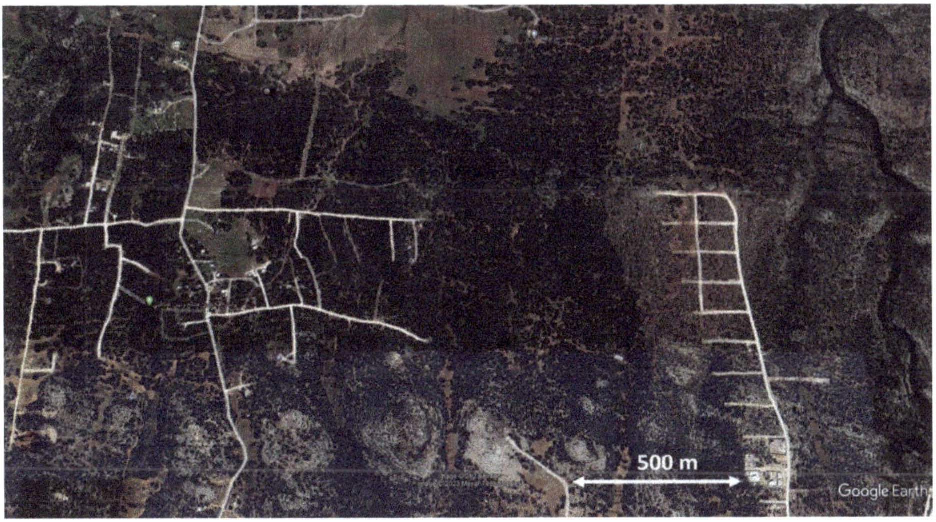

Photo 13.1. Forest (mostly *Quercus calliprinos* (*ballut*)) and matorral slowly cleared for new roads, farms and buildings in the plain (N 32.864 E 21.869) between Shahat and Marsa Suza. (From Google Earth 2023)

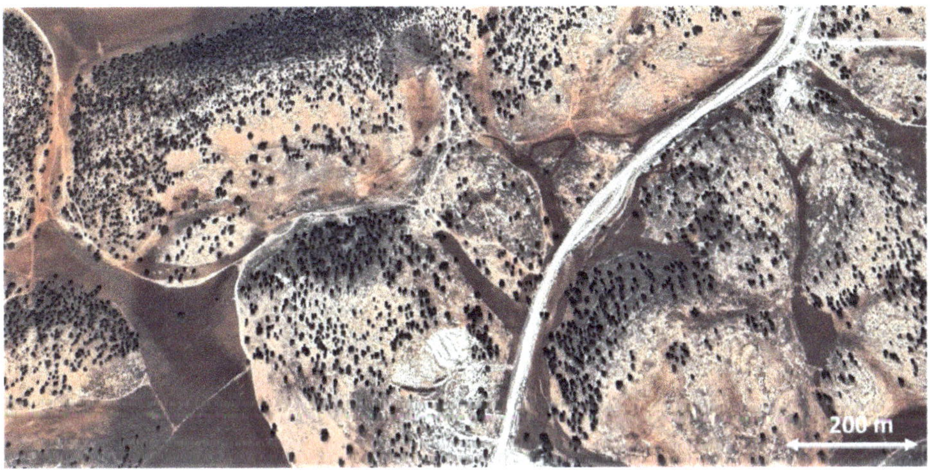

Photo 13.2. Residual patches of open forest of *Juniperus phoenicea* (south Gendoula, N 32.472 E 21.573). (From Google Earth 2023)

are the *Calligonum azel* and *Calligonum arich* Isolated trees now rare in the sand dunes of the Oriental *erg* across the Algerian-Tunisian-Libyan borders. These by no means could be qualified as forests but would be witness to past denser stands of trees and tall shrubs.

Common native trees

Native common and current Libyan trees/tree-like are listed in Table 13.1. Proper native forests often degraded into matorral or garrigues still cover much of the high-rainfall

Table 13.1. List of Libyan native trees (in alphabetical order) as part of a clear or open-forested landscape in northern Libya.

	Trunk diam. at 1.5 m (m)	Max height (m)	Rainfall zone (mm/year)	Soil type	Most present in	Endemism	Wood quality	Wood use	Other use
Acacia tortilis (*talha*)	0.5–1.0	8–15 m	= or below 150	Alluvial-colluvial deposits in wadis	West and Sirte	n.a.	Good – hard wood	Occasional, trunks as pillars in old home in Nafusa, charcoal	Pods as livestock feed (and wildlife)
Arbutus pavarii (*sh'meri*)	0.10–0.30	3–5 m	Above 300	*Terra rossa* on karst	East	Cyrenaica	Good – hard wood, flexible without breaking	Speers, javelins in old days, wood point hardened with fire, tent pickets, good charcoal	Wood used for turning, fresh fruits rich in Vit. C, roots used as red dye
Calligonum sp. (*azel, arich*)	0.10–0.5	1–5	50–150	Sand (mobile and fixed)	West	South-west, Oriental *erg*	Good to medium	?	Fuelwood, charcoal
Ceratonia siliqua (*k'harub*)	0.5–1.0	3–8 m	Above 300	*Terra rossa* on karst	East	n.a.	Good – hard wood	Building material, part of plough *Mihrath*	Beans as livestock feed, flour, roasted seeds used as coffee substitute
Cupressus sempervirens var. pyramidalis (*? 'Erez' ?*)	0.5–1.5	10–30 m	Above 400	Deep organic soils in wadis	East	Cyrenaica	Good – hard wood/ fragrant	Building material	Insect-resistant wood
Cupressus sempervirens var. sempervirens (*'Arz'*)	0.5–1.5	10–15 m	Above 400	*Terra rossa* on karst	East	Cyrenaica	Good – Hard wood/ fragrant	Building material	Insect-resistant wood

Continued

Juniperus phoenicea (*ar'har*)	0.5–1.0	5–12 m	Above 250	Red and brown soils on limestone	East (extinct in west)	n.a.	Good –hard wood, breakable	Occasional, pillars in old home in Nafusa, small tools, superior charcoal	Wood distillation for tar *gatran*, bark as tannin to treat leather
Periploca laevigata (*haleb*)	0.10–0.25	3–5 m	Below 150	Shallow calcareous soils (calcareous crust) and alluvium-colluvium	West and Sirt	n.a.	?	Occasional building material for *zeriba*	Poisonous milky sap?
Pinus halepensis (*snouber*)	0.5–1.0 m	8–15 m	Above 250	Brown – chestnuts soils on limestone	West and east – not in Sirt	n.a.	Medium to poor	Building material, beams, rafters, crates etc., poor-quality charcoal	Wood and resin distillates to produce turpentine
Pistacia atlantica (*batum*)	0.5–1.0 m	5–12 m	Above 150	Deep brown to grey desertic soils	West, not in Sirte, rare in the east	n.a.	Good to medium	Much used for good-quality charcoal, occasional trunks as pillars in old home in Nafusa, carved for large dishes and plates, tools, parts of local plough	Gum, mastic from resin, edible oil once extracted from dry fruits, black ink and dye from galls on leaves

Continued

Table 13.1. Continued.

	Trunk diam. at 1.5 m (m)	Max height (m)	Rainfall zone (mm/year)	Soil type	Most present in	Endemism	Wood quality	Wood use	Other use
Quercus calliprinos (ballut)	0.5–1.0 m	5–10 m	Above 400	Brown –chestnuts soils on limestone	East	Cyrenaica	Good to medium	Building material, small tools, superior charcoal	Acorns as livestock feed, flour from acorns used to mix with wheat flour to make bread in time of famine, bark used for tanning leather, black dye from bark
Rhus tripartita (jdari)	0.10–0.30	3–5 m	Above 150	Shallow calcareous soils (calcareous crust) and alluvium-colluvium	West and Sirt	n.a.	?	Occasional building for *zeriba*, charcoal	Root and bark used for tanning leather
Tamarix sp. (ethel)	0.10–0.30	3–5 m	Azonal with high water table to deep into the desert wadis	Sandy soils and *sebkha*	Sandy – saline depressions	n.a.	Medium to poor	Speers, javelins in old days, wood point hardened with fire, tent pickets, good charcoal occasional building for *zeriba*, charcoal	Root and bark used for tanning leather

None of these trees would withstand soil salinity or gypsum with exception of *Tamarix* sp. isolated trees that would not constitute forest.

zones in the Jebel Al-Akhdar. Some are growing at the fringe of the desert, relic of open forests or savannah-like formations (Keith, 1965; Le Houérou, 1969).

Forest degradation/fuelwood, charcoal and fire

Since antiquity, Libyan forests would have been providing coastal settlements with building material (home building and boat construction) and mostly fuelwood and charcoal for bakeries, pottery (Photo 13.3) and glass kilns, Roman baths and *hammam* for centuries. Ruins of antique ceramic kilns were discovered near Tarhuna and Leptis Magna (Capelli and Leitch, 2011).

During the Roman Empire (by 164 CE), the population of Libya was estimated at about 600,000 occupying approximately 1,500,000 ha in western Libya (Tripolitania) and eastern Libya (Cyrenaica and Marmarica). A guestimate would point to some 12–15,000 ha of forest native stand (30 cm diameter, 8 m height, 100 trees/ha) being felled in Roman Libya per year to provide only fuelwood at 1 kg dry wood/person/day (= 600 t/day) to its entire population. Even if this would have been supplemented with wood coming from olive tree plantations and olive press residues, it certainly largely contributed to clearing Libyan native forests up to the Arab conquest by 643 (Cyrenaica/Barqa) and 647 (western Libya).

The pressure on native wood resources continued unabated and would have been unavoidable up to the time when petroleum oil and gas became available by the early 1960s.

We do have an indication from experts' estimates of the shrinking of Libyan native forested areas in the Jebel Al-Akhdar for major trees (Table 13.2).

No data on native forest areas are available from the west, but we hypothesize that wood collection during the Carthaginian and Roman periods from native forests would have eradicated most trees from the eastern part of the Jebel Nafusa from Gharyan/Ksar Daoun to Msallata and the coastal zone, and hilly back country from the actual Homs to Misurata region. All this region is nowadays covered with *Stipa tenacissima* (*Alfa gedima*) and other (*Globularia alypum* (*Zerga*), *Oryzopsis miliacea*, *Genista microcephala*, etc.), all relict plants always associated with 'open' or 'clear' forests in central and southern Tunisia with more than 200 mm MAR (Le Houérou, 1969). We also hypothesize that during the Roman and Byzantine colonization,

Photo 13.3. Roman kiln at Shiarchara north of Tarhuna (N 32.467 E 13.620).

Table 13.2. Evolution of forested areas (ha) in Jebel Al-Akhdar (1959–1979). (Adapted from Al-Idrissi *et al.*, 1996)

Dominant species	1959	1979	% decrease
Cupressus sempervirens	8000 ha	500 ha	93
Juniperus phoenicea	20,000 ha	15,000 ha	25
Pinus halepensis	7000 ha	5000 ha	29
Quercus coccifera	12,000 ha	1000 ha	92

forested sites with deep soils in the valleys and gently undulating country of the eastern Jebel Nafusa would have been cleared for olive tree plantations, and other fruit trees and vineyards.

Despois (1935) is reporting *Pistacia atlantica* (*batum*) 17 km NE of Nalut in wadi Tirekt (N 31.944 E 11.134) and up along wadi Ech-Chiekh hanging up on the Jebel cliffs. By 1975–1979, 40 years later, our field surveys failed to find any *batum* on these sites. We suppose that they were totally eradicated to provide fuelwood to Nalut. Once oil and gas became easily available to the population, fuelwood harvesting would have slowly dwindled, and large-scale wood cutting would have ceased though charcoal production and use continued to be practised all over the country. However, other forest destruction soon appeared more frequently due to unrest or accident, such as the large bushfires that engulfed the lowland between Shahat and Suza.

Fuelwood production and use

According to FAOSTAT, there is no record of fuelwood production from coniferous species (*Pinus halepensis* or *Juniperus* sp.) in Libya. The estimated production (m^3) by FAO from non-coniferous fuelwood, without mentioning which ones, is presented in Fig. 13.1. It appears to be reconstructed data from Libyan population numbers and wood production estimates[1] from 1982 and later FAO forestry reports. These FAO data do not mention whether this fuelwood originates from native forests or from afforestation projects. We should regard these figures with caution as they are FAO estimates. If we consider that 1 m^3 of non-coniferous fuelwood weighs about 750 kg/m^3, 1,000,000 m^3 represents approximately 750,000 tonnes of fuelwood harvested in 2015. An experiment in Mediterranean Catalonia, north-eastern Spain (Ogaya and Peñuelas, 2020) measured the wood production of a *Quercus ilex* – *Phillyrea latifolia* forest formation receiving an average 600 mm MAR over 21 years (350–900 mm MAR variation). This formation in good condition is reported to have an above-ground net primary productivity of about 2–3 tonnes DM/ha during the year, receiving 300–400 mm/year with about 30% converted to harvestable wood. This is a conversion rate similar to the *Juniperus* – *Quercus* – *Pistacia* sp. matorral from Jebel Al-Akhdar. Considering that the standing wood biomass in the east is ranging from 3–6 tonnes/ha, or 1–2 tonnes harvestable wood/ha, 750,000 tonnes of reported fuelwood wood harvested in 2015 would equate to 375,000–750,000 ha of forest matorral totally cleared in the east. It is clear that these FAO figures of Libyan non-coniferous fuelwood production, even for all of Libya, are grossly overestimated when checked against Table 13.2 or checking Libyan forested areas seen on Google Earth. There is a pressing need to accurately map and determine the Libyan forested areas and evolution. The fact remains that adding bushfire devastation and land clearing for urbanization, the slow, relentless, illegal and unobstructed destruction of the Libyan forests and matorral is happening these days. Even in the west, modern irrigation projects in the 100–150 mm rainfall wadi zones (wadis Merdum, Bir Dufan, Sof-el-Jin, Bey-el-Kebir etc.) keep illegally bulldozering relict slow-growing open formations of mature and young *Acacia tortilis* and *Pistacia atlantica* for private fuelwood collection and charcoal manufacturing.

Fuelwood and charcoal

Fuelwood collection and charcoal making are ancestral ways to provide fire for cooking and heating. To this day, despite being oil- and gas-rich, Libyans, like many other populations from arid zones, still collect fuelwood and manufacture large quantities of charcoal. Fuelwood is regularly collected from rangelands by shepherds and rangeland dwellers. During our field trips with colleagues, we all sinned every day by gathering branches and twigs, and often shamelessly eradicating shrubs (Photo 13.4) for cooking our *Ambakbaka* or *macarona* (Photo 13.5) and for a few cups of reinvigorating sweet tea, or simply gathering and chatting around a campfire at night.

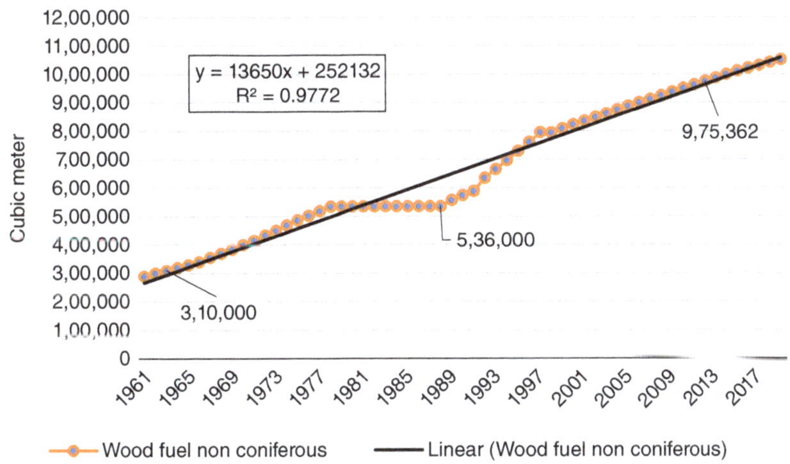

Fig. 13.1. FAOSTAT report of estimated production of non-coniferous fuelwood (m^3) in Libya. (After FAOSTAT)

Photo 13.4. One of the authors (GG) shamelessly collecting fuelwood for the night camp (Hamada, 6 April 2010).

Photo 13.5. Cooking our *ambakbaka* at wadi Arandal (Hamada, 8 April 2010).

While this occasional fuelwood harvest may appear innocent and often necessary, it certainly largely contributes to land degradation and desertification by systematic eradication and collection of arid-zone bushes and soil disturbance. In the mid-1970s, while camping and surveying vegetation on the El-Assah *sebkha*, we used to collect a few branches from *Limoniastrum monopetalum* (*zeita*) large bushes growing on *nebka*. This black solid wood is excellent for fuelwood and its embers last a long time. When surveying the same area by 2010, this plant had nearly disappeared from the Al-Assah region.

How Damaging Is Eradicating and Collecting Fuelwood from Arid Rangelands?

In the mid-1960s, an average figure of about 1 kg of dry wood per person per day is quoted to be collected in the steppe of North Africa (Le Houérou, 1969), the Middle-East (Thalen, 1979) and more in cold Central Asia (Gintzburger *et al*., 2003). It means that a family of ten would have collected and burned about four tonnes of wood per year, possibly more in colder climate in arid zones. If we consider that an *Artemisia herba-alba* rangeland in good condition, common in North Africa to Central Asia at the time, would have a dry wood biomass of aerial and roots of about 500–750 kg/ha, it means that the same family would clear about 5–8 ha of vegetation per year or all vegetation within a radius of 130–160 m around the collection site. Repeated over years, the damage on rangeland vegetation would be considerable and often beyond rehabilitation, and more so in the lowest-rainfall zones (Photo 13.6). This is by no means a criticism of the people quite understandably very much in need of fuelwood, but simply raises the question of other available alternatives.

Photo 13.6. Fuelwood eradication and collection of *Anabasis articulata* (*ajram*) in a 100–150 mm MAR rangeland. (From Flora of Libya Facebook site, 27 November 2021)

Likewise, illegal fuelwood cutting and harvesting in native wooded areas in eastern Libya or on afforestation projects all over northern Libya remain a never-ending problem.

Charcoal Production and Consumption

Using charcoal either for cooking or heating is part of the Libyan culture and traditions.

Charcoal production

There are no accurate data on charcoal production and amount sold in Libya. FAOSTAT provides charcoal production estimates of 5000 tonnes in 1961 climbing steadily to 115,000 tonnes in 2019 (Fig. 13.2).

Annual charcoal consumption per person

From FAOSTAT Libyan population data and charcoal production estimates, the charcoal use went from about 3.4 kg in 1961 steadily rising to 17–18 kg per person and per year (Fig. 13.3). This is no wonder that illegal charcoal production is soaring with the demand. It means that these days, to manufacture this charcoal, each person in Libya contributes to mostly illegal harvesting of about 70–75 kg of dry wood per year or about 115,000 tonnes for the whole Libyan population in 2019. Roughly estimating the fuelwood harvested from native forests or afforestation to be 3–7 tonnes/ha, it points to 20,000–40,000 ha cleared in Libya in 2019, mostly in the east. These estimates need to be refined with accurate surveys at legal and illegal logging sites and mapping by processing temporal series of satellite imagery.

Charcoal manufacturing and markets are open to anyone despite forestry regulations. The demand for local charcoal is high and can be seen on marketplaces from west to east. Charcoal price (Table 13.3) varies according to the region and the type of wood harvested from government forestry plantations, native forests or abandoned orchards.

Potential solutions

Controlling through agricultural police squads is the common solution with officers acting on

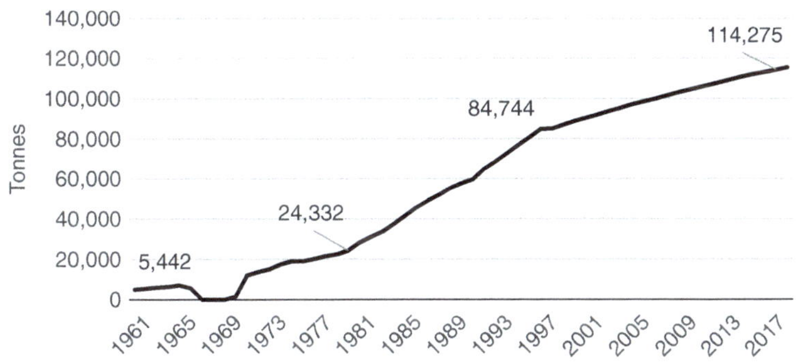

Fig. 13.2. Charcoal production in Libya (tonnes per year estimate, FAOSTAT).

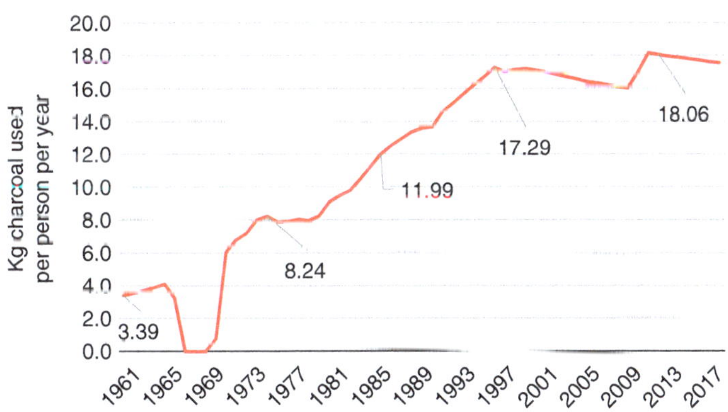

Fig. 13.3. Charcoal consumption in Libya (kg per person and per year estimate, FAOSTAT).

Table 13.3. Current charcoal price in northern Libya (January 2022).

In western Libya (in 25–30 kg bag)		
Eng. D. Amrani, Tripoli, January 2022, personal communication		
Orange tree	100 LD	Best quality
Acacia karroo	80 LD	
Olive tree	50–60 LD	
Casuarina sp.	50 LD	
Acacia saligna	40 LD	
Eucalyptus sp.	30 LD	Lowest quality
In eastern Libya (in 50 kg bag)		
Dr M. Saaed, Benghazi, January 2022, personal communication		
Tamarix sp.	60 LD	
Eucalyptus and *Pistacia lentiscus*	70 LD	
Juniperus phoenicea	75 LD	
Pistacia lentiscus	25 LD	

[1]Libyan dinar = 0.207 US$ (2022)

tips, or on haphazard findings, leading to fines or confiscation of equipment (tractors, trucks, trailers, chainsaws etc). While this may have some temporary effect on a few individuals, this is far from efficient in stopping illicit logging and fuelwood harvesting. Are there other solutions?

Need for proper policy and legislative framework for harvesting fuelwood

Two cases may be considered:

For household use

It refers to small quantities of fuelwood exclusively collected for household use. It is called *affouage* in France or 'common of estovers' in England. It is an ancestral right of taking wood for fire, no further than necessity, given to the users of the land. This right is granted by the shire authorities to collect wood only by registered residents for a small fee, and from predetermined sites on common land. It has evolved over centuries to limit the collection only of dead wood to an acceptable quantity, in some cases limited to a small trailer per family and per collection and limited to certain times of the year. In modern-day Europe, the use of chainsaws is forbidden; only hand saws are allowed and dead dry branches not thicker than 8–10 cm diameter may be collected. While in Libya it may be difficult to apply to seasonal encampments, it may be possible to register permanent residents of villages to collect fuelwood on their *watan*. It also should be that wood harvesting should be only from aerial plant parts and, under no circumstances, should plants be uprooted.

Any fuelwood collected by rangeland residents should not be sold.

At the same time, residents should be encouraged to initiate small tree nurseries to plant trees around their villages on common land assigned to fuelwood tree plantation with carefully selected species (see Table 13.3). Most of the time, *Eucalyptus* sp. and *Tamarix* sp. would be preferred as a source of fuelwood as they are fast-growing and coppice easily. These villagers' plantations should be protected at least for 5–8 years before harvesting above 1 m high, leaving the stumps to regrow and coppice. Harvesting should be decided by resident villagers by common agreement. Villagers should also consider using any wastewater to water the trees, whatever small quantity is available. Schoolchildren must be taught and encouraged to develop school tree nurseries under supervision, participate in tree plantation campaigns, and water and respect trees on the common land. These laudable community or individual initiatives are now flourishing in Libya, especially the Friends of the Trees Association and many others in the west and the east.

Other government incentives such as waiving land tax for ten years and free access to tree seedlings from government nurseries for plantations on private or communal land are in place.

For large-scale fuelwood collectors

Libyan social media regularly display large illegal tree logging on government plantations and bulldozing native trees (*Acacia tortilis, Pistacia atlantica*) on irrigation projects in western wadis. It appears that current laws (if any) and agricultural police actions are inefficient despite well publicized stunts (Photo 13.7).

This is because there are no proper and legal processes to contract harvesting fuelwood from plantations. Fuelwood is a common necessity that is not going to disappear soon in Libya. Instead of using random and inefficient police actions, it may be better for the forestry department supervising native and exotic trees plantations to organize legal fuelwood harvesting from mature trees and on selected sites. This is supposed to:

- train forestry technicians to evaluate fuelwood volumes available from government tree plantations;
- evaluate wood volumes available from maturing trees aged 5–8 years using forestry techniques (see https://www.fao.org/3/ap353e/ap353e00.pdf [accessed 12 March 2022]);
- develop a database and inventory of government plantations with GPS location and boundaries reported on GIS maps, rainfall zone, soil type, year of plantation, area planted (ha), species planted, tree density, survival rate after 1, 2 and 5 years, success or failure of establishment, if failure, record reason; and
- protect the areas planted, prohibiting any grazing or wood cutting, and patrolling the plantations.

At the same time, a legal charter must be developed and implemented by the forestry department to offer legal licences and supervised contracts for fuelwood harvesting produced on mature government plantations. This is suppose to:

- precisely locate the area to be harvested (with plantation sectors properly identified) and mature trees to be properly marked (red paint or other);
- evaluate the volume/weight of wood to be harvested;
- assess the monetary value of the wood volume/weight to be harvested and contract price on delimited site;

Photo 13.7. Illegal *Eucalyptus* fuelwood harvesting, and charcoal manufacturing interrupted by Agricultural Police patrols. (Photo: Libyan Ecological Society Facebook site, 17 July 2021, likely in Jeffara plain)

- establish legal contracts and licences with private forestry firms or contractors wishing to harvest the wood, against reasonable fees, date of beginning and end of operation, machinery and equipment to use on site, penalties in case contract not respected (height of cutting to create coppice for further harvest, stumps and refuse to be removed/burned on site or not, etc.);
- visit the site with the contractor before the operation starts to precisely identify and mark the limit of the site and/or trees to fell;
- have forestry department officers randomly inspect the site during harvesting operations to make sure that terms of contract are respected;
- remove fuelwood from site, properly weighed on calibrated government weighbridge, and recorded with the contract reference and the registration of trucks or vehicles;
- Data and wood volume/weight from site must be entered into the forestry database to compare with the volume/weight of wood evaluated before the operation; adjust techniques used for standing fuelwood volume evaluation per tree species and for future contracts;
- visit by forestry officers with the contractor to verify that contract terms are respected and the site is not damaged (soil removed or heavily disturbed), access tracks respected by heavy machinery and trucks;
- close the harvested site to allow for coppice regeneration (*Eucalyptus, Tamarix*, etc.); and
- monitor tree regrowth to plan for future harvesting contracts.

Similarly, forestry department licences must be organized and implemented for charcoal manufacturing with proper manufacturers' registration. They must keep record of the provenance of the wood (referring to harvesting contract mentioned above), the weight of wood turned into charcoal, and the amount of charcoal produced and sold (Photo 13.8).

Large amounts of illegally manufactured charcoal are regularly transported to main settlements. The charcoal load on the semi-trailer (Photo 13.9) could be around 40–45 tonnes. Considering that charcoal making with

Photo 13.8. Charcoal and fuelwood street vendor. (Photo: Libyan Ecological Society Facebook site, 5 August 2021).

Eucalyptus camaldulensis and *E. gomphocephala* wood is around 25% by weight, it represents approximately 160–180 tonnes of dry wood harvested. A plantation of these trees would produce about two to three tonnes of dry wood/ha/year in the 150–200 mm MAR zone (Aziziya, Suani), hence about 16–24 tonnes in eight years. It means that the charcoal loaded on the truck may come from 8–10 ha of *Eucalyptus* felled. Repeated year after year, the damage on plantations is considerable.

The destruction of tree plantations is catastrophic (Photos 13.10–13.12). On Google Earth, in September 2021, the same site (N 32.114 E 20.468 – El-Abyar region) and surrounding areas are nearly totally cleared of outstanding *Pinus halepensis* plantations (Photo 13.10) and it shows two large new illegal charcoal manufacturing sites. The area cleared in 17 years is approximately 10 ha, not counting neighbouring areas.

The deforestation of this large area continues unabated, and the Google Earth satellite image from September 2021 shows over 11 earth kilns sited on this forested plantation area of approximately 2.5 km^2 x 1.5 km^2 (Photo 13.12).

By 2024, looking at the Google Earth site, even more areas were totally cleared of planted *Pinus*.

Most importantly, wood issued from native Libyan trees and forestry plantations must be strictly prohibited from being harvested, transported or sold without proper governmental forestry contract. Appropriate sanctions and fines must be advertised and implemented.

Photo 13.9. A semi-trailer truck load of charcoal from a charcoal manufacturing site in the Jeffara plain. (Photo: Libyan Wildlife Trust Facebook site, 15 July 2021)

While this may take time to enforce, it may be the only way to make lawful and good use of the many remaining government and private tree plantations established for decades at great cost in Libya and to preserve Libyan native forests, tree plantations and even individual remarkable old trees.

The need for fuelwood in afforestation projects

At the same time, afforestation projects must be a priority from the Ministry of Agriculture and the Forestry Department. Modern nurseries must be established and maintained all over the country. Fuelwood tree plantations must relentlessly be established on every piece of government land available using appropriate fast-growing species already recognized as successful in northern Libya (Keith, 1965): i.e. *Pinus halepensis, Eucalyptus camaldulensis, Eucalyptus gomphocephala, Australian Acacia* sp., *Tamarix* sp., *Calligonum* sp. and *Casuarina cunninghamiana*, with other species that showed potential such as the Central Asian *Haloxylon* sp. and *Elaeagnus* sp. (Table 13.4). Considering the poor soils usually valuable for tree plantations, wood production will be, at best, in the order of 500–1000 kg/ha/year with *Pinus halepensis* with 1000 trees/ha. *Eucalyptus* plantations

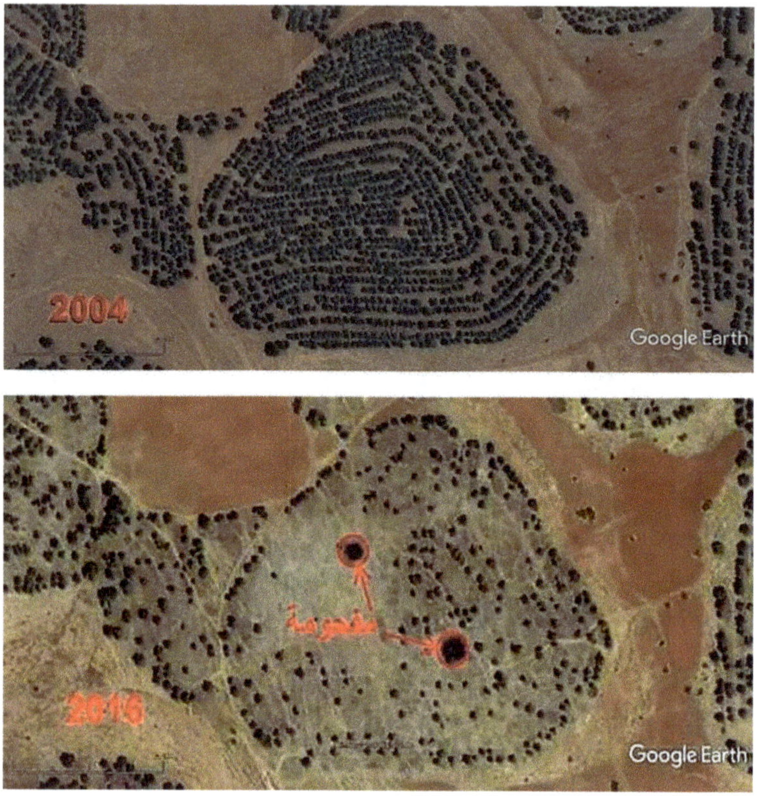

Photo 13.10a, b. The impact of illegal wood harvesting and charcoal manufacturing on government forestry plantation in the El-Abyar region south-west of Bu Maryam in Jebel Al-Akhdar between 2004 and 2016. (From *Libya the Future*, 06/18/2017 at 15:19; https://www.libya-al-mostakbal.org/ [accessed 8/8/24])

may be more productive. The most productive plantations would be on land with high water table with *Tamarix* sp., *Casuarina* sp., *Elaeagnus angustifolia* and *Haloxylon aphyllum*, all four being also salt-tolerant.

Haloxylon aphyllum and *Calligonum* sp. from the steppe and deserts of Central Asia are recognized fuelwood sources and livestock feed (Gintzburger *et al.*, 2003). Flowering *Ceratonia*, *Eucalyptus*, Australian *Acacia*, *Gleditsia* and *Elaeagnus* are also very valuable trees for bees and honey production. Moreover, pods produced by the *Gleditsia* and *Ceratonia* are valuable feed for all livestock. Other exotic species were introduced in Libya (*Parkinsonia aculeata*, and some exotic shrubs) but persistently failed in producing significant quantity of fuelwood per ha.

Studies related to improving charcoal manufacturing needed

The average ratio of dry wood to charcoal produced by weight is approximately 3–4:1. It means that 1000 kg of dry wood would produce approximately 250–300 kg of charcoal in the best conditions. However, this must be calibrated for specific species and charcoal making in Libyan settings. This could be the subject assigned for a Master's thesis. In Libya, traditional charcoal manufacturing is mostly done with the earth kiln technique (Photos 13.13 and 13.14), stockpiling and arranging wood in dome shapes and covered with grass, earth or wet sand before firing it. It requires skill and experience to control the two to three weeks' carbonization process. It often produces

Photo 13.11. The same site (N 32.114 E 20.468), in September 2021, is still cleared of trees and two more active kilns (*mardoma*) are added despite the article published in 2017.

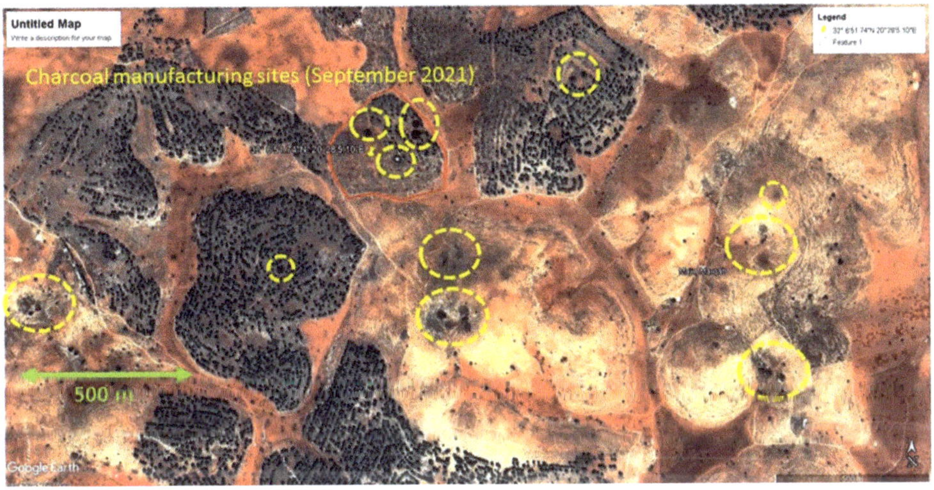

Photo 13.12. New earth kiln sites for charcoal manufacturing near Majin Marqab (N 32.110 E 20.467, south-west El-Abyar, September 2021). (Google Earth)

poor-quality charcoal and significant amounts of refuse, and it needs to be rebuilt after each cycle.

Other charcoal manufacturing easier and more efficient techniques are available in many parts of the world with brick or steel kilns. Both methods produce faster and better-quality charcoal with little unburnt wood. While the brick kiln is static, the steel kiln can be moved to different sites and allows fast manufacturing, usually

Table 13.4. Main native and exotic trees for fuelwood and charcoal in Libya.

LIBYAN NATIVE TREES (GENUS AND SPECIES)	Soil type	Rainfall zones when rainfed (mm MAR)	Growth (1 = slow, 5 = fast)	Potential use	Fuelwood (1 = poor, 5 = good)	Charcoal	Building material	Salt tolerance (0 = nil, 5 = high)	Estimated dry wood production (kg/ha/year)	Suggested protection status
Acacia tortilis ssp. *raddiana*	Loamy to heavy	100–300	1	Afforestation	5	High quality	High quality	0	500	High
Calligonum sp. (*C. comosum, C. azel, C. arich*)	Sand	100–300	2	Sand dune fixation	4	High quality	Poor quality	0	500	High
Ceratonia siliqua	Loamy to heavy	300–600	1	Afforestation, bees	5	High quality	High quality	0	500	High
Cupressus sempervirens	Loamy to heavy	500–600	1	Afforestation	5	High quality	High quality	0	500	High
Juniperus phoenicea	Loamy to heavy	300–600	1	Afforestation	5	High quality	Medium quality	0	500	High
Olea europaea var. *oleaster*	Sandy loam to heavy	300–600	1	Afforestation	5	High quality	High quality	0	500	High
Pinus halepensis	Loamy to heavy	200–600	4	Afforestation and fuelwood	3	Poor	Poor	0	1000	Medium
Pistacia atlantica	Loamy to heavy	200–400	1	Afforestation	4	High quality	High quality	0	500	High
Tamarix sp.	Phreatophyte, halophile	100–600	5	Afforestation and fuelwood	3	Medium	Medium quality	5	5000	Low
Calotropis procera	Sand	100–300	5	Fuelwood	1	High quality (used for gunpowder)	Poor quality	0	?	Low
Periploca laevigata	Loam on calcareous crust	100–300	3	Fuelwood	4	High quality	Poor quality	0	?	Medium
Pistacia lentiscus	Red soils	200–500	3	Fuelwood	3	Medium quality	Poor quality	0	?	Low
Retama raetam	Sandy loam to heavy	100–400	4	Afforestation and fuelwood	4	Medium quality	Poor quality	0	?	Low
Rhus tripartita	Loam on calcareous crust	100–300	2	Fuelwood	4	High quality	Poor quality	0	?	Medium
Arbutus unedo	Red soils	200–500	2	Fuelwood	4	Medium quality	Poor quality	0	?	Medium

EXOTIC (GENUS AND SPECIES)	Soil type	Rainfall zones when rainfed (mm MAR)	Growth (1 = slow, 5 = fast)	Potential use	Fuelwood (1 = poor, 5 = good)	Charcoal	Building material	Salt tolerance (0 = nil, 5 = high)	Estimated dry wood production (kg/ha/year)	From
Acacia karroo	Deep sandy soils, not on calcareous crust	150–400	3	Spiny hedges	5	High quality	High quality	1	1000	South Africa
Acacia saligna (*A. cyanophylla*)	Deep sandy soils, not on calcareous crust	150–400	5	Afforestation, fuelwood, bees, sand dunes fixation	3	Medium	Medium quality	0	5000	Australia
Casuarina sp.	Heavy soil, phreatophyte, halophile	150–400	3	Afforestation, fuelwood	3	Medium	Good quality	4	10,000	Australia
Elaeagnus angustifolia	Loamy to heavy, Phreatophyte	150–400	5	Afforestation, fuelwood, bees, garden hedges	3	Medium	Poor quality	3	2000	Uzbekistan, Turkmenistan
Eucalyptus camaldulensis	All soils, even on fractured calcareous crust	250–500	5	Afforestation, fuelwood, bees	3	Medium	Poor quality	0	10,000	Australia
Eucalyptus gomphocephala	All soils, even on fractured calcareous crust	250–500	5	Afforestation, fuelwood, bees	3	Medium	Poor quality	0	10,000	Australia
Gleditsia triacanthos	loamy to heavy soils, even on fractured calcareous crust	300–500	3	Afforestation, fuelwood, bees, garden hedges	4	Medium	Medium quality	0	2000	Mexico, south USA,

Continued

Table 13.4. Continued.

EXOTIC (GENUS AND SPECIES)	Soil type	Rainfall zones when rainfed (mm MAR)	Growth (1 = slow, 5 = fast)	Potential use	Fuelwood (1 = poor, 5 = good)	Charcoal	Building material	Salt tolerance (0 = nil, 5 = high)	Estimated dry wood production (kg/ha/year)	From
Haloxylon aphyllum	Loamy to heavy, Phreatophyte, halophile	150–400	3	Afforestation, fuelwood	5	Medium	Poor quality	4	5000	Uzbekistan, Turkmenistan
Haloxylon persicum	Sand	150–400	3	Afforestation, fuelwood	5	Medium	Poor quality	2	1000	Middle East
FUELWOOD ISSUED FROM FRUIT TREES										
Citrus sp., almonds, figs, apricot, etc.	Sandy-loam to heavy	Irrigated	4	Fruit tree	5	High quality	Wood inlay on high-quality furniture	0	?	South-east and Central Asia
Olea europaea	Sandy-loam to heavy	Mostly rainfed	2	Fruit tree	4	High quality	High quality	0	?	Old Mediterranean regions

in less than a week. Their efficiency is higher (30–35%) than the earth kilns. These latter techniques should be explored for Libyan conditions. The advantage of the industrial large steel kilns is that they could be registered and controlled by the government.

In Photo 13.14, we can count 28 kilns (average dimensions 3 m high x 10 m diameter base), each with a wood volume of about 80 m^3, i.e. a total of 2240 m^3 of wood, or about 1700 tonnes of wood extracted from 300–500 ha of forestry plantations now totally cleared. This wood will then be turned into a total of 600 tonnes of charcoal. It ultimately represents 20,000 30 kg bags of charcoal at a street value of about 600,000–700,000 Libyan dinars (US$ 120,000–140,000 [2023 prices]). Supposing the wood is illegally harvested from long-established government forestry plantations from Sidi Bahnur sandy area, the return on investment to the Forestry Department and the Libyan government is nil, not counting the extensive damage caused to the once-fixed sand dunes.

Photo 13.13. Wood from *Acacia tortilis* (*talah*) arranged for charcoal making as a traditional earth kiln (*mardoma*) likely at wadi Nefad, before being covered with sand. (Photo: Libyan Ecological Society Facebook site, August 2021)

Other options as sources of fuel

Oil and gas

Libya is rich with gas and oil resources. Fuelwood collection from rangelands is time-consuming and contributes to land degradation and desertification. Fuelwood required by communities is still essential in Libya. It will be extremely difficult to halt fuelwood harvesting by individuals and herders roaming the rangelands. However, for city dwellers, one option would be to ban bakeries and restaurants from using large quantities of fuelwood or charcoal and use fuel or gas operated bread ovens within a reasonable time. This may not apply for registered small restaurants and small vendors offering barbecued meat so much appreciated by Libyans.

Gas cylinders

As for rangeland dwellers and small villages, a government campaign to supply subsidized gas cylinders at a frequency to be decided by the authorities could help. This was successfully implemented in Algeria, a country rich in gas and oil, to stop fuelwood harvesting from the steppe and forested areas. These gas cylinders could be obtained

Photo 13.14. Large fuelwood harvesting and charcoal manufacturing camp from *Eucalyptus* sp. and *Acacia cyanophylla* forestry plantations at Sidi Bahnur (south-east Tajura, 1981).

from vendors by resident families against a governmental coupon delivered by the municipality. This may not completely solve the problem of fuelwood collection on forests and rangelands. People would still occasionally collect and use fuelwood and charcoal. Any further action would be to actively pursue government fuelwood plantations, and control the fuelwood harvest and charcoal manufacturing by means of legal contracts.

Wood and olive cake waste

Libya grows a large quantity of olive trees. Thousands of tonnes of pruned branches are usually burned on site or traditionally used for cooking at household level. Olive-pressing residues (pulp and pits, pumice stones) usually accumulate near olive presses and are ultimately wasted. One should consider the use of these olive tree wastes. Olive cake could be industrially processed by briquetting or pelleting and sold as alternative fuel for households and bakeries. Olive cake was first used in Spain in 1995 to fuel a power plant for electricity production.

Wood waste from old fruit-tree plantations and forestry

Fruit trees are pruned every year. Large quantities of wood, branches and leaves can also be sourced from old and decommissioned fruit-tree plantations (olive, orange, apple, apricot, almond, etc.) and would make a valuable source of fuelwood and quality charcoal.

As for refuse issued from pruning, instead of being burned on site or wasted, it could be processed through a woodchipper/shredder, then sun-dried to less than 12% humidity and ultimately turned into pellets or wood briquettes for sale in local markets. In many parts of the world (China, India, Europe, etc.), these wood briquettes and pellets make a substantial addition to household and industrial alternative fuel supplies issued from forestry industries.

Exotic Rangeland Trees and Shrubs Introduced to Libya

Exotic plants to Libya (Alzerbia *et al.*, 2020; Mahklouf, 2019) were introduced over the past millenaries or recently by travellers, colonizers or foresters. The purpose of their introduction could have been accidental with seeds, forage or luggage from travellers or invading armies. More likely, they were introduced as vegetables (tomatoes, capsicum, zucchini, melons, potatoes etc.), forage (alfalfa), fruit trees (orange, mandarins, apples, pears, peaches, cherries, bananas etc.), as ornamentals, as a source of fuelwood or as construction material (*Eucalyptus*) on irrigated farms and gardens. Very few of them found their way onto rangelands due to the unique arid Libyan conditions. It is only when foresters and land managers took an interest on rangeland improvement that some of them, first as a source of fuelwood and windbreaks, then as fodder shrubs originating from other comparable arid zones, were evaluated in the late 19th century by Turkish foresters (Keith, 1965). Over years of experimenting in northern Libya, some were successfully introduced on large-scale rangeland sites and government projects during 1975–1990. Among the many tested, the main tree and shrub species more or less successfully introduced in Libya since the beginning of the 20th century are listed in Table 13.5. Many more fodder shrubs and trees were evaluated in Libya at one stage or another on Libyan rangeland projects by the mid-1970s following long-term field evaluations in North Africa and other Mediterranean arid regions (Gintzburger *et al.*, 2000; Le Houérou, 2002; Le Houérou and Franclet, 1971). Le Floc'h and Aronson (2013) reviewed many trees from the deserts of the world, some of which could be assessed in Libya.

Exotic plants introduced and assessed on Libyan rangelands – success and failures

Remarkably, none of the exotic species of trees (*Eucalyptus*, Australian *Acacia*) or shrubs (Australian or American *Atriplex* sp.) and *Opuntia* sp. intentionally or accidentally introduced in Libya became invasive. A few exceptions are *Ricinus communis (Kharwa)* and *Nicotinia glauca (Akus Musa)* colonizing roadsides and wasteland near dwellings but never taking root on rangelands. The thorny *Prosopis glandulosa* and *Prosopis juliflora* and other *Prosopis* sp. that became pests in arid sub-Saharan Africa, South Africa and

Table 13.5. Exotic trees and fodder shrubs introduced on Libyan rangelands since beginning of 20th century.

Genus and species	Family	From	Evaluated in Libya	Used in rainfall zone	Use	Comment
Acacia aneura F. Muell. ex Benth.	Mimosaceae, shrub or small tree	Australia	Oushtata, Bir-el-Ghanem	MAR ≥ 150 mm	Limited	Useless, poor palatability
Acacia cyclops A. Cunn. ex G. Don	Mimosaceae shrub or small tree	Australia	Coast from Zuara to Misurata	MAR ≥ 200 mm	Afforestation, dune stabilization close to seashore only	Successful, poor palatability
Acacia farnesiana (L.) Willd.	Mimosaceae, shrub or small tree	Australia	Jeffara and Sirte	MAR ≥ 150 mm	Limited	Useless, poor palatability
Acacia karroo Hayne.	Mimosaceae, small tree	South Africa	West and Sirte coastal zone	MAR ≥ 150 mm	Edges, charcoal	Successful, prickly and thorns
Acacia ligulata A. Cunn. ex Benth.	Mimosaceae, shrub or small tree	Australia	Jeffara	MAR ≥ 150 mm	Limited	Useless, poor palatability
Acacia saligna (Labill.) H.L. Wendl.	Mimosaceae, small tree	Australia	Jeffara and Sirte only	MAR ≥ 150 mm	Afforestation, rangeland plantations, fodder shrubs, fuelwood	Very successful on sandy soils, not suitable in the east on heavy soils
Atriplex canescens (Pursh) Nutt.	Chenopodiaceae, shrub	NW USA	In the Jebel Nafusa and in the east	MAR ≥ 150 mm	Rangeland plantations, fodder shrubs	Limited, poor palatability, two cultivars one for sandy soils and one for heavy soils
Atriplex leucoclada Boiss.	Chenopodiaceae, shrub	Middle-East, Central Asia	In the east only/ Benghazi plain project, Deriana	MAR ≥ 250 mm	Rangeland plantations, fodder shrubs	On heavy red soils, no advantage compared to *A. nummularia*
Atriplex nummularia Lindl.	Chenopodiaceae, shrub	Australia	From west to east	MAR ≥ 150 mm	Rangeland plantations, fodder shrubs	Very successful on loamy-heavy soils, highly palatable but subject to rodents infestations
Atriplex undulata De Dietr.	Chenopodiaceae, shrub	South America	Oushtata	MAR ≥ 150 mm	Rangeland plantations, fodder shrubs	Successful but no advantage compared to *A. nummularia*
Casuarina cunninghamiana Miq.	Casuarinaceae, tree	Australia	From West to East	MAR ≥ 150 mm	Wind breaks, fuel wood	Very successful on loamy-heavy soils,

Continued

Table 13.5. Continued.

Genus and species	Family	From	Evaluated in Libya	Used in rainfall zone	Use	Comment
Casuarina decaisneana F. Muell.	Casuarinaceae, tree	Australia	From west to east	MAR ≥ 150 mm	Wind breaks, fuelwood	Very successful on loamy-heavy soils
Casuarina equisetifolia Forst.	Casuarinaceae, tree	South-east Asia – Australia	From west to east	MAR ≥ 150 mm	Wind breaks, fuelwood	Very successful on loamy-heavy soils
Casuarina glauca Sieb. ex Spreng.	Casuarinaceae, tree	Australia	From west to east	MAR ≥ 150 mm	Wind breaks, fuelwood	Very successful on loamy-heavy soils
Casuarina stricta Ait.	Casuarinaceae, tree	Australia	From west to east	MAR ≥ 150 mm	Wind breaks, fuelwood	Very successful on loamy-heavy soils
Casuarina torulosa Ait.	Casuarinaceae, tree	Australia	From west to east	MAR ≥ 150 mm	Wind breaks, fuelwood	Very successful on loamy-heavy soils
Eucalyptus camaldulensis Dehnhardt.	Myrtaceae, tree	West Australia	From west to east	MAR ≥ 150 mm	Afforestation, roadside, fuelwood, charcoal	Very successful on sandy/ sandy-loam soils
Eucalyptus gomphocephala DC.	Myrtaceae, tree	West Australia	From west to east	MAR ≥ 150 mm	Afforestation, roadside, fuelwood, charcoal	Very successful on sandy/ sandy-loam soils
Eucalyptus rudris Endl.	Myrtaceae, tree	West Australia	From west to east	MAR ≥ 150 mm	Afforestation, roadside, fuelwood, charcoal	Very successful on sandy/ sandy-loam soils
Opuntia ficus-indica (L.) Mill.	Cactaceae, shrub	SW North America	Mostly in the west	MAR ≥ 150 mm	Rangeland plantations, fodder shrubs	Very successful on sandy/ sandy-loam soils, not on heavy soils in the east
Parkinsonia aculeata L.	Caesalpiniaceae, small tree	South and North America	In the west and Sirte	MAR ≥ 150 mm	Rangeland plantations, fodder shrubs	Useless
Schinus molle L.	Anacardiaceae, tree	South America	In the west and Sirte	Ubiquitous	Garden, roadside	Fuelwood, limited
Schinus terebinthifolius Raddi.	Anacardiaceae, tree	South America	In the west and Sirte	Ubiquitous	Garden, roadside	Fuelwood, limited

Australia (Shackleton *et al.*, 2014) were introduced in Libya as early as 1930 at Sidi-el-Mesri, Bu-Slim and near Misurata (Keith, 1965). Fortunately, these introductions were never successful.

The main reasons for unsuccessful expansion of many introduced species on Libyan rangelands are their germination and ecological requirements. Besides some rare germination of Australian *Acacia saligna* observed on sand dune plantations near Sidi-Benhur-Massjed/Tajura (MAR = 250 mm), on good rainy years, the seedlings never got properly established and rapidly faded away. While native species readily germinate and grow, though with difficulties, under Libyan soils and arid climatic conditions, all exotic species introduced on Libyan rangelands have to be raised under nursery conditions, under irrigation in the early stage, transplanted in the field with abundant initial irrigation to secure proper establishment.

In terms of ecological requirements, several valuable fodder shrubs from Australia proved worthless (*Maireana sedifolia*, Chenopodiaceae, Australia) or of limited interest (*Cassia sturtii* = *Senna artemisioides* subsp. *sturtii*, Fabaceae, Australia), and *Medicago arborea* L. and *Medicago arborea* ssp. *citrina* (Fabaceae, from Spain) also due to difficult grazing management.

Kochia sp., possibly *Kochia indica* (= *Bassia indica*) from India or by *Kochia scoparia* (= *Bassia scoparia*) from the Middle East and Central Asia, colonized large areas of the Marmarica coast by the end of WWII. It appears to be an accidental introduction, probably by the British troops in their crates of armaments coming from Egypt or the Middle East. This *Kochia* invasion did not last and disappeared fast from the region by 1950. *Kochia prostrata* and other grasses originating from Russian and Central Asia rangelands kept being assessed by American scientists on rangelands of Morocco and Jordan (Bailey *et al.*, 2010), but never succeeded with proper and long-term establishment below 300 mm MAR in Mediterranean regions.

A Sapotaceae, *Argania spinosa* bushes (= *Sideroxylon spinosum* [http://www.theplantlist.org/tpl1.1/record/kew-14735 (accessed 9/8/24)]) were introduced in the Sirte region in 1985 (El-Gadi and Enayet Hossain, 1986). It has apparently been again assessed and grown at Guettis (Kadem el Jebel – wadi Abu-Shayba) in Jeffara, near Gharyan in the Jebel Nafusa and in Fezzan (wadi Al-Shati). Photos occasionally appeared on the Libyan Ecological Society (LES) Facebook site since 2021 (Ali Agosh and Omar Jally, Tripoli, February 2023, personal communication). Argan tree is a drought-resistant small endemic tree from arid south-western Morocco (Benchekroun and Buttoud, 1989; López-Sáez and Alba-Sánchez, 2009). The argan tree is famous for producing kernels from which oil is extracted either as a local condiment or as a base for industrial cosmetics. As *Argania* is an endemic from Atlantic Morocco benefiting from high air humidity, while it raises questions about its introduction in low-ambient humidity in arid Libyan conditions, it appears to be successfully established from nursery seedlings, growing and fruiting in the field at a few western and southern Libyan sites and must be confirmed with refereed publications.

Exotic plants appearing in Libya is a continuous process. Recently, *Pluchea dioscoridis*, a small bush from the Asteraceae family, was recorded on saline *sebkha* around Benghazi (Altera *et al.*, 2022). Common along the Nile in Egypt and from East Africa, this small bush is moderately invasive and does not present a risk as its peculiar ecology limits its expansion.

Potential new trees for Libya

With recent access to Central Asian countries (Gintzburger *et al.*, 2003), several drought- and/or salinity-resistant trees could be evaluated in Libya.

- *Haloxylon aphyllum* (Minkw.) Iljin (http://www.theplantlist.org/tpl1.1/record/tro-7201217 [accessed 9/8/24]) (Syn. *Haloxylon ammodendron* (C.A.Mey.) Bunge ex Fenzl). This is the most promising and grows as a large tree (4–10 m) on saline land and high saline water table in the steppe of Central Asia in the Kyzyl-Kum and Karakum desert (Gintzburger *et al.*, 2003). It was briefly introduced (1957–63) from Uzbekistan at the Haschan forestry station, Ain Zara, and Agheilat sites (Keith, 1965). It did well on deep sandy sites in Libya; however, Keith reports attacks by birds, mice and rats, damaging young seedlings. This *Haloxylon aphyllum* would do better around *sebkha* with high water table (Al-Assah – Jbebina, from Zuara, along the western coast to Taworga, Sirte, Ajdabiya). It is palatable to livestock and a good source of fuelwood. It was successfully introduced in the Aleppo Syrian steppe in gypseous/saline depressions at Maragha (N 35.563 E 37.658, MAR about 150 mm).

- *Haloxylon persicum* Bunge (http://www.theplantlist.org/tpl1.1/record/kew-2838918 [accessed 9/8/24]), a large low shrub (2–5 m diameter, 1.5 m high) is more suitable for sandy sites and dunes. Common in the Middle East steppe and deserts, it was introduced into southern Tunisia.
- The many desert *Calligonum* sp. from Central Asia and northern China (Gintzburger *et al.*, 2003) are also very diverse and could be suitable for Libyan sandy sites.
- Two large (1–4 m high) woody *Salsola* sp. bushes from Turkmenistan (*Salsola paletzkiana* Litv. [http://www.theplantlist.org/tpl1.1/record/kew-2481468] and *Salsola richteri* (Moq.) Karel ex Litv. [http://www.theplantlist.org/tpl1.1/record/tro-7201329]) may also be suitable for sandy sites (Gintzburger *et al.*, 2003).

All the above are good sources of fuelwood and moderately palatable with seeds available from the Karakul Sheep and Desert Ecology Institute (Samarkand, Uzbekistan) and the Desert Institute (Ashgabat, Turkmenistan).

National Parks and Nature Reserves

Large rangeland and forests reserves were wisely established (Table 13.6) by the government in Libya over many years to protect fauna and flora biodiversity (Bouras, 2013; Elkahwage and Jdeidi, 2018; Khatabi, 1993). They were first initiated in 1949 under the forestry department laws, with subsequent additions in 1970, and establishing the first national park at wadi Kuf in 1978. A Technical Committee of Wildlife and National Parks was instituted in 1990 to oversee and manage the Libyan national parks. Bouras (2013) indicates that protection and management of these reserves lack integration in national policies, lack funds, are weak in legislation, poor understanding from decision makers and public about biodiversity and ecosystems value, and show little research on these areas. There is little governmental policing and control of these national and regional reserves where poaching, hunting, illegal logging and charcoal manufacturing are prevalent. One should note that Libyan NGOs, university researchers and local communities and shires decided to create their own nature reserves and protected areas mostly along the coast and also on rangelands and forested areas. By 2010, a 25 ha exclosure (N 31.689 E 14.958) was established to protect a rare stand of several large *Cocculus pendulus* climbers growing in *Acacia tortilis*. Local shepherds soon destroyed the fences and kept grazing in the protected areas. Unfortunately, without support and approval from heads of tribes and

Table 13.6. Government national parks and nature reserves in Libya.

Biodiversity reserves	Location	Region	Hectares	From	Mean annual rainfall (mm)	Status
Wadi Kuf	N 32.74 E 21.20	East	100,000	1978	300–400	National park
Hisha Jadida	N 31.51 E 15.23	Sirte	160,000	1984	100–150	Nature reserve
Abu-Ghilan	N 32.27 E 13.02	West	4000	1992	150–200	National park
Bir Ayyad	N 32.11 E 12.40	West	12,000	1992	150–175	Nature reserve
Garabulli	N 32.78 E 13.78	West	8000	1992	200–250	National park
Surman	N 32.72 E 12.56	West	400	1992	150–200	National park
El-Naggaza	N 32.77 E 13.87	West	4000	1993	200–250	National park
Sabratha	N 32.76 E 12.44	West	500	1995	200–250	National park
Msallata – Alshafien	N 32.61 E 13.83	West	1800	1998	250–300	Nature reserve
Nalut	N 31.87 E 10.87	West	200	1998	100–150	Nature reserve
Zaltan	N 32.95 E 11.81	West	1000	1998	150–200	Nature reserve
Total (ha)			**291,900**			

local communities, most locally established small natural reserves are doomed.

Conclusions

Libyan relict dense and open forests in Libya, old remarkable trees are national treasures under constant threat from illegal logging and charcoal manufacturing even in the lowest-rainfall zones. It urgently needs effective governmental protection, forestry surveys, academic evaluations, ecological characterization and accurate mapping. To date, there are no reliable data about Libyan forests and wooded areas, only dubious old guestimates. Accurately mapping and monitoring the relict forests and afforestation projects is a necessary first step using modern satellite imagery, up-to-date techniques and GIS. Without this information, the Libyan forest resources will continue to be illegally plundered for fuelwood and charcoal, and relentlessly cleared for unrestrained urbanization.

The demand for fuelwood and charcoal is part of the Libyan culture and is not going to disappear soon. Photos of illegal, wild and large fuelwood harvesting and charcoal manufacturing all over Libya appear regularly on Libyan social media.

The current legislation to forcibly control tree felling, fuelwood harvesting, and charcoal manufacturing appears to be inadequate and inefficient. It will remain so as long as a contractual system regulating these activities is not established and forcibly implemented. A legal framework and forestry contracts for all industrial fuelwood harvesting and charcoal manufacturing on government plantations and national forests are needed to control and curb abusive and illegal operations. Concurrently, transportation and sale of industrial quantity of fuelwood and charcoal must be registered with industrial transactions legally recorded and charcoal manufacturing must be government-registered.

Alternatives to the use of fuelwood and charcoal must be explored and implemented by converting bakeries and public baths to fuel or gas burners, making charcoal manufacturing more efficient by using government-registered movable steel kilns, making better use of olive cake and fuelwood wastes from decommissioned fruit-tree plantations, offering government-subsidized gas cylinders to households in remote

Photo 13.15. Our campfire during a cold winter night at Bir-el-Ghanem (January 1978) with Hedi H'diden, Muamar Asid (ARC technicians) and Haj Ali (shepherd at BGII).

rangeland areas, encouraging native and appropriate exotic fuelwood tree plantations by government organizations and the private sector with appropriate incentives, training forestry technicians to evaluate fuelwood volume and economic output to be legally industrially extracted by private or government contractors, and controlling contracted logging and charcoal-making operations. There is an urgent need for establishing a National Forestry Plantation Monitoring Database accurately recording GPS location with mapping, date of plantation, areas planted, species planted, planting density, survival rate, date of logging, volume extracted, name of contractors, etc. Without immediate actions to remedy rampant illegal and unsustainable logging and charcoal manufacturing, Libya will fast see forestry plantations and unique native forests and trees disappear, hence increasing already rampant desertification.

Overall, many exotic trees and shrubs from different parts of the world and ecologically analogous environments were introduced and assessed in Libya over the past 100 years. Only a few of them have been successfully transplanted from nurseries to rangelands afforestation projects or fodder shrubs large-scale plantations. None of the exotic species introduced in Libya became invasive due to harsh environment conditions.

It remains that despite all legislation and government actions, nothing would replace a woodfire (Photo 13.15) during cold nights while camping on rangelands.

Note

[1] http://www.fao.org/3/AM472T/AM472T.pdf (accessed 8 March 2022)

References

Al-Idrissi, M., Sbeita, A., Jebriel, A., Zintani, A., Shreidi, A. and Ghawawi, H. (1996) Libya country report. *FAO International Technical Conference on Plant Genetic Resources*, 29.

Altera, F.M., Mukassabi, T.A., Elmgasapi, A.M. and Thomas, P.A. (2022) First record of *Pluchea dioscoridis* (Asteraceae) in Northern Libya. *The Scientific Journal of University of Benghazi* 35(2), 157–160.

Alzerbia, A.K., Alaibb, M.A. and Omar, N.O. (2020) Introduced species in flora of Libya. *Libyan Journal of Science and Technology* 11(2), 65–72. DOI: 10.37376/ljst.v11i2.2404

Bailey, D.W., Tabini, R.A., Waldron, B.L., Libbin, J.D., Al-Khalidi, K. *et al*. (2010) Potential of *Kochia prostrata* and perennial grasses for rangeland restoration in Jordan. *Rangeland Ecology & Management* 63(6), 707–711. DOI: 10.2111/REM-D-09-00195.1

Benchekroun, F. and Buttoud, G. (1989) L'arganeraie dans l'économie rurale du sud-ouest marocain. *Forêt Méditerranéenne* 11(2), 127–136. DOI: hal-03555978

Bouras, E.M. (2013) Libya: action plan for implementing the programme of work on protected areas of the convention on biological diversity. Available at: https://www.cbd.int/doc/world/ly/ly-nbsap-powpa-en.pdf (accessed 9 August 2024).

Capelli, C. and Leitch, V. (2011) A Roman amphora production site near Lepcis Magna: petrographic analyses of the fabrics. *Libyan Studies* 42, 69–72. DOI: 10.1017/S0263718900004817

Despois, J. (1935) *Le Djebel Nefousa (Tripolitaine), Etude Géographique*. Larousse, Paris.

El-Gadi, A.A. and Enayet Hossain, A.B.M. (1986) *Poisonous Plants of Libya*. Dar Al-Kitab al-Watani Benghasi, Benghazi, Libya.

Elkahwage, E.I. and Jdeidi, T. (2018) Survey of the wild mammalian species in Misallatah nature reserve and national park. In: *Atti e Memorie dell'ente Fauna Siciliana Society, 4–6 September 2015. Noto-Vendicari (Italy) 3rd International Congress Biodiversity*, pp. 101–112.

FAO (2005) *Global Forest Resources Assessment, Country Reports, Libyan Arab Jamahiriya*. Food and Agriculture Organization of the United Nations. Available at: https://www.fao.org/3/ai886E/ai886E00.pdf (accessed 9 August 2024).

Gintzburger, G., Bounejmate, M. and Nefzaoui, A. (2000) *Fodder Shrub Development in Arid and Semi-Arid Zones*. ICARDA Proceedings Hammamet Fodder Shrubs Workshop, Vols 1, 2. International Center for Agricultural Research in the Dry Areas, Hammamet, Tunisia.

Gintzburger, G., Toderich, K., Mardonov, B. and Mahmudov, M. (2003) *Rangelands of the Arid and Semi-Arid Zones in Uzbekistan.* CIRAD-ICARDA, Montpellier, France.

Keith, H.G. (1965) *A Preliminary Check List of Libyan Flora, Volumes* 1 and 2. United Middle East Executives Ltd, London.

Khatabi, K. (1993) The national report about biodiversity in Libya. *Conference on Biodiversity in North Africa Countries.*

Le Floc'h, E. and Aronson, J. (2013) *Les Arbres des Déserts: Enjeux et Promesses.* Actes Sud Ed.

Le Houérou, H.-N. (1969) *La Végétation de la Tunisie Steppique (Avec Référence aux Végétations Analogues d'Algérie, de Libye et du Maroc)* Vols 1–2, 42–45, Annales de l'Institut National de la Recherche Agronomique de Tunisie, Tunis.

Le Houérou, H.-N. (2002) *Multipurpose Germplasm of Fodder Shrubs and Trees for the Rehabilitation of Arid and Semi-arid Land in the Mediterranean Isoclimatic Zone: A Photographic Catalogue.* CIHEAM, Zaragoza, Spain. Available at: https://om.ciheam.org/om/pdf/b37/02600003.pdf (accessed 9 August 2024).

Le Houérou, H.-N. and Franclet, A. (1971) *Les Atriplex en Tunisie et en Afrique du Nord* (Technical Report 7). Food and Agriculture Organization and US Department of Agriculture, Rome.

López-Sáez, J.A. and Alba-Sánchez, F. (2009) Ecology, ethnobotany and ethnopharmacology of Argan tree (*Argania spinosa*). *Boletin Latinoamericano y del Caribe de Plantas Medicinales y Aromaticas* 8(5), 323–341.

Mahklouf, M. (2019) Invasive alien plant species in Libya. *Journal of Advanced Botany and Zoology* 7(1), 1–4.

Ogaya, R. and Peñuelas, J. (2020) Wood vs. canopy allocation of aboveground net primary productivity in a Mediterranean forest during 21 years of experimental rainfall exclusion. *Forests* 11(10), 1094. DOI: 10.3390/f11101094

Shackleton, R.T., Le Maitre, D.C., Pasiecznik, N.M. and Richardson, D.M. (2014) Prosopis: a global assessment of the biogeography, benefits, impacts and management of one of the world's worst woody invasive plant taxa. *AoB PLANTS* 6. DOI: 10.1093/aobpla/plu027

Thalen, D.C.P. (1979) *Ecology and Utilization of Desert Shrub Rangelands in Iraq.* Springer, Dordrecht, The Netherlands. DOI: 10.1007/978-94-009-9622-9

14

Past and Present Wildlife on Libyan Rangelands

Abstract

Wildlife is an essential element of rangelands and forests. A rich wildlife maintaining a functional food chain is a sign of good health of the ecosystem. These days, in Libya, even in the most remote desert areas, native herbivores, carnivores and birds are mercilessly hunted, decimated and exterminated. Hunting is not any more a vital necessity. It is replaced with indiscriminate illegal sport hunting due to firearms availability leading to wildlife massacres. We review, describe and illustrate the Libyan wildlife from prehistoric to modern times with their current status.

Introduction

The presence of abundant wildlife on rangelands and forests is a sign of good health of the ecosystem. Libya has a long history of rich wildlife from geologic and prehistoric eras proven from fossils, prehistoric rock engravings and paintings, and historical records. Soon came the time when a small population of men and their livestock started roaming rangelands and forests. Hunting was a common and vital activity. Flocks had to be protected against predators. Competition for pasture resources between many resident herbivores and herded flocks escalated. By the 2nd to 3rd centuries BCE, the last North African forest elephants were captured and tamed as tough warfare machines. Roman colonizers accelerated wildlife decimation by paying attractive bounty for live animals captured from their North African colonies, especially carnivores, for their endless circus games in Rome. Unstoppable wildlife extermination continued into modern times.

Long-past Libyan Fauna

Fossilized bones from the Cenozoic geological era were collected from many sites, especially from caves in the Jebel Al-Akhdar at Haua Fteah, Hagfet-ed-Dabba, Hagfet-et-Tera, etc. (Klein and Scott, 1986; McBurney, 1967) and deep into the desert from the Akakus to Kufra/Uweinat, also from south of Sirte – Marsa Brega, along the long-past coastlines of the early Miocene and middle Eocene sea, deep into central and southern Libya (Table 14.1) (Hufnagl, 1972; Le Houérou, 1997).

The Cenozoic era had tropical climatic conditions, warm and damp, and the fauna was mostly tropical. It is also called the 'Age of the Mammals' opening space for Quaternary and modern fauna when climatic conditions during the Pleistocene, about −3 million years to early Quaternary era, slowly changed to cooler and wet climate, adjusting to the Ice Age of the European glaciations. The Pleistocene ended about 12,000 years ago with the start of the Holocene

DOI: 10.1079/9781800627154.0014

Table 14.1. Cenozoic fauna in Libya. (Adapted from Hufnagl, 1972)

Million years	Era	Period	Locality	Fauna
– 3	**CENOZOIC**	Recent, Holocene and Pleistocene (cooler and wet periods in Libya, glaciations in Europe)	Uweinat, wadi Derna, Haua Fteah Cave, Hagfet-ed-Dabba(?), Hagfet-et-Tera(?), Hagfet Kreyiem(?)	Lion, hyena, zebra, rhinoceros, elephant, hippopotamus, giraffe, gazelle, buffalo, sheep
– 12		Pliocene	Qasr Shahabi (near Daryanah(?))	Cetaceans, carnivores, mastodon, deinothere, rhinoceros, horse, pig, anthracothere, bovids, hippopotamus
– 25		Miocene	Jebel Zaltan (south Marsa Brega)	Hyaenodon, carnivores, mastodon, deinothere, sirenians, rhinoceros, pig, anthracothere, giraffid, other ruminants
– 38		Oligocene	Zella (south Sirte)	Carnivores, mastodon, hyracoid, anthracothere
– 55		Eocene	Dor-el-Thala (Tazerbo(?))	Rodents, cetaceans, hyaenodontid, sirenians, proboscideans
– 65		Palaeocene		

era with what we would consider the era of wild animal hunting and domestication by humans. This is documented all over Libya with many rock engravings (petroglyphs) and rock paintings with the most famous being along the Algerian border (Tassili N'ajjer) and in the Jebel Akakus (east of Ghat, south-west Libya) and 100 km east, on the Messak Settafet/wadi Matkhandoush (N 25.762 E 12.170), north-west of the Murzuk Edeyen. There, thousands of rock engravings figuring elephant, giraffe, hippopotamus, Bubalus, lion, cheetah, ostrich, jackal, antelope, crocodile, horse etc. were surveyed by the French Army (3rd *Compagnie Saharienne – Légion Etrangère*) stationed in Fezzan (Diolé, 1955, 1956). These petroglyphs (Photos 14.1–14.6) dated 10,000–12,000 to 4000 years old, depict the large fauna under an environment close to a sub-tropical wooded savannah with increasing aridity.

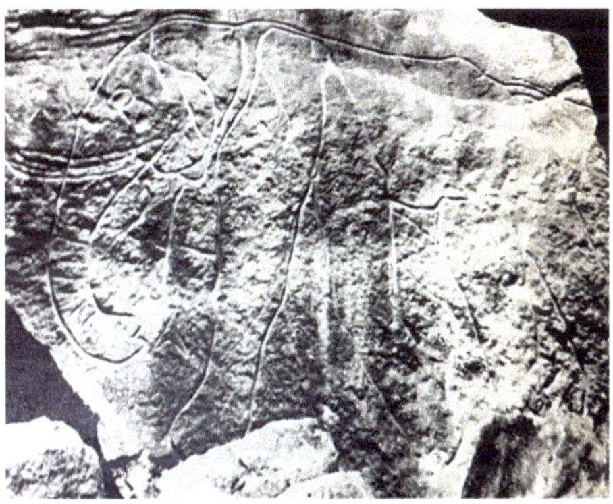

Photo 14.1. Rock engraving of an elephant at wadi Matkhandoush (From Diolé, 1955, 1956)

Are these elephants depicted on petroglyphs in southern Libya different from the North African elephant (*Loxodonta africana* ssp. *pharaohensis*) or the *Elephas atlanticus* (Pomel, 1995) described from the Algerian steppe (Sagne, 1950)? Most likely: the North African elephants used during the second Punic war (218–201 BCE[1]) by Hannibal's armies were reported smaller in size (2–2.5 m at shoulder) with smaller ears and much more difficult to tame compared to the African bush elephant (*Loxodonta africana*, 3–4 m at shoulder). In any case, the small North African elephant used to roam the North African wooded grassland, the pseudo-forests and steppe country from Morocco, Algeria to the Matmata in southern Tunisia to the Jebel Nafusa and likely in Cyrenaica up to the 3rd century CE when they became extinct due to

Photo 14.2. Rock engraving of another elephant at wadi Matkhandoush (From Diolé, 1955, 1956)

Photo 14.3. Rock engraving of giraffes at wadi Matkhandoush. (From Diolé, 1955, 1956)

unremitting hunting and capture for Roman games and lucrative ivory trade.

Wild and domesticated horses were also common. Many rock painting representations of light two-wheeled war carts drawn by horses and driven by antique warriors with spears or bows

Photo 14.4. Rock engraving of ostriches at wadi Matkhandoush. (From Diolé, 1955, 1956)

can be found in the now desertic Tassili and Akakus witnessing long-past climate change.

Other Proof of Prehistoric Wildlife in Northern Libya

The prehistoric and archeological research conducted in several caves (Haua Fteah, Hagfet-ed-Dabba, Hagfet-et-tera, Haj Kreyiem) in the Jebel Al-Akhdar confirm the presence over the past 150,000 years of large and small mammalians, birds, other vertebrates (Table 14.2), snakes, fish, snails, seashells, and humans (Douka *et al.*, 2014; Jacobs *et al.*, 2017; Klein and Scott, 1986; McBurney, 1967). Besides most abundant artefacts from Barbary sheep, goats, and bovines found in these caves, the presence of sub-Saharan wildlife (white rhinoceros, Burchell's zebra, Eland (or Kudu), Hartebeest, long-horned giant buffalo) indicates warmer and wetter interglaciation conditions with abundant vegetation allowing herbivorous wildlife and human expansion from south to north across what is now inhospitable desert. This sub-Saharan wildlife disappeared from northern Libya by the middle-late Holocene (Neolithic) due to increasing dryer conditions, less favourable vegetation cover and increased hunting (Table 14.3). This does confirm the rock engraving from near Shiarchara etched by

Photo 14.5. Rock engraving of a crocodile at wadi Matkhandoush. (From Diolé, 1955, 1956)

Photo 14.6. Rock engraving of an ox or a Bubalus at wadi Matkhandoush. (From Diolé, 1955, 1956)

hunter-gatherers (Fathallah, 1970) who must have witnessed their presence (Barich *et al.*, 2006) in the west and on the plateau as north as actual Tarhuna and Jebel Nafusa. Most remarkable is the presence in eastern Libyan caves of bones and teeth of North African elephants still present in North Africa during the Punic wars (264–146 BCE), of the Kirchberg's rhinoceros, a Eurasiatic forest rhinoceros, and the long-horned giant buffalo (Rossouw, 2001), the two latter that went extinct in northern Libya, Europe and south-west Asia by the end of the last glaciation (12,000–10,000 years BP[2]).

Remarkable are the most isolated 27 petroglyphs figuring Bovidae, lion, rhinoceros, Bubalus(?) located near Shiarchara (Photos 14.7, 14.8, 14.9, 14.10) on the Tarhuna plateau near Ras Sidi-el-Gharib (Fathallah, 1970). When visiting the site in 1976, we counted only 13 exposed engravings, some already damaged by 'gold diggers'(?). All these petroglyphs are artistically and vividly depicted. They are dated from the Neolithic/bovine/pastoral era (5000–2000 BCE) during which occurrences of alternating long wet and long arid episodes that were scientifically identified. The rhinoceros is quite odd in this environment and so far north in Libya, and may be only from herdsmen's memory. The lion is reasonably portrayed as historically contemporary on the Tarhuna plateau. Most interesting are the Bovidae, indicating that cattle were already grazing the land and possibly domesticated when the engraving was rock-cut.

Remnants and artefacts of these large mammalian fauna were found and identified in the Jebel Al-Akhdar caves at Haua Fteah (Klein and Scott, 1986), specifically the common white rhinoceros (*Ceratotherium simum*) at Hagfet-ed-Dabba near wadi Kuf and at Haj Krehiem near Derna, and also the now-extinct small forest rhinoceros (*Stephanorhinus kirchbergensis*) that disappeared by the mid-Holocene (8000–4000 BCE).

All large mammals once present in northern Libya are now extinct due to long-past climate changes, European glaciations and post-glaciations resulting in wet humid to dry long periods impacting the vegetation.

It is remarkable to see a rhinoceros engraving so far north in Libya.

The Bubalus (*Alcelaphus buselaphus buselaphus*), also named *Antilope bubalis*, was a large herbivore roaming the North African steppe to the edge of the Sahara. It is easily recognized with his back hump and the erected horns (Photo 14.8). Now extinct due to relentless

Table 14.2. Main animal artefacts (bones, teeth etc.) found at the Haua Fteah and other caves in Jebel Al-Akhdar. (Adapted from Klein and Scott, 1986)

Common name	Scientific name	Status in Libya	
Wild goat	*Ammotragus lervia*	To present – threatened	Heavily hunted nowadays
Boar	*Sus scrofa*	To present	
Cape hare	*Lepus capensis*	To present	
Crested porcupine	*Hystrix cristata*	To present	
Domestic cattle	*Bos taurus*	To present	
Domestic dog	*Canis familiaris*	To present	
Domestic goat	*Capra hircus*	To present	
Domestic sheep	*Ovis aries*	To present	
Dorcas gazelle	*Gazella dorcas*	To present – threatened	Heavily hunted nowadays
European rabbit	*Oryctolagus cuniculus*	To present	
Gazelle	*Gazella* sp.	To present – threatened	Heavily hunted nowadays
Golden jackal	*Canis aureus*	To present	
Red fox	*Vulpes vulpes*	To present	
Striped hyena	*Crocuta crocuta*	To present	
Greek tortoise	*Testudo graeca*	To present	Heavily collected nowadays
Wildcat	*Felis libyca*	To present	
African elephant	*Loxodonta africana*	Extinct by early 4th century CE in North Africa	Excessive hunting and Roman games
Aurochs	*Bos primigenius*	Extinct by early CE in North Africa	Climate change – post-glaciation era to Roman period
Bear	*Ursus arctos*	Extinct by early CE in North Africa	Excessive hunting and Roman games
Bighorn sheep	*Ovis canadensis*	Extinct by Neolithic	Climate change – up to beginning of Würm glaciation
Burchell's zebra	*Equus burchelli*	Extinct by Neolithic	Climate and vegetation change - post-glaciation era
Dall sheep	*Ovis dalli*	Extinct	Climate change – up to beginning of Würm glaciation
Egyptian mongoose	*Herpestes ichneumon*	Extinct	Excessive hunting
Eland	*Taurotragus oryx*	Extinct by Neolithic	Climate and vegetation change – post-glaciation era
Forest rhinoceros	*Dicerorhinus kirchbergensis*	Extinct by Neolithic	Climate change – up to beginning of Würm glaciation
Greater kudu	*Tragelaphus strepsiceros*	Extinct by Neolithic	Climate and vegetation change - post-glaciation era
Hartebeest	*Alcelaphus buselaphus*	Extinct by early 20th century	Excessive hunting
Lion	*Panthera leo*	Extinct by early 20th century	Roman games and excessive hunting
Long-horned giant buffalo	*Pelorovis antiquus*	Extinct by early-mid-Holocene	End of European Würm glaciation (*c.* –8000 to –10,000 BP)
Ostrich	*Struthio camelus*	Extinct by mid-20th century	Excessive hunting
White rhinoceros	*Ceratotherium simum*	Extinct by end-Neolithic	Climate and vegetation change - post-glaciation era

hunting, they were common and reported in North Africa, Morocco (in the Atlas mountains up to 1917), Algeria (near Geryville in 1920), Tunisia (last occurrence near Tataouine in 1902) and Libya, in the Tassili (*c.*1850).

The lion (*Panthera leo*) was a common occurrence and hunted during the Roman/Byzantine period (5th–6th century AD) all over northern Libya. The North African lion (Photo 14.9) was still present and occasionally hunted (Gérard,

Table 14.3. Suggested dating and climatic conditions at the Haua Fteah (Table adapted from (a) (Jacobs *et al.*, 2017) and (b) (Klein and Scott, 1986)

Haua Fteah layers	Cultural units and age (1) (1000 years)	Period	Estimated years BP (2)	Probable climate near Haua Fteah (1, 2)
I–V	Historic	HOLOCENE	–4700	Warm, relatively moist, but with increasing aridity
VI–VIII	Neolithic (–9.3 to –5.4)		–7000	
IX–X	Libyco-Caspian (–12.7 to –7.9)		–10,000	
XI–XV	Ibero-maurusian (or Oranian) (–7.2 to –12.5)		–12,000	
		LAST GLACIATION	–18 to –32,000	Cold , dry, wetter towards the end of the European Würm glaciation ending about 12,000 years ago
XVI–XXV	Dabban (also from Hagfet-ed-Dabba cave) (–43.5 to –17.1)		–64,000	Mainly cold and dry, but with warmer wetter intervals
XXVI–XXXV	Mousterian (–73.3 to –43.5)		–75,000	Cool and relatively wet
		LAST INTER-GLACIATION	< –127,000	Warm, relatively wet, with cool-dry episodes (first artefacts of human occupation at Haua Fteah by –150,000 years BP)
Deep sounding	Pre-Aurignacian (–101.6 to –67.7)			

1856) in Algeria up to 1890. The North African lion is nowadays extinct (Lakhdar, 2019).

Fathallah (1970) describes three types of Bovidae in the survey he did: (i) with forward horns (the most represented – Photo 14.10); (ii) with sideways horns; and (iii) with horns turned backwards.

One may notice that this Bovidae (Photo 14.10) has a collar of some sort indicating domestication. The partition on body may represent the different colours of its hide, a common representation on wall paintings in the Tassili and the Akakus.

The Bovidae represented with forward horns is from *Bos primigenius* (Muzzolini, 1981), once common in North Africa, also a common figuration on many southern Libya petroglyphs, mixed with *Bos africanus*, having thin lyre shaped erect horns from sub-Saharan Africa (Fig. 14.1). All *Bos* types are represented on rock engravings and paintings in the Tassili and the Akakus.

Photo 14.7. Rock engraving of a rhinoceros near Shiarchara – Tarhuna, 2009.

Photo 14.8. Rock engraving (near Shiarchara – Tarhuna, 2009) of a Bubalus (*Alcelaphus buselaphus*) or of an Eland (*Taurotragus oryx*).

Photo 14.9. Rock engraving of a North African lion near Shiarchara – Tarhuna, 2009.

Photo 14.10. Rock engraving of a Bovidae (*Bos primigenus*), with horns turned forwards (near Shiarchara – Tarhuna, 2009).

All this rock art proved that rich pastures and feed resources would have allowed hunters and herders to bloom in northern and southern Libya during the hunter period (–7000 to –5000 BCE) followed by the pastoral period (–5000 to –2000 BCE), ending during the Garamantian period (–2000 to –1000 BCE), on a land that is now steppic and desertic.

Libyan Fauna in Antiquity

During Antiquity, as early as the 5th century BCE, large predators such as lions, leopards, cheetahs, bears and elephants were hunted to make room for the new Phoenician, Greek and Roman colonizers expanding in Libya. Wandering flocks and crops had to be protected and secured. Many scenes were

Fig. 14.1. Art rock painting of a herd of *Bos primigenus* and *Bos africanus* (After Muzzolini, 1981), from Ouan-Bender – Tassili, south-east Algeria)

represented on mosaics (Photo 14.11) and coloured wall paintings of hunting scenes from villas along the coast were displayed in the museum in Tripoli, Leptis Magna, Shahat and other North African countries. The most disastrous mass-extermination of the native fauna occurred during the Roman Empire (27 BCE–476 CE) to supply the endless circus games offering mock-hunting (in Latin: *Venationes)* and deadly combats between wild beasts or against *bestiarii* and gladiators. Thousands of wild beasts captured in Roman North Africa were exported to and slaughtered in Rome during these popular tournaments. The small North African elephant disappeared from North Africa and Libya, captured and exterminated for the Roman games. It is reported that some 3500 North African elephants were slaughtered during the reign of Caesar Augustus (27 BCE–14 CE). The native Bovidae also became extinct, likely due to increasing aridity, conspicuous land clearing, competition for grazing lands and hunting.

What Is Left of this Fauna these Days?

We list the wildlife we have seen while surveying rangelands or been reported in Libya over the past century. A rangeland in good condition hosts an array of wildlife. Unfortunately, due to range overgrazing and illegal or abusive hunting, wildlife deserted most Libyan rangeland deep into the Hamada and the southern desert. Sighting of large wildlife these days is a rare occurrence deserving attention and recording. Unfortunately, despite strict Libyan hunting laws, Libyan wildlife is under extreme and unremitting extermination.

Mammals

The review of Libyan mammals gives us some insight (Elkahwage and Jdeidi, 2018; Hufnagl, 1972) and neighbouring southern Tunisia (Louis, 1979) completed with our own field observation and notes.

Mammals on rangelands used to be common up to the early 1950s when rangelands were still healthy. Nowadays, due to the availability of firearms and four-wheel-drive cars, illegal and unrestrained hunting is widespread all over Libya, resulting in pointless killing of once-abundant wildlife.

Herbivores

GAZELLES. The most common Libyan gazelle (Photo 14.12) is the small and gracile *Gazella*

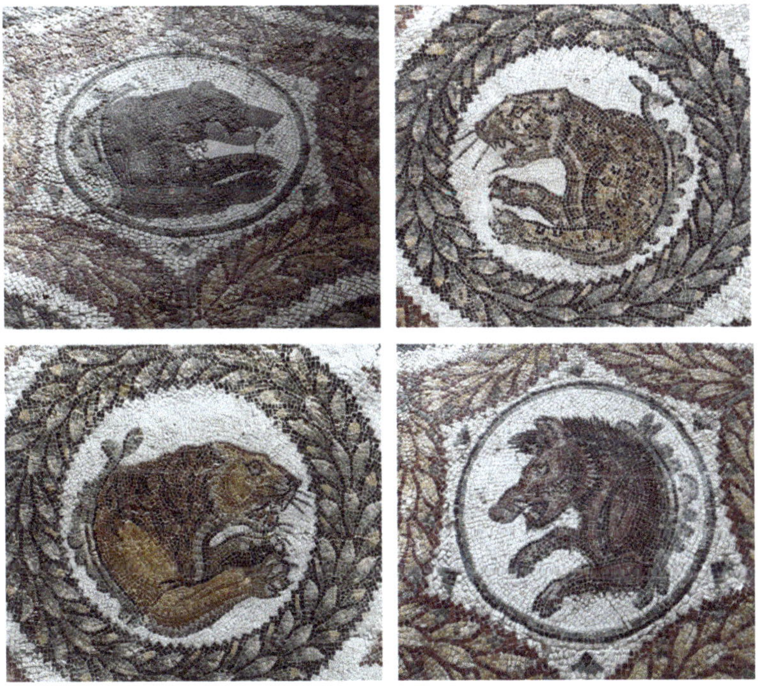

Photo 14.11. Mosaics from the Bardo museum (Tunis) illustrating the wild beasts on the land during Roman times. *Top left to bottom right*: bear, cheetah or panthera, lion and wild boar.

Photo 14.12. *Gazella dorcas* – hunting scene from the early 1970s, wadi Sassu – Bir Dufan.

dorcas (*Ghazal*) that used to roam the steppe to the border of the Sahara from west to east in small flocks of six to ten females. Sighting of these 'red' gazelles were still occasional in the Jeffara plain feeding on the pods and low leaves of the *Talha* at Bir-Ayyad, and along the *sebkha* south of Al-Assah near Hebiliyah and Shabunyah. They were common sights south of Misurata – wadi Sassu to Seddada, in Sirte – south Nofilia and south Jebel Al-Akhdar. Occasionally, captured females are kept as pets.
Status in Libya: threatened/vulnerable

Deep south, along the border of Chad, *Gazella dama* (*mahar* or *riel*) and *Gazella leptoceros* (*Ghazela abiod* or *reem*) are extremely rare. The latter used to be common on the Dahar and the Hamada.
Status in Libya: threatened/vulnerable

Ammotragus lervia (*waddan* or *eawdad*) is a North African endemic wild goat related to ibex. It is not a wild sheep and wrongly named Barbary sheep. Weighing up to 130–140 kg for the male (Photo 14.13), it has a characteristic long-haired, whitish beard, also covering the throat, chest and inside the front limbs. Its powerful horns bend outwards and curl backwards. It was reported to be present near Gharyan in the 1930s (Lepri, 1930). A remarkable climber, it is now only present in the most desertic mountains of central Libya (Sokhna, Jebel Waddan, Jebel Uweinat, Akakus). It is regularly illegally hunted.
Status in Libya: threatened/vulnerable

Photo 14.13. *Ammotragus lervia* (waddan) illegally hunted in Libya. (Photo from Facebook, The Libyan Wildlife Trust, 2017)

Addax nasomaculatus (Baqara mahaliya?) – a large desert antelope weighing up to 80–100 kg, with long (50–60 cm) spiralled and ridged horns, a black tuft between horns, and white strips around the eyes. It used to be seen on the Hamada, Fezzan – Murzuk, Kufra – Jebel Uweinat.
Status in Libya: threatened – extinct in northern Libya, extremely rare in southern Libya

Oryx dammah or scimitar oryx (*Abu harba* or *Baqara barri* (?)) (Photo 14.14) is another larger desert antelope weighing up to 120–150 kg, with long (up to 1 m) backward-curved horns. Large herds used to be common in the Libyan desert, up to the coast of Sirte (Hufnagl, 1972), but mostly deep into the southern desert (Kufra (Bagnold, 1933), Murzuk Edeyen, (Diolé, 1955, 1956), in flat and sand dunes country.
Status in Libya: threatened – extinct in Northern Libya, extremely rare in southern Libya

Bubalus (*Alcelaphus buselaphus*) (figured on Neolithic rock engraving near Shiarchara (see Photo 14.8). A large antelope (100–150 kg) which used to roam in large herds on the rocky steppe and desert of North Africa from Morocco to Egypt. Easily identified with its upright U-shaped set of horns (30–40 cm) with backward-curved ends, and a small hump on the shoulders. Exterminated by hunters during the colonization and later. Last sighted early 1900 in eastern Morocco, southern Algeria and southern Tunisia, possibly also in Libya. Still reported in Tripolitania in 1898; possible sighting by Diolé (1956) in the remote part of the Murzuk Edeyen, doubtfully and simply named by the local guide as *Bagara barri*.
Status in Libya: extinct (Harper, 1945)

Carnivores and omnivores

Golden jackal or *Canis aureus* (*ibn awa*, often called *dhib* or *thib*), the jackal (Photo 14.15) is common all over the Libyan rangelands scavenging food

Photo 14.14. Scimitar oryx (*Oryx dammah*), deep south in Kufra (Bagnold, 1933). (Photo: The Libyan Wildlife Trust. Used with permission from the Royal Geographical Society).

near dwellings, easily interbreeding with domestic dogs, both reputed to be a carrier of rabies; diurnal, very cunning. Shepherds from Al-Assah report that the jackal, in summer, crawls and lies 'dead' close to flocks of sheep, goats or camels grazing during the night. The jackal would jump, catch its prey by the snout and suffocate it.
Status in Libya: common

Photo 14.15. Jackal (*Canis aureus*) in Jebel Al-Akhdar. (Photo: Salem Essaker)

The North African sloughi (or *saluki, slughi maghribi*) is a lean, tall, muscular greyhound with drooping ears and smooth fur. A quiet dog, it rarely barks. The sloughi is a sighthound and not a scent dog. Fast-running, it sights and runs after hares, gazelles, foxes or jackals. To avoid the sloughi being wounded when catching a fox or fighting with a jackal, the owner often cuts and reduces the length of its ears. Three hot-iron brands were applied to the inner side of the sloughi's forearms and lower thighs to reinforce the tendons. Sloughi used to be a common sight in the Jeffara plain hunting with members of the Sian and Nouail tribes. Nomads consider the sloughi a noble dog, with a given individual name contrary to everyday guard dogs; hence, the sloughi was always freely accompanying its master, allowed to enter the master's tent and eat with him. Sloughis are nowadays rarely sighted in Libya.
Status in Libya: vulnerable/threatened

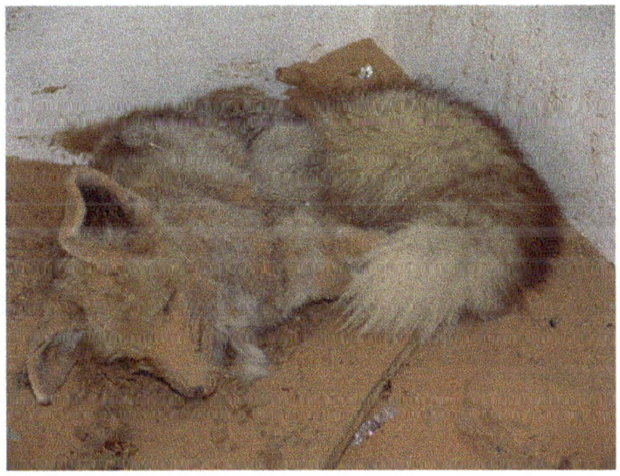

Photo 14.16. Mummified sand fox (*Fox rüppelli*) in an abandoned dwelling at Bir-el-Kelb, Hamada, 2009.

The sand fox or *Fox rüppelli* (Photo 14.16) from the Hamada and rocky desert is quite rare. It is different from the common red fox (*Vulpes vulpes, thaelab*; Photo 14.17), the latter common along the coast and rural dwellings. The sand fox has

Photo 14.17. A young common red fox (*Vulpes vulpes*) sighted in front of its burrow in the Jebel Al-Akhdar (Photo: Salem Essaker)

sand-coloured ears and grey back, while the red fox has the back of the ears black or darkened. Rather shy, it is nocturnal, feeding on insects, lizards, small rodents and roots. Reported by Elkahwage and Jdeidi (2018) in Msallata National Park (eastern Jebel Nafusa).
Status in Libya: vulnerable

Other fox relatives are present such as the fennec (*Fennecus zerda*), a small crepuscular or nocturnal fox with very large ears, from sandy desert in the northern part of Libya (south Sirte), occasionally kept as a pet.
Status in Libya: vulnerable

The striped hyena (*Hyaena hyaena*; *dabba*) (Photo 14.18), a powerful, shy and nocturnal predator still present in the Jebel Nafusa (Tarhuna, Gharyan, Yeffren), hiding during the day in caves, in wadis on the northern slope of the Jebel (Bir Touta, Bir Rabtah(?)). Scavengers, they would prey on young mammals, foxes and small ruminants. In 1978, Chadi shepherds from the Bir-el-Ghanem sheep project killed one big male hyena. Sighting of hyenas in the western Jebel and Jebel Al-Akhdar is common. Unlike wild dogs, the hyena kills its prey, usually only one at a time, and often moves it to deep bushes or a nearby cave. Reported common around Gialo-Awjila (south Ajdabiya) and frequently captured or killed in Jebel Al-Akhdar (see Facebook, Libyan Wildlife Trust).
Status in Libya: vulnerable/threatened

The cheetah (*Acinonyx jubatus; fa'had*) (Photo 14.19), the last large carnivore present in Libya has long legs and tail, small head with characteristic black 'tear' lines. It used to be sighted on the open Hamada (Diolé, 1956) and in the southern desert but also reported along the coast as north as Misurata (1961), south Sirte (1968), Mizda (1969), according to Hufnagl (1972), and south Jebel Al-Akhdar. The last recorded kill, photographed, was near Ajdabiya as late as 1971 (Facebook Libyan Ecological Society post, 10 February 2023).
Status in Libya: extremely rare, threatened/endangered – likely extinct in northern Libya

The caracal (*Felis caracal*; *anag-al-ard* or *nemes*). The name caracal may come from a Turkish contraction of *kara kulacklar* or 'black ears'. The caracal is often confused, when sighting from far away, with the cheetah. The caracal is the North African relative of the European lynx (*Lynx pardinus*); both have the distinctive tufted ears and a short tail. However, while the lynx has black-spotted fur, the caracal has a uniform reddish-brown coat sometimes dotted with vague brown spots on the abdomen. Very shy and stealthy, it is extremely rare to see one as reported in the Jebel Nafusa and wadi Ethel (Bir Ayyad) in the 1960s. It may also still be present in Libya these days in the Jebel Al-Akhdar, only confined to deep matorral/forest country

Photo 14.18. A young hyena in captivity on a farm at Tajura, 2009.

Photo 14.19. A cheetah (*Acinonyx jubatus*) killed in south Jebel Al-Akhdar (Al-Jahishiyah?, c.1966)

and rocky environments far away from humans and flocks. Desperate and hungry caracals are reported to play dead, like the jackal, flicking its ears' tufts to attract the attention of unsuspecting small ruminants or young camels.

Status in Libya: extremely rare; threatened/endangered

Gintzburger's grandfather told him long ago that a black panther was killed in the Kroumirie forests (north-eastern Tunisia) in the early 1900s. It is quite possible as panthers (*Panthera pardus*) were commonly pictured on Roman mosaics and hunted in Algeria in the mid-1850s (Bauer *et al.*, 2017; Gérard, 1856), and one was reported killed in 1913 in northern Algeria. We may also imagine that the black panther from Kroumirie forests could have been a rare and fully black caracal, a known melanistic variation on animals living in deep forests. In any case, we doubt that any panther would still exist in the rare Libyan forests, even deep in the Jebel Al-Akhdar. There, the presence of caracals, though extremely rare, is likely.

Cats: two small wild cats are likely to be found in Libya: the stealthy and dishevelled African wild cat (*Felis libyca* ssp. *cyrenaicum*) with a thick coat and tail is rarely reported present in the forests of the Jebel Al-Akhdar. It looks like a powerful domestic cat but with clear black stripes across front and back legs and tail (Photo 14.20).
Status in Libya: vulnerable

The sand cat (*Felis margarita*), similar in size to a domestic cat but with long hair, with a round, bulky head and short ears, is extremely rare, and nocturnal. Possible occurrences in sandy or rocky desert were recorded long ago along the Algerian border. A sighting was recently reported between Al Saih – Al Damran/Beni Walid and Wadi Zamzam.
Status in Libya: rare/endangered

Genet (*Geneta geneta*; *shafshah*(?)) (Photo 14.21) – a secretive nocturnal small carnivore, looking like a stretched-out cat, it has a pointed face, an elongated body (40–50 cm) with black dots longitudinally aligned and a long tail with black rings. Short-legged, very agile and silent, lives in bushes, in gardens, present in the Jebel Nafusa and the Jebel Al-Akhdar; reported raiding poultry in gardens along the Tripoli coast.
Close to *Geneta genata* is the Libyan striped weasel (*Ictonyx libycus*) present in more arid environments, in the desert and southern oases.
Status in Libya: endangered

Photo 14.20. African wild cat (*Felis libyca* ssp. *cyrenaicum*) in Jebel Al-Akhdar. (Photo: Salem Essaker)

Photo 14.21. Genet (*Geneta geneta*) from Al-Abraq – Jebel Al-Akhdar. (Photo: Idis Al-Osalib)

Photo 14.22. *Poecilictis libyca* from south-east Jebel Al-Akhdar. (Photo: Ayoub Akbar El-Ghaithi, courtesy Salem Essaker)

Poecilictis libyca (Syn: *Ictonyx libycus*; *shihshafa*) (Photo 14.22) is the Libyan stripped weasel, also sighted in the Algerian Sahara, southern Tunisia and western Egypt (Moussouni and Scaravelli, 2024). It is a nocturnal carnivorous mammal feeding on insects, rodents, snakes and small birds. It is 40–60 cm long with a black mask, longitudinal black-and-white body stripes and small round ears with white tips. Rarely encountered, when confronted it will raise its hairs all over its body, turn its tail towards the intruder and squirt a powerful, foul-smelling liquid from glands located at the root of its tail. Swift and agile, it can be confused with *Geneta geneta.*
Status in Libya: endangered

The North African wild boar (*Sus Scrofa* ssp. *Algira*) is a smaller subspecies of the European *Sus scrofa* ssp. *scrofa*. It is omnivorous, feeding both on plants and animals, alive or carrion. Extremely resilient due to its omnivory, fast-moving over long distances, stealthy, producing up to five piglets per litter/year, roaming in forested areas, matorral to oases all over North Africa, it is nowadays rarely seen in Libya though still occasionally sighted from west to east, but absent from Sirte region. Rightly considered as an invasive pest, it is systematically hunted and eliminated. In North Africa, *Sus scrofa*'s meat is not consumed by local populations for religious reasons, based on health justifications, as it is a

known vector of a long list of lethal diseases such as *toxoplasmosis, brucellosis, tularemia*, trichinellids, swine influenza, *Salmonella*, hepatitis, leptospirosis, and pathogenic *E. coli*.
Status in Libya: not threatened

Rodents

Hare and rabbit. There are no wild rabbits in Libya except the ones escaped from farms. But hares (*Lepus capensis; arnab*) (Photo 14.23) are common in the west, in the east and even further south in Fezzan (Oum-el-Arnab). They have very long ears and a light ring around the eyes. The hare population appears to be different in the west, with a clear coat and living in the open steppe or desert, from the one from the east, which has a darker coat and lives in the bushes and forested areas. In the old days, hare would be hunted by riders with Sloughi (*saluki*) in the west while using falcons in Solul el Khoder, Ajdabiya region and in the east.
Status in Libya: common

Crested porcupine (*Hystrix cristata; nis*) are nocturnal rodents, found in the west and the east, likely not in Sirte and the desert. They are common in the coastal zones (Homs to Misurata) and the northern slope of the Jebel Nafusa in *halfa* country (but not in the Jeffara plain), and in the high-rainfall zones and scrubland in the east. They have long, sharp, black-and-white banded spines (up to 40 cm). They are occasionally hunted and barbecued. An extremely rare albino specimen was recently captured in the Jebel Al-Akhdar.
Status in Libya: common

Hedgehog. There are two hedgehogs found in the high-rainfall zones of Libya, mostly in gardens and not in the steppe or the desert: (i) the common Algerian hedgehog (*Atelerix algirus*) with small ears (mostly in western zone); (ii) the long-eared hedgehog (*Hemiechinus auritus*) in the eastern zone. Both are often confused and called *gunfud*. They often bear a lot of ticks. Hedgehogs are occasionally barbecued whole.
Status in Libya: common

Gundis (*Ctenodactylus gundi*) look like guinea pigs. Herbivorous rodents living in groups on rocky slopes of the Jebel Nafusa and deep south, they may also be present in eastern Libya. They used to be caught by villagers and eaten. Gundis are reported vectors of toxoplasmosis (Elsalem, 2024).
Status in Libya: common

Jerboa (*Jaculus jaculus* ssp. *desertii*), a small nocturnal herbivorous/seed-collecting rodent with a slender tail as long as its body and ending with a black tuft with a white tip. Rear legs are long and allow long jumps and look like a very small kangaroo. Common on mobile sand and digging

Photo 14.23. Hare (*Lepus capensis; arnab*) hiding in a tuft of *remeth* in the steppe. (From Facebook, the Libyan Wildlife Trust)

Photo 14.24. Fat-tailed rat (*Psammomys obesus; jird*) in a bush of Suaeda, Magdar-es-Sbot, Al-Assah, 1978).

multi-entry tunnels in compacted sand in *sbot* or *bu rokba* clumps.
Status in Libya: common

Fat-tailed rats (*Psammomys obesus*; *jird*) are small diurnal rodents with short, fat tails, living in colonies on sandy-loam/alluvial soils, and along *sebkha* (Photo 14.24). They feed on vegetation and seeds that they store in their burrows with multiple entries and exits. Extremely prolific, populations would explode when feed is bountiful. That was the case in the mid-1970s, at wadi Abu-Shayba, when annual medic pastures were successfully established (see Chapter 9).
Status in Libya: common

Both the above should not be confused with *Meriones libycus* (also called *jird*), a common small desert rat with a long tail ending with a black tuft (Photo 14.25). They reproduce fast with gestation in 25–29 days producing 3–4 young per litter; they swarm fast and systematically damage the rainfed crops (Photo 14.26) and *Atriplex* plantations (Photo 14.27) in the steppe. In Jebel Al-Akhdar, *Microtus guentheri*, a shy Middle-Eastern vole with a short tail is reported present (Salem Essaker, El Marj, January 2020, personal communication).
All these steppe and desert rodents are vectors of diseases and ticks that may affect humans and wild and domestic mammals.

Reptiles

A comprehensive survey of all Libyan reptiles is available from Bauer *et al.* (2017).

Photo 14.25. *Meriones libycus* (*Jird*) at Bir-el-Ghanem, 1978.

Photo 14.26. Rainfed barley damaged by *Meriones libycus*, Cabao, 1978.

Photo 14.27. *Atriplex nummularia* plantation severely damaged by *Meriones libycus* (Bir-Ayyad, 1979).

Photo 14.28. The head of a cobra (*Naja haje*) crawling in *Lygeum spartum* tuffs, Jbobina, Al-Assah, 2009 – 1.55 m long.

Photo 14.29. A cobra (*Naja haje*) posturing before attacking (Jebel Al-Akhdar). (Photo: Salem Essaker)

Photo 14.30. Huge specimen of cobra (*Naja haje*) killed near Derna. (Photo from Facebook site 'Save Jabal Al-Akhdar from desertification' accessed 6/8/21)

Venomous reptiles

Cobra (*Naja haje*; *thueban*(?)), also sometimes *l'afeaa* (Photos 14.28–14.30), is a venomous snake common all along the coast from west to east, on sandy land, on rocky slopes, keeping to its own territory, hiding in disused burrows from rodents and foxes, in *nebkha*, preying day or night on rodents, hares, birds etc. Mostly fully black (cf. *Naja haje* ssp. *legionis*), in the west and

Photo 14.31. Horned viper (*Cerastes cerastes*) on sandstone, Takut, 1978.

Sirte they are much lighter with yellowish-beige bodies in the Cyrenaica and Marmarica. When in aggressive mode, *naja* raises its head about a quarter of its body length above the ground, with the lower part of the head characteristically expanding as a hood formed from the cervical ribs. *Naja haje* can reach an impressive length, up to 2.5–3 m (6–8 cm in diameter) like the huge specimen over 2 m long killed near Derna (1990?), which the locals called *Konda*, possibly referring to an anaconda, a species not present in Libya.
Status in Libya: common

Vipers are common in the sandy and rocky steppe sites. They are easily recognized by their short and stocky body ending abruptly with a short tail, and by their typical triangular head. With the body coiled in a circle, they are at rest, but, beware, when threatened, they adopt a fast-striking position with the body in an S-shape. All are highly venomous and an untreated clear bite results in death. A simple graze or scratch by a missed bite results in tissue necrosis and a long-standing wound. The symptoms and effect of a *Cerastes*' poisonous bite was documented by Domergue (1960).

Horned viper (*Cerastes cerastes* = Syn. *Aspis cerastes*; *lefha*, *um-qorin*) (Photo 14.31) is a short (10–30 cm long), shy, stumpy venomous snake with a tail ending abruptly. It bears two supraocular horns on a triangular head. The horned viper's colour matches the environment. Mostly found in sandy country, its presence is revealed by sidewinding imprints on sand. It sinks itself in sand, coiled, ready to strike, leaving just the two horns emerging from the sand. When detecting and locating soil vibration (*jerboa*, *jird* or small bird), the horned viper strikes, jumping out of its hiding-place.
Status in Libya: common

OTHER HIGHLY VENOMOUS SNAKES ENCOUNTERED IN LIBYA.

- *Cerastes vipera* (*um-qorin*) is a stumpy viper (30–50 cm long, body diameter 4–6 cm) without horns. More frequent in the desert and sandy places. Its colour closely matches the substate making it difficult to spot.
  Status in Libya: common
- *Daboia deserti* (Syn: *Vipera lebetina*; the lebetine viper), present in southern Algeria and Tunisia, related to *Daboia mauritanica* (Chedad *et al.*, 2024) from western North Africa. *Daboia deserti* is a very venomous fat snake (up to 1 m long) apparently to its most eastern distribution limit, though it may have been sighted in Marmarica (Bauer *et al.*, 2017).
  Status in Libya: threatened (?)
  Serum for these two vipers is available from the Pasteur Institute in Tunis (http://www.pasteur.tn/index.php?option=com_content&view=article&id=161&Itemid=199; accessed 14 January 2024).
- *Echis pyramidum*, up to 60–90 cm long, recorded in the coastal west (Tripoli to Homs), in Marmarica (Tobruk) and in Sebha, mostly on rocky sites.
  Status in Libya: common

Innocuous reptiles and lizards

Malpolon monspessulanus (Syn: *Malpolon insignus*), a harmless diurnal coluber with dark back and yellow belly. Slender with a thin body (up to 1–1.5 m long) and a narrow head, it occurs from coastal west in gardens to the eastern matorral but also sighted in the southern oases hunting rodents and small birds. It does not appear in the arid steppe or the desert.
Status in Libya: common

Varans (*Varanus griseus*; *l'oural*) (Photo 14.32) is a large (up to 70–100 cm long) carnivorous lizard preying on other lizards, snakes, birds and small mammals. It uses its long flexible tail as a whip when caught or fighting. It is present all over the coastal area and deep into the desert. It is fast-running and impressively swells its whole body when cornered. Reported to be innocuous, but bites may result in long-standing infection. Shepherds report that bites from a varan would cause male sterility.
Status in Libya: common

Chamaeleo chamaeleon (*Al-Harba*) (Photo 14.33); a slow-moving lizard-like reptile with large triangular head with globulus, independently moveable eyes. It has a prehensile tail and legs with 2 x 2 'fingers', for grasping vegetation. It perches on shrubs or small trees, remaining motionless, remarkable for its ability to change skin colour to match its environment or during reproduction time. When hunting, it projects its long (as long as his full body) sticky tongue to catch insects. Often kept as a pet in gardens along the coast.

Status in Libya: common

Agama (*Trapelus mutabilis*) (Photo 14.34): small (8–10 cm) innocuous diurnal squat lizard with a large triangular head, common on the *reg* on the Hamada or rocky places, rarely seen running, usually moves slowly, with an ability to mimic the colour of the stones.

Status in Libya: common

Uromastyx acanthinura (*dhobb*) is a diurnal Saharan squat lizard, usually black with white dots, and black, orange, green and yellow back, about 20–30 cm long, with a remarkable short spiny tail that is used as a whip when caught or fighting. It is rather slow-moving, usually basks in the sun in rocky places, and takes shelter in crevices and burrows. It is innocuous, easy to catch in the open but difficult to extract from rock crevices. It swells its body, hence its local

Photo 14.32. *Varanus griseus* (*l'oural*) in a barley field at Bir- El-Ghanem sheep project, 1978.

Photo 14.33. *Chamaeleo chamaeleon* (*Al-Harba*), Sirte, March 1979)

Photo 14.34. A small agama (*Trapelus mutabilis*) on the *reg* on Hamada, near Bir-el-Kelb, 2009.

Arabic name. Common in rocky places in the Jebel Nafusa, on Hamada, and caught in large numbers (Photo 14.35) in the south (Akakus, Messak Settafet) for food or medicine. It does not appear to be present in the Jebel Al-Akhdar, but possibly present in Jebel Uweinat.
Status in Libya: common but possibly threatened

Tortoises

Testudo graeca; fakroun) (Photo 14.36). Found in the Jebel Nafusa (*Testudo graeca* ssp. *Nabeulensis*), in the *Stipa tenacissima* steppe and in the Jebel Al-Akhdar (*Testudo graeca* ssp. *Cyrenaica*) in the scrublands of Cyrenaica. About 15–20 cm long. Prey birds would collect them and drop them from the air to crush the carapace to eat them easily. Tortoises mass-collected in Jebel Al-Akhdar are commonly sold at local markets (Photo 14.37) as pets or illegally exported to nearby Egypt.
Status in Libya: common but threatened.

Photo 14.35. *Uromastyx acanthinura* (*dhobb*) caught for food and medicine. (From Facebook, The Libyan Wildlife Trust)

Birds

Birds' identification is from Etchécopar and Hüe (1964) and Hollom *et al.* (1988).

Common birds

Larks – small birds (10–15 cm); the term covers many different types, either having crested caps (*Galerida theklae*, *Galerida cristata*, the most common in northern Libya), or without crests (*Rhamphocorys clot-bey* on the Hamada, *Calandrella rufescens* (Photo 14.38)); usually found in small bands. Take-off is to low altitude (20–40 m) and then dropping back to ground in jerky motion while singing loudly. Mostly granivorous, all larks nest on the ground.

Grey shrike (*Lanius excubitor*) (Photo 14.39) – a medium-sized bird (15–20 cm) with a grey back and a black band from the beak to the neck; commonly perches on shrubs in the steppe and sites with *Ziziphus lotus*. It has a strong conical beak, slightly curved at the end. It fiercely hunts insects, lizards, small snakes, rodents, even young birds. Nests in bushes, often in *nebkha*.

The European bee-eater (*Merops apiaster*) is a common sight along the coast in isolated couples or large flocks; 25–30 cm including tail, very brightly coloured; blue-green on forehead, yellow throat, with brown back turning yellow on the rump; all other upper parts are dark green with brown underwings; thin, elongated, slightly curved beak; feeds on insects. It is often reported feeding on beehives, because bees are numerous and easily available. It will feed on grasshoppers, dragonflies, wasps, *Scolopendra*,

Photo 14.36. Tortoise (*Testudo graeca* ssp. *Cyrenaica*) from Jebel Al-Akhdar, in Toknis, 2009.

Photo 14.37. Hundreds of tortoises (*Testudo graeca* ssp. *Cyrenaica*) sold at a local market, 2020.

Photo 14.38. A group of larks, *Calandrella rufescens* (?), sheltering under our car during a very hot day (43°C) near El-Watyah (south Al-Assah, August 1976).

spiders, etc. Colonies are nesting in burrows along wadi embankments and loess deposits (Jebel Nafusa). They do not appear to venture into the very arid Libyan zones but migrate south to Sahelian Africa.

Status in Libya: common

Persian bee-eater (*Merops superciliosus*). A medium-sized bird (25–30 cm including two long tail feathers). It is rarely observed, yet colourful, with green-blue back, yellow chin and brown throat. It is often confused with *Merops apiaster*. Mostly present in the arid environments, occasionally

Photo 14.39. *Lanius excubitor* perched on a *remeth* shrublet, Touil-er-Reschiada, south Al-Assah, 2009.

seen perched on isolated *talha* and *batum* in south Sirte (wadi Wishkah, wadi Bey-el-Kebir). Swift flight accelerating to catch flying insects and bees. Nests in old burrows. Usually migrates to the west, to Egypt and further south.
Status in Libya: rare and possibly threatened

Partridge (*Alectoris barbara*) (Photo 14.40) – a commonly sighted 25–30 cm plump grey bird with red legs and around the eyes; fawn neck with white spots. Found in rocky sites, in wadis, in Jebel Nafusa, in matorral in the east. It runs fast and takes off with a heavy noisy flight. It makes a rudimentary nest on the ground under shrubs and is regularly hunted from west to east.
Status in Libya: common

Photo 14.40. *Alectoris barbara*, Jebel Al-Akhdar. (Photo: Salem Essaker)

Ducks and freshwater birds. The ponds from artesian wells near Al-Assah seasonally host a rich diversity of migrating and resident water birds, as well at Taworga (El-Haysah). Common are ducks (*Anas* sp.), grebes (*Podciceps* sp.), coots (*Fulica* sp.), herons (*Ardea* sp., *Ardeola* sp., *Nictycorax* sp., *Ixobrychus* sp., *Bautorus* sp., *Egretta* sp.), some flamingo (*Phoenicopterus roseus*), even the tall and elegant cranes (*Grus* sp.) and some rare white storks (*Ciconia* sp.). All are illegally hunted in large numbers despite local or customary laws that are supposed to protect them. By early 2010, with colleagues from Zuara, we established a sanctuary for migrating birds (Photo 14.41) with a 70 ha fence around the large 15 ha pond at Jbebina (13 km north-west of Al-Assah). The idea was to develop ecotourism and birdwatching on site with

Photo 14.41. The Jbebina Bird Sanctuary (N 32.910 E 11.545) established by A.N. Sultan and G. Gintzburger, October 2010.

the support of the local communities and the Al-Assah camel project. This did not prevent hunters from randomly and indiscriminately shooting resident and migratory birds present in the reserve.

Little owl (*Athene noctua*) – a small (10–20 cm), mostly diurnal, owl often seen perched on rocks, on abandoned dwellings and ruins in the Jebel Nafusa, and deep into the steppe or the desert. Colours (light brown) match the environment. Nests in holes in rocky sites, walls, in abandoned burrows. Hunts insects, rodents, small snakes and birds.
Status in Libya: common

Rare or endangered birds

Chlamydotis houbara ssp. *undulata* (*houbara*) (Photo 14.42) is a frequent sight in the steppe in the Jeffara plain and all over desertic North Africa to Egypt. A bird with long yellow-green legs, more running than flying, 50–60 cm tall with speckled reddish-brown back feathers, head-capped with short white plume and black stripe along both sides of its long neck. Houbara nests on the ground. I observed a houbara's courtship near Bir-Ayyad in 1976, with the noisy male jumping and dancing in a circle around the female. Houbaras are regularly hunted despite being protected by law.
Status in Libya: common but threatened

Peregrine falcon, *Falco peregrinus* (*Al-Sak'r*) (Photo 14.43), is a medium-sized (30–50 cm) resident prey bird with pointed wings and breast striated brown or black, which flies swiftly with a fast wingbeat and long gliding flights. It dives fast onto its prey (hare, small rodent, bird, lizard, snake), knocking them down. It is frequent in coastal zone and could be confused with the Lanner falcon (*Falco biarmicus*), which is the same size as the peregrine falcon but has a white forehead and brownish-red neck, white breast with small black speckles. It is more common in the steppe than on the coastal zone and nests in trees

Photo 14.42. *Chlamydotis houbara* ssp. *undulata* (*houbara*) in *sbot* country (south Sorman), 2009.

Photo 14.43. *Falco peregrinus* (*Al-Sak'r*) – from a falconer at wadi Sassu, 2009.

(on *talha* and *batum*) or on cliffs. Both birds are still caught young and trained locally for falconry. They attract a large price on the illegal market. Occasionally, local falconers are seen hunting in the Jeffara plain, Solul-el-Khoder, Sirte and eastern Libya, in open steppe country to the edge of the desert.
Status in Libya: common but threatened

Large flocks of migratory cranes, white storks and pink flamingos are often spotted in northern Libya from two migratory fluxes coming either from western Europe towards North Africa and sub-Saharan West Africa through western Libya, or from eastern Europe (Turkey) towards eastern Libya, Egypt and further south (Etchécopar and Hüe, 1964; Hollom *et al.*, 1988).

Grus Grus (*alkarki*) in the west (1.15 m tall) and *algarnoog* in the east, are grey all over with a black neck and red cap. *Anthropoides virgo* (1 m tall) is also grey but with white feathers behind eyes. They are migratory birds wintering along the western coast at Farwah Islands near Zuara or on the Jbebina ponds (Al-Assah, spotted in 2009), where they feed on small fish and crustaceans. They fly in V-formation or in line with recurved neck and extended legs, and trumpeting loudly. They are migratory birds coming from western Europe.
Status in Libya: threatened

Phoenicopterus roseus (*Al-Bsharosh*) is the tall (1.6 m), long-legged pink flamingo, common, wading in large noisy colonies on shallow coastal lagoons and *sebkha* in northern Libya. They feed on insects, small freshwater crustaceans and algae and plankton, which are rich in carotenoids, giving them their pink-reddish colour. In flight, their legs are extended and straight. They are never seen inland, in steppe or desert.
Status in Libya: threatened.

Ciconia Ciconia, the *lak-lak*, referring to its loud bill-clattering, is a tall white bird with black wings and long red beak, well known and respected by the locals. Long ago, they used to nest in trees, on tall buildings or electric poles in the Jebel Nafusa. They feed on insects, snakes, frogs and small rodents. Migratory (rarely resident), the Libyan storks migrate from Europe to Africa in November/December. Excellent gliders making use of thermal updrafts, they fly in packs with extended necks and legs, in straight lines. They are never seen inland, in steppe or desert.
Status in Libya: threatened

Unfortunately, these large and superb birds are often gratuitously shot and killed. Not only are these birds considered holy and mentioned in the holy scriptures, but they are also protected by Libyan law.

Extinct

Ostrich (*Struthio camelus*; *naeama*) (Photos 14.44, 14.45) – a large flightless fast-running bird with long legs and long neck. The adult weighs up to 80–120 kg, lays large eggs (15 cm long, up to 1.3 kg in weight), laid in common or bare ground or slightly buried in the steppe or the desert. It used to be hunted for meat and feathers, its eggs consumed; shell fragments are often found in sand dunes (Sidi Benhur – Tajura), in the steppe (Hamada, south Sorman –Jeffara, Sirte), as perforated egg shards shaped and arranged as parts of prehistoric necklaces.
Status in Libya: extremely rare even in deep south; extinct in northern Libya

In 1976, Gintzburger suggested reintroducing ostriches in the Jeffara western rangelands along the Tunisian border. Subsequently, an ostrich farm was established at Al-Assah in the late 1980s producing large quantities of eggs and feathers which were sold in the local market and nearby countries, but this farm was terminated in 2007 with many eggs still on site until 2009.

Photo 14.44. Young ostriches (*Struthio camelus*) kept in captivity, Tajura, 2009.

Photo 14.45. A flock of ostriches (*Struthio camelus*), deep south in Kufra (after Bagnold, 1933). (Photo: used with permission from Royal Geographical Society)

Fish

It may be strange to mention fish in an arid environment where there is no permanent body of water. Fish bones have often been found in prehistoric sites deep in the steppe or the desert (Lévêque, 1990), proof that fish were present and consumed by the locals. They are, however, rarely mentioned these days in southern Libya (Trape, 2016) despite the fact that they are occasionally sighted in temporary freshwater ponds and lakes as mentioned in a video captured south of El-Abyar (on the LES Facebook site – 1 August 2023 – under *Fish in the Desert*). These ponds are resting places for migrating birds crossing the Sahara, feeding on fish, amphibians and small crustaceans and ingesting eggs. The eggs ingested by migrating birds, but not digested, are defecated into water, and survive long droughts. The fish shown on the LES video appear to be of a bream/perch type with an oval body (70–90 mm long) with a long dorsal fin. They are present in freshwater, low-salinity artesian wells around Al-Assah. *Tilapia* are sometimes introduced by farmers in ponds or resurgences as, for example, in Sokhna. Most of these small fish are good to eat despite their many bones. To our knowledge Libyan dwellers do not catch them for human consumption.

Crustaceans – horseshoe crabs

Due to the Daniel deluge (12–13 September 2023) with about 400 mm rainfall in 24 hours in the Derna region and further south, large amounts of water were collected into large and small depressions and cisterns deep into low-rainfall zones resurrecting unsuspected aquatic life. Most remarkable is the finding and photo of a small, 2–3 cm-long, horseshoe crab, a small crustacean (possibly *Triops* sp., *Triops longicaudatus?*) multiplying in the water of the Ablat Al-Ramla near Mekhili.

Molluscs (Helicidae)

Only represented by snails (*belbouch* or *halzun*), they used to be a diet of Mesolithic and Neolithic populations in southern Tunisia during the Caspian era (8000–2700 BCE). Large *escargotières* mounds (10 m high x 50 m x 150 m) of ashes (*rammadiya*) and shells of snails are still found in Gafsa (El-Mekta) and region (Oum El Araies) to Tebessa in Algeria (Lakhdar, 2019). The most common snail species were the small white shell (*Helix melanostoma*) and the large brown-band shell (*Eobania vermiculata* and *Cornu aspersum*). Consumption of snails (*Eremina* sp.) is also reported (Pöllath and Rieger, 2011) during the Greco-Roman era in eastern Marmarica (Marsa-Matruh region). To date, due to desertification and drought, snails are less common in semi-arid zones, though the small white-shell congregates *en masse* on stems of thistles on coastal wasteland, on fences and on stems of spiny acacias. They are still collected and consumed in some parts of North Africa but it does not appear to be the case in Libya.

Arachnids and Insects

Scorpions

Scorpions (*aeqrab*) are a common annoying and dangerous occurrence all over Libya. Hiding during the day, they hunt at night, abounding during *gibli* and hot nights. Some live in the steppe and the desert, others in the south, in oases, along irrigation channels, often climbing and dropping from palm trees.

These were studied by Zourgui *et al.* (2008) who identified nine Libyan types:

- *Buthus occitanus*, the most common yellow scorpion (50–60 mm long) only occurring in the coastal region

Photo 14.46. Possibly *Buthacus leptochelys* (60 mm) common at Ghirza, 1976, on Hamada and all over northern Libya.

- *Androctonus bicolor* (mostly black and up to 80 mm long), *Androctonus australis*, and *Androctonus amoreuxi*, all with fat tails and keeled segments, present all over Libya, common at Al-Assah, Hamada and Sirte
- *Buthacus leptochelys* (Photo 14.46) and *Buthacus arenicola*, both with thin pincers, found in Sirte, Hamada and in the south-east (Awjilah, Jalo, Kufra)
- *Orthochirus innesi* in the south (wadi Atbah, near Murzuk)
- *Scorpio maurus*, small with globulus pincers, in the western coastal regions of Libya; painful sting but rarely lethal
- *Leiurus quinquestriatus* found only in the deep south

Scorpions' stings are always painful and, without proper serum injection, could be lethal. Serums are specific and available from the Pasteur Institute in Tunis.

Centipedes

Scolopendra cingulate (*hamousam*) (Photo 14.47). A common nocturnal, 8–12 cm-long, centipede with yellow and brown stripes on the body. Hiding under rocks and crawling slowly, it is often sighted in derelict dwellings and in the steppe, sometimes under carpets in tents. It has two big, curved, dark venomous fangs (forcipules). It hunts insects, small lizards and snakes. Their bite is painful and may result in strong allergy, fever, swelling, with wounds easily infected, possibly resulting in septicaemia and long-lasting tissue necrosis.

Photo 14.47. *Scolopendra cingulata* (*Hamousam*), 80 mm. (Photo: *https://owlcation.com/stem/Bug-Identification-Scolopendra*; accessed 3/07/22)

Solifugae (spider-like)

Solifugae ('spider-like') (*Eaqrab alriyh*) (Photo 14.48), possibly *Galeodes* sp.: a fast-running large, spider-like (50–60 mm), often sighted in the steppe. They have two large fangs and long hairy legs. They are not reported to be venomous.

Ticks

Ticks (*Hyalomma dromedarii*) or *ougrad* or *alqarad* (Photo 14.49) commonly parasitize dromedaries. They are quite large (10–12 mm) when engorged with blood. They mostly attach to adult dromedaries' inner thighs, udders, scrotum and nostrils. Adult ticks are particularly abundant near watering points and camps (Bir Ougradia, south Al-Assah), in faeces carrying larvae and the nymphs, or under the shade of *talah* or *batum* where flocks gather during hot days. These ticks

Photo 14.48. A common spider-like (*Aqrab-er-riyah*) common in the steppe (at Al-Assah, Magdar-es-Sbot, 1976).

Photo 14.49. *Hyalomma dromedarii* (*ougrad* or *alqarad*), a common large tick parasitizing a dromedary. (Photo: Faye *et al.*, 2022)

would also infest rodents, hares, hedgehogs and horses all over Libya (Abdulsalam *et al.*, 2022), the North African steppe and desert. They occasionally attach to humans with their bite causing itching, inflammation and swelling, and may evolve into long-lasting itching, infection or tissue necrosis.

Hunting in Libya

Long ago, pristine North African rangelands and forests abounded with diverse and rich wildlife. Hunting in Libya and in arid zones is an ancient activity that provides locals with a source of meat or to protect property against predators. Hunters would wisely kill only for their family needs and no more, as there was little need for meat conservation. North African elephants were captured, tamed and used during wars by the Carthaginians (260 BCE). Large carnivorous beasts (lions, panthers, bears, cheetahs) in Libya and Tunisia were caught in huge numbers for popular circus games during Roman and Byzantine times. Using larges nets, javelins, and bows and arrows, hunting large carnivorous beasts was a dangerous exercise. All these large animals were relentlessly hunted and exterminated, often and simply to protect the flocks. Other small and large mammals (hare, gazelle, oryx, waddan, addax, Bubalus), birds (partridge, houbara, ostrich) were traditionally and freely hunted by horse-riders using prey birds or *sloughis*, pursuit dogs. The availability of firearms progressively made hunting much easier and far less dangerous, contributing to the eradication of the last large carnivorous beasts in North Africa and Libya by the early 1900s, with the exception of the cheetah, rarely seen up to the early 1970s in northern Libya and now only extremely rarely observed in the deep south of Libya.

Nowadays, motor vehicles and 4x4 vehicles give unlimited access to the most remote desertic areas where shy wildlife would retreat. Combined with the use of omnipresent powerful military firearms, hunting is not any more a reasonable quest for meat but a reckless sporting activity leading to gratuitous massacre of large numbers of endangered wildlife. This is repeatedly witnessed on Libyan social media photos and videos. Photo 14.50, from the Facebook page of Libyan Wildlife Trust exemplifies one of these senseless massacres. Note that the Libyan Wildlife Trust publishes these shocking photos on its Facebook site to raise awareness of these shameful actions. Current Libyan hunting laws are totally ignored and trampled without any penalties for the perpetrators. The ruthless destruction of the once-rich wildlife is today at an unprecedented level, through irresponsible hunting, unrelenting cultivation, land degradation, rangeland overgrazing and forest resources devastation.

Many years ago, unrestrained hunting in Libya (Fig. 14.2) was already eloquently illustrated by Mohamed Al-Zawawi (Zwawi, 1979).

Conclusions

Wildlife is an essential element of rangelands and forests. A rich wildlife is a sign of a healthy ecosystem. These days, with the exception of remote areas, most large herbivores and carnivores are mercilessly hunted, decimated and exterminated. Large carnivores like the cheetah disappeared from Libyan rangelands by the 1970s. Large native herbivores like oryx, waddan wild goats, addax and Bubalus once frequent in the south retreated to the deep southern desert. Gazelles, occasionally spotted on the Hamada and south Sirte, are hunted on a large scale. Nocturnal and feared hyenas still take refuge deep in the Jebel Al-Akhdar or in rocky places of the Jebel Nafusa where rare stealthy caracal may still be present. Cunning jackals and foxes keep

Photo 14.50. Senseless massacre of gazelles and two waddan. (Photo: Libyan Wildlife Trust, January 2019)

Fig. 14.2. A 1976 illustration of hunting by the Libyan cartoonist Mohamed Zwawi (1979).

pestering livestock owners while smaller carnivores are becoming rare sightings. Native rodents are common and keep decimating crops and fodder shrub plantations. Shifty reptiles like cobras and vipers, and peaceful tortoises, are common, though the latter are collected in large numbers and sold as pets. Birds are not spared: ostriches have been slaughtered since the 1930s and have disappeared from Libya. Large migrating birds like storks, in earlier times wintering in the Jebel Nafusa, are now rarely encountered. Cranes, flamingos and other large migrating birds are needlessly shot. Aquatic birds, still numerous, are hunted in all seasons. Bustards, present in the steppe, are actively hunted despite being protected. The list of wildlife slowly disappearing from Libyan rangelands and forests is long. Hunting for necessity is long-gone. With unabated overgrazing and slow destruction of native forests, encroaching urbanization and hobby hunting, the once-rich Libyan wildlife is inexorably eliminated. The food chain is broken. Rangeland and forest health and fauna diversity are at their lowest point.

Photographic acknowledgements

While most photos are from G. Gintzburger, we acknowledge the photos of Libyan wildlife kindly provided by Mr Salem Essaker, Mr Idris Al-Osalibi, Mr Ayoub Akbar El-Ghaithi, Mr Bernard Faye, and from various and unknown contributors posting photos on Facebook sites with the Libyan Ecological Society site from Dr M. Saaed, the Libyan Wildlife site from Mr Salem Essaker, the Libyan Wildlife Trust site and Save the Jebel Al-Akhdar from Desertification. We regularly accessed these well documented sites from 2015 to 2024.

Notes

[1] BCE = before Christian era; CE = Christian era

[2] BP = before present years, a dating system used by geologists and archaeologists referring to events that occurred pre-radiocarbon dating c.1950. It is also noted as RCYBP for radiocarbon years before present.

References

Abdulsalam, A.M., Saadawi, W.K., Kharwat, H.I. and Shaibi, T. (2022) Species diversity of ticks (Acari: Ixodidae) in Tarhuna, Libya. *Open Veterinary Journal* 12(3), 370–374. DOI: 10.5455/OVJ.2022.v12.i3.10

Bagnold, R.A. (1933) A further journey through the Libyan desert. *The Geographical Journal* 82(2), 103–126. DOI: 10.2307/1785658

Barich, B.E., Garcea, E.A.A. and Giraudi, C. (2006) Between the Mediterranean and the Sahara: geoarchaeological reconnaissance in the Jebel Gharbi, Libya. *Antiquity* 80(309), 567–582. DOI: 10.1017/S0003598X00094047

Bauer, A.M., DeBoer, J.C. and Taylor, D.J. (2017) Atlas of the reptiles of Libya. *Proceedings of the California Academy of Sciences, United States* 64, 155–318.

Chedad, A., Bouam, I., El Bouhissi, M., Dahmani, W., Ait Hammou, M., Mebarki, M.T., Mezzi, M. and Sadine, S.E. (2024) Moorish Viper, *Daboia mauritanica* (Gray, 1849) (Squamata, Viperidae), in Algeria: new provincial records, range extension, and an update on its distribution. *Check List – The Journal of Biodiversity Data* 20(2), 536–543. DOI: 10.15560/20.2.536

Diolé, P. (1955) *Le Plus Beau Désert du Monde.* Edition Albin Michel, Paris.

Diolé, P. (1956) *Dans le Fezzan Inconnu.* Edition Albin Michel, Paris.

Domergue, C.A. (1960) Auto-observation de la morsure par *Aspis cerastes* (vipère à corne). *Bulletin de La Société Sciences Naturelles Tunisie* 9–10, 97–101.

Douka, K., Jacobs, Z., Lane, C., Grün, R., Farr, L. *et al.* (2014) The chronostratigraphy of the Haua Fteah cave (Cyrenaica, northeast Libya). *Journal of Human Evolution* 66, 39–63. DOI: 10.1016/j.jhevol.2013.10.001

Elkahwage, E.I. and Jdeidi, T. (2018) Survey of the wild mammalian species in Misallatah nature reserve and national park. In: *Atti e Memorie dell'ente Fauna Siciliana, 2018.* 3rd International Congress on Biodiversity, 4–6 September 2015, Noto-Vendicari (Italy), pp. 101–112.

Elsalem, R.M.A. (2024) A review of toxoplasmosis in humans and animals in Libya. *Journal of Pure & Applied Sciences* 23(1), 64–71. DOI: 10.51984/jopas.v23i1.2447

Etchécopar, R.D. and Hüe, F. (1964) *Les Oiseaux du Nord de l'Afrique de la Mer Rouge aux Canaries.* Editions N. Boubee et Cie, Paris.

Fathallah, E.E. (1970) Tarhuna zone – prehistoric rupestral pictures. *Archaeological News*, 146–149.

Faye, B., Konuspayeva, K. and Magnan, C. (2022) *L'Élevage des Grands Camélidés.* Editions Quae.

Gérard, J. (1856) *La Chasse au Lion* (Vol. 292). La Nouvelle Librairie, Paris.

Harper, F. (1945) *Extinct and Vanishing Mammals of the Old World.* American Committee for International Wildlife Protection New York Zoological, New York.

Hollom, P.A.D., Porter, R.F., Christensen, S. and Willis, I. (1988) *Birds of the Middle East and North Africa: A Companion Guide.* T. & A. Poyser Ltd, Carlton, UK.

Hufnagl, E. (1972) *Libyan Mammals.* The Oleander Press, Cambridge, UK.

Jacobs, S., Li, B., Farr, L., Hill, E., Hunt, C. (2017) The chronostratigraphy of the Haua Fteah cave (Cyrenaica, northeast Libya): optical dating of early human occupation during marine isotope stages 4, 5 and 6. *Journal of Human Evolution* 105. DOI: 10.1016/j.jhevol.2017.01.008

Klein, R.G. and Scott, K. (1986) Re-analysis of faunal assemblages from the Haua Fteah and other late Quaternary archaeological sites in Cyrenaican Libya. *Journal of Archaeological Science* 13(6), 515–542. DOI: 10.1016/0305-4403(86)90038-5

Lakhdar, F. (2019) Considérations anthropologiques sur les Capsiens, peuple épipaléolithique du Maghreb oriental. *Dialogues Méditerranéen* 3–4, 4–23.

Le Houérou, H.-N. (1997) Climate, flora and fauna changes in the Sahara over the past 500 million years. *Journal of Arid Environments* 37(4), 619–647. DOI: 10.1006/jare.1997.0315

Lepri, G. (1930) Sopra una nuova sottospecie del genere. *Ammotragus. Atti Della Pontificia Accademia Delle Scienze Nuovi* 83, 271.

Lévêque, C. (1990) Relict tropical fish fauna in Central Sahara. *Ichthyological Exploration of Freshwaters* 1(1), 39–48.

Louis, A. (1979) *Nomades d'hier et d'aujoud'hui dans le sud Tunisien, Collection Mondes Méditerranéens.* EDISUD, Aix-en-Provence, France.

McBurney, C.B.M. (1967) *The Haua Fteah in Cyrenaica and the Stone Age of the South-East Mediterranean.* Cambridge University Press, Cambridge, UK.

Moussouni, A. and Scaravelli, D. (2024) New records of Libyan striped weasel *Ictonyx libycus* and common genet *Genetta genetta* from the Algerian Sahara. *Small Carnivore Conservation* 62. Available at: https://smallcarnivoreconservation.com/index.php/sccg/article/view/4247 (accessed 19 August 2024).

Muzzolini, A. (1981) Les premiers boeufs domestiques au Sahara Central. *Bulletin de la Société Méridionale de Spéléologie et Préhistoire* XXI, 19–30.

Pöllath, N. and Rieger, A.-K. (2011) Insights in diet and economy of the East (Greco-Roman sites in North-Western Egypt). *Mittleunden des Deutschen Archäologishens Instituts – Abteilung Kairo*, 163–180.

Pomel, A. (1995) *Les éléphants quaternaires, Paléontologie-Monographie.* Impr. de P. Fontana, Algiers. Available at: https://gallica.bnf.fr/ark:/12148/bpt6k62754098/f88.item# (accessed 19 August 2024).

Rossouw, L. (2001) The extinct giant long-horned buffalo of Africa (*Pelorovis antiquus*). *Culna*, 14–15 September.

Sagne, J. (1950) *L'Algérie pastorale, ses origines, sa formation, son passé, son présent, son avenir.* Imprimerie Fontana, Algiers.

Trape, S. (2010) A new cichlid fish in the Sahara: the Ounianga Serir lakes (Chad), a biodiversity hotspot in the desert. *Comptes Rendus Biologies* 339, 529–536. DOI: 10.1016/j.crvi.2016.08.003

Zourgui, L., Maammar, M. and Emotris, R. (2008) Taxonomical and geographical occurrence of Libyans scorpions. *Archives de l'Institut Pasteur de Tunis* 85(1–4), 81–89.

Zwawi, M. (1979) *The Other Face.* Tripoli, Libya.

15

Assessing Rangeland Recovery, Degradation and Desertification

Abstract

Rangeland biophysical recovery, degradation and desertification are azonal and affect all arid and semi-arid Libyan rangelands. The anthropogenic reasons are well known with climatic high variability and possible climate changes. We review the methods and tools to assess rangeland conditions and health from simple site- and date-specific biomass and vegetation cover measurements to the most advanced complex satellite imagery time-series processing allowing monitoring and mapping trends over large areas. Field calibration and ground-truthing remain essential to verify laboratory satellite image processing. Rangeland degradation, often overemphasized in the media, is not inescapable as many examples confirm rangeland recovery. Powerful satellite-imagery processing tools are operational in arid Australia on large extensive livestock commercial operations led by professional rangeland managers. Libyan rangeland managers, decision makers and scientists could benefit from these technologies to accurately identify and locate rangelands recovering and to prioritize treatments of degraded areas under threat of desertification.

Introduction

From time immemorial, users, developers and managers of natural resources have had to have a deep knowledge and understanding of their environment to make the best use of it without external input. Be it primitive hunters, roaming nomads or settled farmers, all of them knew their local climatic conditions, territory limits, soils, water resources, vegetation and wildlife, and also the support, constraints or antagonism they may expect from neighbours when using the resources available on their land.

Conquerors', explorers' and travellers' accounts from all ethnic groups and cultural background added their contribution to new regions visited and what could be found, along with local agricultural practices and traditions (De Agostini, 1917a, b; Forskål, 1775; Ibn Khaldūn, 1852). The advent of writing and printing accelerated this trend with knowledge becoming available to a larger audience extending their control over territories led by increasing demand for food from an expanding and wealthier population. With ensuing agricultural and rangeland degradation in the Mediterranean regions, especially in arid zones, it became necessary to develop vegetation inventory and later monitoring and mapping methods to better manage the scanty and fragile resources available (Floret *et al.*, 1976, 1977; Floret and Le Floc'h, 1973).

Land degradation and desertification has long been recognized by international and national organizations under the UN with myriad international committees, panels, conferences and publications. Most UN actions by the UN Convention to Combat Desertification (UNCCD) remained general and qualitative with little relevance and practical use to properly quantify and accurately map the actual recovery, degradation

DOI: 10.1079/9781800627154.0015

and desertification, and trends, at management scale. Fortunately, rangeland managers supported by ecologists from national programmes in North Africa, Europe, Australia, sub-Saharan Africa, South Africa and the USA developed methods to identify, quantify and map land condition trends either as stable, recovering, degrading or desertified.

Vegetation Formations, Successions and Climax

As early as the beginning of the 20th century, Clements (1916), studying the ecology and vegetation of the US Great Plains rangelands, formalized the concept of vegetation formation successions with the climax theory. Each vegetation formation is different according to the rainfall and soil type and evolves towards a specific climax. He states that each formation, 'as an organism, ...arises, grows, matures and dies' after reaching a climax, before being replaced by another formation (or plant succession) under the same station ecological conditions. He anticipates that the response to disturbance (grazing, fire, clearing, etc.) of a vegetation formation naturally evolving to its climax is linear and reversible. When the disturbance is moderate, the formation reverses to the pioneer stage and slowly and eventually moves towards its climax or pristine condition. When the disturbance is extreme (overgrazing or long drought), it leads to irreversible degradation. This was the foundation of US rangeland management (Heady, 1975; National Research Council, 1994; Stoddart *et al.*, 1975) postulating that there was an equilibrium between livestock numbers and rangelands resources. This is based on the concept of climax vegetation where carrying capacity is driven by precipitations and temperatures, soil types and floristic diversity driving consistent primary production when the rangeland system is held at equilibrium as long as the stocking rate does not exceed the rangeland carrying capacity. This US concept was largely and unsuccessfully applied to government rangeland projects in North Africa and in Libya, ignoring the local Mediterranean herding practices and socio-economics. Climax vegetation does not exist any more in North Africa due to anthropogenic actions.

New concepts emerged from researchers working in Africa with pastoral communities living with uncertainty (Scoones, 1994). The concept of rangelands at non-equilibrium (Behnke *et al.*, 1993; Vetter, 2005) was also more appropriate to Mediterranean arid and semi-arid rangelands. This new concept considers the necessary flocks' mobility over large rangeland spatial heterogeneity (geomorphology, soils and vegetation) driven by high spatial inter- and intra-rainfall variability (see Chapter 3) and highly fluctuating stocking rates by local herders.

The US equilibrium rangeland management model focuses on system stability by manipulating the stocking rate while the African non-equilibrium model is based on herds' mobility and variable stocking rates according to rainfall (Vetter, 2005). Manipulating stocking rates and flock off-take are possible on privately owned and run (USA, Australia, South Africa, South America), or government-controlled large rangeland areas (Ex-Soviet Union, China) but does not apply to nomadic or semi-nomadic herders where land tenure is often insecure and where keeping the highest number of animals is the only insurance against recurring droughts, defective banking systems and conflicts – common reasons current in Mediterranean arid and semi-arid zones to this day. In these conditions, keeping as many animals as possible leads livestock owners to, *de facto*, ignore adjusting stocking rates to the vegetation resources and practice 'first come, first served'; it leads to rangeland degradation and elimination of the feed resource. Furthermore, oil-rich countries, under political pressure from increasing population's wealth demanding meat, divert their hard currencies to import large amounts of supplementary feed subsidized to livestock owners, hence further stimulating rangeland degradation and desertification.

Defining Rangeland Degradation and Desertification Trends

Degradation

Rangeland degradation is a slow trend reducing vegetation and wildlife biodiversity and accelerated with wind and water erosion slowly abrading soils. Scars appear on the soil with increasing marks of sheet, rill and gully erosion. Soil surface may become compacted and hard-crusted reducing water infiltration and plant germination and recruitment. It results in the steady loss of productivity of an ecosystem compared to its pristine and healthy condition under a set rainfall regime and soil type. Rangeland degradation can be identified with the disappearance of palatable plants replaced by dense stands of unpalatable plants such as thistles (*Atractylis* sp., *Carduncellus* sp., *Carthamus* sp., *Centaurea* sp., *Circium* sp., *Cynara* sp., *Onopordum* sp., *Scolymus* sp.), *Asphodelus microcarpus*, *Ballota* sp., *Marrubium* sp., *Neurada procumbens*, *Peganum harmala*, *Phlomis* sp., *Stipa retorta*, *Thymelaea* sp., etc. It is a spiralling downwards trend leading to desertification. Degradation can be reversed providing the ecosystem is still resilient, as long as the soil seed stock is not totally depleted despite a poor vegetation cover of annuals and perennials. Rangeland protection from grazing and fuelwood collection is the simplest and cheapest method to reverse degradation (see Chapter 16).

Desertification

Desertification is the ultimate stage of land degradation once the irreversibility threshold is reached (Aronson *et al.*, 1995; Holling, 1973; Pontanier *et al.*, 1995). Desertification is defined as the 'land degradation in arid, semi-arid, and dry sub-humid areas resulting from various factors, including climatic variations and human activities' (UNCED, 1992). While desertification often occurs at the margin of the desert, it affects nowadays large swaths of semi-arid lands (Nahal, 2004). Desertification is the expansion of desert-like areas which were in good and healthy condition in men's memory (Aronson *et al.*, 1995; Floret *et al.*, 1986; Le Houérou, 1996; Winslow *et al.*, 2004). It is the result of two main factors operating either singly or in combination in arid Mediterranean zones. These are: man's overexploitation of arid lands (Al-Bukhari, 2019; Le Houérou, 1959, 1969, 1993, 1996, 2005; Le Houérou and Rapp, 1976; Manière and Chamignon, 1986); and, to a lesser extent, periods of prolonged drought. For the past century, arid rangeland conditions rapidly deteriorated due to mismanagement (relentless overgrazing, hazardous rainfed cropping, fuelwood collection, snowballing watering places with new wells combined with escalating use of trucks carting livestock, feed and water to remote rangeland areas, and disastrous irrigation projects with saline water) and confusion of agricultural policies and failed governance. Climate change possibly affecting arid zones may be another factor that appears far less important than current human mismanagement. Deteriorating rangeland conditions, for whatever reason, have catastrophic global impacts on the survival of impoverished livestock owners, herders and nomads slowly deserting their degraded rangelands for city slums (Saïdi *et al.*, 2016).

Deserts and desertification should not be confused. Deserts in the Mediterranean regions are hyper-arid zones where biological activities are limited by current climatic conditions commonly occurring below the 100 mm isohyets and extreme temperatures (see Chapter 3).

Is rangeland degradation inescapable and biological recovery possible?

Recovery, degradation and ultimate desertification trends are azonal as they could affect any ecosystem under any ecological condition, under high or low rainfall, under good soil or poor soils, the difference being the speed at which the degradation or biological recovery processes take place due to changing anthropogenic or climatic factors. Rangeland degradation and desertification are obvious all over the arid zones. But less visible is that degradation can be reversed (Pontanier *et al.*, 1995) when grazing pressure, fuelwood collection, marginal cropping, supplementary feeding, livestock watering control are relieved leading to biological recovery (see Chapter 16).

What are the methods and tools available to identify and assess recovery, degradation and desertification trends? We may distinguish rangeland site- and date-specific surveys from mapping conditions or trends over large areas.

Surveys and Monitoring to Assess Rangeland Positive or Negative Trends

For a long time, evaluation of rangeland conditions were mostly qualitative, highly subjective and based on visual assessment during field trips. However, quantitative scientific assessment of rangeland condition was called for by rangeland managers to optimize commercial ranching operations. In the USA, due to open access to public rangelands to overall commercial grazing, degradation and overgrazing was rife until the Bureau of Land Management introduced grazing rules and assessment of rangeland condition (National Research Council, 1994). In Australia, the ground-based West Australian Rangeland Monitoring System (WARMS) (Watson *et al.*, 2007) was established at the request of pastoralists to have reliable assessments of rangeland conditions and changes affecting their operations. Qualitative methods to assess rangeland conditions slowly emerged, initially being site- and date-specific. By the early 1980s, with the availability of time series of satellite imagery, innovative monitoring methods were developed to quantitatively and accurately rate and locate rangeland status and trends over large areas and over many years (Beutel *et al.*, 2005; Gintzburger and Saïdi, 2009; Karfs *et al.*, 2004) (Figure 15.1).

Site-specific rangeland inventories and surveys

Vegetation surveys are either qualitative or quantitative, or both combined.

Qualitative

The rangeland status, either recovering or degrading, was initially assessed with surveys comparing the floristic composition from nearby sites (same rainfall, same soil types), one open to grazing and the other protected. These qualitative surveys were a one-off study at a set date characterizing the environment and the vegetation with different methods (Emberger and Godron, 1968; Raunkiaer, 1934), ultimately producing phytoecological maps (Le Houérou, 1959, 1969). These maps present a photograph of the status of vegetated sites, describing if degraded or in good condition, but without indicating any trends, either as stable, degrading or improving over time. Vegetation quantitative measurements complemented these qualitative surveys.

Quantitative

There is an abundant literature about quantitative methods to evaluate the arid rangeland vegetation composition, cover, biomass and soil conditions (Emberger and Godron, 1968; Gounot, 1969; National Research Council, 1994).

ABOVE-GROUND RANGELAND BIOMASS MEASUREMENT. It is based on the quadrat method (Gintzburger, 1986; Gintzburger *et al.*, 2003; Gintzburger and Saïdi, 2009) also providing information on plant species diversity, density and cover. The above-ground biomass harvested inside quadrats is sorted by green parts and woody stems to approach the annual production (Floret and Le Floc'h, 1973), the grazeable portion and to estimate the proper range-use factor (see Chapter 8).

LINE-INTERCEPT MEASUREMENTS. The simple step method (Canfield, 1941) and more elaborate line-intercept measurements (LIM) (Brun and Box, 1963; Gintzburger *et al.*, 2009; Gounot, 1969) allow quantification of absolute or relative plant species frequency along a linear path or narrow strip on site complemented with basal or crown cover dimensions, plant structure description per species, biomass, soil surface conditions, extent and distribution of bare soil patches, and photos of sites as implemented by the WARMS (Watson *et al.*, 2007). It is often coupled over time with qualitative and quantitative evaluation of soil and landscape conditions supported by the landscape function analysis (LFA) (Mahmoud *et al.*, 2014, 2021; Saaed *et al.*, 2022; Tongway and Hindley, 2000).

Shortcomings of site- and date-specific assessments and measurements

All these measurements and observations issued from site- or date-specific rangeland surveys,

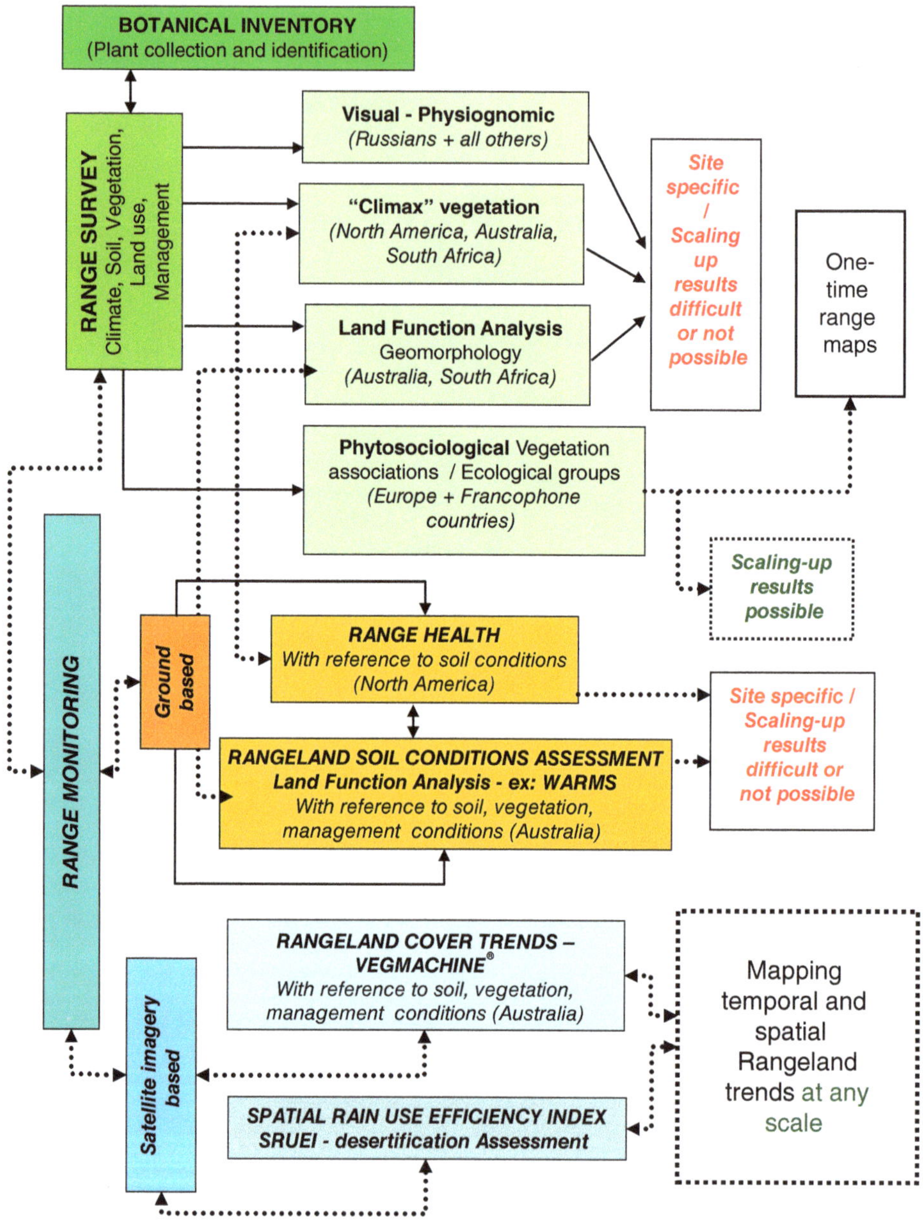

Fig. 15.1. A synthetic view of rangeland survey, analysis and monitoring approaches. (From Gintzburger and Saïdi, 2009)

sometimes repeated over seasons and years, identify elements supporting ground-based rangeland long-term monitoring. It allows establishing the rangeland health that may be 'stable', 'degrading' or 'improving' (National Research Council, 1994) according to a set of pertinent indicators related to vegetation composition and structure, soil condition or management options. These site-specific ground-based surveys would take several years to identify vegetation trends especially on arid zones vegetation, and difficult to scale up to large areas.

These cumbersome and time-consuming methods must be repeated year after year, in spring when the vegetation is at its peak development or in summer when the vegetation is at rest. The results, while revealing, remain site-specific, climate-dependent, difficult to statistically analyse, and challenging to scale up to larger areas (Ludwig *et al.*, 2004; Pringle *et al.*, 2006) unless sites and plant associations are well identified with phytoecological surveys and mapping (Floret, 1974; Le Houérou, 1959, 1969). Furthermore, successions of vegetation associations, vegetation cover changes in Mediterranean arid zones under constant grazing pressure or affected by drought or recovering under protection, may occur over a long period, often 10–20 years depending on the soil type (Floret *et al.*, 1981, 1986; Floret and Le Floc'h, 1973; Mahmoud *et al.*, 2021; Telahigue *et al.*, 1987). National or international projects carrying out biomass or LIM have short financing, labour support and limited lifespan (3–5 years at best) and may therefore be unable to detect significant changes in arid vegetation cover, composition or biomass. Additionally, exceptionally wet years, prolonged droughts or catastrophic events (flood, fire, short-term localized heavy grazing, occasional cropping) may locally and swiftly alter the successions of vegetation types and biodiversity rendering site monitoring useless.

Is floristic composition an indicator of rangeland recovery, degradation or desertification?

Floristic listings may be useful when comparing nearby protected rangelands with open-to-grazing ones, but they remain qualitative and open to endless interpretations. However, the extent of bare soil with abundance of unpalatable species is a clear indicator of degradation as seen around watering stations. It must be complemented with quantitative line-intercept measurements (LIM) or wheel transect recording genus and species occurrence and frequency along paths of different length according to the vegetation density and structure (Photos 15.1 and 15.2). It may also be complemented with measurements of vegetation cover by genus and species by recording each individual plant dimension, height, projection of the crown on soil inside randomly placed quadrats of various size, again

Photo 15.1. Line-intercept measurements in the Karakum desert (April 2002, Ravnina, Turkmenistan). (From Gintzburger *et al.*, 2009).

Photo 15.2. Using the wheel transect method on a steppe vegetation in central Türkiye (Polatli region). (Photo: Saïdi *et al.*)

according to the vegetation overall density and structure (Gintzburger, 1986; Gintzburger *et al.*, 2009). These different line-intercept methods initiated by Canfield (1941) and quadrat measurements are tedious, time-consuming, and site- and date-specific, and results are difficult or impossible to scale up to larger areas.

Are standing biomass measurements reliable indicators of rangeland evolution?

Spring cover of annual plants, though contributing significantly to herbage biomass and forage supply in winter/spring on Mediterranean arid rangelands, are subject to inter- and intra-annual variations with localized precipitation, and hence are poor indicators of degradation/desertification. The perennial vegetation cover, mostly tall and small shrubs, controls wind and water erosion and remains a better indicator of vegetation evolution, but it has its limitations. It is used in monitoring studies displaying and locating the trend in perennial cover (Wallace *et al.*, 2004; Watson *et al.*, 2007), indicating cover changes at property scale but without quantifying the current level of herbage biomass and productivity. Such systems do not indicate whether the environment is in good or poor condition. It assumes that the vegetation cover is related to the photosynthetically active above-ground herbage biomass, but in arid zones, the vegetation cover is poorly related to the productivity. The spatial changes in arid rangeland herbage biomass and seasonal productivity are primarily affected by the seasonal amount and distribution of rainfall received on site, which can be highly variable within and between years (Le Houérou, 1959; Von Wehrden *et al.*, 2010; see Chapter 3) and cannot by itself be an absolute measurement of vegetation trends.

Methods to Spatially Assess Recovery, Degradation or Desertification

Rangeland degradation is not new in Libya as it was already identified as a rampant curse as early as 1935 by Despois (1935) during his travels in the Jebel Nafusa. While at the time rangeland degradation was circumscribed around settlements, it slowly but relentlessly kept affecting arid and semi-arid rangelands over larger and larger areas from west to east, deep into the steppe to the margins of the desert as well as in neighbouring southern Tunisia (Le Houérou, 1959, 1968, 1969, 1982, 1993, 1996). We all witness rangeland degradation by simple sites visual assessment during field trips (Photo 15.3). Due to accepted perception that Libyan rangelands were degrading fast, Libyan academics started assessing the problem (Al-Bukhari, 2019; El-Barasi *et al.*, 2013; El-Tantawi, 2005; Mahmoud *et al.*, 2014, 2016, 2021; Saaed, 2008) with qualitative floristic approaches, however lacking quantitative approach.

While the causes of desertification were known, it was difficult to accurately map recovery, degradation and desertification at a scale practical for land management purposes. Nowadays, quantitative rangeland degradation or recovery assessments on small areas (Floret *et al.*, 1981) and mapping on large areas are available using different methods with satellite images processing supported with GIS software with their pros and cons. We present here three mapping examples of approaching rangeland degradation/desertification or recovery on large areas.

The 'bare soil' approach at Ravnina (Karakum desert, Turkmenistan)

The Ravnina rangelands

It is a sandy country with high fixed sand dunes ridges and interdunes flats comparable to the Bir-el-Ghanem rangelands (Gintzburger, 1986) and receiving an average of 150–175 mm MAR precipitation as rain or snow (Gintzburger *et al.*, 2009). The original Ravnina Soviet sovkhoz covers about 560,000 ha exploited with a flock of about 30,000–40,000 sheep equivalent divided into sovkhoz and small private flocks grazing their territory around 47 assigned wells. The flocks are year-long shepherded and grazing on 3000–7000 ha within 5–7 km of the wells on a total of 215,000 ha during the spring, summer and early autumn. They retreat to sheds near the wells during the cold and snowy winter and usually feed with stored forage provided by the sovkhoz administration. By the late 1980s, the gradual collapse of the centralized Soviet system replaced by the flocks' privatization induced drastic transformations in grazing management on the Ravnina rangelands reflected in vegetation cover evolution.

Why the bare soils approach?

We hypothesize that long-term changes of 'bare soil' cover due to degradation or recovery would be easier to identify than photosynthetic activity from vegetation. For this approach, after geo-correcting and atmospheric corrections of the satellite images, we extract only the bare soil by processing the PVI (Perpendicular Vegetation Index) value pixel by pixel from end-of-summer satellite images time-series when the annuals and ephemeroids have disappeared, and when perennials are at their lowest photosynthetic activity. Hence, late-summer satellite images processing allows us to identify the maximum extent of bare soil. Changes in areas with bare soil over years would indicate by default the vegetation cover trends as temporary degradation, recovery or stable (Fig. 15.2).

Interpretation of the bare soil evolution on the Ravnina rangelands

The change between 1987 and 2000 in bare-soil evolution or recovery on the Ravnina sovkhoz may be partly explained by the inter-annual precipitations (rain and snow). It is nevertheless

Photo 15.3. Visual qualitative assessment of rangeland degradation inside and outside protected area (wadi Sassu protected area, northern boundary).

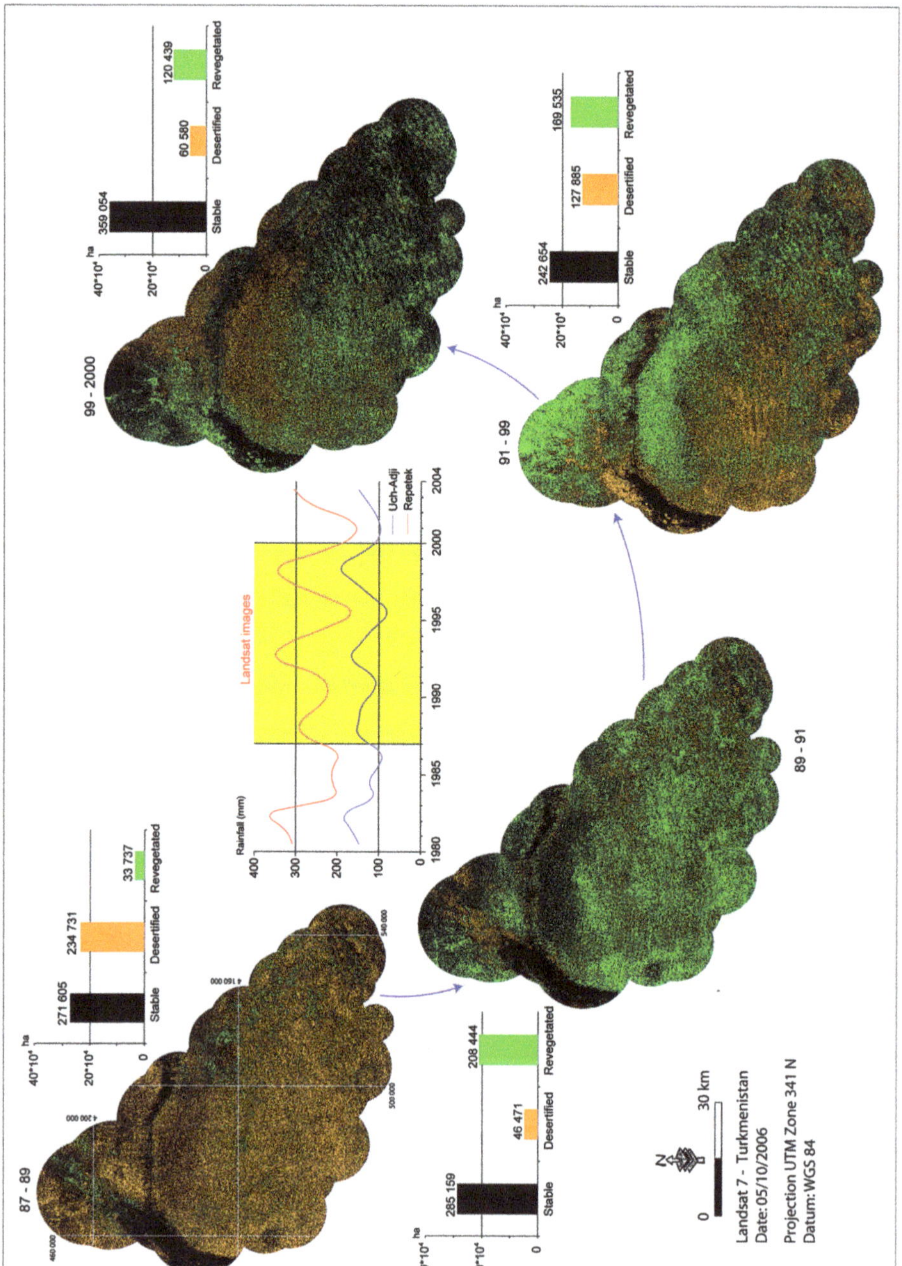

Fig. 15.2. The 'bare soil approach' trends at Ravnina (Karakum desert, Turkmenistan). (From *Gintzburger et al.*, 2009)

mostly explained by the slow collapse of the Soviet system ending in 1991 and impacting the livestock numbers on the Ravnina rangelands. Pre-1991, the Ravnina sovkhoz was supported with funds and state directives directly from Moscow, supplying equipment, financing wells and watering-sites maintenance, and essential winter forage to the large government flocks. The Ravnina rangelands were heavily exploited by the sovkhoz flocks all over, as seen on the 1987–1989 satellite image analysis. By 1989/1990, all support from the Soviet authorities ended, the flocks disbanded or privatized with sovkhoz rangeland grazing disrupted and temporally deserted. The livestock numbers sharply declined allowing rangeland regeneration (1989–1991) throughout except around Ravnina village where the remaining flocks retreated. The period 1991–1999 saw a slow reorganization of the grazing from the Ravnina private flocks, expanding south-west from Ravnina village, and from the east of the Zahmet and Iolotan villages, now having free access to all livestock owners and fuelwood collectors gradually reoccupying the south-western ex-sovkhoz rangelands. As the western rangelands were slowly exhausted, the private flocks moved after 1999 to the north-east where the rangelands were in better condition.

In conclusion, this simple method of identifying the area and kind of change was quite useful and may be used to identify the spatial and chronological degradation and recovery trends in arid and semi-arid areas using the PVI or TSAVI (Transformed Soil Adjustment Vegetation Index). We reiterate that we had no means for fully explaining these vegetation changes.

The Spatial Rain Use Efficiency Index (SRUEI) to assess the rangeland status in the Karakum desert (Turkmenistan)

What is the Rain Use Efficiency Index?

The Rain Use Efficiency (RUE) Index provides a unifying concept of rangeland condition, tied to aridity and ecosystem functioning. It indicates the level of plant production per unit of received rainfall on site (Gintzburger *et al.*, 2009; Huxman *et al.*, 2004; Le Houérou, 1984, 1996). It integrates the precipitation and temperature regimes of the year, soil infiltration characteristics and soil water budget, soil fertility, vegetation cover and structure, plant density and floristic composition. It synthesizes the ability of a vegetation association on a rangeland site to produce dry matter during the year. The RUE, expressed in kg DM /ha/year and per mm of rain, is the ratio between the vegetation production and the annual precipitation received or calculated on site. When interpreting data from about 1000 pairs of rainfall/production sites, Le Houérou (1996) concluded that RUE values of 0.5–2.0 indicate a degraded or desertified environment while non-degraded ecosystems have an RUE ranging from 4 to 6. Thus, RUE is, in essence, an index of 'rangeland health', defined as 'the degree to which the integrity of the soil and the ecological processes of rangeland ecosystems are sustained' (National Research Council, 1994). It is the status of the vegetation condition on a rangeland site compared to its optimal condition, i.e. when the vegetation composition, cover and production are maximized, with little or no soil erosion scars, and where water use and nutrient cycling are boosted in relation to the local climatic and soil characteristics. The more intact all these factors are, the higher is the RUE and, therefore, the better is the rangeland health and the lower the degradation/desertification status. The RUE applied to large areas or SRUEI could be used as a precise synthetic and integrating rangeland condition/desertification indicator, provided that it is calculated with field biomass and productivity measurements (see Chapter 8) in relation to high-resolution satellite imagery and rainfall data generated from a digital elevation model combined with precipitation data from nearby meteorological stations.

Implementing the SRUEI

To this effect, we computed, tested and mapped (Map 15.1) the SRUEI on 530,000 ha of arid rangelands from the Ravnina farm (N 37.95 E 62.66) located about 85 km east of Mary (N 37.61 E 61.83) in the Karakum desert in Turkmenistan (Gintzburger *et al.*, 2009). The average rainfall at Ravnina is 120–150 mm/year increasing with the elevation, climbing from 200 m to the north at Ravnina village to about 350 m to the south. All over the area, the soils are homogenous fixed sand dunes with a vegetation similar to that of the Bir-el-Ghanem rangelands (Gintzburger, 1986). The Ravnina rangeland productivity varies between 0.1 and 0.8 t DM/ha/year, depending on annual rainfall, vegetation cover and plant diversity.

The Ravnina farm SRUEI map (Map 15.1) displays RUE values for each pixel at 30 m resolution. Our interpretation is as follows:

- 0–1.0: bare soil, no production (interdune flats and takyrs)
- 1.0–1.5: very poor condition and low productivity (salt pans)
- 1.5–2.0: very poor condition (degraded vegetation around wells – long-term open grazing)
- 2.0–2.5: poor condition (degraded vegetation – long-term and continuous open grazing)
- 2.5–3.0: medium condition (frequent grazing on recently accessible territory)
- 3.0–3.5: medium to good condition (new grazing territories, frequent grazing)
- 3.5–4.0: good condition (frequent seasonal grazing)
- 4.0–4.5: good condition (occasional seasonal grazing)
- > 4.5: very good condition (rare grazing, territory rarely accessed, or irrigated crops or well-watered areas along the Karakum canal)

The lower RUE values (1 to 3) indicate a poor to medium condition of the rangelands in the northern part within 25–30 km from the main Ravnina settlement where the most accessible pastures are, except for the lush vegetation and irrigation benefiting from the leaks of the Karakum canal. Further south, the pastures being too far away and less accessible to flocks, the RUE values climb from 3 to 4 revealing better status. The satellite imagery 30 m pixel resolution allows us even to identify the dunes crests (RUE = 2.5–3.0) in better conditions than the degraded interdunes pastures (RUE = 2.0–2.5) and pinpointing the degradation rings around wells and watering points with RUE values falling between 1 and 2, indicating utter degradation compared to surrounding rangelands. This was checked with ground-truthing. The SRUEI Ravnina map is different from the Ravnina biomass map displayed in Chapter 8.

We used the same SRUEI method to detect the rangeland status on 250,000 ha (Map 15.2) on the high plateau of Algeria, in the Mecheria (N 33.544 E –0.282) – Naama (N 33.267 E –0.311) region (Saïdi and Gintzburger, 2013). This Mediterranean steppe region receives an average 200 mm MAR. The extent of rangeland in desertified and poor condition covers about 19% of this area.

A similar RUE approach could be used on large areas of Libyan rangeland with homogeneous soils to detect hotspots of recovery, degradation or desertification providing that vegetation biomass and productivity are actually measured, and that accurate rainfall data are available.

Shortcomings of the RUE and the SRUEI mapping

Our Ravnina study covers only one season and provides us with the distribution of the rangeland condition for a specific rainfall season. We anticipate that the SRUEI would evolve over the years with areas remaining stable or becoming degraded or improving according to flock management and displacement. Is the RUE stable for a specific vegetation type and condition over years under different Mediterranean pluviometry? Gamoun (2016) measured the rangeland productivity of four desert rangeland associations in arid southern Tunisia and calculated their RUE evolution per site over ten years. He demonstrated that RUE in all plant associations is not stable but higher during dry years compared to wet years, the reason being that rainfall in wet years is wasted by increased run-off. Therefore, the mapping of RUE values computed pixel by pixel over many successive seasons would not be of any use. The SRUEI calculation and mapping of an ecologically homogeneous region during one single season remain valid to assess the relative rangeland status from desertified to good condition.

Finally, the RUE defined by Le Houérou (1984, 1996) is only valid for Mediterranean regions, i.e. with winter rainfall. It may not be valid in arid sub-Saharan Africa receiving summer rainfall that is less effective due to intense evapotranspiration (Hein *et al.*, 2011).

None of these above methods and techniques provides information on why the range condition or range health is different from one site to another. The rangeland manager or ecologist

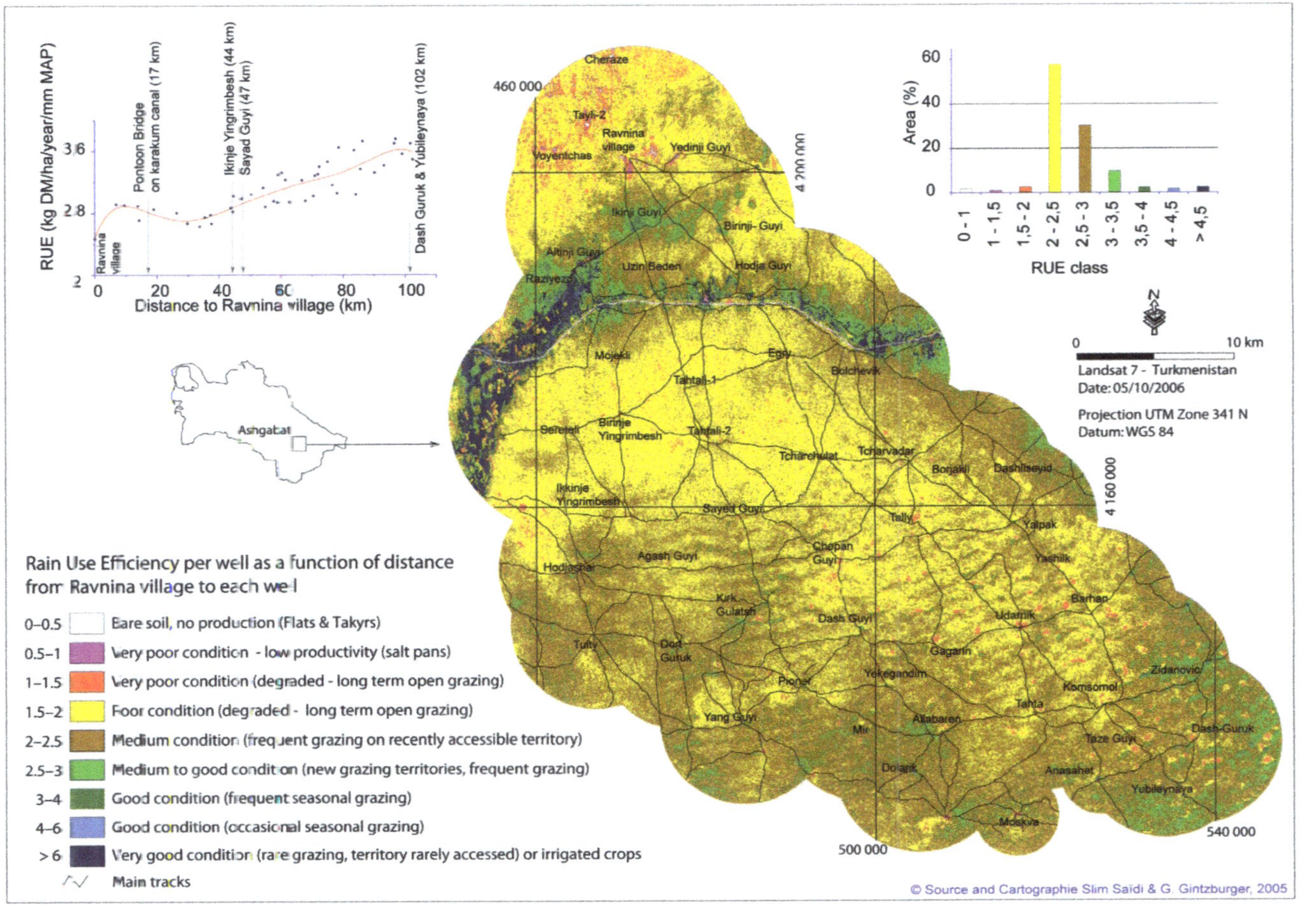

Map 15.1. The SRUEI map from the Ravnina farm (560,000 ha, Karakum desert, Turkmenistan, 2002/03). (From Gintzburger *et al.*, 2009)

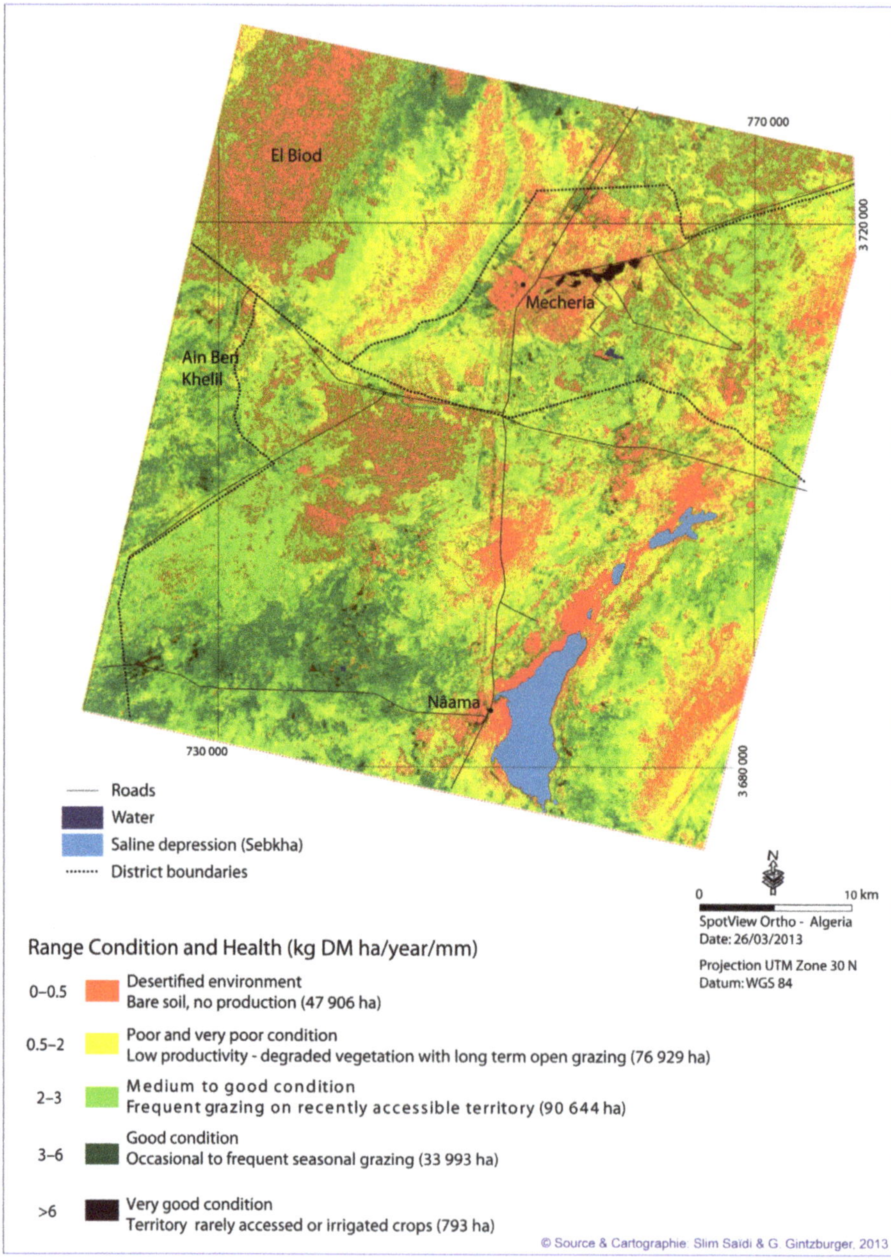

Map 15.2. The SRUEI map from the Mecheria – Naama region (250,000 ha, Algeria. (From Saïdi and Gintzburger, 2013)

may be able to identify the causes of vegetation status and changes. All these methods only define and geographically locate the current rangeland condition. In the existing context of man-made desertification and potential climate change, range managers and decision makers may use the SRUEI to pinpoint priority areas where desertification occurs and where actions on rangeland regeneration may be necessary. Only the range managers familiar with the region surveyed and its rangeland history may usefully interpret the SRUEI map and possibly explain the reasons behind the differences in SRUEI between sites and offer solutions acceptable to communities.

Assessing the vegetation cover and land use with the TSAVI in the Bir-el-Ghanem region 1986 and 2022

Another way to assess the trends in vegetation cover on Libyan rangelands is to compare the spectral signature of satellite images from the same region at different dates. We give here an example of the vegetation cover and land-use changes affecting about 330,000 ha from the Bir-el-Ghanem area. This area has seen drastic land-use changes from the traditional grazing system all over, up to the mid-1970s when the government rangeland projects (Gintzburger, 1986) at BGI, BGII and BGIII and the Australian ley farming system were implemented by the Jeffara Plain Authority at wadi Abu-Shayba and wadi Ethel north of Bir Ayyad (see Chapter 9). These government projects disrupted the traditional grazing system by limiting access to customary rangeland and rainfed cropping areas. Progressively, by 1984–1986, these projects were abandoned due to new land-use policies. It opened the way to small private farms encroaching with their small ruminant flocks and drilling numerous legal and illegal wells for irrigation. The large BGI grazing projects site north of the Aziziya-Bir-el-Ghanem road gave way to a multitude of irrigated small farms invading rangelands south from Sorman to Zawia. The good flat rangelands from wadi Abu-Shayba, turned into rainfed cereal-sheep production from 1974 to 1984, were soon converted to small irrigation farms. The dry wadi Ethel site was partly abandoned and reverted to rangelands. The pressure on rangelands all over the Bir-el-Ghanem region evolved from being well controlled up to the mid-1980s, to fully open to private flocks up to 2011 when the combats around Bir-el-Ghanem made grazing rangelands too dangerous, with many rangeland areas simply deserted except those close to settlements.

The 1986 and 2022 satellite images processed show these changes in rangeland vegetation cover and land use in the Bir-el-Ghanem region.

We selected two satellite images at the spring peak standing biomass: a Landsat 5 TM dated 16 March 1986 and a Landsat 8 OLI dated 29 March 2022 Multispectral Sensor, with high spatial resolution (30 m), were selected and processed to estimate the TSAVI values (Baret *et al.*, 1989), computed from the NIR and the red channels brightness of the image.

The Normalized Difference Vegetation Index (NDVI), often used on arid rangelands (Al-Bukhari, 2019) and by most UN organizations, is not an appropriate vegetation index for arid-zone vegetation due to signal alteration from the bare soil reflectance. The NDVI wrongly returns a bare-soil index and does not differentiate between bare soil and vegetation when the vegetation cover is less than 20–25% (Choudhury and Tucker, 1987; Diouf and Lambin, 2001; Gintzburger *et al.*, 2009; Kennedy, 1989; Saïdi and Gintzburger, 2013; Vonder and Clevers, 1998), a common feature of arid Libyan rangelands.

Preliminary satellite image corrections

The selected Bir-el-Ghanem satellite images were corrected for geometric, radiometric and atmospheric discrepancies before assessing the vegetation cover. These pre-processing steps of satellite imagery are essential to restore or compensate image geometric distortions, eliminate or reduce atmosphere effects on the spectral reflectance values of images (Bonn and Rochon, 1992; Kergomard, 2000). Geometric correction of satellite images involves modelling the relationship between the image and ground co-ordinate systems. Data pixels were linked to exact ground locations by precisely identifying ground control points (GCP) at road intersections or wadi junctions. The root mean square error (RMSE) of all points is less than 0.3 pixel with field information. The spectral reflectance depends on the degree of illumination of the objects. Spectral reflectance is unitless and expressed as a decimal fraction between 0 and 1 or between 0% and 100%. Converting the spectral radiance to exo-atmospheric (Top-of-Atmosphere = TOA) reflectance in-band planetary albedo is performed (Fig. 15.3, step 1). The TOA reflectance does not take into consideration the reflectance at the soil surface. It only removes atmospheric effects such as aerial dust from satellite images and is an important pre-processing step of satellite imagery (Fig. 15.3, step 2). This operation is important when comparing multitemporal images in arid zones. For increased precision, the ATCOR (ATmospheric CORrection) Program, available in ERDAS Imagine software, is used to determine true surface reflectance values measured on the ground. The radiative transfer in ATCOR is based on the MODTRAN4 code (Berk *et al.*, 1998, 2003).

Considering the soil reflectance, an important element in arid area where bare soil is common, we computed the TSAVI for each field centroid to discriminate the soil reflectance from the vegetation cover. From experience, the TSAVI is more appropriate in differentiating bare sandy soils from vegetation than the PVI (Perpendicular Vegetation Index). The TSAVI assumes that the perpendicular distance of the pixel from the soil line is linearly related to the vegetation cover (Richardson and Wiegand, 1977) and herbage mass. It is computed as follows:

$$TSAVI = \frac{a(\mathrm{NIR} - a\mathrm{R} - b)}{\mathrm{R} + a\mathrm{NIR} - ab + 0.08(1 + a^2)}$$

where a = the slope of the soil line and b = intercept point of the soil line.

As a result, the further a given pixel is distant from the soil line, the higher is its TSAVI index value and the vegetation cover. When TSAVI is greater than 0, there is vegetation cover over the pixel; when TSAVI equals 0 there is bare soil on the pixel; when TSAVI is inferior to zero, it indicates that the pixel is either free water or with very low mineral content or low photosynthetic activity (Baret *et al.*, 1989).

Extracting the TSAVI values from the 1986 and 2022 satellite imagery

To check the uniformity between the atmospheric profile of the two images from March 1986 and March 2022, several objects common to the two images were defined. The points chosen are mainly areas without vegetation such as invariants of dark shades (rocky outcrops or tarmac surfaces) and light shades (bare soils or bare dune areas). In this type of arid landscape, the selected vegetation index must be insensitive to the contamination of the soil signal on the vegetation's signal especially when the bare soils are dominant. The TSAVI (Baret *et al.*, 1989) was chosen as it has meaningful dynamics with values close to 0 on bare soils. The same TSAVI radiometric thresholds were applied to identify the status of the vegetation cover (Table 15.1) on the March 2022 and March 1986 satellite images (Fig. 15.4). The geometric shape of cropped areas was identified with the TSAVI thresholds completed with photointerpretation.

Preparing the synthetic image of the vegetation cover evolution 1986–2022

An algorithm with 16 trajectories was developed and applied to synthesize and produce the vegetation cover trend image from 1986 to 2022 (Fig. 15.4).

Interpretation of the two satellite images and the trend synthesis

The rainfall moving average (four years) from the Bir-el-Ghanem meteorological station (1980–2022, Fig. 15.5) may partly explain the vegetation cover changes. It displays four periods:

- 1981 to 1994 with an average of 197 mm and short droughts from 1987 to 1993;
- a long dry period between 1995 and 2009 receiving only 176 mm on average;
- a wetter period from 2010 to 2018 with an average 209 mm; and
- beyond 2018 – fluctuations resume within the normalized interannual rainfall.

The statistical analysis of the Bir-el-Ghanem rainfall data between 1980 and 2022 (see Chapter 3) applying the Z-score test does not detect a significant trend. The annual rainfall irregularities between 1980 and 2022 fall within the normal interannual variability of the lower arid bioclimate. It remains that the better annual rainfall period (2010–2018) may have contributed helping the BGI rangeland recovery

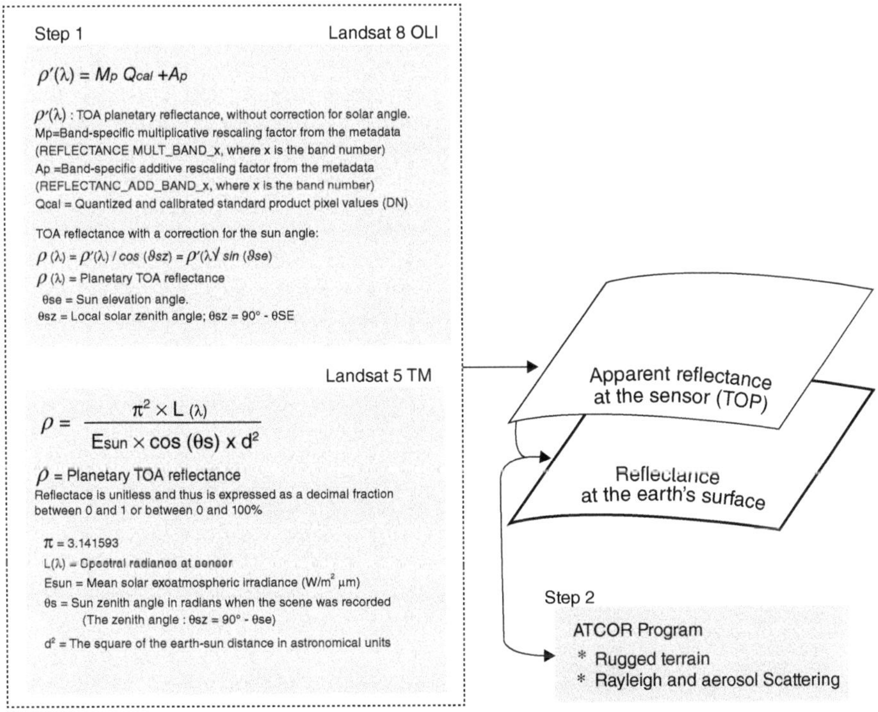

Fig. 15.3. Step-by-step process for atmospheric correction on satellite imagery, 1986 and 2022.

Table 15.1. TSAVI radiometric threshold and code applied to the corrected satellite images.

Code	Interval values TSAVI (μm)	Vegetation cover interpretation
1	[–0.462 to –0.014]	Free water
2	[–0.014 to +0.015]	Bare soil
3	[+0.015 to +0.045]	Low
4	[+0045 to 0.166]	Medium
5	[0.166 to 1]	High

north of the Aziziya Bir el Ghanem road (see image March 2022) but does not explain the degradation on the eastern rangelands from wadi Ethel to south Sorman.

Overall, the March 1986 interpreted image shows abundant rangeland vegetation to the west of Bir-el-Ghanem extending from Bir Ayyad to Jenoub Sorman with positive TSAVI-86 values. This vegetation cover totally disappeared by 2022, possibly resulting from postwar resumption of heavy rangeland grazing from the livestock owners from Qasr-el-Haj/Shakshuk, firming up their customary grazing rights. The wadi Ethel cropping and the Bir Ayyad projects barely survived by 1986. The good and high vegetation cover of protected BGI project (Gintzburger, 1986) in the mid-1970s totally disappeared by 1986 due to irrigated private farms and livestock encroachments. The Wadi Abu-Shayba rainfed ley farming project and the Bir-el-Ghanem sheep project with good *arfej* rangelands (Gintzburger and Alidrissi, 1976) were abandoned by 1984 and turned into a multitude of small irrigated private farms building up their own flocks.

On the 2022 image, the vegetation cover from BGI area and all around wadi Abu-Shayba appears improving conceivably due to war post-2011 and ensuing unrest limiting flocks' movements. A heavily degraded strip appears along the western boundary of wadi Abu-Shayba likely due to safe grazing from flocks from the nearby wadi Bu-Shaba private farms. The rangeland north-west and south-east close to Bir-el-Ghanem town has been heavily degraded. To the west, by 2022, the wadi Ethel project appears to have regained control of the land (starting from 2009) with new crops and tree plantations and 2200 ha rangelands under effective exclosure. The rest of the areas west of wadi Ethel – Bir Ayad to south Sorman – Saniet-el-hood display a heavy degradation and trend towards bare soil confirmed on the synthetic map 1986–2022.

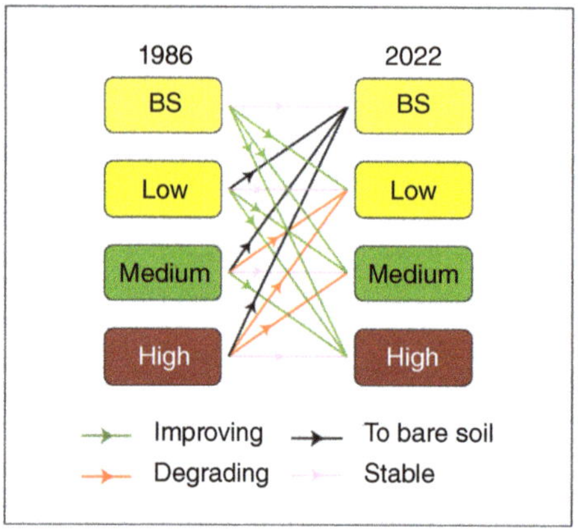

Fig. 15.4. The algorithm synthetizing the vegetation cover trend from the two images 1986 and 2022.

In short, the synthetic vegetation cover changes image 1986–2022 (Fig. 15.6) shows that about 96,000 ha turned to bare soil and became desertified, some 14,000 ha became degraded, while 88,000 ha appeared to have unexpectedly improved. The cropped areas dropped from about 56,000 ha in 1986 to 41,000 ha in 2022. It could be due to the ongoing unrest post the 2011 war affecting most of the Bir-el-Ghanem region. Only field surveys and ground-truthing performed at the two dates may validate. It is worth noting that the spatial vegetation cover changes detected in these areas over the past 36 years are affecting different zones. Some rangelands are becoming accessible to flocks and being degraded while some others are deserted and recovering. The reasons for these changes remain hypothetical and could only be explained by shepherds and farmers knowing the local land-use history.

The vegetation status figured as 'stable' on the synthetic image 1986–2022 means that it has not changed; hence, it could be either unchanged bare soil, low vegetation cover or high cover. It would require ground-truthing to confirm its actual status.

The Australian VegMachine®: Long-term mapping monitoring and accurately detecting rangeland trends

VegMachine® is a monitoring software package based on data-sets of satellite time-series imagery.

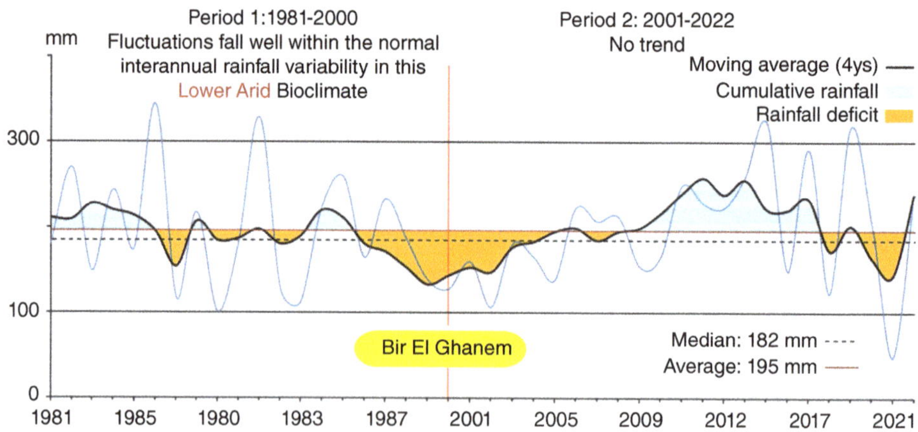

Fig. 15.5. Annual rainfall distribution at Bir-el-Ghanem station (1981–2022).

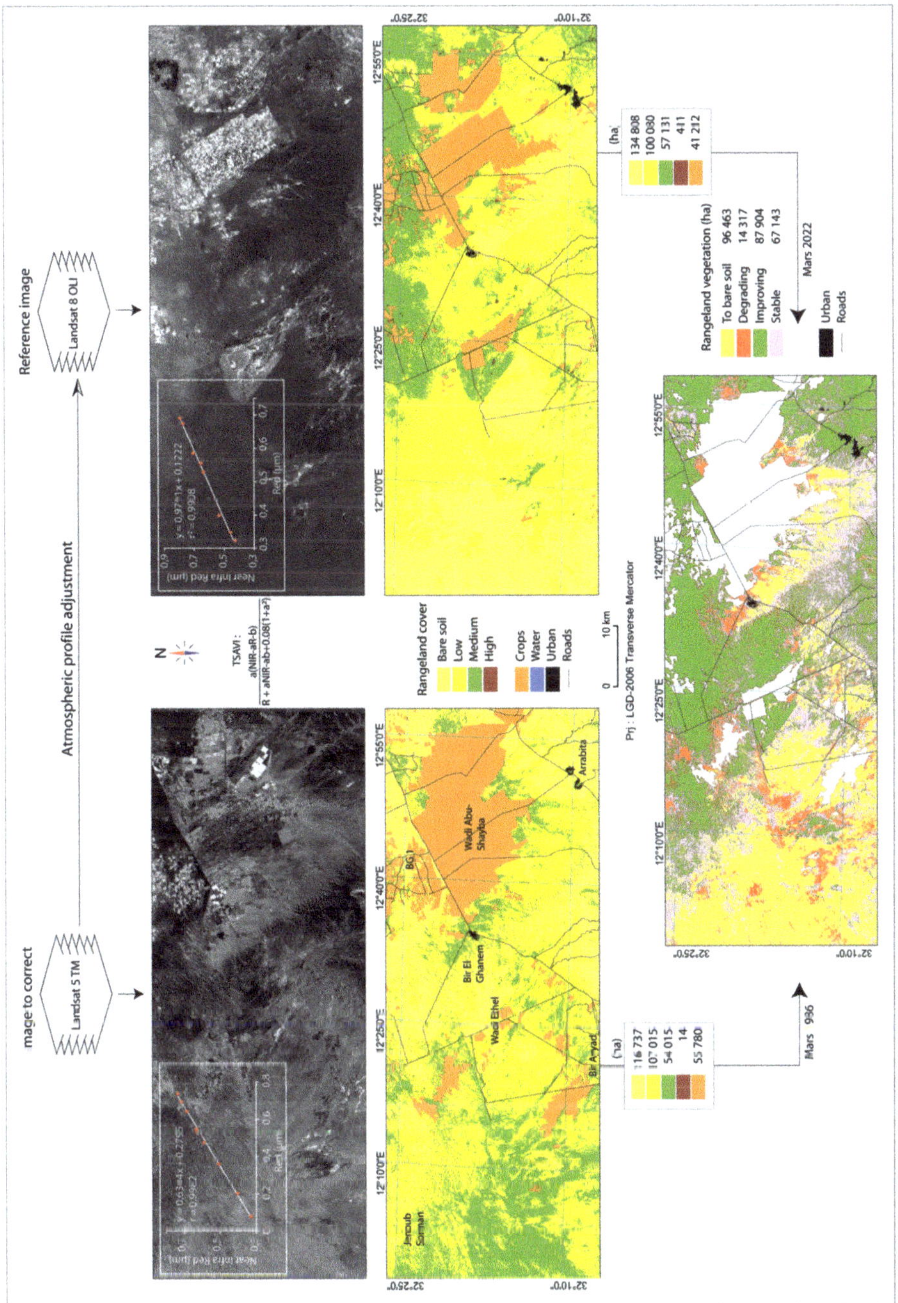

Fig. 15.6. Vegetation cover and land-use trends, and synthesis in the Bir-el-Ghanem region between March 1986 and March 2022.

Developed by CSIRO for commercial grazing operations in Northern Australia, southern Queensland and northern western Australia. It assists graziers to identify, locate, report and manage rangeland ecological changes (degrading, improving, stable) over time, at pixel level and over large areas (Karfs *et al.*, 2004; Wallace *et al.*, 2006; Wallace & Thomas, 1999).

VegMachine® tested in Libya

The test of VegMachine in Libya was supported by the Libyan Rangeland Development Project (Suani) to implement, demonstrate and evaluate the software in a Libyan arid environment at the wadi Sassu protected project (Wallace, 2009). The study area was chosen from the Landsat 5 scene (USGS and Eurimage) Path 187 Row 38 with ten July/August dates between 1998 and 2008. The summer scenes were selected to provide the greatest contrast between perennial vegetation cover and soil, with minimal interference from unpredictable annual vegetation, hence allowing monitoring long-term perennial vegetation cover and bare soil fluctuations. All Landsat images were geo-corrected and calibrated to a common spectral base and resampled to a 30 m pixel size.

The vegetation index used by VegMachine

A unique and robust cover index was calculated out of sophisticated statistical analysis derived from satellite imagery contrast between vegetation cover and bare soil on the Sassu rangelands and surrounding areas. The definition of such a specific index is based on regression and discriminant analysis using ground sites where the vegetation parameter of interest has been measured or estimated (Lehmann *et al.*, 2013; Wallace *et al.*, 2006). Therefore, the specific 'Sassu' VegMachine cover index is specific to the area studied with sandy-loam light-coloured soils. It has nothing to do either with NDVI, TSAVI, PVI and other published vegetation indices and would not be suitable on dark rocky soils as, for example, in the wadi Nefad region further south. The VegMachine cover index has to be recalculated and redefined when exploring larger heterogeneous areas with different vegetation and soil-colour types. It does not disclose the actual percentage of vegetation cover, but simply a relative cover index value inside the mapped areas.

The VegMachine® map and facilities

The legend of the colour-coded map interpreted and produced by VegMachine® is as follows:

Red = large negative trends indicating a rangeland cover degradation over time
Orange = small negative trend
Grey = Stable, no detected trend
Blue = Small positive trend
Green = Large positive trend indicating a rangeland cover improvement

By drawing and selecting polygons drawn on the map, the VegMachine software displays a graph of the vegetation cover index values and trends over years with different colours (Fig. 15.7). The software allows displaying GPS-located photos of the rangeland condition.

One should note that the grey-stable condition does not indicate that the rangeland is in good or degraded condition; it simply indicates that no significant trend is detected. It means that the grey area could be anything from degraded to very healthy. Only ground-truthing with biomass, LIM or LFAnalysis performed on GPS-located sites by rangeland managers or scientists could determine the local rangeland status of these areas.

Conclusions

Rangeland degradation in arid zones and especially in Libya is common and well recognized. Many areas are ultimately desertified. It can be ascertained and accurately located with different methods from basic visual assessment to qualitative surveys and/or quantitative measurements that are site- and date-specific. These field methods are time-consuming and cannot easily be scaled up to large areas rendering long-term monitoring close to impossible. The availability of satellite imagery and GIS software opened innovative monitoring approaches. Satellite imagery processing is a complex procedure, especially on arid-zone vegetation due to the importance of atmospheric dust and bare-soil parasitizing the spectral signals at ground level.

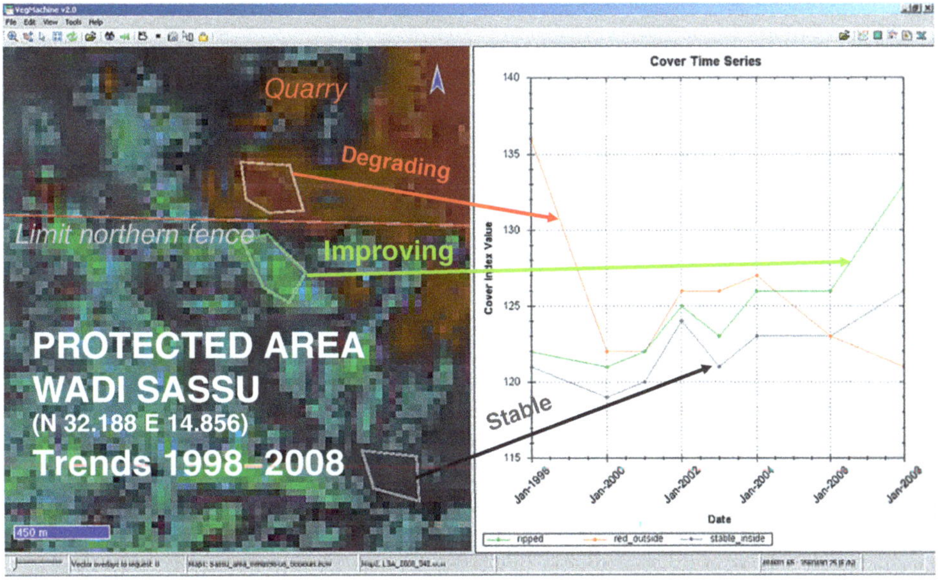

Fig. 15.7. The CSIRO VegMachine® screen display of the cover trends image and graph (1998–2008) on wadi Sassu selected polygons. (From J. Wallace)

Once these corrections are performed, the selection of an appropriate vegetation index such as the TSAVI, the PVI or similar ones eliminating the bare-soil reflectance is paramount to assess the vegetation presence and cover. The easily available NDVI is not suitable for arid-zone vegetation as it returns bare-soil results when the vegetation cover is less than 20–25%, hence exaggerating the extent of degradation and desertification. Once the proper vegetation index is used, it is possible to process satellite imagery and map the status of the vegetation cover during a unique season or monitor long-term vegetation trends either as stable, degrading or recovering using long-term series. Satellite image processing is a fast-evolving technology, and its use must be regularly updated with new commercial software and modules. Long-term monitoring mapping does not give any indication of the reasons why some areas are degraded or recovering. The mapping results must always be checked in the field. Information and explanations of the current vegetation conditions and trends must be collected from shepherds and farmers knowing the climatic and management history of their rangelands.

Rangeland degradation is not definitive and can be reversed with different methods.

References

Al-Bukhari, A.A. (2019) The use of remote sensing and GIS for monitoring rangeland degradation in Libya. PhD dissertation, Cranfield University, Bedford, UK.

Aronson, J., Floret, C., Le Floc'h, E., Ovalle, C. and Pontanier, R. (1995) Restauration et réhabilitation des écosystèmes dégradés en zones arides et semi-arides. Le vocabulaire et les concepts. In: Pontanier, R. (ed.) *L'homme peut-il refaire ce qu'il a défait?* John Libbey Eurotext, Paris, pp. 11–29.

Baret, F., Guyot, G. and Major, D.J. (1989) *TSAVI: A Vegetation Index which Minimizes Soil Brightness Effects on LAI and APAR Estimation. 12th Canadian Symposium on Remote Sensing Geoscience and Remote Sensing Symposium* 3, 1355–1358. DOI: 10.1109/IGARSS.1989.576128

Behnke, R.H., Scoones, I. and Kerven, C. (eds) (1993) *Range Ecology at Disequilibrium: New Models of Natural Variability and Pastoral Adaptation in African Savannas.* Overseas Development Institute, London.

Berk, A., Bernstein, L.S., Anderson, G.P., Acharya, P.K., Robertson, D.C., Chetwynd, J.H. and Adler-Golden, S.M. (1998) MODTRAN Cloud and Multiple Scattering Upgrades with Application to AVIRIS. *Remote Sensing of Environment* 65(3), 367–375. DOI: 10.1016/S0034-4257(98)00045-5

Berk, A., Anderson, G.P., Acharya, P.K., Hoke, M.L., Chetwynd, J.H., Bernstein, L.S., Shettle, E.P., Matthew, M.W. and Adler-Golden, S.M. (2003) *MODTRAN4 Version 3 Revision 1 User's Manual.* Air Force Research Laboratory, Hanscom, Massachusetts.

Beutel, T.S., Karfs, R.A., Bull, A.L., Peel, L. and Wallace, J.F. (2005) VegMachine® – putting pastoralists in the picture. *Agricultural Science* 17(3), 18–20.

Bonn, F.J. and Rochon, G. (1992) *Précis de Télédétection. Volume 1: Principes et Méthodes.* Presses de l'Université du Québec, Sillery, Québec.

Brun, J.M. and Box, T.W. (1963) A comparison of line intercepts and random point frames for sampling desert shrub vegetation. *Journal of Range Management* 16(1), 21–25. DOI: 10.2307/3895032

Canfield, R.H. (1941) Application of the line interception method in sampling range vegetation. *Journal of Forestry* 39(4), 388–394. DOI: 10.1093/jof/39.4.388

Choudhury, B.J. and Tucker, C.J. (1987) Satellite observed seasonal and inter-annual variation of vegetation over the Kalahari, the Great Victoria Desert, and the Great Sandy Desert: 1979–1984. *Remote Sensing of Environment* 23(2), 233–241. DOI: 10.1016/0034-4257(87)90039-3

Clements, F.E. (1916) *Plant Succession: An Analysis of the Development of Vegetation.* Carnegie Institution of Washington.

De Agostini, E. (1917a) *La popolazioni della Tripolitania, Governo della Tripolitania, Parte I.* Tipografia Pirrotta et Breschiano, Tripoli, Libya. Available at: http://archive.org/details/DeAgostiniPopolazioniDellaTripolitaniaParteUno (accessed 17 August 2024).

De Agostini, E. (1917b)= *La popolazioni della Tripolitania, Governo della Tripolitania, Parte II.* Tipografia Pirrotta et Breschiano, Tripoli, Libya.

Despois, J. (1935) *Le Djebel Nefousa (Tripolitaine), Etude Géographique*. Larousse, Paris.

Diouf, A. and Lambin, E.F. (2001) Monitoring land-cover changes in semi-arid regions: remote sensing data and field observations in the Ferlo, Senegal. *Journal of Arid Environments* 48(2), 129–148. DOI: 10.1006/jare.2000.0744

El-Barasi, Y.M., Saaed, M.W.B. and Al Tajoury, R.O. (2013) Land deterioration of a semidesert grazing area in the north-eastern zone of Libya (Cyrenaica). *Journal of Environmental Science and Engineering B* 2, 357–373.

El-Tantawi, A.M.M. (2005) Climate change in Libya and desertification of Jifara Plain using geographical information system and remote sensing techniques. PhD dissertation, Johannes Gutenberg Universität, Mainz, Germany.

Emberger, L. and Godron, M. (1968) *Code pour le Relevé Méthodique de la Végétation et du Milieu: Principes et Transcription sur Cartes Perforées.* Editions du Centre National de la Recherche Scientifique, Paris.

Floret, C. (1974) *Carte phytoécologique et des ressources pastorales d'Hadj Gacem.* (FAO Project TUN/69/001). Institut Nationale Recherche Agronomique, Tunisia.

Floret, C. and Le Floc'h, E. (1973) Production, sensibilité et évolution de la végétation et du milieu en Tunisie présaharienne, Projet – Parcours Sud, Doc. *CEPE* 71.

Floret, C., Le Floc'h, E. and Pontanier, R. (1976) Carte de la sensibilité à la désertisation : Tunisie centrale et méridionale (processus de dégradation en cours des sols et de la végétation) échelle 1:1.000.000. *Sols de Tunisie* 8, 1–68.

Floret, C., Le Floc'h, E., Pontanier, R. and Romane, F. (1977) *Contribution à l'étude de Cas sur la Désertification Région d'Oglat Merteba, Tunisie,* pp. 3–93.

Floret, C., Le Floc'h, E. and Romane, F. (1981) Dynamique des systèmes écologiques de la zone aride: application à l'aménagement sur des bases écologiques d'une zone de la Tunisie Présaharienne. *Acta Œcologica: Œcologia Applicata* 2(3), 195–214.

Floret, C., Le Floc'h, É. and Pontanier, R. (1986) La désertisation en Tunisie présaharienne. *Revue des Mondes Musulmans et de la Méditerranée* 41(1), 291–326. DOI: 10.3406/remmm.1986.2464

Forskål, P. (1775) *Flora Aegyptiaco-Arabica : sive descriptiones plantarum quas per Aegyptum inferiorem et Arabiam felicem.* Hauniae, Ex officina Mlleri.

Gamoun, M. (2016) Rain use efficiency, primary production and rainfall relationships in desert rangelands of Tunisia. *Land Degradation & Development* 27(3), 738–747. DOI: 10.1002/ldr.2418

Gintzburger, G. (1986) Seasonal variation in above-ground annual and perennial phytomass of an arid rangeland in Libya. *Journal of Range Management* 39(4), 348–353.

Gintzburger, G. and Alidrissi, M. (1976) *Range study on Jeffara Plain: Sheep Project at Bir-El-Ghanem* (No. 83/76). FAO and Agricultural Research Center, Tripoli.

Gintzburger, G. and Saïdi, S. (2009) From inventory to monitoring in semi-arid and arid rangelands. In: Squires, V.R. (ed.) *Range and Animal Sciences and Resources Management. Vol. 2: Encyclopedia of Life Support Systems*. EOLSS.

Gintzburger, G., Toderich, K., Mardonov, B. and Mahmudov, M. (2003) *Rangelands of the Arid and Semi-Arid zones in Uzbekistan.* CIRAD-ICARDA, Montpellier, France.

Gintzburger, G., Saïdi, S. and Soti, V. (2009) Rangelands of the Ravnina region in the Kara-Kum Desert of Turkmenistan: current condition and utilisation. In: *Sustainable Agriculture in Central Asia and the Caucasus, N° 2. CGIAR PFU.* Tashkent, Uzbekistan.

Gounot, M. (1969) *Méthodes D'étude Quantitative de la Végétation.* Editorial Masson, Paris.

Heady, H.F. (1975) *Rangeland Management.* McGraw-Hill, New York.

Hein, L., de Ridder, N., Hiernaux, P., Leemans, R., de Wit, A. and Schaepman, M. (2011) Desertification in the Sahel: towards better accounting for ecosystem dynamics in the interpretation of remote sensing images. *Journal of Arid Environments* 75(11), 1164–1172. DOI: 10.1016/j.jaridenv.2011.05.002

Holling, C.S. (1973) Resilience and stability of ecological systems. *Annual Review of Ecology and Systematics* 4, 1–23.

Huxman, T.E., Smith, M.D., Fay, P.A., Knapp, A.K., Shaw, M.R. *et al.* (2004) Convergence across biomes to a common rain-use efficiency. *Nature* 429(6992), 651–654. DOI: 10.1038/nature02561

Ibn Khaldūn (1852) *Histoire des Berbères et des Dynasties Musulmanes de l'Afrique Septentrionale* (trans. B. de Slane). Vol. 1. Imprimerie du Gouvernement, Algiers.

Karfs, R.A., Daly, C., Beutel, T.S., Peel, L. and Wallace, J.F. (2004) VegMachine – delivering monitoring information to Northern Australia's pastoral industry. In: *Proceedings of the 12th Australasian Remote Sensing and Photogrammetry Conference, Fremantle, Australia*.

Kennedy, P.J. (1989) Monitoring the phenology of Tunisian grazing lands. *International Journal of Remote Sensing* 10(4–5), 835–845. DOI: 10.1080/01431168908903924

Kergomard, C. (2000) Pratique des corrections atmosphériques en télédétection: utilisation du logiciel 5S-PC. *Cybergeo: European Journal of Geography*. DOI: 10.4000/cybergeo.1679

Le Houérou, H.-N. (1959) Recherches écologiques et floristiques sur la végétation de la Tunisie méridionale. PhD dissertation. Université Alger Institut de Recherches Sahariennes, Algiers.

Le Houérou, H.-N. (1968) Deux espèces du genre *Erodium* L.'Hérit découvertes en Libye. *Bulletin de la Société Botanique de France* 115, 591–600.

Le Houérou, H.-N. (1969) *La Végétation de la Tunisie Steppique (Avec Référence aux Végétations Analogues d'Algérie, de Libye et du Maroc)* 1–2, 42–45. Annales de l'Institut National de la Recherche Agronomique de Tunisie, Tunis.

Le Houérou, H.-N. (1982) *The Impact of Climate on Pastoralism* (UNTF/Lib 018, Technical Paper No 42). FAO and Agricultural Research Center, Tripoli.

Le Houérou, H.-N. (1984) Rain use efficiency: a unifying concept in arid-land ecology. *Journal of Arid Environments* 7(3), 213–247. DOI: 10.1016/S0140-1963(18)31362-4

Le Houérou, H.-N. (1993) Changements climatiques et désertisation. *Sécheresse* 4(2), 95–111.

Le Houérou, H.-N. (1996) The role of cacti (*Opuntia* spp.) in erosion control, land reclamation, rehabilitation and agricultural development in the Mediterranean Basin. *Journal of Arid Environments* 33(2), 135–159. DOI: 10.1006/jare.1996.0053

Le Houérou, H.-N. (2005) *The Isoclimatic Mediterranean Biomes: Bioclimatology, Diversity and Phytogeography* (Vol. 2). Copymania, Montpellier, France.

Le Houérou, H.-N. and Rapp, A. (1976) Can desertification be stopped? Summary and recommendations. *Ecological Bulletins* 24, 237–240.

Lehmann, E.A., Wallace, J.F., Caccetta, P.A., Furby, S.L. and Zdunic, K. (2013) Forest cover trends from time series Landsat data for the Australian continent. *International Journal of Applied Earth Observation and Geoinformation* 21, 453–462. DOI: 10.1016/j.jag.2012.06.005

Ludwig, J.A., Tongway, D.J., Bastin, G.N. and James, C.D. (2004) Monitoring ecological indicators of rangeland functional integrity and their relation to biodiversity at local to regional scales. *Austral Ecology* 29(1), 108–120. DOI: 10.1111/j.1442-9993.2004.01349.x

Mahmoud, A.M.A., Hasmadi, I.M., Alias, M.S. and Azani, A.M. (2014) Reviews of Landscape Function Analysis (LFA) applications in rangeland ecosystems and its links with vegetation indices (VI's). *World Applied Sciences Journal* 32(5), 986–991.

Mahmoud, A.M.A., Hasmadi, I.M., Alias, M.S. and Azani, A.M. (2016) Rangeland degradation assessment in the south slope of the Al-Jabal Al-Akhdar, Northeast Libya using remote sensing technology. *Journal of Rangeland Science* 6(1), 73–80.

Mahmoud, A.M.A., Hasmadi, I.M., Alias, M.S. and Azani, A.M. (2021) Rangeland restoration analysis on the south slope of Al-Jabal Al Akhdar, Northeast Libya. *Journal of Rangeland Science* 11(1), 26–43.

Manière, R. and Chamignon, C. (1986) Cartographie de l'occupation des terres en zone aride méditerranéenne par télédétection spatiale: exemple d'application sur les Hauts-Plateaux sud Oranais: Méchéria au 1/200 000. *Ecologia Mediterranea* 12(1), 159–185. DOI: 10.3406/ecmed.1986.1123

Nahal, I. (2004) *La Desertification Dans le Monde: Causes, Processus, Conséquences, Lutte*. Editions l'Harmattan, Paris.

National Research Council (1994) *Rangeland Health: New Methods to Classify, Inventory, and Monitor Rangelands.* National Academies Press, Washington, DC. DOI: 10.17226/2212

Pontanier, R., M'Hiri, A., Akrimi, N., Aronson, J., Le Floc'h, E. and Kallala, A. (eds) (1995). *L'homme peut-il refaire ce qu'il a défait?* John Libbey Eurotext, Paris.

Pringle, H.J.R., Watson, I.W. and Tinley, K.L. (2006) Landscape improvement, or ongoing degradation – reconciling apparent contradictions from the arid rangelands of Western Australia. *Landscape Ecology* 21(8), 1267–1279. DOI: 10.1007/s10980-006-0015-x

Raunkiaer, C. (1934) *Life Forms and Statistical Plant Geography.* Clarendon Press, Oxford, UK.

Richardson, A.J. and Wiegand, C.L. (1977) Distinguishing vegetation from soil background information. *Photogrammetric Engineering and Remote Sensing* 43, 1541–1552.

Saaed, M.W.B. (2008) Study of vegetation and land degradation factors in Daphna zone (north-eastern part of Libyan coast-Tobruk). MSc thesis, Department of Environmental and Engineering Science, Libyan Academy of Postgraduate Studies, Benghazi, Libya.

Saaed, M.W.B., Ali, A.B., El-Barasi, Y.M. and Rahil, R.O. (2022) Potential and perspectives for establishing protected areas in El-Jabal El-Akhdar region, northeast Libya: an overview and example from 'El-Abyar area'. *Journal of Degraded and Mining Lands Management* 9(3), 3453–3464. DOI: 10.15243/jdmlm.2022.093.3453

Saïdi, S. and Gintzburger, G. (2013) A spatial desertification indicator for Mediterranean arid rangelands: a case study in Algeria. *The Rangeland Journal* 35(1), 47–62. DOI: 10.1071/RJ12021

Saïdi, S., Gintzburger, G., Bonnet, P., Daoud, I. and Alary, V. (2016) GIS-modelling of land-use trends: impact of drought in the Naghamish Basin (North-Western Egypt). *The Rangeland Journal* 38(6), 605–618. DOI: 10.1071/RJ16062

Scoones, I. (ed.) (1994) *Living with Uncertainty: New Directions in Pastoral Development in Africa.* Intermediate Technology Publications, London.

Stoddart, L.A., Smith, A.D. and Box, T.D. (1975) *Range Management*, 3rd edn. McGraw-Hill College, Blacklick, Ohio.

Telahigue, T., Floret, C. and Le Floc'h, E. (1987) Succession post-culturale en zone aride de Tunisie. *Acta Oecologica: Oecologia Plantarum* 8(22), 45–58.

Tongway, D. and Hindley, N. (2000) Assessing and monitoring desertification with soil indicators. In: Arnalds, O. and Archer, S. (eds) *Rangeland Desertification*. Springer, Dordrecht, The Netherlands. DOI: 10.1007/978-94-015-9602-2_8

UNCED (1992) *Earth Summit Agenda 21: Programme of Action for Sustainable Development.* UN Department of Public Information, New York.

Vetter, S. (2005) Rangelands at equilibrium and non-equilibrium: recent developments in the debate. *Journal of Arid Environments* 62(2), 321–341. DOI: 10.1016/j.jaridenv.2004.11.015

Von Wehrden, H., Hanspach, J., Ronnenberg, K. and Wesche, K. (2010) Inter-annual rainfall variability in Central Asia – a contribution to the discussion on the importance of environmental stochasticity in drylands. *Journal of Arid Environments* 74(10), 1212–1215. DOI: 10.1016/j.jaridenv.2010.03.011

Vonder, O.W. and Clevers, J.G.P.W. (1998) Multisensor remote sensing capabilities: application of present and future optical remote satellite sensors. Wageningen Agricultural University.

Wallace, J.F. (2009) *Implementation, field testing and interpretation of the VegMachine® system in the Libyan rangeland environment at Wadi Sassu.* Report to the Rangeland Development Project (Suani).

Wallace, J.F. and Thomas, P. (1999) Monitoring and summarizing rangeland changes using sequences of Landsat data. In: Eldridge, D. and Freudenberger, D. (eds) *People and Rangelands: Building the Future. Proceedings of the VI International Rangeland Congress, Townsville, Australia*, pp. 19–23, 777–778.

Wallace, J.F., Caccetta, P.A. and Kiiveri, H.T. (2004) Recent developments in analysis of spatial and temporal data for landscape qualities and monitoring. *Austral Ecology* 29(1), 100–107. DOI: 10.1111/j.1442-9993.2004.01356.x

Wallace, J.F., Behn, G. and Furby, S. (2006) Vegetation condition assessment and monitoring from sequences of satellite imagery. *Ecological Management and Restoration* 7(s1), 31–36. DOI: 10.1111/j.1442-8903.2006.00289.x

Watson, I.W., Novelly, P.E. and Thomas, P.W.E. (2007) Monitoring changes in pastoral rangelands – the Western Australian Rangeland Monitoring System (WARMS). *The Rangeland Journal* 29(2), 191–205. DOI: 10.1071/RJ07008

Winslow, M., Shapiro, B.I., Thomas, R. and Shetty, S.V.R. (2004) *Desertification, Drought, Poverty and Agriculture: Research Lessons and Opportunities.* International Center for Agricultural Research in the Dry Areas (ICARDA), the International Crops Research Institute for the Semi-Arid Tropics (ICRISAT), and the UNCCD Global Mechanism (GM), Aleppo, Patancheru, India, and Rome, Italy.

16

Recovery, Restoration and Rehabilitation of Degraded Rangeland

Abstract

Rangeland restoration and rehabilitation techniques were successfully tested on tens of thousands of hectares in North Africa and Northern Libya. Long-term rangeland passive protection and deferred grazing remain the cheapest and most efficient techniques well understood by livestock owners. Active means of rehabilitation, such as reseeding and multipurpose shrub plantations combined with water harvesting techniques and phosphate fertilization may be necessary. Reseeding with native rangeland species is a valid option to be explored, especially with annual *Medicago* sp. (medics) pods ubiquitous on arid Libyan soils and rainfall zones. However, the main constraint is medics' pods supply in quantity. Reallocating the best rangelands to fodder shrubs plantations is complicated, labour-intensive, and expensive, and is neither valued nor accepted by rangeland communities. Rangeland protection, multipurpose shrubs plantations, reseeding, low-cost water harvesting techniques, and minimal fertilization produced satisfactory technical results on Libyan rangelands. They remain doomed to repeated failure unless supported by local communities.

Introduction

Libyan rangelands are in poor condition for many reasons already known (El-Barasi *et al.*, 2013; El-Barasi and Saaed, 2013; Gintzburger and Bayoumi, 1977; Mahmoud *et al.*, 2016; Saaed, 2008; Saaed *et al.*, 2022; Shawesh *et al.*, 1996). The main culprit is man, by relentless overgrazing, cutting and harvesting fuelwood, land clearing for hazardous rainfed cereal cultivation, drilling new wells and cisterns allowing grazing in remote areas, providing endless supplementary feed and water trucking to flocks (Chatty, 2013) in the middle of nowhere, lack of rangeland legislation and policing, unclear land tenure, forestry legislation being ignored, lack of legislation controlling drilling of new wells, ever-increasing amounts of feed imported, etc.

Libyan rangelands degradation and desertification have been long recognized by the Libyan authorities. Many rangeland improvement projects were implemented over many years since the early 1970s at very high cost (GEFLI, 1974a, b) and with, unfortunately, ultimately poor or unsustainable results in arid zones (Gintzburger *et al.*, 2000) due to lack of community participation or support.

Restoration versus Rehabilitation

When rangeland is degraded, it could be at a stage when it could self-repair itself. This is if and when the biological irreversibility threshold is not yet reached, when there is still a reasonable taxonomic diversity present on-site, with a reasonable

DOI: 10.1079/9781800627154.0016

seed stock of native species present, and wind and water erosion is minimal. Restoration is the capacity of the rangeland to repair itself with minimal human intervention. In this case, restoration is possible simply with rangeland protection and exclosures. Rangeland restoration or regeneration aims at re-establishing the taxonomic and structural biodiversity, and the ecological dynamic of the native vegetation as a pre-existing historical environment. Rangeland restoration is usually achieved through passive means. Nature does the work. This was discussed at length in Aronson *et al.* (1995) and Le Floc'h *et al.* (1995).

Rangeland rehabilitation is an enforced human action when the rangelands are utterly degraded beyond the biological irreversibility threshold, with poor taxonomic diversity, lack of seed stock and evident erosion scars. Rangeland rehabilitation necessitates active means to unlock and repair as fast as possible the vegetation dynamic and productivity. This is done with heavy and costly inputs (reseeding, water harvesting techniques, fertilization, fodder shrubs plantations etc.).

We could therefore distinguish two possible actions:

- The rangeland is degraded and overgrazed, or occasionally grazed but still with a good vegetation diversity. In this case, we will consider natural rangeland regeneration by passive means.
- The rangeland is utterly overgrazed and degraded with a very low vegetation cover and poor biodiversity. Here, active means of rehabilitation with necessary inputs (energy, reseeding, soil disturbances, water harvesting, plantations etc.) may be necessary.

Passive Means of Rangeland Restoration

Rangeland protection

Rangeland protection by banning temporary or permanent access for cultivation, grazing, fuelwood collection or hunting is the easiest and cheapest way to allow vegetation regeneration and recovery. It is the oldest and the most efficient method used by communities all over arid rangelands. In the Arab world, in Arabia and the near-East, it was known as the *hima* system (Draz, 1974, 1978, 1985; Zahran and Younes, 1990). In North Africa, from Morocco (Auclair and Alifriqui, 2012) to Tunisia (Nasr, 1993), and also in Libya in the Jebel Nafusa, some rangeland, forested areas or large orchards were traditionally kept as *aguedal, agdal g'dal, gdel or tagdielt* by communities as protected pastures or fodder reserve or, simply, as holy places (*marabout*, graveyards) where neither trespassing nor grazing were provisionally or lastingly allowed. In Libya, it appears to have been long-forgotten and replaced with protected areas (*mahmiyat*) by government authorities dealing with rangeland developments (Range Development projects – RDP, Suani). Some 80 of these protected rangelands areas, each covering from 2000 ha to 5000–6000 ha, were established between 2005 and 2010, from west to east. These *mahmiyat* were fenced off at great cost and sometimes guarded by local communities' factions or hired guards.

During colonial times in North Africa and the Middle East, the military authorities controlling the rangeland areas used to temporarily cap and lock the watering wells and cisterns, just allowing access to water for human consumption through a small trap on wells and cisterns, rendering livestock watering impossible. While these tough measures were once efficient, banning access to degraded rangelands or areas under unrest, they are no longer considered as nowadays water can be carted to any remote area.

The effect of rangeland protection

The beneficial effect on native vegetation growth and regeneration can be seen inside the protected *mahmiyat* at Jenoub Sorman – Saniet-el-Hod compared to the surrounding open grazing area (Google Earth satellite – Fig. 16.1) in a sandy/sandy-loam area in the Jeffara plain despite receiving only an average 150 mm MAR. This area has been protected from grazing for the past 15–20 years and was occasionally open for a short grazing period to make use of the accumulated biomass of dried annuals (*guech*). In 2009, with RDP colleagues, we measured the

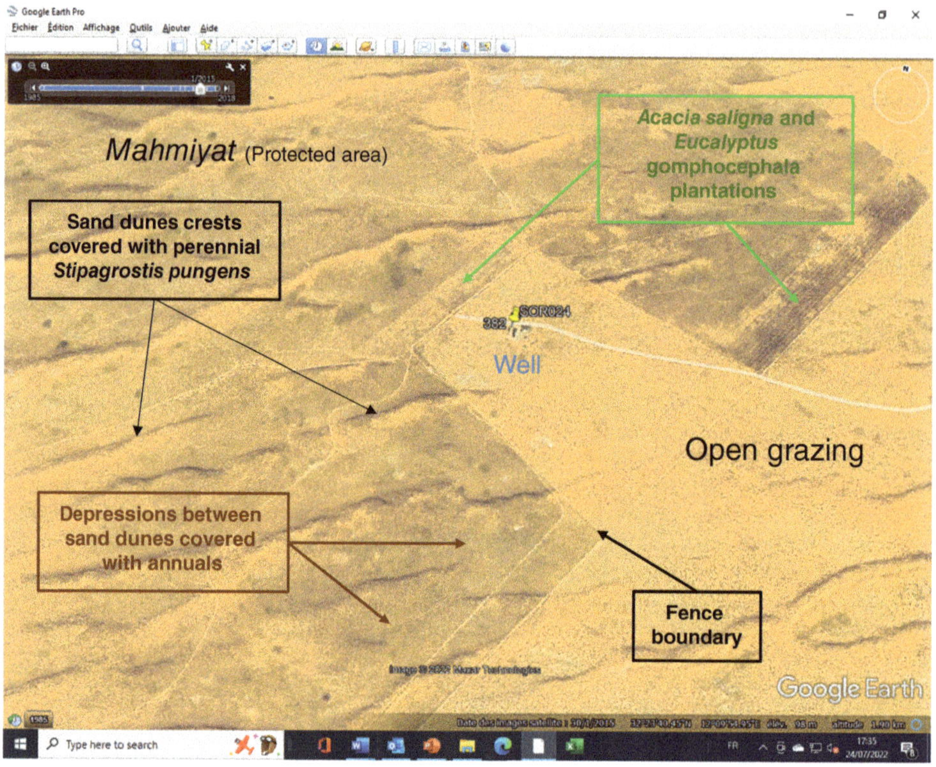

Fig. 16.1. A Google Earth view of a *mahmiyat* at south Sorman (N 32.389 E 12.161), January 2015.

standing biomass of annuals. We then calculated the sustainable number of grazing days for a set number of small ruminants considering a conservative stocking rate removing about 50–60% of the standing biomass of the dry annuals. In consultation with the registered local livestock owners, it was decided to open the *mahmiyat* for a set total number of small ruminants and for a limited period of time at the end of the spring. When implemented, not only the registered flocks showed up, but also many more other unregistered flocks from surrounding rangelands forced their access into the protected areas as the guards and authorities could not control them. That was a total disaster with four to five times more animals entering and grazing the protected area with ensuing serious arguments between the authorities and livestock owners and difficulties to forcefully expel all registered and unregistered flocks. When done, there was no annual vegetation biomass or litter left in depressions between sand dune crests.

A glaring example of an efficient rangeland protection is the community-controlled 1400 ha *mahmiyat* implemented early in 2000 at Al-Bidhan (30 km north-east of Ajdabiya, Fig. 16.2) in an area receiving about 125–150 mm MAR. Despite the low rainfall and very arid conditions, here again, the native vegetation on sandy soils recovered fast with wildlife (birds, hares etc.) reappearing. In this case, the Al-Bidhan community's control, resolution and strength made this possible.

Very long-term rangeland protection and limited grazing can also be seen on Google Earth around the Al-Assah settlement (N 32.83 E 11.63, 150 mm MAR) since the mid-1970s. Similarly, the Al-Watyah/Okba-Ibn-Nafa 5200 ha military airbase rangelands (N 32.48 E 11.89, 100–125 mm MAR, 50 km south-east of Al-Assah, Fig. 16.3) remained strictly closed and

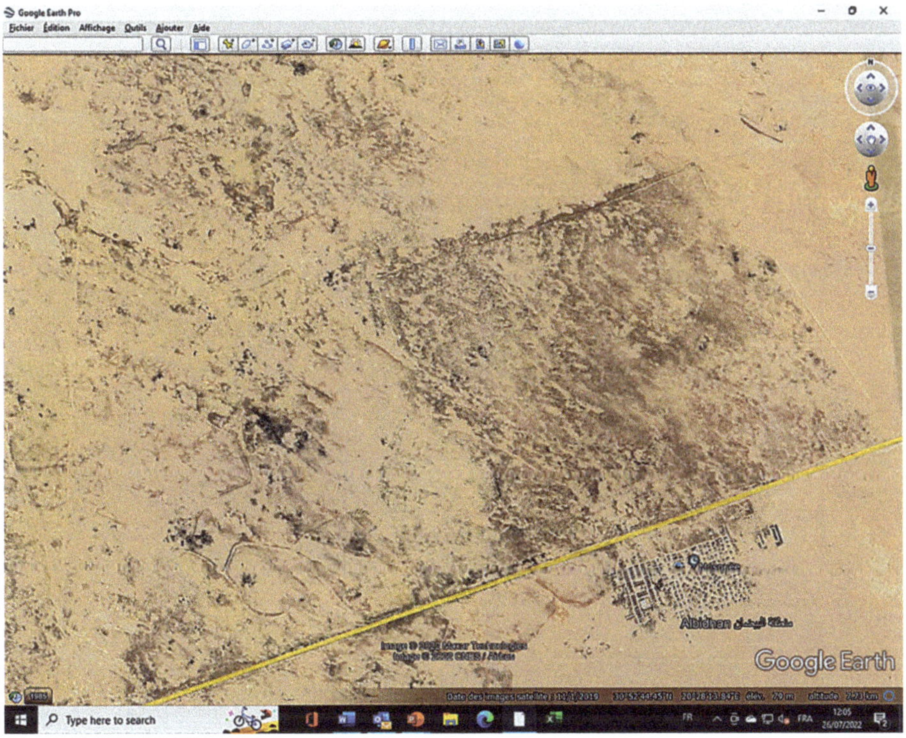

Fig. 16.2. A Google Earth view of the Al-Bidhan (N 30.850 E 20.510) *mahmiyat*, 30 km north-east of Ajdabiya, January 2019.

rangeland vegetation has been ungrazed for the past 50 years. All the above sites are located on predominantly sandy soil.

However, rangeland regeneration by simply banning grazing is not always sufficient due to soil types. In similar rainfall zones, sandy soils respond much faster to protection than rangelands on heavy soils. A study carried out on clay-silt soils on the southern slope of the Jebel Al-Akhdar (Mahmoud *et al.*, 2016) on four sites along a rainfall gradient (250 mm–50 mm MAR) indicates that vegetation regeneration over eight years is slow and would require ten years or more of protection to display some positive results, an impossible prerequisite considering the continuous trespassing by local flocks. Vegetation regeneration is much slower on heavy clay/silty soils than on sandy/loamy soils.

Worth noting are places that were mined during WWII to prevent progress from German, Italian or British troops. These landmines were still present and active long after the end of the war. These mined rangelands presented a constant danger to shepherds and wandering flocks. Some of these properly identified minefields were slowly cleared, but not all; some were simply fenced off to prevent deadly accidents. These areas where access and grazing were *de facto* prohibited for many years allowed remarkable vegetation regeneration (Photo 16.1).

Note that:

- this passive rangeland regeneration (Photo 16.1) can only be attributed to grazing prohibition and not to climate change as rainfall is strictly the same on both sides of the fence; and
- rangeland regeneration is faster on sandy/sandy-loam soils than on heavy soils. This is due to better plant access to water in sandy soils than in heavy soils despite the latter having a higher water-holding capacity.

Fig. 16.3. A Google Earth view of the Al-Watyah/Okba-Ibn-Nafa airbase protected area (N 32.481 E 11.869), 50 km south-east of Al-Assah, January 2019.

Photo 16.1. A view of a WWII landmine field where access and grazing were prohibited showing passive rangeland vegetation regeneration compared to open-access grazing (coastal Sirte, between Nofilia and Ben Jawad, N 30.847 E 17.913, 1984).

Is temporary grazing of protected areas detrimental to regeneration?

Total grazing prohibition for a long period is not necessarily beneficial as, over time, the topsoil gets crusted and slowly covered with biogenic crusts (lichens and moss). Both would bind unstable topsoil, either sandy or heavier, even in the driest environments, and, initially, may limit wind and water erosion. In long-term protected rangelands, the lichen crust (Orlovsky *et al.*, 2004), plant growth and accumulated litter rich in tannins and allelopathic compounds would reduce and even inhibit annual and perennial plant germination (Arroyo *et al.*, 2018; Omar *et al.*, 2017). Seeds produced remain on the topsoil and are easily gathered by ants and birds, hence lost. The *guech* is slowly pulverized during the summer. When autumn rains come, rare seedlings have difficulty emerging through the hard topsoil crust. It leaves the topsoil with large bare surfaces stimulating potential run-off and water erosion. Water storage in the deep soil is considerably reduced with perennials suffering and even slowly drying up. Taking this into consideration, totally banning grazing in protected rangelands for long periods would initially induce a dynamic development of new plant recruitment during the first few years, then followed by a vegetation dynamic stagnation (see the case at Al-Watyah). The only way to start the vegetation dynamic and new germination again is to break the topsoil crust. This is easily achieved by grazing with livestock's hooves stomping the ground, fragmenting the topsoil crust, and facilitating again water infiltration and germination.

In short, long-term banned rangeland grazing is detrimental to rangeland regeneration. Seasonal controlled grazing, also called deferment grazing, would stimulate water infiltration, annual and perennial plant germinations, and promote native shrubs' growth. It supposes that grazing management rules are followed by the livestock owners.

When and where to maximize the benefits of seasonal grazing in protected rangeland areas?

There are two cases when dealing with sandy soils or with loamy-heavy soils:

On sandy light soils, as in the Jeffara, the coastal Sirte and Ajdabiya regions, the leading idea is to replenish the seed stock in the soil. This is achieved at the very end of the spring when the annuals have completed their cycle of flowering and fruiting. At this stage, the vegetation is dry and becomes brittle (*guech*). Grazing the *guech* and trampling it buries seeds and litter in the soil. Many seeds ingested by small ruminants are broken and destroyed, but a large number of seeds remain intact and viable and are scarified during the intestinal transit and eventually pelleted in droppings. This is the case with seeds of perennial legumes (*Retama raetam*, *Anthyllis henoniana*, *Genista* sp., *Calicotome* sp., *Argyrolobium* sp. etc.) and small legumes (*Lotus* sp., *Astragalus* sp., *Vicia* sp., *Lathyrus* sp., *Medicago* sp., *Trigonella* sp., *Hedysarum* sp., *Hippocrepis* sp., *Onobrychis* sp., *Trifolium* sp.). Grazing must be limited to still keep some litter on the soil against erosion. The communities' temptation and pressure is to introduce animals too early in spring while the winter-spring vegetation is green, but it would critically reduce flowering and fruiting. Keeping the flocks on *guech* too long into the summer would lead to the soil being denuded and prone to wind and water erosion. A rule of thumb on protected areas with sandy soil is to graze between the end of the spring and mid-summer, removing a maximum of 50–70% of the *guech* biomass and moving the flocks away. This is easier said than done!

On loamy-heavy soils, as with the case of protected areas where perennial vegetation such as *Artemisia herba-alba* is dominant (foothills wadi Ethel, Jebel Nafusa, wadi Sassu, Solul-el-Khoder), *Anthyllis henoniana* (Haraba, Takut, Kikla, Nalut), the most beneficial time for temporary grazing is late summer/early autumn before these perennials are budding. Shepherds know very well and would avoid grazing green *Artemisia herba-alba* in spring as it would induce diarrhoea and abortion in small ruminants. With late summer and early autumn grazing, small ruminants would prune bushes, remove dry litter and leaves, and soft twigs, hence encouraging winter-spring bushes' growth, flowering and fruiting. This late-summer/ autumn grazing would also reduce allelopathic impact (Omar *et al.*, 2017) of aromatic steppic bushes

(*Artemisia herba-alba*, *Thymus* sp., *Rosmarinus officinalis* etc.) on germination of seeds and plant recruitment and growth.

In conclusion, seasonal grazing of protected areas must be guided by the dominant vegetation and soil types. It is challenging as it rests on controlling the livestock owners and their flocks accessing the protected areas. This is not going to happen without many difficulties and open conflicts with livestock owners or farmers being denied access to their ancestral *watan* or seeing important tracks going through protected areas and being closed (cf. wadi Sassu) without community consultation. This often results in fences being cut or kilometres pulled out (cf. south Sorman, Saniet-el-Hod – Photo 16.2) with encroaching flocks leisurely grazing inside the *mahmiyat* most nights. Hired guards and agricultural police are neither able nor willing to get involved, and, anyway, would not venture on rangelands chasing unruly shepherds and flocks at night.

It is remarkable that the rangeland restoration inside the south Sorman *mahmiyat* clearly visible on Google Earth in 2015 has completely disappeared by 2023 (project area abandoned?), while the community-controlled area at El-Bidhan and the military exclusion zone at Al-Watyah are still clearly visible in 2023.

Photo 16.2. *Mahmiyat* fences and barbed wire destroyed by upset livestock owners and farmers, south Sorman, Saniet-el-Houd (N 32.302 E 12.094), January 2009.

Active Means of Rangeland Rehabilitation

Two non-exclusive active options are available for rangeland regeneration: reseeding and fodder shrub plantations often combined with water harvesting techniques and/or fertilization. Much has been published on the subject in North Africa, from the early reseeding operations attempted in Morocco (Foury, 1950; Villax, 1963) and reviews about fodder shrub plantations (Al-Jadouyi *et al.*, 1984; Dumancic and Le Houérou, 1981; Gintzburger *et al.*, 2000; Le Houérou, 1965, 1980a, b, 1982, 1996, 2002a, b; Le Houérou *et al.*, 1982b; Le Houérou and Barghati, 1982; Le Houérou and Franclet, 1971; Le Houérou and Pontanier, 1987; Nefzaoui *et al.*, 2000; Zaafouri *et al.*, 1994, 1995; Ziani, 1970).

Reseeding degraded rangelands

This was attempted before and after WWII in Morocco using native and exotic plant material, mostly Mediterranean and temperate perennial grasses (Brichambaut, 1950a, b) under rainfed conditions on rangelands. Killian (1961) reported his 1948 observations at El-Mergueb in the steppe of the high plateaux in Algeria. In any case, rare attempts to establish perennial grasses on North African rangelands always failed below the 350–400 mm isohyets (Le Houérou, Montpellier, 2001, personal communication) even with *Agropyron* sp. selected in the USA from collections from arid Central Asia and China (Johnson *et al.*, 2006). The main problem remains with the availability of large quantities of seed of native plants either collected locally or industrially produced.

Potential native plants that could be used for rangeland rehabilitation

PERENNIAL GRASSES. In Libya, the native perennial grasses with drought resistance that could present some interest for rangeland reseeding are *Oryzopsis miliacea* (Syn. *Piptatherum miliaceum*), *Panicum turgidum* and *Pennisetum dichotomum* and *Cenchrus ciliaris*, *Stipa lagascae* (Chaïeb *et al.*, 1990).

- *Oryzopsis miliacea*, a medium-sized bunchgrass, can still be found in the Jebel Nafusa and in the Jebel Al-Akhdar in secluded areas on loamy soils in rainfall zones between 250 mm and 400 mm MAR. It is winter-dormant with strong spring vegetative growth. Seed production and collection are easy.
- *Cenchrus ciliaris*, a small bunchgrass, was common along the northern foothills of the Jebel Nafusa from Bir Touta/Bir-el-Ghanem flats and wadi banks to Bir Ayyad/Nalut, to Misurata/Kararim to Bu-Gren/south Sirte on fixed sandy/sandy-loam soils in rainfall zones between 150 mm and 300 mm MAR. Seeds remain dormant for many years. Dormancy is lifted by exposure to high daily temperatures and sunlight. Ripe fluffy seeds are carried away by the wind (anemochory) and facilitate propagation. *Cenchrus ciliaris* is winter-dormant but extends its growth late spring and early summer as it is a C4 grass. It benefits from rare and late-spring rainfall. Seed production and collection are easy.
- *Stipa lagascae*, a small bunchgrass on sandy-loam soils, re-establishes and regenerates fast in protected areas, common in the 150–400 mm MAR. Seed production is difficult.
- *Panicum turgidum* and *Pennisetum dichotomum* large tufts occur in *Acacia tortilis* formations in sandy/stony wadis and mobile sand (wadi Nefad – Seddada, Bu Gren south Sirte to South Jebel Al-Akhdar) in rainfall zones between 50 mm and 150 mm MAR. Seed production and collection are easy.

The last three perennial Libyan grasses are extremely drought-resistant. Seeds from these grasses could easily be locally collected and used for rangeland reseeding in appropriate environments and rainfall zones. They are all well grazed by livestock.

What is the potential of exotic grasses on Libyan rangelands? Potential exotic Gramineae from other parts of North Africa (Foury, 1950; Villax, 1963), arid China (Jigjidsuren and Johnson, 2003) and USA, mostly *Agropyron desertorum*, *Agropyron cristatum*, were collected, evaluated, selected and used for rangeland reseeding in USA (Johnson *et al.*, 2006). Attempts to use them on rangelands in Morocco, Algeria, Tunisia and the Middle East were never successful. Rare introduction tests in pastoretum in North Africa proved to be useless due to poor field germination, seed production difficulties, and failing rangeland establishment by direct seeding. Le Houérou (Montpellier, 2001, personal communication) remarked that all these exotic perennial grasses would need at least 300–350 mm MAR, an ecological requirement that would prevent using them in most of Libya, not counting challenging grazing management post-establishment.

Perennial and mostly annual legumes are another useful source of native plant material to be potentially used for rangeland reseeding.

PERENNIAL LEGUMES

- *Argyrolobium uniflorum*, a small shrub common on sandy-loam soils of the Jeffara plain, Jebel Nafusa to coastal Sirte under 150–300 mm MAR.
- *Anthyllis henoniana*, a medium-sized shrub present on shallow soils with a loamy matrix under 150–200 mm MAR. Mostly south-west on the southern slopes and northern foothills of the Jebel Nafusa (Rojban to Nalut, Giado to Tiji-Takut).
- *Retama raetam*, a tall shrub present on sandy soils with 100–300 mm MAR (Gintzburger, 1976) from west to east.

So far, the potential of these three plants has not yet been explored in arid North Africa. Seed collection and production of these native perennial legumes should not present any difficulties.

ANNUAL LEGUMES. The Australian annual *Medicago* sp. cultivars[1] or medics were used in the medic-cereal ley farming system (see Chapter 9) tested in the Jeffara plain by the Western Australian agricultural team (Allen and Ewing, 1978; Ewing, 1978, 1979;

Ewing *et al.*, 1979; Halse, 1978, 1982). Seed production techniques are well known and specialized medic seeds harvesting machinery (Horwood Bagshaw Clover Harvester, Photo 16.3) (Moss *et al.*, 2021) from Australia is available and has been used in Libya since the mid-1970s.

However, the Australian cultivars currently available are not suitable for rangeland regeneration operations as they are late flowering.

Indigenous Libyan annual *Medicago* sp. (Table 16.1) offer a better potential for rangeland reseeding and rehabilitation providing that their ecological requirements are respected. Some 1800 native ecotypes of annual medics (Gachet and Elmir, 1972; Heyn, 1963; Lesins and Lesins, 1979; Nègre, 1959; Small and Jomphe, 1989) were collected with ecological details of collection sites and identified in the mid-1970s/early 1980s all over northern Libya from west to east (CSIRO, 1985; Francis, 1978; Gintzburger, 1978a, b, c, 1980, 1983, 1984; Gintzburger *et al.*, 1983; Gintzburger *et al.*, 1984a, b; Gintzburger and Blesing, 1979; Khalil *et al.*, 1977; Webb *et al.*, 1984). The Libyan Medic Evaluation Program (LMEP) was then implemented between Western Australia and Libya (1980–1984) to agronomically evaluate and select the best annual medics species for Libyan arid/semi-arid conditions as most of the Australian cultivars lack early-flowering characteristics to escape the end-of-spring drought, hence low seed production and poor regeneration.

CHARACTERISTICS OF ANNUAL MEDICS FOR LIBYAN ARID RANGELANDS REHABILITATION. The following attributes are essential:

1. Early flowering and abundant seed production
2. High number of seeds or propagules (pods) production
3. Seed dormancy and delayed germination mechanisms to avoid false start due to inter- and intra-annual rainfall variations and to evade years-long droughts
4. Prostrate growth habits to:
    a. minimize overgrazing by livestock
    b. protect soil against wind and water erosion

Photo 16.3. Medic seeds harvesting using the Horwood-Bagshaw Australian Harvester (Bir-el-Ghanem, 2009).

Table 16.1. Potential Libyan annual medics that could be used for rangeland rehabilitation in Libya.

Annual *Medicago* sp.	Soil requirement	Rainfall zone (mm)
Medicago truncatula ssp. *tricycla*	Sandy-loam/inland	150–300
Medicago polymorpha var. *brevispina*	Heavy-soil loam/in depressions	150–300
Medicago laciniata var. *brachyacantha*	Heavy-loamy soils with top sand veil	50–150
Medicago laciniata var. *laciniata*	Heavy-loamy soils with top sand veil	100–300
Medicago littoralis	Sandy soils/coastal	200–300
Medicago tornata	Sandy soils	200–300
Medicago truncatula ssp. *truncatula*	Heavy soils/loam/Jebel	250–400

5. Small seeds produced to minimize destruction during ingestion and transit by livestock
6. High biomass production, with good palatability and feed value
7. Contribute to soils improvement by enriching the soil with nitrogen and organic matter
8. Capacity for spiny pods (propagules) caught in wool and hair to be dispersed by livestock
9. Known commercial seed production and harvesting techniques

From the plant families present in the native Libyan flora, the most abundant annual Poaceae and Brassicaceae have biological mechanisms to evade early drought. However, they lack several of the listed attributes, while the annual *Medicago* sp. (Leguminosae/Fabaceae) possess all attributes listed above.

Early flowering. Early flowering and fruit ripening appear essential to evade the end-of-spring drought. Row evaluations of the 1800 LMEP medic ecotypes were assessed in Western Australia and in western Libya under similar ecological conditions (Gintzburger *et al.*, 1984b). About 200 of the best-performing ecotypes were tested along with Australian cultivars controls, under rainfed conditions, at the ARC Tajoura farm (sandy soils), at Bir-el-Ghanem, wadi Abu-Shayba (sandy-loam soil) and at Wishtata (loamy soil). During the tests, the wadi Abu-Shayba site (MAR = 175–180 mm) received an above-average rainfall with 303 mm in about 30 days, with the last rain in March. The Wishtata site received 137 mm in 17 days with the last rain in late January, hence a poor spring finish. From the 189 Libyan annual medics tested, 75 obtained equivalent or better scores than the Australian cultivars Harbinger and Cyprus on both sites in terms of flowering earliness, winter and spring vigour and burrs/seeds production (Gintzburger *et al.*, 1984b, Gintzburger and Skinner, 1985a, b, c). Quite remarkably, 46 of these best-performing ecotypes came from the Sirte/Solul-el-Khoder region where the average rainfall ranges from 100 mm to 150 mm/year and not much from the Jeffara plain/Jebel Nafusa, which benefit from higher rainfall.

Why do annual medics from the Sirte/Solul-el-Khoder region yield outstandingly vigorous and early medics flowering ecotypes? The possible reason may be the traditional grazing patterns and low stocking rates of this region during the collection. Livestock owners in the Sirte region follow the seasons and adjust to rangeland types. In late autumn and winter, flocks swiftly move south to the inland desert making full use of the winter rain and flush of desert annual vegetation. Thus, the pressure on the Sirte coastal rangeland is released and allows the vegetation to develop fully and set seeds. Being only sparingly grazed, early medics ecotypes can express their potential without being destroyed. By the end of spring/early summer, when watering becomes necessary, the flocks are moving back north, towards the coast where the vegetation is lightly grazed and where water is available. This Sirte grazing pattern has very likely spared all early annual ecotypes. This is not the case of the Jeffara plain and Jebel Nafusa where the population and animal density are very high, inducing all-year-round overgrazing with little flocks' transhumance. It is very likely that early vigorous medics ecotypes would have promptly disappeared from the native vegetation in the Jeffara plain and Jebel Nafusa. A better understanding of traditional and present grazing patterns would be of invaluable help in locating regions of high probability of collecting early maturing plant material.

Is flowering earliness the ultimate attribute for selecting medics for arid zones rangelands? Our tests at Wishtata and wadi Abu-Shayba showed that the *Medicago laciniata* (var. *laciniata* and var. *brachyacantha*) collected from the lowest-rainfall zones in the desert (MAR ≥ 50 mm) flowered marginally later (105–110 days from germination) than the best Australian cultivars for Libyan conditions (Harbinger or Cyprus) usually flowering in 95–100 days. We hypothesize that desert *M. laciniata* developed a fast pod[2] maturity mechanism immediately after flowering. This appears to be the case for Libyan medics tested in low-rainfall zones in Western Australia at Carnarvon (110 mm MAR) and Kalgoorlie (110 mm MAR) during the 1983 season (Gintzburger *et al.*, 1984b). One should notice that *M. laciniata*, tested in Libya, never displayed outstanding winter or spring vigour, hence dodging winter or early spring grazing. *M. laciniata*, nevertheless, always produced pods with many more seeds per pod, up to 8–10 small seeds per pod, with the highest level of seed dormancy compared to other medics from high-rainfall zones. This latter characteristic would be their survival warranty in case of false start or of an early dry-spring finish in the desert. Furthermore, we often noticed that *M. laciniata*, nearly dry, would turn green again with rare and late-spring rainfalls. Therefore, late-flowering species originating from low-rainfall zones may well have a role to play in a seeding mixture to use in arid rangeland revegetation projects.

Pod of annual medics, number of seeds/pods and seed dormancy. One of the most important attributes from the annual medics is their seed dormancy mechanisms (Quinlivan, 1971). Each medic fruit (pod) contains from two to ten seeds according to the species. These seeds remain enclosed in the pod, season after season, and neither shatters nor scatters the seeds like many other annual legumes (*Lotus* sp., *Vicia* sp., *Lathyrus* sp.). All the medic seeds in the pod have different levels of dormancy. The seed dormancy is sequentially and timely relaxed, seed after seed in the pod, by exposure to alternating day and night temperatures on topsoil over many years. It means that all seeds in a single pod would not all germinate at the same time, with the first rain after ripening. In case of a good first rainy season, the first seed would germinate and complete its cycle until fruits ripening, increasing the seed stock on site. In case of false start with insufficient rain followed by drought, the first seed germinating would fail. But, quite remarkably, the next seed, the second in line in the pod, would have its dormancy slowly removed during the following summer, ready for the next autumnal rains to germinate and possibly produce more pods if the season is good. The third and remnant seeds in the pod remain dormant for an extra year, avoiding a possible spring drought, and waiting for their turn to germinate. In case of prolonged drought over several years, live medic seeds remain dormant in the pods for years waiting for a favourable season to germinate, grow, flower and fruit, contributing to increasing the seed stock on site. Obviously, the more seeds per pod, the more chance the medic would evade years of drought. The most drought-tolerant species, *Medicago laciniata*, common in ultra-low-rainfall zones in Libya (MAR ≥ 50 mm into the Hamada), often contains up to ten tiny seeds per pod. These are noteworthy survival attributes and a drought-avoiding mechanism that may be also found in annual legumes with non-shattering pods (*Hippocrepis* sp., *Hedysarum* sp., *Hymenocarpos* sp., *Scorpiurus* sp., *Trigonella* sp.), all present on arid Libyan rangelands. This remains to be studied.

Small seed size of medics better survives ingestion and transit. The seed size is important for the survival of the annual legumes stands as small seeds will escape being destroyed when grinded and ingested by livestock (Ewing, 1995). *Medicago laciniata* var. *laciniata*, *M. laciniata* var. *brachyacantha* and, to a lesser extent, *M. littoralis* var. *littoralis* (Libyan ecotype LIB 1471.6, 725 seeds/g), *M. truncatula* var. *tricycla* (ecotype LIB 1470.3, 336 seeds/g), are present on the most arid Libyan rangelands (MAR ≥ 50–100 mm). They produce smaller seeds (Gintzburger *et al.*, 1984b) compared to the ones present on higher-rainfall zones (MAR ≥ 200–400 mm) such as *M. truncatula* var. *truncatula* (cv. Cyprus, 260 seeds/g), *M. tornata* (ecotype LIB 1484, 160–180 seeds/g; cv. Swani, 190 seeds/g), *M. polymorpha* (ecotype LIB 793.2, 180 seeds/g), *M. scutellata* (cv. Robinson, 66 seeds/g).

Is rhizobium inoculation necessary on Libyan medics? Medics, like many other legumes, have nodules containing nitrogen-fixing bacteria

(*Rhizobium meliloti*, Syn. *Sinorhizobium meliloti*, now *Ensifer meliloti*) on their rootlets. When medics are introduced on new agricultural land, they do need to be specifically inoculated as is the case in Australia in the ley farming system (Puckridge and French, 1983). Fortunately, medics species are ubiquitous in northern Libya and North Africa as well as their specific rhizobium, hence reintroducing medics in arid rangelands rehabilitation programmes would not necessitate any inoculation before seeding operations. Annual legumes effectively contribute to enriching the soil with organic matter and nitrogen.

In conclusion, the rich Libyan annual *Medicago* collection (LMEP) should be explored for future Libyan rangeland reseeding programmes. Broadcast pod seeding, and not naked scarified seeds, must be used to maximize the chance of proper establishment due to inherent seed dormancy characteristics. The best Libyan medic candidates are *M. laciniata*, *M. truncatula* var. *tricycla* and *M. littoralis*, all of them collected all over northern Libyan rangelands and even deep into the 50 mm rainfall zones.

Fodder shrub plantations

Fodder shrub plantations are another means of rangeland rehabilitation by modifying and reallocating an ecosystem to different vegetation types and managements (Aronson *et al.*, 1995). This is the most expensive way to rehabilitate rangelands with multipurpose shrub plantations (Le Houérou, 1974, 1996, 2002b) as it requires outsourcing appropriate exotic seed supply, the establishment of extremely large nurseries capable of producing half to one million seedlings/year close to plantation sites, qualified managers, an army of workers, and good-quality water from new wells or carted by trucks. Implementing the plantation operations requires heavy equipment to prepare the site with deep ripping of the soil with bulldozers, trucks to transport the tens of thousands of seedlings from nurseries to plantation sites, a mass of labourers to transplant the seedlings, and follow-up watering of individual seedlings during the following summer. The fodder shrub plantation then needs to be protected from grazing for 3–5 years with fences and guards. Fodder shrubs grazing management, when not properly controlled, remains a serious issue (Gintzburger *et al.*, 2000; Le Houérou, 2002b; Telahique, 1981). Despite all this shortfall, fodder shrub plantations on rangelands constitute useful fodder reserve for dry periods or years of drought providing they are protected and not used all year long to feed desperate livestock wandering on degraded rangelands.

Most successful shrubs planted on Libyan rangelands

A variety of native and exotic multipurpose shrubs are available and were tested in arid zones of the world and especially in the Mediterranean semi-arid and arid zones (Gintzburger *et al.*, 2000; Le Houérou, 2002b). A few were used in Libya for rangeland rehabilitation as drought fodder reserves during 1975–2000. We detail only the ones that were successfully established on largescale plantations in Libya, ignoring the anecdotal ones.

NATIVE SHRUBS

- *Atriplex halimus* (Chenopodiaceae/Amaranthaceae) is a dense vigorous monoecious shrub occurring all over North Africa and northern Libya (MAR > 150 mm) mostly in wadis and around *sebkha* (*Atriplex halimus* ssp. *halimus* – erect type, leafy with soft stems, phreatophyte, salt-tolerant) and in a more arid environment receiving 100–150 mm MAR. *Atriplex halimus* ssp. *schweinfurthii* is a hemispherical bush, intricate, with stiff bare branches ending with pointed twigs preventing browsing from small ruminants, but avidly consumed by dromedaries. Both are self-reseeding with MAR > 150 mm. Seed sources are important as genetic variability is large, inducing woodiness and palatability differences (Al-Jadouyi *et al.*, 1984; Elhamrouni and Sarson, 1975; Le Houérou, 1981, 2002b; Le Houérou and Franclet, 1971; Ziani, 1970). Both *Atriplex halimus* subspecies were established on tens of thousands of hectares on predominantly sandy loam/heavy soils in north-western Libya (1975–1990) and in wadis and depressions (*gdama*) under the Jeffara Plain Authority. Attempts

Photo 16.4. Alley farming with rainfed barley in-between double rows of transplanted *Atriplex halimus* (south Zliten, wadi Um el-Jdari, 2007).

to initiate alley farming with double rows of *Atriplex halimus* separated by large strips of rainfed barley were established south of Zliten (Photos 16.4 and 16.5). *Atriplex* sp. transplanted on deep-ripped calcareous crust always showed poor establishment, low survival and minimal shrub growth.

- *Calligonum comosum* (Polygonaceae), a medium-sized desert shrub occurring on mobile sand (south El-Watyah-Takut and eastern Hamada Derj-Sinaween, MAR > 50–100 mm), abundantly used for sand dunes fixation at BGI and BGII (from 1975, Bir-el-Ghanem rangeland projects); fruits and seeds abundant and easy to collect once established; fluffy fruits are disseminated by wind. *Calligonum* palatability is poor and highly drought-tolerant.
- *Periploca laevigata* (Asclepiadaceae), a medium-sized woody/twining shrub, highly drought- and grazing-resistant, common on shallow soils over calcareous crust (MAR > 75–150 mm) in southern Tunisia (Ferchichi, 1995; Ferchichi *et al.*, 1991) and from northern and southern foothills of Jebel Nafusa, Hamada, Sirte to Marmarica along wadi lines. *Periploca laevigata* is frost-sensitive and occurs in places with 'm'[3] > 3°C. Not really convincing when transplanted as it grows very slowly. Better growth observed on protected sites with native *Periploca* formations. Once established it is very resistant to browsing. Mostly browsed by dromedaries. Often collected for fuelwood.

EXOTIC SHRUBS

- *Atriplex nummularia* (Chenopodiaceae/Amaranthaceae, from Australian Mediterranean rangelands), a medium-sized erect dioecious shrub, which may reach 3 m in height and 4–5 m in diameter at Deriana in Benghazi plain project on *Terra rossa* when in favourable conditions (Le Houérou, 2002b; Le Houérou and Barghati, 1982). Transplanted with success on large areas on rangeland projects in Jebel Nafusa, Wishtata – Tarhuna, south Zliten region (MAR > 150–200 mm) over deep-ripped calcareous crust over loam-marl/heavy soils. Tolerates salinity and cold ('m' > 0°C). Male *A. nummularia* plants have larger fleshy-silvery leaves compared to female plants. Usually well grazed by all livestock (Salem *et al.*, 2010) but highly susceptible to damage from native rodents. In terms of browsing management, *Atriplex nummularia* branches are fragile and easily broken by browsing livestock. The leaves are rich in sodium chloride, and small ruminants feeding on *Atriplex*, especially in summer, must have access to fresh water.

- *Opuntia ficus-indica* (Cactaceae) coming from South and Central America were long ago introduced into Spain and Portugal by the 17th century, and then into North Africa (Elhamrouni and Sarson, 1973; Gouhis *et al.*, 2018; Nefzaoui and El Mourid, 2008). The spineless form *Opuntia ficus-indica* var. *inermis* is most exclusively established on sandy/sandy-loam deep soils, low tolerance to cold ('m' > + 2°C) and good drought tolerance (MAR > 150 mm); does not withstand weed competition in low-rainfall zones; used as supplementary feed and source of fruits; does not tolerate direct grazing, usually used as cut-and-carry supplementary feed chopped and mixed with straw-cereal bran (Nefzaoui, 2000). The pads are rich in water, but nitrogen-deficient; feeding over long summer periods leads to strong diarrhoea with small ruminants.
- *Acacia saligna* (= *Acacia cyanophylla*, Fabaceae/Caesalpinioideae from Western Australia), a small phyllodineous, thornless, multi-trunked tree (4–8 m high), high genetic heterogeneity, young seedlings only to be transplanted on deep fixed sandy/ sandy-loam soils, fast-growing, widely used for sand dune fixation (Leone, 1924) in western Libya (Magdares-Sbot, Sidi-Benhur, Bir-el-Ghanem, Sirte); cold-sensitive ('m' > + 2–3°C) and drought-resistant (MAR > 150 mm); short-lived (15–20 years), with wide superficial root systems with suckers; fresh foliage and twigs avidly browsed by dromedaries, but often rejected by small ruminants (Dumancic and Le Houérou, 1981) because of high tannin content; cut-and-carry feed to small ruminants after leaves and twigs air-desiccated for a couple of days; abundant seed production; seeds used to feed poultry (Keith, 1965), wood collected for firewood,
- *Atriplex canescens* (Chenopodiaceae/Amaranthaceae, seeds from USA), a medium-sized shrub with elongated leaves, very cold- ('m' > 0°C) and drought-resistant (MAR > 125–150 mm), two subspecies are recognized: (i) ssp. *Canescens* to plant on loamy/heavy soils, much used in northern USA, Syria, Jordan, Iran, northern Pakistan, in Libya on the Jebel Nafusa; and (ii) ssp. *Linearis* for sandy soils, much planted on the BG (Bir-el-Ghanem) rangeland projects. Well grazed by all livestock. Flexible branches not as fragile as *A. nummularia*. Selection of *Atriplex* sp. was done in Tunisia (Elhamrouni and Sarson, 1975).

OTHER SHRUBS OCCASIONALLY PLANTED ON LIBYAN RANGELANDS. The following shrubs or small trees were tested on a few hectares but were either unsuitable due to their poor production and survival, and poor acceptance by livestock. All originate from semi-arid /arid zones and are drought tolerant.

- *Acacia aneura*, Fabaceae, from all inland Australian Mulga rangeland, a small tree with needle-like phyllodes, very drought resistant, unpalatable
- *Acacia cyclops*, Fabaceae, from Western Australia, a phyllodineous globe-shaped small tree, only suitable on coastal dunes and close to seashore, highly salt spray-tolerant; unpalatable.
- *Acacia karoo*, Fabaceae, from South Africa, small thorny tree use for fences, fuelwood and apiary.
- *Acacia ligulata*, Fabaceae, from inland arid Australia, medium-sized shrub, very drought-tolerant, sand binder (Bir-el-Ghanem projects), but unpalatable.
- *Acacia pycnantha*, Fabaceae, from semi-arid Australia, small tree with lanceolate curved leaves, poor palatability.
- *Acacia salicina*, Fabaceae, from eastern and central Australia, small phyllodineous tree with pendulous branches and leaves, cases of stock poisoning reported in Australia, gypsum-tolerant.
- *Acacia victoriae*, Fabaceae, from arid Australia, medium-sized thorny shrub/tree used for fences and fuelwood in Libya.
- *Argania spinosa*, Sapotaceae, from south Morocco/Mauritania, a small tree only recently grown in nurseries and some adult specimens implanted near Gharyan in undisclosed location.
- *Atriplex amnicola*, Chenopodiaceae, endemic to salt land in Western Australia, a small

woody multi-branched shrub, prostrate/ slightly erect, waterlogging- and grazing-tolerant, low production.

- *Atriplex glauca*, Chenopodiaceae, peri-Mediterranean to Australia and USA, on saline lands, prostrate with good ground cover.
- *Atriplex lentiformis*, Chenopodiaceae, from south-west USA, a small erect, woody and leafy shrub, well grazed.
- *Atriplex repanda*, Chenopodiaceae, endemic from Chile.
- *Atriplex semibaccata*, Chenopodiaceae, from arid Australia, a small prostrate hemispherical shrub with red berries, highly palatable, low production.
- *Atriplex undulata*, Chenopodiaceae, from Argentina, a small hemispherical shrub with wavy leaves, often established by direct seeding, turns woody with age.
- *Cassia sturtii* and other *Cassia*/*Senna* sp., Fabaceae, from arid Australia, small erect shrubs, susceptible to overgrazing, tested successfully at Wishtata, palatable.
- *Colutea istria*, Fabaceae, from west Asia and Middle East, a medium-sized shrub, tested successfully at Wishtata, root system has suckers, susceptible to overgrazing.
- *Coronilla glauca*, Fabaceae, peri-Mediterranean, native on red soils in Jebel Al-Akhdar, a small leafy erect shrub, susceptible to overgrazing.
- *Gleditsia triacanthos*, Fabaceae, from arid USA, a small thorny tree in Libya, produces valuable pods as livestock feed, fuelwood and apiary, also used as live fences.
- *Haloxylon persicum*, Chenopodiaceae, from Egypt to Middle East and Central Asia, a medium-sized multi-branched tree, psammophyte, phreatophyte, poorly palatable, good source or fuelwood.
- *Kochia prostrata* and *K. indica*, Chenopodiaceae, from Egypt to Middle East – Central Asia – Mongolia, highly palatable but toxicity reported due to high calcium oxalate, accidentally introduced in Marmarica by Australian soldiers during WWII, poor survival.
- *Medicago arborea*, Fabaceae, northern Mediterranean countries, a medium-sized leafy erect hemispherical shrub tested in Wishtata, very palatable, very susceptible to overgrazing, would shed leaves during summer or prolonged drought.
- *Medicago citrina*, Fabaceae, from Spain and Baleares, similar to *M. arborea* but with horizontal branches and light-yellow flowers, poor drought resistance.
- *Opuntia fusicaulis*, Cactaceae, from Central America/Mexico, with elongated pads, cold-resistant, rare in Libya.
- *Parkinsonia aculeata*, Fabaceae, from south-west USA to South America, a multi-branched small tree, planted along road from Aziziya to Jebel Nafusa in Jeffara plain, soft spines and small leaves, poor source of feed and wood, useful to apiarists.
- *Prosopis glandulosa*, Fabaceae, from Central and South America, a thorny small multi-branched tree, poor palatability, invasive in India, Mauritania but not in Libya.
- *Rhamnus oleoides*, Rhamnaceae, peri-Mediterranean to Iran, medium-sized woody shrub.
- *Rhus tripartita*, Anacardiaceae, from North Africa, sub-Sahara to Middle East, native in Libya, a very drought-tolerant medium-sized woody shrub, avidly grazed by all livestock despite strong and acute thorns, occasional source of fuelwood.
- *Salsola vermiculata*, Chenopodiaceae, a woody grey-pubescent intricate small shrub (50–100 cm), spinescent at end of stems, polymorphic, abundant white to pinkish flowering, common from Al-Assah to Tobruk on loamy-clayey soils, very palatable, widely used on rangeland plantations in Syria, very resilient to overgrazing, self-reseeding with MAR > 150 mm, seeds viability is short (about 3–4 months).
- *Sesbania* sp., a Fabaceae introduced from coastal south-east Asia into Egypt on irrigated land, tested at Al-Assah, slender fast-growing tree, needs irrigation, useless on rangelands.
- *Ziziphus lotus*, Rhamnaceae, from Spain, North Africa to Middle East, very valuable source of nectar for apiarists. *Ziziphus* honey produced is highly priced in Libya.

Nursery production and plantation techniques

Multipurpose shrubs are raised in nurseries before being transplanted. Germination and establishment by direct seeding on semi-arid and arid rangeland sites always failed due to low and irregular rainfalls.

NURSERY PREPARATION. The shrubs' seedlings are individually raised in black polyethylene pouches in nurseries from seed germination to the time they are ready for field transplanting. Seed scarification (*Acacia* sp.) or water leaching (*Atriplex* sp.) is necessary to facilitate germination. *Atriplex* seedlings may also be produced from fresh cuttings. The pouches must be high enough (8 cm diameter x 25 cm high), perforated to avoid waterlogging or open-ended to prevent coiled roots or root strangulation detrimental to proper growth post-transplantation. The planting media, slightly fertilized with old sheep manure, must be porous but firm enough to protect the rooting systems at transplanting. Regular and moderate watering is essential for proper seedlings development under shade or in a plastic house. Nursery operation must be properly timed so that seedlings must be less than 6–9 months old when ready for transplantation on-site. When the time for field planting comes close, seedlings must be pruned to remove excess leaves and rootlets protruding from pouches. They must then be hardened by progressive exposure to direct sun and watering gradually reduced. Unused and old seedlings must be destroyed and not used for plantations as survival prospects on rangeland sites are extremely poor.

SOIL PREPARATION ON PLANTATION SITES. Establishing a fodder shrub plantation requires soil preparation that varies according to the soil type. Soil furrowing or ripping (Photo 16.5) must be done before the rainy season starts to store as much water as possible in the soil. All seedlings transplantation operations must be done during the winter and ended by mid-spring, and in any case, far before the rangeland annuals plants are wilting, that is by mid-April in northern Libya. Any shrub plantations done after early/mid-April would have a low survival rate and likely fail. On deep sandy soils, as common in the Jeffara plain, deep furrows (50–75 cm) are cut along contour lines. Then, plantation holes are hand-dug with a shovel or using an auger. On sites with a calcareous or gypsum crust, soil preparation is done by sub-soiling or ripping on contour lines. Seedlings, separated from plastic raising pouches, are placed into the planting hole and foot-pressed for a firm contact between the nursery media and the soil and immediately watered on site.

In Photo 16.5 is seen a common plantation mistake in arid zones of the region. The elevated

Photo 16.5. Double rows of Australian *Atriplex nummularia* plantations, south Zliten plateau (MAR = 175 mm), 1977.

earth bonds both sides of the ripped line prevent natural water harvesting into the plantation holes. These bonds must be cut open on both sides at regular intervals for the planted shrubs to benefit from water run-off from surrounding areas.

Spineless cactus pads (or cladodes) are cleanly cut, collected from old plantations and trucked to plantation sites (Elhamrouni and Sarson, 1973; Gouhis *et al.*, 2018; Nefzaoui and El-Mourid, 2009). Each cactus to be transplanted must be undamaged and composed of three to four pads (Anonymous, 1972a, b). The two lower pads are then buried into the sandy soil and will soon take root. It is important to note that the rooting system of cactus is superficial and would suffer from weed competition, hence the need, from time to time, to scarify the topsoil on both sides of the cactus plantation line.

SEEDLINGS OR CACTUS PADS' DENSITY AT TRANSPLANTATION. *Atriplex* seedlings must be hardened in nursery before transplanting. After hardening, they must then be abundantly watered just before being transported to plantation sites. A suitable number of seedlings must be available on site for plantation by manual labour available during the day. On Libyan rangelands in the 200–300 mm MAR, small shrubs' density plantation is about 1 m between shrubs and 5 m between rows (1500–2000 shrubs or cactus pads/ha) at plantations. In alley farming, *Atriplex* are planted on widely spaced double rows (2 m in-between rows) with 1 m between shrubs and 10–12 m rainfed barley strips resulting in an overall shrub density of about 1200–1500 shrubs/ha. This plantation density is reduced by half in the 150–200 mm MAR.

Acacia cyanophylla is usually densely planted at about 1000–1500/ha.

WATERING AND SURVIVAL. All transplanted seedlings are immediately watered once with 10–15 litres of fresh water after planting to facilitate early establishment and growth. Hand watering must be done two or three times during the first summer each transplanted individual seedlings. With four waterings during the first summer, each ha of plantation at 1500 shrubs/ha requires about 60–90 m^3 of water from new wells on site or carted from far away. While *Atriplex* would withstand moderately saline water, *Opuntia* sp. needs salt-free water.

During the following winter, seedling survival is monitored and, when necessary, dead seedlings may be replaced and watered accordingly. Survival rates vary with soil types and depth, and rainfall zones. Deep soils and high MAR have a 60–80% survival rate while it drops to 30–40% on shallow soils and low-rainfall zones unless planted in wadis or supplemented with run-off water in depressions.

PROTECTION AND MANAGEMENT. It is necessary to protect the fodder shrubs plantation for at least three to five years before being used either as direct grazing (*Atriplex* sp.) or as cut-and-carry (*Acacia cyanophylla*, spineless *Opuntia*). Both latter shrubs do not tolerate direct grazing, especially by dromedaries. While the shrubs are growing, the native annual vegetation is recovering and contributes to the biomass and fodder quality available per ha. *Acacia cyanophylla* displays a good capacity to regrow (coppicing) fast after back-cutting at 60–70 cm above ground level as tested at Bir-el-Ghanem (Le Houérou, 2002b). Despite demonstrating excellent technical results even in low-rainfall zones, the sustainable management of protected, restored or rehabilitated areas is rarely achieved unless communities are actively involved and participating (Gintzburger *et al.*, 2000).

After a few years of total protection and if the plantation is successful, it is open for grazing during summer to livestock owners belonging to a local community or co-operative against a grazing fee for short periods and controlled stocking rates. In Syria, the grazing fee to access the Maragha fodder shrubs plantation (N 35.562 E 37.660) was about US$ 2.5/ha with a maximum stocking rate of three sheep/ha and for one month. Each contract holder also has to pay, up front, a refundable 20% deposit to prevent them camping inside the plantation and collecting fuelwood (Nordblom *et al.*, 1997). This was not without problems as tents and large flocks gathered around the fodder shrubs plantation were waiting for their turn to enter the plantations, hence degrading the rangeland around the plantations.

Expected production from multipurpose shrubs and survival

Production figures on 200 mm MAR sites are about 0.3–1.5 kg DM/*Atriplex nummularia* shrub/year on shallow soils and deep soils collecting water run-off, respectively (Le Houérou, 2002b; Le Houérou and Barghati, 1982). *Acacia cyanophylla* would produce up to 3–10 kg DM/ small tree/year, with half of it available for browsing (Zaafouri *et al.*, 1995). Survival of *Acacia cyanophylla* three years after plantation ranges from 20–90% in the Jeffara in southern Tunisia with the best results on deep sandy-loam soils. *Acacia cyanophylla* gets well established on deep sandy over saline areas for four to five years, but as soon as their roots go deep into salty soil or water, they perish fast, as seen along the Zuara-Sabratha *sebkha*.

Soil improvement due to shrubs plantation

All the above shrubs contribute little to soil improvement as they drop very little organic matter as litter, except from the *Acacia* rooting system that would nodulate and fix nitrogen. *Atriplex* leaves, rich in sodium chloride, would tend to sterilize the soils around the base of the shrubs.

Mixed-shrub plantations

Mixed multipurpose shrub plantations are not recommended as they would have different levels of palatability according to grazing season, livestock and other uses (firewood, sand fixation etc.). 'Cafeteria' browsing experiments at Wishtata have demonstrated that small ruminants would consume some shrub species of diverse origin and ignore others (Le Houérou *et al.*, 1982a) rendering management extremely difficult. It is therefore preferable to establish mono-specific plantations and, when possible, from identical origin, though other problems may arise such as stimulating diseases and rodent damage on high-density mono-specific plantations.

Economics of fodder shrub plantations

The economics (return on investment) of Libyan fodder shrub plantations were not evaluated. The costs of the fodder shrub plantations in North Africa, and likely valid for Libya, are colossal (Le Houérou, 2002b; Le Houérou and Pontanier, 1987). The land preparation costs effected by commercial contractors (Bouchamaoui[4]) in Libya – furrowing or ripping, manual seedling plantation (1000–2000 shrubs/ha) or cactus pads (400–600/ha), and subsequent watering – are about US$ 200–500/ha, not considering the nursery seedlings production, cactus pads supply and fences erection post-plantation. In the Jeffara plain, by the mid-1970s, five rows of barbed wire and concrete posts were erected to protect rangelands or fodder shrub plantations at a cost of about US$ 5–7/linear metre. It does not consider the side-benefits from erosion control, biodiversity improvement, return of wildlife and the fuelwood that could be collected from the mature *Atriplex nummularia* plantations evaluated to 0.3–5 kg dry wood/shrub/year and possibly up to 10–15 kg/tree/year from the *Acacia cyanophylla*. An economic valuation of governmental fodder shrub plantations at Obeisan (MAR ≥ 175 mm, Aleppo steppe, 110 km south-east of Aleppo) on 7325 ha of loamy/saline-gypseous soils with about 1000 *Atriplex* sp. or 400 *Salsola vermiculata*/ha was US$ 32.7/ha in 1996 (Table 16.2) by Rae *et al.* (2000). The site preparation for planting, and seedling production and transport to sites, respectively, represents 35% and 48% of the total costs.

Table 16.2. Establishment costs of governmental fodder shrub plantations at Obeisan (Aleppo steppe, Syria) in 1996. (Adapted from Rae *et al.*, 2000)

Operations	Details	US $/ha	
Site selection	Wages and fuel	0.08	0.24%
Site protection	Border trenches around plantation sites	2.46	7.53%
Site preparation for planting	Bulldozer ripping and furrowing	11.44	35.01%
Seedling production	Nursery operations and transport to site	15.54	47.55%
Planting on site	Manual labour and fuel costs	2.24	6.85%
Guards' wages	Four years of site protection	0.92	2.82%
Total plantation costs (US$/ha)		**32.68**	100%

Table 16.3. Establishment costs of cactus, *Acacia cyanophylla* and *Atriplex* sp. in central and southern Tunisia. (Adapted from Nefzaoui, 2000)

	Cactus	% cost	Other shrubs	% cost
Land preparation	27	3.6	27	2.6
Plant supplies	80	10.6	50	4.8
Transport to site	60	8.0	40	3.8
Labour for planting	100	13.3	125	11.9
Dead plant replacement	12	1.6	30	2.9
Labour for replacement	25	3.3	25	2.4
Watering	–	–	300	28.7
Early protection (60 days)	300	39.8	300	28.7
Subsidies (3 years)	150	19.9	150	14.3
Total TND/ha	**754**	100	**1047**	100
Conversion TND to US$ (1994)/ha	*228*		*316*	

In central and southern Tunisia, establishment costs of cactus, *Acacia cyanophylla* and *Atriplex* sp. by a governmental agency are shown in Table 16.3.

Costs of multipurpose shrub plantations vary from country to country among semi-arid and arid countries according to local socio-economics and labour costs (Gintzburger *et al.*, 2000).

Combining different restoration and rehabilitation techniques

Soil fertility and water are the two dominant limiting factors of vegetation growth on arid rangelands. Fertilizing rangelands or harvesting and concentrating rainwater onto sites under shrub plantations or regeneration sites have been attempted on Libyan rangelands.

Fertilization

The Libyan arid rangeland soils, similar to the one in southern Tunisia (Le Houérou, 1959, 1960, 1969) are always poor in organic matter, nitrogen and especially phosphate. The response of perennial rangeland plants to fertilization is difficult to discern as it takes many years to obtain results due to intra- and inter-annual rainfall variability. Native annuals have a higher relative growth rate (Gintzburger, 1986) and respond faster to protection and phosphate fertilization even under low and irregular rainfall (Floret *et al.*, 1981; Gintzburger, 1979; Waechter, 1978).

Nitrogen fertilization is essentially used to stimulate vegetative growth. In 1976, large-scale aerial application of hundreds of tonnes of finely crushed urea (a nitrogen-release fertilizer) was carried out at the BGIII site (MAR = 200–175 mm) covered with a native vegetation of *Stipagrostis pungens – Retama raetam*. The aerial spreading was done with small agricultural aircraft[5]. Aerial fertilizer application was combined with aerial reseeding of naked perennial grasses seed (*Dactylis glomerata, Lolium perenne*). The choice of these temperate species was obviously erroneous for this arid bioclimate. Moreover, the aerial seeding technique was a waste as the seeds would be exposed on topsoil and not buried, hence lost to collecting ants, rodents and birds. Logistical problems, delayed application and rain turned the dumped urea into hard mounds of unworkable fertilizer. That was a total and very expensive failure and was soon abandoned. Small-scale experiments at wadi Abu-Shayba using 250 kg/ha of compound fertilizer (N-P-K, 16-48-0) with rainfall harvesting boosted the native annual grass stand (*Cutandia dichotoma*) from nearly nothing to about 320 g DM/m^2 in one single season (Gintzburger and Huxley, 1979). While this experiment is economically irrelevant, it showed the hidden production potential of annual native grasses when fertilized.

Phosphate is an essential element stimulating flowering and fruiting of annual legumes. A rangeland fertilizer experiment was established in September 1978 at Bir Ougradia (25 km south-west Al-Assah (MAR = 125–150 mm)

(Gintzburger, 1979). An early-autumn unique application of single superphosphate (SSP, 18–20% P_2O_5) at 75, 150, 250 and 400 kg/ha on a sandy/sandy-loam protected area proved to stimulate the frequency (transect and point quadrat methods (Canfield, 1941; Gounot, 1969), growth and seed production of annual medics and other annual legumes (*Lotus pusillus*, *Hippocrepis* sp., *Onobrychis* sp.) and *Plantago* sp. (*P. albicans*, *P. ovata*, *P. ciliata*) when compared to ungrazed, unfertilized control (Gintzburger, 1979). The hit frequency on annual plants on sandy protected plots increases from 20–30% on unfertilized plots to 40–60% on plots fertilized with 40–80 kg P_2O_5/ha after one season receiving only 172 mm of rain. Most important is that phosphate fertilization on semi-arid and arid rangeland must be done exclusively with single superphosphate that contains sulphur and microelements promoting legumes' root growth, flowering and fruiting. The single superphosphate must be broadcast at the end of the summer before the first autumnal rain and on topsoil so that the annuals rooting system have easy access to it. This is supported by similar rangeland fertilization experiments in Syria (Osman *et al.*, 1990, 1994) and in Israel (Henkin *et al.*, 1996)

In any case, rangeland fertilization in low-rainfall zones is always considered uneconomical, more so in the Libyan context with an 'open grazing for all' on collective land in the absence of a recognized land tenure system.

Micro water harvesting techniques

On arid rangelands, water is the other main factor limiting vegetation growth. Long ago, desert dwellers and communities learned to efficiently harness the meagre water resources for survival and rainfed farming (see Chapter 11).

In the desert, on the Hamada, which is mostly a bare and rocky gravelly plateau receiving, on average, 50–75 mm/year, it is common to observe small 100–200 m^2 shallow depressions (Photo 16.6) where scanty rainfall and fine sand is collected, allowing the growth of desert plants like *Aristida plumosa*, *Aristida obtusa*, *Medicago laciniata* var. *brachyacantha* and some *Farsetia aegyptiaca* and *Oudneya Africana*, while there is no vegetation around. This is also a common sight with different plants in south Sirte and south Jebel Al-Akhdar.

Other examples of micro-topographic water harvesting facilitating plant establishment in arid environments are common sightings in the steppe (Photo 16.7).

Photo 16.6. Shallow depressions collecting water on the rocky flat desert Hamada.

Photo 16.7. Micro water harvesting in a shrub plantation hole facilitating the establishment of native *Medicago truncatula* in an *Artemisia herba-alba* steppe (Gharyan-Gdama, 175 mm MAR, 1976).

Water harvesting at macro and micro level, in small furrows, ditches, small embankments and bunds, planting holes, or even on car tracks on loamy soils on rangelands, mimics small ruminants' hoof imprints in the soil where plants take refuge and benefit from extra water available for completing their life cycle. The same micro-topographic alterations could be used to increase water availability at the plant level when reseeding rangeland plants or transplanting multipurpose shrubs. Several simple techniques are available (Johnston and Mann, 2024). Some are economical and could be applied on large rangeland areas. One is the pitting technique with a water harvesting/seeding machine based on these observations.

THE AUSTRALIAN PITTING / SEEDING MACHINE. Pitting is a simple water harvesting technique consisting of creating lines of interrupted shallow depressions scooped from the topsoil. The pitting machine must be exclusively used on loamy/heavy soils and not on sandy soils. Each shallow pit collects water, seeds and organic matter facilitating germination and growth. The pitting machine comprises a two-wheeled trailer with the pitting implement articulated behind. The modified 66 cm diameter plough disc has welded radiating pins extending (Photo 16.8) on the cut-out periphery of the disc. The tilt and side angle of the modified disc may be adjusted to change the width of the pits' cut. The unique rear support wheel may be lowered or raised to adjust the pits' depth. While rolling, the cut-out part of the disc alternatively digs the pits while the pins keep the disc turning between pits.

A specialized pitting and seeding machine was developed in Australia for revegetation operations on degraded arid rangelands (Gintzburger and Skinner, 1985a). It is registered under an international patent[6] (Gintzburger *et al.*, 1986) (Fig. 16.4). This machine creates a series of intermittent shallow pits in which seeds or medics' burrs are dropped and rainwater is trapped. The disc directly activates a seed box that distributes the seeds, burrs and, possibly, superphosphate fertilizers into the pits. Intact soil, 1–1.5 m width, separates each pit to avoid triggering water erosion. The interval between pits can be altered by the disc cut-out and pin numbers. The depth of the pits must be only a few centimetres (up to 10–12 cm) to avoid digging into the sterile subsoil (Fig. 16.5 and see Chapter 5). The pitting lines must be spaced by three or five metres of intact soil to allow for water run-off, organic matter and native seeds accumulating into the pits. The pitting implement is articulated on a mobile frame enabling adjustment of the depth of the pit, and freely rolled on soil, rocks or calcareous crust.

The pitting machine was successfully used in arid Australian rangelands at several sites for reseeding scarified seeds and medics' burrs (Gintzburger and Skinner, 1985a, b, c). On Syrian rangelands, fresh seeds of *Atriplex halimus*, *A. canescens*, *A. leucoclada*, *Haloxylon aphyllum*, *Salsola vermiculata* and *Artemisia herba-alba* were sown. It was implemented at Adame, Maragha, Obeisan and Delboah stations (MAR = 150 mm, Aleppo province) in an experiment comparing seedlings establishment by broadcasting seeds without soil cultivation, harrowing after seed

Photo 16.8. The principle of the simple pitting machine mounted on a two-wheeled trailer with details of the modified plough disc with radiating pins, manufactured at ICARDA workshop, Aleppo, 1997.

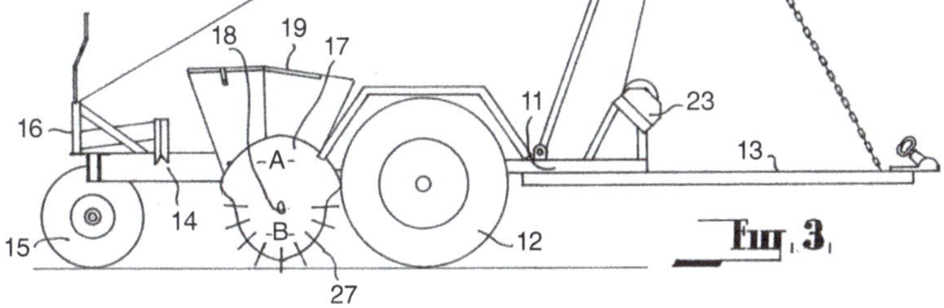

Fig. 16.4. Diagram of the original Australian pitting and seeding machine. (11: support frame of the trailer; 12: transport wheels; 13: draw bar; 15: support wheel to control depth of pits; 16: pit's depth control; 17, 18: modified pitting plough disc at angle to the longitudinal axis of the implement; 27: radial pins; 19: seed-dispensing box; 23: winch to raise or lower the pitting implement into transport position or into operation).

broadcasting, and pitting with seeds in the pits. The site received 70 mm during the 1996/97 season. Only pit seeding showed successful seedlings' emergence and establishment of *Artemisia herba-alba* (Photo 16.9) and *Salsola vermiculata* while the two other treatments proved useless (Gintzburger, 1998). Pitting was also employed at Tallila FAO rangeland project (Tadmor/Palmyra, Syrian steppe) and on the wadi Sassu plateau (MAR = 150 mm, 40 km south-west of Misurata) with Libyan medics (Photos 16.10 and 16.11) on a degraded vegetation of *Arthrophytum scoparium*/*Artemisia herba-alba*.

Pitting operations usually are done in lines, but they were often operated in large spirals

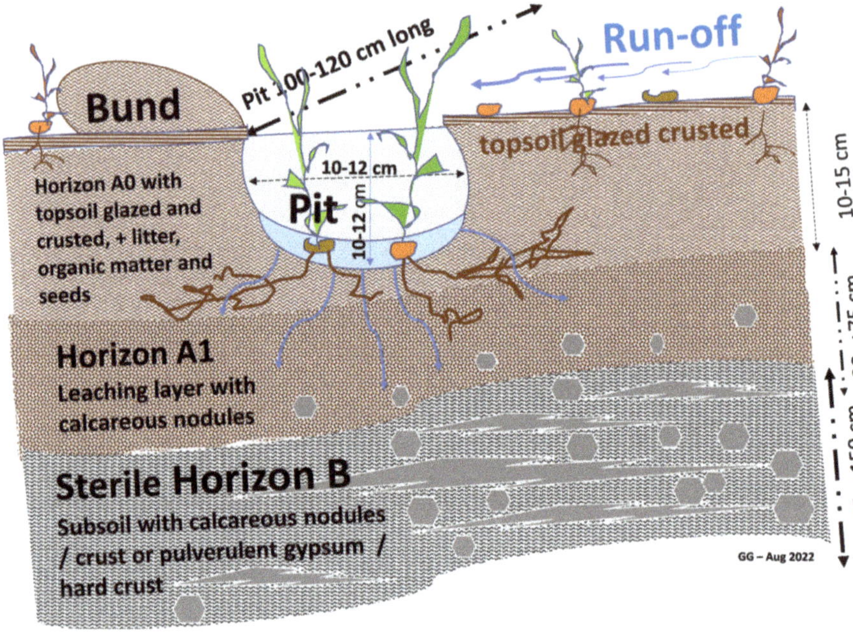

Fig. 16.5. Schematic profile of a shallow pit from the Australian pitting and seeding machine on rangeland.

Photo 16.9. *Artemisia herba-alba* re-establishment with only 70 mm rainfall when sown in shallow pits using the pitting technique at Maragha (N 35.575 E 37.647), Aleppo steppe, Syria, 1997.

Photo 16.10. Pitting and medic seeding operations at wadi Sassu with the KimSeed double pitting-reseeding implements® on *Arthrophytum scoparium/Artemisia herba-alba* degraded flats, wadi Sassu (N 32.150 E 14.808), October 2009, with Engineer Ali Nezdef.

Photo 16.11. Establishment of Libyan annual medics burrs sown into pits on the wadi Sassu plateau rangelands with 130 mm rainfall, by the end of November 2009).

(snail pitting) on sites to avoid triggering water erosion and preferential grazing by flocks on pitted lines.

This pitting machine is low-cost (less than US$ 1000) and easy to construct in a village workshop with simple tools and reformed plough discs. It requires neither a hydraulic system nor a motor to operate the pitting disc and the seed box and is maintenance-free. It is fast-operating (4–8 km/h) and rolls over rocks or hard crust. Pulled behind a car, without the need for a tractor, it is easily moved between rangeland sites with the pitting-seeding implement winched up.

Syrian farmers and livestock owners from Khanasser/Maragha (Syria), after watching our ICARDA reseeding rangeland operations, surreptitiously borrowed our pitting machine at night to sow barley in shallow pits. They refused to reveal the location of their attempt or if it was successful, as seeding rainfed barley on rangelands below the 200 mm isohyets was forbidden in Syria.

It has been argued that pits may become filled with soil over time; however, our snail-pitting site established at wadi Sassu in 2009 covering 1 ha is still visible on the 2024 Google Earth image

(N 32.150 E 14.808). Pits established on other rangeland sites north of Al-Assah (N 32.911 E 11.579) in 2009 are still seen too. It means that small pits retain their micro water harvesting function, soil particles and native seeds capture over many years.

SHALL WE USE A STANDARD SEED DRILL FOR RANGELAND RESEEDING OPERATIONS?. Seeding rangeland native species using standard agricultural (discs or tines) seed drills (Photo 16.12), many times attempted on Libyan rangelands, always failed due to water deficit at root level. Standard seed drills are only to be used on well-prepared cropping land and, beyond doubt, not on arid rangeland with shallow and rocky soils and sparse native vegetation.

THE NARDI/VALLERANI SYSTEM. A water harvesting machine was manufactured by Nardi/Vallerani[7]. Pulled by a powerful 130–210 hp tractor, the Vallerani/Delfino 3 system is activated by a complex hydraulics system (Photo 16.13) to move a heavy reversible plough up and down.

The Nardi/Vallerani machine digs intermittent semi-circular large macro-basins (L: 1.6–5 m, W: 50–60 cm, D: 50–60 cm). Each macro-basin is separated by 0.7– 2.3 m ripped soil. The operating speed of the tractor and implement is sluggish according to the soil type and varies from 2 km/h to 4 km/h due to the slow plough-lifting hydraulic mechanism. The spacing between lines of macro-basins tested on Jordanian flat alluvial rangelands was 4 m, 8 m and 12 m, allowing treatment of 1–6.5 ha/h according to the tractor speed (Gammoh and Oweis, 2011). While capable of trapping large amounts of water (up to 0.75–1 m^3) on loamy/heavy soils, it would lead to waterlogging in the basins, a detrimental impact on shrubs' growth during the rainy season. Digging to 50–60 cm deep into the rangeland soils would expose the sterile subsoil containing calcareous concretions or gypsum (Fig. 16.5) in which native or exotic shrubs would have difficulties in getting established and surviving. Considering that each macro-basin disturbs the soils on 5 x 1 m, and 2.3 m ripped on each line (14 macro-basins + interval/00 m linear), with line spacing of 4 m, 8 m and 12 m, this is 26%, 13% and 8%, respectively, of the area treated where the native vegetation is totally destroyed, and the soil seed stock buried and wasted. The severely excavated area may appear minimal but native perennial plants will take many years to germinate and re-establish in waterlogged macro-basins and ripped lines, more so in lower-rainfall zones receiving less than 200 mm MAR.

The Nardi machine has no seeding implement attached to it, therefore each macro-pit must be hand seeded, adding to labour costs on sites.

Each unit of the Nardi/Vallerani system represents a high up-front investment (tractor, machine and spare parts about €170,000 in 2014). It necessitates skilled operators and mechanics for maintenance, notwithstanding limited availability and high cost of spare parts

Photo 16.12. A standard seed drill used for crop seeding and not to be used for rangeland reseeding operations (South Zliten, 1978).

Photo 16.13. The complex water-harvesting Nardi / Vallerani system (Delfino3s Plow - https://www.Vallerani.Com/en/vallerani-system-en/equipment/ (accessed 29 June 2023)).

for the machinery in developing countries. The operating cost, with 5 m inter-basin lines, is about €75–170/ha when used on large areas in Morocco (Boutaleb, 2013) and €30–130/ha in Sahelian Africa, the Middle East and inner Mongolia (Vallerani System; accessed 20/8/22). In Morocco, the rangeland co-operatives intended to charge a fee to members for use of the potentially regenerated rangelands to fund the cost of future range rehabilitation work. Even ignoring the biological shortcomings, lack of understanding of rangeland soils and ecology, and socio-economical context of Vallerani's advocates, one wonders which community or government could afford such expensive machinery, maintenance expenses and high rehabilitation costs on rangelands barely carrying an average 1–2 sheep unit/ha. Biological expectations in arid environments from this complex and expensive machinery are high but have yet to be confirmed with tangible results.

LAND IMPRINTING. The land-imprinting idea sprung from the 'sheepsfoot' roller used for plastic soil compaction of road base or airfield; however, the rangeland imprinter does not compact the soil. The land imprinter is a roller on which parallel large corner irons are welded (Photo 16.14). Weight may be adjusted by filling the roller with water. Due to its weight (over 1 tonne), it must be pulled behind a tractor. It embosses micro-topographic alternating indentations on sandy/sandy-loam soils exclusively. It does not disturb the soils and does not damage the native vegetation as it simply rolls over. It could be fitted with a seedbox dispenser dropping appropriate seeds in front of the roller when operating on dry soils or behind the roller during wet events to avoid seeds sticking to the implement (Gintzburger, 1999). Seeds dispersed in front of the land imprinter would be pressed into the soil. If dispersed behind the roller, a chain curtain dragged behind would profitably cover the seeds with soil. The land imprinter technique could be tested in Libya on degraded rangeland areas with a thin sand veil (Bir-el-Ghanem, Magdar-es-Sbot (El-Assah), Sirte coast) with local seeds of *Rhanterium suaveolens*, *Plantago albicans*, *Argyrolobium uniflorum*, *Cenchrus ciliaris* and pods of Libyan annual medics. It could also be worth testing rolling

Photo 16.14. The land imprinter built at the ICARDA workshop (Tel-Hadia, Syria, 1998)

over medium-sized sand dunes with reseeding of *Calligonum comosum*, *Panicum turgidum* and *Pennisetum dichotomum*.

As far as using pitting or imprinting techniques, the idea is to trigger a biological recovery by creating islands of useful plants that would disperse their seeds around. Hence, pitting or imprinting would only be performed on widely spaced strips affecting only 20–30% of the overall protected area allowing native vegetation to recover by simple rest-restoration and getting perennial plants re-established.

BUNDS AND EMBANKMENTS. The last water harvesting technique for promoting growth of multipurpose shrubs and trees is to construct bunds or embankments. This is a millenary technique (Vetter *et al.*, 2009; Vetter and Rieger, 2019a, b) used in arid environments to concentrate run-off on farming areas (see Chapter 9). This technique is still used in the Jebel Nafusa, in southern Tunisia (Matmata) and Egyptian Marmarica to support olive and fig trees or palm tree growth on terraces in 150 mm country. It is manual, labour-intensive farming consisting of building and maintaining stone terraces on contour lines with hand constructed small basins to concentrate and hold captured water from scanty rainfall on individual fruit trees. While private farmers would invest in such work for high-value fruit trees or perhaps for establishing multipurpose shrubs (Delaroque, 1998), it is doubtful it would be adopted by livestock owners for establishing multipurpose shrubs and trees on large-scale community-owned rangeland regeneration operations.

Land Tenure and Community Perceptions on Rangeland Rehabilitation

The main problem with rangeland regeneration/rehabilitation operations and overall management remains with the communities and individuals who are not inclined to invest in money and labour on land they do not fully control (Gintzburger *et al.*, 2000). One of the main concerns of rangeland farmers and livestock owners regarding governmental rangeland plantations is that it would be a land grab in disguise. Without their initial involvement and participation and long-term support from the rangeland communities, and without a transparent and accepted land tenure system, any rangeland improvement is doomed to fail. Progress with leading rangeland farmers, livestock owners and communities in Algeria (Redjel and Boukheloua, 2000), central and southern Tunisia (Nefzaoui, 2002; Nefzaoui *et al.*, 2000, 2014), northern Syria (El-Fikiki *et al.*, 2000; Ghassali *et al.*, 1999; Rae *et al.*, 2000) and tribal lands in northern Pakistan (Nawaz, 2000) would be the examples to follow.

This is not without problems. Blunt opinions relating the pros and cons from the users' point of view on the establishment of private and governmental protected areas and fodder shrub plantations in the Syrian steppe were powerfully expressed by elders of the Eneza, Abraz and Bu-Kurdi tribes (El-Fikiki *et al.*, 2000). The main points they raised *in extenso* and *verbatim* were:

- When government rangeland plantations are established, alternative grazing areas should be provided to livestock owners deprived of grazing areas.
- Protecting rangelands is cheaper than establishing fodder shrub plantations.
- Prohibit fuelwood harvesting from good rangelands to allow for perennials' growth.
- *Atriplex halimus* and *Salsola vermiculata* are the best shrubs in the Syrian steppe as they are palatable and self-reseeding.
- Plantations should be a mixture of different shrubs for proper livestock feeding. 'Man can't survive by eating bread only.'
- Private livestock owners should be allowed to camp and water their flocks inside governmental plantations instead of gathering around the plantation perimeter waiting for their turn to enter the plantation, wasting time and energy to reach their allocated browsing sites often 4–6 km away from the camp site.
- Grazing only *Atriplex* is not sufficient for proper small ruminants' feeding and fertility; milk production (from Awassi sheep) increases while spring annuals are green in-between *Atriplex* rows, but falls back when annuals dry off when livestock are only browsing *Atriplex*.
- *Atriplex* should be grazed as early as 2–3 years after plantation to prevent them from getting woody and too large as sheep could not reach leaves and green twigs that are high up or deep inside bushes.
- Hungry sheep browse *Atriplex* or barley crop residues for 1.5 hours only. Hungry again, they return browsing *Atriplex* after about two hours while they would rest for 8–10 h before grazing barley residues again.
- Browsing salty *Atriplex* planted on salty land doubles up watering to 8–9 litres of fresh water per head of small ruminants per day, with water trucked to our camp, hence adding expenditure to flock maintenance (costing up to the value of one head of small ruminant per day for a flock of 250 heads).
- Barley cropping in the steppe (below 200 mm MAR) should not be prohibited as it is sustainable, providing that it is located in depressions and wadis. 'When wadis flow, we consider them as our irrigated land.' 'Our tribal rules ... respect the diversity of natural resources, soils and topography.'
- Any new fodder shrubs reserve should be established in close co-operation and in agreement with the traditional users or plantations will fail in the future. The local people can be effective in managing their rangelands, controlling access and animal numbers if they are allowed to do so.

Even though this is from the Syrian experience, a lot has to be learned from their clear and practical contributions as well as from rangeland managers, farmers and NGOs' experiences of planting and using fodder shrubs in Morocco, Algeria, Tunisia, Jordan, Baluchistan, South Africa and Australia. Diverse rangeland co-operative agreements were implemented in Morocco, Algeria, Tunisia, Syria and Jordan. To our knowledge, such arrangements were never considered on Libyan government-protected areas and fodder shrub plantations.

Conclusions

Restoration and rehabilitation techniques and their implementation have been successfully tested on tens of thousands of hectares in North Africa and northern Libya. Regeneration and rehabilitation techniques are well known. Appropriate native plant material has been long ago identified and tested. Their seeds can be locally collected and produced in quantity.

Long-term rangeland protection and deferred grazing remain the cheapest and the most efficient rangeland restoration techniques, providing that the seed stock in the soil is still sufficient. It is well understood and accepted by rangeland communities, providing that it remains under their exclusive control. Rangeland protection works better and faster on light sandy soils even on the most arid rangelands while vegetation recovery takes a long time, over ten years of total protection, or never, on rangelands with glazed heavy soils and depleted native seed stock. Expensive active means of rehabilitation such as specific rangeland reseeding techniques and/or multipurpose shrub plantations, combined with water harvesting techniques and phosphate fertilization, may then be necessary.

Rangeland rehabilitation by reseeding with native rangeland species is a valid option to be explored but the main constraint is the appropriate seed supply. Short-term reseeding experiments have always failed due to intra- and inter-annual rainfall variability and inappropriate species, often irrelevant exotic species being tested. Most promising are the native Libyan annual *Medicago* sp. (medics), all perfectly adapted to different types of soil and rainfall zones. Industrial seed production of medics is well known and does not present any technical problems. Reseeding arid rangelands with entire medic burrs is the most promising technique taking advantage of hard seed dormancy and sequential seed germination over several years. Reseeding with other common arid rangeland annual legumes with pods (*Hippocrepis* sp., Onobrychis sp., *Hedysarum* sp., *Lotus* sp., *Scorpiurus* sp., *Trigonella* sp.) and other drought-resistant perennials (*Plantago albicans*, *Argyrolobium uniflorum*) should be explored. Reseeding with annuals should be considered as a triggering regeneration mechanism for revegetation by producing large numbers of seeds, catching fine soils particles, adding organic matter, facilitating water infiltration and allowing perennials to slowly return.

Rangeland rehabilitation with fodder shrub plantations is complicated, labour-intensive and expensive. It should only be assigned where shrubs establishment and fodder production are secure with a proper land tenure system and production maximized in depressions receiving run-in and with the best soils. However, it will compete with rainfed barley cropping land and therefore is not well accepted by rangeland communities. The choice of successful fodder shrubs and multipurpose small trees is limited but years of successful experiences in Libya may justify their use.

Combining protection, multipurpose shrub plantations, reseeding, low-cost water harvesting techniques and fertilization are all possible and give satisfactory and proven results on Libyan rangelands. However, over years, expensive and technically successful large areas of protected areas and multipurpose shrub plantation have been destroyed due to traditional grazing practices (first come, first served), excessive stocking rates, and lack of organized rangeland management. The main problem with rangeland restoration and rehabilitation remains unsolved with the current grazing and browsing management as long as the land tenure systems and community agreement and support are not concurrently considered.

Arid rangeland restoration or rehabilitation is a long and patient process requiring a wide understanding of plant taxonomy, plant ecology, environmental and socio-economic conditions. It necessitates years of critical field experience and observations, and, most important, learning from the many past successes and failures from the Mediterranean and other arid regions of the world.

Notes

[1] Australian cultivars of annual medics are available from CSIRO publications at: Register of Australian Herbage Plant Cultivars – Grasses and legumes (csiro.au) (accessed 21 August 2024).

[2] The medic's pods are also commonly named 'burrs' in Australia.

[3] 'm' is the average daily minimum of the coldest month, usually January in Libya (see Chapter 3).

[4] See the Hedi Bouchamaoui Group (hbg-holding.com)

[5] The Embraer EMB 202 Ipanema is an agricultural aircraft with a 260 hp Lycoming engine, McCauley propeller and 550 kg capacity. It was used worldwide for aerial application on crops.
[6] The international patent of the original Australian pitting machine is available at WO1986004771A1.pdf (storage.googleapis.com) (accessed 19 August 2022). An improved version is manufactured and marketed by KimSeed Australia Pty Ltd: Microsoft Word - Kimseed Camel Pitter Brochure - Australia 2014 (accessed 19 August 2022).
[7] The Nardi/Vallerani Fiat system: https://nardigroup.com/en/vallerani-system/ (accessed 19 August 2022); or at: https://qcat.wocat.net/en/wocat/technologies/view/technologies_1528/ (accessed 29 June 2023).

References

Al-Jadouyi, J.M., Shawesh, O.M. and Al-Hamrouni, A.M. (1984) Atriplex *Species and Their Importance in Range Improvement in the North African Countries of the Green Belt Project.* National Office for Agriculture Advice and Studies, Tripoli.

Allen, J. and Ewing, M.A. (1978) *Report on 4 Years' Experience (1974–78) Growing Annual Medics in the Jeffara Region.* Western Australian Advisory Group (Cereal Project – Jeffara Plain Authority), Libya.

Anonymous (1972a) *Cactus Sans Epines* (FAO Leaflet WI/C5540/2.72). Food and Agriculture Organization, Rome.

Anonymous (1972b) *Fodder Shrubs* (FAO Leaflet, PI/D6572). Food and Agriculture Organization, Rome.

Aronson, J., Floret, C., Le Floc'h, E., Ovalle, C. and Pontanier, R. (1995) Restauration et réhabilitation des écosystèmes dégradés en zones arides et semi-arides. Le vocabulaire et les concepts. In: *L'homme peut-il refaire ce qu'il a défait?* John Libbey Eurotext, Paris, pp. 11–29.

Arroyo, A.I., Pueyo, Y., Giner, M.L., Foronda, A., Sanchez-Navarrete, P., Saiz, H. and Alados, C.L. (2018) Evidence for chemical interference effect of an allelopathic plant on neighboring plant species: a field study. *PLOS ONE* 13(2), e0193421. DOI: 10.1371/journal.pone.0193421

Auclair, L. and Alifriqui, M. (2012) *Agdal: Patrimoine socio-écologique de l'atlas marocain* (Vol. 29). IRCAM-IRD, Publications de l'Institut Royal de la Culture Amazighe, Centre des Etudes Historiques et Environnementales, Rabat.

Boutaleb, A. (2013) Water harvesting increases productivity of degraded range in Morocco, MENARID Gateway. (Knowledge Fact Sheet 17).

Brichambaut, G. (1950a) Espèces introduites et spontanées du genre *Pennisetum*. *Les Cahiers de la Recherche Agronomique* 3, 339–400.

Brichambaut, G. (1950b) Un genre groupant de nombreuses espèces fourragères importantes: le genre panicum. *Les Cahiers de la Recherche Agronomique* 3, 332–337.

Canfield, R.H. (1941) Application of the line interception method in sampling range vegetation. *Journal of Forestry* 39(4), 388–394. DOI: 10.1093/jof/39.4.388

Chaïeb, M., Floret, C., Le Floc'h, E. and Pontanier, R. (1990) Les graminées pérennes, un recours pour la réhabilitation des terres de parcours dégradés en zone aride tunisienne? *Ecologia Mediterranea* 16(1), 415–425. DOI: 10.3406/ecmed.1990.1681

Chatty, D. (2013) *From Camel to Truck: The Bedouin in the Modern World.* White Horse Press. Available at: https://www.jstor.org/stable/j.ctv289dtw3 (accessed 24 August 2024).

CSIRO (1985) Australian plant introduction review. *Australian Plant Introduction Review* 18(1), 182.

Delaroque, F. (1998) *Use of a Geographical Information System to Identify Potentialities of Range Improvement in the Wadi Tips – Region of Marsa Matrouh – Northwest Coast of Egypt.* Mémoire de Maitrise IGARUN, Nantes, France.

Draz, O. (1974) The Hema system of range reserves in the Arabian Peninsula: its possibilities in range improvement and conservation projects in the Near East, Appendix 2. In: *Range Management and Fodder Development: Report to the Government of the Syrian Arab Republic*. FAO, Rome.

Draz, O. (1978) Revival of the Hema system of range reserves as a basis for the Syrian range development program. In: Hyder, D.N. (ed.) *Proceedings of the 1st International Rangeland Congress*, pp. 100–103. SRM.

Draz, O. (1985) The Hema system of range reserves in the Arabian Peninsula. In: McNeely, J.A. and Pitt, D. (eds) *Culture and Conservation: The Human Dimensions of Environmental Planning.* Croom-Helm, London, pp. 109–121.

Dumancic, D. and Le Houérou, H.-N. (1981) *Acacia cyanophylla* Lindl. as supplementary feed for small stock in Libya. *Journal of Arid Environments* 4(2), 161–167. DOI: 10.1016/S0140-1963(18)31581-7

El-Barasi, Y.M. and Saaed, M.W.B. (2013) Threats to plant diversity in the northeastern part of Libya (El-Jabal El-Akhdar and Marmarica Plateau). *Journal of Environmental Science and Engineering* 2, 41–58.

El-Barasi, Y.M., Saaed, M.W.B. and Al Tajoury, R.O. (2013) Land deterioration of a semidesert grazing area in the north-eastern zone of Libya (Cyrenaica). *Journal of Environmental Science and Engineering B* 2, 357–373.

El-Fikiki, A.R., Ibn Nuri, F.A. and Ibn Jassim, A.M. (2000) Agropastoral community experiences with fodder shrubs in Syria. In: Gintzburger, G., Bounejmate, M. and Nefzaoui, A. (eds) *Fodder Shrub Development in Arid and Semi-Arid Zones* Vol. 1. International Center for Agricultural Research in the Dry Areas, Hammamet, Tunisia, pp. 239–243.

Elhamrouni, A. and Sarson, M. (1973) *Cactus locaux ou introduits en Tunisie, INRF – Tunisie* (Note techniques N°16, FAO TUN/71/540). INRF-Tunisia.

Elhamrouni, A. and Sarson, M. (1975) Appétabilité de certains *Atriplex* spontanés ou introduits en Tunisie, INRF. *Note de Recherche, Food and Agriculture Organization* 8(540), 10.

Ewing, M.A. (1978) *Report on Experience with Cereals and Medic Pastures on the Cereals Project – Neffusa Region.* Western Australian Advisory Group (Cereal Project – Jeffara Plain Authority), Libya.

Ewing, M.A. (1979) *Progress with Annual Medics* (Medicago *sp.*) *in the Jeffara Region of the Socialist People's Libyan Arab Jamahiriya.*

Ewing, M.A. (1995) Pâturage à légumineuses annuelles: un élément essentiel pour la restauration et la réhabilitation des écosystèmes méditerrannéens sous faible pluviométrie. In: *L'homme peut-il refaire ce qu'il a défait?* John Libbey Eurotext, Paris, pp. 161–186.

Ewing, M.A., Standing, W., Fituri, M. and El-Taash, S. (1979) *Medic Experiments in the Cereal Project – Jeffara Plain.*

Ferchichi, A. (1995) Caractérisation morpho-biologique et écologique d'une espèce pastorale de la Tunisie présaharienne (*Periploca angustifolia* Labill.) = Implications pour l'amélioration pastorale. *Cahier Options Méditerranéennes* 12.

Ferchichi, A., Neffati, M. and Le Floc'h, E. (1991) Comportement comparé d'espèces pastorals de diverses provenances de la zone aride Tunisienne installées en pastoretum. *Proceedings of the International Rangeland Congress* 4, 364–367.

Floret, C., Le Floc'h, E. and Romane, F. (1981) Dynamique des systèmes écologiques de la zone aride: application à l'aménagement sur des bases écologiques d'une zone de la Tunisie Présaharienne. *Acta Œcologica: Œcologia Applicata* 2(3), 195–214.

Foury, A. (1950) Les légumineuses et graminées fourragères au Maroc. *Les Cahiers de la Recherche Agronomique* 3(3), 25–317.

Francis, C.M. (1978) *Distribution and Ecology of Annual* Medicago *sp. in Northwest Libya – A Report Based on a Plant Collection Tour 1978.* Western Australia Department of Agriculture, Perth, Australia.

Gachet, J.P. and Elmir, A. (1972) *Etude Monographique des* Medicago annuelles. *Annales de l'Institut National Agronomique de Tunisie* (Vol. 45–1). INRAT, Tunisia.

Gammoh, I.A. and Oweis, T.Y. (2011) Performance and adaptation of the Vallerani mechanized water harvesting system in degraded Badia rangelands. *Journal of Environmental Science and Engineering* 5, 1370–1380.

GEFLI (1974a) *Programme de mise en valeur de la zone pastorale développement de l'élevage ovin. Plaine de Gefara et djebel Nefusa. Rapport Agrostologique.* Groupement D'Etude Française en Libye.

GEFLI. (1974b) *Survey for Agricultural Development (Wadi Hira, Bu Shaiba and Gharyan Projects).* Groupement D'Etude Française en Libye.

Ghassali, F., Cocks, P., Osman, A. and Leybourne, M. (1999) Rehabilitation of degraded grassland in north Syria: use of farmer participatory research to encourage the sowing of annual pasture legumes, 1999. *Experimental Agriculture* 35(4), 489–506. DOI: 10.1017/S0014479799354089

Gintzburger, G. (1976) *Utilization of a Shrub Legume* Retama raetam *(R'tem)* (No. 82/76). FAO and Agricultural Research Center, Tripoli.

Gintzburger, G. (1978a) *A Prospection and Collecting Tour in the Middle Zone and the Eastern Part of Zone* 7 (No. 231/78). FAO Agricultural Research Centre, Tripoli, pp. 118–151).

Gintzburger, G. (1978b) *Collection of Seeds and Shrubs (Bir-el-Ghanem, Bir Ayyat and Nalut Area)* (No. 183/78). FAO Agricultural Research Centre). Tripoli.

Gintzburger, G. (1978c) *Plant Genetic Conservation, Forage Legume Collection in Libya, Part I: Organization and Preliminary Results* (No. 171/78). FAO Agricultural Research Centre, Tripoli.

Gintzburger, G. (1979) *Rangeland Regeneration in Libya – Protection and Fertilization: Preliminary Test at the Assah Camel Project* (No. 266/80). FAO and Agricultural Research Center, Tripoli.

Gintzburger, G. (1980) Forage legumes in Libya. *Plant Genetic Resources* 43, 7–11.

Gintzburger, G. (1983) *Libyan Medic Evaluation Project.* Mission report. Western Australian Overseas Project Authority, Perth, Australia.

Gintzburger, G. (1984) Annual *Medicago* spp. for the Mediterranean semi-arid and arid rangelands. Proceedings, 2nd International Rangeland Congress, Adelaïde, pp. 319–320.

Gintzburger, G. (1986) Seasonal variation in above-ground annual and perennial phytomass of an arid rangeland in Libya. *Journal of Range Management* 39(4), 348–353.

Gintzburger, G. (1998) *ICARDA Annual Report to Japan. Improvement of Native Pastures and Rangeland.* International Center for Agricultural Research in the Dry Areas, Aleppo.

Gintzburger, G. (1999) Range roller spirals to steppe seed success. *CARAVAN (ICARDA, Aleppo)* 10, 13–15.

Gintzburger, G. and Bayoumi, M. (1977) *Survey of the Present Situation and Production of the Libyan Rangelands* (No. 118/77). FAO and Agricultural Research Center, Tripoli.

Gintzburger, G. and Blesing, L. (1979) *Genetic Conservation in Libya – Indigenous Forage Legumes Collection in Northern Libya, Distribution and Ecology of* Medicago *sp.* Food and Agriculture Organization, Rome.

Gintzburger, G. and Huxley, P.A. (1979) *Surface Water Micro-Catchment and Range Plants Establishment – Preliminary Investigation* (No. 234/79). FAO and Agricultural Research Center, Tripoli.

Gintzburger, G. and Skinner, P. (1985a) A simple single disc pitting and seeding machine for rangeland revegetation. *Australian Rangeland Journal* 7–1, 29–31.

Gintzburger, G. and Skinner, P. (1985b) Annual medics species and pitting for rangeland regeneration. *Range Management Newsletter of the Australian Rangeland Society* 85(1), 13–15.

Gintzburger, G. and Skinner, P. (1985c) *Annual Medics Species for WA Rangelands: Selection and Establishment Techniques/Experimental Results 1984–85.* Western Australia Department of Agriculture Division of Plant Research, Perth, Australia.

Gintzburger, G., Sbeita, A., Rojbani, H., Fituri, M. and Francis, C.M. (1983) Annual *Medicago* sp. collected from the Sirte Region of Libya. *Plant Genetic Resources* 56, 23–29.

Gintzburger, G., Francis, C.H., Sbeita, A., Rojbani, H.R., Fituri, M. and Skinner, P.W. (1984a) *LMEP – Catalogue of the Annual* Medicago *sp. from the Western and Central zones of Northern Libya* (Vol. 2). Western Australian Overseas Project Authority, Perth, Australia.

Gintzburger, G., Sbeita, A. and Francis, C.M. (1984b) *Libyan Medic Evaluation Project – Final Report.* WAOPA-Western Australian Department of Agriculture, Perth, Australia.

Gintzburger, G., Skinner, P. and Liedel, L. (1986) A simple single disc pitting and seeding machine for rangeland revegetation operation (Spain Patent No. 551912 1987). Available at: https://patentimages.storage.googleapis.com/55/f6/7c/a4a2828ef73870/WO1986004771A1.pdf (accessed 24 August 2024).

Gintzburger, G., Bounejmate, M. and Nefzaoui, A. (2000) *Fodder Shrub Development in Arid and Semi-Arid Zones* (ICARDA Proceedings Hammamet Fodder Shrubs Workshop, Vols 1, 2). International Center for Agricultural Research in the Dry Areas, Hammamet, Tunisia.

Gouhis, F., Louhaichi, M. and Nefzaoui, A. (2018) Cactus crop (*Opuntia ficus-indica*) to rehabilitate rangelands in semi-arid regions of Tunisia. In: Sáenz, C., Dubeux, J.C.B., Prat, L. and Louhaichi, M. (eds) *IX International Congress on Cactus Pear and Cochineal CAM Crops for a Hotter and Drier World,* Coquimbo, Chile, 26–30 March 2017.

Gounot, M. (1969) *Méthodes D'étude Quantitative de la Végétation.* Editorial Masson, Paris.

Halse, N.J. (1978) *Land Use Recommendations for the Jeffara Plain,Tripoli, Libya.* Western Australia Department of Agriculture, Perth, Australia.

Halse, N.J. (1982) *Annual Inspection Visits to Cereal Project – Libya.* Western Australian Advisory Group Cereal Project – Jeffara Plain Authority.

Henkin, Z., Noy-Meir, I., Kafkafi, U., Gutman, M. and Seligman, N. (1990) Phosphate fertilization primes production of rangeland on brown rendzina soils in the Galilee, Israel. *Agriculture, Ecosystems and Environment* 59(1–2), 43–53.

Heyn, C.C. (1963) The annual species of *Medicago*. *Scripta Hierosolymitana* 12(1), 1–154.

Jigjidsuren, S. and Johnson, D.A. (2003) *Forage Plants in Mongolia.* Admon Press.

Johnson, D.A., Jigjidsuren, S., Sheehy, D.P., Majerus, M.E., Winslow, S.R. and Holzworth, L.K. (2006) Collection and evaluation of forage germplasm indigenous to Mongolia. *USDA Forest Service Proceedings RMRS* P-39, 50–61.

Johnston, D.R. and Mann, R.K. (2024) Rangeland pitting for revegetation and annual weed control. *Rangelands* 46(1), 23–30. DOI: 10.1016/j.rala.2023.11.002

Keith, H.G. (1965) *A Preliminary Check List of Libyan Flora, Volumes 1 and 2.* United Middle East Executives Ltd, London.

Khalil, H., Francis, C.M. and Halse, N.J. (1977) *Annual* Medicago *sp. Ecotypes from Northwest Libya. I – Preliminary Agronomic Evaluation* (No. 37; Technical Bulletin). Western Australia Department of Agriculture.

Killian, C. (1961) *Amélioration naturelle et artificielle d'un pâturage dans une réserve algérienne 'Le Mergueb'.* Imprimeries La Typo-Litho et Jules Carbonel réunies.

Le Floc'h, E., Neffati, M., Chaïeb, M. and Pontanier, R. (1995) Un essai de réhabilitation en zone aride: le cas de Menzel Habib (Tunisie). In: *L'homme peut-il refaire ce qu'il a défait?* John Libbey Eurotext, Paris, pp. 139–160.

Le Houérou, H.-N. (1959) Recherches écologiques et floristiques sur la végétation de la Tunisie méridionale. PhD dissertation. Université Alger Institut de Recherches Sahariennes, Algiers.

Le Houérou, H.-N. (1960) Contribution à la connaissance de la flore phanérogamique de la Libye. *Bulletin de la Société d'Histoire Naturelle de l'Afrique du Nord* 51, 175–200.

Le Houérou, H.-N. (1965) *Improvement of Natural Pastures and Fodder Resources.* Report to the government of Libya. Food and Agriculture Organization, Rome.

Le Houérou, H.-N. (1969) *La Végétation de la Tunisie Steppique (Avec Référence aux Végétations Analogues d'Algérie, de Libye et du Maroc)* (Vol. 1–2), 42–45. Annales de l'Institut National de la Recherche Agronomique de Tunisie, Tunis.

Le Houérou, H.-N. (1974) Principles, methods and techniques for range and fodder crops development in Tunisia and the Mediterranean. In: *Pasture and Fodder Crops Studies No. 2,* 3rd edn. Plant Production and Protection Division, Food and Agriculture Organization, Rome.

Le Houérou, H.-N. (1980a) L'*acacia cyanophylla* Lindl. comme fourrage complémentaire pour les petits ruminants en Libye. In: Le Houérou, H.-N. (ed.) *Les Fourrages Ligneux en Afrique.* CIPEA/ILCA, Addis Ababa, pp. 315–322.

Le Houérou, H.-N. (1980b) Les plantations d'arbres et d'arbustes fourragers: techniques d'implantation et de gestion. In: Le Houérou, H.-N. (ed.) *Les Fourrages Ligneux en Afrique*, pp. 345–352. CIPEA/ILCA, Addis Ababa.

Le Houérou, H.-N. (1981) *The Feed Value of* Atriplex *spp.* (Technical Paper No. 13, UNTF/Lib 18). FAO and Agricultural Research Center, Tripoli.

Le Houérou, H.-N. (1982) *The Impact of Climate on Pastoralism* (UNTF/Lib 018, Technical Paper no. 42). FAO and Agricultural Research Center, Tripoli.

Le Houérou, H.-N. (1996) The role of cacti (*Opuntia* spp.) in erosion control, land reclamation, rehabilitation and agricultural development in the Mediterranean Basin. *Journal of Arid Environments* 33(2), 135–159. DOI: 10.1006/jare.1996.0053

Le Houérou, H.-N. (2002a) Cacti as a fodder crop for marginal lands in the Mediterranean Basin. In: Nefzaoui, A. and Inglese, P. (eds) *Proceedings of the IVth International Congress of Cactus Pear and Cochinel.* Acta Horticultura.

Le Houérou, H.-N. (2002b) *Multipurpose Germplasm of Fodder Shrubs and Trees for the Rehabilitation of Arid and Semi-arid Land in the Mediterranean Isoclimatic Zone: A Photographic Catalogue.* CIHEAM, Zaragoza, Spain. Available at: https://om.ciheam.org/om/pdf/b37/02600003.pdf (accessed 24 August 2024).

Le Houérou, H.-N. and Barghati, M.S. (1982) *An Evaluation of Fodder Shrubs Plantations in the Benghazi Plains* (UNTF/Lib 018, Technical Paper no. 45). FAO and Agricultural Research Center, Tripoli.

Le Houérou, H.-N. and Franclet, A. (1971) *Les triplex en Tunisie et en Afrique du Nord* (Technical Report 7). Food and Agriculture Organization and US Department of Agriculture, Rome and Washington, DC.

Le Houérou, H.-N. and Pontanier, R. (1987) *Les plantations sylvo-pastorales dans la zone aride de Tunisie.* Man and Biosphere Technical Notes no. 18. UNESCO.

Le Houérou, H.-N., Dumancic, D. and Eskileh, M. (1982a) *Study of Fodder Shrubs Consumption by Sheep in the Wishtata Range Development Project* (UNTF/LIB 018, Technical Paper no. 43). FAO and Agricultural Research Center, Tripoli.

Le Houérou, H.-N., Dumancic, D., Eskileh, M., Schweisguth, D. and Telahigue, T. (1982b) *Anatomy and Physiology of a Browsing Trial: A Methodological Approach to Fodder Shrubs Evaluation* (Technical Paper 28). FAO and Agricultural Research Center, Tripoli.

Leone, G. (1924) Consolidamento ed imboschimento delle zone dunose della Tripolitania. *L 'Agric. Colon. Firenze* XVIII, 9, 299–308.

Lesins, K.A. and Lesins, I. (1979) Genus Medicago *(Leguminosae): A Taxogenetic Study*. Dr W. Junk Publishers, The Hague, Boston and London.

Mahmoud, A.M.A., Hasmadi, I.M., Alias, M.S. and Azani, A.M. (2016) Rangeland degradation assessment in the south slope of the Al-Jabal Al-Akhdar, Northeast Libya using remote sensing technology. *Journal of Rangeland Science* 6(1), 73–80.

Moss, W.M., Guzzomi, A.L., Foster, K.J., Ryan, M.H., Erskine, W. and Nichols, P.G.H. (2021) Vacuum harvesting sucks: improving the Horwood Bagshaw Clover Harvester. In: Cullen, B. (ed.) *Proceedings Australian Grassland Association Inc. Virtual Symposium*. Australian Grassland Association, pp. 15–19.

Nasr, N. (1993) *Systèmes agraires et organisations spatiales en milieu aride: Cas d'el Ferch et du Dahar de Chenini et de Guermessa (Sud-Est Tunisien).* Geography thesis, University Paul Valéry, Montpellier, France.

Nawaz, K. (2000) Rangeland rehabilitation through plantation of fodder shrubs in Kanak Valley, Baluchistan, Pakistan: community participation. In: Gintzburger, G., Bounejmate, M. and Nefzaoui, A. (eds) *Fodder Shrub Development in Arid and Semi-Arid Zones.* ICARDA Proceedings Hammamet Fodder Shrubs Workshop, Vol. 1, pp. 253–260. International Center for Agricultural Research in the Dry Areas, Hammamet, Tunisia.

Nefzaoui, A. (2000) Nutritive value of spineless cactus (*Opuntia ficus-indica* var. *inermis*) and *Atriplex* (*Atriplex nummularia*) based diets for sheep. In: Gintzburger, G., Bounejmate, M. and Nefzaoui, A. (eds) *Fodder Shrub Development in Arid and Semi-Arid Zones.* ICARDA Proceedings Hammamet Fodder Shrubs Workshop, Vol. 2, pp. 518–523. International Center for Agricultural Research in the Dry Areas, Hammamet, Tunisia.

Nefzaoui, A. (2002) Rangeland management options and individual and community strategies of agropastoralists in central and Southern Tunisia. In: Ngaido, T., McCarthy, N. and Gregorio, M. (eds) *International Conference on Policy and Institutional Options for the Management of Rangelands in Dry Areas.* Vol. 23. CAPRI Working Paper.

Nefzaoui, A. and El Mourid, M. (2008) Rangeland improvement and management in arid and semi-arid Environments of West Asia and North Africa. Available at: http://icarda.org/wspublications//Regional_program_reports_archive/North_Africa/Rangeland_improvement.pdf (accessed 24 August 2024).

Nefzaoui, A. and El Mourid, M. (2009) Cacti: a key-stone crop for the development of marginal lands and to combat desertification. *Acta Horticulturae* 811, 365–372. DOI: 10.17660/ActaHortic.2009.811.50

Nefzaoui, A., Ben Salem, H., Zaafouri, M. and Chouki, S. (2000) Tunisian experience with fodder shrubs. In: Gintzburger, G., Bounejmate, M. and Nefzaoui, A. (eds) *Fodder Shrub Development in Arid and Semi-Arid Zones* ICARDA Proceedings Hammamet Fodder Shrubs Workshop, Vol. 1, pp. 210–228. International Center for Agricultural Research in the Dry Areas, Hammamet, Tunisia.

Nefzaoui, A., El-Mourid, M. and El-Ouachi, M. (2014) The tribe – lpatform of participatory local development and management of communal rangeland resources. *Journal of Arid Land Studies* 24–1, 57–60.

Nègre, R. (1959) *Révision de* Medicago *d'Afrique du Nord, Extrait de la Société d'histoire Naturelle de l'Afrique du Nord*, pp. 267–314.

Nordblom, T.L., Arab, G., Gintzburger, G. and Osman, A.E. (1997) Results of a survey of Bedouin groups with contracts to graze the government rangeland plantation at Maragha, Aleppo province, Syria. In: Hadad, N., Tutwiller, R. and Thomson, E. (eds) *Improvements of Crop-Livestock Integration Systems In Dry Areas of West Asia and North Africa.* International Center for Agricultural Research In the Dry Areas, Aleppo, pp. 437–445.

Omar, M.K., Masoud, M. and Ali, H. (2017) The allelopathic effects of *Artemisia herba-alba* Asso: aqueous extracts on seed germination and seedling development of *Ceratonia siliqua* L. *Journal of Marine Sciences and Environmental Technologies* 3(2).

Orlovsky, L., Dourikov, M. and Babaev, A. (2004) Temporal dynamics and productivity of biogenic soil crusts in the central Karakum Desert, Turkmenistan. *Journal of Arid Environments* 56(4), 579–601. DOI: 10.1016/S0140-1963(03)00075-2

Osman, A.E., Pagnotta, M., Russi, L., Cocks, P.S. and Falcinelli, M. (1990) The role of legumes in improving marginal lands. In: Osman, A.E., Ibrahim, M.H. and Jones, M.A. (eds) *The Role of Legumes in the Farming Systems of the Mediterranean Areas.* Proceedings of a workshop on the Role of Legumes in the Farming Systems of the Mediterranean Areas UNDP/ICARDA, Tunis, 20–24 June 1988, pp. 205–216. Springer, Dordrecht. DOI: 10.1007/978-94-009-1019-5_18

Osman, A.E., Cocks, P.S. and Bahhady, F. (1994) Response of Mediterranean grassland to phosphate and stocking rates: livestock production. *The Journal of Agricultural Science* 123(3), 319–326. DOI: 10.1017/S0021859600070313

Puckridge, D.W. and French, R.J. (1983) The annual legume pasture in cereal ley-farming system of Southern Australia: a review. *Agricultural Ecosystem and Environment* 91, 229–237.

Quinlivan, B.J. (1971) Seed coat impermeability of legumes. *Journal of the Australian Institute of Agricultural Science* 37, 283–295.

Rae, J., Arab, G., Jani, K., Murad, N., Ngaido, T., Gintzburger, G. and Nordblom, T.L. (2000) Socioeconomics of shrub plantations in Syria. In: Gintzburger, G., Bounejmate, M. and Nefzaoui, A. (eds) *Fodder Shrub Development in Arid and Semi-Arid Zones*. ICARDA, Proceedings Hammamet Fodder Shrubs Workshop, Vol. 2, pp. 627–648. International Center for Agricultural Research in the Dry Areas, Hammamet, Tunisia.

Redjel, N. and Boukheloua, J. (2000) Algerian experience with fodder shrub plantation. In: Gintzburger, G., Bounejmate, M. and Nefzaoui, A. (eds) *Fodder Shrub Development in Arid and Semi-Arid Zones*. ICARDA, Proceedings Hammamet Fodder Shrubs Workshop, Vol. 1, pp. 203–209. International Center for Agricultural Research in the Dry Areas, Hammamet, Tunisia.

Saaed, M.W.B. (2008) Study of vegetation and land degradation factors in Daphna zone (north-eastern part of Libyan coast-Tobruk). MSc thesis, Department of Environmental and Engineering Science, the Libyan Academy of Postgraduate Studies, Benghazi, Libya.

Saaed, M.W.B., Ali, A.B., El-Barasi, Y.M. and Rahil, R.O. (2022) Potential and perspectives for establishing protected areas in El-Jabal El-Akhdar region, northeast Libya: an overview and example from 'El-Abyar area'. *Journal of Degraded and Mining Lands Management* 9(3), 3453–3464. DOI: 10.15243/jdmlm.2022.093.3453

Salem, B., Norman, H.C., Nefzaoui, A., Mayberry, D.E., Pearce, K.L. and Revell, D.K. (2010) Potential use of old man saltbush (*Atriplex nummularia* Lindl.) in sheep and goat feeding. *Small Ruminant Research* 91(1), 13–28. DOI: 10.1016/j.smallrumres.2009.10.017

Shawesh, O.M., Suleiman, K.A. and Yahia, T.A. (1996) *Aridity and Desertification and Their Effects on Food Security*. National Council for Scientific Research, Tripoli.

Small, E. and Jomphe, M. (1989) A synopsis of the genus *Medicago* (Leguminosae). *Canadian Journal of Botany* 67(11), 3260–3294. DOI: 10.1139/b89-405

Telahique, T. (1981) *Wishtata Grazing Perimeter*. Internal Report, FAO TF/LIB 10. Tripoli.

Vetter, T. and Rieger, A.-K. (2019a) Ancient water harvesting in the Old-World dry belt – synopsis and outlook. *Journal of Arid Environments* 169, 42–53. DOI: 10.1016/j.jaridenv.2019.103996

Vetter, T. and Rieger, A.-K. (2019b) Water harvesting as a key for understanding adapted ancient livelihoods in an arid environment – approaches to dryland archaeology in the Eastern Marmarica (NW Egypt). *Journal of Arid Environments* 171, 102940. DOI: 10.1016/j.jaridenv.2019.02.009

Vetter, T., Rieger, A.-K. and Nicolay, A. (2009) Ancient water harvesting system in the north Eastern Marmarica (North Western Egypt). *Libyan Studies* 40, 9–23. DOI: 10.1017/S0263718900004489

Villax, E.J. (1963) La culture des plantes fourragères dans la région Méditerranéenne Occidentale. *Les Cahiers de la Recherche Agronomique* 17.

Waechter, P. (1978) Essai d'orientation sur les possibilités d'accroitre la production végétale et d'améliorer les conditions d'integration des systèmes culturaux et d'élevage. Rapport, Institut des Régions Arides, Médénine, Tunisie 31.

Webb, L.G., Gintzburger, G. and d'Antuono, M.F. (1984) *The Libyan Medic Data Base: A Computerised Database for Plant Genetic Resources Using INFO HENC01 Software*. Western Australia Department of Agriculture, Perth, Australia.

Zaafouri, M.S., Akrimi, N., Floret, C., Le Floch, E. and Pontanier, R. (1994) Les plantations sylvo-pastorales en Tunisie pré-saharienne. *Sécheresse* 5, 265–275.

Zaafouri, M.S., Akrimi, N., Floret, Ch., Le Floc'h, E. and Pontanier, R. (1995) Les arbustes fourragers exotiques. In: Pontanier, R. (ed.) *L'homme peut-il refaire ce qu'il a défait?* John Libbey Eurotext, Paris, pp. 211–229.

Zahran, M.A. and Younes, H.A. (1990) Hema system: traditional conservation of plant life in Saudi Arabia. *Journal of King Abdulaziz University-Science* 2(1), 19–41. DOI: 10.4197/Sci.2-1.2

Ziani, P. (1970) *Atriplex halimus*: exploitation des formations naturelles et des plantations. Technical note. PNUD, Tunisie, INRF Tunisie.

17

Agro-ecological Zoning of Libya: The Case of Barley and the Olive Tree

Abstract

Considering the barley grain deficit for livestock feed and plans to expand olive tree production, we research if there is still some agricultural land available for these rainfed crops in Libya. We developed an agro-ecological zoning methodology based on algorithms and GIS mapping filtering limiting climatic, soil, and barley and olive tree specific agro-ecological and agro-biological factors. We produced maps of western and eastern Libya excluding current land use and accurately locating suitable land for the two crops. We identified 78,200 ha in the west for rainfed barley possibly yielding an extra 39,000 tonnes of grain/year. In the east, about 116,700 ha may produce 70,000 tonnes of grain/year. As for olive trees, 6,900 ha (240,000 trees) and 26,600 ha (1,330,000 trees) may be available in the west and in the east, respectively. The methodology could be further improved providing an updated land-use map and refined limiting factors. Cropping, planting new trees and harvesting on additional suitable land would have to remain with the identified actors and communities, their property rights, their traditional laws and operational rules.

Introduction

The need for sustainable strategic choices and management regarding land use is a major challenge for the development of agriculture in Libya. Libya faces a considerable shortage of locally produced livestock feed and therefore needs to import massive and expensive amounts of it (see Chapter 10). One wonders whether there would still be suitable land available in northern Libya for producing livestock feed, especially barley grain. Similarly, one wonders if there would still be available land for rainfed olive tree production mostly cornered in western Libya and rarely practised these days in eastern Libya. Would it be possible to find ancillary vacant and suitable land for additional rainfed barley and expand olive tree production without resorting to irrigation and, most importantly, without encroaching rangeland areas that are already under threat and degradation?

Combining agro-ecological zoning and specific agro-biological limiting factors for barley and olive trees offers solutions to these questions.

Agro-ecological Zoning

Agro-ecological Zoning (AEZ) defines homogeneous areas based on the interaction of climate, soils and topography combined with current land use. It assists in evaluating, locating and mapping suitable and available agricultural land. It was implemented long ago in Libya by land surveyors and managers to make the best use of available land resources (Le Houérou, 1984, 1989; Rowland, 1945) combining environmental

DOI: 10.1079/9781800627154.0017

information from hand-drawn maps and aerial photography (Allan *et al.*, 1973). It is now done much faster with geospatial analysis with GIS software and algorithms capable of filtering many layers of information to produce extremely accurate maps over large areas (Gintzburger *et al.*, 2009; Saïdi and Gintzburger, 2013) and even at a country scale.

Two types of AEZ potential are sequentially assessed:

- the pedoclimatic potential or theoretical potential defining the AEZ suitable or capable to sustain rainfed agricultural activities in broad terms (Batidzirai *et al.*, 2012). It is the land where the soil and climatic conditions meet the requirements for rainfed agriculture, i.e. rainfall, sunshine (duration and intensity), derived parameters (evapotranspiration, length of growing season, soil-water deficit), and topography (altitude and slopes); and
- the available or technical potential which is a fraction of the pedoclimatic potential. It identifies rainfed areas available for any possible crops combined, excluding all areas already under agricultural activities, urban development, natural reserves, forests etc.

Approach to agro-ecological zoning

The process to evaluate the soil and climate potential available for agricultural rainfed activities is presented in Fig. 17.1.

Assessing the pedoclimatic potential

Land potential according to climate

In arid Mediterranean regions, conditions are too dry for much of the year to allow maximizing crops' growth without irrigation. During the winter period, receiving annual rains, low temperatures may limit the growth and production of vegetation and crops. Calculating the length of the growing period (FAO, 1996) makes it possible to define the period of the year when the appropriate temperature and available water in the soil are met for vegetation growth and agricultural production. Climatic variables allow us to calculate the length of the growing periods and then compare them to the needs of crops. This concept is essential to AEZ, bringing in the seasonality in the assessment of land resources and agricultural capabilities. The parameters considered are as follow:

- LGP = length of growing period
- Mwb = monthly water balance
- Mr = cumulative rainfall for the month (in mm)
- Etpm = evapotranspiration of the month (in mm)
- UWRm = useful water reserve at the start of the month (in mm)
- WR = water reserve

The length of the growing period model

The length of the growing period (LGP) is a synthetic parameter affecting crops and integrating periods of drought, evapotranspiration and months with low or high temperatures. It defines the periods when the monthly average temperature (Tavg-m) is greater than 5°C, with 'm' (mean minimum temperature of the coldest month) and when the Mwb is positive. The data used to calculate the Mwb come from the Worldclim Version-2 database (Fick and Hijmans, 2017) and given by:

$$\text{Mwb} = (\text{Mr} - \text{Etpm}) + \text{UWRm}$$

The soil water reserve for each month (UWRm) depends on the Mwb of the previous month and computed as follows:

Condition [Mwb-1 > 0, UWRm = Mwb-1]
Condition [Mwb-1 ≤ 0, UWRm = 0)]

The Mwb is calculated month after month. It starts when the soil is saturated with water, generally when the rainy season starts at the end of the dry summer. In this case, the UWRm equals the initial soil-water reserve (WR) depending on the depth and type of soil. In the case of Libya, the starting date for calculating the Mwb, the monthly water reserve (UWRm) and the monthly growth period (GP) is the month of November.

Calculating the soil water reserve

The useful soil water reserve (WR) is the quantity of water the soil may store and supply to plants.

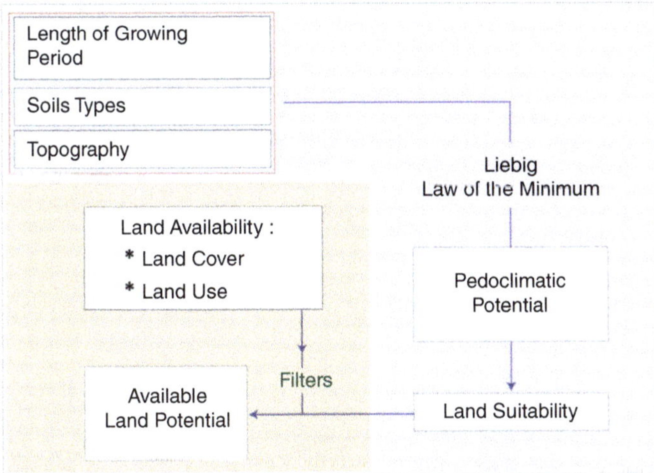

Fig. 17.1. Agro-ecological zoning process.

WR is computed for each type of soil according to our soil map (see Chapter 5) and ISRIC data (Hengl *et al.*, 2014) with:

- initialization of UWR (October) with WR, the period when the soil is saturated with water;
- initialization of the number of months of growth with GP = 0; and
- for each month (e.g. November to the preceding October, and so on):

UWRm = UWRm-1
Mwb = (Mr – Etpm) + UWRm
Condition [Mwb >0, UWRm = min (WR, Mwb)]
Condition [Mwb ≤0, UWRm = max (0, Mwb)]
Condition [Mwb > 0, GP = 1]
Condition [Mwb ≤ 0, GP = 0]
LGP tot = Σ(1 to 12) GP

The length of the growing period (Map 17.1) is labelled as follows according to FAO classification (FAO, 1996):

- Class 5 (S1): highly suitable, longest growth period (about 150 days/5 consecutive months), appropriate for all cereals and fruits trees;
- Class 4 (S2): moderately suitable, growth period (about 90–120 days/3–4 consecutive months, appropriate for cereals;
- Class 3 (S3): marginally suitable, short growth period (about 60 days/2 consecutive months), appropriate for cereals and rangelands;
- Class 2 (N1): marginally not suitable, very short growth period (about 30 days/one month after first rain), opportunistic rainfed barley cropping, mostly rangelands; and
- Class 1 (N2): permanently not suitable, no growth period (0 days – mostly desert), seasonal rangelands in exceptional years.

Land potential according to soil types

The soil types (depth, texture, physical and chemical characteristics) define the land potential for agriculture purpose based on the soil map of Libya (see Chapter 5). This is translated into terms of soil potential agricultural classes (Map 17.2) as follows:

- Class 4: highly suitable, soils with high agricultural potential (deep sandy-loam, loam, loess, *Terra rossa*, brown steppic soil);
- Class 3: suitable, soils with marginal agricultural potential (depth less than 1 m, some rock or stone content);
- Class 2: moderately suitable, soils with limited agricultural potential (mobile sand,

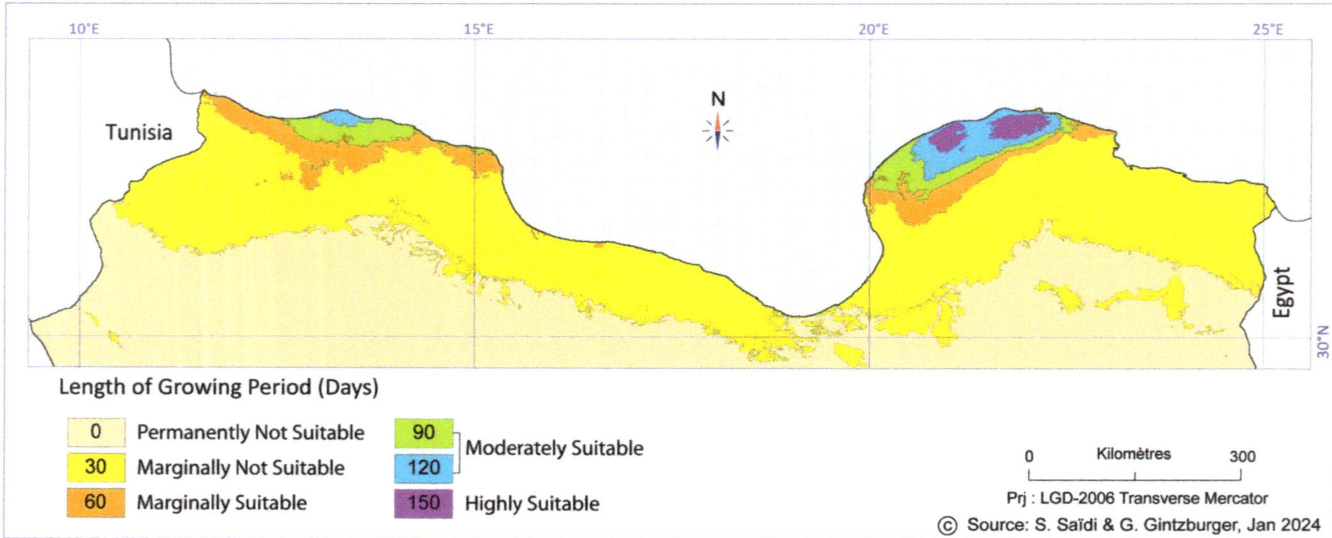

Map 17.1. The length of growing period (LGP) in northern Libya.

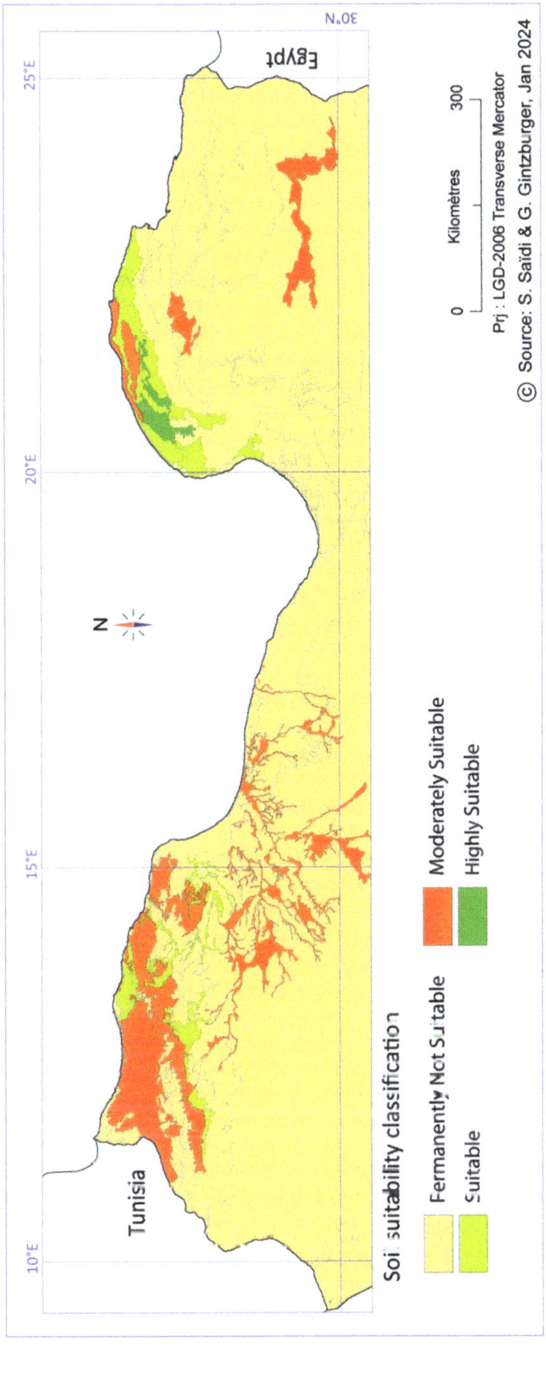

Map 17.2. The soil agronomic suitability in northern Libya.

fixed sand dune, heavy soil, rocky soils); and

- Class 1: soils permanently not suitable (all shallow soils, rock outcrop, exposed calcareous or gypseous crust, saline depressions).

We define agricultural potential when a soil is appropriate for rainfed cropping, fruit trees etc., mostly from class 4 and 3. Some of the class 2 may be used for rainfed cropping and occasional irrigation projects providing heavy input. Class 1 would be mostly rangelands.

Land potential according to topography

The topography defines the capability of the land for agriculture activities according to its position on the digital terrain model: summit, slope, lowland, foothills etc. The slopes' classification (Map 17.3) is extracted from the SRTM (USGS, 1996) with a 90 m resolution using the 'slope' function of ArcGIS as follows:

- Class A: 0:3%, strongly suitable, mechanized cultivation without major constraints
- Class B: 3:6%, suitable, mechanized cultivation may require special equipment
- Class C: 6:6559%, marginally, suitable requiring intensive manual labour (terracing)
- Class D: > 9%, permanently not suitable, limiting any cultivation due to rugged terrain and erosion risks

Mapping the Pedoclimatic and the Technical Potential

Each growth period, soil-water reserve, soil type and topography layers are rasterized and resampled at 100-m resolution to obtain the pedoclimatic agricultural potential maps (Map 17.4 and Map 17.6).

The technical agricultural potential (Map 17.5 and Map 17.7) is then identified by removing the urban and infrastructure zones, the current agricultural land use, and the conservation of biodiversity and natural habitats (IUCN-UNEP, 2019) from forested and afforested areas. In the absence of an official Libyan national up-to-date land-use map, we used the one published by FAO and UNDP (2006). Subsequently, each raster is reclassified into classes from 1 to 4. Finally, the ultimate layers of land availability are mapped by retaining the most limiting factor for each pixel according to Liebig law of the minimum (Liebig, 1840).

The resulting pedoclimatic agricultural suitable potential is mapped according to the FAO recommendations (FAO, 1983):

- Class S1: highly suitable land having few limitations not significantly reducing productivity. The expected yield is greater than 60% of the rainfed potential yield.
- Class S2: moderately suitable land with some limitations reducing productivity and possibly increasing inputs. The expected yield is greater than 40% of the potential yield achievable without irrigation.
- Class S3: marginally suitable land with serious limitations reducing productivity with input hardly justified. The expected yield is less than 20% of the rainfed potential yield.
- Class N2: permanently not suitable land having physical limitations prohibiting any possible cropping except rangelands grazing.

The suitable and the available agricultural potentials are mapped according to the different steps described above with a 1 km resolution.

Discussion and Preliminary Conclusions on the General AEZ

Considering all pedoclimatic and topographic limitations and data available from the current Libyan land cover (Nwer *et al.*, 2013), the theoretical and finally available rainfed potential agricultural land in Libya are summarized in Table 17.1.

In the east (Map 17.5), a scanty 8,000 ha of high rainfed potential land may be available in the Jebel Al-Akhdar close to El-Abyar, though these may already be occupied with afforestation or occasionally cropped. Available areas with moderately rainfed potential covers some 300,000 ha, but it raises doubts about their ease of access as these are mostly rugged karstic

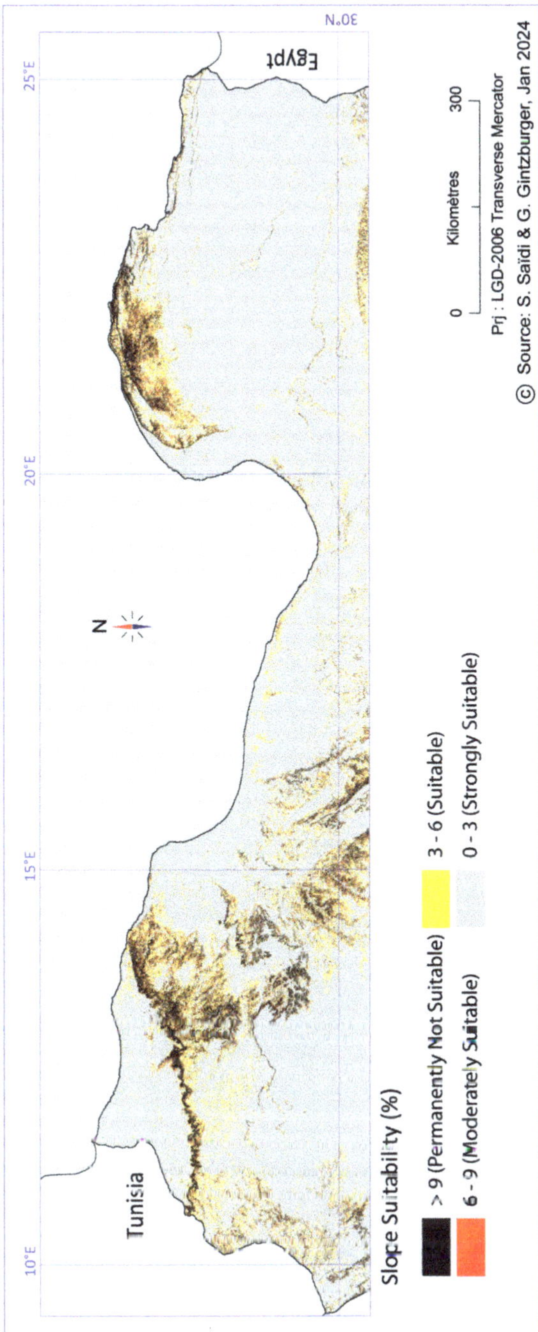

Map 17.3. The slope classification in northern Libya.

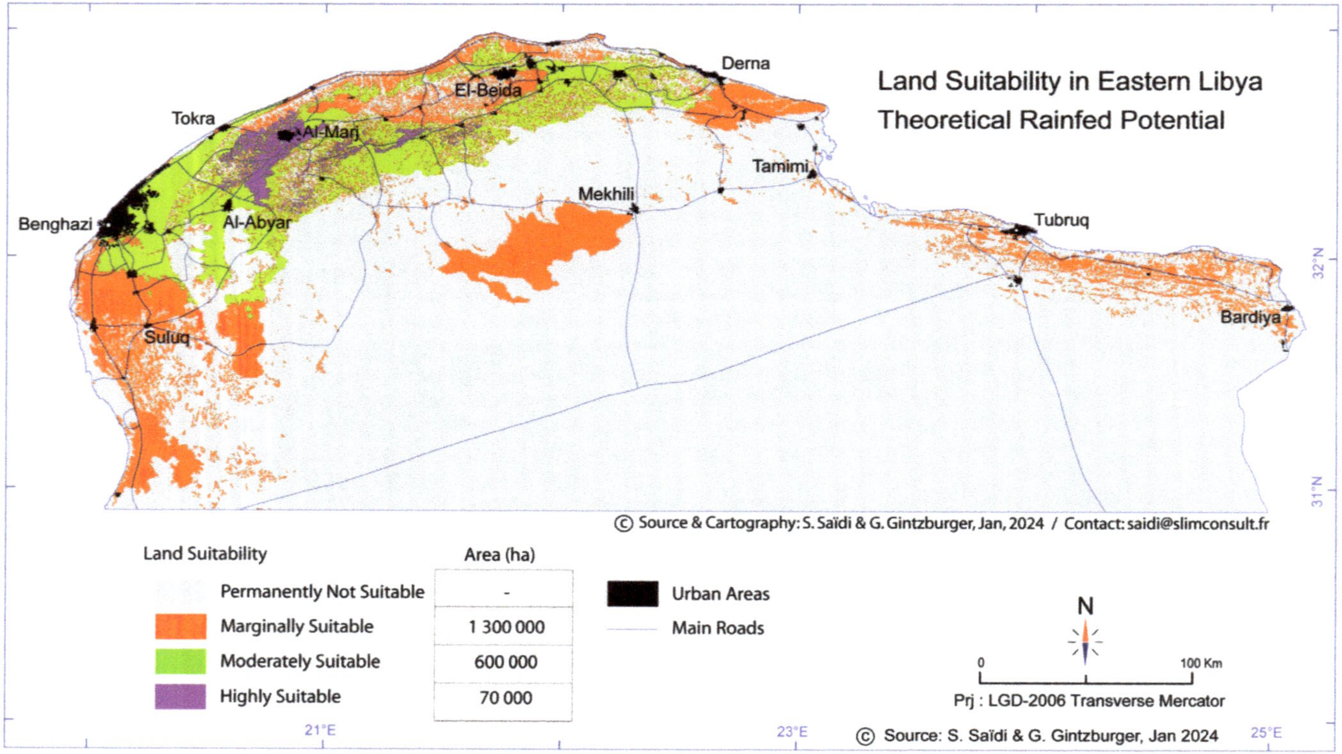

Map 17.4. Theoretical rainfed potential (east).

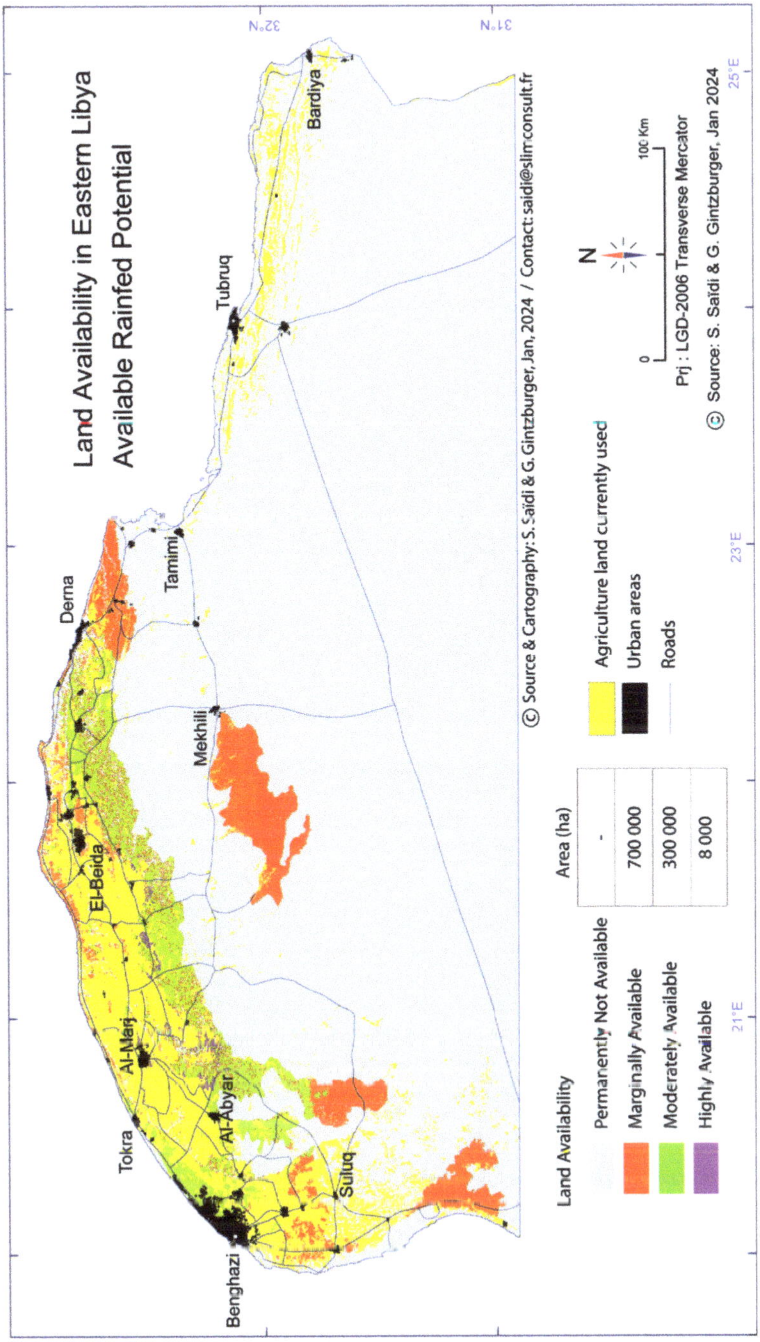

Map 17.5. Available rainfed potential (east).

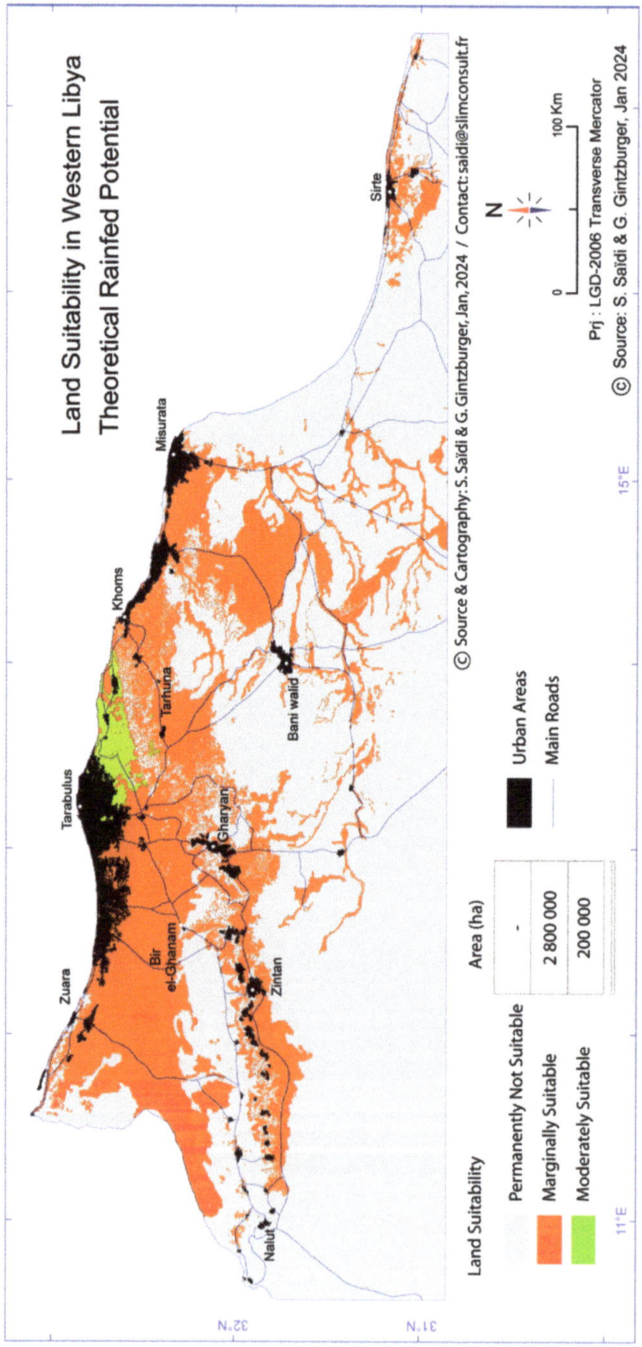

Map 17.6. Theoretical rainfed potential (west).

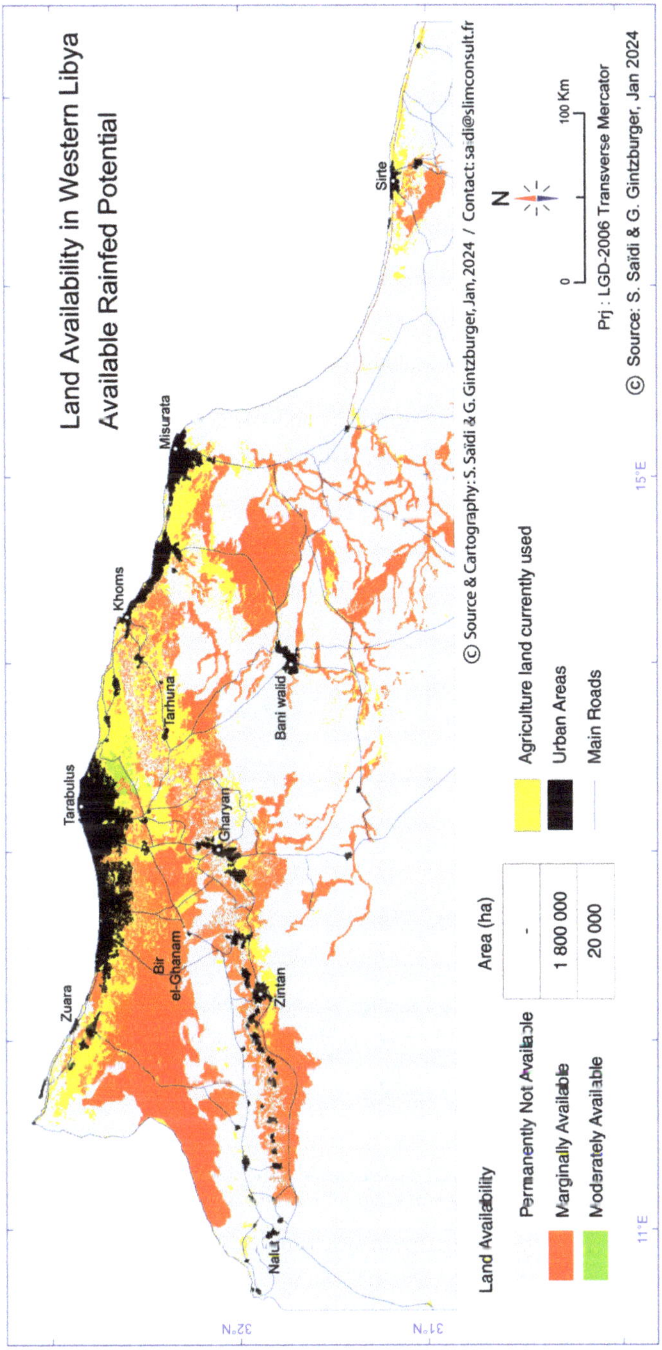

Map 17.7. Available rainfed potential (west).

Table 17.1. Agricultural theoretical and available rainfed potential in northern Libya.

(1) Theoretical rainfed potential		(2) Available rainfed potential		
East (Map 17.4)	ha	**East** (Map 17.5)	ha	% (2/1)
Highly suitable	70,000	Highly available	8000	11%
Moderately suitable	600,000	Moderately available	300,000	50%
Marginally suitable	1,300,000	Marginally available	700,000	54%
West (Map 17.6)	ha	**West** (Map 17.7)	ha	% (2/1)
Highly suitable	0	Highly available	0	0%
Moderately suitable	200,000	Moderately available	20,000	10%
Marginally suitable	2,800,000	Marginally available	1,800,000	64%

rangeland countries with forest and garrigues. It must be checked and corrected once the mountainous forested areas in the east are accurately mapped. The marginally available rainfed potential areas, with about 700,000 ha, are located in low-rainfall areas in the *baltat* zones, wadis Samalus and Mekhili, south of the Jebel Al-Akhdar, sporadically cropped with barley in good rainy years and definitely not suitable for any other crops. The exception to these marginal potential areas is east to wadi Derna to the north of wadi Al-Khabta north of Martuba, where many new private irrigated farms are now expanding. It is also the case around Suluq – Jardinah supplied with water from the Omar Al-Mukhtar reservoir.

In the west (Map 17.6), all the highly suitable rainfed land is already used and exhausted. Only about 20,000 ha of moderately available areas may be accessible close to Tripoli, south-east of Tajoura and Sidi Ben Hur. However, these areas are covered with mobile sand dunes where forestry plantations have been implemented over many years and where irrigated farms are steadily encroaching. The remaining 1,800,000 ha with marginal potential, mostly rangelands, offer little prospect of rainfed development as arid and semi-arid rangelands are increasingly infringed by private irrigated farms as seen south of Zuara – Sorman toward Bir-el-Ghanem (see Chapter 15). Further south, wadis and depressions are occasionally cropped with rainfed barley, or coming under pressure from new irrigation projects.

Our rainfed theoretical and available rainfed maps were produced with currently accessible climatic, soil, digital elevation models and the FAO 2004–06 land-use data that must be updated. Our aim is to describe a GIS procedure that could be refined, if and when updated information and accurate land-use maps become available.

It is clear that Libya is already occupying all the best soils receiving sufficient rainfall for rainfed cropping. Moderate- and marginal-potential lands and flats and traditional rangelands are now the target of hazardous cropping and numerous new private and governmental irrigation projects, the latter as long as water remains available in quantity and quality.

Potential Agricultural Land Available for Additional Rainfed Barley and Olive Trees in Northern Libya

In a context of food insecurity combined with climate change, it becomes necessary to identify where it may still be possible to hypothetically extend local crops or new crops. In most cases, the implementation of new crops on new land may involve a change in land use such as shifting from one crop to another or even converting rangelands or forests into cultivated land, an option we would not encourage. Using an advanced AEZ agro-biological methodology, our goal is to identify potential additional agricultural areas that are not yet allocated in the current Libyan land use, but which may be suitable and available for cropping rainfed barley or growing olive trees. The challenge is enormous as it supposed that the land-use map available is up-to-date and accurate. The emphasis is on crop sustainability and the means of achieving it through existing production systems.

Our approach is important for Libya in terms of food security and economic development by combining agro-ecological zoning and agro-biological limiting factors. It presents a comprehensive assessment to locate and estimate the potential additional land available for two crops, rainfed barley and olive, both of major economic importance in Libya. This modelling is based on the most recent scientific, geospatial techniques, and practical and operational knowledge, providing solid support to Libyan agricultural authorities aiming at changing current investments and redirecting them towards new arable lands.

Materials and methods

The first step consists in identifying and mapping the potential theoretical and available rainfed land potential (Maps 17.4, 17.5, 17.6 and 17.7).

The second step considers the limiting agro-biological factors permitting locating which part of the land is available for sustainably growing a specific crop or tree. It is based on a multi-criteria raster-data spatial analysis of the agro-ecological zoning with limiting agro-biological factors specific to the intended crops or trees. We chose to test rainfed barley and olive trees, both traditional and essential crops grown in Libya either on small family farms or on larger areas by livestock owners with low level of agricultural input and equipment.

Barley

Barley is commonly cropped in Libya and from Maghreb to Mashreq to support the mixed crop-livestock rangeland system. It is easily cropped in the most difficult dry but risky environments with minimum farming equipment. It is a multiple-use cereal for the local farmers as coarse bread or other typical Libyan staple food. It is also the most common supplementary livestock feed used for rangeland livestock as grain (see Chapter 10), but also as green forage or dry crop residues, all with high nutritive value (Goodchild *et al.*, 2022).

The cultivated barley (*Hordeum vulgare*) is likely the oldest domesticated cereal (Parry and Parry, 1993; Snir *et al.*, 2015). The centre of biodiversity of the *Hordeum* genus are the Middle East and the rim of the Mediterranean basin (von Bothmer *et al.*, 2003) with *Hordeum vulgare* ssp. *spontaneum*, its wild ancestor, still present in the Jebel Al-Akhdar (Keith, 1965). Over centuries, Libyan farmers have selected many local drought-resistant barley varieties (Al-Jibouri, 1966; Keith, 1965) slowly replaced by modern commercial varieties.

The agro-biological factors relating to rainfed barley cultivation are presented in Table 17.2.

Olive trees

Olea europea ssp. *sylvestris* is the wild form of olive tree. It is still present, though rare, in the Jebel Al-Akhdar (wadi Saaîdi, N 32.518 E 22.646, 14 km south-west of Martuba). The domesticated olive tree (*Olea europaea* L.) dates back approximately 6000 years (Amouretti and Comet, 2000) in Asia Minor and appeared to expand further from Syria to Greece via Anatolia between 3800 BC and 3200 BC, then extend to the west by the 6th century BC with the Phoenicians throughout the Mediterranean basin into North Africa.

Dating back to the Phoenician-Roman time until today (Mattingly, 1988, 1993, 1994, 1995), olive groves are mostly present in the north-western regions of Libya in the Jebel Nafusa, from Khoms to Misurata-Zliten, Msallata, Tarhuna – Gharyan – Yefren, from Zintan to Nalut and on the coastal zone of the Jeffara plain from Zuara to Garabulli. In Cyrenaica, many antique olive presses are still present (Buzaian, 2022). There are reports of olive orchards as early as the 3rd century BC by the Greek and Roman colonizers. Past the 7th century BC, however, olive production faded in Cyrenaica. It was gradually abandoned by the 11th century BC when the Banu Hilal and Banu Sulaym nomads colonized the country. By the early 19th century, explorers noted that Cyrenaican dwellers were using *ghee* butter from sheep milk for cooking and not olive oil. Olive drupes from long-abandoned eastern orchards were fed to sheep and cattle. Remarkably, during the Ottoman Empire, by the end of the 19th century, European travellers mentioned 200,000 olive trees, i.e. approximately 2000–4000 ha of olive orchards between

Table 17.2. Limiting agro-biological factors for barley.

Limiting factors for barley	Factors	Data source	Interval	Threshold	Score	Notes
Length of growing period (LGP) from Nov. to Feb. (NDJF)	LGP	Fick and Hijmans (2017)	3–5 months	> 4	5	Highly suitable
				2 < LGP < 4	4	Suitable
				< 2	1	Permanently unsuitable
Seasonal water deficit (SWD)	SWD	Bennani *et al*. (1992); Fick and Hijmans (2017); Trabucco and Zomer (2018)	0– 262 mm	≤ 50	5	Highly suitable
				50 < SWD < 150	4	Suitable
				150 < SWD < 250	3	Moderately suitable
				250 < SWD < 350	2	Marginally suitable
				> 350	1	Permanently unsuitable
Growing season temperature (NDJF)	Tmax	Fick and Hijmans (2017); Najar *et al*. (2017)	15– 35°C	> 35	1	Permanently unsuitable
				25 < Tmax < 35	3	Moderately suitable
				20 < Tmax < 25	4	Suitable
				15 < Tmax < 20	5	Highly suitable
	Tmin	Fick and Hijmans (2017); Najar *et al*. (2017)	1–25°C	≥ 25	3	Moderately suitable
				1 < Tmin < 10	5	Highly suitable
				< 1	1	Permanently unsuitable
	Tavrg	Bensemane (2018); Fick and Hijmans (2017)	18–30°C	≥ 30	2	Marginally suitable
				25 < Tavrg < 30	4	Suitable
				18 < Tavrg < 25	5	Highly suitable
				< 18	1	Permanently unsuitable
Topography	Slope	USGS (1996)	%	≤ 6	5	Highly suitable
				6 < Slope < 12	4	Suitable
				12 < Slope < 24	2	Marginally suitable
				> 24	1	Permanently unsuitable
Soils, if MAR > 300 mm/year	Texture	See Chapter 5 Hengl *et al*. (2014)	Particle size	Clay	4	Suitable
				Sebkha	1	Permanently unsuitable
				Sandy clay	4	Suitable
				Clay loam	4	Suitable
				Sandy-clay loam	5	Strongly suitable
				Loam with 40% sand, 40% loam and 20% clay	5	Highly suitable

				Sandy loam	5	Highly suitable
				Loamy sand	4	Suitable
				Mobile sand	1	Permanently unsuitable
	Depth	See Chapter 5; Hengl *et al.* (2014)	cm	≥ 100	5	Highly suitable
				75–100	5	Highly suitable
				50–75	4	Suitable
				25–50	2	Marginally suitable
				≤ 25	1	Permanently unsuitable
	Coarse elements	See Chapter 5; Hengl *et al.* (2014)	%	≤ 25	5	Highly suitable
				25–50	3	moderately suitable
				> 50	1	Permanently unsuitable
Soils, if 175 < MAR < 300 mm/year	Texture	See Chapter 5; Hengl *et al.* (2014)	Particle size	Clay	1	Permanently unsuitable
				Sebkha	1	Permanently unsuitable
				Sandy clay	2	Marginally suitable
				Clay loam	3	Moderately suitable
				Sandy-clay loam	5	Highly suitable
				Loam with 40% sand, 40% loam and 20% clay	5	Highly suitable
				Sandy loam	5	Highly suitable
				Loamy sand	4	Suitable
				Mobile sand	1	Permanently unsuitable
	Depth	See Chapter 5; Hengl *et al.* (2014)	cm	≥ 100	5	Highly suitable
				75–100	5	Highly suitable
				50–75	4	Suitable
				25–50	2	Marginally suitable
				≤ 25	1	Permanently unsuitable
	Coarse elements	See Chapter 5; Hengl *et al.* (2014)	%	≤ 25	5	Highly suitable
				25–50	3	Moderately suitable
				> 50	1	Permanently unsuitable

Benghazi and Derna (Buzaian, 2022), and between Gubba and Ain Mara. Early in the 20th century, travellers observed the wild abandoned olive trees along the road from El-Marj to Shahat and wadi Kuf. The Italian colonization attempted to revive olive plantations in Cyrenaica, later supported by the British Military Administration post-WWII, but without success, seeing the local livestock owners systematically grazing in abandoned orchards and cutting trees for fuelwood. Later, post Libyan independence in 1951, the Ministry of Agriculture promoted the plantation of olive trees free of charge, and even financed a government olive press in Massah, the latter finally closing down in 1965 due to the lack of olive supply as Cyrenaican dwellers were not interested in harvesting olives (Buzaian, 2022). To date, it is reported that Libya has roughly 8–9 million olive trees in production on about 205,000 ha at an average density of 40–45 olive trees/ha (Abuseif *et al.*, 2024).

The agro-biological factors relating to rainfed cultivation of olive trees are presented in Table 17.3.

Principle of the method

Deciding about land accessibility for barley and olive trees is a two-step approach: the first step is determining and mapping the theoretical suitable rainfed potential (Maps 17.4 and 17.6); the second step is similar to the method to produce the land availability (Maps 17.5 and 17.7) but with processing the GIS AEZ algorithms with the specific agro-biological parameters of the two crops that can be biologically and technically farmed.

Assessing the agro-biological potential

The agro-biological factors relating to rainfed barley cultivation and olive trees are presented in Tables 17.2 and 17.3, respectively.

TEMPERATURES. During the winter, receiving most annual rains, low temperatures limit crops and tree growth, and production. High temperatures and *ghibli* also limit yield. The thermal requirements of barley and olive trees are presented in tables 17.2 and 17.3.

RAINFALL AND SOIL-WATER DEFICIT. The soil-water deficit is the main factor limiting the yield of cereal crops and olive trees in arid and semi-arid zones. It results from the daily or monthly availability of rainfall accumulated in the soil supplying the crop minus the evapotranspiration. In Libya, the seeding and growing season starts by late October/early November with the first winter rain after the dry summer and autumn. It is the usual barley seeding time for the region when the soil reaches a useful soil-water reserve estimated at 120 mm (Trabucco and Zomer, 2018). The soil-water reserve represents the water stress that barley or olive trees may be exposed to. The crop yield is reduced when the soil-water reserve is insufficient, and, if exhausted, the crop yield will fail. Jouve (1984) assesses the impact of the soil-water deficit on potential production with a linear regression between the average monthly water deficit and the barley yield. This water deficit is converted into theoretical barley yield values ranging from 0 q/ha to 28 q/ha. According to Bennani *et al.* (1992), wheat crops in arid zones may accept an average water deficit ranging from 21 mm/month for an early variety up to a total of 262 mm per cropping season for a late variety. Beyond that, the area is considered unsuitable for cereal cropping. It confirms Le Houérou's (1959, 1969) North African field observations indicating that barley will fail with less than 200 mm MAR and 300 mm for wheat. We used these values to scale the ability of barley to produce or fail during a season.

Olive trees lose between 2.7 mm and 8.5 mm/day (Villalobos *et al.*, 2000). Habaieb and Masmoudi-Charfi (2003) indicate that olive tree cultivation becomes profitable when the total evapotranspiration does not exceed 500 mm/year.

In order to find the areas limiting the spatial distribution of this crop (olive or barley) according to the ETP, we calculate the monthly water balance. The latter shows that in Libya the dry months extend from April to October, i.e. about 210 days of drought. To establish the ability of the olive tree to grow, depending on seasonal dry period, the maximum ETP value retained is 1800 mm/year.

Table 17.3. Limiting agro-biological factors for olive trees.

Limiting factors for olive trees	Factors	Data source	Interval	Threshold	Score	Note
Length of growing period (LGP)	LGP	(Fick and Hijmans (2017); Le Houérou (1959)	2–5 months	≥ 4	5	Highly suitable
				2 < LGP < 4	4	Suitable
				< 2	1	Permanently unsuitable
Annual water deficit (AWD)	AWD	(Fick and Hijmans (2017); Habaieb and Masmoudi-Charfi (2003); Trabucco and Zomer (2018); Villalobos *et al.* (2000)	0–1800 mm	≤ 1200	5	Highly suitable
				1000 < AWD < 1200	5	Highly suitable
				1200 < AWD < 1500	4	Suitable
				1500 < AWD < 1800	2	Marginally suitable
				> 1800	1	Permanently unsuitable
Temperature	Tmax	Baldy (1990); Fick and Hijmans (2017)	15–35°C	> 35	1	Permanently unsuitable
				30 < Tmax < 35	2	Marginally suitable
				20 < Tmax < 30	4	Suitable
				15 < Tmax < 20	5	Highly suitable
				< 15	2	Marginally suitable
	Tmin	Baldy (1990); Fick and Hijmans (2017)	1–25°C	≥ 25	3	Moderately suitable
				15 < Tmin < 25	4	Suitable
				1 < Tmin < 15	5	Highly suitable
				< 1	1	Permanently unsuitable
	Tavrg	Baldy (1990); Fick and Hijmans (2017)	15–40°C	> 35	1	Permanently unsuitable
				20 < Tavrg < 35	4	Suitable
				15 < Tavrg < 20	5	Highly suitable
				< 15	2	Marginally suitable
Topography	Slope	(Loukili and Ihissou (2016); USGS (1996)	%	≤ 12	5	Highly suitable
				12 < Slope < 25	4	Suitable
				> 25	1	Permanently unsuitable

Continued

Table 17.3. Continued.

Limiting factors for olive trees	Factors	Data source	Interval	Threshold	Score	Note
Soils, if MAR > 300 mm/year	Texture	See Chapter 5; Hengl *et al.* (2014)	Particle size	Clay	4	Suitable
				Sebkha	1	Permanently unsuitable
				Sandy clay	4	Suitable
				Clay loam	4	Suitable
				Sandy-clay loam	5	Highly suitable
				Loam with 40% sand, 40% loam and 20% clay	5	Highly suitable
				Sandy loam	5	Highly suitable
				Loamy sand	4	Suitable
				Mobile sand	1	Permanently unsuitable
	Depth	See Chapter 5; Hengl *et al.* (2014)	cm	≥ 100	5	Highly suitable
				75–100	3	Moderately suitable
				50–75	1	Permanently unsuitable
				0–50	1	Permanently unsuitable
				≤ 25	1	Permanently unsuitable
	Coarse elements	See Chapter 5; Hengl *et al.* (2014)	%	≤ 25	5	Highly suitable
				> 25	1	Permanently unsuitable
Soils, if 175 < MAR < 300 mm/year	Texture	See Chapter 5; Hengl *et al.* (2014)	Particle size	Clay	1	Permanently unsuitable
				Sebkha	1	Permanently unsuitable
				Sandy clay	2	Marginally suitable
				Clay loam	3	Moderately suitable
				Sandy-clay loam	5	Highly suitable
				Loam with 40% sand, 40% loam and 20% clay	5	Highly suitable
				Sandy loam	5	Highly suitable
				Loamy sand	4	Suitable
				Mobile sand	1	Permanently unsuitable
	Depth	See Chapter 5; Hengl *et al.* (2014)	cm	≥ 100	5	Highly suitable
				75–100	3	Moderately suitable
				50–75	4	Suitable
				25–50	3	Moderately suitable
				≤ 25	1	Permanently unsuitable
	Coarse elements	See Chapter 5; Hengl *et al.* (2014)	%	≤ 25	5	Highly suitable
				> 25	1	Permanently unsuitable

COMPUTING THE WATER DEFICIT. The water deficit (Wd) is calculated using the IRHO method (Brunin and De Berchoux, 1994) based on the calculation of the monthly water balance (Mwb) where:

Mwb = (Mr – Metp) + UWRm and mapped (Maps 17.8 and 17.9)

The water deficit (Wd) is calculated month by month starting with a saturated IRW in November set at 120 mm (Trabucco and Zomer, 2018). Subsequently, we iteratively calculate the WR of each month. It stands for the reserve from the end of the previous month. It takes the value of the previous month if it is greater than zero. Otherwise, IRW is set to 0. The Wd is the maximum between zero and the absolute value of Mwb when negative. Tables 17.2 and 17.3 reflect the ability of barley and olive trees to produce.

LAND SUITABILITY ACCORDING TO SOIL TYPES. The agricultural suitability map is based on the characteristics of its soil (texture, depth, physical and chemical characteristics) from the Libyan soil map (see Chapter 5) and the ISRIC database (Hengl *et al.*, 2014), then translated in terms of agronomic suitability classes in Tables 17.2 and 17.3.

LAND SUITABILITY ACCORDING TO TOPOGRAPHY/LANDFORMS. The topography impacts the land suitability according to local agriculture practices and position and slope of the potential land on the landform extracted from a digital elevation model: summit, lowland, foothills and slope. The slope classification is scored using the Shuttle Radar Topography Mission SRTM (USGS, 1996) classification at 90-metre intervals in Table 17.2 and 17.3.

ASSESSING AND MAPPING OF POTENTIAL ADDITIONAL LAND SUITABILITY. For each crop type, once the suitable geographical areas are mapped, the potentially additional available land for a specific crop is identified, excluding urban areas, protected areas, current forests and rangelands, *sebkha*, water surface, rocky outcrop, unsuitable for cultivation (Maps 17.5 and 17.7). These exclusion criteria allow the delimitation of potentially available land.

Results

We then produce the additional land availability maps by processing the agro-biological parameters of barley and olive trees in the AEZ GIS algorithms.

Mapping additional land availability using limiting agro-biological factors for barley

The limiting agro-biological factors for rainfed barley cropping are detailed in Table 17.2.

Mapping additional land availability using limiting agro-biological factors for olive trees

The limiting agro-biological factors for rainfed olive trees are detailed in Table 17.3.

Discussion and Conclusions

This two-step geospatial theoretical modelling exercise considers first environmental factors, allowing us to map the land capability, and then land suitability, with agro-biological limiting factors to identify additional land areas for cropping rainfed barley or olive trees. This is without encroaching on existing land use including rangelands. It can be applied to any other crops and trees. It is still a speculative mapping application that must be refined with updated soil and land use correct mapping, and more accurate agro-biological crops parameters.

We named three potential availability classes (high, moderate and marginal) for both rainfed barley and olive trees. We shall only consider the 'high' class offering the highest agricultural sustainability as the two other classes may present too many risks of failure, more so in the 'marginal' class, due to short growth period, unreliable or low rainfall, poor soils or unsuitable topography that may affect sustainable production for both crops.

It is, nevertheless, Libyan farmers and livestock owners who may sporadically elect to grow these crops in less suitable land receiving higher localized rainfall and in low-lying areas, wadis and depressions collecting large quantities of run-on water.

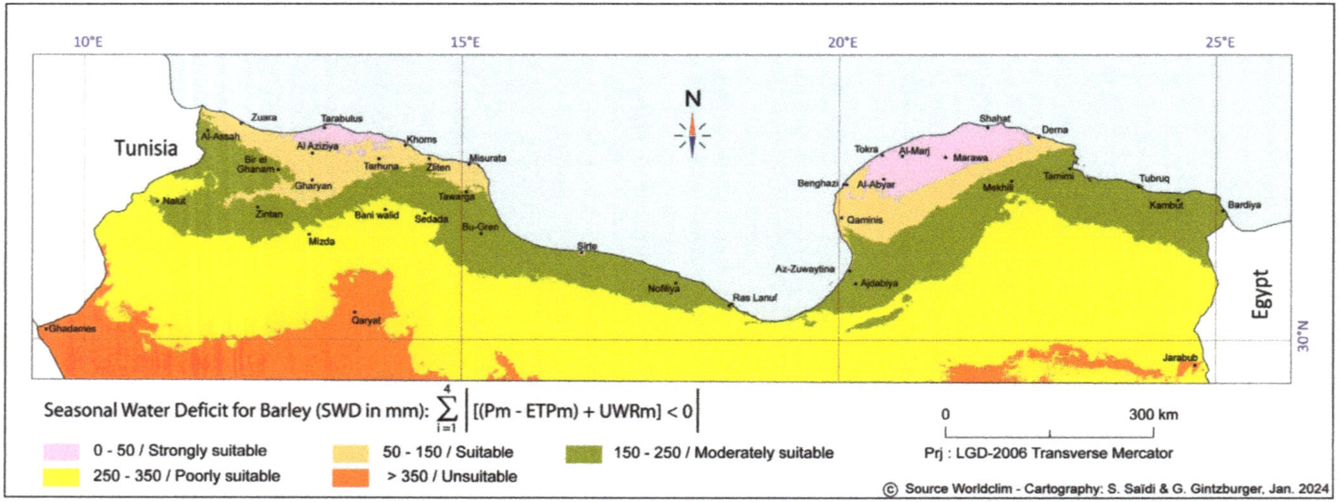

Map 17.8. Seasonal water deficit for Barley.

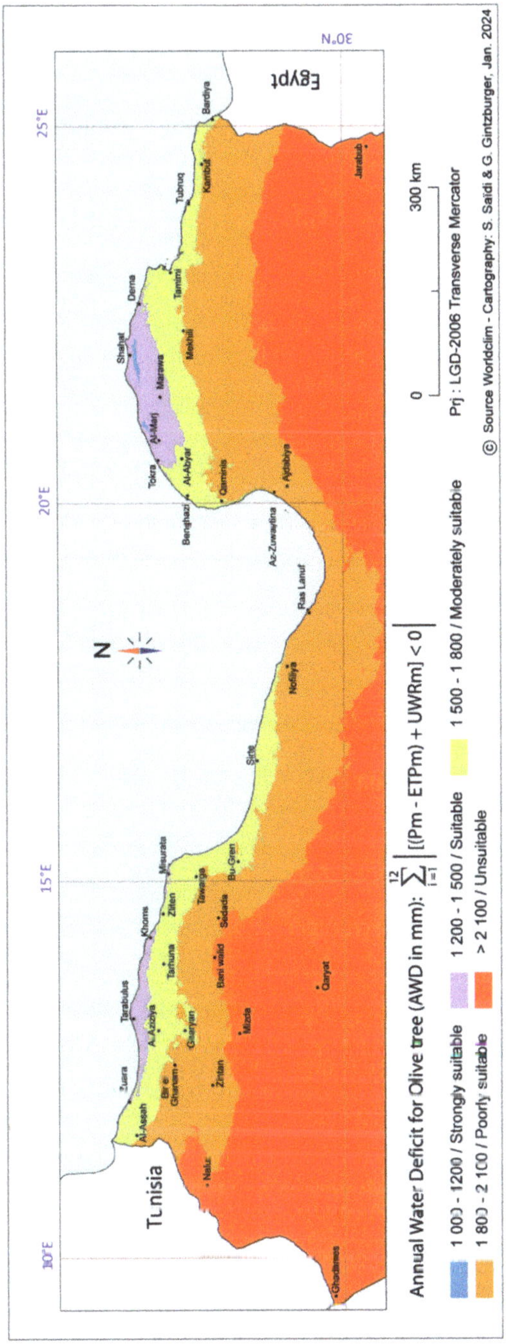

Map 17.9. Annual water deficit for olive tree.

Potential additional land for rainfed barley cropping

Considering only the extra highly suitable and available land for additional rainfed barley cropping, about 116,700 ha (Map 17.10) may still be available and suitable for rainfed barley in the east, in the triangle Tokra–El-Abyar–Soluk/Qaminis with an expected production of about 70,000 tonnes/year of barley grain at an average 0.6 tonnes grain/ha/year.

In the west, it may be possible to further expand about 78,200 ha (Map 17.11) of rainfed barley between Msallata-Khoms and close to Misurata-Zliten, possibly yielding an average 0.5 tonnes grain/ha/year, hence an extra 39,000 tonnes/year of barley grain. It remains that farmers and livestock owners may still risk cropping rainfed barley on marginally suitable land, especially in the wadis and depressions collecting extra rainwater, a common unreliable practice.

These theoretical figures appear small considering the enormous gap of the current Libyan livestock feed deficit (see Chapter 10). It may, however, locally ease the grazing pressure on some rangeland.

Potential additional land for rainfed olive plantations

It confirms that the east has more potential than the west for new olive plantations (Maps 17.12 and 17.13). In fact, when looking at the current land use on Google Earth, we hardly find orchards, especially olive, in the east as expected, while these are ubiquitous in the west in the Jebel Nafusa and the Jeffara plain.

In the east (Map 17.12), olive tree plantations are uncommon these days. However, about 26,600 ha of available land could still be planted in the highly suitable zone with 1,330,000 rainfed olives at a density of 59 trees/ha on *Terra rossa* with about 12 m x 14 m spacing, mostly around Tokra and Jardas-el-Abid and further towards Taknis-Marawa and south Gandoula, in the many inter-hills flats actually exclusively cropped with rainfed barley.

In the west, where rainfed olive tree plantations are prevalent since antiquity, only 6900 ha (Map 17.13) for new rainfed olive trees could still be available in the highly suitable zone close to and west of Khoms (Bani Hasan – Naqqazah), amounting to a maximum of 240,000 trees at a locally practicsd density of 35 rainfed olive trees/ha with a standard 17 m x 17 m spacing. However, this area already bears many sparse olive trees on small private farms under pressure from urbanization.

The maps produced to locate potential additional land for barley cropping or olive tree groves are accurate. Exact coordinates of each polygon identified as suitable can be extracted from the original maps. Using the 'Combine' ArcGIS function, it would further be possible to identify the new land either exclusively suitable for barley or for olive trees, or both.

The social aspects

The sustainability of these new crops as well as their social acceptance is an essential condition for realizing the potential (Elbehri *et al.*, 2013; Sims *et al.*, 2006; Wüstenhagen *et al.*, 2007). The right to crop, plant trees and harvest on additional suitable land depends on the identified actors and communities, their property rights, their traditional laws and operational rules. There is little doubt that local farmers and livestock owners know precisely where and what their land can offer in terms of cropping or planting fruit trees. If some of the suitable land identified is not currently cropped, it means that other constraints or cultural choices exist and are not considered in our modelling at this stage, especially social, cultural or practical constraints, as to whether to accept or reject our proposals. It also does not anticipate governmental options and necessary private or state support and investments to implement our suggestions.

The agro-ecological zoning and modelling is a rational and powerful land management tool providing that it is developed based upon solid agro-climatic and agro-biological data and accurate updated land-use maps. Our AEZ model can certainly be further improved.

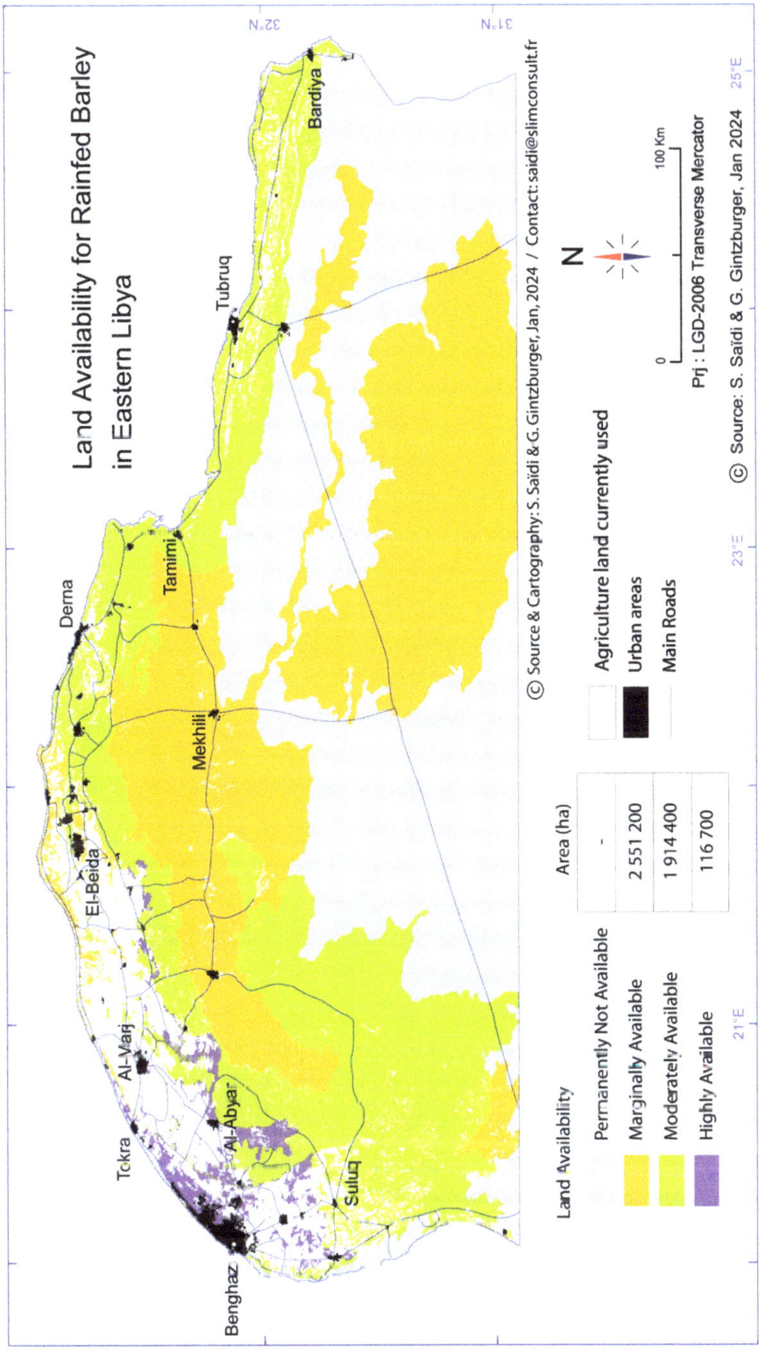

Map 17.10. Potential additional land availability for rainfed barley in eastern Libya.

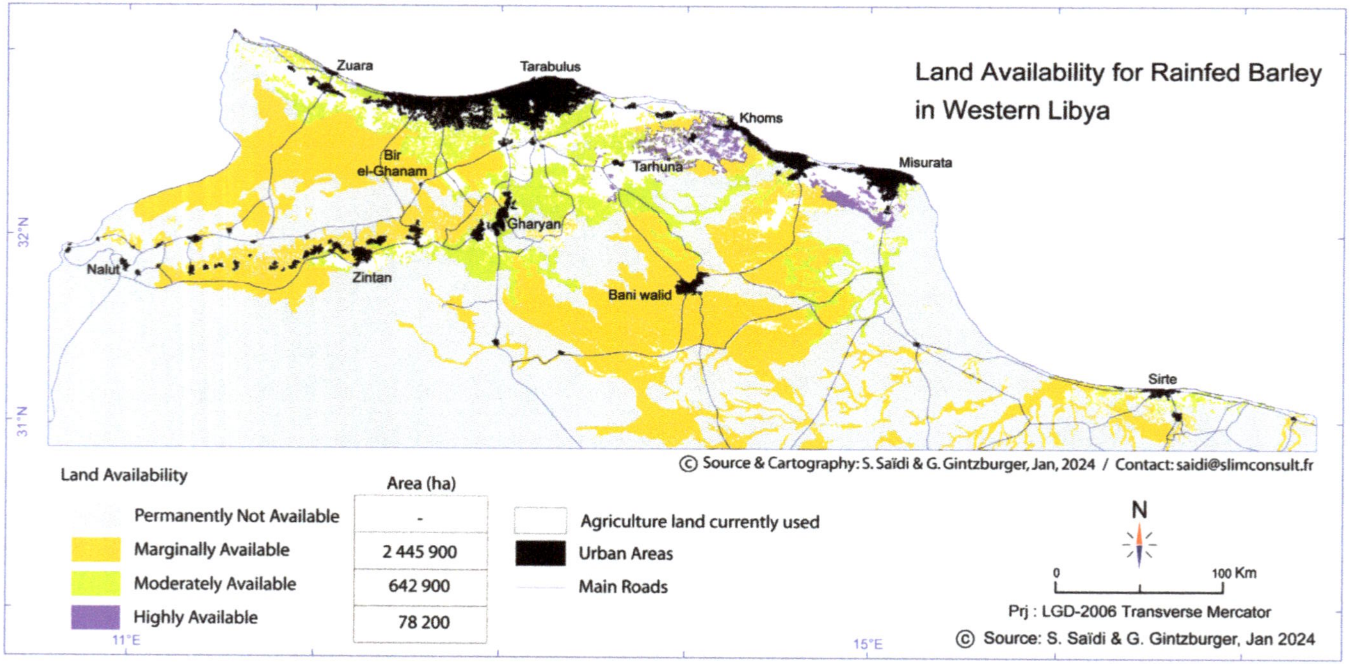

Map 17.11. Potential additional land availability for rainfed barley in western Libya.

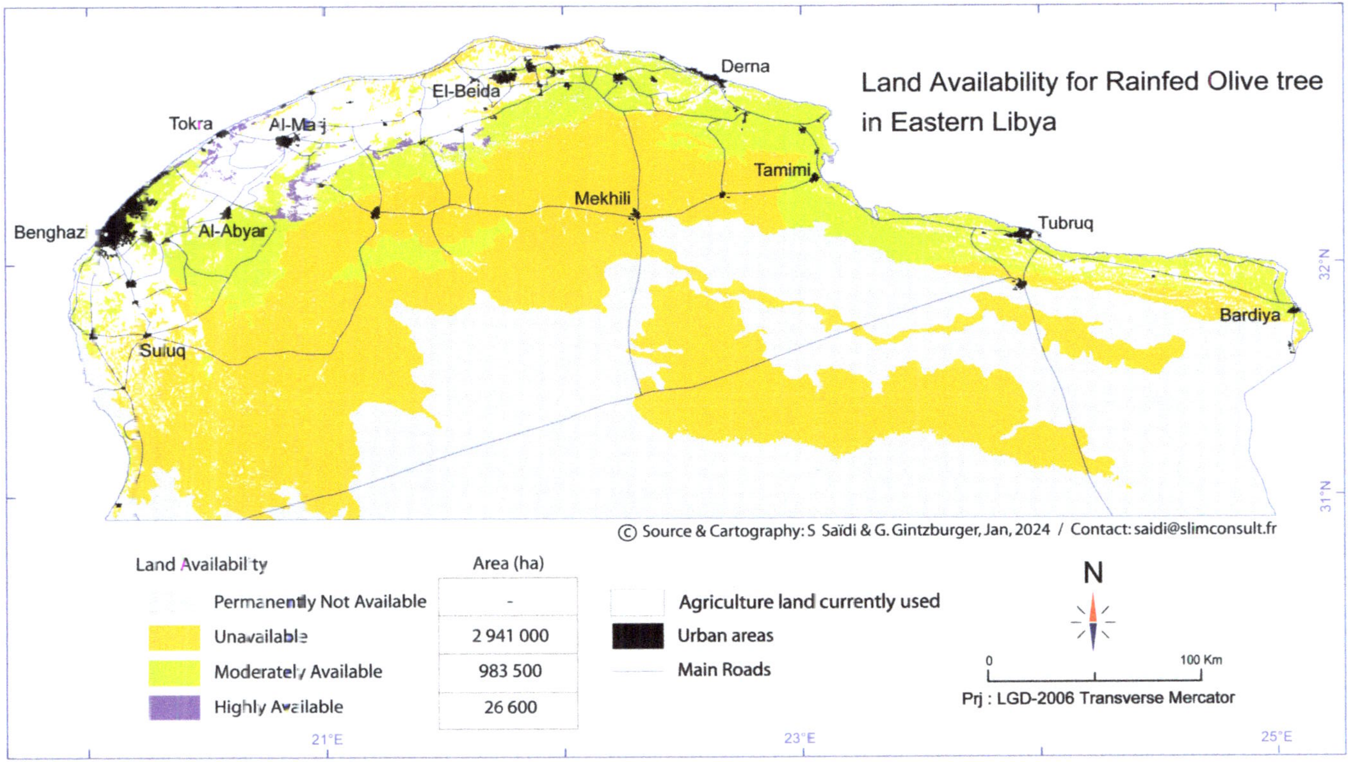

Map 17.12. Potential additional land availability for rainfed olive trees in eastern Libya.

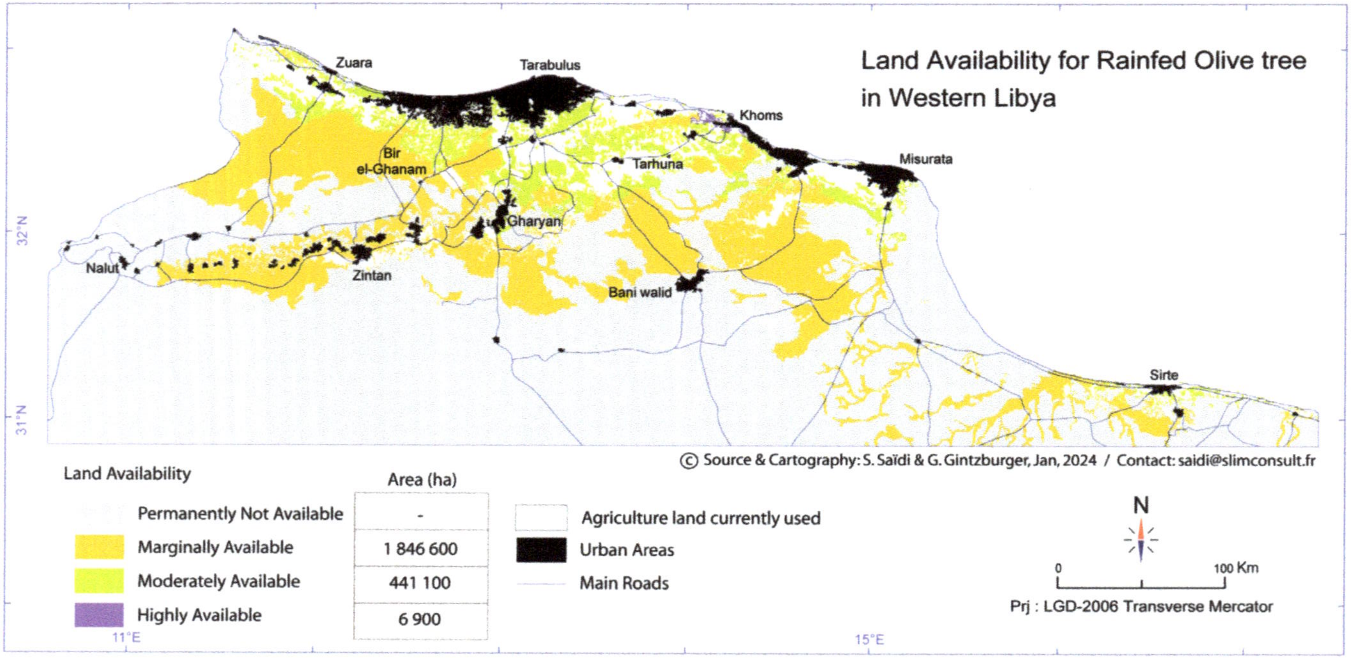

Map 17.13. Potential additional land suitability for rainfed olive trees in western Libya.

References

Abuseif, M.I., Amrajaa, B.N. and Mohamed, E.O. (2024) Assessment of the quantitative genetic diversity among some imported olive cultivars at Benghazi Governorate – Libya. *Journal of Plant Production* 15(3), 119–122. DOI: 10.21608/jpp.2024.275166.1315

Al-Jibouri, H.A. (1966) *Description and Distribution of Wheat and Barley Varieties Grown in Libya.* Food and Agriculture Organization of the United Nations, Rome.

Allan, J.A., McLachlan, K.S. and Penrose, E.T. (eds) *Libya: Agriculture and Economic Development.* Frank Cass, *London.*

Amouretti, M.C. and Comet, G. (2000) *Le Livre de l'Olivier.* Édisud, La Calade (Bouches-du-Rhône).

Baldy, C. (1990) Le climat de l'Olivier (*Olea europaea* L.). *Ecologia Mediterranea* 16(1), 113–121. DOI: 10.3406/ecmed.1990.1656

Batidzirai, B., Smeets, E.M.W. and Faaij, A.P.C. (2012) Harmonising bioenergy resource potentials – methodological lessons from review of state of the art bioenergy potential assessments. *Renewable and Sustainable Energy Reviews* 16(9), 6598–6630. DOI: 10.1016/j.rser.2012.09.002

Bennani, S., Ouattar, S. and Ismaîli, M. (1992) Modèle de quantification des déficits hydriques subis par des blés de différentes durées de cycle en climat méditerranéen sub-humide (Saïss, Maroc). *Cahiers Agricultures* 1(5), Article 5.

Bensemane, L. (2018) *Analyse de l'adaptation de l'orge (*Hordeum vulgare *L.) sous conditions semi-arides.* PhD dissertation, Université Ferhat Abbas, Sétif, Algéria. Available at: http://dspace.univ-setif.dz:8888/jspui/handle/123456789/1612 (accessed 26 August 2024).

Brunin, C. andDe Berchoux, C. (1994) IRHO: 50 ans au service des oléagineux tropicaux. In: *Les 50 ans de l'AFECG.* AFECG, Neuilly-sur-Seine, France. Available at: https://agritrop.cirad.fr/463894/(accessed 26 August 2024).

Buzaian, A. (2022) *Ancient Olive Presses and Oil Production in Cyrenaica (North-East Libya).* The British Institute for Libyan and Northern African Studies, London. Available at: https://www.jstor.org/stable/10.2307/jj.4350573 (accessed 26 August 2024).

Elbehri, A., Segerstedt, A. and Liu, P. (2013) *Biofuels and the Sustainability Challenge: A Global Assessment of Sustainability Issues, Trends and Policies for Biofuels and Related Feedstocks.* Food and Agriculture Organization, Rome.

FAO (1983) *Guidelines: Land Evaluation for Rainfed Agriculture* (FAO Soils Bulletin. No. 52). Food and Agriculture Organization, Rome.

FAO (1996) *Agroecological Zoning Guidelines.* FAO Soils Bulletin. Food and Agriculture Organization, Rome.

FAO and UNDP (2006) *Land Cover/Vegetation Map of Libya.*

Fick, S.E. and Hijmans, R.J. (2017) WorldClim 2: new 1-km spatial resolution climate surfaces for global land areas. *International Journal of Climatology* 37(12), 4302–4315. DOI: 10.1002/joc.5086

Gintzburger, G., Saïdi, S. and Valerie, S. (2009) *Rangelands of the Ravnina Region in the Karakum Desert of Turkmenistan: Current Condition and Utilisation.* Sustainable Agriculture in Central Asia and the Caucasus Serie, N° 2. CGIAR PFU. Tashkent, Uzbekistan.

Goodchild, A.V., Grando, S., Thomson, E.F., Haylani, M. and Ceccarelli, S. (2022) Heritability and genetic gain of digestible organic matter intake of barley straw genotypes by sheep, using repeatable in situ and laboratory-based indicator traits. *The Journal of Agricultural Science*, 1–11. DOI: 10.1017/S0021859622000399

Habaieb, H. and Masmoudi-Charfi, C. (2003) Calcul des besoins en eau des principales cultures exploitées au nord de la Tunisie: estimation de l'évapotranspiration de référence par différentes formules empiriques (cas des régions de Tunis, Béja et Bizerte). *Sécheresse* 14, 1–9.

Hengl, T., Jesus, J.M., MacMillan, R.A., Batjes, N.H., Heuvelink, G.B.M. and Ribeiro, E. (2014) *Global Soil Information Based on Automated Mapping. SoilGrids1km.* DOI: 10.1371/journal.pone.0105992

IUCN-UNEP (2019) Base des données mondiales sur les aires protégées. Available at: http://www.iucn.org/ (accessed 29 August 2024).

Jouve, P. (1984) Relation entre déficit hydrique et rendement des céreales (blé tendre et orge) en milieu aride. *L'Agronomie Tropicale* 39(4), 308–334.

Keith, H.G. (1965) *A Preliminary Check List of Libyan Flora,* Volume 2. United Middle East Executives Ltd, London.

Le Houérou, H.-N. (1959) Recherches écologiques et floristiques sur la végétation de la Tunisie méridionale. PhD dissertation, Université Alger Institut de Recherches Sahariennes, Algiers.

Le Houérou, H.-N. (1969) *La Végétation de la Tunisie Steppique (Avec Référence aux Végétations Analogues d'Algérie, de Libye et du Maroc) 1–2, 42–45*. Annales de l'Institut National de la Recherche Agronomique de Tunisie, Tunis.

Le Houérou, H.-N. (1984) An outline of the bioclimatology of Libya. *Actualités Botaniques* 131(2–4), 157–178. DOI: 10.1080/01811789.1984.10826657

Le Houérou, H.-N. (1989) Classification écoclimatique des zones arides (s.l.) de l'Afrique du Nord. *Ecologia Mediterranea* 15(3–4), 95–144. DOI: 10.3406/ecmed.1989.1643

Liebig, J. (1840) *Organic Chemistry and Its Application to Agriculture and Physiology*. Taylor and Walton, London.

Loukili, M. and Ihissou, Y. (2016) L'analyse intégrée du milieu à l'aide du SIG en vue d'une planification rationnelle de la gestion conservatoire des terres: application dans la région de Dhar Souk (Province de Taounate, Maroc). *Revue Marocaine Des Sciences Agronomiques et Vétérinaires* 4(1), 9–20.

Mattingly, D.J. (1988) The olive boom: oil surpluses, wealth and power in Roman Tripolitania. *Libyan Studies* 19, 21–41. DOI: 10.1017/S0263718900001060

Mattingly, D.J. (1993) Maximum figures and maximizing strategies of oil production? Further thoughts on the processing capacity of Roman olive presses. In: Amouretti, M.-C. and Brun, J.-P. (eds) *La Production du Vin et l' Huile en Mediterranée*. Actes du Symposium International, pp. 483–498. Ecole Francaise d'Athènes. https://dialnet.unirioja.es/servlet/articulo?codigo=592657

Mattingly, D.J. (1994) Regional variation in Roman oleoculture: some problems of comparability. In: Carlsen, J., Ørsted, P. and Skydsgaard, J.E. (eds) *Landuse in the Roman Empire*. L'Erma di Bretschneider, Rome.

Mattingly, D.J. (1995) *Tripolitania*. Routledge, New York. DOI: 10.4324/9780203481011

Najar, A., Ben Ghanem, H., Kumari, S., Sayed, H., Rezgui, S. and Baum, M. (2017) Sélection de lignées d'orge (*Hordeum vulgare* L.) pour leur résistance au barley yellow dwarf virus (BYDV) et évaluation de leurs performances agronomiques. *Canadian Journal of Plant Science* 97(2), 277–285. DOI: 10.1139/cjps-2016-0170

Nwer, B.A., Zurqani, H.A. and Rhoma, E.A. (2013) The use of remote sensing and geographic information system for soil salinity monitoring in Libya. *GSTF Journal of Geological Sciences* 1(1). DOI: 10.5176/2335-6774_1.1.5

Parry, M.L. and Parry, C.J. (1993) Agricultural geography of barley in the agrometeorology of rainfed barley-based farming systems. In: Jones, M., Marthys, G., and Rijks, D. (eds) *Proceedings of the International Symposium*, pp. 15–31.

Rowland, J.W. (1945) *Survey of Land Resources in Tripolitania*. British Administration, Department of Agriculture.

Saïdi, S. and Gintzburger, G. (2013) A spatial desertification indicator for Mediterranean arid rangelands: a case study in Algeria. *The Rangeland Journal* 35(1), 47–62. DOI: 10.1071/RJ12021

Sims, R.E.H., Hastings, A., Schlamadinger, B., Taylor, G. and Smith, P. (2006) Energy crops: current status andfutureprospects.*GlobalChangeBiology*12(11),2054–2076.DOI:10.1111/j.1365-2486.2006.01163.x

Snir, A., Nadel, D., Groman-Yaroslavski, I., Melamed, Y., Sternberg, M., Bar-Yosef, O. and Weiss, E. (2015) The origin of cultivation and proto-weeds, long before Neolithic farming. *PLOS ONE* 10(7). DOI: doi.org/10.1371/journal.pone.0131422

Trabucco, A. and Zomer, R.J. (2018) *Global Aridity Index and Potential Evapotranspiration*. ET0 Climate Database: Version 2. Data set. DOI: 10.6084/m9.figshare.7504448.v3

USGS (1996) *USGS 30 ARC-second Global Elevation Data, GTOPO30* (Gtopo30 ed.). Data set. U.S. Geological Survey's Center for Earth Resources Observation and Science (EROS). DOI: 10.5065/A1Z4-EE71

Villalobos, F.J., Orgaz, F., Testi, L. and Fereres, E. (2000) Measurement and modeling of evapotranspiration of olive (*Olea europaea* L.) orchards. *European Journal of Agronomy* 13(2), 155–163. DOI: 10.1016/S1161-0301(00)00071-X

von Bothmer, R., Sato, K., Knupffer, H. andvan Hintum, T. (2003) Barley diversity – an introduction. In:von Bothmer, R.,van Hintum, T., Knüpffer, H. and Sato, K. (eds) *Developments in Plant Genetics and Breeding* 7, 3–8. Elsevier. DOI: 10.1016/S0168-7972(03)80003-8

Wüstenhagen, R., Wolsink, M. and Bürer, M.J. (2007) Social acceptance of renewable energy innovation: an introduction to the concept. *Energy Policy* 35(5), 2683–2691. DOI: 10.1016/j.enpol.2006.12.001

18

Overall Conclusions

Despite their apparent monotony, the Libyan rangelands are ecologically extremely diverse. This is due to a rich geology and to climate shaping the landscapes supporting various soil types, in turn defining distinctive plant associations. For centuries, these landscapes and plant associations provided free feed and wildlife resources to rangeland dwellers and farmers. There was, for a long time, a fragile equilibrium between resources and demand. This equilibrium, often broken by droughts or unrest, tilted towards creeping land degradation after WWII and especially accelerated in the early 1970s with the discovery of oil, bringing increased human and livestock population, wealth and rampant urbanization.

The Libyan rangelands are, these days, in crisis, unable to sustain the extraordinary current livestock numbers. This is unsustainable without importing vast quantities of complementary and subsidized feed, illegally drilling wells and constructing countless new watering places, all contributing to more rangeland desertification. The culprit is not climate change, but simply humans.

Climatic data, mainly rainfall and temperature, from all points of the compass, helped us discriminate different bioclimates controlling plant diversity and distribution. Specific plant associations, ecological groups and climatic variants are closely defined by matrices of bioclimates, soil types and grazing regimes. These original plant associations have fast evolved and been degraded over the years and especially during the last 50–60 years due to overuse and abuse, overgrazing with unsustainable numbers of livestock, fuelwood uprooting, charcoal manufacturing, expanding hazardous rainfed barley cropping, and confiscating the best rangelands for fodder shrub plantations. Rangeland degradation in Libya is obvious and ubiquitous, sometimes irreparable. While climate change may be incriminated, our study on meteorological data over the past 70 years does not allow us to detect a clear climatic evolution over northern Libya, though there are some drying trends detected mainly on the coastal stations. Climate change and oscillations between wet and dry eras in arid zones over past million of years is not new as ascertain by paleoclimatic studies. The current cause of rangeland degradation rests with increased demand for meat, mainly sheep.

The Libyan flock livestock fed on rangelands was around 1–1.5 million sheep-equivalent in the 1975 rising to about 8–9 million by 2010. In the mid-1970s, the Libyan rangelands were supplying about 40–50% of the yearly feed and were already overstocked by 200%. Rangeland annual feed supply dropped to its lowest with less than 5% by 2010. This rangeland overstocking combined with unsustainable rangeland management is continuously fuelled with ever-increasing use of costly imported feed and

DOI: 10.1079/9781800627154.0018

subsidized stale bread. Furthermore, traditional transhumance patterns are totally disrupted with accessibility to new legal and illegal wells, cisterns, watering points and water-carting to the most remote arid rangelands and landscapes. The large Libyan livestock flock is now mostly sedentary and artificially fed on open-air industrial or small feedlot operations.

Rangeland degradation is reaching irreversibility on many sites. Desertification is looming. The Libyan rangelands situation is grim if not desperate unless drastic actions are soon implemented. Since the mid-1970s, many technically well-planned rangeland and rainfed cropping projects were developed all over northern Libya with billions of dollars invested. These are soon to be abandoned for many reasons, the major one being lack of preliminary consultation and support from the traditional rangeland communities. Who better know their territory and their ancestral rangeland resources than them?

Reversing arid rangeland degradation is still possible and demonstrated mostly on fixed sandy loam soils, but it remains much more difficult on heavier soils. We do have the technical expertise. Many examples of efficient rangeland natural restoration (governmental or communities' *mahmiyat*) are visible all over the northern Libyan rangelands. Fodder shrub plantations with known appropriate species are a valuable option providing that they are solely accepted, established, controlled and managed by the local communities. Rangeland reseeding must be similarly attempted with native species, mostly Libyan annual legumes of which many have been already collected long ago and safeguarded as Libyan plant genetic resources in Libya and Australia. They are available with known industrial native seed production already performed in Libya. This must be a priority in rangeland research and development.

Water on rangelands is another essential and problematic issue. New illegal or governmental watering points, deep wells and cisterns, may only be curbed by appropriate governmental legal control and implementations. This has been efficiently applied on rangelands in other arid countries of the region as simple measures to limit rangeland access to remote areas out of season.

Forests and afforestation in the east and the west are under heavy and persistent threats by illegal fuelwood harvesters and uncontrolled charcoal manufacturers. Afforestation with strictly protected native species is necessary before it is too late. Locals should be encouraged to establish fuelwood plantations on private land with exotic species such as fast-growing and coppicing *Eucalyptus* to lessen the pressure on native forests. Implementation of legal registration and control of fuelwood harvesting and industrial charcoal manufacturing is the only solution to end the destruction of irreplaceable native trees and forest, and pilfering governmental and private reforested areas.

Agro-ecological zoning mapping based on environmental factors and agro-biological information on crops or trees would help identify agronomically suitable and available vacant land without encroaching on rangelands already under much pressure.

Degraded rangelands offer little feed and protection to wildlife. Once diverse and abundant on healthy rangelands and a valuable source of meat to the locals, wildlife is relentlessly decimated by illegal and out-of-season hunting, documented on countless gory posts on Libyan social media.

Is it too late to act to restore the richness, diversity and usefulness of Libyan rangelands? What would be the cost and consequences of doing nothing? It is obviously time to act without any further delay and always in consultation with and support from the local communities. Persistently promoting education and information to younger generations of Libyans at school about the value of a healthy environment, rangelands and forests would help in securing a better future for all.

Annex 1 – Collecting plants, photos and herbarium specimens

Collecting Plant Specimens: Yes or No?

The purpose of our field trips is to collect information on rangeland plants and their ecology. Occasionally, we may be able to observe and describe soil profiles.

The question remains: Do we collect plant specimens to keep in our herbarium? While there is no problem in collecting common plants, we still must be mindful not to contribute to destroying plant biodiversity or rare plants. We can take detailed photos and macro-photos when we properly name specimens *in situ*. However, a photo, as good and as detailed as it is, may not always be enough to properly name the plant with reference to detailed description available in various flora. Often, there is a need for more careful examination under a magnifying lens. This can be done in the field consulting the flora we carry with us: cumbersome and difficult when the wind is blowing dust, the sun is cooking you and you and you friends are impatiently waiting to proceed to the next stop or set up camp before nightfall. So, hurriedly, we capture a few photos, collect samples, take ecological notes on-site (possibly with GPS coordinates), vernacular names and, quickly, arrange the samples in our herbarium press or, if in a rush, keep them in a large plastic bag until we reach the night camp. At the camp, we arrange our plants in a herbarium press with each individual plant labelled and placed between blotting paper; butcher's paper is the best and cheapest, or newspaper. We change the wet papers every day. Usually, during the field trip, we keep our many presses stockpiled on the roof rack – our specimens dry much faster, but beware – make sure all presses are well secured!

There is a problem: we just collected a few mysterious plants, but back at the lab the specimen has neither flowers nor fruits, no bulb or corm. There is no option but to go back to the field, to the very site where we collected the specimen, hopeful that we will find the site (re- GPS coordinates) and the enigmatic plant, possibly having benefited from a good season, flowering and fruiting, if not totally grazed by desperate livestock. It is a perpetual challenge. But remember, do NOT collect or uproot rare plants; leave the orchids, and the like, in place!

At the lab we have now properly identified our specimen. We must make sure that they are properly desiccated, accurately labelled, annotated (flower colour when fresh, date of collection, location with toponymy, taxonomic Latin name, family, vernacular name if available, soil type, rainfall zone, name of the collector(s) preceded by 'Leg.' (*Legit* = collected by...), name of the person who did the determination preceded by 'Det.' (*Determinavit* = identified by ...), etc.) and sorted by family. You may poison the specimen, but the process is quite awkward and dangerous. Instead, we process to an occasional toxic fumigation of the properly sealed herbarium storeroom. We must also regularly check that our specimen is not parasitized with silverfish, borers and various insects. Another possibility is to seal the desiccated specimen in herbarium paper folders in a large plastic bag and place them in a freezer (–18°C to –20°C) for a few days. The cold will efficiently kill all insects and their eggs.

Properly Naming and Labelling Plant Specimens

Back at the lab starts the determination work using florae (we say 'several florae', if possible). Many florae and scientific publications are available covering North Africa, Libya, Egypt and further to the east, dating back to the 18th century. In our book, for naming the Libyan plants, we referred to the *Flora of Libya* (El-Gadi and Jafri, 1976) and to *A Preliminary Checklist of Libyan Flora* (Keith, 1965). Both publications are readily available in Libya. Though they need updating with revised genus and species names, and recently discovered plants in Libya, they are still easily accessible to most Libyan botanists and environmentalists. An update of the *Flora of Libya* with synonyms is available from Essokne and Mahklouf (2023) and Essokne *et al.* (2024). If you need further updating of scientific names and to check the taxonomy, several on-line sites are available (See Chapter 7).

You need to use a good low-power stereo microscope (5x – 25x) and a lot of patience. Once you have narrowed down to the family of your specimen, start using the family dichotomic key that will eventually lead you to the genus and possibly the species. Do not hesitate to consult with several florae as you may not have the required plants' parts (flower, style, ovary, fruits, young or old leaves or branches etc.) on your specimen. Compare the descriptions between different florae. They may complement each other and give you some useful evidence. Misidentification is still possible and nobody is immune to wrongly identify and name a plant or an animal.

You may find that the plant you are trying to determine has different names in different flora; it may have been described initially and later by different botanists. Therefore, when naming the specimen on your herbarium label with the genus and the species, you must also write down the name of the first author (usually his initials) who first discovered and named this plant. If the author's name is in brackets followed with another author's name, it means that the plant's name was reviewed and updated by the second author. In any case, check the possible synonyms of the plant's name.

Remember that the genus name starts with a capital letter, while the species name usually starts with a lower case (in fact, not always, as, in the old days, when the name of the species was attributed to a person (a collector or a famous botanist etc.), the species' name would start with a capital too, but not any more).

Both *Genus* and *species* names are always italicized in publications.

A few final notes on plants' names:

- Internationally documented and recognized plants' names are written in Latin using Latin alphabet. Check and recheck the correct spelling!
- Refrain from attaching an English foreign name to your specimen: it is completely irrelevant as a foreign name attributed to a plant in a country may be ascribed to other genus/species in other countries. Often we hear English-speaking colleagues using the name *Sagebush* in the field or in publications for the mediterranean *Artemisia herba-alba* (*Shih*) or even for *Artemisia campestris* (*Chahal* or *Tegouft*). *Sagebush* is a generic name used in the USA for west-American *Artemisia* sp. Irrelevant and wrong! Why not use the vernacular North African or Libyan name transcribed from Arabic or Berber?
- Find out the vernacular name from local farmers or shepherds. On your herbarium label, it should be written in Arabic, if possible, and/or transcribed in Latin letters. Transcription/transliteration of vernacular names from Arabic into Latin alphabet presents a lot of difficulties as it may be pronounced by young or old people with different accents in different regions of Libya. The same plant may have different vernacular names from different regions, though we found that Arabic names of many steppic or desert plants, long-since attributed by Bedouins, are nearly always the same from the Middle East to Algeria.

- Sometimes, the vernacular name simply refers to different plants with the same local name, for example, and especially due to the colour of the flowers (Ex: *Zerga*, *Zriga*, *l'Azreg* for plants with blue flowers ... and there are many of them in the Libyan steppe!). The vernacular name may also give you clues about the plant's characteristics: for example *Bu Rokba* (the one with a knee) for *Panicum turgidum* that has geniculated – knee-like stems, or *Hallab*/*Bu Halibah*/*Dahar-el-halab* (referring to milk) for *Periploca laevigata* or *Euphorbia* sp., that have a milky sap, or *Sarat-al-kabsh* (the ram's navel) for *Anacyclus cyrtolepidioides*, etc. The list is long.

Photography

Nowadays, everyone uses a sophisticated camera (or a multi-tools mobile phone) during field trips to capture photos of plants and landscapes. The camera technology has fast evolved since the early 1970s, evolving from analogue to digital photography.

In this book, some photos date back to 1975 when using an Exakta Varex IIa analogue reflex camera fitted with a Schacht 50 mm f2.8 M lens. This lens was specifically designed for capturing macro-photography to a 1:1 magnification. We used at the time Kodak Ektachrome F 35 mm film. Taking a photo with this equipment required some ability, setting the exposure and diaphragm to get the proper exposure and depth of field. It comes with experience. The films were developed in Europe meaning that it would take about four to six weeks before we could view our photos as slides, often a bit too late to return to faraway sites near Nalut or wadi Sof-el-Jin, and, in any case, too late to capture new photos of plants that had already disappeared or faded away. We also used black-and-white negatives processed and printed at home. (A lengthy process!)

For the purpose of illustrating this book, we scanned some of these old colour slides and B&W negatives with a Nikon Slide Scanner Cool Scan V ED driven by a powerful scanning software (ViewScan x64) at 4200 DPI and which automatically corrects scratches and dust and offers many more possibilities.

Later, by 1995, while working in Central Asia, we started using our first digital camera, a Sony Cyber-Shot DSC (1995) for illustrating the book on rangelands of Uzbekistan; no more fiddling with the speed, the diaphragm, all automatic focusing and light-adjusting. Much faster operation but still having sometimes to adjust both for a proper depth of field.

We then shifted to a superior Sony DSC-F828 (late 2003), a digital camera that has never failed us. This camera is fitted with an outstanding Carl Zeiss T28 – 200 mm – F2.0–F2.8 lens and a mechanical zoom (7x) allowing superb macro-photos. It has many possibilities, including night shoot in total darkness using infra-red detection and focusing. But this camera was still quite bulky and collected a lot of fine dust during our incessant field trips to the Libyan steppe and rugged country. While we were still using it in Libya (2007–2010), we ended up also carrying the smaller Olympus Stylus µTough-6000 (10 megapixels), a dust-,water-, shock-proof small camera packed with optical capacities: a reliable companion during dusty/bumpy/searingly hot field trips to the Hamada, to wadi Wishkah, to Martuba, to Mekhili, and more.

All these digital cameras captured images easily stored on computer. Their 3–10 megapixel sensors were large enough to produce quality photos (up to 13 cm x 18 cm format) in 300 dpi minimum resolution: plenty for illustrations in a book or on a website.

Capturing a digital photo is cheap and easy, so do not hesitate to take many photos of the same plant at different angles, exposures, distances, general views, details of flowers, fruits, leaves, stems, etc. You may not be able to come back to the same place or find the same plant.

In the field, it is better to operate under natural light. The best light is mid-morning or mid-afternoon. Avoid using a flashlight as it often gives poor results and flattens details on macro-photos. Use a contrasting background to your subject. Once you take a photo in the field or in the lab, at once check that you properly focused on the detail you want to capture. Use the manual command of the diaphragm to change the depth of field of your photo. If not focused, shoot another one and recheck.

Often, we do regret having neglected taking photos of the most common or modest plants, thinking that we will take photos or specimens later. This later never occurs, so do it now!

Remember that your camera may be able to geotag your photos with the GPS position. If this option is available on your camera, activate it. Record your photos identifier and GPS position in your notebook. You may want to come back to the same remote spot to complement your photos with leaves, flowers, fruits or other important details you missed.

While taking photos of plants is a valuable tool, it does not replace collecting herbarium specimens whenever possible to refine or confirm plant determination and take as many morphological, ecological and biological notes as possible on-site. Do not wait until the night camp as details fade away fast from your tired memory after a long day jogging cross-country.

References

El-Gadi, A.A. and Jafri, S.M.H. (eds.) (1976) *Flora of Libya*. El Fateh University of Tripoli/Arab Development Institute Tripoli.

Essokne, R. and Mahklouf, M. (2023) Index to new updated names of the Flora of Libya. Volume I. *Journal of Science* 16, 41–52.

Essokne, R., Mahklouf, M. and Isweiri, H. (2024) Index to new updated names of the Flora of Libya. Volume II. *Journal of Science* 17, 70–78.

Keith, H.G. (1965) *A Preliminary Check List of Libyan Flora, Volumes 1 and 2.* (Vol. 2). United Middle East Executives Ltd, London.

Annex 2 – Glossary

The most common Arabic vernacular plant names and items brought centuries ago by the Bedouins Banu Sulaym and Bani Hilal tribes survived to this day but are slowly being eroded and forgotten with modernity and abandonment of living on rangelands.

During our fieldwork, we collected the vernacular names of plants, landscapes and common items used by shepherds and farmers living on rangelands. We phonetically transcribed them in Latin letters, a difficult task as these words may be pronounced by young or old persons with their own local western or eastern Libyan dialect and accent, specific phonemes or graphemes, from dialectal Arabic or Berber; hence, several words for the same item may appear in the list. Libyan words for plants, wildlife or common items may be different in the west and the east.

Dr Manam Saaed (manam.saaed@uob.edu.ly, Associate Professor, Rangeland and Landscape Ecology, University of Benghazi) kindly went over the list and completed the Libyan Arabic names in this glossary. We are certainly missing or misspelling some names of plants, wildlife or items. We leave it to our Libyan colleagues to add, amend and correct our glossary, possibly compiling western or eastern local Arabic or Berber names with their description or meaning.

Transliterated Libyan names		Vernacular names
Plants	**Scientific names**	
Talah	*Acacia tortilis* ssp. *raddiana*	طلح
Ba'athran/Quisoom	*Achillea santolina*	قيصوم/بعثران
Ain-el-Buma/Sgemera/Telma	*Adonis dentata*	عين البومة
Akrosh	*Aeluropus littoralis*	عكريش
Gassul	*Aizoon hispanicum*	غسول
Chendgoura	*Ajuga iva*	شندقوره
Agol	*Alhagi maurorum*	عاقول
Quzul	*Allium* sp.	قازول
Ajram/Baguel/Jill, Kiddad/Taza	*Anabasis articulata*	عجرم
Seret-el-kebsch	*Anacyclus cyrtolepidioides*	سرة الكبش
Akaraba	*Anastatica hierochuntica*	
Nesrim/karaba (?)	*Anastatica hierochuntica*	
Chaicut/Afahlehle	*Androcymbium punctatum*	

Transliterated Libyan names		Vernacular names
Rezdaïr/Rouzdir/Hossir	*Anthyllis henoniana*	رزدير
Chabrom	*Anvillea australis*	
Nogud	*Anvillea garcinii*	نقد
Sch'meri	*Arbutus pavarii*	شماري
Herga/Kherta	*Argyrolobium uniflorum*	
Hanzaab	*Arisarum* sp.	حنزاب
Noussia	*Aristida obtusa*	
Erer/Henna-el-agreb	*Arnebia decumbens*	
Tegoufet/Chahal	*Artemisia campestris*	تقفوت
Shih	*Artemisia herba-alba*	شيح
Balbal Jmal	*Arthrocnemum glaucum*	بلبال جمل
Baguel	*Arthrophytum schmitianum*	باقول
Remeth	*Arthrophytum scoparium*	رمث
Nerisc/Rhinesh	*Arum cyrenaicum*	رينش
Gasba/Hagnah	*Arundo donax*	قصبة/حجنة
Anseil/Bellouz	*Asphodelus fistulosus*	عنصيل
Ansil/Belouz	*Asphodelus microcarpus*	عنصل
Izian/Lahiat-el-tes/Tazia	*Asphodelus tenuifolius*	لحية التيس
Dagees	*Asteriscus pygmaeus*	دقيس
En Hanesh	*Astragalus cruciatus*	عين الحنش
Gundal	*Astragalus spinosus*	قندول
'Ser	*Atractylis cancellata*	صر
'Ser	*Atractylis serratuloides*	صر
Qatf	*Atriplex coriacea*	قطف
Qatf	*Atriplex glauca*	قطف
Gu'taf/Guetaf	*Atriplex halimus*	قطف
Gtaf/Zell	*Atriplex mollis*	قطف السبخ
Rubia (?)	*Ballota hirsuta*	
Ouhas/Choubta	*Bassia muricata*	
Bo Shtertah	*Bromus mollis*	بوشترتة
Bu-sh'renta	*Bromus rubens*	بوشرنته
Rassol	*Calendula aegyptiaca*	
Harta/Arta	*Calligonum comosum*	
Brumbakh	*Callotropis procera*	
Gundul	*Calycotome intermedia*	قندول
Gundul	*Calycotome villosa*	قندول
Kabr	*Capparis spinosa*	كبار
Bin-Awma/Hussina	*Carrichtera annua*	بنوما
Quartom/Ser	*Carthamus lanatus*	صر
Ghazel	*Catananche arenaria*	
Regma	*Cenchrus ciliaris*	رقمة
Belala/Blal	*Centaurea dimorpha*	
Sooret-en-Nagga	*Centaurea glomerata*	صرة الناقة

Transliterated Libyan names		Vernacular names
Kbeshia	*Centranthus ruber/calcitrapa*	
Kharoub	*Ceratonia siliqua*	خروب
Dum	*Chamaerops humilis*	
Oualeglis	*Chrysanthemum macrocarpum*	
Tarthooth asfar	*Cistanche phelypaea*	طرثوث أصفر
Tarthooth (?)/*babouch-el-kelb/ Zeb-el-Ard*	*Cistanche violacea*	طرثوث
Berbesh (-el-abiod) (?)	*Cistus parviflorus*	بربش أبيض
Arfina/Um-el-Djeladjel	*Cleome arabica*	
Hanthal	*Colocynthis vulgaris*	الحنظل
Jeld /-el-had/Zeri	*Cornulaca monacantha*	
Arz/Erz/Saraw	*Cupressus sempervirens* var. *horizontalis*	الأرز/سرو
Arz/Erz/Saraw	*Cupressus sempervirens* var. *sempervirens*	الأرز/سرو
Hamul	*Cuscuta* sp.	حامول
Ziwan/Buruk'ba	*Cutandia dichotoma*	زيوان
Besbes (?)	*Cutandia memphitica*	
Rukaf	*Cyclamen rohlfsianum*	ركف
Laberint/Lemmed	*Cymbopogon schoenanthus*	
Shawk-el-bel	*Cynara cardunculus*	شوك البل
Nedjem/Nageelah	*Cynodon dactylon*	نجم/نجيله
Tarthooth Kelb/Zeb-el-Ard	*Cynomorium coccineum*	طرثوث
Abasis	*Cyperus conglomeratus*	
Nishia	*Dactylis hispanica*	
Sharai	*Danthonia forskahlii*	
Khadrat-el-Hamir	*Daucus syrticus*	خضرة الحمار
Grunfel	*Dianthus caryophylus*	قرنفل
Harra	*Diplotaxis harra*	الحارة
Fakous himar	*Ecballium elaterium*	فقوس حمير
Bu-Zerga/Z'riqa	*Echiochilon fruticosum*	زريقة
Henit-el-Agreb/Masais	*Echium angustifolium*	مصيص
Alenda	*Ephedra alata*	علنده
Lesles	*Eremobium aegyptiacum*	لسلس
Hamarah	*Erica multiflora*	حمره
Khelret-el-Ghoul/Ragma	*Erodium glaucophyllum/E. hirtum*	
Dehma	*Erodium guttatum*	
Halablab	*Euphorbia characias*	حلبلب
Seret-el-kebsch	*Evax desertorum*	سرة الكبش
M'tarba or Mutaraba	*Fagonia glutinosa*	
Um-terba	*Fagonia microphylla*	
Shejera adub	*Farsetia aegyptiaca*	
Kalkh	*Ferula* sp.	كلخ
Sfishifa/Dbebiga	*Frankenia laevis*	
Nefel	*Generic for Medicago* sp., *Trigonella* sp., *Lotus* sp., *all small legumes with yellow flowers*	نفل

Transliterated Libyan names		Vernacular names
Jilban	*Generic for Vicia* sp., *Lathyrus* sp.	جلبان
Shedida/Layet-el-Sheikh/Kesma	*Genista capitella*	الشديدة
Seif-el-Grab/Seif-el-Khalouf	*Gladiolus segetum*	
Zriga/Zariqa	*Globularia alypum*	زريقة
Guejroud/Bugreid/Krashaïa	*Gymnocarpos decander*	قجرود
Hdidat/Scenin/Guthaam	*Halocnemum strobilaceum*	غذام
Shegeret-Er'rih	*Haplophyllum tuberculatum*	شجيرة الريح
Geshroud	*Helianthemum desertorum/E. ruficomum*	
Erga/rega	*Helianthemum kahiricum*	
Rguiga/Zefzef	*Helianthemum lippii*	زفزف
Aud/Reguig/Semhari/Zefzef	*Helianthemum sessiliflorum*	
Hasheb-el-Rnab	*Helichrysum stoechas*	عشب الارنب
Tahenna	*Heliotropium undulatum*	
Um labeid	*Herniaria fontanesii*	
Neffel	*Hippocrepis bicontorta*	
Hashishat El-Agrab	*Hippocrepis cyclocarpa*	حشيشة العقرب
Ziwan	*Hordeum murinum* or *Hordeum maritimum*	زيوان
Gangeet	*Hyoscyamus albus*	قنقيط
Hamra	*Hyparrhenia hirta*	
Handakuk	*Hypecoum aequilobium*	حندقوق
Bugrad	*Hypericum triquetrifolium*	بوقرط
Heulma	*Ifloga spicata*	
Diss	*Imperata cylindrica*	
Smar	*Juncus acutus*	
Ar' Aaar Bahree	*Juniperus macrocarpa*	عرعر بحري
Ar' Aaar	*Juniperus phoenicea*	عرعار
Ghiz-el-ghazal	*Koelpinia linearis*	
G'dom/Rezedid	*Lasiurus (Rottboellia) hirsutus*	
Jilban/Ahmara	*Lathyrus cicera*	
Huthaan	*Launaea nudicaulis*	حوذان
Azim/Lahyet-Jeddi	*Launaea resedifolia*	لحية الجدي
Kamun-el-Jmel	*Lavandula coronopifolia*	كمون الجمل
Gamah-el-Watan	*Lecanora esculenta* (?)	قمح الوطن
Gababbgab	*Leontice leontopetalum*	
Zeita	*Limoniastrum monopetalum*	زيتة
Rhadem-el-Ghazel	*Limonium pruinosum*	
Helga	*Linaria aegyptiaca*	
El Samah	*Lolium rigidum*	الصامة
Gudrat	*Lotus corniculatus*	قرط
Kheta	*Lotus creticus*	خيطة
Amranann	*Lotus glinoides*	
Kheta	*Lotus pusillus* or *Lotus creticus*	خيطة
Auwsedj/Sekkum	*Lycium arabicum*	عوسج

Transliterated Libyan names		Vernacular names
Halfa mahboula	*Lygeum spartum*	حلفا مهبولة
El-hahma	*Malcolmia aegyptiaca*	
Khoubeïsa	*Malva parviflora*	خبيزة
Rubia/Marubia	*Marrubium vulgare*	روبيا
Babounej/Flefla	*Matricaria aurea*	
Shgara	*Matthiola aegyptiaca*	الشقارة
Jaddah/Zgara	*Matthiola oxyceras*	
Aoiunet-el-Hnesh	*Medicago laciniata*	عوينة الحنش
Kerachet-el-Arneb	*Medicago minima*	كريشة الأرنب
Shankurah	*Melilotus alba*	
Ghazul	*Mesembryanthemum cristallinum*	غسول
Ghazul/tefel	*Mesembryanthemum nodiflorum*	غسول
Alkhiass/Hemin (?)	*Moricandia arvensis*	
Kltul	*Muscari comosum*	
Marsin	*Myrtus communis*	مرسين
Kaleel	*Narcissus elegans*	
Defla	*Nerium oleander*	دفله
Saadan	*Neurada procumbens*	
Akkuz Musa	*Nicotinia glauca*	عكوز موسى
Ghordog	*Nitraria retusa*	غردق
Shouk-el-hanesh	*Noaea mucronata (= N. spinosissima)*	شوك الحنش
Aefej-el-Ghazel	*Nolletia chrysocomoides*	
Tassak, el-Had, Askaf	*Nucularia perrini*	
Zek'eter/Zebbous/Kerkeita	*Olea oleaster*	كركيطة
Adeda	*Onobrychis caput-galli*	سن العجوز
Sin-el-Ajus	*Onobrychis crista-galli*	سن العجوز
Sheedadah	*Ononis natrix*	شديده
Hindi/Karmus-el-Hindi	*Opuntia ficus-indica*	هندي
Gousmir	*Oryzopsis miliacea*	
Zuwetina	*Oudneya africana*	زويتينة
Hammad	*Oxalis cernua*	حامض
Bakhur/Schook-es-sera/Chabroum	*Pallenis spinosa*	شبرم
Abu rukba	*Panicum turgidum*	أبو ركبة
Bi qareoun	*Papaver rheas*	بو اقرعون
Saysaban	*Parkinsonia aculeata*	سيسبان
Tifun/Theia-el-arab/Theia el-Jbel/ Ez-Zbada	*Paronychia argentea*	
Harmel	*Peganum harmala*	حرمل
Ater	*Pelargonium odoratum*	
Bu'rokba/Mekhamla	*Pennisetum dichotomum*	
Harsha	*Pennisetum elatum*	
Hafan/Harsha/Samma/Tman	*Pennisetum setaceum*	
Taam El Naser	*Pergularia tomentosa*	طعم النسر
Hallab/Bu Halibah/Dahar-el-halab	*Periploca laevigata*	حلاب

Transliterated Libyan names		Vernacular names
Ta'am-el-Arneb	*Phagnalon rupestre*	طعمة الارنب
Chetam/Ketab/Sakabh	*Phillyrea media*	
Sahaina/Shera	*Phlomis floccosa*	
K'ssab	*Phragmites communis*	قصبه
Houdilan	*Picris coronopifolia*	
Snouber	*Pinus halepensis*	صنوبر
Betum/Batoum/Tizert	*Pistacia atlantica*	بطوم
Patum (?)	*Pistacia lentiscus*	بطوم
Alajzan/Guezah	*Pituranthos chloranthus*	قزاح
Taghat/Aljian	*Pituranthos scoparius*	
Taghat/Gouzah	*Pituranthos tortuosus*	قزاح
'Ihnem	*Plantago albicans*	إينم
Sbouss	*Poa bulbosa*	
Ghardab/Gourdab	*Polygonum equisetiforme*	قرضاب
Bilebsha/Bender Akresh	*Portulaca oleracea*	بليبشه
Gatroun	*Psoralea bituminosa*	
Ballout	*Quercus calliprinos*	بلوط
Belbal	*Randonia africana*	
Gula/harir/Saghil	*Ranunculus asiaticus*	
Umm neda	*Reaumuria vermiculata*	أم الندى
M'sawiya	*Reseda alba*	
Babush-el-kharoof	*Reseda arabica*	
R'tem	*Retama raetam*	رتم
Sellof	*Rhamnus lycioides*	سّلوف
Arfedj	*Rhanterium suaveolens*	عرفج
Tafga	*Rhaponticum acaule*	
Summac/Argh/Djederi/Jdari	*Rhus tripartitum*	جداري
Karrwah	*Ricinus communis*	خروع
Mungar	*Roemeria hybrida*	
Klill	*Rosmarinus officinalis*	إكليل
Fokhia/Fuha	*Rubia tinctorum*	فوه
Hummaid	*Rumex vesicarius*	حميض
Shajarat-el-Riyah/fessel/Faijel	*Ruta graveolens*	فيجل
Ghedem	*Salicornia fruticosa*	
Lebed	*Salsola kali*	
Shadeedah	*Salsola longifolia*	شديدة
Gel	*Salsola tetragona*	جل
Feres/Gel	*Salsola tetrandra*	فرس/جل
Rhadam/Malika-el-Badia	*Salsola vermiculata*	
Bu Fetha/Ejiz	*Salvia aegyptiaca*	
Sag-en-Naga	*Salvia verbenaca*	ساق الناقة
Sharuneah	*Sanguisorba minor*	شرونية
Shibrig	*Sarcopoterium spinosum*	شبرق
Shabraq/Shabreg	*Sarcopoterium spinosum*	شبرق

Transliterated Libyan names		Vernacular names
Goulglane/Sarrah	*Savignya aegyptiaca/S. parviflora*	
Eyin tishisma	*Scabiosa arenaria*	
Zabbal	*Schismus barbatus*	
Semmid	*Scirpus holoshoenus*	
Sunnaryiah/Guernina	*Scolymus hispanicus*	
Losga	*Scorpiurus muricatus*	
Guizz/Salsafil-el-Aswad/Dehheba	*Scorzonera undulata*	ذباح
Acheb-el-Husan	*Scrophularia* sp.	عشبة الحصان
Dag'aa/daraeita	*Senecio coronopifolius*	
Um-magrain	*Silene colorata*	
Derias	*Silphium cyrenaicum* (?)	درياس
Alayiq (?)	*Smilax aspera*	عليق
Haneb-el-dib	*Solanum nigrum*	عنب الديب
Olik	*Spartium junceum*	
Ziwan	*Sphenopus divaricatus*	
Gerara	*Statice pruinosa*	
Shenne	*Statice thouinii*	
Adem-el-dib	*Stipa barbata*	
Adem	*Stipa lagascae*	
Behmah	*Stipa retorta/S. tortilis*	بهما
Halfa gedima/Gedimah	*Stipa tenacissima*	حلفة قديمة
Sbot	*Stipagrostis (Aristida) pungens*	
Souida	*Suaeda fruticosa*	
Souad	*Suaeda pruinosa*	
Ethel or Tarfa (?)	*Tamarix* sp.	أثل/طرفة
Terfez	*Terfezia boudieri*	ترقاس
Garambush	*Tetragonolobus purpureus*	غرمبوش
Jaada	*Teucrium polium or Teucrium pilosum*	جعدة
Derrias	*Thapsia garganica*	درياس
Methnan	*Thymelaea hirsuta*	مثنان
Methnan-el-Abiod	*Thymelaea microphylla*	المثنان الأبيض
Zaater	*Thymus algeriensis*	زعتر
Za'atar	*Thymus capitatus*	زعتر
Damrane/Sereef	*Traganum nudatum*	ضمران/ضريف
Kutrab/Shouk Aziz	*Tribulus terrestris*	
Nefel (?)	*Trifolium tomentosum*	نفل
Handgoug	*Trigonella* sp.	نفل
Nefel	*Trigonella stellata*	نفل
Faraoun	*Urginea maritima*	بصل الفرعون
Chabrom Shouk/Kadad	*Zilla spinosa*	شبرم/كداد
Seder/Sedr	*Ziziphus lotus*	سدر
Bugreiba/Aggaia/Belbal	*Zygophyllum album*	بلبال

Transliterated Libyan names		Vernacular names
LIVESTOCK	**Description**	
Abel Maghrebia	Most common dromedary breed in Libya and North Africa	أبل مغربية
Alluch	Lamb (in the west)	علوش
Atrus	Mature ram	عتروس
Awassi	Fat tail sheep from the Middle East	عواسي
Barbary	Fat-tail sheep/95% of Libyan sheep/ north-western and Sirte	بربري
Barki	Barbary sheep from Jebel-Akhdar and Marmarica (see also Dernawi)	برقي
Bell/Ebel	A flock of dromedaries	بل/إبل
Dernawi	Barbary sheep from Jebel Akhdar and Marmarica = Barki	درناوي
Dubasi	Thin-tail sheep from southern Libya (Fezzan)	
Galem	Standard Libyan flock of small ruminants	غلم
Gaoud	Young dromedary from half a year to 4 years old	قعود
Gatha'a	Male or female sheep at the age 8–12 months	جذع
Ghanam	Generic for sheep	غنم
Ghimi	Barbary sheep breed from Fezzan (Hon, Zella) (?)	
Himliya	Dromedaries' flock free grazing without shepherd's supervision	هاملة/هملية
H'war	Young dromedary up to 6 months old	حوار
Im'Azba	Herd does not return at tent for night	أمعزبة
Jamel	Mature male dromedary	جمل
Kebs'h	Leading ram	كبش
Kharf'i	Lamb born in winter	خرفي
Khruf	Weaned lamb	خروف
Ma'az	A group of 20–25 goats/a flock of goats	ماعز
Ma'sz Mahali	Breed of goats in northern Libya	ماعز محلي
Mehari	Slim dromedary breed from the Sahara and southern Libya	مهاري
Na'aga	Female sheep at the age of more than 30 months, the most productive age	نعجة
Naga	Mature female dromedary	ناقة
Rabah	Male or female sheep at the age 24–30 months	رباع/رباعية
Rahma	Sheep, goats and dromedaries disease	رهمة
Sadas	Male or female sheep at the age 30 months	سداس/سدسة
Saluki or Slouhgi	Tall fast-running sight-hunting dogs used by nomads' riders – considered as noble dogs allowed to enter the tent	سلوقي
Serk	Incision on sheep's ear to identify	
Thany	Male sheep at the age 12–24 months	ثني
Wabar	Dromedary's hair or wool	الوبر
Warwara	Soft palate projection from mature male dromedary in rut	ورورة
Watish	Thin-tail sheep from southern Libya (Fezzan)	

Transliterated Libyan names		Vernacular names
WILDLIFE	**Scientific names or description**	
Abn Awaa	*Canis aureus*, the common Libyan jackal (See also *Dhib*)	ذيب
Aeqrab	Generic for scorpions	عقرب
Al-Bsharosh/Amgawakh	*Phoenicopterus roseus*	البشروش/أمجوخ
Alkarki	*Grus grus*	الكركي
Al-qarad/Garad/Ougrad	*Hyalomma dromedarii*	قراد
Al-Sak'r	*Falco peregrinus*	صقر
Al-washq/Washaqa	Lynx sp. (?)	وشق
Anaq-al-Ard	*Felis caracal*	عناق الارض
Arnab	*Lepus capensis*	أرنب
Baouda/Namoos	Generic for mosquito	بعوض/ناموس
Baqara barria/Abu harba	*Oryx dammah*	بقرة برية/أبو حربة
Baqara mahaliya	*Addax nasomaculatus*	بقرة محلية
Belbouch	*Helix melanostoma*, small white snail	بلبوش
Bu Aqraban	*Scolopendra cingulata* (in the east)	بوعقربان
Bubalus	*Alcelaphus buselaphus*	
Chuka/Chees	Spine from porcupine	شوكة/شيص
Dabba	*Hyaena hyaena*	ضبع
Dhib	*Canis aureus*, not occurring in Libya, often confused with Jackal (*Abn Awaa*)	ذئب
Dhobb	*Uromastyx acanthinura*	
Eaqrab al-riyh	*Galeodes* sp., Solifugae (spider-like)	عقرب الريح
Fa'had	*Acinonyx jubatus*	الفهد/الشيتا
Fakrouna	*Testudo graeca*	فكرونة
Fennec	*Fennecus zerda*	الفنك
Garad	*Hyalomma dromedarii*	قراد
Garnoog	*Grus grus*	غرنوق
Gata	*Pterocles alchata*, large flocks of long pin-tail red partridges from the desert	قطا
Ghazal	*Gazella dorcas*	غزال
Ghazela Abiod	*Gazella leptoceros*	غزال
Gundi	*Ctenodactylus gundi*	كوندي
Gunfud	*Atelerix algirus* or *Hemiechinus auritus*	قنفود
Hagaal	*Alectoris barbara*	حجل
Halzun/Bo Zwia	*Eobania vermiculata* and *Cornu aspersum*, large brown snails	حلزون/بوزوية
Hamousam	*Scolopendra cingulata* (in the west)	
Harba	*Chamaeleo chamaeleon*	حربة
Houbara	*Chlamydotis undulata*, protected bird (Ahabbara (female) , Horb (male))	حبارة
Jorboa	*Jaculus jaculus*, desert rodent common on sandy places, with a long tail ending black	الجربوع
Jerboa	*Jaculus jaculus* ssp. *desertii*	جربوع
Jirth	*Psammomys obesus* or *Meriones libycus*	جرذ
Jrad	Generic for grasshoppers	جراد
La' Riel/Mahar (?)	*Gazella dama*	الأريل

Transliterated Libyan names		Vernacular names
Lak-Lak	*Ciconia ciconia*	لق لق/أعنزه
Lefha/Um-qoroon	*Cerastes cerastes* = Syn. *Aspis cerastes* or *Vipera lebetina* /(?) *Daboia* sp. (?)	أفعى أم قرون
Lefha'	*Cerastes cerastes* = Syn. *Aspis cerastes*	لفعة
Naeama	*Struthio camelus*	نعامة
Nahal	Bee	نحل
Nemer	*Panthera* sp.	نمر
Nemes Al-Rabah/Shfasha	*Geneta geneta*	نمس الرباح
Nis/Said Al'Lail	*Hystrix cristata*	صيد الليل
Rabdda	Female ostrich	الربدا
Reem	*Gazella leptoceros*	الريم
Shifsha	*Poecilictis libyca (Syn: Ictonyx libycus)*	شفشه
Sloughi/Saluki	A sight hunting noble dog once common with nomads and hunters all over North African arid rangelands	سلوقي
Taban	Venomous snake – Cobra (?)	ثعبان الكوبرا
Thaelab	*Vulpes vulpes*	ثعلب
Thueban	*Naja haje*	ثعبان
Waddan	*Ammotragus lervia*	الودان
Woural/Oural	*Varanus griseus*	ورل
RANGELANDS	**Description**	
Aacheb	Green grazable spring vegetation	عشب
Abiyyar Hukumiyii	Government cisterns/wells	أبيار أو صهاريج حكومية
Abyar	A place with many wells (sing. = bir)	أبيار
Adar	The place where a family or tribe camps and builds their tents, it is always associated with memories and nostalgia	الدار
Aguedal/Guedal	Protected area by private owner or community	
Ahjaar	A rope that connects the animal's front leg with the back leg on the same side together to prevent it from walking a long distance	أهجار
Aila (Pl: 'Ailaat) = Aït	Descent group, family	عائلة (جمعها عائلات)
Aïn	A well providing water to dwellers and livestock, also used for artesian well	عين
Al-barr	The steppe in the east, open rangelands (see *Jisha*)	البر
Al-batnan	The western part of Marmarica plateau west of Tobruk to Bomba Gulf extending from the coast to south	البطنان
Alef	Forage	علف
Al-ghaba	The forest zones in the east	الغابة
Al-Hisha	Best-watered floodplain	الهيشة
Al-jabal	The mountains zones	الجبل
Al-Jabaliya	Berbers tribes from the Jebel Nafusa	الجبالية

Transliterated Libyan names		Vernacular names
Al-maksura	Long strips of woven material often decorated (*Margum*) used to separate the side of men from women and kids in a tent in the west	
Al-sirwal	The low-rainfall steppe country extending into the desert in the east	السروال
Al-suhub	The steppe country (neither cropped, nor desert) in all Arab countries	السهوب
Al-thahar	The second plateau in Cyrenaica	الظهر
Amazigh/or Imazighen	The name the Berbers give to themselves	الأمازيغية
Ambakbaka	Meat browned in a pot in olive oil with onions first and then hot water and macarona with tomato paste and seasoning, slow-cooked; the name *Ambakbaka* comes from the onomatopoeia of the clanging of the pot cover during cooking macarona	امباكباكا
Amgalaam	Shearing season is the end of spring, and people usually volunteer to help in exchange for a sumptuous local meal with lamb meat	أمجلم
Ard Gabila	Tribal land = descent group land (see *watan*)	أرض القبيلة
Ard Melak	Privately own land	أرض ملك
Arqub	The high area between two valleys	عرقوب
Asabiyyah lildam	'Esprit de corps', solidarity within and between social groups as tribes, clans or families	العصبية للدم
B'sat	Small carpet or rug	بساط
Bait (Pl: Buyuut)	Bedouin tent/all those who live in the tent (see *khayma*)	بيت (جمعها بيوت)
Bait-es-sha'r	The common Bedouin tent (made mainly of goat's hair) used in winter as it is more suitable to rainy season - see *khayma*	بيت شعر
Balta (Pl: Balat)	Large desert depressions (south Jebel Al-Akhdar on the fringe of the desert) fed with run-off water, opportunistically cropped with rainfed barley	بلطة (جمع بلط)
Barqua/Barce/Barke	The El-Marj plain	برقة/بارشي
Barraka	A simple hut or house, usually made of zinc	براكة
Bartha'a	A thick protective layer of fabric and burlap sewn together, placed on the donkey's back to facilitate riding and carrying things on it	بردعة
Batha	Garrigue/maquis in the Jebel Al-Akhdar	باثا
Bawadi	The people of the tents (in the east) = the Bedouin	بوادي
Bazin	Meal made with roasted and ground barley with olive oil or milk	بازين
Bialsabq	'First come, first served', by first occupancy	بالسبق
Bir (Pl: Abiyyaar)	A well providing water to dwellers and livestock	بير

Transliterated Libyan names		Vernacular names
Bir Malik or Bir Khassa	Private well	بير ملك/بير خاصة
Bsissa	Meal – roasted barley flour and other seeds, hand-mixed with olive oil	بسيسة
Budairi	Autumn (Sept./Oct.) – first rain	البداري
Burkan	Volcano	بركان
Busat/B'sat	Rug or carpet, *Kilim* woven	بساط
Chayïr	Barley	شعير
Chemla	Knitted udder-bag deterring dromedaries' calf from suckling	شملة
Daffa	Plateau with shallow soil	دفة
Damus/Domus	Troglodyte dwellings dug by Berbers into the deep loess in the Jebel Nafusa	
Danah As-Siid (Pl: Danah al-Lassiyaad)	Cousins	ضنا السيد جمعها (ضنا الأسياد)
Dârâa	Site or place protected from wind	ذرا
Dar-ech-shte	Winter camp or tent	دار الشتاء
Dar-es-Sif	Summer camp or tent	دار الصيف
Dellou or dalu	Bucket (leather or other) to draw water from well	دلو
Eassal	Honey	عسل
Emidat alkhayma	Wooden poles to support the roof of the tent	أعميدات الخيمة
Erg	Large mobile sand dune areas	عرق
Fasquiya	Antique or modern cistern holding fresh water usually from natural springs	فاسكية
Feidah	Large depression feed with run-off water, usually cropped with rainfed barley (in the west)	
Fitna Barbaryya	Intertribal conflicts and wars	فتنة بربرية
Fitura	Residues from pressed olives	فيتورة
Flij	Long and narrow tent strip made of woven wool and hairs from sheep, dromedaries and goats	فلج
Ga'aa	The place where ears of grain are collected after being harvested for threshing	قاعة
Gab	When sheep are watered once every two days or every three days	غب
Gabaila (Pl: Gubayil)	Tribe	قبيلة (جمعها قبائل)
Gadab	The year when the land receives very little rain in the rainy season	جدب
Gaid	A rope that connects the animal's front legs together to prevent it from walking a long distance	قيد
Galaia	Small pieces of lamb meat fried and cooked in oil	قلايا
Galam	Mixed flock of sheep and goats	غلم
Gammah	Wheat	قمح
Gar'â	Desert depression sporadically flooded, saline or not, sometimes partially cropped	قارة

Transliterated Libyan names		Vernacular names
Gatran	Tar extracted from *Juniperus* wood distillation, used to sanitize *shakwa*	قطران
Gerret el-Anz	Mid-February when goats are allowed to get back on the range after the cold and wet weather	غرة العنز
Ghedir/R'dir	A large pool of water after rain or after a flood in wadi	غدير
Ghrara	Bag (approx. 1 m wide × 1.4 m high) to store grain, made of a *flij* cut to size, folded and sewn	غرارة
Grisha	A small enclosure (see *zeriba*)	
Guech	Dry grazable annual vegetation	قش
Guelta	Pond/water accumulation usually in mountainous country/watering place	
Guerba	Goat skin container for saving or carrying water	قربة
Haajja Jumla	Collective/communal belonging	حاجة جملة
Hadd (Pl: Huddud)	Border/limits of territory	حد (جمعها حدود)
Hadur	The people of the towns (in the east)	حضور
Hamel	Flat carpet, blanket (*Kilim* woven), long tent strips made of woven sheep wool, goats and dromedary hair	
Hammam	Public bath	حمام
Hatiyah	Shallow depression, sometimes cultivated	
Hawia	A thick layer of protective fabric is placed over the dromedary's back so that it is not harmed by the weight of the load on its back	حوية
Hima	Protected area by local community	حِمَى
Holy or Huli	A traditional Libyan summer costume for men, made of a single length (4–6 × 1.5) m of white wool (see *jard*)	حولي
Imazighen	The Berbers (the free men)	الأمازيغ
Ishta	Winter (Sept./Oct. to Feb.)	الشتاء
Jabeh	Beehive	جبح
Jalim	Special scissors used to shear sheep wool, goat hair, or dromedary hair	جلم
Jard or Jerd	A traditional Libyan costume for men, made of a single length (4–6 × 1.5) m of white wool (see *holy*)	جرد
Jarraya	Low stone walls on gentle slopes to collect and direct to crops harvested water in low-rainfall zones	جراية
Jebbad	Labourers working in the palm groves in the south	جباد
Jisha	Steppe/treeless, rocky, and difficult terrain (see *barr*)	الجشة
Kaaf	Cave used to store grain (in the east)	كاف
Kailaa	A sack of grain weighs about 80 kg	كيلة
Khayma	The common Bedouin tent (made of woven sheep wool and dromedary and goats hairs) (see *Bait-es-sha'r*)	خيمة

Transliterated Libyan names		Vernacular names
Leyyah or Layeh	Fat from the tail of Barbary sheep	اللّيّة
Lusaita or Al-wasaita	The first plateau between the sea and the second plateau in Cyrenaica	الوسيطة
Luwaiyia	Late spring (April/May)	اللوايا
Macarona	Popular Libyan pasta meal	مكرونة
Madmusa	A flat unleavened bread made by shepherd on the hot sand	مدموسة
Mahmiyat	Protected area by government authority	محميات
Malik (Pl: Amlak)	Immobile/land property	جمعها أملاك) ملك)
Manehaa	A sheep or several sheep given on loan, so that they can feed them and benefit from their milk and wool	جمعها منايح) منيحة)
Maraei	Rangeland/pasture	مراعي
Mardoumat/Mafhuma	Traditional earth or sand covered kiln to manufacture charcoal	مردومة/مفحومة
Margum	Wool woven hangings or flat carpets brightly coloured – red dominant, without pile and decorated with geometric patterns	مرقوم
Masda	Horizontal loom traditionally used by Bedouin to weave long tent strips	مسدى
Maslach/Garbah	Water container made of raw goat skin	قربة
Matmura	Grain storage in soils (in the east)	مطمورة
Menfes	A spillway in a tabia to evacuate excess water run-off to the next tabia downstream	منفس
Mensez	Vertical loom	منسج
Mesbut	Rangeland with dominant *Stipagrostis pungens*	
Mihrath/Maharat	A traditional wooden plough	محراث
Moklah	Bag carried by shepherd, or muzzle for dromedary, or feed bag	مخلاه
Mortier de tuilleau	Roman-Byzantine very long-lasting waterproof lime mortar in antique cisterns	
Moya quibrity	Sulphurous water from well	مياه كبريتية
Murabin	Animal flocks owners	مربين
Murtafa	Hill	مرتفع
Nagab (Pl: Anguuba)	Cluster of tents or households (common ancestor)/extended family	نقب
Naja'	Bedouin camps (in the east)	نجع
Nebkha	Soil accumulation around a shrub, forming an accrescent mound	نبخة
Nool	Common name for a vertical loom (at home or in weaving shops)	النول
Quibrity	Sulphurous water from artesian wells, also matches (*Quibreet*)	كبريت
Rabii'	Early spring (Feb./Mar. – early Apr.)	ربيع
Raïe	Shepherd	راعي
Ramla	Sandy area/sandy soils	رملة

Transliterated Libyan names		Vernacular names
Rammadiya	Mounds of ashes and snail shells from prehistoric settlements in southern Tunisia	رماضية
Reba'i	Lambs born in Feb./Mar. and later	رباعي
Reg	Large desert flat areas covered with stones and pebbles	رق
Riwaq	Pieces of burlap and cloth sewn together, rectangular in shape, closing the sides and opening of the Bedouin tent	رواق
Rizik	Movable property (herd)	
Sa'ada	The common female ancestor of all the eastern Libyan tribes	سعدة
Saba	When the vegetation and the cultivated grain produced in abundance due to favourable rainfalls	صابة
Sabkha (or *Sebkha*)	A saline depression	سبخة
Saïf	Summer (May–August)	الصيف
Sajada	Carpet or rug – floor covering in Bedouin tent	سجادة
Samen	Sour salted and fermented butter from ewes' milk	سمن
Sarj	A thick, luxurious protective layer of textile is placed on the horse's back to make it easier to ride	سرج
Sekkah	Section of cropping land, part of the *watan* allotted to a clan or a family (in the west, possibly a Turkish word)	
Sha'ara	Juniper tree (in Jebel Al-Akhdar)	شجرة الشعرة
Shakwa	Tanned goatskin to keep milk	الشكوة
Shawaia	Pieces of meat grilled directly over the fire	شوايا
Silbiyya	High and cold crestline in Jebel Akhdar	الصلبية
Silik	Lower section of floodplain/less watered	السلك
Sima	Project's or owner's hot-iron brand on dromedary; each tribe has its own brand	السيمة
Sqifa/Teqaq	Tent partition between men and women and kids	الطقاق
Tabia	Stone embankments or low dams across wadis to create terraces and slow down water flow	الطابية
Tamazight	The language shared by the Berbers	اللغة الأمازيغية
Tanur	Cylindrical vertical oven for cooking bread	تنور
Taqsium alma'/Muah alma	'Where the mountains divide the water', often the watershed limit between tribal territories	ميوح الماء
Targi	Local breed of goats from southern Libya	تارقي
Tariq al-Nahr	The 'road of the river' referring to the Great Man-Made River Project	طريق النهر
Tawila/Al-Basaat	Plateau (landform)	الطويلة/البساط
Tbikha	Meal – soup cooked with potato, chickpeas, lentils or broad beans and any type of meat	طبيخة

Transliterated Libyan names		Vernacular names
Tebagi	Local breed of goats from southern Libya	
Tebben	Dry cereal crop residues fed to livestock	تبن
Tefelah	White clay used to degrease wool	طفلة
Tobel or Tbol	Drum, used in festivals and war as well	طبل
Usadah	Cushions or bags common in bedouin tents (see *ghrara*)	وسادة
Washuun (Pl: Washnaat)	The ladies and kids members of the family (mother, daughters, kids)	واشون (جمعها وشنات)
Wassath/Jaber	Middle pole supporting the roof of the Bedouin tent	وساطة/جابر
Watan	Customary territory including arable land, rangeland and wells belonging to a tribe, clan or family	وطن
Wazra	Winter traditional Libyan dark wool cloth (dromedary and sheep's wool) rectangular 5 m × 1.5 m, wrapped around body	وزرة
Zawiya	Building and institutions associated with Sufi brotherhood	زاوية
Zeriba/Zriba	Enclosure made of branches to keep small young ruminants near the tent at night (see *grisha*)	زريبة
Zumeita	Barley flour mixed with olive oil or milk as simple meal; usually cumin and coriander seeds are added after grinding barley grain	زميتة
Zwariyya	Berbers tribes from Zuara along the western coast	زوارية

Index

Note: The page numbers in *italics* and **bold** represents *figures* and **tables** respectively.

www.ingramcontent.com/pod-product-compliance
Lightning Source LLC
LaVergne TN
LVHW070903300926
849162LV00011B/1846